Site Characterization in Karst and Pseudokarst Terraines

Richard C. Benson • Lynn B. Yuhr

Site Characterization in Karst and Pseudokarst Terraines

Practical Strategies and Technology for Practicing Engineers, Hydrologists and Geologists

Springer

Richard C. Benson
Technos, Inc.
Miami, FL, USA

Lynn B. Yuhr
Technos, Inc.
Miami, FL, USA

ISBN 978-94-024-0425-8 ISBN 978-94-017-9924-9 (eBook)
DOI 10.1007/978-94-017-9924-9

Springer Dordrecht Heidelberg New York London

Softcover re-print of the Hardcover 1st edition 2016

Printed on acid-free paper

Springer Science+Business Media B.V. Dordrecht is part of Springer Science+Business Media (www.springer.com)

Preface

The purpose of this book is to present both an insight into the complexities of karst and pseudokarst conditions and a broad strategy and practical approach needed to carry out an effective site characterization within these or any other complex geologic conditions. Our intent is to provide a basic understanding of these topics and introduce the vast array of tools available to solve the geologic puzzle. We believe that the geology is the key to a site characterization, since the geology is what controls the engineering and hydrologic conditions at a site. If the geology is properly understood, then all subsequent work can be done with more confidence, including design, construction, contaminant assessment, remediation, modeling, and long-term operation.

This book is divided into three parts:

- Part I provides a brief introduction to some of the many karst and pseudokarst conditions that may be encountered. It provides a brief overview of karst development, various conditions that may be present, and their wide range of scale. While there are many benefits to a karst landscape (groundwater resources, minerals, and recreation), our focus is on the damaging impacts to both engineered structures and groundwater resources. By developing an understanding of the nature of karst and appreciating its potential impact, we have a starting point for an effective site characterization effort.
- Part II is the core of the book and presents a strategy to carry out an effective site characterization. The strategy emphasizes the use of a wide range of measurements and technology, over a range of scales. We introduce many of the off-the-shelf technologies that are presently available to help characterize karst conditions. This section includes numerous examples and mini-case histories to illustrate the strategies and methods presented.
- Part III presents three site characterization case histories that include:
  - A landfill that was being developed over an abandoned limestone mine in the Kansas City area, which had experienced a major mine-roof collapse due to a deep-seated paleo-collapse feature
  - A geotechnical assessment for a new bridge into the Florida Keys in an area of suspected karst
  - An EPA Superfund site in west central Florida that required assessment of geologic, hydrologic, and karst conditions prior to initiating remediation efforts

The Authors

This book is, for the most part, based upon the authors' combined experience of more than 80 years as consulting geologists. The authors are a father and daughter team, who has worked together since 1978 providing consulting services through their company Technos, Inc. Richard Benson pioneered many of the early applications of various geophysical methods in the 1960's and 1970's. He founded Technos Inc. in 1971 specializing in applied earth sciences. Their hands-on experience in site characterization, with both karst and pseudokarst, is the basis for the strategies and array of methods presented in this book. This personal experience is represented by the many technical examples included in the book. These examples are from sites in the USA, the Bahamas, much of the Caribbean, and Guam.

The authors have developed all tables, diagrams, and photos, unless specified. Most of the figures shown are simplified. We recognize that the real-world, especially in karst and psuedokarst, is much more complex. In many cases, the examples do not contain the specific location or name of the owner in order to avoid concerns of liability.

Miami, FL, USA

Richard C. Benson
Lynn B. Yuhr

Acknowledgments

We dedicate this book to our families and thank them for their unconditional patience and support. With a special dedication to Bonnie Benson (wife of Richard), Eric Yuhr (husband of Lynn) and Marisol Sanchez (adopted family member and office manager for over 20 years). To our friends and colleagues, we thank you for the continuous positive nods, pats on the back, and review of chapters that were often incomplete and incoherent. To our publisher, thank you Petra and Hermine for your patience, support, and allowing us to check one off of our bucket list.

A special thank you is extended to our reviewers and contributors who include:

David Carrier, Argila Enterprises
Charles Finkl, Professor Emeritus of Geosciences, Florida Atlantic University, and editor in chief of the Journal of Coastal Research
Thomas Aley, Ozark Laboratories
Tony Waltham, Geophotos, www.geophotos.co.uk
Chris Anderson, www.darklightimagery.net
Jim Lococo, Mount Sopris
Peter Annan and Steve Cosway, Sensors & Software, Ltd.
Dwain Butler, (retired) US Army Waterway Experiment Station
Peter Fenning, (retired) consulting geophysicist in UK
Devraj Sharma, founder Principia Mathematica
Gary Olhoeft, (retired) Colorado School of Mines
Paul Beam, Dept. of Energy
Richard Woods, University of Michigan
Lewis Land, NCKRI
George Veni, NCKRI

Authors can be contacted at:

Richard C. Benson, CPG, PG
richardbenson65@gmail.com

Lynn B. Yuhr, PG
lynnyuhr@gmail.com

Contents

Part I

A Brief Overview Karst and Pseudokarst

Part I of this book provides the reader with a brief overview of some of the wide range of karst and pseudokarst conditions that can exist along with their benefits and damaging effects. The scale of these features that can impact engineering works or environmental issues can range from small voids or permeable zones of centimeters to caverns with dimensions of hundreds of meters or more. Most karst conditions will have evolved over thousands of years or more while those involving mans activities (pseudokarst) may be formed over much shorter periods of time from a few years to a few hundred years. Having an understanding of the type of karst or pseudokarst features that can develop or are already present in a particular setting will provide a starting point for an effective site characterization effort.

Some Common Terms 1

Abstract

There is a wide range of terminology encountered in the field of karst. The following terms will set the stage for the topics to be covered. Additional terms will be presented throughout the book. The two key divisions are karst and pseudokarst. Karst conditions are formed due to a natural process of dissolution of soluble rock. Pseudokarst may look or act like karst, but it may be formed naturally in non-soluble rock or may be caused by man's activities. In addition, the term paleokarst is used to indicate those karst features that are very old, in geologic terms, and are typically inactive. For a complete discussion of karst terminology see Monroe (A glossary of karst terminology. US Geologic Survey Water Supply Paper 1899K, 1970), Lowe and Waltham (Dictionary of karst and caves. In: Judson D (ed) British Cave Research Association cave studies series 10. British Cave Research Association, Malvern, 2002), and Fields (A lexicon of cave and karst terminology with special reference to environmental karst hydrogeology. US EPA/600/R-02/003, 2002).

1.1 Karst

The word karst stems from the meaning stone, and its derivatives are found in many languages of Europe and the Middle East (Ford and Williams 2007). The term karst is commonly used to describe a wide range of surface landforms that have developed from the dissolution of soluble rock. These surface features may include bare rock, closed depressions or collapsed sinkholes. These features result in unique hydrologic conditions such as sinking streams, springs, or conduit flow providing high porosity. White (1988) and LaMoreaux (1995) provide a brief review of historical development of karst literature.

The most common karst images are those of sinkholes in the news or from our visits into caves. In a broader sense, karst conditions include both surface and subsurface features including sinking streams, springs, highly irregular top of rock profile, the epikarst, dissolutionally enlarged joints or bedding planes, conduits and caves (Fig. 1.1). Sediments or younger rock strata may cover karst features with no visible evidence of their presence at the surface.

While subsidence, sinkholes, and collapse can occur on a human time scale, the dissolution of carbonate rock and void forming process has been occurring over much longer periods of many thousands of years or more. However, certain conditions in evaporate rocks, such as gypsum or salt and some pseudokarst conditions can result in subsidence, sinkholes, and collapse that can develop over a human time scale.

Traditionally karst refers to the landforms that result from the dissolution of soluble rock, these landforms are result from subsurface dissolution features. For this book, the term karst will be used to include both surface and subsurface features, which are an interconnected system. The presence of the surface landforms requires the presence of the subsurface features. Therefore, we are using the term in a broader sense to include both surface and subsurface features when referring to karst terrains.

1.2 Paleokarst

Paleokarst refers to old or mature karst (sinkholes or caves) that has developed over geologic periods of time and has become inactive. That is, paleokarst has been uncoupled

R.C. Benson, L.B. Yuhr, *Site Characterization in Karst and Pseudokarst Terraines*,
DOI 10.1007/978-94-017-9924-9_1

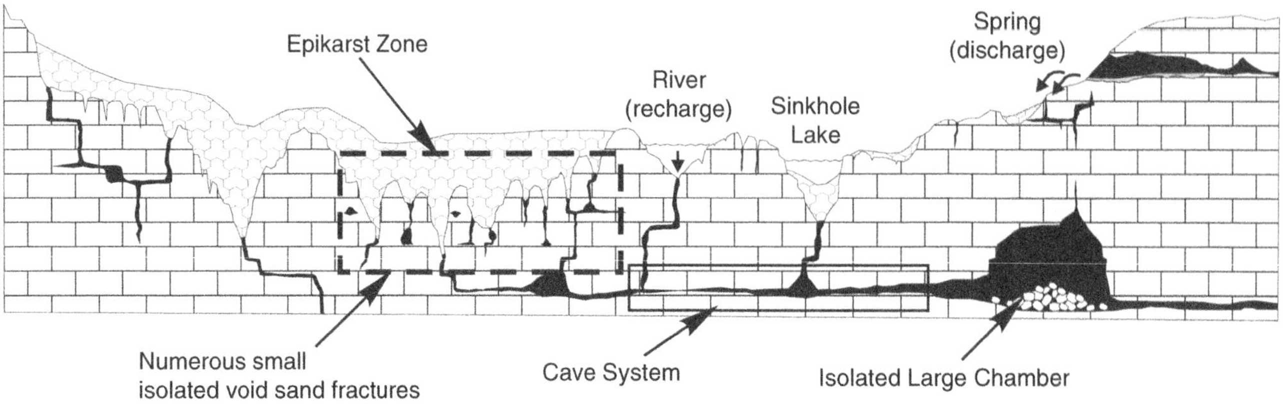

Fig. 1.1 A conceptual sketch of a karst system showing some of the many geologic and hydrologic complexities

from the dissolution process and as a result there is little, if any, further dissolutioning taking place. However, subsidence or collapse may still occur if there are large open caverns present.

Paleokarst may have undergone tectonic changes with uplifts or subsidence and/or sea-level changes. While typically buried, it may be exposed at the surface or at road cuts or high in the mountains. Paleokarst can sometimes be reactivated by natural causes or man's actions. Paleokarst is of importance because of its association with deposits of oil and gas. In addition, minerals such as silver, lead, zinc and bauxite are found in zones of paleokarst and breccias pipes (circular zones of collapse filled with broken rock fragments).

Palmer and Palmer (1989) discuss the extent of paleokarst within the United States. Bosak et al. (1989) and James and Choquette (1988) provide a general discussion of paleokarst terminology, its presence worldwide, along with associated mineral deposits, hydrogeology and engineering hazards.

1.3 Pseudokarst

Pseudokarst conditions are those that resemble or act like karst but were not formed by the natural dissolution of rock. They are also known as "false-karst" or are sometimes referred to as "analogous karst". Pseudokarst may be associated with natural conditions such as caves in lava tubes, sea caves, and ice caves. Pseudokarst may be associated with fissures and caves in sandstone, or even granite caused by structural changes or erosion. In some cases, it may be associated with man's activities such as subsidence or collapse due to underground mines, fissures and subsidence due to withdrawal of groundwater, oil or gas or even acid leaks from industrial sites. Many of the sinkholes reported in the evening news are commonly pseudokarst "sinkholes" caused by leaky water pipes or sewer lines.

While karst alone can present large problems, pseudokarst covers a broader range of conditions and is probably as pervasive as, if not more than, karst, particularly in developed areas. The number of pseudokarst events is probably equal to those of karst and may be of greater impact and cost than that of karst alone. Palmer (2007) provides a brief discussion of pseudokarst related to non-soluble rocks. Borchers (1998), Holzer (1984), and Johnson and Neal (2003) all provide papers on pseudokarst related to fluid withdrawal, hydro-compaction, earth fissures, ground cracks and mining.

References

Borchers JW (ed) (1998) Land subsidence: case studies and current research. In: Proceedings of the Dr. Joseph F. Poland symposium on land subsidence, Association of Engineering Geologists, special publication no 8. Star Publishing Co., Belmont

Bosak P et al (eds) (1989) Paleokarst a systematic and regional review. Developments in earth surface processes 1. Elsevier, New York

Fields MS (2002) A lexicon of cave and karst terminology with special reference to environmental karst hydrogeology. US EPA/600/R-02/003

Ford D, Williams P (2007) Karst hydrogeology and geomorphology. Wiley, New York

Holzer TL (1984) Ground failure induced by ground water withdrawal from unconsolidated sediment. In: Holzer T (ed) Man-induced land subsidence, The Geologic Society of America, Reviews in engineering geology VI. , Geologic Society of America, Boulder, pp 67–105

James NP, Choquette PW (eds) (1988) Paleokarst. Springer, New York

Johnson KS, Neal JT (eds) (2003) Evaporate karst and engineering/environmental problems in the United States, Oklahoma Geological Survey, Circular 109

LaMoreaux PE (1995) Historical references to karst studies In: Beck BF (ed.) Proceedings of the 5th multidisciplinary conference on sinkholes and the engineering and environmental impacts of karst: karst geohazards, Gatlinburg, Tennessee. Balkema, Rotterdam, pp 11–16

Lowe D, Waltham T (2002) Dictionary of karst and caves. In: Judson D (ed) British Cave Research Association cave studies series 10. British Cave Research Association, Lancaster

Monroe WH (1970) A glossary of karst terminology. US Geologic Survey Water Supply Paper 1899K

Palmer AN (2007) Cave geology. Cave Books, an Affiliate of the Cave Research Foundation, Dayton

Palmer MV, Palmer AN (1989) Paleokarst of the United States. In: Bosak P, Ford D, Glazek J, Horacek I (eds) Paleokarst, a systematic and regional review. Developments in earth surface processes 1. Elsevier, New York, pp 337–363

White WB (1988) Geomorphology and hydrology of karst terrains. Oxford University Press, New York

The Development of Karst Conditions

2

Abstract

The following is a brief introduction to the development of karst conditions from deposition to the variety of post-depositional processes that have created what we see today. This is a very broad topic and is specific to the area in which it occurs as well as the maturity of the process. This includes the types of soluble rock, rates of dissolution and geomorphic processes that may affect the dissolution process. The benefit of this information lies in the ability to understand what karst conditions may exist or be expected at a particular site. Often the various geologic settings provide clues as the type, depth, shape or relationship of expected karst features that may be present.

2.1 Carbonates and Other Soluble Rock

Carbonate rocks include limestone, dolomites, and chalks. Other readily soluble rocks include evaporates such as gypsum, anhydrite and salt. All of which are found in a wide range of environments worldwide. Nearly all carbonate sediment is derived from calcareous algae and calcium carbonate ($CaCO_3$) producing animals living at or near the site of deposition. Deposition of carbonates typically occurs in shallow, marine environments in tropical and subtropical climates. On land, both carbonate and evaporate deposits can also occur in lakes (Braithwaite 2005).

The term limestone is used for rocks in which the carbonate fraction is primarily calcite ($CaCO_3$). The term dolomite is used for rocks in which the carbonate fraction is primarily magnesium ($MgCa(CO_3)^2$). The original deposition of limestone is as an unconsolidated mud. It is then altered to a consolidated rock by compaction, cementation, dissolution and replacement.

Today active deposition is taking place over the Bahamas platform, which extends more than 966 km from Florida to the island of Hispaniola. This area has developed along the subsiding continental margin of North America and consists of a series of shallow-water carbonate banks. It is the largest carbonate platform in the world and is also one of the thickest ever developed with a possible thickness of 10 km (Purdy 1963). The islands of the Bahamas contain a wide range of karst conditions in terms of size and depth. The well known and spectacular Blue Holes found in the Bahamas are deep solution conduits as much as 100 m in diameter and more than 100 m deep. Dean's Blue Hole on Long Island is the deepest known at about 200 m (Wilson 1994).

Chalk is a marine limestone consisting of fine grain carbonates. It is often not well cemented, friable, and is highly porous. Because of its high porosity and low strength large caves do not generally develop in chalk. As a result, chalk is not strong enough to give rise to typical karst landforms. Waltham et al. (2005) provide further details on chalk.

Evaporate rocks include gypsum ($CaSO_4\ 2H_2O$), or anhydrite ($CaSO_4$), and salt (NaCl). Both salt and gypsum are much more soluble than limestone. These evaporates are found in 32 of the 48 contiguous states (Johnson 2003) and over 25 % of the continental surface of the world (Ford and Williams 2007).

2.2 Post Deposition Processes

2.2.1 Dissolution of Limestone

While limestone is nearly insoluble in distilled water, groundwater will contain dissolved carbon dioxide that

R.C. Benson, L.B. Yuhr, *Site Characterization in Karst and Pseudokarst Terraines*,
DOI 10.1007/978-94-017-9924-9_2

forms a weak carbonic acid (H_2CO_3) in which limestone is slightly soluble. While some carbon dioxide is derived from the atmosphere during rainfall resulting in a weak carbonic acid, most of the carbon dioxide is derived from organics within the soil and by decaying organic material. A detailed discussion of the chemical reactions and the complex dissolution process can be found in White (1988), Ford and Williams (2007), and Palmer (2007).

The rock mass is initially formed with some degree of primary porosity as it is deposited and modified by compaction and cementation. It is then modified with cracks, joints, fractures, and bedding planes (secondary porosity) over time by a variety of stresses including tectonic stress, loading and rebound from glaciations and the constant small flexure of daily earth tides (about 2 cm) over geologic time (Fett 1992, personal communication). The resulting openings (secondary porosity) in the rock serve as preferential means for water to flow through the rock system.

When the flow of slightly acidic water is concentrated through the zones of secondary porosity, it leads to an increase in local porosity (tertiary porosity) through dissolution. Dissolution begins with laminar flow of water through small cracks. Over time the crack widens and flow becomes more turbulent and the process of dissolution accelerates. Selective zones of secondary porosity then become dominant flow paths becoming a self-perpetuating means of dissolution resulting in a network of connected conduits. These tertiary pathways can be significantly enlarged by a combination of dissolution, mechanical erosion and transport of rock. Their size can range from a few centimeters to tens of meters in width. Large volumes of rapidly moving fresh water are required to create these tertiary porosity features. The effectiveness of the acidic water is greatest when it first comes into contact with rock and is reduced as the water flows deeper and becomes saturated with dissolved calcite.

2.2.1.1 Rate of Dissolution in Limestone

The rate of dissolution (limestone removal) is a function of the chemical dissolution (or corrosion) and the mechanical erosion (corrasion) process. Both the dissolution and the mechanical erosion processes are a function of a number of factors that make the exact rates difficult to determine. Many environmental factors play an important role in the dissolution of carbonate rocks.

Dissolution rates of limestone have been measured at a number of locations throughout the world based upon the amount of surface denudation or by the chemistry of springs. The results are commonly reported in units of centimeters per thousand years (cm/ka^{-1}). While the natural dissolution of carbonate rock is highly variable, 1–3 cm/1,000 years is an average range of values (Ford and Williams 2007).

In general, the natural dissolution of limestone is a very slow geologic process, so slow that significant new cavities or enlargement of caves or subsidence cannot occur within the 100 year lifetime of most engineered structures (Sowers 1996; Waltham et al. 2005). However, dissolution and subsidence may be accelerated by man-made changes such as:

- Turbulent flow (due to pumping or reservoir leakage at dams).
- Changes or concentration of surface water runoff from roads, roofs and paved parking lots (primarily add to the erosional effects).
- Acids that have leaked from industrial facilities can result in rapid dissolution creating cavities over tens of years.

Based upon the very slow process of natural dissolution for limestone and dolomites, the subsurface voids, conduits, cavities or caves that may impact a site are already present and have taken more than 10,000–100,000 years to develop (White 1988). It is these well-developed features that become the focus of many of our site characterization projects in karst settings.

2.2.1.2 Two Models for Dissolution

The classic model for dissolution of carbonate rocks is that of surface water with elevated levels of carbonic acid that infiltrate into the rock resulting in dissolution and karst. This is referred to as epigenic karst (Fig. 2.1a). However, in the last few decades, an alternate model of karst development has become recognized. The alternate model has dissolution dominated by upward flow and gradients from deeper waters which are referred to as hypogenic karst (Fig. 2.1b). Hypogenic karst develops from the upward flow of deeper groundwater systems that are independent of recharge from the surface.

Additional models have suggested a combination of processes involving upward migration of fluids of higher temperatures, or carrying sulfuric acid in addition to carbonic acid (Palmer 2007). The concepts of hypogene karst development are summarized by Klimchouk (2009) and discussed at two symposiums:

- National Cave and Karst Symposium 1: Advances in Hypogene Karst held in 2009 and published by National Cave and Karst Research Institute (Stafford et al. 2009).
- Hypogene Speleogenesis and Karst Hydrogeology of Artesian Basins held in 2009 and published by Ukrainian Institute of Speleology and Karstology: Special Paper 1 (Klimchouk and Ford 2009).

The depth of active dissolution in the epigenic model is generally thought of as being shallow. Water will accomplish 50–80 % of its solutional work within about 10 m of the surface (Ford and Williams 2007). Milanovic (2004) suggests that the most significant dissolution from a hydrogeologic

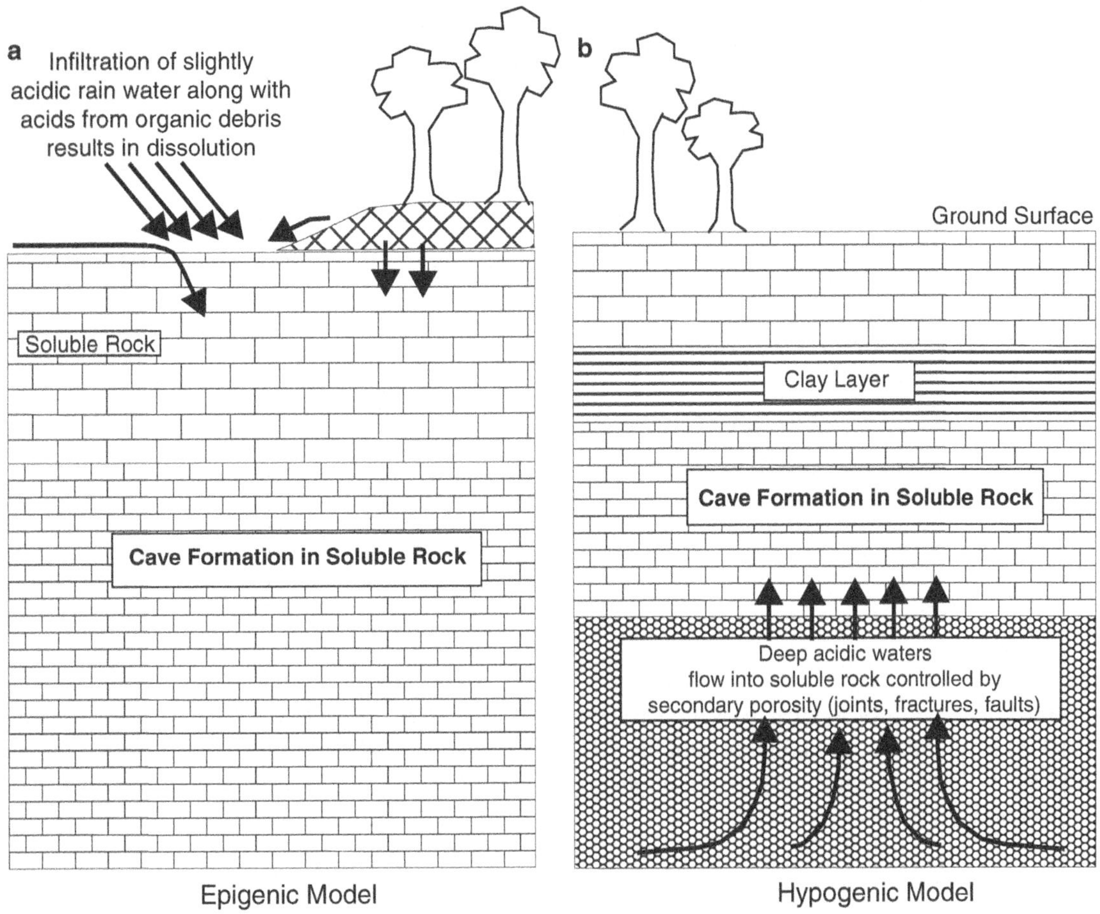

Fig. 2.1 Two models of karst development, the classic model of epigenic karst development results from the infiltration of surface waters (**a**) and a more recent model of hypogenic karst development results from the upward flow of deeper waters (**b**)

standpoint will occur in the shallow environment, generally less than 100 m deep. The depth of active dissolution in the hypogene model is much deeper. Deep-water circulation systems have been recently recognized with karst water circulation detected to depths greater than 2 km (Glazek 1989).

The Carlsbad Caverns and nearby Lechuguilla Cave in New Mexico are two of the largest cave systems which have been formed, in part, by circulating deep waters and the presence of hydrogen sulfide. The origin of the acids is from much deeper oil deposits developed by bacteria activity. The resulting hydrogen sulfide gas migrates upward into the thick overlaying limestone where it comes in contact with groundwater and produces sulfuric acid. Dissolution of the limestone by sulfuric acid is about eight times faster than that due to carbonic acid, which can account for the large cave systems encountered in these two cave systems.

With the development of hypogenic karst model, it is clear that the karst system is much more complex than originally thought. In fact, some of the existing karst systems are beginning to be reinterpreted to include both epigenic and hypogenic karst conditions (Barr et al. 2008).

Biologic microbes in certain caves have been found to produce acids that also result in dissolution of limestone. Such microbial activity may occur at much greater depths in carbonate aquifers. Palmer (2007) provides further discussion of microbial effects on karst chemistry.

2.2.1.3 The Fresh Water Lens and Mixing Dissolution

The fresh water lens is an important source of drinking water within coastal areas and on islands (Fig. 2.2). The thickness of the fresh water lens is approximated by the Gheyben-Herzberg equation, that is, for every meter of fresh water head, (above sea level) the fresh water lens is approximately 40 m thick (a 1:40 ratio due to density differences between fresh and salt water) (Palmer 2007). This can account for large amounts of fresh water on islands and coastal areas.

Dissolution is associated with fresh waters flowing from the surface (Fig. 2.2a), but when infiltrating fresh water becomes saturated with CO_2 further dissolution ceases. However, where fresh water and saline waters come in contact at the fresh water/salt water interface, further dissolution

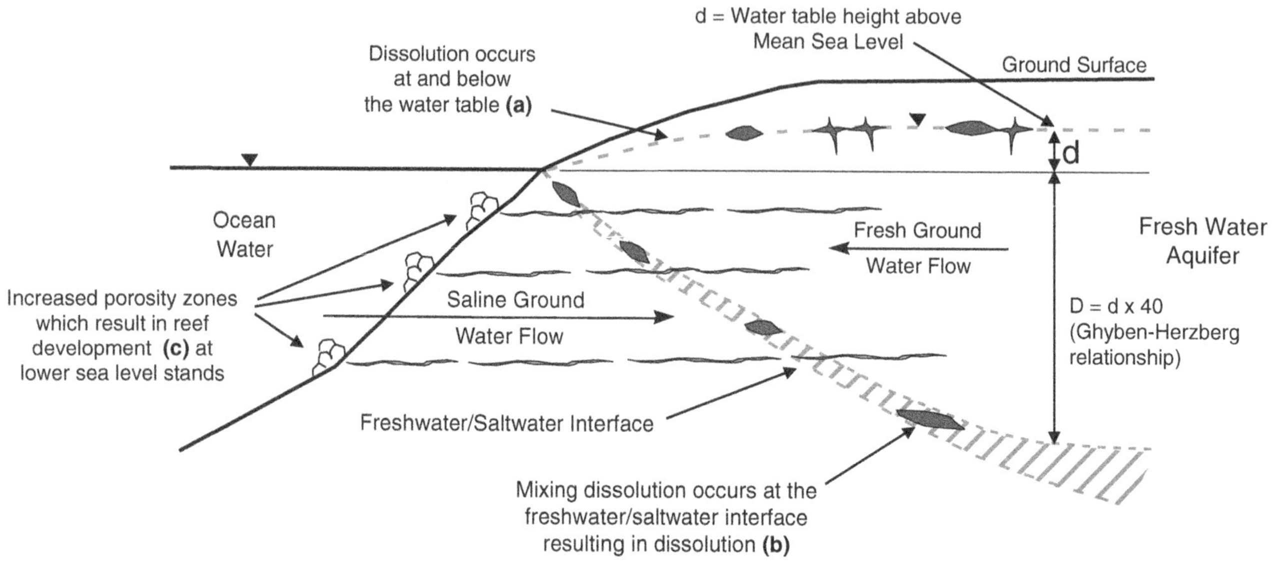

Fig. 2.2 A fresh water lens is found along the coast and on islands and its thickness is determined by the Ghyben-Herzberg relationship. Dissolution of limestone occurs near the surface of fresh water (**a**) and at the freshwater saltwater interface (**b**)

can occur even when the fresh water has become saturated (Fig. 2.2b). This is referred to as mixing dissolution. White (1988) and Palmer (2007) provide a discussion on the process by which mixing dissolution can take place.

The depth of the saltwater interface is a function of fresh water level above sea level at a 1:40 ratio. As the fresh water lens becomes thicker, the interface will become deeper, resulting in a deeper mixing dissolution. The freshwater/saltwater interface can vary from a very sharp boundary to a broad diffuse boundary depending upon site-specific conditions and can change overtime (Yuhr and Benson 1995).

2.2.2 Dissolution of Other Soluble Rocks

Evaporites, such as halite (salt) and gypsum, are much more soluble in contrast to carbonates such as limestone, dolomites and chalks. Halite is the most soluble of karst rock, about three orders of magnitude more soluble than limestone. Large cavities within salt deposits have been known to develop over a few years to tens of years associated with extraction of salt water for oil and gas drilling (Johnson 1998).

The dissolution of gypsum is approximately two orders of magnitude greater than limestone. Under high groundwater gradients such as beneath dams, dissolution rates for gypsum may be as high as 10 mm/year (Dreybrodt et al. 2002). Because gypsum dissolves faster, sinkholes and caves have been known to form within a few years where gypsum is located beneath dams (Brune 1965).

The many aspects of dissolution and cave development in other soluble rocks are covered by Klimchouk et al. (2000).

2.2.3 Dissolution of "Non-soluble Rocks"

Development of karst conditions in non-soluble rock such as sandstone and granite are unusual, but due occur (Ford and Williams 2007). Work by Alexander et al. (2005) have identified karst conditions including sinkholes and rapid groundwater flow by dye traces in the Hinckley sandstone in east central Minnesota. A number of caves in the St. Peter Sandstone occur under the streets and buildings of downtown Minneapolis, Minnesota. Some of these caves are thought to be pseudokarst, created by the piping of the poorly cemented St. Peter Sandstone into a main sanitary sewer line for the downtown area (Brick 2000).

2.2.4 Mechanical Erosion and Transport

Mechanical erosion (corrasion) and transport of rock occurs under rapid flow conditions which is common in karst. Rapidly flowing water can mechanically remove bedrock by abrasion and could exceed rates of dissolution (Palmer 2007). Under flood conditions the increased buoyancy provided by water reduces the weight of the rock and enables rapidly flowing flood waters to move large blocks of rock. As a result, the mechanical erosion and transport of rock may add to the removal of limestone by dissolution alone by an order of magnitude or more.

2.2.5 Geomorphology

Karst development occurs over a geologic time scale and can be modified over that time by the dynamics of the earth

(Fookes et al. 2005). This includes changes in sea level, by uplift of a land mass and by climate change. Recognizing these past processes will aid in understanding the karst conditions that may be present.

2.2.5.1 Changes in Sea Level

Sea levels have changed considerably over geologic time. Due to multiple glacial and interglacial stages, sea level stands have ranged from about 60 m above current sea level to at least 130 m below. Recent analysis of sea-level data from the northern Gulf of Mexico by Balsillie and Donoghue (2004) used carbon-14 (C-14) dating along with other historic data. These changes in sea levels are tied to the last period of glaciation. Their findings suggest that sea level has been as much as 130 m below present day sea level. These results from the northern Gulf of Mexico have significance beyond the local region since the data compare favorably with data from the Red Sea.

Both Florida and the Bahamas, which have not been impacted by uplift, have deeper karst development as a result of the lower sea levels (Fig. 2.3). Stalactites found in some underwater caves are evidence for a water table (sea level) that was once much lower than today.

Throughout the last episode of sea level rise there were intermediate intervals of stability. During these times, as the fresh water flowed seaward dissolution took place creating local horizontal zones of tertiary porosity (Fig. 2.2c). As the fresh water exited the sea floor, it carried nutrients that created ideal conditions for reefs to grow. Evidence for this is seen on Florida's southeast coast. As sea level rose, three main reefs are found at approximately 6, 12 and 24 m below MSL. On-shore there are horizontal zones of increased tertiary porosity found at approximately 6, 12 and 24 m below MSL, (Finkl and Esteves 1997; Esteves and Finkl 1999; Finkl and Krupa 2003). Today these horizontal zones of tertiary porosity are often encountered when pouring cement piles for foundations resulting in excessive quantities of cement (Berkowitz 1994 personal communication).

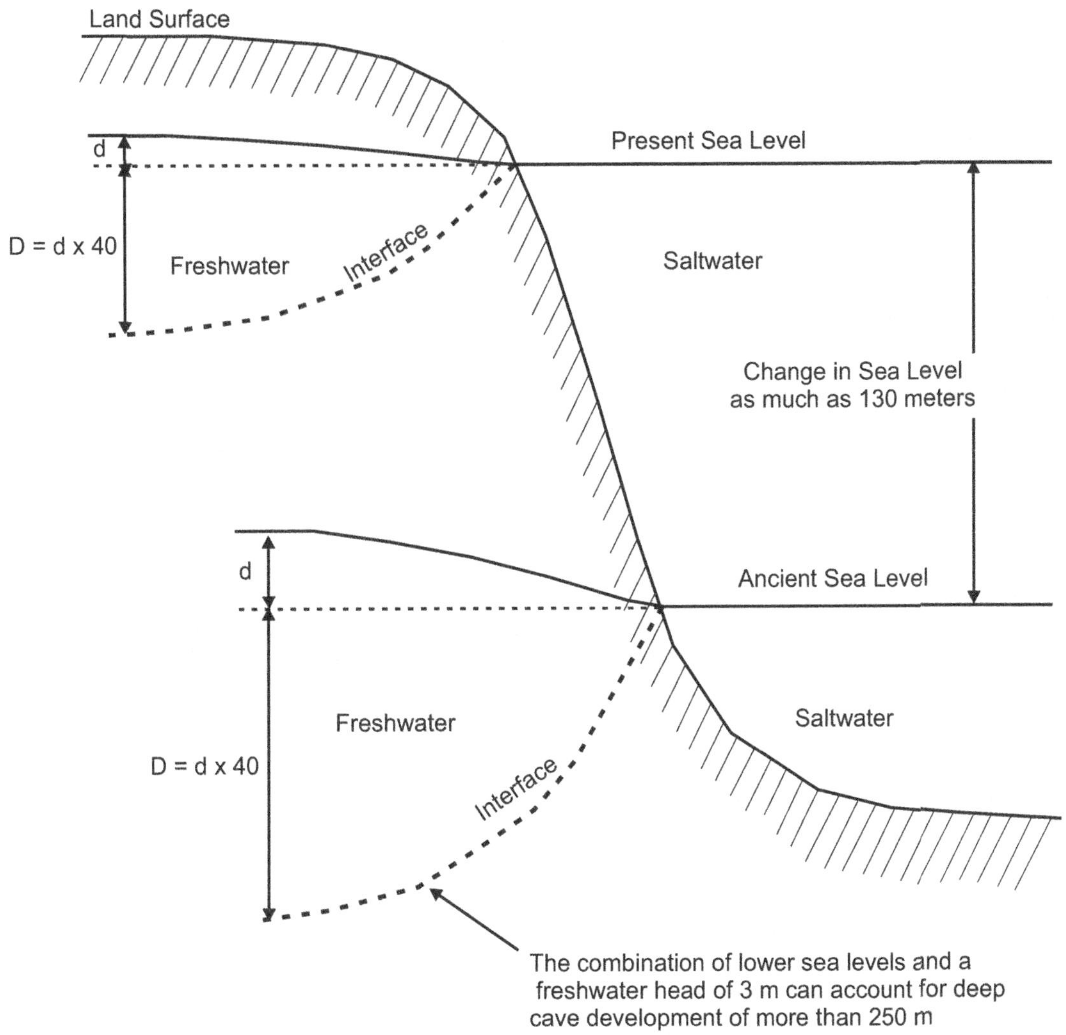

Fig. 2.3 During periods of lower sea level stands, with a thick freshwater lens, there can be extensive dissolution deep within the limestone

The combination of lower sea levels and a thick fresh water lenses (Fig. 2.3) can account for the deeper zones of dissolution in Florida and the Bahamas. A few examples include:

- Key Largo, Florida – A sinkhole found by United States Geological Survey (USGS) off of Key Largo in the reefs was more than 55 m deep with a C-14 date of 15 ka (Shinn et al. 1996);
- Tarpon springs, Florida – A paleosinkhole more than 49 m deep with a C-14 date of more than 40 ka (Yuhr et al. 2003);
- Bahamas – the deep Blue holes up to 200 m deep (Wilson 1994);
- Fort Lauderdale, Florida – Two sinkholes with diameters of 365 and 670 m were observed in the Gulf Stream approximately 9 and 13 km offshore from Fort Lauderdale at depths of approximately 250–300 m (US Coast Guard 2007).

Similar conditions of large paleosinkholes and underwater cave systems can be expected at other locations with soluble rock throughout the world.

2.2.5.2 Tectonic Uplift of Land Masses

Some land masses and islands have been uplifted by tectonic forces essentially resulting in an apparent lowering of sea level. The effects of uplift will provide greater potential energy and increased hydraulic gradients within an aquifer resulting in a lowering of the base level of karst development. For example, limestone cliffs up to 180 m high exist around the perimeter of northern Guam (Taborosi 2004). These cliffs have been uplifted and contain flank margin caves that were developed when the land was at a lower level (Fig. 2.4). Similar results can be seen on the Isla de Mona off of Puerto Rico (Mylroie 2005; Palmer 2007).

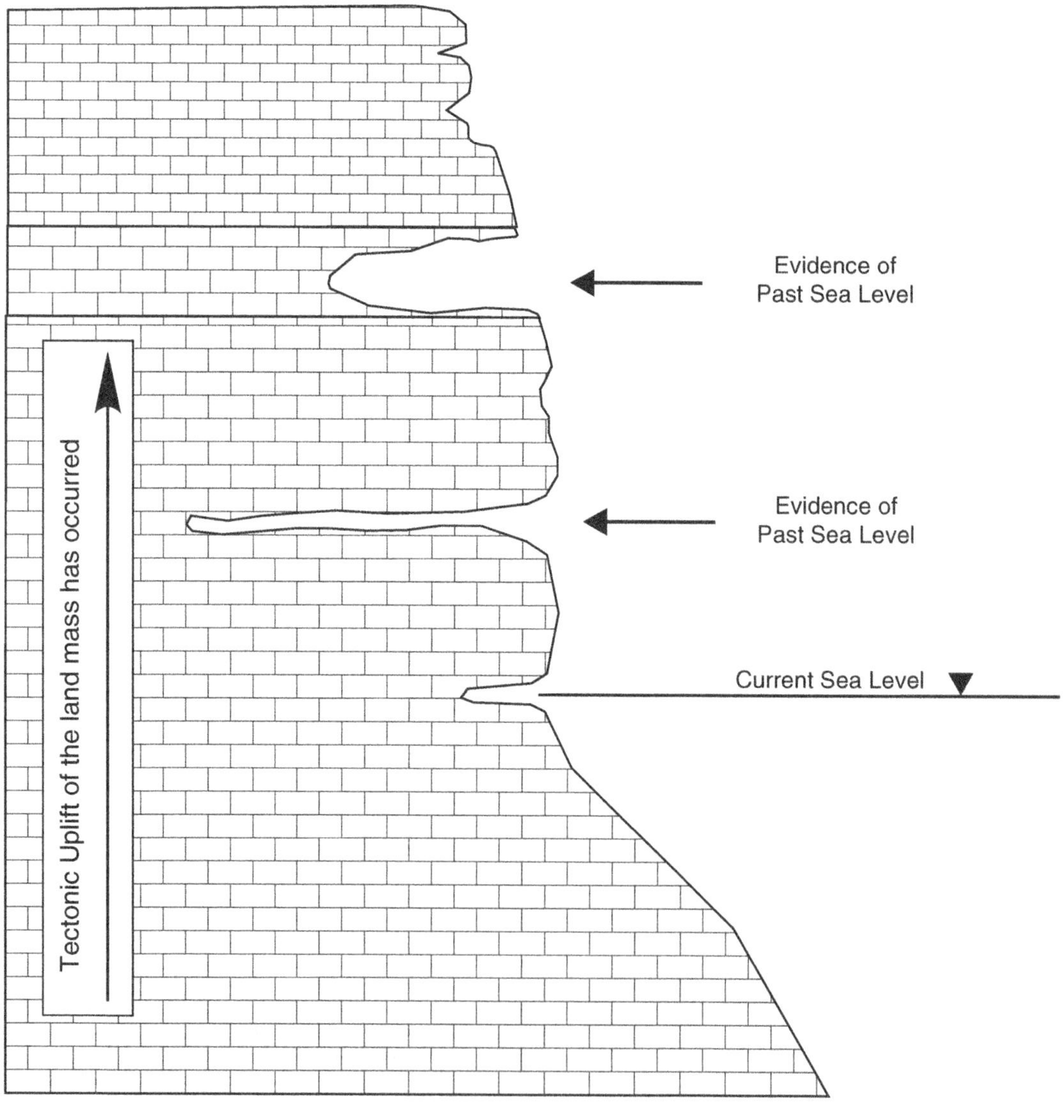

Fig. 2.4 Sea caves originally developed at sea level can been seen at higher elevations due to tectonic uplift

2.2.5.3 Climate

Regional climate has a strong influence on the development of karst landforms. The most mature karst occurs in wet tropical environments. The principal climatic factor that impacts the development of karst appears to be the mean annual runoff rate (Whitaker and Smart 1997). As the earth's climate has evolved, some areas that were once optimum for development of karst conditions are now dry deserts and mountains that contain the presence of paleokarst.

Even local changes in climate and associated vegetation are cited for the differences in hydraulic conductivity in the Bahamian limestone. The average hydraulic conductivity is two orders of magnitude greater in the northern Bahamas (with rainfall of 1,550 mm) than in the relatively more arid southern islands (with a rainfall of 810 mm). The greater rainfall to the north results in greater vegetation with an increase in soil CO_2 providing conditions for shallow dissolution. In addition, the combination of greater rainfall, increased vegetation, along with larger islands to the north result in a larger fresh water lens provides conditions for deeper bedrock dissolution at the fresh/saltwater interface (Whitaker and Smart 1997).

2.3 Some Properties of Karst Rock

The physical properties vary widely for limestones and dolomites. "Although there are general correlations between age, bulk density, compressive strength and permeability, these may vary within the scale of samples" (Braithwaite 2005). The range of limestone strength and its properties can vary significantly: As an example:

- The Bethany Falls limestone from the Kansas City area is an older (Pennsylvania Period), massive, low porosity and hard limestone (1,229 psi, 139.3 MPa) (Fig. 2.5a),

Fig. 2.5 Variations in limestone range from an older, massive strong limestone Kansas City, Kansas (**a**), a younger, highly weathered weak, almost friable limestone, in Puerto Rico (**b**) and a very young weaker, porous, Miami limestone (**c**) in south Florida

- The Aymamon Limestone from Puerto Rico is a younger (early Miocene), and at some surface exposures it is highly weathered and almost friable limestone (Fig. 2.5b), and
- The Miami Limestone from Southeast Florida is a very young (Pleistocene age), extremely porous and relatively weaker limestone (Fig. 2.5c).

Limestone strength in unconfined compression can range between 1.5 MPa, the lower limit for rock by its engineering definition, to more than 150 MPa, greater than concrete (Sowers 1996). Sowers' describes the young, porous Miami Limestone as having very different porosities above and below the water table. Above the water table porosity is about 15 % versus below the water table where circulation is the greatest may be as high as 75 % and appears sponge-like (Fig. 2.5c). Mean values for compressive strength of limestones are on the order of 100 MPa, with a few ranging up to 200 MPa. General values for limestone and dolomite strength characteristics and bearing capacity of cave roofs can be found in Deere and Miller (1966), Bell (2004), Sowers (1996), and Waltham et al. (2005).

When the primary porosity of the rock is high, such as in chalk or the Miami Limestone, the flow of groundwater will tend to be widely diffuse and tertiary porosity development will be minimized (Sowers 1996). Cave systems do not commonly form in weaker carbonate rocks and those that do form will not persist. These weaker rocks lack the strength to span large cavities and result in limited caves and surface karst features.

In strong rocks that are well cemented with low porosity, flow is restricted to fractures that then become enlarged by dissolution and erosion resulting in a cave system. Most caves form in the stronger rocks with low primary porosities. The greatest karst development is generally seen in massive thick-bedded limestone (Waltham et al. 2005).

References

Alexander EC Jr et al (2005) Dye tracing sewage lagoon discharge in a sandstone karst, Askov, Minnesota. In: Beck B (ed) Proceedings of the 10th multidisciplinary conference on sinkholes and the engineering and environmental impacts of karst, San Antonio, Texas Sept 2005, vol 144. ASCE Geo-Institute Geotechnical Special Publication, Reston, Virginia, pp 449–458

Balsillie JH, Donoghue JF (2004) High resolution sea-level history for the Gulf of Mexico since the last glacial maximum, vol 103, Report of investigation. Florida Geologic Survey, Tallahassee

Barr K et al (2008) Hypogenic karst and its implications for Minnesota hydrogeology. In: Yuhr L, Alexander Jr CE, and Beck BF (eds) Proceedings of the 11th multidisciplinary conference on sinkholes and the engineering and environmental impacts of karst, ASCE Geo-Institute, Geotechnical Special Publication No. 183:43–53

Bell FG (2004) Engineering geology and construction. Spoon Press, London

Braithwaite CJR (2005) Carbonate sediments and rocks, a manual for earth scientists and engineers. Whittles Publishing, Dunbeath, Scotland

Brick G (2000) Subterranean twin cities. University of Minnesota Press, Minnesota

Brune G (1965) Anhydrite and gypsum problems in engineering geology. Eng Geol 2(1):26–38

Deere DU, Miller RP (1966) Engineering classification and index properties for intact rock. Air Force Weapons Laboratory Technical Report AFWL-TR-65-116

Dreybrodt W et al (2002) Karstification below dam sites: a model of increasing leakage from reservoirs. Environ Geol 42(5):518–524

Esteves LS, Finkl CW (1999) Late Cenozoic depositional paleoenvironments of Southeast Florida interpreted from core logs. Rev Bras Geocienc 29(2):129–134

Finkl CW, Esteves LS (1997) A preliminary analysis of neogene deposital environments in Broward County, Florida. Fla Sci 60(2):94–103

Finkl CW, Krupa SL (2003) Environmental impacts and coastal plain activities on sandy beach systems: hazards, perception and mitigation. J Coast Res Spec Issue 35:132–150

Fookes PG et al (eds) (2005) Geomorphology for engineers. Whittles Publishing, Dunbeath, Scotland

Ford D, Williams P (2007) Karst hydrogeology and geomorphology. Wiley, New York

Glazek J (1989) Tectonic conditions for karst origin and preservation. In: Bosak P, Ford DC, Glazek J, Horacek I (eds) Paleokarst, a systematic and regional review. Elsevier Academia, Amsterdam, pp 569–574

Johnson KS (1998) Land subsidence above man-made salt dissolution cavities. In: Borchers JW (ed) Proceedings of the Poland symposium on land subsidence, Land Subsidence Case Studies and Current Research. AEG Special Pub. No. 8, Star Pub. Belmont, California, pp 385–392

Johnson KS (2003) Evaporite-karst problems in the United States. In: Johnson KS, Neal JT (eds) Evaporate karst and engineering/environmental problems in the United States, Oklahoma Geological Survey, Circular 109:1–20

Klimchouk AB (ed) (2000) Speleogenesis evolution of karst aquifers. National Speleological Society, Huntsville, Alabama

Klimchouk AB (2009) Principle features of hypogene speleogenesis. In: Klimchouk AB, Ford D (eds) Hypogene speleogenesis and karst hydrogeology of artesian basins. Ukrainian Institute of Speleology and Karstology: Special Paper 1, Simferopol, Ukraine, pp 7–16

Klimchouk AB, Ford D (eds) (2009) Hypogene speleogenesis and karst hydrogeology of artesian basins. Ukrainian Institute of Speleology and Karstology: Special Paper 1, Simferopol, Ukraine

Milanovic T (2004) Water resources engineering in karst. CRC Press, Boca Raton

Mylroie JE (2005) Coastal caves. In: David CC, White WB (eds) Encyclopedia of caves. Elsevier, New York, pp 122–125

Palmer AN (2007) Cave geology. Cave Books, an Affiliate of the Cave Research Foundation, Dayton

Purdy EG (1963) Recent calcium carbonate facies of the Great Bahamas Bank. Sediment Facies J Geol 71:472–497

Shinn EA et al (1996) A giant sediment trap in the Florida Keys. J Coast Res 12:953–959

Sowers GF (1996) Building on sinkholes, design and construction of foundations in karst terrain. ASCE Press, Reston (out of print)

Stafford KW et al (eds) (2009) National cave and karst symposium 1: advances in hypogene karst. National Cave and Karst Research Institute, Carlsbad, New Mexico

Taborosi D (2004) Field guide to caves and karst of Guam. Bess Press, Honolulu

US Coast Guard (2007) Draft environmental impact statement calypso LNG deepwater port license application, Volume I impact analysis US Coast Guard, Washington, DC

Waltham AC et al (2005) Sinkholes and subsidence, karst and cavernous rocks in engineering and construction. Springer-Praxis, Chichester, UK

Whitaker FF, Smart PL (1997) Climatic control of hydraulic conductivity of Bahamian limestones. Groundwater 35(5):859–868

White WB (1988) Geomorphology and hydrology of karst terrains. Oxford University Press, New York

Wilson WL (1994) Morphology and hydrology of the deepest know cave in the Bahamas: Dean's Blue Hole, Long Island. Bahamas J Sci 2(1):10–14

Yuhr L, Benson RC (1995) Saltwater intrusion: concepts for measurements and regional characterization for Broward County, Florida. In: Proceedings of the symposium for the application of geophysics to environmental and engineering projects, EEGS Annual Meeting, Orlando, FL

Yuhr L et al (2003) A case history of a large karst investigation. Presented at the 3rd international conference on applied geophysics, Orlando, FL

3 Types of Karst Features

Abstract

There are a wide variety of surface and subsurface features associated with various karst terrains. They occur at a wide range of scale due to processes both past and present. However, not all features will be found at all locations. In general, the types of karst features can be divided into surface landforms and subsurface features. The surface landforms include sinkholes (dolines), areas of subsidence, sinking streams, springs and cave entrances. The subsurface features are the areas of dissolution within the rock. These subsurface landforms include tertiary porosity (enlarged fracture systems), open conduits some with flowing water and large caverns as well as buried sinkholes. Some of these subsurface landforms may have already caused subsidence or collapse features at the surface. The following chapter discusses the typical karst features that may be encountered. Understanding the type of karst features that can exist in a particular geologic setting will provide insights that can be used as a basis to guide the site characterization efforts.

3.1 Sinkholes

Sinkholes are one of the most common features we think of when working in an area of karst geology. The terms doline and sinkhole are generally used interchangeably with sinkhole dominantly used in the United States. They are the visible evidence of the impact of karst and are often the topic of the evening news and newspapers. Big catastrophic collapse and loss of life are rare, however, even a small depression of a few centimeters on a high speed expressway can result in devastating vehicle accidents. Most sinkholes develop slow enough (typically over a few hours or even days) so that people have time to avoid or escape them, but on occasion they have occurred almost instantaneously. Property damage ranges from inconvenience to catastrophic. As late as the 1970s sinkholes were commonly considered "Acts of God" and their prediction and assessment were thought to be beyond the geologic and engineering capabilities of the day.

Waltham et al. (2005) have characterized six types of sinkholes (Fig. 3.1). This classification of sinkholes provides a common basis of terminology and a means of characterizing the collapse features depending upon their geologic conditions and the nature or mechanism of their collapse. There has also been a wide range of terminology used to describe the variety of sinkholes (Table 3.1). See Waltham et al. (2005) for a more complete discussion of sinkhole terminology used by others. In the United States, the term sinkhole is widely used to describe any event of subsidence or ground collapse whether truly associated with dissolution of rock or not. The term doline is used by much of the karst community. For all practical purposes the terms sinkhole and doline are the same. We will use the term sinkhole to mean an area of obvious physical surface collapse (Fig. 3.2a). The term subsidence will be used to describe an area of gentle depression (Fig. 3.2b).

Sinkholes can evolve over time. For example, a sinkhole may remain open and dry providing access into the cave system for exploration by cavers (Fig. 3.3a) or it may be filled with water limiting access to cave divers (Fig. 3.3b). The throat of the sinkhole may be closed off with sediments allowing the sinkhole to be filled with water resulting in a sinkhole lake (Fig. 3.3c). In some cases, a sinkhole lake will periodically drain and refill. Depending upon the age and location of the collapse, it could also be naturally filled by

R.C. Benson, L.B. Yuhr, *Site Characterization in Karst and Pseudokarst Terraines*,
DOI 10.1007/978-94-017-9924-9_3

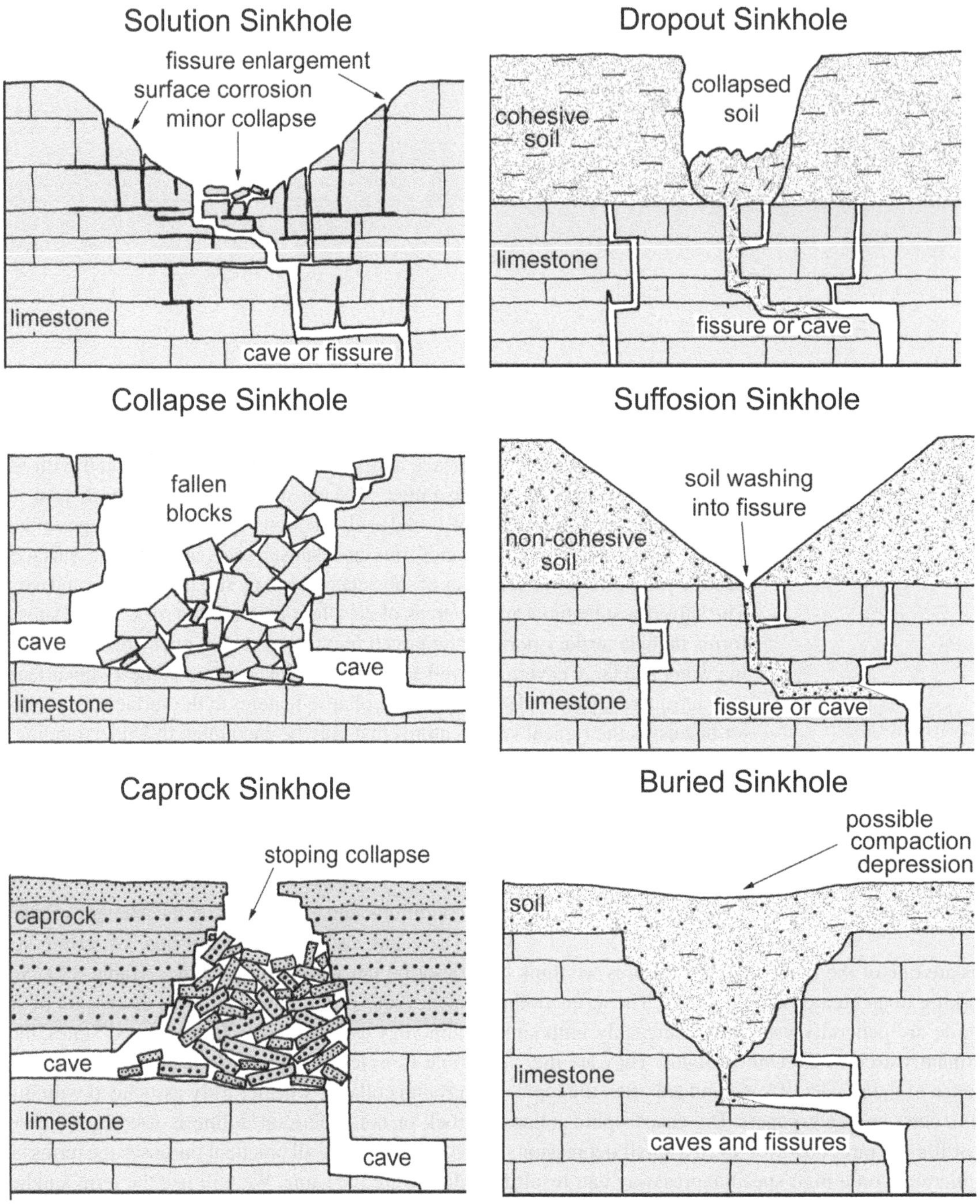

Fig. 3.1 Six types of sinkholes have been classified by Waltham and Fookes (2003) (Courtesy of T Waltham Geophoto)

sediments and buried so that there is little, if any, surface evidence remaining (Fig. 3.3d).

3.1.1 A Wide Range of Sizes

Sinkholes can range in size from almost insignificant to gigantic in both width and depth. The smaller subsidence or sinkholes are often simply a nuisance and are typically remediated using a small amount of backfill material. Figure 3.4 shows a quarry in northern Florida where the rock surface has been exposed. These numerous small diameter pits were created by dissolution due to vegetation. Similar pits are seen along canal banks in south Florida and in the easternmost Everglades. When these small local voids are present and soil cover is thin, material easily settles into the voids and small sinkholes develop. These conditions often result in small nuisance-

Table 3.1 Various terms used to describe sinkholes (After Waltham et al. 2005)

Sinkhole type (Fig. 3.1)	Other terms in use
Solution sinkhole	Dissolution sinkhole, doline, cockpit, or subsidence
Dropout sinkhole	Subsidence sinkhole, cover collapse sinkhole, alluvial sinkhole
Collapse sinkhole	Cave collapse sinkhole, cenote
Suffosion sinkhole	Subsidence sinkhole, cover subsidence sinkhole, alluvial sinkhole
Caprock sinkhole	Subjacent collapse sinkhole, interstratal collapse, breccia pipe, caprock collapse
Buried sinkhole	Filled sinkhole, compaction sinkhole, paleosinkhole

a

b

Fig. 3.2 A sinkhole is commonly thought of as an obvious collapse (**a**) while a gentile depression can be thought of as subsidence (**b**) (Photo (**a**) courtesy of T Scott, Florida Geological Survey)

type sinkholes. Figure 3.5 shows two examples of small nuisance sinkholes that are generally backfilled for remediation.

When the sediment cover becomes thicker, (6–15 m or more) the sinkholes become larger and can reach tens of meters across with losses up to 100,000 m^3 or more. Figure 3.6 shows an old large sinkhole in England. Note the person in the photo for scale reference. New sinkholes in thick soil mantle constitute the most wide spread karst hazard (Waltham et al. 2005) and become a major hazard when they occur in populated areas.

The Winter Park Sinkhole in central Florida (Fig. 3.7a) is a very large sinkhole with dimensions of 100×106 m in diameter and 30 m deep. It consumed more than 228,000 m^3 of sediments along with a house, large trees, a few vehicles, and half of an Olympic swimming pool. This is an example of a large drop-out or cover collapse sinkhole and can be seen in Fig. 3.7b.

There are also examples of much larger sinkholes. While encountering such conditions would be a very rare event, they are mentioned here only to illustrate that such extreme conditions exist.

- Investigators described "Crveno Jezero" the Red Lake located in the Croatian Coastal area as the world's largest sinkhole. It is 518 m deep from the upper rim to water levels with a diameter of 300 m at lake level. Besides its huge dimensions, the investigation of this great sinkhole is a fascinating story of perseverance, logistics, rigging, diving technology and adventure (Aspacher et al. 2000).
- Li (2004) described extremely large areas of karst in China and indicated that there are about 50 known large

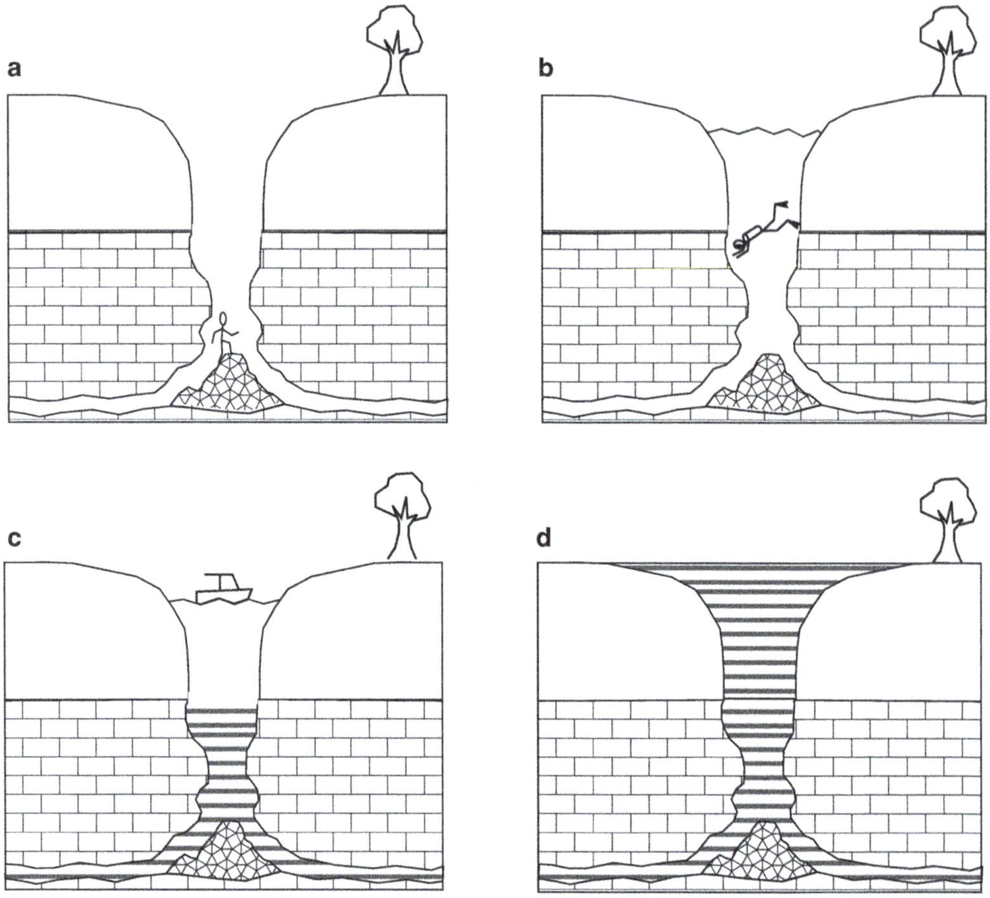

Fig. 3.3 Sinkholes can remain open or evolve over time by filling with water or sediment. (**a**) Open accessible cave (**b**) Water-filled cave accessible by cave divers (**c**) Sinkhole lake (**d**) Sediment-filled "buried sinkhole"

Fig. 3.4 The top of rock in a North Florida quarry was cleared of sediments exposing extensive small pits (Photo courtesy of B. Wisner 1972 Florida Department of Transportation)

Fig. 3.5 Examples of small sinkholes are due to small pits or isolated cavities at or near the top of shallow rock (**a**) Sinkholes along roadway in Tennessee (**b**) Sinkhole along railroad in south Florida

Fig. 3.6 A very large sinkhole in England estimated to be 100 m in diameter (note person for scale)

Fig. 3.7 The Winter Park Sinkhole in central Florida is typical of a very large sinkhole (**a**) Senior author at edge of collapse (**b**) Aerial view of collapse (Photo courtesy of T. Scott, Florida Geological Survey)

areas of karst referred to as tiankeng in China. These are giant sinkholes, three of which are more than 500 m deep and 500 m in diameter.

3.1.2 Sinkhole Densities and Linear Trends

The presence of sinkholes is a clear indicator of a karst terrain. The Mitchell Plain in southern Indiana, where local sinkhole densities are over 380 per km^2 is a classic example of a sinkhole plain. Figure 3.8 is an aerial photo over the Mitchell Plain showing the pervasiveness of sinkholes (closed depressions) throughout the area. However, sinkholes may be rare to absent in many karst areas.

The number of sinkholes present and their densities are tied to site-specific geologic conditions such as thickness of soil cover, type of underlying rock, depth to water table, the degree of dissolution or karst maturity. In many cases, where sinkholes are present they tend to line up as a linear trend or other geometric patterns. These linear trends or other geometric patterns often follow main joints that control groundwater flow resulting in preferential dissolution.

Fig. 3.8 Sinkhole densities can be quite high such as the Mitchell Plain in southern Indiana (Courtesy of T Waltham Geophoto)

For example a topographic map from central Florida (Fig. 3.9) shows extensive sinkholes and sinkhole lakes. The sinkholes form very distinct linear trends. The linear trends are associated with the subsurface geologic conditions such as preferential dissolution. There are also major linear trends associated with topographic changes that range up to 24 m. A known north-south fault runs parallel less than 1 km to the west of the main road.

These sinkhole patterns and densities may provide initial clues regarding areas of dissolution and therefore areas of higher risk. If a site is to be located within an area of higher sinkhole density or on a linear pattern of sinkholes it is not a guarantee of a major cave system at depth being present. However, the likelihood of some dissolution feature being present is much higher.

3.1.3 Sinkhole Susceptibility Maps and Databases

Sinkhole susceptibility or risk maps have been developed for some karst areas. An early sinkhole risk map of Florida (Fig. 3.10) was developed by Sinclair and Stewart (1985) and is based upon sediment cover thickness and type of sinkholes. Many county, states and regions now have GIS-based sinkhole information and maps which are readily available for use. Sinkhole susceptibility or risk maps are generally regional in nature, identifying areas of sinkhole concentration or frequency of occurrence.

The sinkhole databases provide information such as the location, date of occurrence, size, shape and depth, along with the circumstances of collapse. They provide an excellent overview of location and concentration of sinkhole occurrence and can be used for risk assessment for collapse or for vulnerability of groundwater contamination. These databases are also used to calculate new sinkholes per km^2 per year (NSH). This term was developed by Wilson (1995) as a means of characterizing the sinkhole activity within an area.

These maps and databases do have limitations such as under-reporting, non-technical observations and errors within the database itself. Estimates of under-reporting sinkhole events range from 2.5 to 22 in two different areas of Florida and from 5 to 8.5 in eastern Tennessee (Wilson 1995). The databases are also biased toward more developed areas where sinkholes are typically of greater concern and reported more frequently. As a result of these limitations such data must be used with caution. However, they are a good place to start when assessing the potential for sinkhole activity.

3.2 Sinking Streams and Springs

Sinking streams and springs represent the recharge and discharge points for groundwater flow. A spring is where groundwater discharges to the surface, while a sinking stream is where surface water enters the subsurface recharging the groundwater system. Sinking streams are also referred to as swallowholes or swallets.

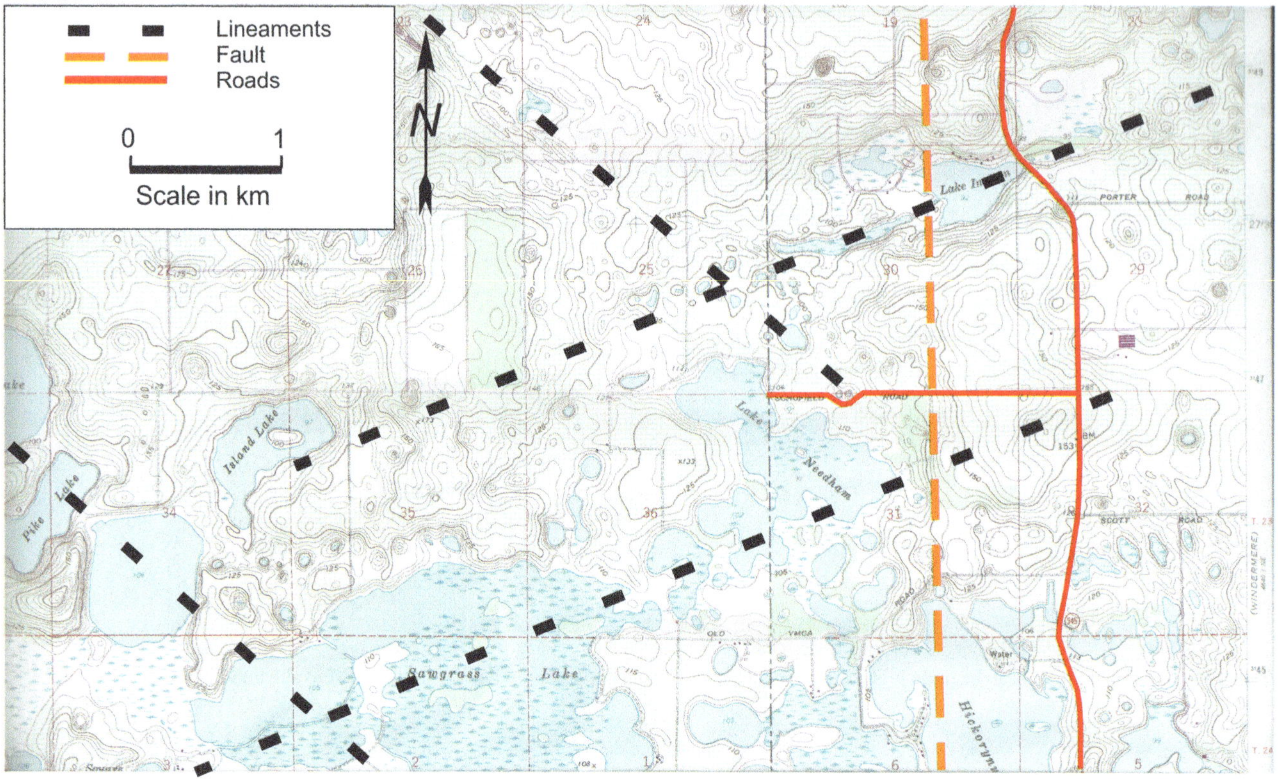

Fig. 3.9 A topographic map from central Florida reveals possible alignment of sinkholes (USGS Lake Louisa Quadrangle, 7.5 min Topographic, revised 1980)

Regional recharge occurs over a sinkhole plain with diffuse infiltration. Local recharge is commonly concentrated into topographic lows or individual depressions, collapse features, and along fractures. Figure 3.11a shows a sinking stream in southeastern Minnesota where stream flow enters a small sinkhole recharging the groundwater.

The majority of springs are located along the perimeter of the erosion base at the outer boundary of the karst basin or at the seacoast (Milanovic 2004). Figure 3.11b shows a spring discharging from a hillside in England. Springs are also known to occur underwater and off-shore at the outer boundary of a karst basin.

In a karst environment, the presence of sinks or springs is controlled by site-specific hydrologic conditions that are simply taking advantage of a flow path along a fracture, joint, conduit, or cave. As hydrologic conditions change (naturally or due to man's influence) the flow into sinks and out of springs will vary, can become intermittent and sometimes dry up all together or move to another location. Identifying and understanding the location and size of sinks and springs provides initial insight to a site's hydrology.

These hydrologic connections between surface and groundwater allow rapid transmission of waters within a karst basin. This allows karst aquifers to rapidly recharge and provide an excellent source of drinking water. This also makes the groundwater within a karst basin vulnerable to contamination. White (1988) discusses the various water resource and contaminant problems in karst.

In the US, springs are common in Alabama, Kentucky, Missouri, Tennessee, Texas, Virginia, West Virginia and Florida. Those states with a large number of springs also have significantly large areas of karst. However, it should be noted that springs are not unique to karst areas but are also found in other geologic settings.

At sea level, fresh water flows from caves and enlarged fractures into the sea. Considerable loss of fresh water occurs by flow from fractures and springs in coastal settings. Taborosi (2004) estimates that some fracture springs in Northern Guam discharge over 8 million l per day. Submarine springs have been discovered at many locations throughout the world. In ancient times, the Phoenicians drew potable water from the springs in the sea bottom (Milanovic 2004). Historically sailing ships would stop in south Florida to fill flasks with fresh water from springs within Biscayne Bay (Parks 1977).

Springs are proportional to the size of the karst water basin and are classified by their flow volume (Scott et al. 2004). Their point of discharge is further classified as vents or seeps. Vents are defined as a larger cave like opening while seeps are small openings with more diffuse discharge. They are also classified as to their location on-shore or off-shore (Scott

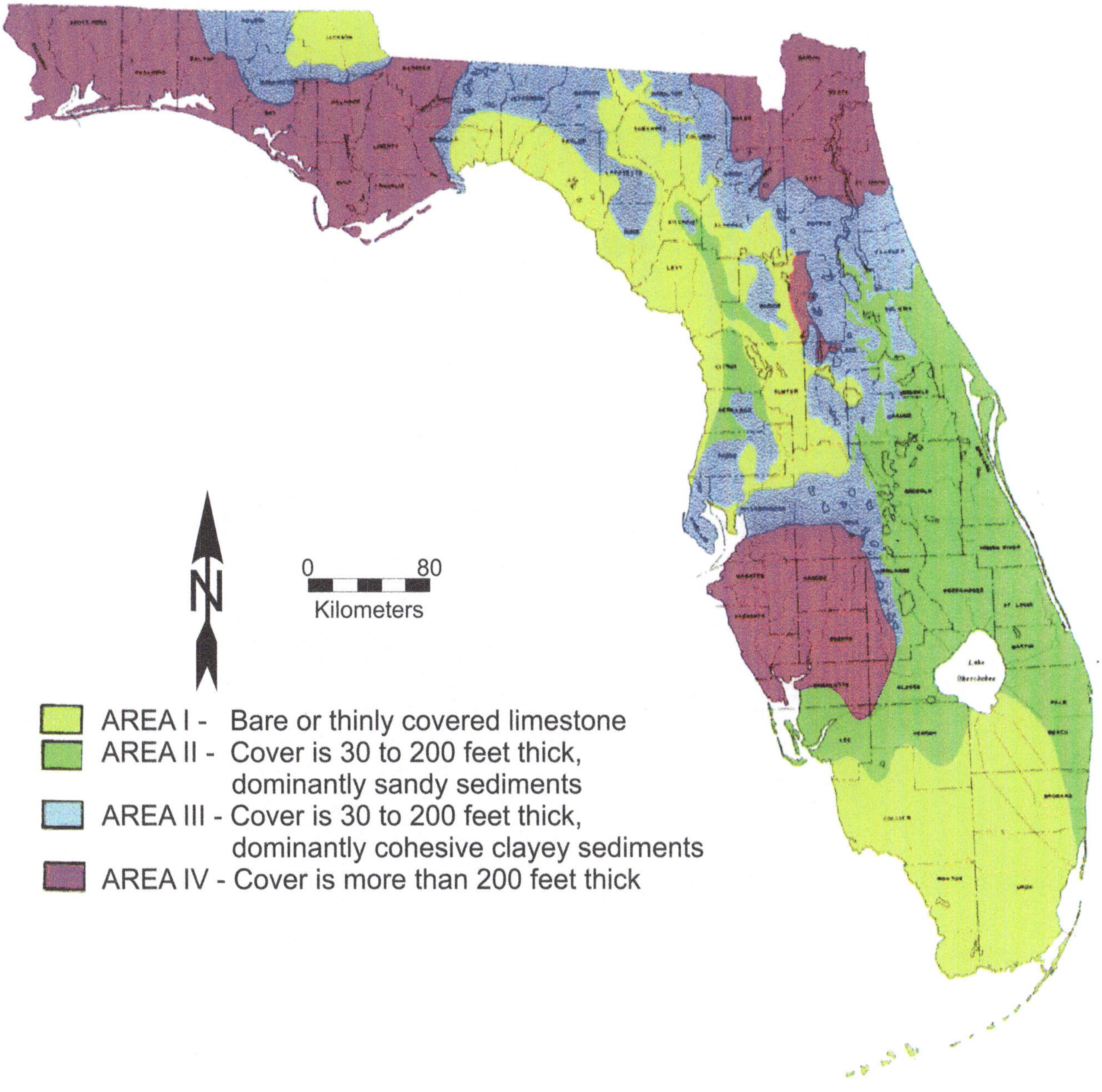

Fig. 3.10 An early sinkhole risk map was developed for the State of Florida (Sinclair and Stewart 1985)

et al. 2004). A very small flow rate would be considered a seep and classified as a magnitude 8 spring having a flow rate of less than 8 ml/s. A first magnitude spring has a very large flow rate of 2,800 l/s. First magnitude springs have been mapped in Florida, Arkansas, Missouri and even Idaho.

3.2.1 Springs in Florida

More than 700 springs have been recognized in the state of Florida, with 33 first magnitude springs, more than any other state (Scott et al. 2004). In the early years of Florida's tourism development, the springs became popular locations for recreation. Today many of Florida's springs continue to be major recreational areas and tourist attractions, including Silver Springs with its glass bottom boats, and Weeki Wachee Springs with its mermaids. Divers have explored many of the springs and sinkholes in the state making major contributions to understanding our karst system (Stamm 1994).

Numerous submarine springs exist off Florida's Atlantic and Gulf of Mexico coastlines (Scott et al. 2004). While most offshore springs today are concentrated in the northern

Fig. 3.11 A small sinking stream in southeastern Minnesota (**a**) and a moderate spring flowing from a hillside in England (**b**)

portion of the state, there is also a history of springs in southeast Florida, where sailing vessels stopped in the Miami area to replenish their fresh water supply. Historic photos have shown sailors filling barrels with fresh water from a spring in the Biscayne Bay in South Florida (Parks 1977). As a result of development in the early twentieth century, water levels were lowered which were the driving force for these springs. All of the springs in the southeastern portion of the state now have little to no flow.

In contrast, one large offshore spring still flows in the Gulf Stream 4 km off of Crescent Beach in northeastern Florida. The spring occurs at a water depth of 18 m with the vent extending 38 m below the bottom. Its discharge is confined to a localized area and flow greatly exceeds that of a first order spring. Recharge occurs from lakes and sinkholes in an area east of Gainesville Florida, (some 120 km inland from the offshore spring) at an elevation of approximately 33 m above sea level (Swarzenski and Holmes 2000).

Unfortunately some of the major spring systems in Florida are being affected by surface water runoff leading to pollution of the spring system. An increase of tenfold in nitrate concentration has occurred in 13 of the first magnitude springs of Florida (Scott et al. 2004). Extensive study of the Wakulla spring system has clearly demonstrated man's impact on the springs of Florida (Kincaid et al. 2012).

An understanding of the presence and location of springs begins to form the basis of characterizing the karst basin, groundwater flow regime and level of risk from potential contamination. While many older springs are no longer flowing, their permeable pathways remain and may impact both geotechnical projects as well as groundwater flow.

3.3 The Epikarst Zone

The term epikarst is used to describe the "skin of the karst". Engineers commonly referred this zone as the "top of rock" or "rockhead". This is the dissolutionally weathered upper portion of the bedrock where the waters are more aggressive and maximum dissolution has taken place.

Exposed rock without sediment cover can be highly dissolutioned and is referred to as karren karst. Figure 3.12 shows two examples of extreme karren karst, one is from Manitoulin Island, Canada and the other is from England. However, many of the areas of dissolutioned rock are commonly covered by sediment and referred to as the epikarst.

The epikarst can range from thin, almost non-existent to tens of meters thick or more. The top of rock conditions can range from hard (older, well consolidated and cemented (Fig. 3.13a) to highly weathered rock (Fig. 3.13b). Typically the epikarst is commonly about 10–15 m thick, and consists of highly-fractured and dissolved bedrock (Fig. 3.14). However, the depth of weathering may exceed 100 m especially in humid tropical climates (Fookes et al. 2005). The amount of rock removed by dissolution within the epikarst varies from less than 1 % to more than 50 %. The percent of the bedrock void volume that is filled with sediment within the epikarst can range from less than 5 % to more than 95 %, with the higher percent values being the most common (Aley 1997).

The epikarst zone typically has lower overall permeability than underlying portions of the bedrock and can function as a perched aquifer providing appreciable water storage. It may be separated from the saturated zone. In contrast, the epikarst zone may also provide a means to convey water laterally over large distances. Flow into the epikarst zone is more rapid than flow out of it. Discharge from the epikarst is by limited highly permeable vertical pathways transmitting water downward (Aley 1997).

Fig. 3.12 Exposed limestone with extensive dissolution along secondary fractures is referred to as limestone pavement or karren fields (**a**) Karren karst in Manatoulin Island, Canada (**b**) Extreme karren karst in England (Photo courtesy of T Waltham Geophoto)

Roadcut in northeast Alabama showing epikarst over a more massive limestone

Roadcut in Puerto Rico showing highly weathered epikarst

Fig. 3.13 Epikarst, top of rock or rockhead can range from massive rock (**a**) to highly weathered rock (**b**)

Industrial sites or roadside spills underlain by complex epikarst zones can have significant difficulties with environmental remediation. The epikarst can provide storage, lateral flow and even multidirectional flow. Therefore, a single point-source of contamination can often result in a pattern of contaminant distribution, that suggests multiple sources of contamination. In contrast, contaminants within the epikarst may also move slowly or remain trapped in pockets.

In some cases, the epikarst zone may function as an aquitard or confining layer (see case history in Chap. 27). At this site, there is a surficial aquifer consisting of sands over a deeper limestone aquifer. The epikarst separating these two aquifers is a layer of clay over the top of highly weathered rock forming an aquitard. At this site, the epikarst has been referred to as the semi-confining layer (SCL). Its presence is critical for limiting the downward migration of contaminants as well as minimizing sinkhole development.

The highly variable geologic and hydrologic conditions found within the epikarst make it an important part of any site characterization effort. Highly variable rock conditions beneath a soil cover inevitably provide greater geohazards because of conditions are obscured and difficult to define by drilling alone (Waltham and Fookes 2003). In addition, the highly variable hydrologic conditions present a complex

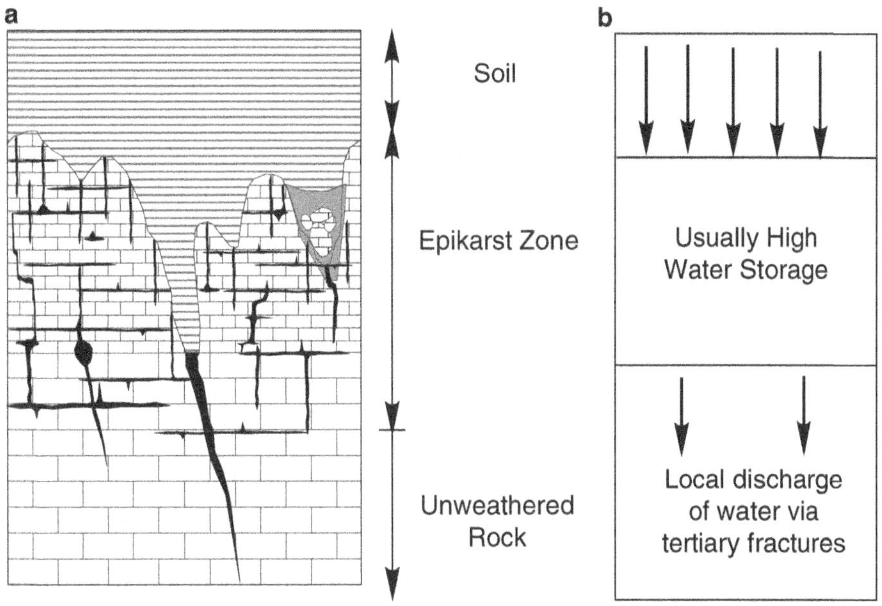

Fig. 3.14 Epikarst conditions are complex both in terms of geologic and hydrologic conditions (**a**) Geologic model of epikarst (**b**) Hydrologic model of epikarst

zone for environmental cleanup of contaminants. In some cases, the true conditions will only be known when the rock is exposed. For a further discussion of the epikarst see Jones et al. (2004).

3.4 Caves

Caves are only explored and mapped where an opening can be found to provide human access. By definition, caves are openings into rock that can be explored by humans for an appreciable distance. This implies that the cave has a minimum dimension of 0.5 m, and often much larger. It should be recognized that this size limitation for direct exploration implies that there are probably many smaller conduits and fissures that are part of the cave network that have not been identified. It is also possible that some larger portions of a cave system cannot be explored because of restricted access or fallen rock blocking a passage (Fig. 3.15). As a result, a cave map may not necessarily represent the complete cave system.

Fig. 3.15 Piles of fallen rock can nearly block off access in a cave and in some cases will completely block off access (Courtesy of T Waltham Geophoto)

Culver et al. (1999) have developed a map of caves in the US that shows the location of nearly 45,000 caves. The location, mapping and characterization of caves provide us with considerable insight to regional karst conditions. These maps can be invaluable in the assessment of local and site-specific karst conditions, if located nearby.

In the US, a great deal of information about caves can be obtained from state agencies who maintain such records and by consultation with the National Speleological Society (NSS) and their caving members and cave divers who have formed local groups called grottos. These groups are actively involved in locating, mapping, and managing caves. The Grotto locations, names and contact information can be found by contacting the NSS on-line at www.caves.org. In addition, there are cave organizations, speleological societies, throughout the world with similar missions.

If available, cave maps can provide considerable information, well beyond its location. Cave development can occur along bedding planes, fractures, or structure and can be a combination of them. The level of information that has come from cave explorations and mapping provides evidence of dissolution patterns, preferred depth of dissolution and preferred formation for dissolution. The cross section of a cave will allow us to make an assessment of minimum and maximum depths of a cave and its relation to the geologic stratigraphy. It will also reveal the location of any large chambers, if they are present.

3.4.1 Cave Geometry and Densities

The development of most caves is influenced by local geology (Palmer 2007). This includes the geologic strata (bedding planes), fractures, faults, folds and structure along with dip and strike which all play a role in cave development and its geometry. Conduit development begins along the secondary porosity zones driven by recharge and groundwater flow. Those conduits with higher flow eventually become dominant and take on various forms. Cave maps can often reveal many characteristics such as whether the cave is fracture or bedding plain controlled, which geologic formation it is in and its minimum depth of rock cover. Cave systems are discussed in detail by Palmer (2007).

For example, Fig. 3.16 is the cross section of Sorcerers Cave in southwest Texas that was mapped by Veni (1980). This cave system has horizontal development at two main levels that are controlled by nearly horizontal bedrock conditions. Figure 3.17 is a partial cave cross section from an early cave map that shows the periodic development of larger chambers in the underwater Alachua Cave System north of Gainesville, Florida. These periodic caverns within the cave have most likely developed at intersections of fractures with zones of weakened rock. One of these enlarged chambers has broken through to the surface providing the divers access. By noting the spacing of the larger cavern development, one might be able to predict where the next large cavern with a higher risk of subsidence might be located. One might also expect to find near surface indications of the presence of these caverns associated with lineaments caused by the fracture zones and vegetation associated with their recharge.

By understanding local geologic fabric or geomorphology of a cave system one can begin to make informed decisions as to the trends of a cave system beyond its mapped extent and identify what possible unique features may be present for a particular site. Invaluable as they are, cave maps only represent those areas of a cave system that are accessible and only those caves that have been explored and mapped.

The density of cave conduits has been estimated to be a relatively small portion of the total surface area, which is known to contain caves. Quinlan (1991, personal communication) suggests that the Mammoth Cave System has approximately 585 km of conduit within an area of approximately 90 km^2. Assuming an average conduit dimension of 7.5 m this implies that about 5 % of the surface has a conduit under it. Worthington et al. (2000) suggest that the cave passages of Mammoth Cave underlie between 0.36 %

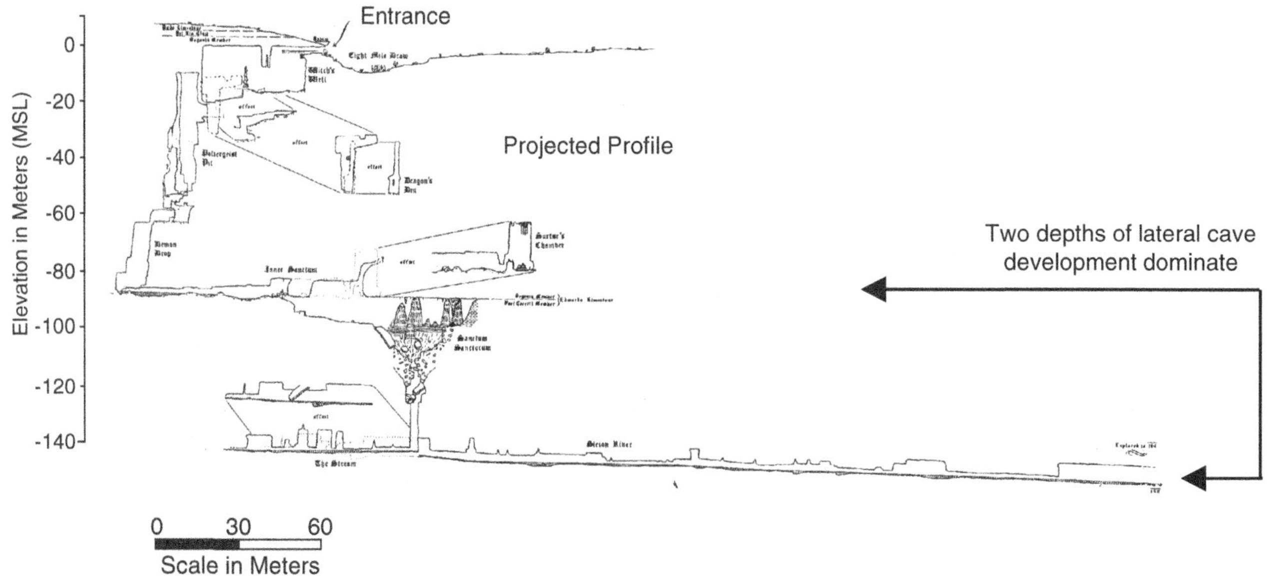

Fig. 3.16 The cave map of Sorcerer's Cave in southwest Texas indicates linear cave development at two levels. Such data provides insight as to other possible areas of development in the area (Veni 1980)

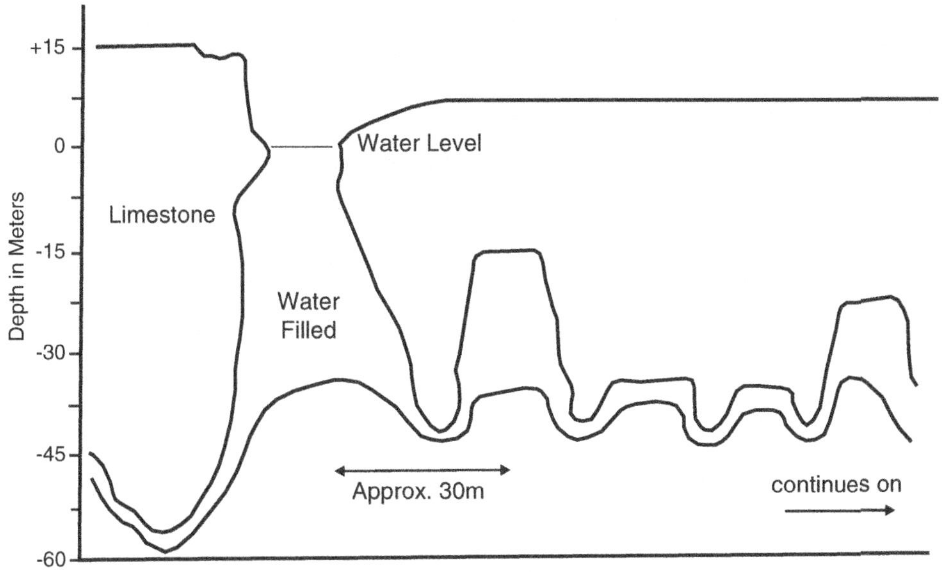

Fig. 3.17 An initial cross section map of Alachua cave developed by cave divers shows enlarged caverns occurring at intervals along the cave system (Mount 1973)

and 3.3 % of the surface area. The extremely enlarged caverns or dome areas within a cave system are limited to an even smaller percentage of area, probably something on the order of 0.1–1 % or less.

3.4.2 Large Cave Systems Develop in Thick Massive Limestone

Large cave systems develop best in massive competent, limestone (Waltham and Fookes 2003). An extensive system of solution cavities and large caverns is known to exist in the thick (over 120 m) Mississippian limestone and dolomite beds throughout the mid-continent of the US (Gentile 1984).

Finding the deepest, longest and largest caves has been the "holy grail" of those adventurers who are pushing the limits of cave knowledge. The following are some examples of extreme dimensions and depths of larger caves. While these large features are rare they are included to illustrate the possible worse case conditions for geotechnical concerns.

- The Longest Cave: The benchmark for the longest cave system has been 100 km. There are currently 14 cave systems longer than 100 km with Mammoth Cave system in Kentucky being the longest, in fact, longer than the next two combined. The Mammoth Cave System includes approximately 643 km of complex maze passageways (Gulden 2012).
- Longest Underwater Caves: As a result of the introduction of SCUBA equipment in mid-1950s, divers began to venture into springs and sinkholes. Exley (2004) provides a listing of longest underwater caves, mostly in Florida and Mexico, some lengths of more than 106 km (combined total passage length) with depths ranging from 8 to 120 m. Cave systems in the Yucatan peninsula exceed 200 km.
- The Deepest Cave: By 2007 there were several caves over 1,500 m deep, with early reports of one 2,000 m deep. The current record for the deepest cave system is Krubera (Voronja) Cave in the Western Caucasus, Republic of Georgia at about 2,191 m (Williams 2012).
- The Largest Chamber: The largest cave chamber is the Sarawak Chamber in Malaysia, which is approximately 700 m long, 400 m wide and at least 70 m high, (Dixon 2011).

3.4.3 Other Types of Caves

While the majority of caves are formed as a result of dissolution (karst), caves can be formed in a number of other ways and in a wide variety of materials. These caves include:

- Lava tube caves which form in volcanic rock as magma is deposited at the surface during an eruption and then is cooled.
- Coastal caves which include both sea caves and flank margin caves. Sea caves are formed in all types of rock and are due to mechanical erosion of a weak zone within the rock. Flank margin caves are formed at the outer edges of the fresh-saltwater interface due to dissolution (Mylroie 2005).

- Structural caves which form due to tectonic movement of rock forming a void or cave system.
- Ice caves which form in glacial areas.
- Caves which form in sandstones, quartzite and granite.

White and Culver (2005) as well as Palmer (2007) describe the various types of caves in more detail. Those caves formed by a process other than dissolution would be considered natural pseudokarst. While all caves are not necessarily formed the same way, the void space they create can potentially impact a geotechnical or environmental project in the same way.

3.4.4 Secrecy and Discretion as a Cave Management Tool

Obtaining cave locations and maps from both state and private groups is becoming more difficult because of concerns regarding damage to the cave systems as their locations become more commonly known. The issues of concern range from destruction of delicate geologic features, plundering archaeologically significant caves, (some of which contain pictographs and burial remains), destruction of unique habitats and contamination. Gookin (1997) provides a summary of some of the threats to caves.

Even as a member of the NSS, the senior author has encountered some problems with secrecy in attempting to determine the presence of caves near a low level radioactive waste site for the Department of Energy in Missouri. While the state agency provided the number of caves within a specified radius of the site they would not provide the locations and even shielded the computer printout from view. On another site in West Virginia the local cavers provide the number of caves and general location within a specified radius of the site but avoided providing any further details.

In general, the caving community is reasonably helpful in providing limited data if one clearly identifies themselves and explains the need for the data and does not become too intrusive with the inquiry. Work closely with the caving community and do not publish cave locations.

Under the Federal Cave Resources Protection Act (1988), federal land managers are required to inventory all known caves so they can be mapped and safeguarded against vandalism and exploitation. Many cavers insist that the best way to protect caves is to keep their locations secret. After years of struggling with the issue, the National Speleological Society established a federal cave management policy that encourages its members to cooperate with the federal inventory (Cave Conservationist Newsletter 1994).

References

Aley T (1997) Groundwater tracing in the epikarst. In: Beck B, Stephenson JB (eds) Proceedings of the 6th multidisciplinary conference on sinkholes and the engineering and environmental impacts of karst, the engineering geology and hydrogeology of karst terranes, Springfield, Missouri. Balkema, Rotterdam, Netherlands, pp 207–211

Aspacher B et al (2000) Beyond the blue. NSS News 58(5):13–20

Cave Conservationist (1994) Newsletter of cave conservation and management. NSS 13(1), February, 24 p.

Culver DC et al (1999) Distribution map of caves and cave animals in the United States. J Cave Karst Stud 61(3):139–140

Dixon K (2011) Measuring Sarawak Chamber. The Mulu caves project: exploring the world's largest caves. www.mulucaves.org. Accessed 21 June 2013

Exley S (2004) The taming of the slough, a comprehensive history of peacock springs. In: Poucher S (ed) National Speleological Society, Huntsville, Alabama, p 170

Fookes PG et al (eds) (2005) Geomorphology for engineers. Whittles Publishing, Dunbeath, Scotland

Gentile RJ (1984) Paleocollapse structures: longview region, Kansas City, Missouri. Bull Assoc Eng Geol 21(2):229–247

Gookin J (1997) Secrecy and discretion as a cave management tool. NSS News 55(4):114–136

Gulden B (2012) Worlds longest caves. www.caverbob.com Accessed 3 June 2006

Jones WK et al (eds) (2004) Proceedings of the 2003 epikarst symposium. Karst Waters Institute Special Publication 9

Kincaid TR et al (2012) Demonstrating interconnection between a wastewater application facility and a first magnitude spring in a karstic watershead: tracer study of the southeast farm wastewater reuse facility, Tallahassee, Florida. Florida Geologic Survey, Report of Investigation 111

Li F (2004) Introduction to karst Tiankeng in China. NSS News 62(8):220–221

Milanovic T (2004) Water resources engineering in karst. CRC Press, Boca Raton

Mount T (1973) Preliminary Alachua cave map. Consultant report

Mylroie JE (2005) Coastal caves. In: David CC, White WB (eds) Encyclopedia of caves. Elsevier, New York, pp 122–125

Palmer AN (2007) Cave geology. Cave Books, an affiliate of the cave research foundation, Dayton, Ohio

Parks AM (1977) The forgotten frontier, Florida through the lens of Ralph Middleton Monroe. Banyan Books, Miami

Scott TM et al (2004) Springs of Florida. Florida Geologic Bulletin No. 66

Sinclair WC, Stewart JW (1985) Sinkhole type, development and distribution in Florida, Florida Geologic Survey, Map Series 110

Stamm D (1994) The springs of Florida. Pineapple Press, Sarasota

Swarzenski PW, Holmes CW (2000) Re-examining the submarine spring at Crescent Beach, Florida. USGS Open File Report 00-158

Taborosi D (2004) Field guide to caves and karst of Guam. Bess Press, Honolulu

Veni G (1980) Sorcerer's cave map. In: The Texas caver, February 25(1)

Waltham AC, Fookes PG (2003) Engineering classification of karst ground conditions. Q J Eng Geol Hydrol 36:101–118

Waltham AC et al (2005) Sinkholes and subsidence, karst and cavernous rocks in engineering and construction. Springer-Praxis, Chichester, UK

White WB (1988) Geomorphology and hydrology of karst terrains. Oxford University Press, New York

White WB, Culver DC (2005) Cave definition of. In: Culver DC, White WB (eds) Encyclopedia of caves. Elsevier, New York, pp 81–84

Williams RA (2012) Braving caves. National Geographic Magazine, National Geographic Society, Washington, DC, 224(2):122–123, August

Wilson WL (1995) Sinkhole and buried sinkhole densities and new sinkhole frequencies in karst of northwest Peninsular Florida. In: Beck BF (ed) Proceedings of the 5th multidisciplinary conference on sinkholes and the engineering and environmental impacts of karst: karst geohazards, Gatlinburg, Tennessee. Balkema, Rotterdam, pp 79–91

Worthington SRH et al (2000) Matrix, fracture and channel components of storage and flow in a Paleozoic limestone aquifer. In: Sasowsky ID, Wicks CM (eds) Groundwater flow and contaminate transport in carbonate aquifers. Balkema, Rotterdam, Netherlands, pp 113–128

4 Karst Maturity and Development

Abstract
This chapter introduces the concepts of maturity and karst development, which can span millions of years. The type of karst features present at a site can provide clues as to where in the process of development it is. Understanding the maturity of a site will aid in understanding the conditions that exist and aid in assessing the degree of their stability or instability. These concepts help us understand this complex geologic system and put it into the context of development from initiation to decay.

4.1 Karst Maturity

Waltham and Fookes (2003) both engineering geologists, have proposed an engineering classification of karst that describes the karst in terms of degree of maturity, complexity and geomorphology along with specific features that may impact engineering projects. Waltham et al. (2005) have continued these ideas in their recent book on sinkholes and subsidence. They have classified karst maturity into five levels, (Table 4.1 and Fig. 4.1). The five levels of karst classification generally increase in number, size and variety of karst features from Juvenile to Extreme. This classification presents the karst as a system illustrating the range features and their development. It provides an excellent focus for one's early thinking about site conditions and also provides a means for effective communication with team members and discussions with the client/owner.

In many cases, without intimate site knowledge and experience, it may be difficult to decide upon one of the five classifications. It may be better to start by using just three levels, Juvenile, Mature, and Extreme. For example, the karst class in southern Indiana (Fig. 3.8) is most likely Juvenile or Youthful Karst. The karst class in Puerto Rico (Fig. 4.2) is most likely Mature Karst. The karst class in Guilin, China (Fig. 4.3) is Extreme Karst. These evolutionary stages represent increasing geologic complexities and concern for engineering projects.

It should also be noted that very mature karst conditions could have a low number of new sinkholes per km^2 per year (NSH) (Table 4.1). Such examples occur in Puerto Rico and the cockpit karst of Jamaica. This does not mean that mature karst is completely stable. Old, existing mature karst, sinkhole lakes or paleokarst can contain open voids that may impact engineering projects and can sometimes be reactivated.

4.2 Karst Development Time Scale

It is important to place the stages of karst development and maturity in context with time. Geologists have divided Earth's history into several eras, which are further partitioned

R.C. Benson, L.B. Yuhr, *Site Characterization in Karst and Pseudokarst Terraines*,
DOI 10.1007/978-94-017-9924-9_4

Table 4.1 Karst classification based upon maturity (Waltham and Fookes 2003)

Classification	Sinkholes	Caves	Rockhead relief	NSH (new sinkholes per km^2 per year)
Juvenile Karst	Few, barely developed	Few, 1 m wide, minimum collapse	Minimal	<0.001
Youthful Karst	Common but <10 m dia	Larger cave development beginning	Some fissures and relief	0.001–0.05
Mature Karst	Frequent, 100 m diameter	Large caves present	Fissures and relief is common	0.05–1.0
Complex Karst	Large sinkhole dominant, 1 km diameter	Large caves common	Fissures and extreme relief	0.5–2.0
Extreme Karst	Largest size and variety of features, only found in wet tropics	Large caves common with cavernous zones and collapse	Extensive fissures and extreme relief	>>1.0

into periods. The most recent periods have been further divided into epochs (Fig. 4.4). Most caves and karst date from the late Cenozoic Era (Neogene Period) and represent less than 0.1 % of the Earth's history. However, the rocks that contain caves can be of almost any age. The oldest rock that contain large solution caves are Precambrian, with ages of up to about 2.3 billion years, although the caves themselves are much younger (Palmer 2007).

A karst system is initiated and evolves by a variety of conditions over time. Karst voids and caverns are created by a combination of dissolution (corrosion), erosion (corrosion) and incasion (collapse). Figure 4.5 illustrates the major events in the life of a karst system from its initiation to decay. Initially laminar flow occurs through secondary fractures. As dissolution occurs the fractures are widened. At some critical threshold turbulent flow begins to occur leading to accelerating rates of enlargement resulting in tertiary fractures. As dissolution continues the fractures become enlarged and selective ones become dominant resulting in large conduits and in some cases larger caverns may develop.

The process proceeds from initiation to enlargement over a period of 3,000–5,000 years. Eventually degradation begins after about 10,000–100,000 years and finally decay after about 1–10 million years (White 1988; Esteban and Wilson 1993) (Fig. 4.5). This corresponds to the evolution of karst stages described by Waltham and Fookes (2003) from Juvenile, Youthful, Mature, Complex and Extreme Karst. These evolutionary stages also represent increasing concern for engineering and environmental projects.

The eventual decay and collapse of a karst system may evolve in various ways:

- After collapse, it may remain partially open as small to large isolated voids.
- Bulking of rock may occur filling the cavern with porous rubble or breccia which may become recemented.
- The system may be eroded from above resulting in a stable valley.
- Or some combination of these may occur.

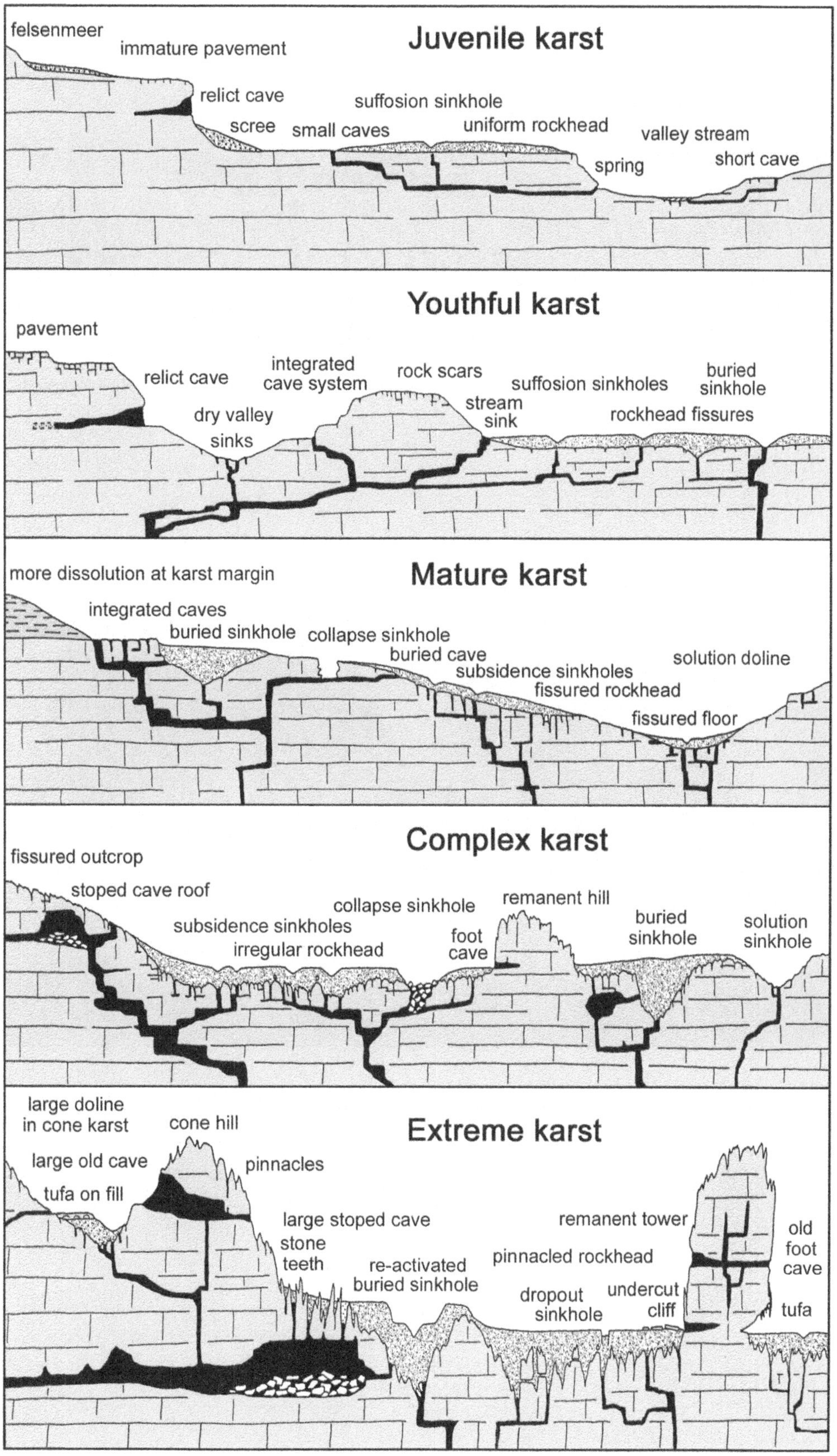

Fig. 4.1 Five conceptual models of karst maturity developed by Waltham and Fookes (2003). These engineering classifications of karst range from simple young juvenile system to a well developed extreme system (Courtesy of T Waltham Geophoto)

Fig. 4.2 The karst of Puerto Rico is likely a Mature Karst. Note two of the magotes (the term used in Puerto Rico to describe these features) have been cut away for road development

Fig. 4.3 The karst of Guilin, China is likely Extreme Karst

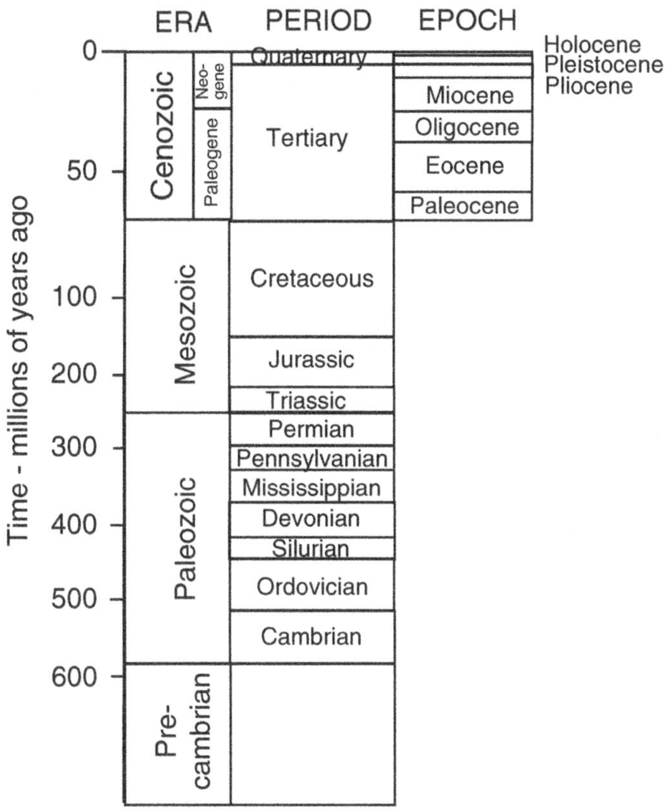

Fig. 4.4 Abbreviate geologic time table

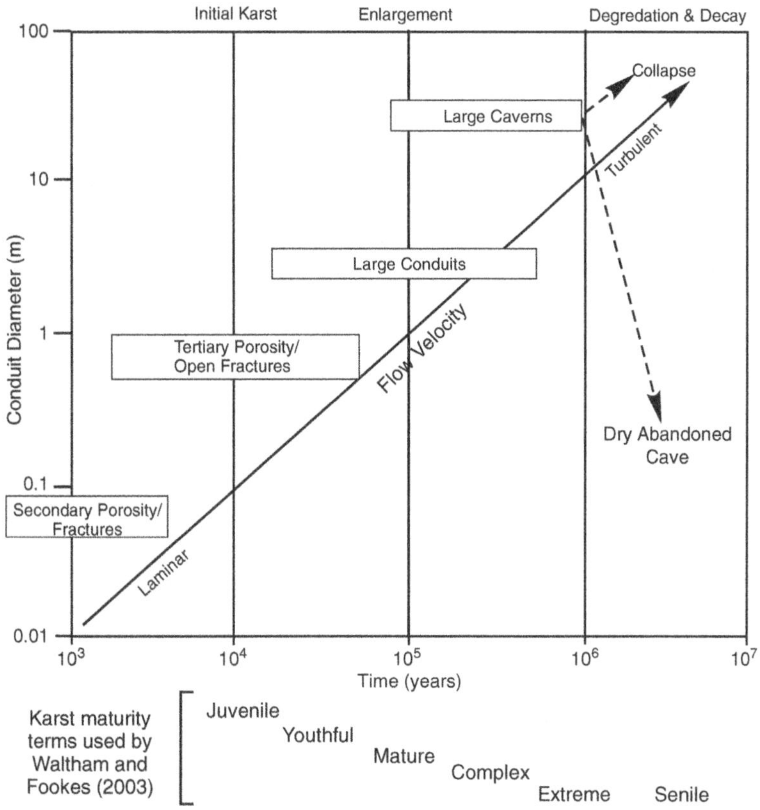

Fig. 4.5 Karst development and its evolution, compiled from information presented by White (1988), Esteban and Wilson (1993), and Waltham et al. (2005)

References

Esteban M, Wilson JL (1993) Introduction to karst systems and paleokarst reservoirs In: Fritz RD, Wilson JL, Yurewicz DA (eds) Paleokarst related hydrocarbon reservoirs, SEPM Core Workshop No. 18:1–9

Palmer AN (2007) Cave geology. Cave Books, an affiliate of the cave research foundation, Dayton

Waltham AC, Fookes PG (2003) Engineering classification of karst ground conditions. Q J Eng Geol Hydrol 36:101–118

Waltham T et al (2005) Sinkholes and subsidence, karst and cavernous rocks in engineering and construction. Springer-Praxis, Chichester

White WB (1988) Geomorphology and hydrology of karst terrains. Oxford University Press, New York

5 Areas Affected by Karst and Pseudokarst

Abstract

This chapter briefly discusses the presence of karst and pseudokarst in the US and worldwide. Soluble rocks exist at or near the surface over approximately 25 % of the US. If pseudokarst and deep conditions are included up to 75 % of the US may be affected. Similar conditions may exist worldwide. Knowing where these conditions exist will help in the planning and management for land use, groundwater resources, mineral resources and unique habitats.

5.1 United States

Davies et al. (1984) developed a map of the US showing areas affected by karst and pioneered the topic of karst and its engineering aspects during the 1960s and 1970s. Based upon his map, limestone or other soluble rock is present at or near the surface over 25 % of the United States. The most extensive karst in the United States occurs in the limestone's of Mississippian age (about 325–345 million years old) (Gentile 1984) which covers the mid-US continent.

Davies (1977) in a keynote address at an early Symposium on Detection of Subsurface Cavities at the US Army Waterways Experiment Station in Vicksburg, Mississippi estimated that:

- Limestone and other soluble rock were present at or near the surface over approximately 25 % of the United States
- Adding natural pseudokarst (such as lava caves) increases the area to about 32 %;
- When withdrawal of fluids (water and oil) and subsidence due to mining are included the area increases to about 54 %; and
- If deeper conditions to 1,000 ft and more (paleokarst) are included, about 75 % of the U.S. is affected.

The 1984 map produced by Davies et al. was updated by Veni et al. (2001) and included both karst (carbonate and evaporate) and pseudokarst. The American Geological Institute (AGI) presented this map in a document *Living with Karst*. Weary and Doctor (2014), with the USGS developed new digital maps of the United States, Puerto Rico and the US Virgin Islands. They identified areas having karst or the potential for karst and pseudokarst using GIS maps developed for various states. Their digital maps are available as downloadable files at various scales from USGS. Figure 5.1 is the 2014 digital map of karst for the contiguous 48 states. These karst maps are an invaluable resource of regional trends for use by engineers, geologists, land planners and others as a starting point for an investigation.

The mapping of karst, paleokarst and pseudokarst conditions is continually being updated. Improvements are being made as terminology is unified, map projections are standardized and more accurate, complete and detailed information is gathered.

5.2 Worldwide

Carbonate rock is widespread and karst occurs to some extent in almost every part of the world. Estimates of the amount of carbonate rocks found at or near the earth's surface range from 15 %, (Sweeting 1973) to 20 % (Ford and Williams 2007). An estimated 25 % of the world's population is supplied water from karst aquifers (Ford and Williams 2007). Ford and Williams (2007) as well as Palmer (2007) discuss the distribution of karst and caves worldwide.

R.C. Benson, L.B. Yuhr, *Site Characterization in Karst and Pseudokarst Terraines*, DOI 10.1007/978-94-017-9924-9_5

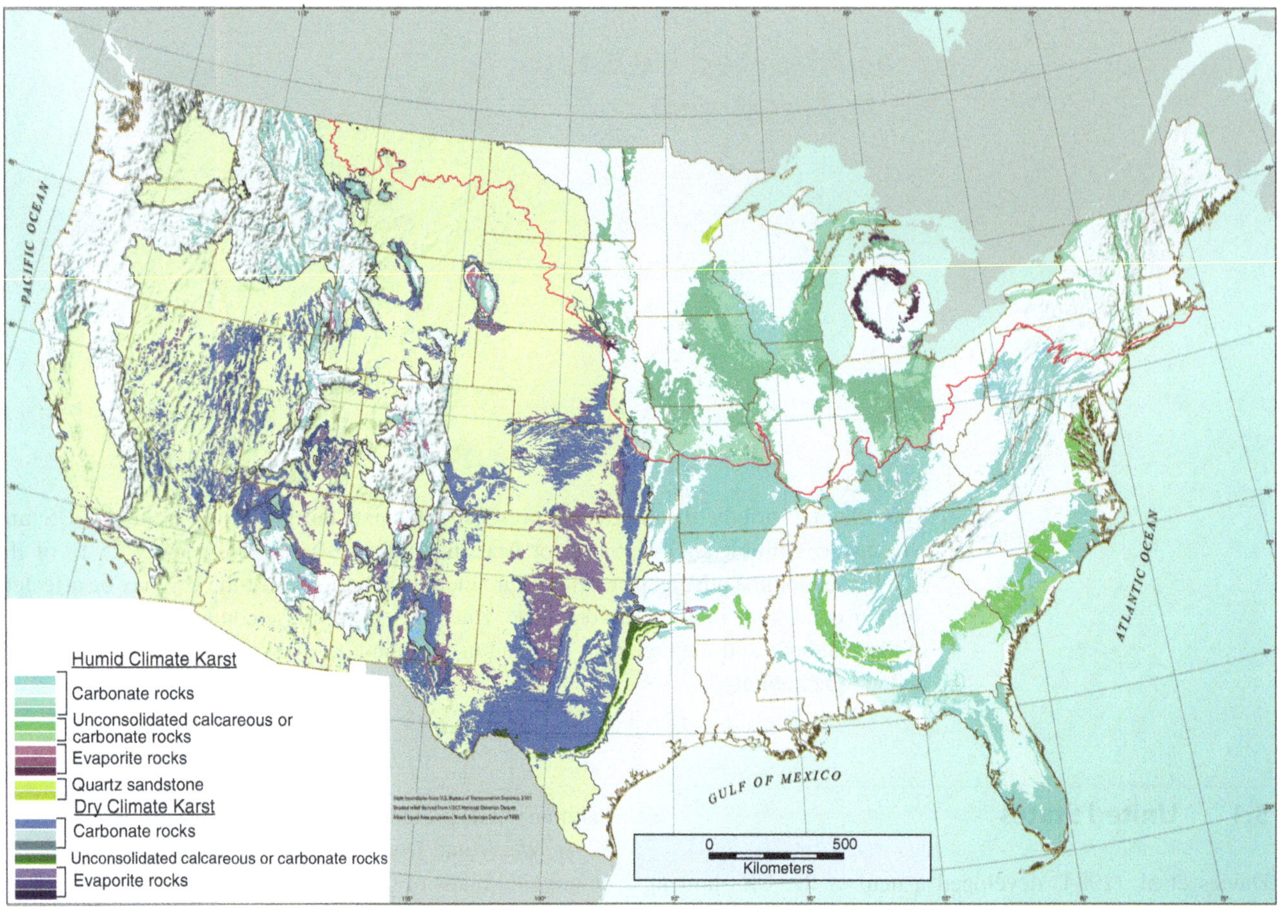

Fig. 5.1 A map of the contiguous United States shows a general representation of karst and pseudokarst areas. Although most karst is exposed at the land surface, some is buried under layers of sediment and rock (Weary and Doctor 2014)

Karst conditions are commonly thought of as occurring in a tropical or semi-tropical environment such as Mexico, Florida, Guam, the Caribbean and Central America, South Pacific, and Asia. However, we also find karst in temperate climates and even in the deserts of the middle east and Afghanistan. In desert environments the sinkholes and caves were formed under historically different climatic conditions when the area had a wetter climate. Paleokarst, occurs over much of the world between 30 and 60° latitude (Bosak et al. 1989). Gypsum, anhydrite and salt underlie more than 20–25 % of the worlds land surface. More than 90 % of evaporates, including all of the salt, is buried under other rocks (Kozary et al. 1968).

Pseudokarst caused by mans activities such as pumping of groundwater or oil and gas as well mining of coal, salt, gypsum, and other resources also extends across much of the world. The impact of mans activities results in a wide range of subsidence and collapse problems.

Efforts are being made to improve the worldwide maps of karst. One such effort was a thesis supported by The Nature Conservancy and the University of Arkansas which began efforts to develop the Karst Regions of the World (KROW) (Hollingsworth et al. 2008). This was an effort to develop and populate an interactive database as well as output maps on a continental and worldwide scale showing areas of near-surface karst (within 100 m of the surface). This work however does not include buried or deep-seated karst and focuses on surface areas of karst that contain fragile ecosystem. This effort is being continued by a team of researchers that include the USGS, the National Cave and Karst Research Institute and the Nature Conservancy as part of the University of South Florida libraries' Karst Information Portal (www.karstportal.org).

References

Bosak P et al (eds) (1989) Paleokarst a systematic and regional review, vol 1, Developments in earth surface processes. Elsevier, New York

Davies WE (1977) Keynote presentation to the symposium on detection of subsurface cavities. US Army Engineer Waterways Experiment Station, Vicksburg (Lecture recorded by Benson RC)

Davies WE et al (1984) Engineering aspects of karst, a map of the karst and pseudokarst of the United States. US Geologic Survey US Geological Survey, National Atlas of the United States of America, scale 1:7,500,00

Ford D, Williams P (2007) Karst hydrogeology and geomorphology. Wiley, New York

Gentile RJ (1984) Paleocollapse structures: longview region, Kansas City, Missouri. Bull Assoc Eng Geol 21(2):229–247

Hollingsworth E et al (2008) Karst regions of the world (KROW): global karst datasets and maps to advance the protection of karst species and habitats worldwide. In: Kuniansky E (ed) US Geological Survey Karst Interest Group Proceedings, Bowling Green, Kentucky May 27–29. USGS Scientific Investigations Report 2008–5023. Reston, Virginia

Kozary MT et al (1968) Incidence of saline deposits in geologic time. Geol Soc Am Spec Pap 88:43–57

Palmer AN (2007) Cave geology. Cave Books (An affiliate of the Cave Research Foundation), Dayton

Sweeting MM (1973) Karst landforms. Columbia University Press, New York

Veni G et al (2001) Living with karst, a fragile foundation, Awareness Series No. 4. American Geologic Institute, Alexandria

Weary DJ, Doctor DH (2014) Karst in the United States: a digital map compilation and database. USGS Open File Report 2014–1156

6 Karst and Its Many Benefits

Abstract

There are a number of benefits to karst, from its abundant source of water to its mineral resources and recreational aspects. Karst landforms provide a wide array of picturesque scenery and conditions providing many opportunities for recreational activities such as swimming, canoeing, fishing and boating. Most of us have experienced these benefits either directly or indirectly by spending recreational time at a spring; visiting a cave as a tourist attraction; utilizing groundwater from a karst spring or aquifer; benefiting from the mineral resources found in some karst terrains.

In addition, some caves and pseudokarst features such as man-made caves, mines and tunnels have all provided benefit to man for shelter or commercial use. This chapter briefly summarizes some of the benefits that have been realized in karst terraines.

6.1 Springs

Springs discharging from karst aquifers and their associated streams and rivers have attracted both animals and humans to their cool refreshing source of fresh water. The vital role of karst groundwater dates back to pre-historic times. Desert people all came to know the critical locations of fresh water. Indians and settlers located their villages at springs and along rivers that provided both food and water.

Today much of the bottled water comes from springs. One bottled-water company in Florida lists five springs as the sources of their water. Kentucky spring water is perfect for bourbon distilling because it is free of minerals that affect taste. Bourbon whiskey has been called "one of America's unique cultural contributions to the world" (Fryar 2009).

Springs have provided resources to pre-historic people and wildlife during a time when sea level was much lower. Archeology finds have been made at many springs, spring streams, and in sinkholes of Florida. Sites have been found along ledges within sinkholes and at springs and streams which are now well below sea level. These sites were occupied during lower sea levels. Little Salt Spring is located near the west coast of Florida some 96 km south of Tampa. The sinkhole has an hour glass shape which is about 73 m diameter at the surface and is as much as 67 m deep. There are ledges at approximately 16 and 26 m deep where preserved ancient artifacts dating back 8,000 to more than 12,000 years have been found. More than 10,000 years ago the water table was much lower and the ledges within the sinkhole were accessible to animals and humans (Gifford 2008 personal communication).

The springs and their rivers of Florida are used extensively for recreation including swimming, fishing, various forms of boating, canoeing and floating. In addition, many of the springs and water-filled sinkholes provide access for cave divers to explore and map these underwater cave systems (Stamm 1994).

R.C. Benson, L.B. Yuhr, *Site Characterization in Karst and Pseudokarst Terraines*,
DOI 10.1007/978-94-017-9924-9_6

6.2 Caves

Tourist caves are open to the public and are found throughout many countries. These caves provide safe access for tourists to view the wonders of the underground network of caverns and their spectacular formations of stalactites and stalagmites, large chambers, and hidden rivers.

Caves have also been used as shelters since the dawn of mankind. Many of the world's greatest archeological sites have been found in caves that have preserved evidence of prehistoric cultures including early art and records of mans early activity. The Dead Sea Scrolls were placed in a cave for safe storage and found by a Bedouin boy in 1947 (Tushingham 1958). To the Maya Indians, the cave was a realm between the supernatural and the source of life. Archeologists have discovered brightly painted clay pottery and stone alters dating back 2,000 years, evidence of cave rituals carried out by the Maya (Roberts 2004).

Natural caves have been used as places of refuge from prosecution in biblical times through World War II. During World War II a group of Jews hid and survived in a cave in the Ukraine. On May 5, 1943 38 Ukrainian Jews started a 344 day underground stay to avoid capture. The cave was called Priest's Grotto and is one of the fourth longest caves (124 km) in the world today. Cris Nicola tells this fascinating story in the NSS News (Steele 2005), where he discusses his experience with the cave and it's few living survivors.

Other uses include caves on islands in the South Pacific which were very effectively used by the Japanese to defend against attach by US forces in World War II. Today some caves continue to be used as emergency shelters during hurricanes. In 2004, residence of the Cayman Islands took shelter in the island caves during a hurricane, as they had in previous storms.

In an era prior to modern refrigeration, the brewing industry used the caves of St Louis, Missouri. In the early 1800s the beer making industry utilized these caves for cold storage. By 1860, there were some 40 breweries operating and using caves for storage (Rother and Rother 2004).

6.3 Sinkholes

Sinkholes have found many uses throughout history. The sinkholes (cenotes) of Yucatan were used as ritual places and it is believed that they were used for human sacrifices. The ancient Maya considered the sinkholes sacred entrances to the underworld (Vesilind 2003).

The Arecibo Observatory in northwest Puerto Rico is part of the National Astronomy and Ionosphere Center (NAIC). It is located in a large circular karst depression in the limestone. The depression provided a natural geometry for the 305 m reflector. Originally built in the early 1960s, with a primary goal to study the ionosphere. Today, with many upgrades, it is the site of the world's largest single unit radio telescope, which employs its reflector to listen for radio signals from possible life in space (NAIC 2014).

6.4 Karst Aquifers and Groundwater Resources

Karst aquifers supply about 25 % of the worlds drinking and irrigation water and are among the most prolific and important aquifers. Many regions including arid coastal areas and island nations of the Caribbean depend almost entirely upon water from karst aquifers (Ford and Williams 2007). The extremely high porosity and often cavernous nature of a karst aquifer allows large volumes of water to be stored underground. In some parts of the world, cave streams are large enough to economically allow damming to store water or to utilize the flow for hydroelectric power (Milanovic 2004).

6.4.1 The Edwards Aquifer

The Edwards aquifer of central Texas is one of the most prolific aquifers in North America. It occupies 1,035 km^2, ranges from 137 to 183 m thick and is the primary source of water for approximately 1.7 million who live in the region and provides much of agricultural and industrial water needs. A well drilled in 1991 is reportedly the worlds greatest flowing well with a discharge of about 95,000 l/min.

The Edwards Aquifer has an extremely high porosity and permeability, characteristic of many karst aquifers. The aquifer has been subjected to several uplifts, major faulting and karstification. Faulting and subsequent dissolution along fractures create a very heterogeneous and anisotropic permeability distribution. This karst system includes sinkholes, sinking streams, caves, springs, and an extensive system of subsurface drainage. The stratigraphic and structural features serve to control the distribution of recharge features along with water chemistry. There are extensive cave systems that support diverse ecosystems (Hovorka et al. 1996).

6.4.2 The Floridan Aquifer

The Floridan aquifer is the primary aquifer in the southeastern United States and extends under all of Florida, and portions of Alabama, Georgia, and South Carolina (Miller

1990). The Floridan aquifer in Florida yields over 950 million l/day to wells. An estimated 8 quadrillion liters of fresh water are contained within the Floridan aquifer system.

Porosity due to dissolution ranges from isolated vugs to caverns several meters across. The aquifer consists of a number of very high permeability zones, which generally conform to bedding planes, which are commonly either solution riddled or fractured. These karst conditions extend deep into the Floridan aquifer system and were formed at lower sea level stands (Miller 1990). In South Florida, the lower Floridan aquifer at depths of 760–900 m is not considered part of the freshwater system. This zone contains saline water and is used extensively for disposal of treated municipal wastes along Florida's southeast coast via injection wells.

6.5 Mineral Resources

Resources have historically been mined from caves such as flint to make tools, minerals for medicine, and paint pigments. Bat guano from caves was also used as fertilizer in the eighteenth and nineteenth centuries. During the Revolutionary and Civil wars extensive mining of saltpeter (nitrates) from caves was used in the production of gun powder.

Bauxite (an aluminum ore) occurs as a clay-like material produced by intense weathering in warm humid climates and can be found in carbonates in numerous locations throughout the world. The largest bauxite region is on the island of Jamaica where vast bauxite deposits are found within paleosinkhole pits 10–30 m deep.

Mineral deposits are often found in zones of paleokarst and breccia. Breccia zones include sinkholes and collapse where angular fragments of carbonate rocks containing lead and zinc deposits are found in a few isolated areas of the United States, Canada and Europe. The Tri-state district of Missouri, Oklahoma, and Arkansas, are one of the best known ore producers. Within the Tri-State District, the breccias zones can be up to 30 m deep, 150 m wide and several kilometers long (Sangster 1988). Some uranium deposits are also found in breccia pipes in the United States.

Many of the oil and gas fields over the world are developed in the high permeability's associated with paleokarst conditions. Hydrocarbon reserves are sometimes found within deep-seated paleokarst collapse zones, which consist of brecciation. This brecciation sometimes has an increased porosity providing a trap for hydrocarbons, (Fritz et al. 1993). Large areas of the United States are affected by paleokarst (Palmer and Palmer 1989). James and Choquette (1988) and Bosak et al. (1989) provide detailed discussions of paleokarst.

References

Bosak P et al (eds) (1989) Paleokarst a systematic and regional review, developments in earth surface processes 1. Elsevier, New York

Ford D, Williams P (2007) Karst hydrogeology and geomorphology. Wiley, New York

Fritz RD et al (1993) Paleokarst related hydrocarbon reservoirs. Soc Econ Paleontol Miner Core Work 18:1–275

Fryar AE (2009) Springs and the origin of bourbon. Ground Water 47(4):605–610 (July–August)

Hovorka SD et al (1996) Edwards aquifer ground-water resources: geologic controls on porosity development in platform carbonates, south Texas. The University of Texas at Austin, Bureau of Economic Geology, Report of Investigation 238

James NP, Choquette PW (eds) (1988) Paleokarst. Springer, New York

Milanovic T (2004) Water resources engineering in karst. CRC Press, Boca Raton

Miller JA (1990) USGS ground water atlas of the United Stated: Floridan aquifer system, Alabama, Florida, Georgia and South Carolina. USGS HA 730-G

National Astronomy and Ionosphere Center (NAIC) (2014) Arecibo observatory, www.NAIC.edu. Accessed 12 May 2014

Palmer MV, Palmer AN (1989) Paleokarst of the United States. In: Bosak P, Ford D, Glazek J, Horacek I (eds) Paleokarst, a systematic and regional review, developments in earth surface processes 1. Elsevier, New York, pp 337–363

Roberts D (2004) Descent into the Maya underworld. Natl Geogr Mag 206(5):36–53

Rother H, Rother C (2004) Lost caves of St Louis: a history of the cities forgotten caves. Virginia Publishing Co, St. Louis, 119p

Sangster DF (1988) Breccia-hosted lead zinc deposits in carbonate rocks. In: James NP, Choquette PV (eds) Paleokarst. Springer, New York, pp 102–116

Stamm D (1994) The springs of Florida. Pineapple Press, Sarasota

Steele B (2005) Spelean spotlight interview with Chris Nicola (discussing the priest's grotto). NSS News, Jan, pp 6–12

Tushingham, AD (1958) The men who hid the dead sea scrolls. Natl Geogr Mag 114(6):785–808

Vesilind PJ (2003) Maya water world. Natl Geogr Mag 204(4):82–100

7 Karst and Its Damaging Impact

Abstract

Karst and pseudokarst can impact sites in a number of ways. The problems can range from small nuisance type of subsidence or collapse that simply require routine maintenance to catastrophic collapse which have major economic impact and in some cases loss of life. The sudden collapse impacting structures (buildings, roads, etc.) are those that get everyone's attention including the attention of the media. Other impacts due to karst which are just as important include flooding and contamination issues. The following chapter highlights a variety of real-life impacts due to karst and psuedokarst related activity. These examples are not meant to be inclusive, but are meant to show a variety of impacts in terms of type of karst feature, size and type of impact.

7.1 Structural Impacts

Nuisance sinkholes are typically small (Fig. 3.5) don't necessarily have a large impact on structures or possess a high risk for injury. They can often be "managed" by simple back-filling measures. This type of sinkhole is often common in surface water retention ponds, parking lots, roads, etc.

However, the size of a sinkhole does not always correlate to its impact, but is typically related to the proximity to structures, infrastructure, and culturally developed areas. For example, a large sinkhole in the middle of a grassy field may have little to no impact. A similar size sinkhole or even a much smaller one within in a developed area may have catastrophic effects. Even a small sinkhole in the middle of a high-speed highway or industrial waste pond could have a significant impact.

The following section illustrates a number of different structural impacts that include houses, landfills, runways, bridges and dams. These examples are not meant to be inclusive but to show a range of serious impacts due to karst activity.

7.1.1 Private Residences

The impact of sinkholes on individual residential structures is quite common in Florida. Insurers, in the state of Florida, have been required to offer coverage for damages resulting from sinkholes since 1981 (Florida Statutes 2014). Sinkhole insurance claims have increased substantially in number and cost over the past decades. The majority of the claims have occurred in central Florida in counties with obvious sinkhole activity. However, even areas not subject to dramatic sinkhole activity, like southeastern Florida, the number of claims has also increased. Insurers have reported claims increased from 2,360 in 2006 to 6,694 in 2010 (Florida Senate 2011).

When there is an obvious sinkhole collapse present, such as a house about to disappear (Fig. 3.2a) or the Winter Park sinkhole (Fig. 3.7), there is little dispute regarding insurance coverage. However, if the evidence of a sinkhole is simply cracks in a walkway or the house walls (Fig. 7.1) without the obvious presence of a collapsed sinkhole there can be considerable differences of opinion as to the cause. Many factors may cause small differential settlement of a structure such as: shrink/swelling clays, post construction settlement, ero-

R.C. Benson, L.B. Yuhr, *Site Characterization in Karst and Pseudokarst Terraines*,
DOI 10.1007/978-94-017-9924-9_7

Fig. 7.1 In some cases the structure may be impacted by settlement resulting in cracks

sion, decomposition of organic materials, faulty construction, and thermal expansion/contraction of construction materials. Even the removal of trees or large shrubs can lead to swelling of clay which can go on for decades after their removal (Clayton et al. 1995). There has been an expanding industry of consultants, remediation firms, and lawyers to deal with these issues. Many other states have also developed regulations for sinkholes or mine related subsidence.

7.1.2 Surface Water Management System

The management of surface water runoff at a landfill included a concrete liner to guide waters off of and away from the landfill into a retention pond. Small leaks in the liner resulted in sinkholes developing beneath the concrete lining impacting and eventually breaching the concrete liner. Once breached, the surface water runoff was concentrated into these locations rapidly enlarging the sinkholes and extending the damage (Fig. 7.2).

7.1.3 Sinkholes on an Airport Runway

Small sinkholes at a local airport runway on Andros Island in the Bahamas became more than a nuisance. Small pits in the surface of the rock are commonly referred to as "banana holes". These small pits, typically on the order of 1 m in diameter or so, had been filled with organic debris and were underlying the asphalt runway. Once the organic debris began to decay, local collapse features occurred along the runway. Their frequency of occurrence increased after a major storm had gone through the islands. Upon landing, an aircraft encountered one of these small sinkholes. Figure 7.3 shows the damage to the propeller of the aircraft after encountering a small sinkhole.

7.1.4 Multiple Collapse at a Housing Development

A new housing development in west central Florida (Hernando County) experienced the formation of 760 sinkholes in less than 24 h over an area of approximately 8 ha (Fig. 7.4). This may be the greatest number of sinkholes to occur in a limited area within this short period of time. The sinkholes ranged from 1 m (Fig. 7.5a) to more than 30 m in diameter (Fig. 7.5b), with the majority of them smaller in size. There was no loss of life or equipment in this case and the incident occurred prior to housing construction.

7.1.5 Seepage and Collapse at Dams

Many of the early well known problems with karst are associated with the failure of dams. There are records of historic dam failures with some dating before 4000 B.C. Legget and Hatheway (1988) provide a summary of dam failures.

Fig. 7.2 Seepage of water through a concrete lining used to transfer surface water away from a landfill resulted in washing away of sediment and ultimately the collapse of the lining

Fig. 7.3 Damage to an airplane propeller that encountered a small sinkhole (banana hole) on the runway at Congo Town airport, Andros, Island Bahamas

- The first masonry dam in the world in Egypt was constructed before 4000 B.C. It failed because it was built on limestone and much of the water flowed underneath the dam.
- The first big problem that occurred in the United States in the early 1890s was a reservoir in New Mexico. The dam failed to hold water because it was built adjacent to a rather large gypsum outcrop (Davies 1977).
- The Austin dam constructed in 1893 started leaking before it was finished and a flood carried away portions of it in 1900. Legget (1962) provides further details.
- More than 54 sinkholes were documented in the reservoir area behind Anchor Dam in Wyoming, one as big as 90 m in diameter. While built, it holds only a fraction of the water that was intended for it (Davies 1977).

Fig. 7.4 A new housing development in west central Florida 760 sinkholes occurred within 24 h as a result of drilling and completing a well for irrigation purposes

In addition, the Hales Bar Dam was constructed on limestone on the Tennessee River. Thousands of cavities (>8,000) were identified in the 2,000 borings and several hundred thousand barrels of cement were injected into these cavities and fissures. Leakage at the dam had become serious a little more than 10 years after completion in 1926. After many attempts to stop the leakage, hot liquid asphalt was used with success. Unfortunately, leakage began again in 1941. TVA then became the new owner and decided on a final remediation for the dam. A continuous curtain of concrete was install across the face of the dam to stop the leaks. When this failed in 1963 the TVA decided to give up and build the Nickajack Dam downstream. The Hales Bar Dam was demolished in 1968 (Legget and Hatheway 1988).

7.1.6 Elevated Expressway Failure

A pile failure occurred during the construction of the new, elevated Tampa Bay Crosstown Expressway on the west coast of Florida in April of 2004. A single massive column supporting a three-lane road over a 30-m span, sank into the ground about 5 m over a 10 s period along with the newly constructed roadway resting on top of the column (Fig. 7.6). The other 60 piers that were in-place showed no signs of any problems.

The columns extended down into a caisson that was founded on strong limestone about 18 m below the surface. Prior to construction, the geotechnical investigation included drilling 22 m deep to test subsurface conditions (Stein et al. 2004). The fatal flaw in the limestone was obviously deeper than the test borings.

Fig. 7.5 The majority of the 760 sinkholes in Fig. 7.4 were 1–3 m in diameter (**a**) along with a few larger ones of 15–30 m in diameter (**b**)

Fig. 7.6 Damage occurred during construction of the Crosstown Expressway in Tampa, Florida when one pile slowly sunk about 6 m

Expressway officials and road builders said the collapse has as much chance of happening again "as someone getting hit by a meteor". Other comments in the press included; "A sinkhole as deep as this is undetectable", "It was just a bizarre event", "A small problem with the soil, something 1–6 m across, is easy to miss. If you try to find everyone, you could never afford the project", and "It was an act of God". These comments showed a complete lack of understanding of karst conditions, (Waltham et al. 2005). The cost of this single catastrophic event was estimated at about $100 million, but the cost of additional investigation (deeper initial borings or employing other methods) prior to construction is negligible in comparison when dealing with possible karst conditions at such critical structures.

7.1.7 Deep Unknown Paleokarst

Buried paleokarst are usually not noticed and can be a surprise in many cases. The Straights of Mackinac Bridge in Michigan was built on an area that had collapsed into the vast solution chambers of evaporates about 450 million years ago. Subsequently new rock was deposited on top of it. Piers for the bridge should have been on solid rock, but the ancient buried karst was encountered leading to problems in construction (Davies 1977). Black (2012) provides further details on this regional problem and its history.

7.1.8 Excessive Grout Quantities

It is quite common to encounter a void, cavity or open fracture in test borings and then remediate with grouting. However, on occasion the amount of grout required will greatly exceed estimates. In these instances, the understanding of subsurface conditions may not be clear. Over the years working in South Florida, stories of excessive grout were continually cropping up. A tall building at a local marina in Ft Lauderdale, Florida encountered foundation problems when borings encountered voids. Unusually large quantities of grout were needed to stabilize conditions.

During expansion of the Macarthur Causeway bridge in Miami, an excessive amount of grout was needed to fill voids encountered (Berkowitz 1994, personal communication). The need for excessive grout was most likely associated with high permeable zones developed at lower sea level stands. These permeable zones are laterally discontinuous and are found at depths of roughly 6, 12 and 24 m below mean sea level associated with older sea level stands (Finkl and Esteves 1997; Esteves and Finkl 1999).

7.1.9 Reactivation of Sinkholes and the Drainage of Sinkhole Lakes

Most sinkhole lakes are stable because the opening or throat of the sinkhole has been filled with sediments. The stability of a sinkhole lake is associated with the thickness and clay content of the sediments filling the throat of the sinkhole as well as the hydraulic head differences between the surficial and deeper aquifers. On occasion however, an old sinkhole may re-activate (or possibly a new sinkhole occurs within an existing sinkhole lake) draining much or all of the water from the lake.

An example from 1999 is that of Lake Jackson, north of Tallahassee, Florida. Lake Jackson is a closed karst lake of approximately 16 km^2 that has no surface outflows (Fig. 7.7a).

a

Lake Jackson (USGS 7.5 Minute Quadrangle Lake Jackson, Florida 1982)

b

Photo of sinkhole in bottom of Lake Jackson (photo courtesy of T. Scott, Florida Geological Survey)

Fig. 7.7 Lake Jackson, a sinkhole lake in north Florida (**a**), has periodically drained allowing geologists to investigate some of the sinkholes (**b**)

The lake had been a valuable recreational resource for northern Florida and southern Georgia, well known for its bass fishing and water sports. The triggering mechanism that reactivated the sinkholes is thought to be drought, resulting in increasing differences between the water table and the potentiometric surface (Scott 2013, personal communication).

In 1999, the lake completely drained. Once the outflow ceased, geologists and biologists found a rare opportunity to investigate the dry lake basin. State geologists entered the sinkhole to explore the system (Fig. 7.7b). State geologists have estimated that there are hundreds of buried sinks in the Lake Jackson Basin. Drilling on the dry lake bottom revealed a highly irregular, karstic surface. Historically, the lake drains into a sinkhole approximately every 25 years (Scott 2013 personal communication).

7.1.10 Problems with Man-Made Lakes

Man-made lakes are sometimes created to improve a new housing development and provide recreation. Cedar Lake was a man-made lake of about 30 ha and up to 17 m deep located 72 km west of Oklahoma City. It was designed as a recreational lake for a privately owned residential community but one night it drained. Initial investigations indicated that the lake had been saturating the underlying gypsum for about 20 years creating a crevice where the water leaked out traveling 0.4 km underground before flowing out of a canyon wall (Etter 1986). This was a major economic impact to the local community.

At a site in North Carolina, numerous small diameter sinkholes (0.3–0.6 m in diameter) were developing at the surface. These sinks were associated with voids and cavities within the upper portion of the limestone that was at a depth of about 12 m below grade. The primary triggering mechanism for these small sinkholes was the installation of a dam that created a lake to enhance property values. The lake artificially increased water levels changing hydraulic gradients and thereby initiating sinkhole activity. In addition to higher water levels, which was the primary cause of sinkhole activation, a secondary factor contributing to the rate of soil piping may have been the low frequency vibrations of the heavily loaded freight trains passing directly over the area of sinkhole development.

7.2 Groundwater Contamination

Karst aquifers are inherently susceptible to contamination. Rapid recharge via fractures or sinkholes provides a direct pathway for contaminants to enter the groundwater system. The flow of karst groundwater through conduits and fractures is often measured in kilometers per day as compared to groundwater movement through a granular (diffuse-flow) aquifer commonly measured in meters per year. Because of the rapid flow, and minimal filtration, karst aquifers are particularly vulnerable to contamination. Contaminants such as nutrients, pesticides, heavy metals, acids, organic solvents, petroleum products and fecal bacteria have been documented in karst aquifers. In addition, there is the potential for accidental spills to occur along roads and railways that pass through karst areas. Once contaminants have reached the groundwater the conduits provide a pathway for contaminants to move quickly over long distances and contaminate large areas.

In many engineering designs, surface runoff from roofs, paved parking lots and roadways are disposed of in nearby sinkholes or within retention ponds. While clearly not the best of environmental practice, it has been a very common practice in order to manage surface water. This storm water runoff is a source of non-point pollution transporting solids, heavy metals, nutrients, bacteria, road salt, herbicides, and hydrocarbons from highways. Stephenson and Beck (1995) provide a detailed discussion of the problem along with extensive references.

The continued rapid development of real estate leads to increased runoff from built up areas. As the runoff from roofs, roads, parking lots and buildings exceeds 15 % of the watershed area, water quality degradation begins to take place (Veni et al. 2001).

Both agricultural and industrial sites underlain by epikarst zones can have significant environmental impacts. Agricultural application of pesticides, fertilizers, farm wastes and their associated pathogens can easily impact the groundwater system. The waste from massive cattle, pig and chicken farms and their point recharge via closed depressions and thin patchy soil cover can have an even greater impact. Leaks and on-site waste disposal from industrial sites present similar long-term contamination issues (Coxon 2011).

Because of the multidirectional nature of water movement in the epikarst, a single point-source of contamination can result in a pattern of contaminant distribution, which may initially suggest that there are multiple sources of contamination (Aley 1997). Because of the extreme variability within the epikarst zone and deeper conduit flow, contaminant flow can be rapid and complex making the characterization, modeling and remediation much more difficult.

7.2.1 Mining Wastes

Waste impoundments are commonly utilized to manage waste streams from mining and industrial facilities. Many of these impoundments are simple excavations into the existing soils and are unlined. While these waste streams are composed of a variety of materials, those that are acidic in nature can be of great concern in karst areas. Release of acids is controlled by neutralization, however, this is often delayed or neglected. While small discharges of acids are quickly neutralized by limestone bedrock, long-term leakage can result in both geotechnical and environmental problems. Sasowsky et al. (1995) provide a discussion of acid mine drainage and its impact within carbonate geology. They found that "acid waters can persist far into the basin, particularly if contact with the limestone is minimized by clastic or precipitated coatings on the rock".

7.2.1.1 Jamaica Bauxite Wastes

Bauxite is found in abundance within the karst depressions of the western part of Jamaica. The refining of bauxite to alumina results in a waste known locally as "red mud", which is disposed of in unsealed mined out karst depressions in the limestone. The "red mud", which is more than 70 % water is highly caustic and infiltrates into the groundwater. Groundwater contaminated by "red mud", shows an increased sodium concentration and increased pH. Two areas have been affected by such contamination resulting in over 20 km^2 of aquifer being unsuited for groundwater development, (Fernandez 1991).

7.2.1.2 Phosphate Mining in Florida

Phosphate is a key ingredient in fertilizer and many other products and has been mined in central Florida since the 1880s. Phosphate deposits are found about 7.5 m below the surface and giant draglines are used to uncover and dig up the matrix, which consists of phosphate, sand and clay.

Gypsum waste slurry (silt to sand size particles) results from the manufacturing process. This waste slurry is deposited in huge diked ponds for dewatering that are referred to as gypstacks. The suspended materials settle out and form a weak porous rock-like mass. The remaining water at the surface is acidic with a pH of 1.5–2.0 (Fuleihan et al. 1997).

The gypstacks range in size from less than 40 to more than 280 ha and range in height up to 60 m above grade. More than 20 gypstacks are located within 40,400 ha in central Florida. All new gypstacks are lined at their base and have water-circulation systems to prevent the escape of waste slurry. With exception of six new stacks, all of the others are unlined (Florida Institute of Phosphate Research Library 1999 personal communication).

Gypstacks have had sinkhole failures (Sowers 1996), the most recent formed in June of 1994. A major collapse occurred at a 168 ha, 60 m high gypstack at the New Wales plant of IMC-Agrico Company. During a routine morning visual inspection, an elongated erosion feature was seen in the surface of the gypstack. That afternoon a sinkhole developed which was approximately 48 m in diameter and tapered to 32 m wide at the base of the gypstack (Fig. 7.8), (Fuleihan et al. 1997).

Six borings including four angle borings along with piezometers were used to determine the dimensions of the collapse and assess hydrologic conditions. Figure 7.9 shows a conceptual cross section of the conditions at the failed gypstack. The nearly vertical shaft tapered from 48 m diameter at the surface to about 32 m diameter at a depth of 18 m and then extended vertically to a depth of 55 m where blocks of gypsum were encountered. The sinkhole extended more than 122 m below the top of the stack. An estimated 113,000 m^3 of phosphogypsum and an undetermined amount of contaminated water disappeared down the sinkhole (Fuleihan et al. 1997).

Fig. 7.8 A large sinkhole developed in a gypstack, a waste material from the phosphate industry in Florida (Tihansky 1999)

Hayward Baker carried out remediation by installing angle borings and grouting with the focus upon sealing of the confining layer (Fig. 7.9). Over 2,890 m^3 of concrete grout were injected through 50 injection points to seal the base of the sinkhole within the confining layer. The remaining sinkhole was then filled with gypsum. Fuleihan et al. (1997) provides a complete discussion of the collapse event, the groundwater contamination and remediation.

Sinkholes are not uncommon in this area of Florida. The original conceptual model showed a narrow throat going into limestone. However, there was almost certainly a pre-existing subsurface karst feature (i.e., an enlarged well interconnected fracture or cave system) that enabled the 113,000 m^3 of materials to rapidly disappear into the underlying limestone (Fig. 7.9). The combination of the additional increased weight of the 60 m high gypsum stack, along with possible water level fluctuations and some of the dilute acid reaching the underlying limestone could all have contributed to the failure below the gypstack. Sowers (1996) described a previous gypstack failure resulting in a collapse feature within the limestone of 30 m in diameter.

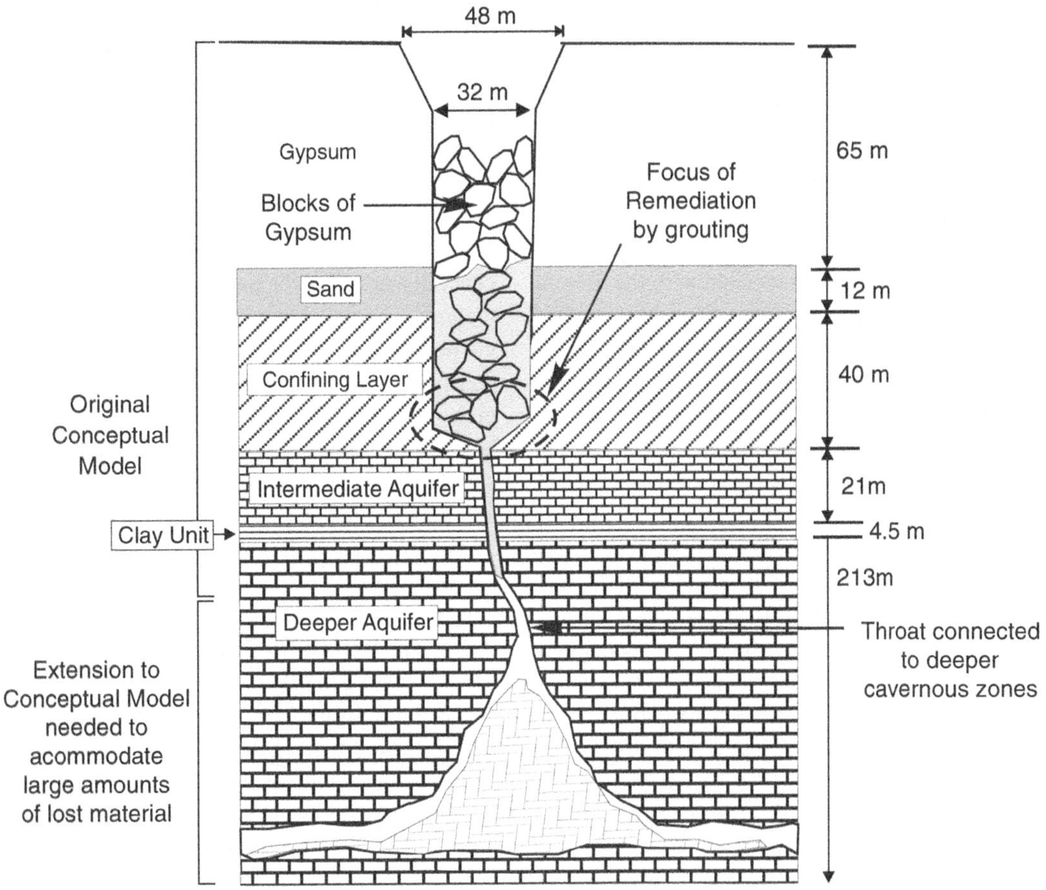

Fig. 7.9 A conceptual cross section of the gypsum stack collapse

7.2.2 Regional Contamination: The Woodville Karst Plain (WKP)

The Woodville Karst Plain (WKP), in Northern Florida, covers approximately 1,200 km^2. Wakulla Spring is the largest inland spring within the WKP and is the third largest spring in Florida. Unfortunately, the water quality has been declining within Wakulla Spring. Nitrates that come from human waste and fertilizer have been blamed for the degradation of the spring waters that includes an increase in hydrilla and algae resulting in a 30 % decrease in water clarity. In the past 30 years the spring has seen a fivefold increase in nitrogen levels (Chelette et al. 2002).

Recharge to the Floridan aquifer in the Woodville Karst Plain occurs by sinking streams, direct infiltration of precipitation through sinkholes, infiltration through the variable thick sands and soils overlaying the aquifer, and groundwater flow into the WKP from the north. An estimated 1,000 sinkholes are located within the WKP. Several of these sinkholes are known to receive water from disappearing surface streams that drain upland regions (Gerami 1984; Davis 1996). This pervasive recharge increases the vulnerability of the groundwater to contamination. Three possible sources of contamination to Wakulla Springs have been recognized: a waste water disposal spray field located south of Tallahassee, Florida; storm water runoff that rapidly recharges the Floridan aquifer through sinkholes; and discharge from private septic tanks.

The Wakulla Springs Woodville Karst Plain Symposium was held in October of 1998 (Schmidt et al. 2000). The Hydro Geo Consortium Group at the Florida Geologic Survey and the Florida State University initiated a regional assessment of the sinkholes, springs and conduit flow. Extensive studies have been completed to characterize the Woodville Karst Plain. Cave divers have made tremendous contributions to the descriptions and connections of the underwater network of caves within the WKP. The results are described by Wisenbaker and McKinlay (2006) and Wisenbaker et al. (2007). Microgravity surveys have been carried out to detect the presence of possible karst conduits not mapped by cave divers between the spray field near Tallahassee and Wakulla Springs a distance of about 16 km (Yuhr et al. 2008). Quantitative dye tracing has also been performed in the WKP in order to gain defensible estimates of the contributions of specific pathways to the discharge at Wakulla Spring and to measure hydraulic parameters for use in numeric modeling (Kincaid et al. 2012).

7.3 Pseudokarst Impacts (Natural and Man-Made)

Besides subsidence and collapse associated with the dissolution of soluble rock, there are a number of conditions which resemble or behave like karst but are not caused by natural dissolution of the rock. These karst-like features (pseudokarst) may occur due to natural processes or man-induced processes and may impact a site as dramatically as natural karst conditions. The pseudokarst that is man-induced tends to be more pervasive in developed areas.

The natural processes creating pseudokarst can occur in soil or rock. Naturally occurring processes in soils include fissures and erosional processes. Karst-like conditions can also occur within insoluble rocks such as lava, sandstone, or even granite.

Man-induced processes creating pseudokarst can be divided into small and large sizes. The small pseudokarst can range from very small, localized problems with dimensions of cm to tens of meters. The smaller scale pseudokarst includes dissolution due to acid spills or leaks, broken water or sewer lines, boring or tunneling and voids behind concrete such as tunnel liners. The large pseudokarst can be quite dramatic and range from tens of meters to regional in nature over several square km. This type of pseudokarst includes subsidence or collapse associated with petroleum activities and underground mining as well as regional subsidence due to withdrawal of groundwater, oil and gas.

7.3.1 Naturally Occurring Pseudokarst

7.3.1.1 Earth Fissures and Erosional Features

The term "earth fissure" is defined as a crack at or near the ground surface, generally occurring in unconsolidated earth material. These fissures develop below the ground and migrate towards the surface. Earth fissures have been widely observed in a number of the western United States. They go undetected before they break through to the surface and can cause problems with foundations (Shlemon 2004).

Fissures are relatively long, curvilinear pattern of discontinuous, relatively narrow, open steep-sided cracks in earth materials. Holzer (1984) suggests that fissures due to subsidence are generally long and may be on the order of 1,000 m, while desiccation fissures or giant polygons are on the order of 10 m. They are typically 1– 2 m wide and 2–3 m deep. Major mechanisms to explain earth fissuring are groundwater depletion and erosion associated with the rapid flow of surface water.

In areas with poorly consolidated sediments and little or no vegetative cover, extremely rapid weathering and mass transport may occur under flash flood conditions. Erosional processes can lead to the development of pipe-like openings within the unconsolidated material such as silt, clay and loess along with certain shales, claystones and volcanic ash and tuff. Soil piping is a major factor in the erosional process. Although the forms resemble karst landforms they are generally not large or expansive in area. These pseudokarst features may develop and change rapidly as a result of a single heavy rainstorm.

Loess usually contains a carbonate cement and frequently gives rise to small sinkholes and caves (Sweeting 1973). In loess, the erosion process may develop to the level where sinks, pipes, and caverns develop looking like classic karst and are known in China as "loess karst" (Fookes et al. 2005). These erosion features present problems to engineering works when they lie undetected just below the surface.

7.3.1.2 Swelling (or Expansive) and Shrinking Soils

Some clay soils undergo slow volume changes with a change in moisture content. A decrease in moisture causes shrinkage and fissuring while an increase in moisture causes swelling and heave. While swelling soils do not present the ultimate hazard like a large sinkhole, they can cause considerable differential settlement, cracking and damage to structures similar to the impact from early phase of natural karst subsidence or collapse.

Factors causing changes in moisture content include:

- Changes in precipitation
- Wet and dry seasons
- Poor drainage
- Leaking underground pipes and
- Tree and vegetation cover and its removal

Transpiration from vegetation cover is a major cause of water loss from soils in semi-arid regions. The removal of trees or large shrubs from a site removes their desiccating effect, therefore a subsequent increase in moisture results in swelling (Bell and Culshaw 1998). The effects from removal of vegetation can go on for decades after their removal (Clayton et al. 1995).

The depth of the active zone in which swelling and shrinkage occurs varies by season and may extend over 6 m deep in some arid regions. In temperate regions such as Britain with a damp climate, changes are typically restricted to the upper 1–1.5 m in clay soils (Bell 2004).

Swelling soils are present in much of the United States as well as the rest of the world. They impact structures resulting in cracking much like slow subsidence and early sinkhole activity. When cracks to structures occur, they can often be interpreted as a sinkhole problem. This is common in the west central area of Florida where expansive soils are present. Often an insurance claim of a sinkhole without the presence of obvious collapse can lead to disputes as to the actual cause. Is the problem due to a sinkhole, swelling soils or poor foundation conditions due to fill and debris? Bell and Culshaw (1998), Noe and Dodson (1999), Bell (2004), and Hunt (2005) provide further discussion on the subject of swelling soils.

7.3.2 Smaller Man-Induced Pseudokarst

7.3.2.1 Dissolution by Release of Acids

Significantly accelerated dissolution of limestone can occur when acids are introduced whether natural or man induced. Here the focus is on the release of acids, which can occur under a variety of conditions such as during transportation, from waste impoundments or through pipelines at industrial facilities. Whether produced by the oxidation of sulfides from mine wastes or rocks containing sulfides, the presence of sulfurous acids will dissolve limestone rapidly (Sowers 1996).

The release of acids is controlled by neutralization. If addressed in a timely manner, the impact can be minimized. In some cases, leaks can go undetected leading to the development of large cavities, subsidence or collapse as well as contamination problems. It is this long-term leakage that can have the greatest impact, creating both geotechnical and environmental problems.

It is not unusual for manufacturing facilities to develop leaks in their processing systems. In some cases, acids leaking from pipelines at commercial facilities have dramatically increased dissolution of limestone under the facility leading to the formation of large cavities and sinkholes. The following examples illustrate some of the problems encountered:

- Cavities had developed within the limestone underlying a power plant chemical laboratory due to an acid leak. Some cavities were up to 2 m in diameter with channels extending latterly as much as 3–6 m. A remedial grouting program was carried out utilizing acid resistant cement to stabilize the structure.
- Electroplating facilities utilize large vats of acids. A plant had been in operation for a number of years with acid leakage occurring. This created significant voids under the concrete floor that resulted in foundation problems for the large tanks and equipment within the building.
- One large manufacturing plant had unknowingly lost a large amount of acids (unknown volume) over the years. This resulted in dissolution of the limestone and the development of a large cavity under the building. The cavity was estimated to be more than 600 m^3 based upon microgravity measurements, borings and geophysical logging data. Extensive remediation grouting was required to stabilize the foundation.

7.3.2.2 Sinkholes Caused by Broken Water and Sewer Lines

Broken water and sewer lines are a common cause of much pseudokarst collapse in developed areas and are commonly reported in the news as "sinkholes". Voids are created by soils washing away due to the broken water or sewer lines. These voids ultimately collapse and look just like traditional sinkholes. While most problems are relatively small and localized there are occasional examples of catastrophic collapse.

In 2005, residents in Guatemala City began to complain about rumbling and shaking in their homes. Because the city is located in a seismic zone and has volcanic activity complaints were ignored until February 2007 when a large sinkhole collapse occurred. The sinkhole was 30 m diameter, and 60 m deep, five homes collapsed and three people died. Then in May of 2010 a second sinkhole collapsed (18 m in diameter and 36 m deep) including a three-story building and the disappearance of at least three people. These were both large, straight-sided collapse features due to a deep sewer system (approximately 40–50 m below ground) with a diameter of 2.25 m (Hermosilla 2012).

In Atlanta, a parking lot at a hotel developed a large sinkhole due to a leaky sewer line. Problems with the sewer line were known and a reinforced fabric liner was placed under the parking lot designed to support a 6 m diameter sinkhole allowing time for evacuation. A number of smaller sinkholes developed which the liner supported until they all coalesced. The large sinkhole, 45 × 60 m, caused two fatalities. A hotel employee was found crushed in a car at the bottom of the sinkhole and another employee was found roughly 3 km away in a combined sanitary and storm water line (Melvin 1993).

7.3.2.3 Settlement and Collapse Due to Horizontal Borings and Tunneling

Horizontal borings are commonly used for the installation of utilities and pipelines. Some of these borings have resulted in significant subsidence and sizeable sinkholes, which can occur during installation or after completion.

A subway tunnel was being dug below existing city streets in Los Angeles. The construction induced a large collapse, about 24 m wide, in the overlaying soils and breaking both water and gas lines. The collapse was stabilized with a low strength cement and construction proceeded. During subsequent modifications to the liner a second collapse occurred at the same location (Gordon and Kennedy 1995).

Unsuspected vertical settlement of up to 12 cm was observed in fractured granitic gneiss several hundred meters over the Gotthard highway tunnel in central Switzerland.

Settlement was probably related to consolidation of fractures resulting from fluid drainage into the tunnel and pore pressure changes after tunnel construction (Zangerl et al. 2003).

7.3.2.4 Voids Behind Concrete

Small voids of a few centimeters to a meter or so are commonly encountered behind many concrete structures. These voids behind concrete are typically small but can have a significant impact on structures such as highways, runways, sea walls and concrete tunnel linings. For example, hydraulic pumping of concrete slabs on highways and airport runways can lead to a thin void space under the concrete slab leaving one side of the slab lower than the other. This causes bumps in our highways and backbreaking shocks to pilots on landing when the nose wheel crosses the variations of up to 2.5 cm in the top of the concrete slabs. The 727 stretch aircraft are known as concrete breakers of runways. Delaminating of concrete in bridge decks is also a problem. A small industry has developed to carry out non-destructive testing to evaluate the presence of such voids.

7.3.3 Larger Man-Induced Pseudokarst

7.3.3.1 Regional Subsidence Due to Withdrawal of Groundwater

Pumping of groundwater has lead to significant regional subsidence in more than 14 areas of the United States (Holzer 1984). Land subsidence of one type or another has been reported in at least 37 of the 50 states. This type of subsidence is expected to increase as the demands for water supply and natural resources increase all over the world (Johnson 1998a).

Subsidence was first recognized in the Houston-Galveston area in 1918 as a result of oil field withdrawals and then subsequently pumping of groundwater covering 9,420 km^2 (Holzschuh 1991). A classic case of subsidence up to 9 m due to groundwater pumping has occurred in the San Joaquin valley of southern California encompassing 25,800 km^2 (Johnson 1998a). This type of subsidence is regional in nature but can have local impacts such as fissures and erosion or impact to infrastructure.

For example, Rogers Dry Lakebed in the Mojave Desert north of Los Angeles consists, in part, of pluvial clay deposited during a period when the area was a permanent lake. A thin layer of silt covers the clay. Below the pluvial clay lies a layer of gravel containing an extensive aquifer. Ward et al. (1995) describe the geologic setting.

The groundwater pumping from the regional valley has exceeded the estimated annual recharge almost every year since the early 1920s. After World War II, pumping expanded driven by the growth in California. Originally groundwater was under artesian flow conditions. In some areas, the water table has declined more than 30 m since the early 1950s (Londquist 1995).

A large portion of Rogers Lakebed is utilized as earthen runways at Edwards Air Force Base (EAFB). One of the runways is 11 km long. By 1990 groundwater levels at EAFB were more than 21 m below the lakebed surface. Near the southern edge of Rogers Lakebed, the land had subsided more than 0.6 m between 1961 and 1989 due to groundwater withdrawal. This contributed to the formation of surface fissures on the runways (Kratochvil 1989).

The fissures would develop below the surface hidden from view. They would eventually break through the surface with polygonal patterns (Fig. 7.10a). They would then gradually become larger and more obvious (Fig. 7.10b) eventually reaching dimensions of about a meter or so deep and a meter wide (Fig. 7.10c). Once obvious they would be excavated and repaired. Undetected, they could be a significant hazard to fighter aircraft, bombers and shuttle landings. On three occasions the shuttle's wheels broke through the desert floor, fortunately without damage (Kratochvil et al. 1992).

7.3.3.2 Pseudokarst Associated with Petroleum Industry Activity

Evaporites are present in 32 of the 48 contiguous states and underlie approximately 35–40 % of the land area. Natural evaporite karst is present in almost all areas in which evaporates are found. The most wide spread and pronounced examples of problems in both gypsum and salt are in the Permian basin of the southwestern United States (Johnson 2003).

Subsidence and collapse features resulting from petroleum activity have been well documented and are considered man-made pseudokarst. Examples include the Wink Sinks in Texas, brine well sites in or near Carlsbad New Mexico and subsidence over the Ekofisk Oil Field in Norway.

Wink Sinks

The Wink Sinks are two very large sinkholes in west Texas that occurred in 1980 and 2002 in the giant Hendrick oil field near the town of Wink, Texas. Wink Sink 1 occurred in 1980 and was 110 m in diameter and 34 m deep with an estimated volume of 159,000 m^3. Wink Sink 2 formed in 2002 around a water supply well. It was approximately 238 × 186 m wide and 61 m deep with an estimated volume of 1,330,000 m^3. In addition, broad subsidence features occurred with sags of 7–8.5 m.

The origin is believed to be from naturally occurring cavities within a salt/anhydrite (gypsum) bed about 400–655 m below the surface. Unconsolidated clastic materials overlie the soluble deposits. Natural dissolution is known to exist from deep fresh water upwelling through fractures and dissolving of the overlaying salts.

Fig. 7.10 Fissures at the Edwards Air Force Base in California extended over many of the earthen runways. A typical fissure early in its development with polygonal patterns at the surface (**a**), a second stage of development displaying open fissures (**b**), and a fully developed fissure with subsidence (**c**)

It is thought that oil production over the area subsequently aided in further dissolution and ultimately to an unstable state. The area has a large oil field including one well within the original sink. Freshwater from surficial aquifers moved down through fractures created by the borehole. Large evaporation ponds added to the downward flow of fresher water that also greatly enhanced dissolution (Johnson 1987; Johnson et al. 2003).

Brine Wells in New Mexico

Numerous brine wells are located in New Mexico. These wells are drilled into massive salt beds. Fresh water is pumped into the well to dissolve the salt. The salt water is recovered from a nearby well and is used as drilling fluid. The process can develop large cavernous zones within the salt bed. Without proper engineering control the cavern can become large enough so that it collapses resulting in large sinkholes.

A site 35 km northeast of Carlsbad, New Mexico was pumping water from a brine well when a worker on-site noted a rumbling noise and quickly left the site. Minutes later a large sinkhole abruptly formed, engulfing the brine well and structures. The initial sinkhole was several tens of meters in diameter and filled with water to within 12 m below the surface. There were large concentric fractures around its perimeter. The sinkhole grew to about 111 m in diameter and 45 m deep and became dry as the water drained from it (Fig. 7.11). An estimated 432,400 m^3 of sediment was lost. A few months later another sinkhole formed about 17 km to the northeast of the initial site (Land 2013).

Fig. 7.11 A large sinkhole north of Carlsbad New Mexico was a result of dissolution mining of salt brine to be used as drilling fluids in drilling for oil (Land 2013)

As a precaution, another brine well within the city of Carlsbad has been shut down and is being monitored. This site is at the intersection of three state highways with adjacent mobile homes, a drainage canal and a church. Signs along the roads warn of the potential danger (Fig. 7.12). Surface geophysical surveys were made to determine the lateral extent of the cavern and the site is being monitored by inclinometers installed within the boreholes with their outputs sent to critical emergency staff phones (Land 2013).

Subsidence Over the Ekofisk Oil Field

The Ekofisk Oil Field located in the North Sea off of Norway withdraws oil from a chalk reservoir. Subsidence due to withdrawal of oil was first recognized in 1984 and was 2.6 m maximum. The subsidence depression was 6 km long and 4 km wide. Subsidence at sea impacts critical drilling rig platforms, bottom mounted storage tanks and piping, (Whittaker and Reddish 1989).

7.3.3.3 Pseudokarst Associated with Underground Mines

Pseudokarst conditions have developed as a result of subsidence and collapse over underground mining operations and mine shafts. There has been subsidence and collapse over abandoned limestone, gypsum and coal mines in the United States. Underground coal mining has occurred beneath 3.2 million hectares of land in the US, almost 1 million hectares of which have been affected by subsidence. Most of the coal mining has taken place in the eastern half of the US where thousands of hectares in urban areas are threatened by subsidence, (Gray and Bruhn 1984).

Room and pillar mining in the US averages 40–60 % extraction but can be as great as 80–90 %. The room and pillar coal mine can result in classic looking sinkholes when the roof span between pillars fails and when the overlaying rock is not thick or strong enough to prevent the collapse from reaching the surface. Subsidence has occurred as early as a decade after mining and as late as a century. More than half the subsidence incidents took place 50 or more years after mining (Gray and Bruhn 1984; Dunrud 1984).

Long-wall mining inherently leads to subsidence as the entire coal seam is removed as mining advances. Immediately after the coal is removed the mine-roof begins to collapse and the collapse begins to propagate slowly to the surface. About 90 % of the subsidence occurs in the first 3 months after extraction, while residual subsidence continues for up to 3 years at the deeper sites (Gray and Bruhn 1984; Dunrud 1984).

In one area of southern Italy, limestone was mined underground to maximize land space for farming. Urban expansion over time resulted in buildings over the mines. Over the decades memory of the existence and lateral extent of the mines was lost. In addition weathering processes and illegal discharge of waste caused a decrease in strength of the limestone resulting in failures. Vattano et al. (2013) describe the history and investigation of these abandoned mines.

Underground Coal Mine Fires

Fires in coal mines are reported from nearly all parts of the world were coal is being mined. Coal mine fires have lead to subsidence as a result of decreasing the volume of coal through burning. Local communities have had to be moved because of environmental issues. Stracher (2007) provides a review of coal fires from around the world.

Fig. 7.12 Another site within the city of Carlsbad, New Mexico where salt was dissolved for drilling purposes has created a large cavity and a road sign alerts drivers to potential of sinkhole occurring in this area

Gold Mining in South Africa

The richest gold mines in the world are located in the "Far West Rand" of South Africa about 64 km west of Johannesburg. Gold is mined from the conglomerates, which underlie the thick (1,000 m or more) dolomites. The mining takes place between thick dykes that divide the dolomite into isolated compartments. By 1960, the mines were deep enough that it was necessary to start extensive pumping and the water table was lowered through the dolomites to provide access to the gold deposits. The impacts were first seen when springs began to dry up and then sinkholes began to occur as the water level was lowered (Brink 1984; De Bruyn and Bell 2001).

The most catastrophic sinkhole event occurred in December of 1962, when a three-story crusher plant along with 29 occupants was swallowed by a sinkhole. More than 200 other sinkholes have destroyed buildings and taken additional lives. These were all very rapid collapses, each occurring in a period of minutes and not allowing time for people to escape. Other sinkholes have occurred in the area over the past 25 years without the loss of life (Brink 1984; De Bruyn and Bell 2001).

7.3.3.4 Pseudokarst Associated with Underground Salt Mines

In the US, salt deposits underlie some part of 25 of the 48 contiguous states. Some of the salt deposits are extensive and rank among the greatest salt deposits of the world (Johnson 2003). Salt was first obtained by surface mining and evaporation ponds, then by underground mining. There are two methods for underground mining of salt, the room and pillar mining where the salt is physically extracted and lifted to the surface and by solution where the salt is dissolved and pumped to the surface as a brine. Collapse of room and pillar mines are rare unless they become flooded. Examples of dry salt mines that have developed sinkholes are the Retsof Mine (New York), Weeks Island (Louisiana) and Jefferson Island Diamond Salt Mine collapse (Louisiana). These mines are now flooded and are closed (Johnson 2003; Neal 1997).

Retsof Mine of New York

The Retsof Mine is owned by Akzo Nobel Salt Inc. and is the largest salt mine in the western hemisphere. It is located south of Rochester in New York. The mine was in operation from 1882 to 1995. The large (26 km^2) room and pillar mine with a mined thickness of 3.5 m is located about 335 m below grade. A complex system of geology and aquifers more than 305 m thick overlies the mine which includes approximately 182 m of shale and limestone with a cover of glacial sediments. In 1994, the mine-roof failed in the eastern portion of the mine that was located beneath a buried glacial valley with a fresh water aquifer. The failure occurred with the inrush of brine and gas, followed by a sustained inflow of fresh water. The mine filled in approximately 21 months with an estimated flow of 113,000,000 l/day (Nieto and Young 1998).

Very little geotechnical drilling data were available prior to 1995. However, the literature contained numerous references describing stress conditions and up-warps beneath sedimentary valleys. Several deep, water wells located 1–5 km from the mine had water level declines of 15 m or more. Initial indicators of impacts on private wells were disregarded until a widespread pattern of surface subsidence and groundwater changes became obvious. Published reports from the nineteenth century reveal that natural brine and gas pools existed in the region long before mining began (Gowan and Trader 2000). The unfavorable geologic and hydrologic conditions within the overburden along with the existing stress regime near the collapse could have been anticipated (Nieto and Young 1998).

Weeks Island

Sinkholes have occurred over the U. S. Strategic Petroleum Reserves storage facility at Weeks Island, Louisiana. The Weeks Island salt dome (a room and pillar salt mine) was one of six original storage sites selected for the US Strategic Petroleum Reserve, which was established after the middle-east oil embargo of 1973–1974. The salt mine was originally opened in 1902 and salt was extracted until 1977 when the mine was converted to oil storage. Petroleum reserves are stored in salt mines more than 152 m below grade (Neal 1997).

In 1992 a sinkhole formed over the outer edge of the former salt mine. A second smaller sinkhole formed early in 1995. These sinkholes were associated with a combination of factors including anomalous geologic conditions, salt creep (the gradual movement of salt), and stress cracks propagated upward leading to collapse. The sinkholes were stabilized and oil reserves were finally moved to another location and the mine was filled with 85 % saturated brine (Neal 1997).

Jefferson Island Diamond Salt Mine Collapse

A room and pillar salt mine was in operation 400 m under the freshwater Lake Peigneur in Louisiana. Oil exploration was taking place within the shallow lake in 1980. A 120 m error in the location of an oil exploration borehole resulted in the borehole hitting the salt mine at the 400 m level. The result was a huge whirlpool that drained the entire lake along with additional water flowing through a canal from the Gulf of Mexico. This water formed a 45 m high waterfall. The five million dollar drill rig vanished into the giant "sinkhole" along with barges and a tugboat. Adjacent property of 26 ha with 150 year old trees were washed away in a landslide. A local fisherman saw the calamity and headed to shore. He had just tied his boat to a tree and walked away when he turned around his boat and the tree were gone. Compressed air from the mine resulted in a huge geyser of water from the mine shaft. Although there were more than 50 workers in the mine along with the drilling crew on the lake, there was no loss of life (Johnson 2003). A commentary and video can be seen on the Modern Marvels Engineering Disasters 5 (A&E Television Networks 2000/2001).

In this case the rapid flow of large quantities of water and sediment quickly created a large diameter hole from a small 35 cm borehole, large enough to allow the drill rig, two barges and a tugboat to be lost down the hole.

Solution Mining of Salt

Cavities developed during solution mining of salt are typically 10–100 m in diameter and are 10–600 m high. Unfortunately the cavity sometimes becomes too large and the roof collapses. Most solution mining collapses have been a result from cavities formed 50–100 years ago before modern-day engineering safe guards were developed. Proper design has virtually eliminated this problem in new facilities. Examples of subsidence resulting from solution mining are provided by Johnson (1998b) and include the Cargill Sink in Kansas, along with the Grand Saline Sinks in Texas. Ege (1984) provides examples of the Meade Salt Well and the Hutchinson sink both in Kansas along with the two sinkholes on Grosse Ile, an island in the Detroit River, Michigan.

References

Aley T (1997) Groundwater tracing in the epikarst. In: Beck B, Stephenson JB (eds) Proceedings of the 6th multidisciplinary conference on sinkholes and the engineering and environmental impacts of karst, The Engineering Geology and Hydrogeology of Karst Terranes. Balkema, Springfield/Rotterdam, pp 207–211

Arts & Entertainment (2000/2001) Modern marvels engineering disasters 5, The History Channel A&E Television Networks DVD, Cat. No. AAE-73136

Bell FG (2004) Engineering geology and construction. Spon Press, London

Bell FG, Culshaw MG (1998) Some geohazards caused by soil mineralogy, chemistry and microfabric: a review. In: Maud JG, Eddleston M (eds) Geohazards in engineering geology. Engineering geology special publication, vol 15. Geological Society, London, pp 427–441

Black TJ (2012) Deep karst system research, Michigan, USA. Carbonate Evaporite 27(2):119–122

Brink ABA (1984) A brief review of the South African sinkhole problem. In: Beck BF (ed) Proceedings of the 1st multidisciplinary conference on sinkholes: their geology, engineering and environmental impact, Orlando, Florida. Balkema, Rotterdam, pp 123–127

Chelette A et al (2002) Nitrate loading as an indicator of nonpoint source pollution in the lower Saint Marks-Wakulla rivers watershed. Northwest Florida Water Management District, Water Resources Special Report 02-1. Havanna, Florida

Clayton CRI et al (1995) Site investigations, a handbook for engineers, 2nd edn. Blackwell Science, Cambridge

Coxon C (2011) Agriculture and karst. In: van Beynen PE (ed) Karst management. Springer, Dordrecht, pp 103–138

Davies WE (1977) Keynote presentation to the symposium on detection of subsurface cavities. US Army Engineer Waterways Experiment Station Vicksburg, Mississippi. (lecture recorded by Benson RC)

Davis H (1996) Hydrogeologic investigation and simulation of groundwater flow in the upper Floridan aquifer of north-central Florida and southwestern Georgia and delineation of contributing areas for selected city of Tallahassee, Florida. Water Supply Wells: USGS Water-Resources Investigation Report 95-4296. Denver, Colorado

DeBruyn IA, Bell FG (2001) The occurrence of sinkholes and subsidence depressions in the far west Rand and Gauteng Province, South Africa, and their engineering implications. Environ Eng Geosci 7(3):281–295

Dunrud CR (1984) Coal mine subsidence – western United States. Geol Soc Am Rev Eng Geol 6:151–194

Ege JR (1984) Mechanisms of surface society of subsidence resulting from solution extraction of salt. In: Holzer TL (ed) Man induced land subsidence. Reviews in engineering geology, vol 6. The Geologic America, Boulder, pp 203-221

Esteves LS, Finkl CW (1999) Late Cenozoic depositional paleoenvironments of southeast Florida interpreted from core logs. Rev Bras Geosci 29(2):129–134

Etter J (1986) Geological mishap drains lake. NewsOK (Oklahoma), 16 Nov 1986

Fernandez BF (1991) Caustic waste contamination of karstic limestone aquifers in two areas of Jamaica. In: Proceedings of the 3rd conference on hydrogeology, ecology, monitoring, and management of ground water in karst terranes. US EPA and Ground Water Scientists and Engineers a division of NGWA, pp 389–402

Finkl CW, Esteves LS (1997) A preliminary analysis of neogene depositalenvironments in Broward County, Florida. Fla Sci 60(2):94–103

Florida Senate (2011) Issues relating to sinkhole insurance. Interim Report 2011-104 Committee on Banking and Insurance Dec 2010

Florida Statutes (2014) Title XXXVII, Insurance, Chapter 627, insurance rates and contracts, 627.706 Sinkhole insurance; catastrophic ground cover collapse. www.leg.state.fl.us/statutes. Accessed 12 Oct 2014

Fookes PG et al (eds) (2005) Geomorphology for engineers. Whittles Publishing, Dunbeath, Scotland

Fuleihan NF et al (1997) The hole story: how a sinkhole in a phosphogypsum pile was explored and remediated. In: Beck BF, Stephenson JB (eds) Proceedings of the 6th multidisciplinary conference on sinkholes and the engineering and environmental impacts of karst, the engineering geology and hydrology of karst terranes. Balkema, Rotterdam, pp 363–369

Gerami A (1984) Hydrogeology of the St. Marks river basin, northwest Florida. M.S. thesis, Florida State University, Tallahassee, FL

Gordon L, Kennedy JM (1995) Street vanishes in subway sinkhole. Los Angeles Times, 23 Jun 1995

Gowan SW, Trader SM (2000) Mine failure associated with a pressurized brine horizon: Retsof salt mine, western New York. Environ Eng Geosci 6(1):57–70

Gray RE, Bruhn RW (1984) Coal mine subsidence–eastern United States. Geol Soc Am Rev Eng Geol 6:123–149

Hermosilla RG (2012) The Guatemala city sinkhole collapses. Carbonate Evaporite 27(2):103–107 (Special issue on integrating science and engineering to solve karst problems)

Holzer TL (1984) Ground failure induced by ground water withdrawal from unconsolidated sediment. In: Holzer T (ed) Man-induced land subsidence. Reviews in engineering geology, vol 6. The Geologic Society of America, Boulder, pp 67–105

Holzschuh JC (1991) Field trip guidebook for the 4th international symposium on land subsidence, 12–17 May, Houston

Hunt RE (2005) Geotechnical engineering investigation handbook, 2nd edn. Taylor & Francis, Boca Raton

Johnson KS (1987) Development of the wink sink in west Texas due to salt dissolution and collapse. In: Beck BF, Wilson WL (eds) Proceedings of the 2nd multidisciplinary conference on sinkholes and the environmental impacts of karst, karst hydrogeology: engineering and environmental applications. Orlando, Florida. Balkema, Rotterdam, pp 127–136

Johnson AI (1998a) Land subsidence due to fluid withdrawal in the United States, an overview. In: Borchers JW (ed) Proceedings of the Dr. Joseph F. Poland symposium on land subsidence, Land subsidence case studies and current research. Star Pub, Belmont, California, pp 51–57

Johnson KS (1998b) Land subsidence above man-made salt dissolution cavities. In: Borchers JW (ed) Proceedings of the Poland symposium on land subsidence, Land subsidence case studies and current research. Star Pub, Belmont, California, pp 385–392 (AEG Special Publication No. 8)

Johnson KS (2003) Evaporite-karst problems in the United States. In: Johnson KS, Neal JT (eds) Evaporate karst and engineering/environmental problems in the United States, Oklahoma Geological Survey, Circular 109:1–20

Johnson KS et al (2003) Sinkholes and land subsidence owing to salt dissolution near wink, west Texas, and other sites in western Texas and New Mexico. In: Johnson KS, Neal JT (eds) Evaporate karst and engineering/environmental problems in the United States, Oklahoma Geological Survey Circular 109:183–195

Kincaid TR et al (2012) Demonstrating interconnection between a wastewater application facility and a first magnitude spring in a karstic watershed: tracer study of the southeast farm wastewater reuse facility. Florida Geologic Survey, Tallahassee, Report of Investigation 111

Kratochvil GL (1989) Geological evaluation of potential hazards to shuttle landing operations on Rogers Dry Lakebed, Edwards Air Force Base, California. Astronaut Office NASA National Aeronautics and Space Administration, Johnson Space Center, Houston

Kratochvil GL et al (1992) Fishing for fissures. Mil Eng 84:66–68

Land L (2013) Geophysical records of anthropogenic sinkhole formation in the Delaware Basin region, southeast New Mexico and West Texas, USA. Carbonate Evaporite 28(1):183–190 (Special Issue on Integrating Science and Engineering to Solve Karst Problems)

Legget RF (1962) Geology and engineering. McGraw-Hill, New York

Legget RF, Hatheway AW (1988) Geology and engineering, 3rd edn. McGraw-Hill, New York

Londquist CJ (1995) Hydrology and land subsidence, Antelope Valley California, In: Prince KR, Galloway DL, Leake SA (eds) US geological survey subsidence interest group conference, Edwards Air Force Base, Antelope Valley, California, 18–19 Nov 1992. US Geologic Survey Open-File Report 94–532:38–39

Melvin D (1993) Storm sewer collapse, killing 1, hole swallows 3 cars outside Atlanta hotel. Sun-Sentinel 15 June 1993. Available online at www.articles.sun-sentinel.com. Accessed April 12, 2012

Neal JT (1997) Mine induced sinkholes over the US strategic petroleum reserve (SPR) storage facility at Weeks Island, Louisiana: geologic mitigation and environmental monitoring. In: Beck BF, Stephenson JB (eds) Proceedings of the 6th multidisciplinary conference on sinkholes and the engineering and environmental impacts of karst, the engineering geology and hydrology of karst terranes. Balkema, Rotterdam, pp 357–361

Nieto AS, Young RA (1998) Retsof salt mine collapse and aquifer dewatering, Genesee Valley, Livingston County, New York. In: Borchers JW (ed) Proceedings of the Dr. Joseph F. Poland symposium on land subsidence, case studies and current research, pp 309–325

Noe DC, Dodson MD (1999) Heaving-bedrock hazards associated with expansive, steeply dipping bedrock in Douglas County, Colorado. Colorado Geological Survey, Denver, Colorado (Special Publication 42)

Sasowsky ID, White WB, Webb JA (1995) Acid mine drainage in karst terranes: geochemical considerations and field observations. In: Beck BF (ed) Proceedings of the 5th multidisciplinary conference on sinkholes and the engineering and environmental impacts of karst: karst geohazards, Gatlinburg, Tennessee. Balkema, Rotterdam, pp 241–247

Schmidt et al (2000) The Wakulla springs Woodville karst plain symposium transactions, 9 Oct 1998. Florida Geological Survey Special Publication 46

Shlemon RJ (2004) Subsidence and ground fissures in southern California: an engineering and geologic perspective. In: The Shlemon conference earth fissures, sponsored by the Association of Engineering Geologists and Engineering Geology Foundation, AEG special publication

Sowers GF (1996) Building on sinkholes, design and construction of foundations in karst terrain. ASCE Press, Reston (out of print)

Stein L et al (2004) Unseen sinkhole downs span. St. Petersburg Times, 14 Apr 2004

Stephenson JB, Beck BF (1995) Management of the discharge quality of highway runoff in karst areas to control impacts to ground water, a review of relevant literature. In: Beck BF (ed) Proceedings of the 5th multidisciplinary conference on sinkholes and the engineering and environmental impacts of karst: karst geohazards, Gatlinburg, Tennessee. Balkema, Rotterdam, pp 297–321

Stracher GB (ed) (2007) Geology of coal fires: case studies from around the world, Reviews in Engineering Geology XVIII. Geological Society of America, Boulder, Colorado

Sweeting MM (1973) Karst landforms. Columbia University Press, New York

Tihansky AB (1999) Sinkholes, west-central Florida. In: Galloway D (ed) Land subsidence in the United States, USGS circular 1182. U.S. Geological Survey, Reston

Vattano M et al (2013) Examples of anthropogenic sinkholes in Sicily and comparison with similar phenomena in southern Italy. In: Land L, Doctor DH, Stephenson BJ (eds) Proceedings of the 13th multidisciplinary conference on sinkholes and the engineering and environmental impacts of karst, National Cave and Karst Research Institute, Symposium 2:263–271

Veni G et al (2001) Living with karst, a fragile foundation, vol 4, Awareness Series. American Geologic Institute, Alexandria

Waltham AC et al (2005) Sinkholes and subsidence, karst and cavernous rocks in engineering and construction. Springer-Praxis, Chichester, UK

Ward AW et al (1995) Geologic setting of east Antelope Basin, with emphasis on fissuring on Rogers Lake, Edwards Air Force Base, Mojave Desert, California. In: Prince KR et al (eds) US Geological Survey Subsidence Interest Group Conference, Edwards Air Force Base, Antelope Valley, California, 18–19 Nov 1992. US Geologic Survey Open-File Report 94–532, p 35–37

Whittaker BN, Reddish DJ (1989) Subsidence – occurrence, prediction and control, vol 56, Developments in Geotechnical Engineering. Elsevier, New York

Wisenbaker M, McKinlay C (2006) A history of exploration in America's longest underwater cave, Part 3. NSS News 64(8):13

Wisenbaker M et al (2007) Explorers connect America's longest underwater cave to Wakulla Springs. NSS News 65(12):4–11

Yuhr L et al (2008) Mapping and drilling of conduits within the Woodville Karst Plain of north Florida. In: Yuhr LB, Alexander EC Jr, Beck BF (eds) Proceedings of the 11th multidisciplinary conference sinkholes and the engineering and environmental impacts of karst. ASCE Geotechnical Special Publication No. 183:340–350

Zangerl C et al (2003) Ground settlements above tunnels in fractured crystalline rock: numerical analysis of coupled hydromechanical mechanisms. Hydrogeol J 11:162–173

Triggering Mechanisms for Sinkholes

8

Abstract

Dealing with the engineering issues associated with karst collapse features requires an understanding of the sinkhole development process including an understanding of the mechanisms that trigger sinkhole collapse. Sinkhole collapse is often thought of as a natural process. However, collapse can be triggered by either natural conditions or man's activities and is often a complex combination of both. If we can gain an understanding of the likely triggering mechanisms, we can possibly avoid areas of potential collapse, design around it or at least minimize its impact.

8.1 Statistics

Statistics have been collected by a number of different groups regarding the triggering mechanisms for collapse. Overwhelmingly heavy rain, drought or other changes in water conditions play a major role. Davies (1977) cites causes of collapse recorded by the Missouri Geological Survey, which had kept excellent records of collapse. Their findings indicate that about 56 % of the collapses were induced by some activity by man. The remaining 44 % were due to natural causes.

Newton (1976, 1986), a geologist with the USGS had reviewed sinkhole phenomena for the Alabama Highway Department. He found that water was the dominant triggering factor and indicated that collapse was accelerated by a decline in the water table that resulted in:

- A loss of buoyant support to unconsolidated deposits,
- An increase in the velocity of movement of water,
- Water-level fluctuations at the base of unconsolidated deposits adding to erosion and soil piping, and
- Induced recharge.

Wisner and Denahan (1985 personal communication) geologists with the Florida Department of Transportation compiled statistics over a 10-year period on sinkhole collapse. The causes of these sinkholes were classified into six categories (Table 8.1). The data show that at least 66 % of the collapses were directly caused by water related factors (low water table and heavy rainfall). While these statistics in Florida generally represent roadway-related occurrences, they indicate that natural and man-made changes in surface water and groundwater are the dominant factors in precipitating collapse. They also tend to agree with independent findings by Newton (1976).

8.2 Water-Related Triggering Mechanisms

Water is a powerful and key component that triggers subsidence and collapse of sinkholes. Changes in water levels can be caused by many factors that include heavy rainfall, drought, pumping and dewatering, along with the concentration of surface water run-off, to name a few. A number of man-made structures can also lead to changes in water levels. Industrial impoundments, cooling ponds, man-made lakes and dams where the hydraulic head has been artificially raised can all trigger sinkhole activity. While a singular event may ultimately trigger a sinkhole it is usually a combination of events that has lead up to developing unstable conditions prior to the final triggering of collapse.

R.C. Benson, L.B. Yuhr, *Site Characterization in Karst and Pseudokarst Terraines*,
DOI 10.1007/978-94-017-9924-9_8

Table 8.1 Causes of collapse

Cause	Percentage (%)
Blasting	5
Drilling	5
Low water table	8
Construction	11
Other	11
Heavy rain fall	58

8.2.1 Changes in Water Levels

Sinkhole development often tends to be seasonal in nature and follow the trends in rainfall. For example, during drought periods, the groundwater levels are lowered by lack of recharge and pumping for drinking water and agriculture purposes. This results in a loss of buoyancy support to both unconsolidated materials and rock. The buoyant force is approximately 1 ton/m^3. Therefore, if the water table were reduced by 30 m the increased load would be approximately 30 tons/m^3 (Hunt 2005). In periods of heavy rainfall, groundwater level is raised. Such cycling of the water level cause changes in hydraulic gradients between surficial and deeper aquifers including artesian head. These conditions create stresses due to drying and wetting as well as changes in head pressure within the unconsolidated materials. The combination of these stresses in an area where dissolution has already created cavities results in a high risk of materials raveling into void spaces and ultimately resulting in sinkhole collapse.

In many areas of Florida, especially along the elevated sandy ridges, sinkhole development peaks when the potentiometric surface of the Floridan is the lowest and groundwater demand is highest usually in April and May. In the Orlando area, 42 % of all new sinkholes occur during the months of April and May during the peak of the rainy season. In the coastal lowlands, a double peak occurs in the seasonal sinkhole frequency. One peak occurs in the dry season, but a second occurs, usually in August, which is the height of the summer thunder-storm season (Wilson 1991).

The heavy rainfall associated with hurricane Frances in September 2004 caused extensive sinkholes to occur along its path through central and the west coast of Florida. Thirty centimeters of heavy rain fell within 8-h near Ocala, Florida which triggered about 200 sinkholes (Wilson 1991). Again in 2012, Tropical Storm Debbie dumped tremendous amounts of rain, at a time that Florida was experiencing a drought (Wilson 1991). The combination of a lowered water table due to the drought and the heavy rainfall triggered numerous sinkholes.

Rapid changes in water levels, sometimes referred to as water hammer, can create both compressional and tensional stress on trapped water and air (Chen et al. 1995). The resulting stress can contribute to breaching the overlying sediments within the voids of the epikarst, creating a pathway for sediments to move downward. These rapid fluctuations in the water table may be natural or man-induced (such as purging of a well).

8.2.2 Pumping

During Florida winters, farmers pump large quantities of groundwater to spray on crops to prevent freezing. This lowers the water table rapidly and commonly results in sinkhole activity. An increased pumping of 11 million liters/day in a well field near Tampa resulted in 64 new sinkholes within 1 month (Beck and Sinclair 1986).

8.2.3 Dewatering

Dewatering associated with quarries, mining operations or construction have been known to cause sinkholes nearby. Again, the triggering mechanism is the change in water level.

A limestone quarry leased land approximately 760 m from a local college in northwest Georgia. The water table was within 1.5–3 m from the surface and rock lay 3–6 m below grade. As the quarry excavated deeper to find better quality rock it became necessary to increase pumping to dewater the quarry and keep it dry. A few sinkholes begin to occur near the quarry and were filled with quarry spoil. With time the sinkhole activity began to approach the college. Eventually, a man-made lake on campus went dry and sinkhole activity increased on campus. At one point, quarry blasting resulted in columns of water spurting from the ground on campus (Sowers 1996).

The December Giant sinkhole (or "Golly Hole") occurred in the isolated woods of Shelby County about 48 km south of Birmingham, Alabama in 1972. A farmer was startled by his house shaking one evening and he heard the sound of breaking trees. The next morning he walked in the direction of the sound and found a giant sinkhole that was more than 91 m in diameter and 36 m deep (about 86,700 m^3). Several obvious fissures were observed around the circumference of the sinkhole. Sowers (1996) reported the presence of a large limestone quarry about 2.5 km from the sinkhole which was more than 30 m below the water table and had been pumping large amounts of water to maintain the quarry dry and operational. This pumping probably triggered the Golly Hole.

8.2.4 Surface Water Run-Off

Changes in drainage patterns that result from construction such as major roads and their associated drainage systems can result in sinkhole activity. Most highway drainage and repairs are (or should be) designed to utilize the natural karst systems recharge. Yet inadequate drainage lines along highways commonly trigger failures (Moore 1987, 1988, 2006).

Concentrated surface water drainage from roofs of large buildings and their associated paved parking lots are also a factor causing sinkholes. Where buildings, parking areas and roadways are present 50–80 % of the surface area may be covered from which surface water is being diverted and concentrated. Retention ponds are often used to collect surface waters and divert this concentration of water away from important structures. However, the retention ponds themselves are frequently plagued with sinkholes.

A commercial office complex near Frederick, Maryland collected rainwater from approximately a few hectares of building roofs and paved parking area that was diverted into a central sump of about 0.5 ha. The top of rock was shallow (about 3 m below grade) and highly irregular with intermediate size voids and small conduits. Within 1 year the concentration of surface water began to take its toll around the perimeter of the sump. Localized subsidence became numerous and continued to cause problems (Fig. 8.1). The initial subsidence and localized collapse were repaired only to have subsequently worse problems. Major remediation was undertaken consisting of excavation to the top of rock and backfilling with gravel. Water was then redirected into the central portion of the drainage area.

8.2.5 Leaky Water Pipes and Sewers

Broken water and sewer lines are the cause of much pseudokarst collapse. A break in a storm sewer or sanitary sewer line provides an opening, allowing unconsolidated materials to flow into the sewer. An alternative is when a water line breaks, the flow of water washes material away resulting in a sinkhole. Many of the sinkholes reported in the news are, in fact, caused by such leaks. While most problems are relatively localized there are a few examples of catastrophic collapse (see Sect. 7.3.2).

Fig. 8.1 A local storm water disposal area near Fredrick, Maryland resulted in settlement and numerous small sinkholes

8.3 A Guideline to Minimize Sinkholes Triggered by Water

In South Africa, surface water is also a major triggering mechanism for sinkhole collapse. So much so that detailed guidelines have been developed to control the impact of water, including leaky water lines and management of surface water runoff. The dominant factor in the guideline focuses upon design installation and monitoring of water services and runoff, (South Africa Institute of Engineering and Environmental Geologists 2003).

The following are some examples from their guidelines.

- Water bearing services must be placed above ground or in sleeves.
- Drainage from rooftops must discharge onto sealed surfaces or lined furrows and then into storm water systems.
- No trees shall be planted within 1.5 times their eventual height from the line of storm-water services.
- All ponds, water courses and road surfaces shall be impervious.
- Criteria are provided for the design of swimming pools. All water pipes, pumps and connections should be installed in the open, on the surface or in service canals where they may be inspected.

Note that the guideline focuses upon controlling a variety of sources of water as well as the potential breaching of pipelines due to the root systems of trees.

Fig. 8.2 The Guest Sinkhole in west central Florida triggered by drilling a monitor well (Tihansky 1999)

8.4 Other Triggering Mechanisms

While water is the major factor in developing of karst and triggering of sinkholes, many other factors also may trigger a sinkhole including:

- Construction activities such as drilling and grouting, and
- Vibrations, blasting or earthquakes

8.4.1 Drilling Operations

Drilling operations can trigger collapse by creating a pathway through a confining layer, between a surficial and deeper groundwater zones. Breaching a confining layer and allowing sudden downward drainage from a surficial aquifer can result in a sinkhole. Alternatively, when an artesian aquifer is breached the upward flow of water can also lead to a collapse.

The Guest Sinkhole in west central Florida occurred when a saltwater monitoring well was being drilled. The drill bit was down about 62 m on the third day of drilling when the ground began to give way. Within minutes, the sinkhole swallowed the well-drilling rig, a water truck, a trailer loaded with drill pipe and a 12 m pine tree. Figure 8.2 shows the beginning of the collapse around the drill rig. The sinkhole was about 91 m in diameter and 7.5 m deep (taking about 200,160 m^3 of sediment). As the sinkhole formed, it was originally dry, however, within 24 h it had filled with water seeping into the sinkhole from the surficial aquifer. Ten small satellite sinks as much as 6 m in diameter formed along a NNW-SSE trend through the main sinkhole (Scott 2009, personal communication, Tihansky 1999).

Although there are obvious sinkholes in the region, there was no surface indication of a problem at this site. No one could have anticipated the event without a more detailed investigation of the area. An example of a large sinkhole located about 16 km to the south is Eagles Nest, which is one of the premiere cave diving spots in Florida. The entrance to Eagles

Nest is typical of a large dome breakdown, a shallow lake at the surface with a narrow opening into a very large chamber. Lateral tunnel development occurs at depths of 60–88 m (Florea 2008). It is most likely that drilling at Guest Sink had encountered one of the large chambers similar to those found at Eagles Nest, which triggered the collapse and allowed the rapid loss of all of the equipment, trees and sediments.

An example of multiple sinkhole collapse at a new housing development in west central Florida (Hernando County) was presented in Sect. 7.1.4. A total of 760 sinkholes occurred in less than 24 h (Fig. 7.4). A deep well was being drilled for irrigation purposes. Air rotary drilling had proceeded below the sand sediments and a clayey epikarst layer into the massive limestone. During the subsequent development of the well (a procedure that involved flushing of the well to obtain maximum production) sinkholes began to occur.

The process of air rotary drilling had probably loosened the clayey sediments within the epikarst zone due to air pressure from below. The repeated flushing pressures during well development continued to aggravate these conditions. The combination of these two activities was sufficient to initiate collapse. As the collapse process continued the head difference between the surficial aquifer and the deeper Floridan aquifer aided in sinkhole development.

The many smaller sinkholes (Fig. 7.5a) were likely associated with small isolated voids near the top of weathered limestone at a depth of about 12 m. The larger sinkholes (Fig. 7.5b) were likely related to cavernous zones that are known to exist deeper within the lower third of the Suwannee Limestone, at depths of approximately 30–60 m below grade (Wetterhall 1964).

8.4.2 Impact of Vibrations, Blasting and Earthquakes

Deeper underground mines, tunnels and cave systems are generally known for the excellent stability during earthquakes. In general, vibrations, blasting and earthquakes will have little, if any, impact to natural cave systems deeper within the rock mass. This is because the energy is dominated by low frequency waves, which travel as surface waves, therefore the deeper caves and mines are much less affected by this energy. Wisner and Denahan (1985 personal communication) have noted that only 5 % of the sinkholes were caused by blasting associated with nearby quarries. In addition, caves often evolve into stable shapes with inherently strong and stable arched roofs. Air raid shelters have commonly been established in tunnels, subways, mines and caves to avoid the effects of bombs. However vibrations, blasting and earthquakes may trigger roof-collapse in caves and mines which are shallow, where the rock is highly fractured and where conditions are already unstable.

Cave divers in north-central Florida reported seeing blocks of rock <1 m in diameter spalling off of the cave roof as a train passed overhead (Mount 1973 personal communication). In contrast, the New Madrid earthquakes of 1811 and 1812 had an estimated magnitude of 7.5–8.0 and occurred within 240 km of Mammoth Cave in Kentucky. Workers present within Mammoth Cave at the time reported no rock failure. Observations in Appalachian caves show no rock falls although local earthquakes have taken place during the period of observations (Davies 1951).

Further details are provided by Bolt (1993) who discusses earthquakes, Bell (2004) who discusses blasting and earthquakes, and Hunt (2005) who discusses vibration and earthquakes.

8.5 Size and Rate of Sinkhole Collapse

8.5.1 The Size of a Sinkhole

The size of a dropout sinkhole or cover collapse sinkhole is related to the thickness of unconsolidated sediments and the size or volume of the void space in the underlying rock and its interconnections. When the sediments are thin and the localized voids in the top of rock are limited in size, the sinkholes are small.

Some dropout sinkhole or cover collapse sinkholes may initially be quite small but commonly enlarge with time. At a site in North Carolina where sediments were about 12 m thick, small (0.3–0.6 m diameter) soil piping sinkholes would slowly migrate to the surface (Fig. 8.3a). Then a slow process of raveling of sediments would take place resulting in a funnel shape sinkhole of a 3–6 m diameter within weeks (Fig. 8.3b). The sinkholes in Figs. 3.5, 7.5a, and 8.3 are all related to small isolated voids within the epikarst.

Larger dropout sinkhole or cover collapse sinkholes are generally associated with thicker overburden, and occur due to the presence of well-developed and interconnected open fracture system or an open cave system within massive rock. The thicker sediments begin to slowly ravel into the fractures or voids creating a void in the sediments that enlarges and migrates toward the surface. At some critical point the soil arch collapses forming a sinkhole. Examples of these sinkholes are seen in Figs. 3.7 and 7.5b.

8.5.2 The Speed of a Sinkhole Collapse

The dissolution of rock, the formation of tertiary porosity and the development of large conduits and cave systems occurs over a geologic time scale (Fig. 4.5), while sinkhole collapse itself occurs on a human time scale. After a

Fig. 8.3 Initial sinkholes at a site in North Carolina were quite small (<1 m in diameter) as a result of piping in sandy soils (**a**). Within a few weeks, these sinkholes would slowly enlarge to 3–6 m in diameter as the sandy soils continued to ravel (**b**)

triggering event, most sinkhole collapses develop over a few hours or a few days, most often providing ample warning for human safety. However, there are a number of examples where collapse has been almost instantaneous. Those unique situations are where both speed of development and size come together resulting in catastrophic collapse. The following illustrate some examples of rapid and large sinkhole development:

- On the west coast of central Florida, at a new housing development 760 sinkholes developed in less than 24 h over an 8-ha area (Fig. 7.4). There was no loss of life or equipment in this case.
- On the west coast of central Florida, the Guest Sink consumed a drill rig and two support trucks, within a matter of 10 min (Fig. 8.2) (Scott 2009, personal communication).
- In the South Africa gold mining district, in December of 1962 an entire Crusher plant and 29 people disappeared almost instantaneously. Dewatering of mine operations 1 km of more below grade were likely responsible for the collapse (Brink 1984).
- In another case in South Africa, Mr. Nortjie sat in a pavilion watching four friends play a game of tennis. The friends heard a loud noise like a pistol shot, when the dust cleared there was nothing to be seen of the pavilion or Mr. Nortjie (Brink 1984).
- March of 2013 in a small town east of Tampa, Florida, without any warning the concrete floor slab of a single story home collapses taking with it the bed and its occupant, all within a few minutes. The house was later demolished and efforts to recover the body were abandoned. In this case the sediments had already flowed

into deeper voids resulting in a large void under the concrete slab, which grew until eventually the slab failed (Altman 2013).

In some cases, there will be precursor indications of impending collapse such as cracks in the soil, pavement, or the structure itself. In others there may be no obvious prior indications of impending collapse. The above examples are illustrations of unsuspected collapse, some of which occurred almost instantaneously.

In these examples, the speed at which a collapse occurs depends on many factors such as the character of the sediments, clay content, partial cementation, hydraulic head differences and the size of the opening into the rock along with the size and interconnection of the void space within the rock. Therefore, the speed at which a collapse takes place is unique to the geologic setting and the temporal conditions (natural and man-made), which are impacting the collapse. If the sediments are loose sands, the process may proceed rapidly. Where the sediments contain a significant amount of clay or are weakly cemented the mature slope may develop quite slowly.

8.5.2.1 The Winter Park Sinkhole

The Winter Park Sinkhole in the Orlando, Florida area is another excellent example of a large sinkhole collapse with thicker overburden (Fig. 3.7). It has been studied and well documented including its time sequence of events. The following chronological summary of the Winter Park Sinkhole has been edited from the Winter Park Sinkhole Report (Jammal 1982).

Friday at 8 pm, May 8th 1981 Mrs. Owens, a homeowner, heard a queer swishing noise and saw a sycamore tree disappear into the ground. She called the police department. An initial cone shaped sinkhole 12 m in diameter and 6 m deep had occurred.

Saturday at 4:00 am – Overnight the sinkhole had slowly enlarged to a diameter of 24 m. Mrs. Owens saw the ground cracking on her property. Latter in the day the sinkhole rapidly expanded and half of Mrs. Owens yard was gone.

Two Porsches parked at the nearby German Car Services were engulfed by the sinkhole. A Winter Park firefighter saw two 18 m trees sucked into the hole just like they were going down the toilet. By noon a pickup truck and another Porsche fell into the sinkhole. The diving board of a large swimming pool was halfway down the 30 m slope. One corner of Mrs. Owens' frame house was starting its descent into the sinkhole and another corner of the house rose off the ground before sliding down the slope.

By 1 pm A fourth Porsche slipped off the edge and settled 6 m below the surface, where it stayed the rest of the afternoon. Mrs. Owens' house was in the pit, about 7.5 m down the slope. A fifth Porsche rolled into the pit.

At 2:40 pm – A portion of a large swimming pool collapsed and a crane had lifted some of the vehicles out of the pit.

At 2:50 pm – There was a reverberating crash, then a swooshing sound as Mrs. Owens' house literally disappeared into the abyss about 38 m below. It was about 19 h after the beginning of the sinkhole before Mrs. Owens' house disappeared below the water. At days end, six vehicles had been swallowed by the sinkhole.

Sunday Morning – The Olympic sized swimming pool, valued at $150,000 was ripped in half.

In this example, there was time to prevent loss of life. While this was not instantaneous, there was nothing to be done to stop the collapse progress.

References

Altman H (2013) Search for man under Seffner sinkhole abandoned. The Tampa Tribune 18 Mar 2013, Tampa, Florida

Beck BF, Sinclair WC (1986) Sinkholes in Florida, an introduction. University of Central Florida, College of Engineering, The Florida Sinkhole Research Institute, Orlando

Bell FG (2004) Engineering geology and construction. Spon Press, London

Bolt BA (1993) Earthquakes. W. H. Freeman and Company, New York

Brink ABA (1984) A brief review of the South African sinkhole problem. In: Beck BF (ed) Proceedings of the 1st multidisciplinary conference on sinkholes: their geology, engineering and environmental impact, Orlando, Florida. Balkema, Rotterdam, pp 123–127

Chen et al (1995) Mechanical impacts of air and water compression in karst conduits. In: Beck BF (ed.) Proceedings of the 5th multidisciplinary conference on sinkholes and the engineering and environmental impacts of karst: karst geohazards, Gatlinburg, Tennessee. Balkema, Rotterdam, pp 17–22

Davies WE (1951) Mechanics of cave breakdown. Natl Speleol Soc Bull 13:36–43

Davies WE (1977) Keynote presentation to the symposium on detection of subsurface cavities. US Army Engineer Waterways Experiment Station Vicksburg, Mississippi. (lecture recorded by Benson RC)

Florea LJ (ed) (2008) Caves and karst of Florida. In: A guidebook for the 2008 NSS National Convention, NSS, Huntsville

Hunt RE (2005) Geotechnical engineering investigation handbook, 2nd edn. Taylor & Francis, Boca Raton

Jammal & Associates, Inc. (1982) The winter park sinkhole, a report of the investigation, findings, evaluation and recommendations. Consultant report for the City Commission of Winter Park Florida, March

Moore HL (1987) Sinkhole development along 'untreated' highway ditches in east Tennessee. In: Beck BF, Wilson WL (eds), Proceedings of the 2nd multidisciplinary conference on sinkholes and the environmental impacts of karst, karst hydrogeology: engineering and environmental applications. Orlando, Florida. Balkema, Rotterdam, pp 115–120

Moore HL (1988) Treatment of karst along Tennessee highways. In: Geotechnical aspects of karst terrains: exploration, foundation design and performance, and remedial measures. ASCE Geotech Special Publ No 14, pp 133–148, ASCE, Reston, Virginia
Moore HL (2006) A proactive approach to planning and designing highways in east Tennessee karst. Environ Eng Geosci 12(2):147–160
Mount T (1973) Preliminary Alachua cave map. Consultant report
Newton JG (1976) Early detection and correlation of sinkhole problems in Alabama with a preliminary evaluation of remote sensing applications. Alabama Highway Research Report 76
Newton JG (1986) Development of sinkholes resulting from man's activities in the eastern United States. US Geologic Survey Circular 968, Reston, Virginia
South Africa Institute of Engineering and Environmental Geologists (2003) Guideline for engineering-geological characterization and development of dolomitic land. Council for Geoscience, Pretoria
Sowers GF (1996) Building on sinkholes, design and construction of foundations in karst terrain. ASCE Press, Reston (out of print)
Tihansky AB (1999) Sinkholes, west-central Florida. In: Galloway D et al (eds) Land subsidence in the United States. USGS Circular 1182, Reston, Virginia
Wetterhall WS (1964) Geohydrologic reconnaissance of Pasco and southern Hernando counties, Florida. Florida Geological Survey Report of Investigation 34
Wilson WL (1991) The new picture of sinkhole development in Florida. The Florida Specifier, Feb, Goldenrod, Florida

Cave and Cavern Collapse

9

Abstract

This chapter presents the concept of large caverns and domes found in many caves. It discusses the process of their development and modes of breakdown. This includes discussion of factors necessary to prevent or minimize the risk of subsidence or collapse at the surface including bulking of rock, which can sometimes restrict further collapse and the thickness of rock needed over a cave or mine to be considered safe. Regardless of the many rules of thumb and safety factors there are cases where collapse has propagated to the surface from great depths.

9.1 Breakdown Domes

Breakdown domes are huge caverns which form as a result of a localized area of collapse within the rock mass. They typically have arched roofs and have a large debris pile on the floor (Fig. 9.1a). High-arched cathedral roofs with large spans are seen in many caves and are part of their spectacular beauty. The largest dome in the eastern United States and the second largest in the US is the Rumble Room in Rumbling Falls cave system in eastern Tennessee (Fig. 9.1b). It covers an area of more than 1.6 ha and is approximately 120 by 210 m and about 48 m high (Davis 2002).

These large caverns with breakdown domes tend to develop at intervals along a cave system that are impacted by a major fracture, the intersection of fractures, a syncline or incised valley where the rock has been stressed. Such areas are often more permeable allowing recharge from the surface over long periods of time, which further weakens the rock. These zones provide an area of rock failure that slowly enlarges upward resulting in a large dome-shaped cavern.

9.1.1 A Conceptual Model of a Large Cavern

In Fig. 9.2a, a fracture is shown where water slowly begins to seep down into the cave weakening the ceiling rock and rock failure begins to occur falls. Over time the localized rock fall develops into a sizeable cavern (Fig. 9.2b). This large cavern may develop without the benefit of significant further dissolution (Waltham et al. 2005).

These breakdown domes tend to become inherently stable when they develop compression arches in the roof rock. If there is a sufficient rock cover, they can remain stable for long periods of time. The risk of surface subsidence or collapse is relatively low and they may constitute little, if any, hazard to normal civil engineering works at the ground surface.

This process of developing a cavern within a cave occurs on a geologic time scale. A time period of 10,000–100,000 years (or more) is needed under most conditions to develop large cavernous conditions (White 1988). As the cavern slowly continues to expand upwards it eventually reaches the top of the rock (Fig. 9.2c). The sediments begin to fall into the cavern (Fig. 9.2c). If the sediment is sandy, a narrow piping feature may slowly work its way toward the surface (Fig. 9.2c). If there is clay in the sediment or it is slightly cemented, a larger spherical void may develop and remain stable for some time (Fig. 9.2e). Eventually the void in the sediment will reach a critical state and a single event or combination of events will act as the final triggering mechanism resulting in a large sinkhole (Fig. 9.2f).

Once the event is triggered it may proceed rapidly in minutes or may occur over a few hours or even days. The speed and volume of material transfer will depend upon the volume of the void space within the rock, the degree to which the void space is open and interconnected, the size of the opening

R.C. Benson, L.B. Yuhr, *Site Characterization in Karst and Pseudokarst Terraines*,
DOI 10.1007/978-94-017-9924-9_9

a

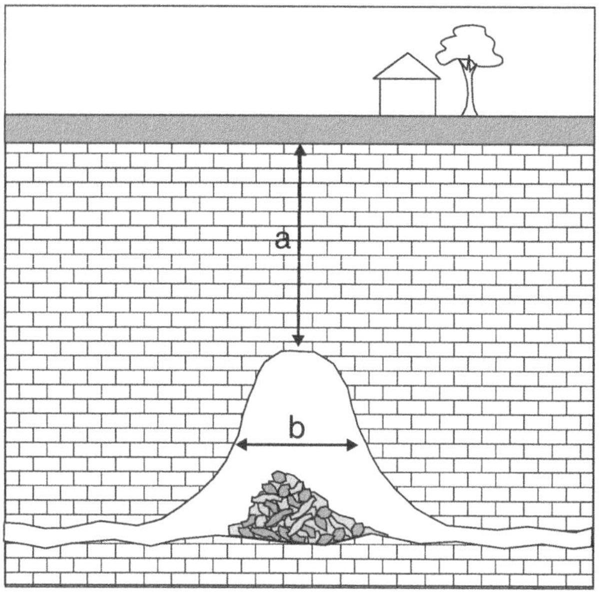

A domed cavern with adequate overlaying rock thickness is inherently stable, a>>b

b

Photo of large cavern Rumble Room in Rumbling Falls Cave, Tennessee (courtesy of Chris Anderson)

Fig. 9.1 Huge caverns can develop arched domes that can be inherently strong and stable for long periods of time (**a**). An example is seen in the Rumble Room of Rumbling Falls Cave, Tennessee (**b**). Note person for scale standing on top of the rock debris and another person on rope (Photo courtesy of Chris Anderson, www.darklightimagery.net)

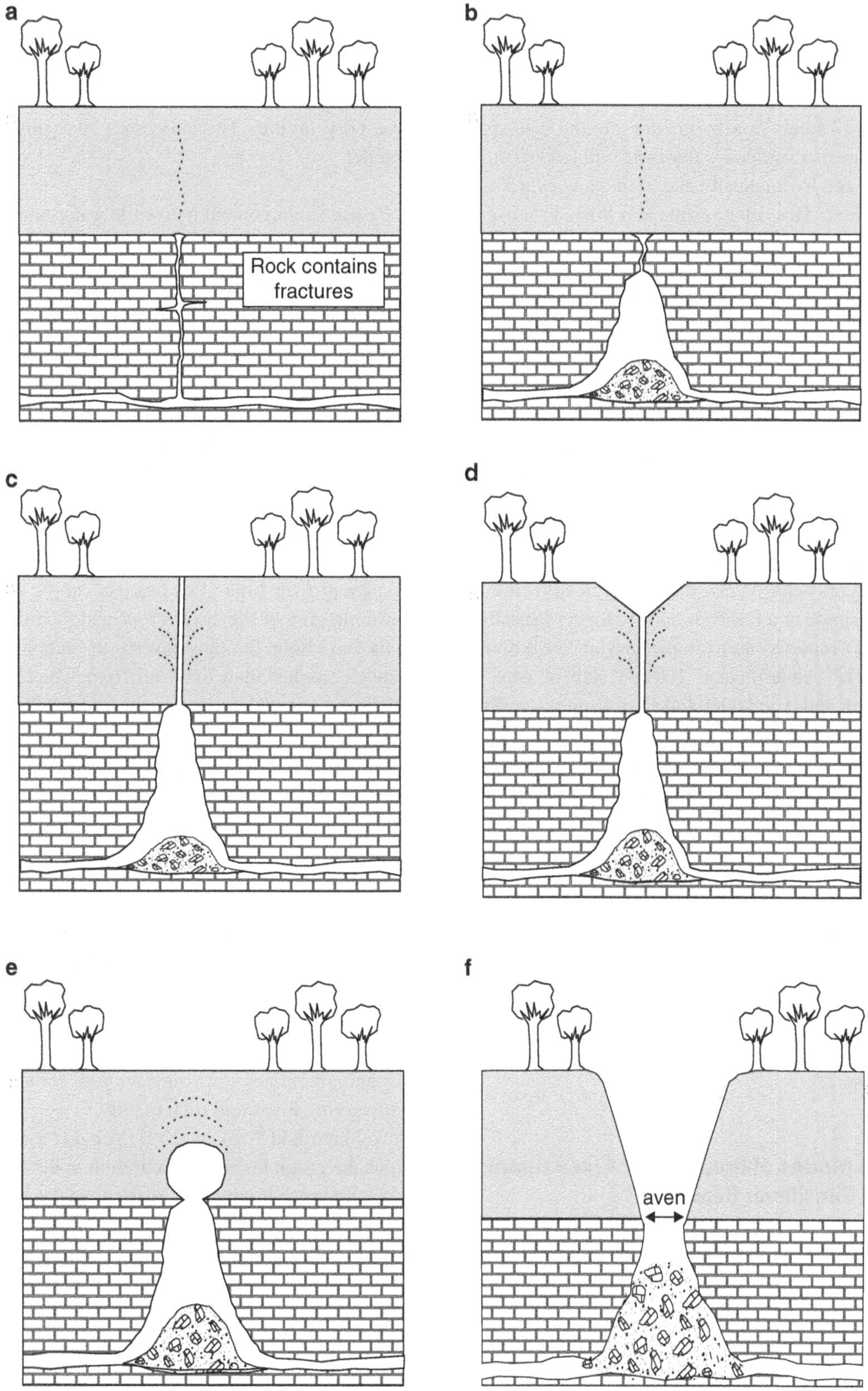

Fig. 9.2 A conceptual model of the development of a very large cover collapse sinkhole over a cavern. (**a**) A cave system develops along bedding plane. (**b**) Enlarged cavern develops along fracture. (**c**) Enlarged cavern breaks through and causes soil piping in sandy unconsolidated materials. (**d**) Enlarged cavern continues to take soils and sinkhole expands. (**e**) Enlarged cavern breaks through and causes void in clayey, cohesive materials. (**f**) Void in soil reaches critical state and collapse into cavern occurs

into the rock, the properties of the sediment and the presence of a hydraulic head. The properties of the sediment (its range of particle size, density, viscosity, cohesiveness and degree of cementation, if any) will affect the rate and amount of the material that will freely flow by gravity into the void space.

The presence of a surficial water table will tend to fluidize the sediments and hydraulically add to the downward movement of sediment. This often results in a flushing noise during the collapse. In many cases, the surficial water table becomes depressed and the sinkhole may be dry immediately after the collapse. The sinkhole eventually fills as the water level slowly recovers and conditions begin to stabilize.

9.2 Mechanics of Cavern Breakdown

In 1951, Davies presented the mechanics of cavern breakdown by brittle fracture. He describes two types of breakdown, block and slab. Block breakdown occurs with blocks that are typically short laterally and have significant thickness. Slab breakdown occurs with blocks that are typically limited in thickness to a single or a few beds and are longer laterally.

Ellis (1994) reported an example of slab breakdown in Mammoth Cave. An estimated 100-ton slab of roof rock about 6 by 21 m and 0.3 m thick fell 12 m from the ceiling of the rotunda near the entrance to Mammoth Cave. It is thought that the extremely cold air caused the slab to pry loose because of wedging effect of freezing moisture. Temperatures outside had dipped as low as −16 °F and were accompanied by heavy winds.

An example of massive block breakdown occurred instantaneously over a 6.8-ha area referred to as the "central collapse area" (CCA) in the Tobin limestone mine in the Kansas City area. The 6.8-ha CCA occurred shortly after the area had been mined in 1973. It's air blast shattered windows of the trucks within the mine and reportedly was recorded on seismographs some 48 km away. Further details are discussed in Chap. 25.

9.2.1 Two Modes of Breakdown: Fixed Beam and Cantilever Beam

Fixed beam breakdown occurs when the rock strata of the cave roof extend from one side of the cavern to the other, supported at both ends with no fractures in between (Fig. 9.3a). Cantilever beams are those supported on one side of the cavern and extend into the cavern some distance unsupported (Fig. 9.3b). Long fixed beam strata will eventually sag and fracture resulting in a cantilever beam (Davies 1951). Failure of roof rock tends to occur gradually a slab at a time as local stress ultimately leads to failure. The simplest approach for roof breakdown is the brittle fracturing model based upon work from the mining literature (Davies 1951). White (2005) provides a further summary of breakdown and their mechanics including simple equations for analysis of fixed and cantilever beams.

White (2005) summarizes the factors that contribute to cave roof failure. The processes that initiate breakdown include:

- Passage enlargement below the water table as water flows through the cave
- Removal of water removes the buoyant support (about 42 %) of the roof rock
- Effects of occasional flooding can destabilize the roof
- Action of vadose water flowing into the cave from above enlarges fractures and changes fixed beams into cantilever beams.
- Ice wedging within joints further fracture the bedrock
- Crystal wedging, replacement by minerals such gypsum can fracture the bedrock

Under these various conditions, collapse will have slowly progressed over time. The process can be observed by the slow enlarging of the cavern roof and the pile of rock debris on the cave floor. In many caverns the debris pile on the floor is much smaller than the void from which the rock came. This is due to rapid flowing water through the conduit system resulting in dissolution of the fallen rock and the transport of rubble and sediment through the conduit system thereby maintaining large open conduits.

9.3 Thickness of Rock Needed to Prevent Surface Subsidence or Collapse

Various rules of thumb have been used to predict safe levels of rock thickness necessary in order to prevent subsidence or collapse. Waltham and Fookes (2003), Waltham et al. (2005), and Waltham and Lu (2007) all provide a number of examples and guidelines showing cave roof stability in limestone with varying loads and rock quality.

Waltham and Fookes (2003) suggest a rule of thumb in which the ground is considered stable if the overlaying thickness of the rock is equal or greater than the roof span excluding any soil cover. This concept is conservative and would apply to a limestone with a normal density of fractures and bedding planes and excludes any heavily fissured limestone.

Waltham et al. (2005) provide a discussion on safe cover thickness over caves. Their rule of thumb suggests that the roof thickness should exceed 70 % of the cave width. Guidelines for safe cover must be increased to accommodate gypsum, chalk, and weak limestone.

Full scale loading tests carried out by Waltham and Swift (2004) within man-made caves in the sandstones of Nottingham provide further insight to the thickness of rock necessary for stability over such caves or mines. They

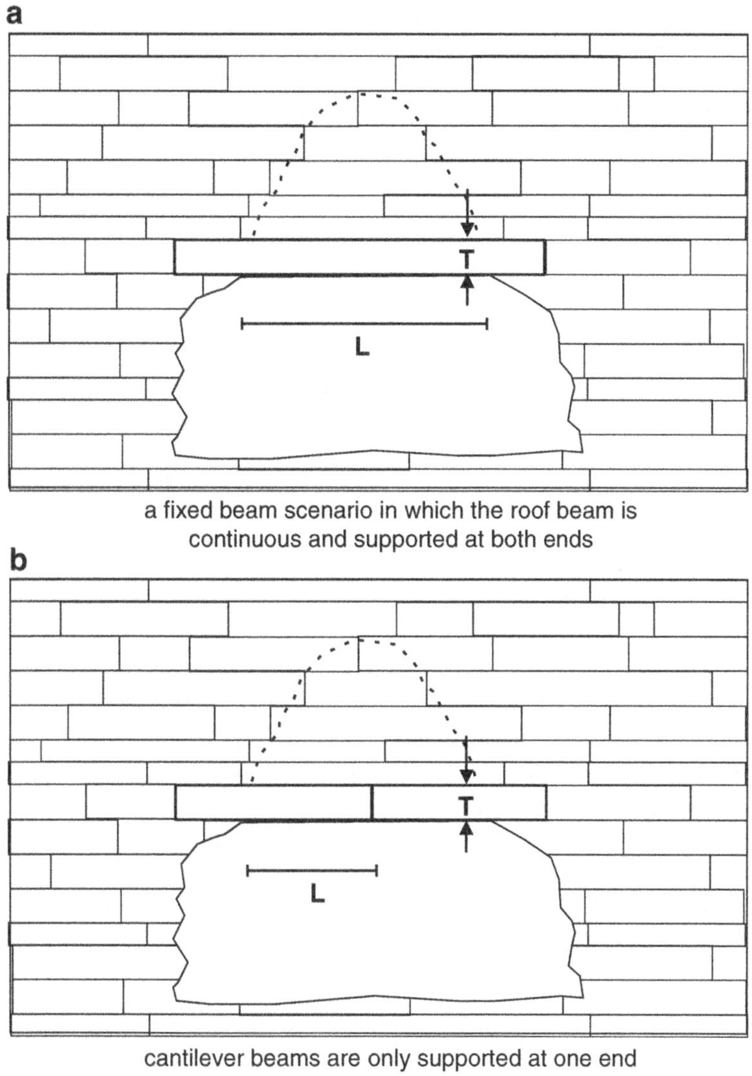

Fig. 9.3 A fixed beam (**a**) and a cantilever beam (**b**). The length (*L*) and thickness (*T*) of the beam as well as its density and flexural strength are factors in roof stability

concluded that thickness of overlaying rock should exceed 50 % of cave width for most projects on typical karst limestone. Where higher loads are imposed, cover thickness would need to exceed 70 % of cave width to be considered stable. Numeric modeling has also been done by Waltham and Lu (2007) to determine the thickness of rock cover needed to avoid subsidence over a cave when additional loading is imposed by engineered structures. An intact rock-cover thickness that exceeds half the cave's width appears to be safe in most karst terrains in strong limestone.

Waltham et al. (2005) discuss the range of roof thickness required for safe conditions with various rock mass ratings (RMR). If the RMR for a typical cavernous limestone is taken conservatively as between 30 and 40, a cover ratio of roof thickness being half the cave width (t=w/2) appears to be adequate for most engineering practices.

All such rules of thumb are only guidelines and must be tempered by site-specific data and observations. Site-specific data should include establishing bed thickness, rock quality, fracture spacing and apertures along with identifying other conditions such as lineaments, major fractures, incised valleys and structure which may indicate weak rock conditions. This data must then be combined with experienced professional judgment.

9.4 Experience from Mine Failures

We can also learn about collapse and subsidence from experience with mines because there is much published literature from the mining industry. Miners, engineers and geologists have often observed roof rock failure in mines shortly after it occurs and have been able to quantify it. However, it is rare that observations and analysis of roof failure in cave systems have been made, although cavers have photographed the results of some failures.

Cavern breakdown and roof failure in a mine are similar with the following exceptions, Davies (1951):

- A lack of blasting in caves
- The mine has a short-term development of 10's of years or so while the cavern has evolved over 100,000's of years or more.
- The stress are fully developed and are in equilibrium within the cavern
- There is an absence of any artificial supports within the cavern

As a result, stress fields around mines tend to be unstable compared to natural cave systems, which have developed over long periods of time where stress tends to be in equilibrium.

Much of the experience that has evolved within the mining industry can be used to help understand the collapse of mines as well as collapse of cave systems. There have been numerous efforts to predict the magnitude and extent of mine subsidence (mostly coal mines). While good mine design and practice minimizes failures, the problems of subsidence remain pervasive.

Gray and Salver (1971) provide an early summary of state-of-the-predictive capabilities for subsidence over coal mines. The National Coal Boards *Subsidence Engineering Handbook* (1975) became the definitive early reference for coal mine subsidence in England and elsewhere. More recent work includes; an excellent review by Whittaker and Reddish (1989) along with a recent summary of Empirical Rock Failure Criteria (Sheorey 1997), which provides an excellent overview of approaches to mine roof failure and the resulting surface subsidence. Kane et al. (1993) provide an excellent review of the geologic factors affecting coal mine roof stability in the eastern United States. These approaches utilize a variety of numeric and empirical methods to predict the lateral extent and depth of surface subsidence.

While most of these authors mention the importance and impact of geologic conditions, they do not discuss them in any detail. For example, Whittaker and Reddish (1989) suggests that local geology and natural strength of the immediate roof (and overlying strata) are important factors. However, the mining dimensions and geometry of workings are of equal importance and should be considered in making assessments of subsidence risks above such mines. Kane et al. (1993) found that geologic factors are a major factor in failure of coal mine roofs. Geologic conditions which contributed to roof failure in underground coal mines include both lithologic and structural discontinuities. Lithologic conditions may include paleochannels, scours, rider seams, clay veins, or crevasse splay deposits. Fractures (joints and faults) also contribute to roof falls. Stress fields both tectonic and mining-induced can determine whether failure will occur, and the mode of such failure. They suggest that paleoenvironmental studies are necessary in determining hazardous roof conditions and should be incorporated into the mine design.

In summary, where mines (and natural cavities) are near the surface with little rock cover, roof collapse and subsidence are common. However, in those situations where there is abundant rock cover, with strong rock beds, the risk of subsidence or collapse is lower, yet it may still occur. Both cavern and mine breakdown are a function of the site-specific geologic conditions, the width of span along with fracture spacing, hydrologic conditions, and the thickness of the overlaying rock.

The authors have spent hundreds of hours in the Tobin Limestone Mine and many hours in other abandoned and modified limestone mines in the Kansas City area. In some cases, stable roof spans of up to 30 m have been identified while shorter spans of 15 m have commonly failed. However, our observations in the Kansas City limestone mines show the typical localized breakdown begins as localized spalling as described by Hasan et al. (1988).

This localized spalling is commonly initiated where paleochannels of rubble rock have breached the mine roof exposing the "rubble zone". The rubble zone is a calcite mudstone and silty claystone (shale). Some of the rubble zone contains swelling clay minerals (smectite-illite mixed layers). When exposed to the moist air within the mine, localized breakdown occurs, typically less than a cubic yard. The zone of breakdown slowly enlarges and may ultimately include up to ½ ha or more of collapse extending upward 3 m and in some cases more. The presence of the rubble zone and its impact on mine collapse is discussed in more detail in Chap. 25. By paying attention to lithologic and structural discontinuities such as paleochannels, clay veins, joints and faults remedial action can then be taken (Kane et al. 1993).

There are three conditions that are cause for concern of roof failure in caves or mines which can result in surface subsidence or collapse (Fig. 9.4):

- a wide roof span at shallow depth (a).
- a roof span at depth with weakened rock cover, such as an incised valley and fractured rock which has allowed infiltration of surface waters resulting in weathering and weakened rock (b).
- the roof of a localized enlarged cavern which is approaching a critical state of stability (c).

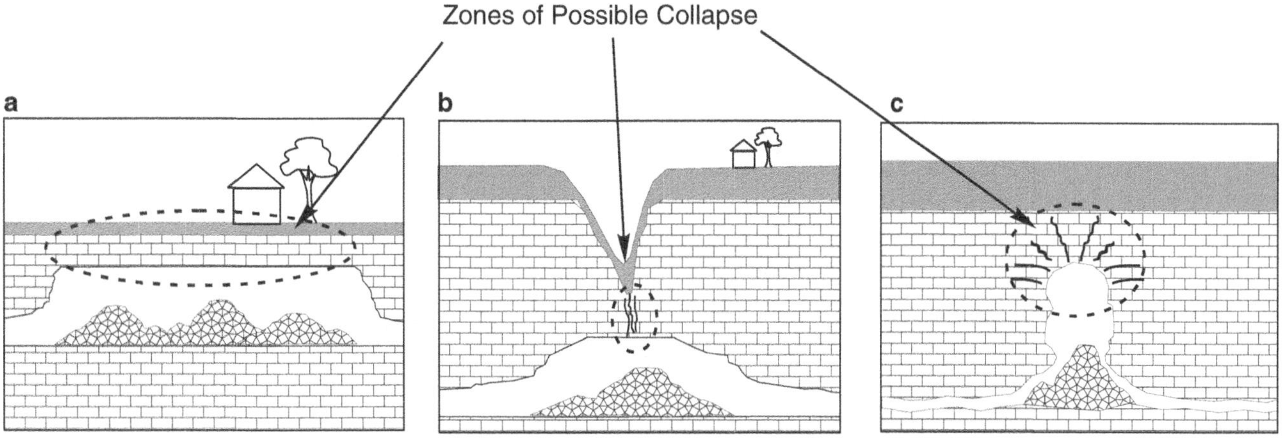

Fig. 9.4 Examples of where roof collapse can result in surface subsidence or collapse. (**a**) Wide span with thin rock cover. (**b**) A thick overlying rock with a localized zone of weakened rock cover such as an incised valley or major fracture. (**c**) Enlarged cavern in a critical state

9.4.1 Bulking of Fallen Roof Rock

As rock falls from the roof of a cave or a mine it commonly falls as small to medium size blocks that rotate while falling and break, ending up in a random pile of rock. This broken rock will occupy a greater volume than they did when the rock was in place (Fig. 9.5a) this is known as bulking factor. As the upward collapse continues, the greater volume of the fallen rock begins to fill the void space (Fig. 9.5b). At some point, the volume of bulked rock will nearly or totally fill up the void space thereby, preventing further roof failure (Fig. 9.5c).

Typical bulking factors from the coal mining industry are 0.3–0.5. That means that the expanded volume of fallen rock is 30–50 % more than its in-situ volume due to its breaking into pieces (Piggott and Eynon 1978). This factor may be as low as 0.2 if claystone and shale are present (Dunrud and Osterwald 1980). Blasted rock will typically have the highest bulking factors of 50–60 % (Colaizzi, 1990, personal communication).

The authors have made extensive direct measurements and photo documentation of bulking in a limestone mine in the Kansas City area. Measurements were made of the volume of the void created by the fallen rock and the volume of the rock rubble pile on the mine floor (Fig. 9.6a). These bulking measurements yielded an average bulking factor of 0.4 (Benson et al. 2000). However, in some cases rock may fall as a massive slab resulting in bulking factor close to zero (Fig. 9.6b). More details from this case history is presented in Chap. 25. Piggott and Eynon (1978) and Bell (2004) provide equations for various geometries and plots of maximum height of collapse for various bulking factors.

Piles of breakdown rock on the floor are random in size and orientation and highly permeable. Massive breakdown (Fig. 3.15), that completely blocks a cave passage, is referred to as terminal breakdown (White 2005). In many cases, the quantity of debris on the cave floor is much less than the volume of rock that has fallen from the roof. This indicates that large quantities of rock have been removed when water was flowing through the cave system. Examples of such conditions are seen in many large caverns and in sinkholes accessed by cave divers where the volume of fallen rock debris on the cavern floor is substantially less than the volume of rock removed from the cavern roof.

There are factors that can reduce bulking of fallen rock that include:

- The bulking factor may decrease if there is flowing water through the rubble leading to further degradation and transport of the fallen rock.
- The bulking factor will decrease as the void space becomes filled reducing the distance the rock falls and its tendency for the rock slabs to rotate and fracture.
- Some compaction and possible degradation can be expected after the fallen rock has been in place for a while. Rock strata with a high shale content will have less bulking factor because of weaker shale crumbling on impact and being compacted into voids within the rubble pile without adding much to the bulking factor.

Consideration of the bulking is another factor which allows us to estimate the risk of surface collapse for a given rock thickness over a cavern or mine. However, without a detailed understanding of the local rock conditions (rock

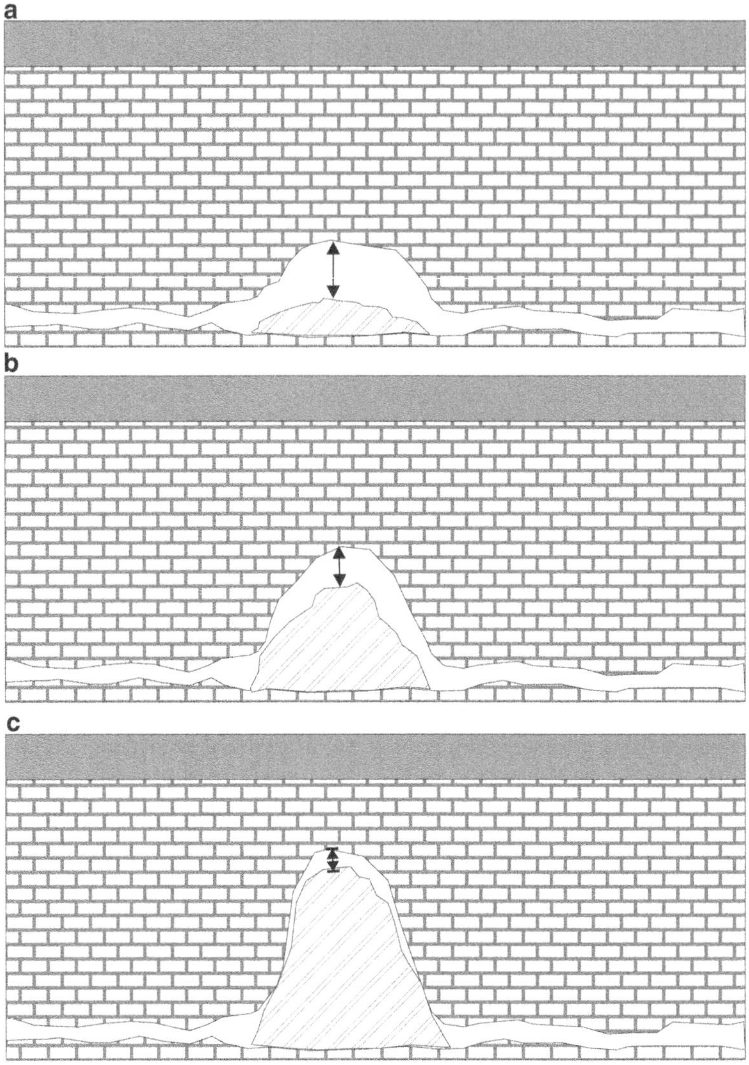

Fig. 9.5 Bulking of fallen rock begins to build up (**a**), slowly filling the void space (**b**), and eventually choking off further collapse (**c**)

strength, fracture patterns, structure and local hydrology) these bulking factors should only be used as guidelines.

9.5 Propagation of Subsidence and Collapse from Great Depths

It is relatively easy to understand the collapse reaching the surface in the presence of thinly bedded and fractured strata over a shallow cave or mine. It is difficult to conceptualize subsidence at the surface due to collapse of very deep natural cavities or mines. Especially when considering the issues of bulking, thick sections of overlaying rock (100's of m or more) with massive strata. Nevertheless, collapse from great depths does occasionally propagate to the surface.

Some of the most extensive karst in the United States occurs in the limestone and dolomite beds of Mississippian age (about 325–345 million years old), which have a combined thickness of over 120 m. Extensive cavern development has occurred in the Mississippian rocks and is recognizable throughout the mid-continent (Fig. 9.7a) (Gentile 1984). The Pennsylvanian System, a section of shale and limestone, about 150–180 m thick lie over the Mississippian limestone and dolomite beds in the Kansas City area. A number of large cavern collapse have occurred within the thick Mississippi Limestone have propagated upward through the Mississippian and Pennsylvanian System (more than 200 m) and have been identified in the Kansas City area road cuts (Fig. 9.7b). There are places where large blocks of bedrock have moved downward a few meters along high angle, normal faults (Gentile 1984).

a

smaller blocks of rock provide greater bulking

b

large blocks of rock provide little bulking

Fig. 9.6 An example of rock bulking that will eventually choke off further collapse (**a**) and an example of a single large slab failure of roof rock approximately 3 by 6 by 2.25 m thick resulting in low bulking (**b**)

Deep mines have also resulted in surface subsidence. A few examples are:

- Dewatering a deep gold mine in South Africa (>1,000 m below grade) resulted in a few rapid catastrophic collapses (Brink 1984).
- A room and pillar coal mine collapse 365 m underground occurred near Birmingham, Alabama, and reached the surface (Colaizzi, 2004, personal communication)
- The Retsof room and pillar salt mine more than 300 m below grade in New York State resulted in surface subsidence (Nieto and Young 1998).

Roof collapse in deep-seated mines or caves can only reach the surface when the overlaying rock has been weakened by some means such as the presence of a major fracture zone, fault or structural weakness. For example an unknown major fracture system may intersect a mine. Rainfall may seep downward in the fracture over long periods of time weakening a zone of rock. Then surface operations require an impoundment be constructed or surface water drainage be diverted concentrating it over the fracture. Or a building and parking lot is built concentrating large amounts of runoff into the fracture contributing to its further weakening and ultimately resulting in a deep cave or mine collapse reaching the surface. Here the presence of either the natural fracture system or the surface modification independently may not have triggered collapse, but when combined, the two, further weaken the rock allowing propagation of the deep collapse to reach the surface.

The structural integrity of natural cave systems and mines are impacted by both geologic conditions and man's activities. The natural geologic and hydrologic conditions above caves or mines combined with certain impacts of mans activities are, in fact, the dominant factors in a collapse propagating to the surface. In many cases when the details are established we find that there is a geologic flaw present, which has lead to the subsidence or collapse.

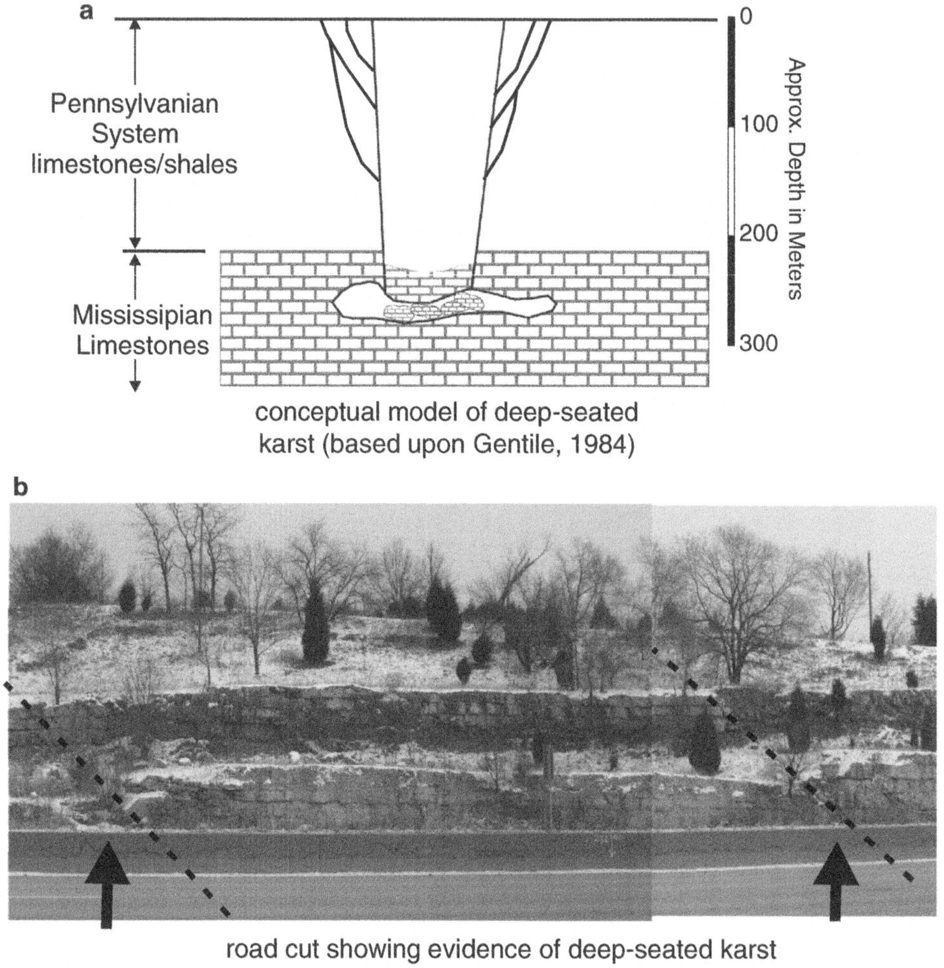

Fig. 9.7 Deep-seated collapse has occurred in the Mississippian Limestone that reached the surface in and around Kansas City area (**a**) (Based on Gentile 1984). Edges of such a paleocollapse along Interstate-470 near Kansas City, Missouri with *arrows* showing the locations of fault scarps (**b**)

References

Bell FG (2004) Engineering geology and construction. Spon Press, London

Benson RC et al (2000) Stabilization against geologic instability-underground documentation of a mine backfilling case history. ASCE Geo-Denver Conference Denver, Colorado, 5–8 August 2000

Brink ABA (1984) A brief review of the South African sinkhole problem. In: Beck BF (ed) Proceedings of the 1st multidisciplinary conference on sinkholes: their geology, engineering and environmental impact, Orlando, Florida. Balkema, Rotterdam, pp 123–127

Davies WE (1951) Mechanics of cave breakdown. Natl Speleol Soc Bull Number 13:36–43

Davis C (2002) Exposing Tennessee's titanic cave chamber. National Geographic Adventure Magazine, March 4(2): on-line extra at www.nationalgeographic.com

Dunrud CR, Osterwald FW (1980) Effects of coal mine subsidence in the Sheridan, Wyoming area, USGS professional paper 1164, Reston, Virginia

Ellis I (1994) Rockfall in Mammoth Cave. NSS News April 52(4):129

Gentile RJ (1984) Paleocollapse structures: longview region, Kansas City, Missouri. Bull Assoc Eng Geol XXI(2):229–247

Gray RE, Salver HA (1971) Discussion of state of predictive art in subsidence engineering. J Soil Mech Found Div, ASCE 97(SM1)

Hasan S et al (1988) Geology of greater Kansas City, Missouri and Kansas, United States of America. Bull Assoc Eng Geol XXV(3):281–341

Kane WF et al (1993) Geologic factors affecting coal mine roof stability in the eastern United States. Bull Assoc Eng Geol 30(2):165–179

National Coal Board (NCB) (1975) Subsidence engineer's handbook. NCB Mining Department, London, 111p

Nieto AS, Young RA (1998) Retsof salt mine collapse and aquifer dewatering, Genesee Valley, Livingston County, New York. In: Borchers JW (ed) Proceedings of the Dr. Joseph F. Poland symposium on land subsidence, case studies and current research, pp 309–325, Star Pub., Belmont, California

Piggott RJ, Eynon P (1978) Ground movements arising from the presence of shallow abandoned mine workings. In: Geddes JD (ed) Proceedings 1st international conference on large ground movements and structures. Pentech Press, London, pp 749–780

Sheorey PR (1997) Empirical rock failure criteria. Balkema, Rotterdam

Waltham AC, Fookes PG (2003) Engineering classification of karst ground conditions. Geol Soc Lond Q J Eng Geol Hydrol 36:101–118
Waltham AC, Lu Z (2007) Natural and anthropogenic rock collapse over open caves. In: Parise M, Gunn J (eds) Natural and anthropogenic hazards in karst areas: recognition, analysis and mitigation. Geologic Society special publications 279. Geologic Society, London, pp 13–21
Waltham AC, Swift GM (2004) Bearing capacity of rock over mined cavities in Nottingham. Eng Geol 75:15–31
Waltham AC et al (2005) Sinkholes and subsidence, karst and cavernous rocks in engineering and construction. Springer-Praxis, Chichester
White WB (1988) Geomorphology and hydrology of karst terrains. Oxford University Press, New York
White EL (2005) Breakdown. In: Culver DC, White WB (eds) Encyclopedia of caves. Elsevier, Burlington, pp 56–60
Whittaker BN, Reddish DJ (1989) Subsidence – occurrence, prediction and control, vol 56, Developments in geotechnical engineering. Elsevier, New York

Insight into the Nature of Cover Collapse Sinkholes

10

Abstract

The focus of this chapter is upon providing insight to the depth, size and nature of the void space in the rock based upon the size and speed of cover collapse sinkholes. Insight is provided by scale model tests of cover collapse sinkholes, mine backfilling as well as data on many large and very large collapse features. Five levels of conceptual models are presented relating the size of the sinkhole to the possible size and nature of void space within the rock. These models can provide a first approximation to the nature of conditions at a site.

10.1 Introduction

Cover collapse sinkholes are also known as dropout sinkholes, subsidence sinkholes, or alluvial sinkholes, (Fig. 3.1 and Table 3.1). These are sinkholes in which sediments have collapsed into the void space within the rock. These sinkholes have a wide range of sizes and occur under a wide range of conditions. Sinkhole dimensions (diameter and depth) along with the volume of sediment lost are commonly measured or estimated. In some cases, we may also know the size of the opening into the rock (aven) along with the time required for the collapse to occur. What we normally do not know is the size, depth, geometry and openness of the void space within the rock, which is often one of the many goals for a site characterization in karst terrain.

Over the years we have had the opportunity to observe the nature and behavior of a variety of cover collapse sinkholes and to investigate the conditions under which these sinkholes have developed. In addition, we have gained some insight into the size and nature of void space in the rock by the amount of sediment lost based upon:

- A review of scale model tests showing the general behavior of cover collapse sinkholes
- Extensive experience in backfilling of mines by others
- Direct observations and measurements of mine backfill materials injected into an abandoned limestone mine

These data have provided us with the ability to develop simplistic conceptual models relating the size of a cover collapse sinkhole to the approximate size and nature of the void space within the rock.

10.2 Insight from Scale Model Sinkhole Tests

Chen and Beck (1989) discuss the use of a 2D sand model to illustrate the development of a cover collapse sinkhole. They used a plexiglass model 132 cm long by 91 cm high with wooden sides and bottom. Its width was adjustable from 6 to 18 cm. Six holes from 0.6 to 5 cm were drilled in the bottom to simulate the opening into the limestone. Twenty-nine different trials were run with a variety of conditions ranging from dry sand to saturated conditions, their results indicated that:

- Gravitational movement of the sediment is a major factor in sinkhole development.
- Very small initial openings (avens) in the limestone can result in large sinkholes if there is sufficient void space deeper within the limestone.
- A more cohesive sediment strata may slow the upward development of the collapse resulting in horizontal enlargement below the more cohesive strata.

R.C. Benson, L.B. Yuhr, *Site Characterization in Karst and Pseudokarst Terraines*,
DOI 10.1007/978-94-017-9924-9_10

- Infiltrating recharge aided the collapse process. The hydraulic head drops discontinuously with the surrounding water table, creating a hydraulic jump. The water flow over the opening is vertically downward with a great scouring effect

Additional sand model results are shown in Whittaker and Reddish (1989) who show the time sequence of collapse with similar results. These scale model test results are in good agreement with many observed sinkholes over known geologic conditions.

10.3 Insight from Mine Backfill Stabilization

Insight into the amount of sediment or material that can flow into a large open void space can be obtained from experience filling abandoned room and pillar coal and limestone mines with fly-ash, sand, and crushed rock. At some point the mine becomes sufficiently full of material so that the system chokes off before the entire mine space is full. This data provides us with insight and practical limits to the amount of material that can be placed into a very large or infinite open space within a mine from a single opening (a borehole).

10.3.1 Comparison Between Sinkholes and Mine Backfilling

The mine backfilling process uses a small diameter borehole (of 20 cm or so). Larger cover collapse sinkholes usually pass through a larger opening in the rock (1–15 m or more). The mine backfilling process also takes place over a few days, longer than most sinkhole collapse that typically occur over a shorter period of time of a few minutes to a day or more. The void space in a mine is laterally uniform and extensive compared to dissolution features like a cave, which are often tubular in extent and highly variable.

The particle sizes of backfill materials (fly ash mixtures, sand and crushed rock) are reasonably similar to the sediment lost into a sinkhole. The backfill material is commonly fluidized similar to the sediment being fluidized by the presence of a surficial water table. In addition, the backfill material is usually under a high head of 15–30 m or more, similar to the sediment in a large cover collapse sinkhole. While there are differences between the two processes there are sufficient similarities to make useable comparisons between natural sinkholes and the mine backfilling process.

Data from mine backfilling has been obtained from three sources:

- From mine backfilling literature which include scale model tests and full scale projects by the Bureau of Mines (Colaizzi et al. 1981)
- Experience backfilling abandoned coal and limestone mines from Goodson Associates (Colaizzi, 1990, personal communication) and
- Direct observations and documentation from mine backfilling of the Tobin limestone mine in Kansas City Kansas (Benson et al. 2000).

10.3.1.1 Scale Model Tests and Experience with Full Scale Projects

Colaizzi et al. (1981) presented the results of two scale model tests of backfilling of mines. The conclusions of these tests were:

- The slurry was generally distributed over an area of approximately 0.4 ha.
- Deposits within the model were not dependent upon the concentration of the sediment or injection velocity, providing the velocity was high enough to transport the slurry through the injection pipe.
- Backfill material extended further from the injection hole in a flooded mine as opposed to a dry mine.
- The extent to which backfill material was transported was more or less independent of the slope of the model.
- Backfill material was transported over areas of roof falls that blocked corridors.
- Backfill material was transported into blind areas as long as a circulation of sediment laden water occurs in and out of the blind areas.
- As resistance to flow of the slurry develops, a new channel is formed in another direction along a line of less resistance.
- At the end of each test a backpressure is developed within the injection system and a final breakout of flow occurs along an unobstructed corridor. The entire flow from the injection pipe then travel down this channel transporting material long distances from the injection pipe to the perimeter of fill material.

In addition, three full-scale backfilling projects were carried out by the Bureau of Mines (Colaizzi et al. 1981). The largest amount of fill placed into a single fillhole at each of the three sites respectively, was 29,800, 53,300 and 63,500 m^3 (Table 10.1).

10.3.1.2 Experience in Backfilling of Mines

Goodson and Associates of Denver, Colorado have had extensive experience stabilizing both coal and limestone mines with fly ash, sand and crushed rock. They have placed up to 54,500 m^3 in a 2.5 m high mine through a single cased

Table 10.1 Summary of mine backfill data

Name	Dimensions	Maximum amount of backfill in one borehole (m^3)
Bureau of Mines Report (Colaizzi et al. 1981)	Two scale model test and three full scale backfilling projects	29,800
		53,400
		63,500
Backfilling of coal mine, Goodson Associates, (Colaizzi, 1990, personal communication)	2.5 m high coal mine with an estimated 60 % extraction	54,500
Backfilling with crushed limestone in the Tobin mine (Benson et al. 2000)	4.25 m high with 85 % extraction	Up to 7,600–15,300[a]

[a]These numbers are on the low side because in all cases fill was being placed against mine walls or adjacent to previous fill on at least one side, thereby limiting the total amount of fill material that could be put into a single fillhole

borehole (20–33 cm) before the system became choked off. Fly-ash mixtures and sand will flow for 30 m or more from the fillhole into an air or water-filled mine. In some cases, flow up to 300 m from the fillhole has been observed (Colaizzi, 1990, personal communication).

10.3.1.3 Direct Observations of Mine Backfilling in a Limestone Mine

A landfill was being constructed over an abandoned limestone mine in the Kansas City area. The Tobin mine was a room and pillar mine with a height of 4.25 m, pillars of approximately 7.5 by 7.5 m and spaced on about 18 m centers. Even though there was 52 m of limestone and shale over the mine, the state required that the mine be filled to 90 % or 3.8 m to prevent possible collapse at the surface.

Goodson Associates was employed to fill the mine using a crushed rock slurry. The uppermost rock at the site (Argentine Limestone) was removed by ripper and crushed to 1.25 cm. The rock was then mixed with water and pumped through a 20 cm pipe to an open borehole into the mine at 5,680 l/min. Water was pumped from the mine to the surface and injected back into the mine as part of the fill process.

Direct observations, photographs and video of the mine backfilling process were made from within the mine as the fill was being injected. Initially we were within a few meters from the injection hole. The fill material would eventually reach the roof of the mine and would then cut a channel for the crushed rock to flow to the perimeter (Fig. 10.1a). At approximately 20 min intervals the channel would choke-off and a new channel would be cut in a different direction allowing crushed rock to flow to the perimeter. This sequence of breakthroughs would continue to occur in a somewhat random pattern, 360° from the injection point until an area of about 23–30 m radius around the fillhole was backfilled to the mine-roof (100 % full) and filled to the 90 % level over a distance up to 68 m from the fillhole (Fig. 10.1b). At this point, the system would become choked-off and the fill process would cease. The slope (angle of repose) around the perimeter of fill ranged from 4 to 9° and averaged 6°.

As part of a QA/QC process to document conditions of backfill for the State of Kansas, both photographic as well as observations and measurements were made of the extent of fill after each fillhole had choked off. Additional details are included in Benson et al. (2000) and in Chap. 25, the first case history.

A spacing of approximately 45–60 m was selected for the fillhole borings to achieve 90–100 % filling of the mine. The amount of fill placed in each hole would vary from 7,600 to more than 15,300 m^3. This volume is on the low side since fill was being placed against mine walls, and or existing mine fill on one side or more. As a result the amount of fill in each borehole would be limited depending upon the location of nearby mine walls and previous fill material. Portions of the mine were dry while other areas of the mine were filled with water. There was little difference between the amount of fill that could be placed in a dry or wet portion of the mine.

10.3.1.4 Conclusions from Mine Backfilling

The experience from backfilling mines (Table 10.1) shows that there are limits to the amount of fill material which can be placed into a large open space through a single opening. The amounts of fill material ranged from 7,600 to 63,500 m^3. These observations and measurements show that:

- There is a maximum amount of backfill material that can be placed in a single fillhole into an open mine space, and
- The void space will generally be greater than the maximum volume of backfill material.

This information can provide insight into understanding of cover collapse sinkholes.

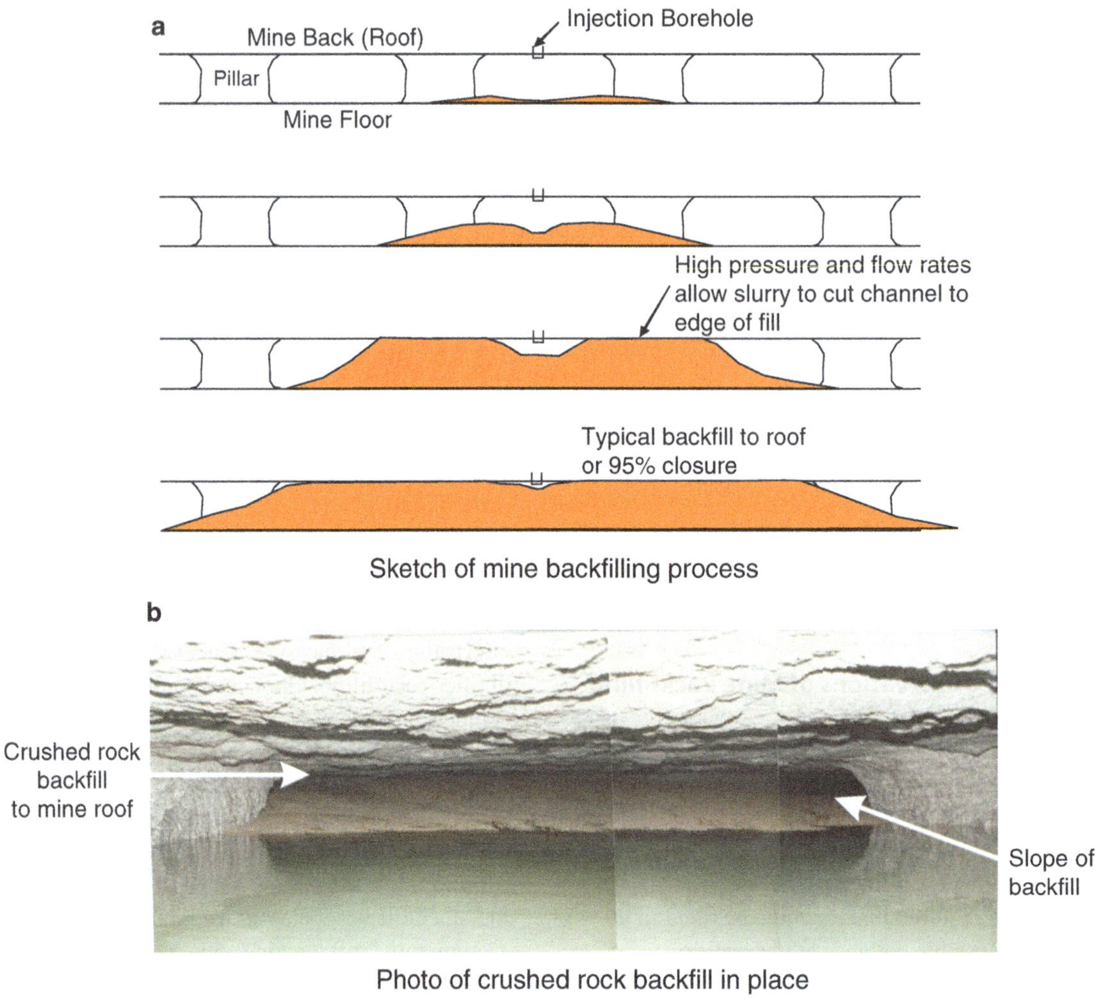

Fig. 10.1 A sketch of the crushed rock fill process showing progressive amounts of fill (**a**). A photo of shows the crushed rock fill to the roof of the mine and the slope of the edge of fill (**b**) at this time, water had filled the 4.25 m mine, more than half full

10.4 Conceptual Models of Cover Collapse Sinkholes

Five generic conceptual models have been developed which relate the size of a sinkhole collapse (volume of material) to possible void space in the subsurface. We have simplified issues and are assuming a one-time collapse event where the sediment is not washed away by subsurface flowing water. These conceptual models are based upon our sinkhole investigation experience, insight from scale model sinkhole tests and experience from backfilling of mines. The models are simplified and assume one dominant type of dissolution feature. The many other variables that may be present have been ignored for the sake of simplicity. The five models include:

- Small cover collapse sinkholes associated with voids in the epikarst zone or top of rock.
- Intermediate size cover collapse sinkholes associated with fractured rock
- Larger cover collapse sinkholes associated with a cave system
- Very large cover collapse sinkholes associated with the presence and collapse of a large cavern
- Extremely large cover collapse sinkholes associated with evaporate karst and some types of pseudokarst

10.4.1 Small Cover Collapse Sinkholes

Smaller cover collapse sinkholes are due to movement of sediments into isolated voids, fissures, and cavities in the epikarst zone or upper surface of the rock. These sinkholes are generally small (on the order of 1–5 m in diameter and volumes of 1–15 m^3) and result in limited amount of sediment lost. While the percentage of dissolved bedrock within the epikarst can range up to 50 % or more, a high percentage of this void space will be filled with sediment (Aley 1997).

This results in small voids, which are commonly isolated or poorly interconnected.

The conceptual model for this size of cover collapse includes a shallow top of rock (about 15 m or so, our guesstimate) and overlying sediment that is dominantly sands with limited clay content or cemented layers. This allows piping or raveling of soils to occur vertically over small isolated void spaces and results in small individual sinkholes.

Two such examples include:

- A site in North Carolina where the limestone was 12 m below grade. A small diameter sinkhole (about 0.3–0.6 m in diameter) with vertical sides would slowly work its way to the surface through mostly sandy soil (Fig. 8.3a). Within a few weeks or so, the sinkhole would enlarge to about 3 m in diameter with more sloped sides (Fig. 8.3b).
- A site in west central Florida had 760 sinkholes develop overnight (Fig. 7.4). The vast majority of them were small, a meter or so in diameter (Fig. 7.5a). These smaller sinkholes were likely associated with small isolated voids within the upper surface of the limestone about 12 m below grade.

While sinkholes originating in the epikarst generally manifest themselves as small individual sinkholes, the individual sinkholes may some times merge and appear as larger ones. In addition, when the sediment thickness is small the dimensions of the cover collapse sinkholes are quite limited. As the sediment thickness increases and the void space in the epikarst increase the dimensions of the sinkholes increase but they are almost always small and isolated.

10.4.2 Intermediate Cover Collapse Sinkholes

Intermediate cover collapse sinkholes are 10–15 m in diameter with a material loss of greater than a few 100 up to 1,000 m^3 and occur when the rock is highly fractured. A tertiary fracture system within the massive rock can accommodate a significant amount of unconsolidated material only if the fracture apertures are wide, well interconnected and are not clogged with sediment. In addition, the sediment moving into the fracture system needs to be in a fluidized state.

Waltham et al. (2005) presents a conceptual model of a tertiary fractured rock system. It consists of a massive limestone, with fissures 0.1 m wide spaced every 10 m on each of three rectilinear orthogonal fracture systems, and with conduits 0.6 m in diameter along each fracture intersection. This results in a mean cavernous porosity of only 4.6 %.

Such a fracture system could result in a cover collapse sinkhole that could consume a volume of sediment on the order of 100–1,000 m^3 or more. The sediment would have to be in a fluidized state with a large hydraulic head to accommodate larger volumes of sediment.

10.4.3 Large Cover Collapse Sinkholes

Large cover collapse sinkholes include those that are 30–50 m in diameter with a material loss of greater than a few 1,000–50,000 m^3. A conceptual model for this size of collapse would be a typical open cave system.

Based upon data from Mammoth Cave its mean cavernous porosity has been estimated by Quinlan (1989, personal communication) at about 5 % and by Worthington et al. (2000) between 0.36 and 3.3 % of the subsurface area. This is very similar to the fractured rock model by Waltham et al. (2005) of 4.6 %. While the fracture systems are generally more spatially prevalent they tend to be smaller and tend to be in-filled thereby limiting the amount of sediment they can accommodate. Caves are less spatially prevalent, but tend to have a larger open cross section that can accept greater amounts of sediment than fractured rock. Therefore, it is reasonable that sinkholes over a cave system would be larger than a fractured rock system.

The open void space of a cave system would be much less than that of an open mine due to the typical linear nature and meandering of a cave system. Since there is a limited amount of fill material that can be placed into an open mine (Table 10.1) this suggests that there is an inherent limit as to the amount of sediments that can be transferred into a cave system with less open space.

10.4.4 Very Large Cover Collapse Sinkholes

Very large cover collapse sinkholes, 100 m in diameter or more, are those that have consumed unusually large quantities of sediment 100,000–200,000 m^3 or more. In many cases these large sinkholes have not only consumed large quantities of sediment, but numerous large objects such as trees, structures, vehicles and even drill rigs.

Many karst texts, professional papers, and reports discuss these very large collapse features. However, their illustrations often show the geologic cross section under a large cover collapse sinkhole with relatively small fissures or voids limited to the uppermost rock. For example, the initial conceptual model of Winter Park Sinkhole (Fig. 10.2a) showed the original geologic cross section with relatively small fissures tapering to zero with no cavernous system. This suggested a limited void space into which large amounts of sediment and other large debris could flow. This sinkhole had quite an appetite and consumed two 18 m trees, a house, part of an Olympic swimming pool, part of a city street, a number of vehicles and more than 229,000 m^3 of sediments.

Based upon experience with mine backfilling (Table 10.1) it would be impossible to lose that much sediment and other debris into small narrow fissures. Figure 10.2b shows a revised conceptual model with a relatively large open cavern within the limestone that would be necessary to have accommodated all of this unconsolidated material as well as the large amount of debris. It is the presence of this large cavern, that eventually failed and lead to this very large sinkhole.

In addition, the opening into the rock must be sufficiently large to enable such large items to pass into the deeper cavity without clogging the opening in the rock. In the case of the Winter Park Sinkhole (Fig. 3.7), the initial opening within the uppermost sands and the deeper clay aquitard may have been only a meter or so. After the surface collapse occurred and sediment began to move downward, the opening in the sands and aquitard are rapidly enlarged by the large volume of water and debris. Even the size of the opening in the limestone is probably enlarged by erosion. The final size of the opening at the Winter Park sinkhole was measured at 16 m (Scott, 2005, personal communication).

Table 10.2 includes examples of very large cover collapse sinkholes that have already been discussed throughout this book. Included are examples where large amounts of material have disappeared into a sinkhole and in some cases large trees, structures, and vehicles have also disappeared. When very large amounts of sediment are lost (100,000–200,000 m^3 or more) along with other large debris, a large cave system along with a cavern must exist within the massive rock. There also must be, or must quickly develop, a sufficiently large opening (aven) in the top of rock to allow large quantities of other debris to flow into the cavernous system. In each of these cases there were large caverns that had developed over geologic time and collapse was finally triggered by some event.

10.4.5 Extremely Large Cover Collapse Sinkholes

Table 10.3 includes examples of extremely large sinkholes, more than 200 m in diameter that have occurred in both evaporate karst and certain pseudokarst conditions. However, this type of collapse must be treated independently because of their wide range of variations from typical karst in carbonate rock. The extremely large cover collapse sinkholes in evaporate karst are often a result of dissolution mining of salt and brine extraction which resulted in a large cavity within the salt (Fig. 7.11). Unless very carefully engineered and managed these large cavities have been known to collapse resulting in large sinkholes. Examples from west Texas and southeastern New Mexico have resulted in sinkholes with dimensions of as much as 240 by 186 m and estimated losses of sediment up to 1,300,000 m^3.

Table 10.3 also includes a few examples of pseudokarst that are noteworthy and have resulted in extremely large sinkholes and losses. For example:

- Two sinkholes in Guatemala were the result of a sewer line of only 2.25 m diameter about 40–50 m below the surface. Yet the sediment lost was more than 42,000 m^3 and included five homes, a three-story building and six people (Hermosilla 2012).
- The three-story crusher plant at a South African Goldmine disappeared down a huge hole (Brink 1984).
- The Jefferson Island salt mine in Louisiana, where the initial opening was a borehole of about 35 cm in diameter. It then became rapidly enlarged as a very large volume of water and the abrasion of sediment scour out and enlarge the opening, which allowed 56,536,000 m^3 of sediment, a large oil drilling rig, barges and a tug boat to disappear (Johnson 2003).

10.4.6 Conceptual Models and Their Limitations

The general features of the five conceptual models are summarized in Table 10.4. These conceptual models provide an estimate of the relative depth, type and size of open void space within the rock based upon the size and depth of the cover collapse sinkhole along with the volume of sediment lost.

The range of dimensions and volume of sediments in Table 10.4 and Fig. 10.3 are estimates and not intended as absolute values, there can be a wide range of overlap in the values. These models are subjective first approximations of conditions to guide further work and there will be exceptions due to the many variables involved. Only after we have obtained additional site-specific data from the site characterization effort will we be able to better characterize the final nature and magnitude of the open space within the rock.

A plot of the volume of material lost versus estimated void space within the rock is shown in Fig. 10.3. In most cases, the void space will often be much greater than the material lost. This is supported by the scale models of mine backfilling and actual mine backfill data that shows that a limited amount of material can be placed in an open mine from a single fillhole. The system then becomes choked off preventing additional material from entering the mine. Sinkholes likely behave similarly where there is a limit on the material that can be moved into the open void space before the system is choked off leaving the void space less than 100 % filled.

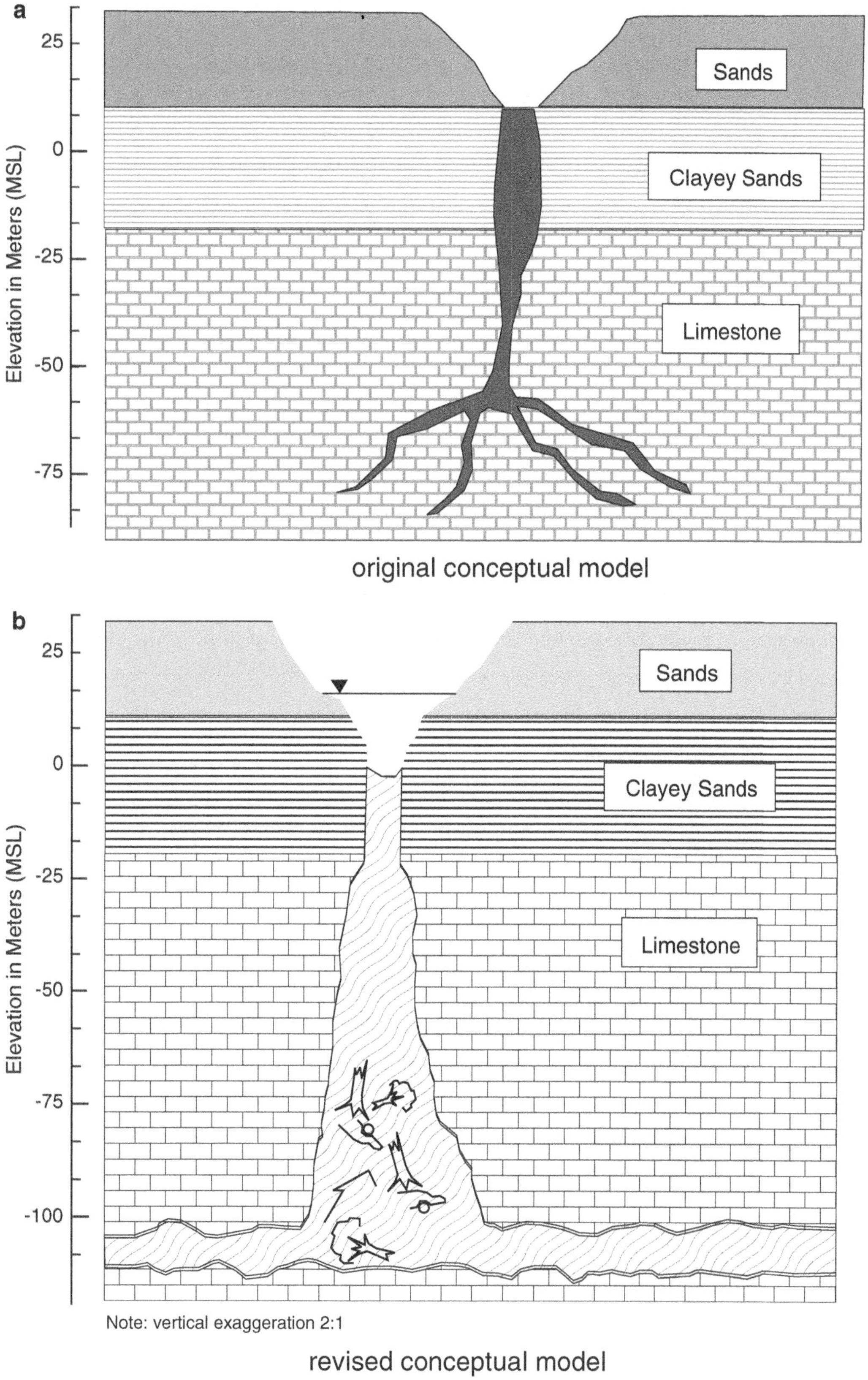

Fig. 10.2 The original conceptual model of the geologic conditions at the Winter Park sinkhole shows limited void space within the rock (**a**). A revised conceptual cross section of the Winter Park sinkhole shows a more realistic geologic condition with a large cavern at depth that has sufficient space to accommodate the large volume of sediment and other debris (**b**)

Table 10.2 Examples of very large cover collapse sinkholes

Site name	Size of sinkhole (m)	Estimated amount of material lost (m^3)	Type of material lost	Rate of collapse
Winter Park Sinkhole, Orlando Florida	100 × 106 m and 30 m deep	>229,000	Sediments, a house, large trees, portion of a large pool and multiple vehicles	Over 36 h
Guest sinkhole west central Florida	About 90 m dia. and 90 m deep	>191,000	Sediment, a drill rig, water truck, pipe truck and large trees	Rapid (within minutes)
Gypstacks in west central Florida	48 m dia	113,200	Phosphogypsum waste	Approx. 6 h

Table 10.3 Examples of extremely large cover collapse sinkholes

Site name	Size of sinkhole (m)	Estimated amount of material lost (m^3)	Type of material lost	Rate of collapse
A sinkhole north of Carlsbad, New Mexico	111 m dia. and 45 m deep	432,400	Sediment	Initial collapse in minutes then expanded for more than a week
Wink Sink 1 in Texas	110 m dia. and 34 m deep	159,000	Sediment	
Wink Sink 2 in Texas	240 × 186 m	1,330,000	Sediment	
Gold Mining in South Africa	55 m dia. and 30 m deep	>500,000 est.	A three story crusher plant along with 29 people	Very rapid
Diamond Crystal Salt Mine, Jefferson Island, Louisiana	Up to 800 m wide	56,536,000	Sediment and water draining Lake Peigneur, an oil drilling rig, barges and a tug boat	Hours
Coal Waste Impoundment, Kentucky	Impoundment failure 30 ha	945,450	Waste coal slurry	Rapid
Sinkholes in Guatemala City	Up to 30 m dia. and 60 m deep	Up to 42,400	Sediment along with buildings and people	Rapid

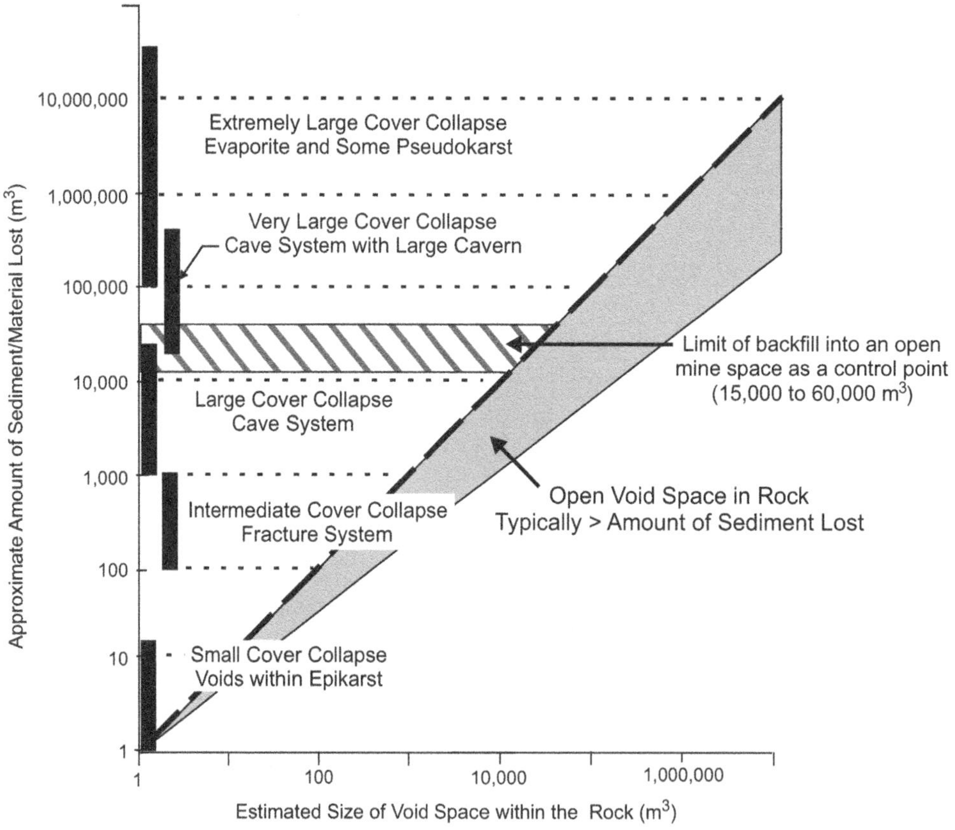

Fig. 10.3 The relationship between sinkhole size (based upon the volume of material lost) and expected void space in subsurface rock for a cover collapse sinkhole

Table 10.4 Summary of cover collapse sinkhole conceptual models

Relative size of the cover collapse sinkhole	Diameter of sinkhole (m)	Volume of sediments lost (m^3)	Nature of the void space within the rock
Small	1–5 m	1–15	Typically associated with isolated voids within the epikarst or top of tock
Intermediate	10–15 m	100–1,000	Open tertiary fracture system
Large	30–50 m	1,000–50,000	Cave system
Very large	100 m or more	100,000–200,000 or more	Large cavern associated with a cave system
Extremely large	More than 200 m	100,000 to more than 50,000,000	Evaporite and pseudokarst

General conditions and assumptions include:

- These conceptual models are based upon a single collapse event ignoring changes with time such as hydrologic transport of sediments.
- Sinkhole size is directly related to sediment thickness along with the size of the void space within the rock.
- If there is a surficial aquifer or hydraulic head to fluidize the sediments they can flow more rapidly and greater quantities can flow into a given void space.
- Evaporite karst and pseudokarst sinkholes are unique cases and must be considered separately.

The data in Fig. 10.3 can be utilized in conjunction with the sinkhole classification (Fig. 3.1), the evolution of karst (Fig. 3.3) and the models of karst maturity (Fig. 4.1) to develop a more complete understanding of the potential karst or pseudokarst conditions that may exist at a particular site.

References

Aley T (1997) Groundwater tracing in the epikarst. In: Beck B, Stephenson JB (eds) Proceedings of the 6th multidisciplinary conference on sinkholes and the engineering and environmental impacts of karst, the engineering geology and hydrogeology of karst terranes, Springfield, Missouri. Balkema, Rotterdam, pp 207–211

Benson RC et al (2000) Stabilization against geologic instability-underground documentation of a mine backfilling case history. In: ASCE Geo-Denver conference Denver, 5–8 August 2000

Brink ABA (1984) A brief review of the South African sinkhole problem. In: Beck BF (ed) Proceedings of the 1st multidisciplinary conference on sinkholes: their geology, engineering and environmental impact, Orlando, Florida. Balkema, Rotterdam, pp 123–127

Chen J, Beck BF (1989) Quantitative modeling of the cover-collapse process. In: Beck BF (ed) Proceedings of the 3rd multidisciplinary conference on sinkholes and the engineering and environmental impacts of karst, St Petersburg Beach, Florida. Balkema, Rotterdam, pp 89–95

Colaizzi GJ et al (1981) Pumped-slurry backfilling of abandoned coal mine workings for subsidence control at Rock Springs, Wyoming. US Bureau of Mines Information Circular 8846

Hermosilla RG (2012) The Guatemala city sinkhole collapses. Carbonates Evaporates Spec Issue Integr Sci Eng Solve Karst Probl 27(2):103–107

Johnson KS (2003) Evaporite-karst problems in the United States. In: Johnson KS, Neal JT (eds) Evaporate karst and engineering/environmental problems in the United States, Oklahoma geological survey, circular 109. Norman, Oklahoma, pp 1–20

Waltham AC et al (2005) Sinkholes and subsidence, karst and cavernous rocks in engineering and construction. Springer-Praxis, Chichester, UK

Whittaker BN, Reddish DJ (1989) Subsidence – occurrence, prediction and control, vol 56, Developments in geotechnical engineering. Elsevier, New York

Worthington SRH et al (2000) Matrix, fracture and channel components of storage and flow in a Paleozoic limestone aquifer. In: Sasowsky ID, Wicks CM (eds) Groundwater flow and contaminate transport in carbonate aquifers. Balkema, Rotterdam, pp 113–128

Part II

The Strategy and Methods for Site Characterization

Part II of this book outlines the site characterization strategies and methods that have been found to be effective by the authors and their associates for the past 40 years in the fields of geotechnical and environmental investigation, much of which was in karst and pseudokarst conditions. The discussion on strategy introduces the concepts, strategies and pitfalls, which need to be considered to achieve a reasonably complete and accurate site characterization. They include obtaining appropriate and adequate data followed by a through assessment of all data. *The focus is upon understanding the sites geologic conditions, since the geology ultimately controls all geotechnical aspects as well as groundwater and contaminant flow*. An understanding of the geologic conditions will enable us to make effective, informed decisions for construction, remediation and the long-term management of the site. To accomplish this requires a dedicated hands-on team of senior professionals with a range of expertise and familiarization with the many methods available for site characterization.

The discussion on methods includes the many existing and diverse tools that are available for our toolbox, which can be used for effective site characterization. Numerous examples and mini-case histories are included to illustrate their applications and successful results. Emphasis is placed upon simple approaches such as old fashion observations (almost a lost art) as well as current technology such as a variety of remote sensing, surface and borehole geophysical methods. However, it is this integrated approach using both traditional methods as well as current technologies that enables us to obtain a more complete and accurate site characterization. While the focus of the book is on karst and pseudokarst, the strategies and methodology used are equally applicable in all geologic environments for both geotechnical and environmental site characterization problems.

No single book, including this one, can possibly provide all of the details related to the methods and their use, necessary to carry out a proper site characterization. The focus of the methods section is to present a brief introduction to the many off the shelf tools we have available for site characterization. The details of how to use them are left for many other texts, short courses and a life long learning experience.

11 What Is Site Characterization

Abstract

Site characterization is the process of developing an understanding of the geologic, hydrologic and engineering properties at the site including the soil, rock, along with groundwater and in many cases, man-modified conditions in the subsurface (e.g. utilities, structures, mines and tunnels) that can impact site conditions. It also includes the spatial and temporal assessment of contaminants when they are present. Various terms such as site investigation, site assessment and site characterization have been used to describe this process and are often used interchangeably. Many case histories have shown that most failures are caused by not properly understanding the site conditions that can impact the project. These failures could have been avoided by focusing attention on the geologic and hydrologic conditions.

11.1 Introduction

Each site characterization is a unique combination of setting, objectives, logistics, technical issues and non-technical issues (budget, politics, etc.). Davies (1977) and Fookes (1997) believe that every site characterization in karst must be treated as unique and the unexpected should be anticipated until proven otherwise.

There are generally, two different times when a site characterization is required for a project:

- Before a problem has occurred, prior to construction of a building, bridge, dam or nuclear power plant, etc. to assess the potential for settlement, collapse, or leakage. This is the easiest and most cost effective time to complete a site characterization since there is usually better site access and problems can be corrected or avoided before construction.
- After a problem has occurred investigating the cause of settlement, collapse or leakage at an existing facility in order to address problems that have already occurred and plan remediation actions. This is usually a more difficult time to complete a site characterization due to the presence of existing structures both above and below ground limiting access, interfering with measurements as well as increasing cost.

In either case, a site characterization is required. If done before construction or development you will save money and time and have the opportunity to incorporate findings into a quality design. Therefore, you may as well do it right the first time (Davies 1977).

Site characterization is done in a number of ways by different professions including: geotechnical engineers, geologists and hydrologists. The category of engineering geologists is a special group who are focused upon the process of site characterization. Culshaw et al. (2008) describe the history of engineering geology along with the leading pioneers in the field. Medley (2009) provides further details on the role of the engineering geologist. Knill (2003) discusses the role of the engineering geologist with respect to larger projects such as dams where a large team of specialists may be involved over a period of several years. Fookes (1997) suggests that the job of the engineering geologist is to get the geology right. It is this focus on geology that lays the foundation for the site characterization efforts.

R.C. Benson, L.B. Yuhr, *Site Characterization in Karst and Pseudokarst Terraines*,
DOI 10.1007/978-94-017-9924-9_11

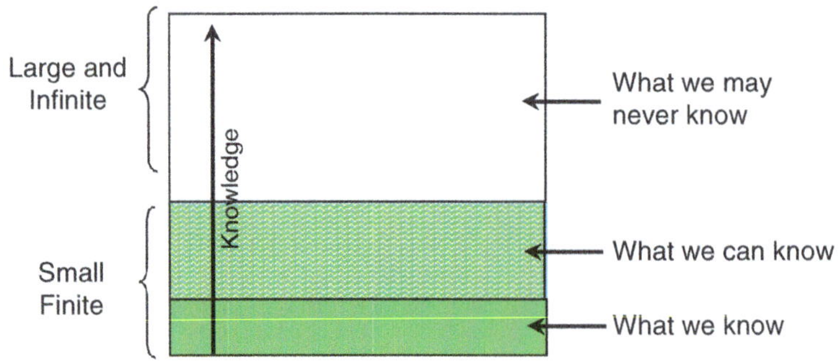

Fig. 11.1 The nature and limitations of knowledge involved in solving a geotechnical problem. This figure is probably representative of many site characterization projects

11.2 Uncertainties in Site Characterization

There are many issues, both technical and non-technical, which can impact the completeness and accuracy of the site characterization process. A fundamental aspect is the inherent limitations of our knowledge when attempting to solve complex geotechnical problems. Figure 11.1 has been adapted from a presentation by George Sowers, who provided a lecture to the Southeast Department of Transportation Engineers on the topic of site characterization. A key point made, was that one of the most difficult tasks is to identify what we don't know. Recognizing what we have missed in the site characterization and what its potential impact to the process is, may be critical. Sowers went on to point out that unfortunately only a small percentage of existing knowledge is commonly used and that most knowledge is acquired by trial and error. A significant part of the site characterization process lies in reducing uncertainties by understanding the variable conditions and by evaluating their consequences to various project decisions, be they geotechnical or environmental in character. In order to minimize the impact of karst, the complex interaction of the soil, rock and groundwater must be understood. Many failures in site characterization are related to misunderstanding of the geology (Sowers 1982).

Others have also cited similar philosophies throughout the years. Glossop (1968) suggests that in order to minimize uncertainties, what you should be looking for will be dictated by the geologic setting and project needs. McMahon (1985) points out in his Davis Memorial Lecture, that uncertainty in geotechnical engineering is often large with enormous economic consequences because the natural subsurface is often highly variable and its sampling is typically limited. The need to remain vigilant and expect surprises throughout a site characterization is critical. Oreskes and Belitz (2001) also stress the need to be aware that there may be many unknowns.

No site investigation will ever be 100 % complete and accurate and there are some facts that may never be known about a site. Based upon over 100 years of cumulative experience, Benson and Sharma (1994, 1995) suggest that most geotechnical, environmental and groundwater site characterization work is significantly less than 50 % complete and accurate (Fig. 11.2). Frequently it is as low as 10–15 % and rarely ever is the completeness and accuracy at the 75–95 % level. These findings tend to support the concepts presented in Fig. 11.1. However, the knowledge and technology is currently available to achieve completeness and accuracy at the 75–95 % confidence levels. These levels are only achieved with a focused level of effort that includes a sound strategy, a team of experienced senior professionals in the field and most important, the owner's commitment to the project.

11.3 The Technical Literature

Site characterization is not new. Modern site characterization began with Carl Terzaghi, the father of soil mechanics, in the early 1930s (Goodman 1999). Hvorslev (1949) compiled an early approach for geotechnical investigations. More recent geotechnical insights are provided by:

- Clayton et al. (2005) Site Investigations, 2nd edition
- Bell (2004) Engineering Geology and Construction
- Hunt (2005) Geotechnical Engineering Investigation Handbook, 2nd edition
- Hencher (2012) Practical Engineering Geology

Recent approaches for environmental site investigation are presented in:

- Sara (2003) Site Assessment and Remediation Handbook, 2nd edition and
- Neilsen (2006) Practical Handbook of Environmental Site Characterization and Ground Water Monitoring, 2nd edition

These six books provide detailed information on a wide range of geologic and hydrologic issues along with technology for geotechnical and environmental site characterization. However, their discussion of karst is quite limited.

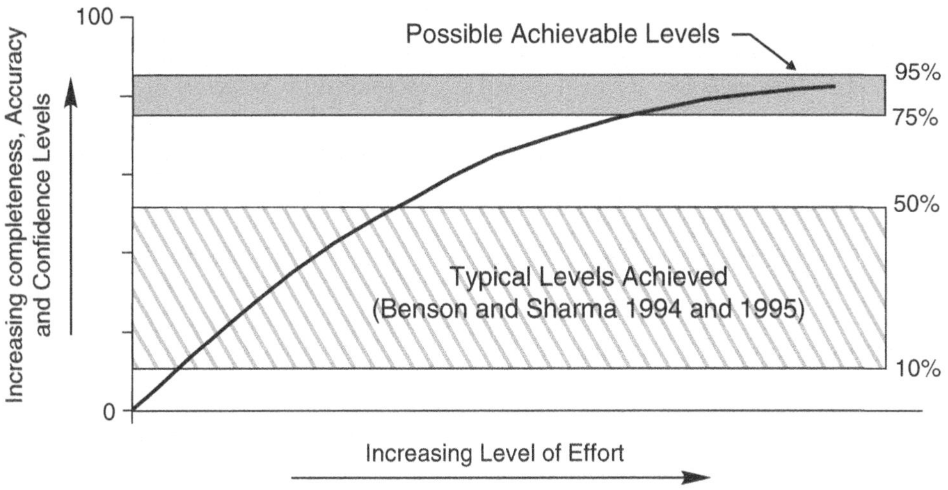

Fig. 11.2 A hypothetical plot showing how confidence levels will increase with the increasing level of effort as the site characterization proceeds

There are many excellent books on karst, paleokarst and pseudokarst. However, the authors have only found three books that deal with karst from a practical engineering perspective. The authors of each of these books have extensive experience in karst.

- Sowers (1996) Building on Sinkholes, Design and Construction of Foundations in Karst Terrain, unfortunately this publication is out of print.
- Milanovic (2004) Water Resources Engineering in Karst. This excellent book deals with the problems of managing water resources within the extreme karst environment of Yugoslavia and other areas.
- Waltham et al. (2005) Sinkholes and Subsidence, Karst and Cavernous Rocks in Engineering and Construction.

In addition, the semi-annual Sinkhole Conferences, initiated by Berry Beck beginning in 1984, have provided extensive information on applied karst issues. These conferences continue today under the direction of the National Karst and Cave Research Institute and PE LaMoreaux Associates. The focus of this conference is upon applied solutions to karst problems. The conference has brought academics, cavers, engineers, geologists and hydrologists together for effective technology transfer. For more information on these conferences see www.sinkholeconference.com.

11.4 Concepts and Strategies for Site Characterization by Others

The strategies presented in this book are not new, they have evolved over time as technology has advanced from the 1930s on, and our understanding of karst conditions have also evolved. Terzaghi's *Observational Method* was introduced in the 1930s and 1940s when there were few, if any, remote sensing, geophysical or minimally invasive means of measurements available. Peck (1969) summarizes the steps of the *Observational Method*. Terzaghi made observations before and throughout the project as excavations and boring data became available. He predicted the most likely geologic scenario and the worse case scenario. As work progressed he was gradually able to converge upon an accurate model of geologic conditions. All this was accomplished without any of the modern tools we have today. Peck (1969) attributed Terzaghi's great success to his:

- unique observational skills;
- understanding of geology and its implications to the specific construction project; and
- insistence upon his presence in the field coupled with his full, personal responsibility and authority on critical jobs.

The observational method still fits today's site characterization needs and remains one of the most powerful tools in our toolbox. However, it is commonly misused with the term "observation method" cited as little more than a marketing tool (Peck 1969).

Site characterization began as a largely observational science. In the past, observations demanded that many hours be spent walking the site, taking notes and making sketches. As technology has advanced, many have relied upon data downloaded from the internet, limited site-specific data and sometimes unfortunately may not have set foot on the site.

Davies (1977) a karst geologist with USGS provided a keynote address at the US Army Corps of Engineers Facility at Vicksburg, Mississippi at a Symposium on Detection of Subsurface Cavities. This was probably one of the earliest geotechnical meetings on karst in the United States. Davies summarized his experience and his recommended approach for dealing with karst. The following is a summary of the key points from his lecture.

- At the beginning of any project in karst, you are either going to avoid the potential problems and find a better site, accept the problems, understand them and correct them before construction, or fix problems later after construction.
- Most consultants are concerned with the corrective action, more than the initial site characterization problem. The site characterization approach is an entirely different route and it's a rare route. It costs money but it certainly does not cost anywhere near as much as unexpected corrective action that will ultimately have to be taken along with political and legal delays.
- If you are working in karst you must do one of two things either accept the risks and go forward without a site characterization and plan on spending a lot of money, or complete a thorough site investigation and then remediation, which will also cost a lot of money. However, in the latter case you will have had a chance to understand the site conditions in greater detail prior to construction and then carry out remediation before the fact when the site is completely open and accessible.

Other engineers and geologist who have worked in karst echo similar ideas. Peter Fookes, a British engineering geologist with world-wide experience, much of it in karst, presents an excellent overview of the site characterization problems from a practicing engineering geologists viewpoint (Fookes 1997). He covers all geologic settings, emphasizes the use of geomorphology and discusses some karst issues. His use of detailed sketches is excellent. This is a must read for anyone practicing site characterization. In his lecture he states that the difficult ground conditions found in karst are often poorly understood by most engineers. Karst conditions are immensely variable, and require a thorough investigation by a team that appreciates the complex characteristics of karst and at many sites, the details of the karst conditions will only be known when the site has been excavated (Waltham and Fookes 2003).

The Tennessee Valley Authority (TVA) began its site characterization of dams in the 1930s and then nuclear power plants in the 1980s with most, if not all, developed in a karst setting. TVA's site characterization strategy (Hopkins 1977) is a phased approach and included:

- Phase I of a site investigation consists of a comprehensive study of regional geologic literature.
- Phase II was a localized site investigation which consisted of a site visit followed by detailed geologic mapping, remote sensing, tectonic studies, and geophysical investigations. Only then, were the percussion drilling and geophysical logging performed along with cross-hole seismic measurements.
- TVA used a zone approach to karst investigations. Most karst problems are caused by a system of relatively small joint-controlled and or bedding plane features, rather than a single large cavity. As such the field investigation would focus upon locating and characterizing these zones.
- TVA had an understanding of the importance of geology and the necessity of a tight working relation between the geologist, geophysicist, and the engineer. TVA would move the team of key personnel to the site and they would be dedicated to that site until the site characterization had been completed.

The Corps of Engineers published a document, "Foundation Considerations in Siting of Nuclear Facilities in Karst Terrains and Other Areas Susceptible to Ground Collapse", (Franklin et al. 1981). This document provided an overview of the nature of karst and some pseudokarst problems along with a strategy for site characterization. They provided extensive details of the probability and difficulty of detecting a solution cavity or channel by borings and concluded that a grid search was more effective than a random search. They were also early proponents of the use of surface and borehole geophysical methods in the site characterization process.

The following comments were taken from Hatheway (1996), Perspectives No.21 "Karstic may not be Karst; When is it Safe for a Landfill" a commentary section published in the Association of Engineering Geologists News.

> Karst, by its very definition, is a condition that represents the potential for the presence of geologic conditions that could compromise the integrity and function of engineered works or environmental facilities. Karst conditions can range from nil to extreme.
>
> It is not enough to spot one or two indications of karst in the regional setting and come to the conclusion that the site itself is unsafe. Just to say that an area is "karst", does not declare it to be unsuitable for sitting and operation of engineered works. Karst features in the area, by themselves, do not indicate that the ground beneath a particular site has been or will be affected by existing karst conditions or activity at any time in the future. Even if karst features are found on the site itself they should not be cited as a project-stopper rationale. A complete site characterization may characterize the existing karst conditions sufficiently well so that they may be avoided or properly remediated avoiding impact to engineered works. It is the engineering geologist who must carry the message to the owner, but it is the owner who must make the decision to commit the funds and time sufficient to characterize the site adequately.

Just because a site is in a karst area, or has karst around it does not mean that the site itself has karst activity or cannot be used for an engineered purpose. Many dams, water works, landfills and other critical structures have been built and successfully operated within karst settings.

The Expedited Site Characterization (ESC) process developed by the Department of Energy (DOE) in the 1990s incorporated many of the early concepts on site characterization by others. DOE had looked at the progress of Environmental Protection Agency (EPA) field investigations and had decided that DOE's areas of contamination were significantly greater than that of EPA. As a result they were searching for a better and more focused strategy for site characterization to more rapidly converge upon the actual conditions at a site.

The ESC strategy emphasizes the use of a variety of methodologies, which will commonly include remote sensing and surface geophysics, followed by minimally invasive push technology. Subsequently, borings and/or trenches and sampling along with geophysical logging and hole-to-hole measurements are made with their locations based upon data, rather than guesses. In all cases, the use of on-site experienced hands-on professionals in the field managing the project and making observations and measurements, (as in the Observational Method) is absolutely essential. Daily field team meetings to review data and change direction and approach, as needed to rapidly converge on site conditions. The ESC process is flexible and is neither site nor technology specific.

Some of the unique features of the ESC method include:

- A small core team of two to six professionals who are highly experienced, hands-on multidisciplinary geologists and engineers.
- The core team members are on-site and are directly involved in the daily field observations, data acquisition, processing and interpretation.
- Emphasis is placed upon understanding of the site geology.
- The work plan is dynamic and can be modified by the field team on-site, as needed.
- A broad variety of measurements and technologies are used.
- A preliminary conceptual model of site conditions is developed early and is tested and improved as new data becomes available.
- The use of on-site laboratories or a minimum turn around time from an off-site lab for geotechnical or chemical analysis is used;
- Data from the various activities are analyzed daily. Each day the team meets to discuss findings and modifies the site characterization program as needed to optimize data acquisition.
- A unified database/graphics system is used to manage data and present results to a common scale with all data points shown;
- The same team is involved throughout the site characterization process from beginning to end.

A more detailed discussion of the ESC process can be found in (Beam et al. 1997; Benson 1997; Benson et al. 1998; Yuhr 1998) and ASTM's Standard Practice for Expedited Site Characterization (ASTM 1998). The key ideas of ESC have been used as part of the authors overall approach and strategy to site characterization for decades. The authors have had the opportunity to participate in the demonstration of the ESC concept at three DOE sites in karst and pseudokarst. On these projects, Ames Laboratory teamed with private consultants to carry out this work.

11.5 The Site Characterization Team

An effective site characterization team of two to six senior hands-on professionals is usually sufficient for most site characterization efforts. They must be sensitive to the issues of geologic uncertainty and possess the skills, experienced and persistence to pursue the details. There is no substitute for good judgment based upon experience and on-site observations. The senior professionals should directly participate in data acquisition and remain involved for the duration of the project in order to provide continuity. This critical asset is often a key to success.

We recognize that a small team of experts, no matter how experienced will not be able to cover all technical issues. That is why it is important to know what you and your small team know and what they don't know and to fill the gaps where necessary. The selection of members of the site characterization team and its supporting members and subcontractors is a critical component for a complete and accurate site characterization effort.

In addition, the site characterization team must be flexible and be able to change direction and activities at any time as new data is obtained, if necessary. All field operations (including all decisions on changes), data assessment, analysis and reporting must be under the control of the core team of senior professionals who are in the field.

If the site characterization effort is compromised due to time, costs or capability; the result are almost certain to have a much higher risk of failure and negative impact to future efforts. Failure to effectively complete an early site characterization will almost always results in greater risks, remediation costs, schedule delays, along with lawsuits and other needless expenses. You will pay for a site characterization eventually whether you want to or not. The most economical approach by far is to carry out an effective site characterization early before design and construction have begun, as part of the site selection process (Davies 1977). By doing so, both time and costs of the long-term project can often be easily reduced by factors of 2–5 or more.

Figure 11.3 shows the geologic ESC team who were working at the Faultless site (an underground nuclear test site) in the Central Nevada Test Area. The team is standing at the surface casing for the 975 m deep boring used to deploy the test bomb. After detonation the ground subsided about 2.5 m. The focus of the ESC effort was to investigate possible contamination

Fig. 11.3 The geologic ESC team at Faultless Site (an underground nuclear test site) in the Central Nevada Test Area. The 2.4 m diameter casing was used to lower the bomb

associated with drilling mud pits and other waste materials. The work was carried out for DOE under the direction of Ames Laboratory with Team members consisting of:

- Technos, who was responsible for location and characterization of mud pits and other waste as well as geologic characterization of the site
- McLaren-Hart, who was responsible for chemical analysis utilizing on-site mobile chemical laboratories.

11.6 Some Pitfalls of Site Characterization

There are a wide variety of pitfalls that can impact the site characterization process. They can be divided into technical and non-technical issues. The following is a brief summary of some of these significant pitfalls and reasons that prevent adequate site characterization from taking place (Benson and Sharma 1994, 1995). The examples cited cover a wide range of issues related to geotechnical and environmental projects. While many of the comments do not apply directly to karst, they do apply to the general process of site characterization.

Throughout the 1990s there has been a notable increase in the number of papers, articles, and editorials on ethics and the degradation of technical work within environmental and engineering practices. The editorial "What has Gone Wrong" by Freeze and Cherry (1989) presented an assessment of the environmental industry and identified problems with the process after a decade of experience including:

- A Lack of Understanding or Appreciating the Impact of Geology.
- Politics Outweighing Science.
- Distrust Between Parties

The problem of inadequate site characterization and lack of attention to geologic conditions is also common in geotechnical work as highlighted by Shuirman and Slosson (1992) in their book Forensic Engineering. They cite five areas of concern in the state of the practice in geotechnical site characterization:

- Frivolous Lawsuits:
- Incompetent Professional Work:
- Outdated Codes and Regulations:
- Educational Shortcoming:
- Most geotechnical related accidents, disasters, and failures are not "acts of God". In almost every instance, the impact event was recognizable, predictable and preventable if only good professional judgment (and good site characterization) had been used.

More recent comments regarding pitfalls were cited in three articles in Geo-Strata by:

- Mitchell and Kavazanjian (2007) "Geoengineering Engineering for the 21st Century"
- DiMaggio (2007) "Commentary: Advancements and Disappointments in GeoEngineering and Geo-Construction – A 30 Year Reflection", and

- Matheson (2009) "Engineering Geology Past, Present and Future-This time it's Different".

They suggest that two major technical pitfalls that have impacted site characterization are the lack of an interdisciplinary approach and the over reliance on the use of computers.

11.6.1 Lack of Interdisciplinary Approach

Applying an interdisciplinary approach to site characterization seems like such a logical approach. However, the push for developing narrow specialties within a given field is stronger than ever within our educational systems as well as in practice. As such, there is a lack of being able to visualize the broader picture of site characterization including the integrated aspects of geology, hydrology and engineering. This lack of understanding and communication occurs between disciplines as well as within each profession viewing the problem with limited insight from narrow specialties. Clearly, greater emphasis is needed on interdisciplinary studies. Most geotechnical engineers have not studied enough applied geology and geologists have not studied enough applied geotechnical engineering.

11.6.2 Impact of Computers

Advancements in technology, computers and software is moving at a rapid pace. One of the impacts of the digital age is that we have the ability to gather and manage vast amounts of data very quickly. While advancements in technology, computers and software have obvious benefits, we also have to recognize the disadvantages.

Since the launch of "Sputnik" in 1957, there has been a shift from the more practical hands-on aspects of engineering, hydrology, and geology towards a more advanced theory and use of computer programs to solve problems. Many geologists now graduate without having the advantage of a field course and geotechnical engineers without having seen a drill rig in operation. The lack of these hands-on experiences ultimately impacts their understanding of the site characterization process.

- There is a need for better geotechnical databases, models and information systems to handle the large data streams.
- Another impact of this rapid change in technology is the lack of critical thinking and practical problem solving abilities by professionals.
- Computer software often lacks reality checks resulting in overconfidence that the computer always yields the right answer.
- The vast amount of data that is available often provides a false sense of confidence in a project.
- Many professionals are becoming too reliant upon technology rather than judgment gained through experience.
- Engineering geology remains in part an observational science. There is no substitute for first-hand observations and on-site experience. Analysis by computer alone cannot replace on-site observations and experience.

There are also many non-technical pitfalls that regularly impact site characterization efforts. Some of the many non-technical pitfalls include:

- Cost, quality control and project management aspects
- Fragmenting of a project into small segments, often over gaps of time, and
- Lack of senior experienced professional involvement.

Cost, quality control, and project management aspects have occasionally become more important than the real technical mission. All too often projects are being administered by personnel with little or no understanding of the technical issues, with little emphasis on qualifications and experience of the consultant being retained. In addition low-bid and fast-track contracts have been promoted in the past few decades which tend to focus upon cost and time as opposed to getting things done right.

Fragmenting of a project is not unusual, with a project broken into small segments spread over months and years by the owners. Multiple consultants and contractors are commonly involved and there are often personnel changes with both the owner and the consultants. When this happens there is a lack of continuity in the project and understanding site conditions is never complete. Critical information is easily lost, and quality suffers.

Lack of direct involvement of senior experienced professionals. Firms commonly will list the most senior staff on the proposals and they show up at the presentations and meetings but are rarely ever involved in the project once awarded. The junior staff is then sent to the field, reserving dollars for billable time of the senior staff in the office. Senior experienced professionals in the field are almost unheard of in today's cost conscious world.

There are a wide variety of technical and non-technical pitfalls, which can significantly impact site characterization process. Most of these limitations and pitfalls have been recognized over the decades, yet they continue to persist. It requires an awareness and commitment by both the site owner and the consultant to overcome these many limitations and pitfalls.

References

American Society for Testing and Materials (1998) Standard practice for expedited site characterization of vadose and ground water contamination at hazardous waste contaminated sites, ASTM D 6235-98a. ASTM International, West Conshohocken

Beam P et al (1997) Lessons learned, a brief history of site characterization. HazWaste World/Superfund XVIII, Washington, DC, 2–4 December 1997

Bell FG (2004) Engineering geology and construction. Spoon Press, London

Benson RC (1997) ESC: just another new acronym or maybe something meaningful. Environ Eng Geosci 3(3, Fall):453–456

Benson RC, Sharma D (1994) The pitfalls in site characterization and how to avoid them. Short course presented by Environmental Education Enterprises, Salt Lake City Utah

Benson RC, Sharma D (1995) Pitfalls of site characterization and how to avoid them. Short course presented by Environmental Education Enterprises, Charlotte, NC, November

Benson RC et al (1998) ESC: how it differs from current state of the practice. In: Symposium on the application of geophysics to environmental and engineering problems, EEGS, Denver, CO, pp 531–540

Clayton CRI et al (2005) Site investigation. Available on line at www.geotechnique.info

Culshaw MG et al (2008) Two hundred years of engineering geology. Geol Soc Lond Q J Eng Geol Hydrol 41(2):137–142

Davies WE (1977) Keynote presentation to the symposium on detection of subsurface cavities. In: Butler DK (ed) US Army Engineer Waterways Experiment Station Vicksburg, Mississippi. (lecture recorded by Benson RC)

DiMaggio JA (2007) Commentary: advancements and disappointments in geo-engineering and geo-construction, a 30–year reflection. Geo-Strata July/August, pp 12, 13, 26

Fookes PG (1997) Geology for engineers: the geologic model, prediction and performance. Geol Soc Lond Q J Eng Geol 30(4):293–431

Franklin AG et al (1981) Foundation considerations in sitting of nuclear facilities in karst terrains and other areas susceptible to ground collapse. Geotechnical Laboratory, U.S. Army Engineers Waterways Experiment Station NUREG/CR- 2062, Vicksburg

Freeze A, Cherry J (1989) What has gone wrong, a guest editorial. Groundwater 27(4):458–464

Glossop R (1968) The rise of geotechnology and its influence on engineering practice. The Institution of Civil Engineers, London. Geotechnique 18(2):107–150

Goodman RE (1999) Karl Terzaghi, the engineer as artist. ASCE Press, Reston

Hatheway A (1996) "Karstic" may not be karst; when is it "safe" for a landfill? Assoc Eng Geol News, Perspectives No.21, 39(2):103–110

Hencher S (2012) Practical engineering geology. Spoon Press, London

Hopkins R (1977) TVA philosophy and methods for site investigation. In: Butler D (ed) Symposium on cavity detection of subsurface cavities. US Army Waterways Experiment Station, Vicksburg, pp 25–26

Hunt RE (2005) Geotechnical engineering investigation handbook, 2nd edn. CRC Press, Boca Raton

Hvorslev MJ (1949) Subsurface exploration and sampling of soils for civil engineering purposes. US Army Waterways Experiment Station, Vicksburg

Knill J (2003) Core values: the first Hans-Cloos lecture. Bull Eng Geol Environ 62:1–34

Matheson GM (2009) Commentary: engineering geology past, present and future-this time it's different! Geo-Strata March/April, pp 14–16

McMahon BK (1985) Geotechnical design in the face of uncertainty. J Aust Geomech Soc 10:7–19

Medley EM (2009) Commentary: engineering geology, a vital phase of geoengineering. ASCE Geo-Inst Geo-Strata 10(2):10–12

Milanovic PT (2004) Water resources engineering in karst. CRC Press, Boca Raton

Mitchell JK, Kavazanjian E Jr (2007) Geoengineering for the 21st century, vol 8(4). ASCE Geo-Institute, Geo-Strata, Reston, pp 14–18

Neilsen DM (ed) (2006) Practical handbook of environmental site characterization and groundwater monitoring, 2nd edn. CRC Press, Boca Raton

Oreskes N, Belitz K (2001) Philosophical issues in model assessments. In: Anderson MG, Bates PD (eds) Model validation: perspectives in hydrological science. Wiley, London, pp 23–41

Peck RB (1969) Advantages and limitations of the observational method in applied soil mechanics. Geotechnique 19(2):171–187

Sara MN (2003) Site assessment and remediation handbook, 2nd edn. Lewis Publishers, New York

Shuirman G, Slosson JE (1992) Forensic engineering, environmental case histories for civil engineers and geologists. Academic, San Diego

Sowers G (1982) Lecture presentation to the southeast department of transportation engineers. Recorded by RC Benson

Sowers GF (1996) Building on sinkholes, design and construction of foundations in karst terrain. ASCE Press, Reston (out of print)

Waltham AC, Fookes PG (2003) Engineering classification of karst ground conditions. Geol Soc Lond Q J Eng Geol Hydrol 36:101–118

Waltham AC et al (2005) Sinkholes and subsidence, karst and cavernous rocks in engineering and construction. Springer-Praxis, Chichester

Yuhr L (1998) Managing data and its integrity in the ESC process, symposium on the application of geophysics to environmental and engineering problems. EEGS, Chicago

The Strategy

12

Abstract

The objective of the site characterization process is to develop an understanding of the geologic conditions (both regional and site-specific), which is sufficiently complete and accurate so that a conceptual geologic model of the subsurface can be created with reasonable confidence level for the intended project needs. The difficulty is one of solving the three-dimensional geologic/hydrologic puzzle with many variables and unknowns. The success of a site characterization effort lies in the ability to locate, describe and quantify the natural geological, and hydrological heterogeneity at a site. Heterogeneity is a natural result of the processes that create and modify the geologic setting, and are sometimes further modified by man's activities. The presence of karst typically complicates the process by introducing a higher level of heterogeneity in the three-dimensional geologic puzzle. It can often be the fatal flaw for a project.

Site characterization must focus upon obtaining a thorough and complete understanding of geologic conditions that will impact the site. While uncertainties, assumptions and opinions are a part of any site characterization (Fig. 12.1a), they should be minimized. Interpretations must be supported in a logical and obvious way by sufficient data, which have been tested and proven to be correct. A solid base of data enables us to carry out subsequent efforts such as modeling, risk assessment, construction, or remediation with greater confidence and accuracy while minimizing uncertainties, assumptions and opinions (Fig. 12.1b) (Benson et al. 1996; Yuhr et al. 1996).

12.1 The Detection Dilemma

One of the biggest problems with site characterization efforts has been our ability to effectively locate and map subsurface geologic conditions especially localized anomalous conditions. This is particularly true in fractured rock and karst conditions due to their limited spatial extent, which represent less than 1–5 % of the subsurface area. As a result traditional approaches to investigation, using borings, have a very low probability of intersecting these features. Beginning in the 1950s the development of geophysical instrumentation and its application to characterization of shallow subsurface conditions began. Since then digital technology has resulted in a greatly expended group of geophysical tools for our site characterization tool box (Olhoeft 1994). This has provided us with three basic approaches that can be used to improve our ability to detect and evaluate subsurface conditions and geologic variability. They are:

- Direct Detection
- Indirect Detection
- Statistical Assessments

In most site characterization efforts all three approaches will be employed to effectively converge upon a complete and accurate conceptual model of site conditions.

R.C. Benson, L.B. Yuhr, *Site Characterization in Karst and Pseudokarst Terraines*,
DOI 10.1007/978-94-017-9924-9_12

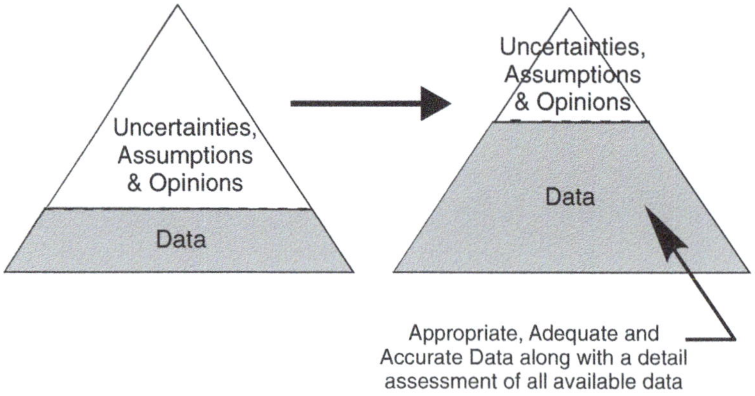

Fig. 12.1 Minimizing uncertainties, assumptions and opinions in site characterization by increasing the amount of use full data

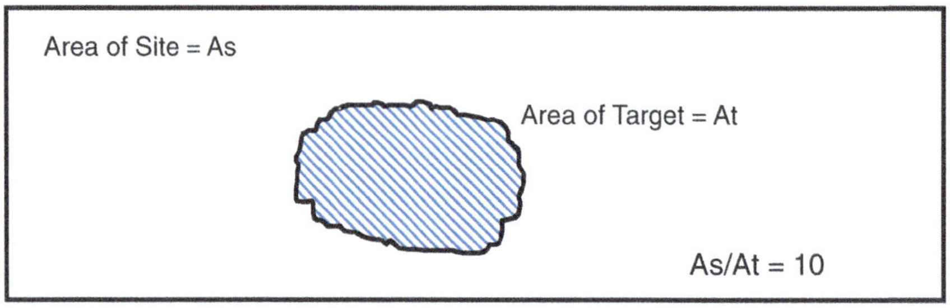

Probability of Detection	As/At = 10	As/At = 100	As/At = 1000
100	16	160	1600
98	13	130	1300
90	10	100	1000
75	8	80	800
50	5	50	500
40	4	40	400

Fig. 12.2 The probability of detecting an isolated target using borings based upon the site to target area ratio. For a given probability of detection, the number of borings increases if the site to target ratio increases

12.1.1 Direct Detection

Direct detection of geologic conditions can be accomplished in a number of ways such as by observations at outcrops, conventional drilling and sampling, or trenching. This provides a very high level of confidence in the detailed data from a local area or the area immediately within the borehole or the trench walls. This type of detection is an integral part of the site characterization process, but needs to be optimized in terms of the location.

Most subsurface geologic assessments are based upon vertical boreholes, commonly made along a survey line or over a grid. In feasibility and preliminary studies, spacing may range from 30 to 90 m or more. Final study programs, depending upon the project type, a borehole spacing of 15–30 m is commonly recommended (Hunt 2005). While the use of boreholes with this type of spacing is a general guideline, in practice, it all too often is treated as a standard and certainly should not be applied when assessing complex geologic conditions.

Benson et al. (1984) has shown that boring densities are commonly inadequate to detect geologic anomalies. For example, Fig. 12.2 shows a target that could be a cavity or buried sinkhole, buried debris, or a contaminant plume. The table provides the number of borings necessary to simply detect the target for different site to target ratios. If the site to target ratio is 10:1 then ten regularly spaced borings will be required to provide a detection probability of 90 %. Note that these statistics only apply to the detection of the target by one boring and are not sufficient to define the shape or depth of the target, which would require many additional borings. Smaller targets, such as widely spaced joints or fractures, would create a much larger site to target ratio and require a 100–1,000 borings to achieve a 90 % probability of detection.

Quinlan (personal communication, 1989), has indicated that the area of caves within Mammoth Cave system underlie less than 5 % of the surface area encompassing the overall boundaries of the cave system. Worthington (1999) and Worthington et al. (2000) have compiled data on the extent of some mapped caves. His data suggests that cave passages occupy 3 % or less of the bedrock in which they are located. Such low percentages of occurrence makes a subsurface investigation using a limited number of boreholes like "looking for a needle in a haystack" and almost assures failure. Not only would it be cost prohibitive to use a large number of borings to accurately resolve conditions, but it would reduce the site to swiss-cheese. If vertical fractures are targeted by vertical borings the probabilities of detection are even worse. See Franklin et al. (1981) for further details on the probability of detection.

A more effective way to utilize the direct detection methods is needed so they can be used in a focused manner on the critical anomalous areas. This can be accomplished by the use of surface geophysical methods (indirect detection) and statistical methods to aid in the detection of anomalous areas so that the direct methods can be located based upon solid data rather than rules of thumb or guesses.

12.1.2 Indirect Detection

Indirect detection is accomplished in a variety of ways using simple observation, remote sensing and surface geophysical methods to obtain indirect evidence of subsurface conditions. Indirect detection may be simple observations of a variety of indicators present at the surface in the form of surface cracks, water puddles and flow, vegetation stress, or regional photo-lineaments (linear trends seen in aerial photos). For example, when we use aerial photos for geologic interpretation, the photo only reveals the conditions at the surface. However, we often interpret aerial photos to provide subsurface information. A photo-lineament may be an indication of possible fracture zone, which may be highly weathered with water seeping downward. A linear trend of sinkholes seen in aerial photos may indicate the presence of a cave system between the sinkholes at depth.

We often observe evidence of deeper-seated karst activity within the shallow soil and rock due to the presence of Near Surface Indicators (NSI). These NSI may be present at the surface such as cracks, or just below the surface (within the upper couple of meters) and show up in some of the shallow geophysical data. For example, ground penetrating radar data may show dipping strata that would be a NSI of possible deeper activity. These NSI can provide an initial insight to the subsurface and become an effective means of focusing our attention on anomalous conditions and locating the direct sampling methods.

12.1.3 Statistical Approach

Statistical data and trends can be used to predict the risk of subsidence and collapse of sinkholes. For example, a sinkhole database or a cave database and cave maps can be utilized to provide a statistical trends for an area in which data is available. This type of data may be used to calculate frequency of occurrence of sinkholes within an area. Wilson (1995) has utilized this approach in Florida to define the number of new sinkholes per year per unit area (NSH). This approach can provide reasonable confidence levels as to what conditions might be expected over a large area. But such methods are not intended to provide reliable site-specific data.

All three approaches (direct and indirect detection along with statistical methods) will commonly be used to evaluate subsurface conditions. Each has its advantages and limitations. Adequate spatial density and coverage over an area is the key to any geologic site characterization to adequately define the background and anomalous subsurface conditions. Once the background and anomalous subsurface conditions have been accurately located, our direct sampling methods (direct push methods, borings and trenches) can then be located with confidence (rather than guesses) to further improve our understanding of subsurface conditions.

12.2 Appropriate, Adequate and Accurate Data

At this point, it should be clear that a reasonably accurate and complete site characterization is all about data. In order to have a sound understanding of the subsurface conditions, we have to have data. In order to minimize uncertainties and assumptions we have to have data. But it's not just any data. One must have appropriate data, adequate data and accurate data followed by a thorough assessment of all data. Determining what data are appropriate, how much data are adequate along with assuring accurate data, are some of the most important steps in carrying out site characterization efforts (Benson et al. 1996; Yuhr et al. 1996). All data associated with any site characterization should meet these three criteria. The less one relies on assumption and opinions (Fig. 12.1a), the more confident we can become in our site characterization because of the solid base of appropriate, adequate and accurate data which we are basing our interpretations on (Fig. 12.1b).

12.2.1 Appropriate Data

There is an extremely wide range of technology that is available in our tool box for site characterization. All of these techniques have their advantages and limitations. It is critical to understand the limitations of each of the many technologies so that the most effective tools can be selected. The selection will be site-specific and based upon the project needs so that the resulting data is appropriate to the problem to be resolved. The following are some concepts for selecting appropriate methods.

12.2.1.1 Keep It Simple, Focus Upon Basics

So often enthusiasm for some new technology by someone who has seen a demo or heard a paper drives the project. We are not adverse to technological advancements, but when a simple measurement or an approach can provide the needed information cost-effectively then it may be the most appropriate. We recommend following the KISS principle (keep it simple stupid) and always work from the simple to the complex.

12.2.1.2 The Sequence of Work

In practice, we should approach a site characterization by first understanding the regional setting and its geomorphology along with its implications to the site itself and then focus upon the site-specific details (Fig. 12.3). The data acquired to assess each of these settings (regional, local and site-specific) needs to be appropriately scaled. In this way, our detailed, site-specific data and interpretations can be based upon and placed into the context of the regional geologic setting and its geomorphology, providing a solid basis for our conceptual model. This concept is stressed by Fookes et al. (2000) who point out that "The present ground conditions, at any site, including the soil and bedrock, are a result of the total geological and geomorphological history of the site, including the stratigraphy, the structure, the former and current geomorphological processes, the past and present climatic conditions over the geologic time scale. It is this total geological history that is responsible for the mass and material characteristics of the ground today".

While this strategy of working from the regional to local investigation is not new, it is seldom followed in practice. All too often the process is reversed and the detailed work of drilling, sampling and laboratory analysis occurs early in the site characterization process, often omitting the regional setting and its geomorphic impact to the sites history. Hoek and Bray (1977) state that the review rock cores are often the starting point in a geotechnical investigation. While this is critical information, they suggest that the regional framework of the geological environment should be developed first so that subsequent information can be placed into perspective.

The scale of observations and measurements can range from the regional setting of many square kilometers to the most detailed site-specific sampling and measurements of soil samples or rock cores (with dimensions of <10 cm^2) (Fig. 12.4). Methods such as satellite data are best applied to regional coverage while aerial photography can be applied to regional, local and site-specific coverage. The surface geophysical measurements are usually limited to local or site-specific coverage. The invasive measurements such as borings are usually limited to site-specific use. In some cases, limited spatial data from geologic mapping of outcrops

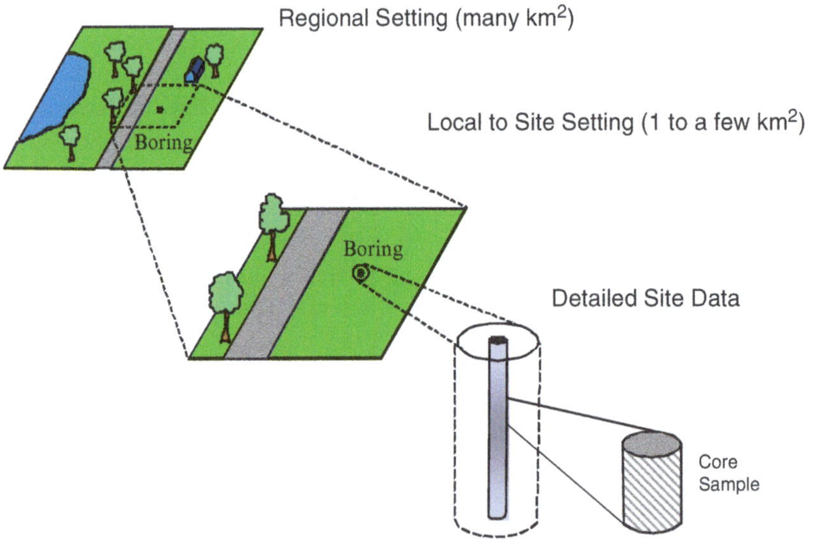

Fig. 12.3 The sequential approach to site characterization should start with the regional setting and work toward the local setting and then the site-specific details

	Regional	Local Setting	The Site	Localized Within the site	Sediment or Core Samples
Method	Many km^2	A few km^2	About 1 km^2	<10 m^2	<10 cm^3
Airborne/Satellite Measurements			Limited		
Aerial Photos					
Geologic Mapping/Observations					
Surface Geophysics					
Invasive Measurements					
Geophysical Logging					
Engineering Measurements					
Hydrologic Measurements					
Dye Tracing					

Fig. 12.4 Typical methods used for site characterization and their range of scale

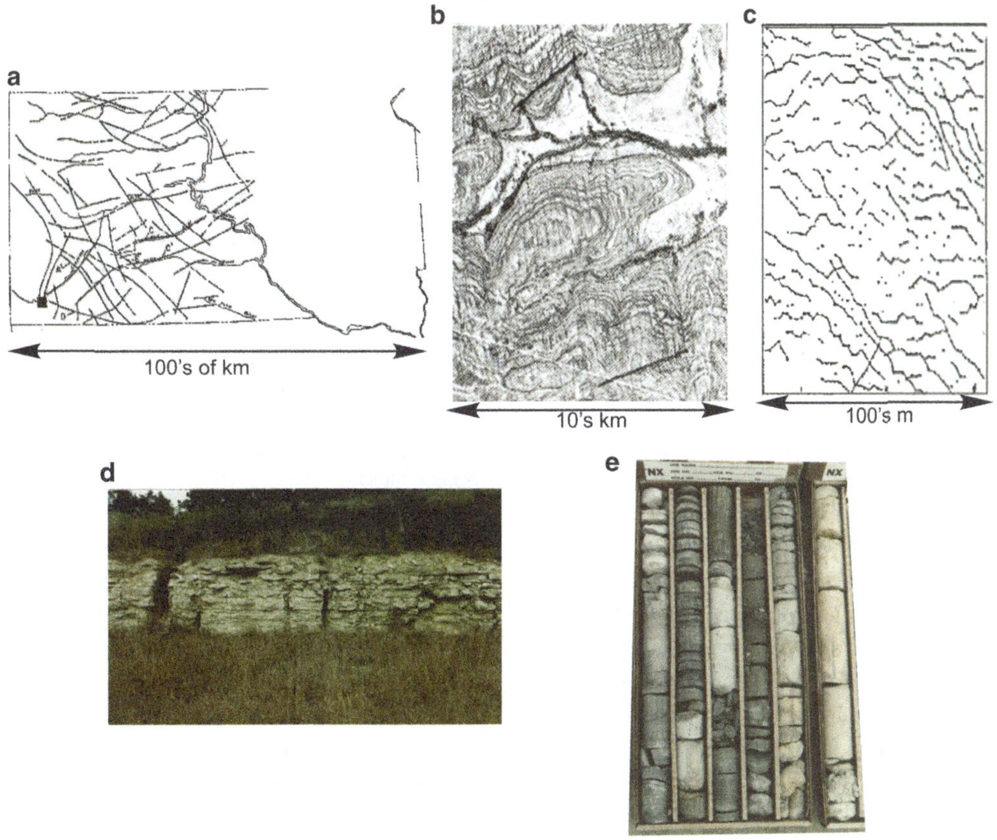

Fig. 12.5 The evaluation of fractures can encompass a wide range of methods and scale of measurements from regional to site-specific data. (**a**) Regional data – Landsat. (**b**) Local aerial photos. (**c**) Site-specific surface geophysical data. (**d**) Site-specific – roadcut. (**e**) Site-specific detailed – cores

or borings can be expanded over larger areas by interpolation and extrapolation. Each method has its own advantages and limitations, which must be understood in order to select the appropriate method for the scale of data needed.

Figure 12.5 illustrates various observations and measurements that can be used when characterizing lineaments, joints and fractures proceeding from the regional to the site itself and finally to the more site-specific details.

- A project might start with a regional lineament assessment based upon Landsat imagery. Figure 12.5a is an example covering more than 82,000 km^2.
- Aerial photos can also be used for lineament analysis Fig. 12.5b. This example covers about 10 km^2. The obvious N10°W joint pattern was observed at both a regional and local setting.
- A lineament assessment was made over 4 ha using surface geophysical measurements (electromagnetics) to map joints in shale (Fig. 12.5c). These data provide detailed coverage over the site itself.
- Geologic mapping along road cuts on or surrounding a site, expose joints and fractures, over 10's–100's of meters (Fig. 12.5d). Sampling is now at the localized site-specific level.
- A core sample (Fig. 12.5e) provides data on bedrock fractures at a very detailed level (7–10 cm diameter sample). However, we cannot assume that they are laterally extensive or are hydraulically interconnected over the scale of the site itself unless verified by other data.

12.2.1.3 The Impact of Scale

It is important to understand the concept of scale effect because it impacts the results of our measurements and affects the appropriateness of data to the site characterization effort. Results from various mechanical and hydrologic tests will vary based upon the area selected for the test and the volume tested. The mechanical properties of the rock mass includes fractures, their orientation, spacing, aperture which vary with scale along with deformability, strength, and internal stress. Hydrologic properties of the rock mass will also vary with the size of the sample. For example, a slug test measures the hydraulic conductivity of the volume near the well screen, a packer tests measures a larger volume of a meter or so around the well and a pump test will measure a volume between wells which may be several meters or more. The relationship between the test results and volume measured is called the "scale effect".

DaCunha (1990a) introduced the concept of Representative Elementary Volume (REV). He shows that for the same rock mass, a sampling volume can be reached where the test results become independent of the sampling size. The REV is the smallest volume that can be considered representative for that property (Fig. 12.6). Nelson (1984) has suggested a rule of thumb that the volume of the sample should be at least ten times that of the fracture spacing in order to have it be representative. Such rules of thumb are reasonable when no other data is available. The hydrologic properties of a rock mass also vary with scale as illustrated by Fig. 12.7. The many issues of scale effects are discussed in Da Cunha (1990b).

When determining appropriate data, it is recommended that:

- The sequence of work is from the regional toward site-specific,
- Simpler measurements or methods should be used when and where appropriate,
- The REV for various properties and their scale effects needs to be considered and,
- The advantages and limitations of each measurement should be understood.

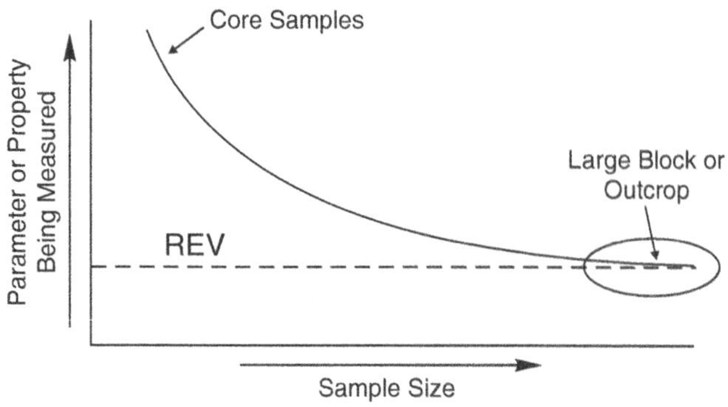

Fig. 12.6 As the volume of the sample increases within the same rock mass the value for the parameter being measured changes. A volume can be reached, beyond which the test results become independent, that volume is the REV. In this example, we have shown the REV to be a large block of rock or rock outcrops

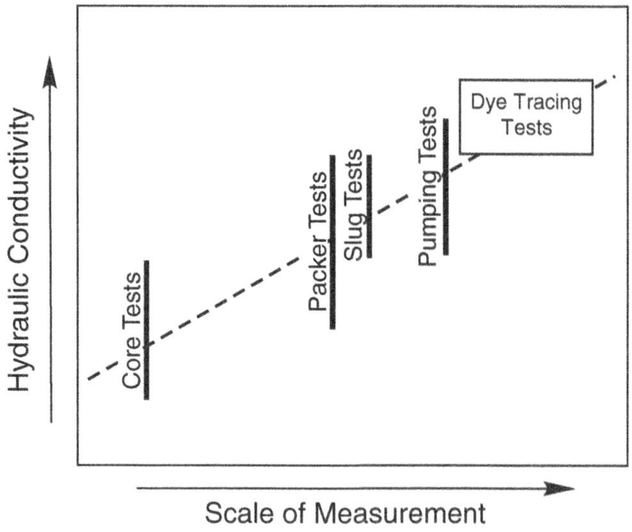

Fig. 12.7 Hydrologic properties also vary with the size or volume being sampled

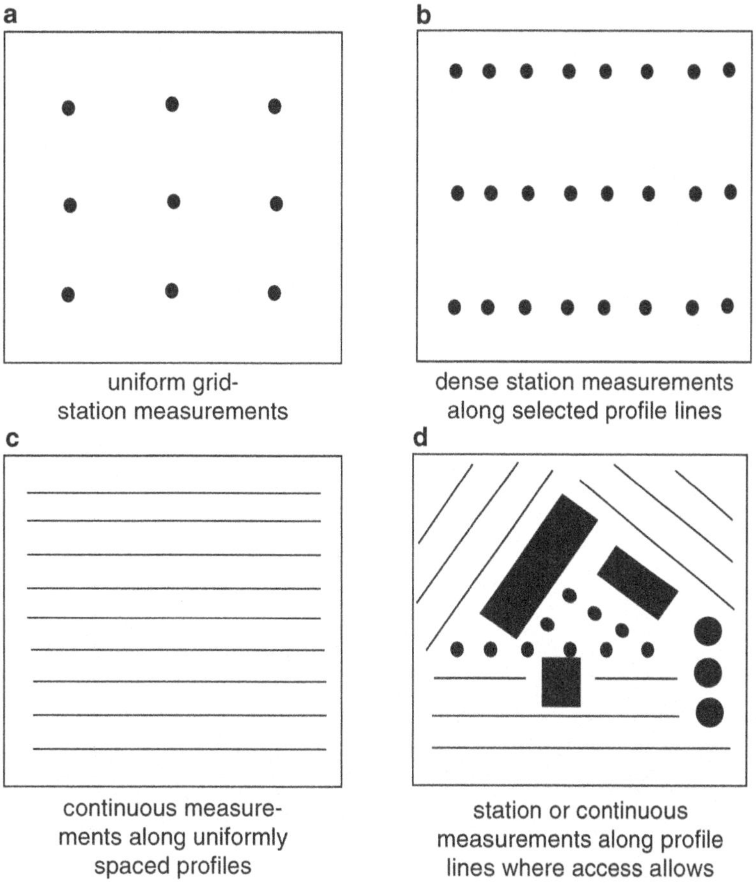

Fig. 12.8 The survey grid used for data acquisition may be a uniform grid of station measurements (point data) over the site (**a**), a series of parallel survey lines using dense station measurements (**b**), "continuous" measurements over parallel survey lines (**c**), or the survey may need to accommodate existing structures and incorporate both station and continuous measurements, where accessible (**d**)

12.2.2 Adequate Data Density

While a very high data density providing 100 % coverage would be optimum, it is not usually possible based upon both technical and economic reasons. Therefore, once appropriate technologies are selected, the next critical decision is how can they be used to provide the best spatial sampling.

Adequate data simply means that there should be sufficient spatial sampling to reduce the uncertainty of variations in geologic and hydrologic conditions. Insufficient spatial sampling can easily miss anomalous conditions leading to an incomplete and erroneous understanding of site conditions.

This is where the use of some remote sensing or surface geophysical method will greatly benefit a project because of their ability to dramatically increase spatial data density. Satellite or aircraft measurements and imaging provide coverage over large areas (regional conditions) and in some applications local data. Surface geophysical measurements can be used to increase data density along survey lines or over a grid on a site-specific basis. If adequate data density is obtained then geologic variability as well as anomalous and background conditions can then be effectively located based upon data rather than guesses.

Adequate data is typically thought of as having good lateral coverage over a site. However a site characterization includes all three dimensions as well as changes over time. Therefore, adequate data density must also address sampling with depth and over time. These factors are particularly critical where hydrologic conditions play a role.

12.2.2.1 Lateral Data Density

Lateral data density can be accomplished by using tightly spaced measurements on a grid or "continuous" (or nearly continuous) measurements on closed spaced parallel lines. Todays digital instruments have fast sample rates, such that the data is almost continuous (see Chap. 16 for further discussion). This approach is used when there is no specific target of interest and the objective is to determine the presence of variable or anomalous geologic conditions. Figure 12.8 illustrates various approaches to optimizing lateral data density whether using station or continuous measurements. The reality is that there may be site access restrictions (steep

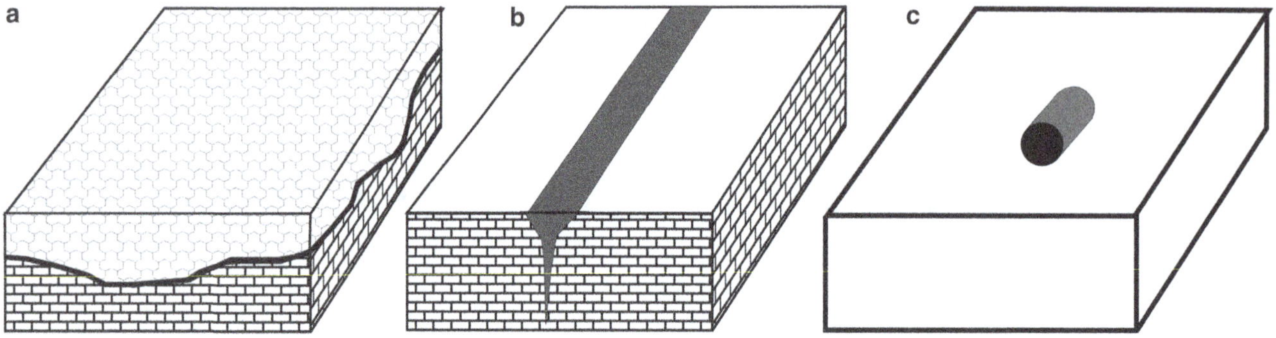

Fig. 12.9 The size and shape of the target will influence the approach to data acquisition and the amount of data required to be adequate. (**a**) Planar features. (**b**) Linear features. (**c**) Discrete features

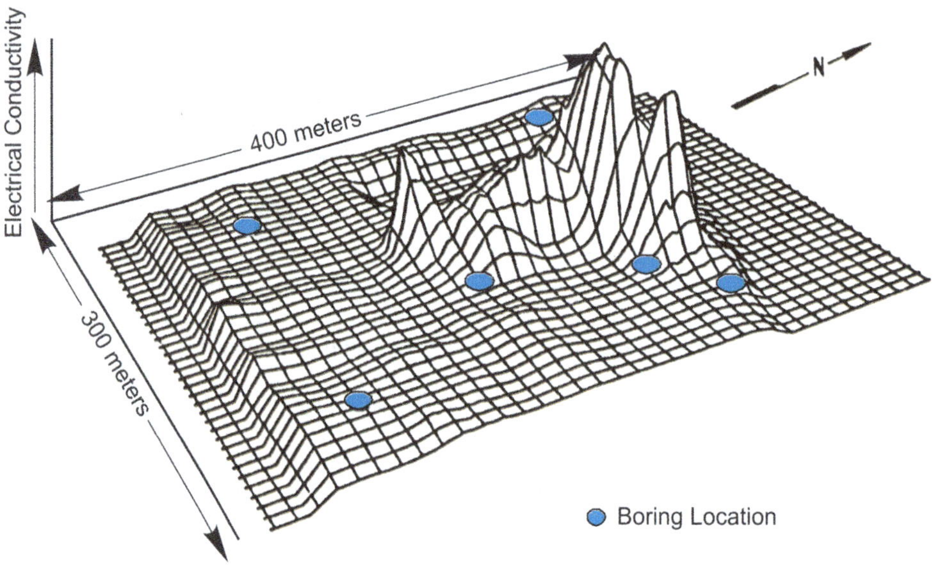

Fig. 12.10 Surface geophysical electromagnetic conductivity data (EM31) from a school yard near Love Canal (1980) providing essentially 100 % site coverage. Boring locations are shown by *dots*

topography, buildings or fencing) that require a more creative, mixed approach. However, data density can also be optimized if something is known about the site or a particular target of interest. For example, the target may be:

- a generally planar surface such as a clay layer or top of rock,
- a linear target such a buried channel in the bedrock, fault or fracture zone; or
- a discrete feature such as an enlarged cavern or buried paleosinkhole.

Given the nature of the target one can optimize a given number of measurements or survey lines (Fig. 12.9). For example:

- If the target is known to be planar, fewer grid points or points along a survey line may be needed.
- If the target is linear, then the survey lines should be oriented to cross as close to perpendicular as possible to the target and should be spaced close enough to provide adequate definition of the target.
- If the target is a discrete feature, then the expected size of the target may impact the spacing of the measurements to improve the probability of detection (Fig. 12.2).

Once an anomalous feature is detected, it is often necessary to add additional data for a more detailed characterization of the shape, boundaries and depth.

While not an example of karst, Fig. 12.10 is a classic example of a site, which clearly shows the impact of inadequate spatial sampling by borings (direct sampling) alone and the benefits of indirect measurements to locate anomalous conditions. There had been some reports of contamination in a schoolyard near Love Canal (Niagara Falls, New York). Five borings had been located, one in each corner and one in the center of the site. A sixth boring was place over an area of dead grass. These boring locations were a reasonable approach given no other information, however they were simply a best guess and none of the six borings encountered any contaminated material.

Surface geophysical measurements were continuously acquired using electromagnetic (EM) measurements. The EM instrument used was an EM31 measuring to an approximate depth of 6 m. Continuous data was acquired along survey lines spaced 3 m apart. This provided virtually 100 % coverage of the site. The resulting data in Fig. 12.10 is a three dimensional perspective of the electrical conductivity values which clearly show the spatial extent and magnitude of a high conductivity area. One new boring was located at the center of the anomalous area and indicated the presence of buried fly ash, which is an electrically conductive contaminant. This example clearly illustrates the benefit of such indirect measurements to locate direct measurements.

12.2.2.2 Data Density Versus Depth

Data density versus depth is often overlooked, yet, applies to all of our measurements. For example, we often encounter boring locations on a map. At first glance it might appear that there is adequate spatial density. However, if we plot the depth of borings we often find few, if any, borings extend to sufficient depths to meet all project objectives.

For example, a low-level radioactive waste site in St. Louis in a karst setting had 80 borings over 67 ha (Fig. 12.11a). Initially, this appears to be adequate spatial coverage across the site. However, the borings and wells were installed by different contractors at various phases of investigation. When the distribution of borings with depth was plotted it revealed that most of the borings were terminated well above the top of rock and only two borings (Fig. 12.11b) had actually penetrated bedrock. Furthermore, they only penetrate the upper 3 m of weathered rock. Clearly, any critical assessment of top of rock, karst conditions, groundwater or possible deeper radioactive contaminants at the site is limited based upon information from only two such borings.

Often the resolution of the data with respect to its intended purpose may determine if it is adequate. For example, drilling data that samples every 1.5 m has a vertical resolution of +/−0.75 m. If the intended purpose is to determine the thickness of an aquifer 30 m thick, a resolution of +/−0.75 m or 5 % of the thickness may be sufficiently adequate. However if the thickness of the aquifer is only 3 m, then the same resolution would be 50 % of the thickness which is clearly not adequate.

Much as with the scale issue, the question of depth of measurements needs to be addressed. Since there are a wide range of measurements whose depths vary from the surface (for example aerial photos) to much greater depths such as borings. We need to select measurements that will provide the appropriate depth of interest for the project needs. Figure 12.12 is an example from a site characterization for a proposed hazardous waste facility in southwest Texas. A wide range of measurements with a wide range of depths was used on that project.

12.2.2.3 Data Density Over Time (Temporal Data)

Temporal adequacy is concerned with sufficient data points to avoid gaps in the data or changes in conditions over time. If natural periodic events, such as seasonal water levels, spring or stream flow, are to be represented, the data must have an adequate sampling over the time to document these changes. Where the water level changes slowly with time (e.g. a decline over periods of months due to seasonal changes) fewer water level measurements may be needed than where water levels change more rapidly (e.g. in response to periodic storm events). In either case, it is important that these temporal changes are adequately represented by the measurements.

An example of temporal changes at a site, include a mountain stream that was but a trickle. One could easily cross it without getting your feet wet and we had done so daily over a 2 week period (Fig. 12.13a). Then one day, a heavy rain in the mountains turned the stream into a raging torrent, which could easily sweep you off of your feet (Fig. 12.13b). Such rapid events in stream or spring flow are critical in evaluating hydrology and contaminant transport. Such conditions can only be monitored by electronic water level gauge since such events occur infrequently and often tend to be of short duration, which are easily missed by personnel manually monitoring a staff gauge.

12.2.3 Accuracy of Data

Accuracy can be defined as “closeness to the truth”. Errors to some degree occur in all measurements and data. Errors will also occur in consultant reports and even our own work. To make an error is easy, to catch it and correct it is much more difficult and requires attention to detail. Some errors may be minor with little, if any, effect on our results. Some errors are bigger than others and can have a major effect upon the project. Once we recognize that all measurements and data will have some degree of error, we need a means to assess if a given piece of data is sufficiently accurate to be used for a given purpose.

As data is being acquired, a preliminary assessment of its quality and interpretation should be made in the field. After all data has been acquired, a thorough assessment must be made so that subsequent interpretations, conceptual models, and risk assessments, are not impacted by errors, gaps or incomplete data. This assessment includes a critical review

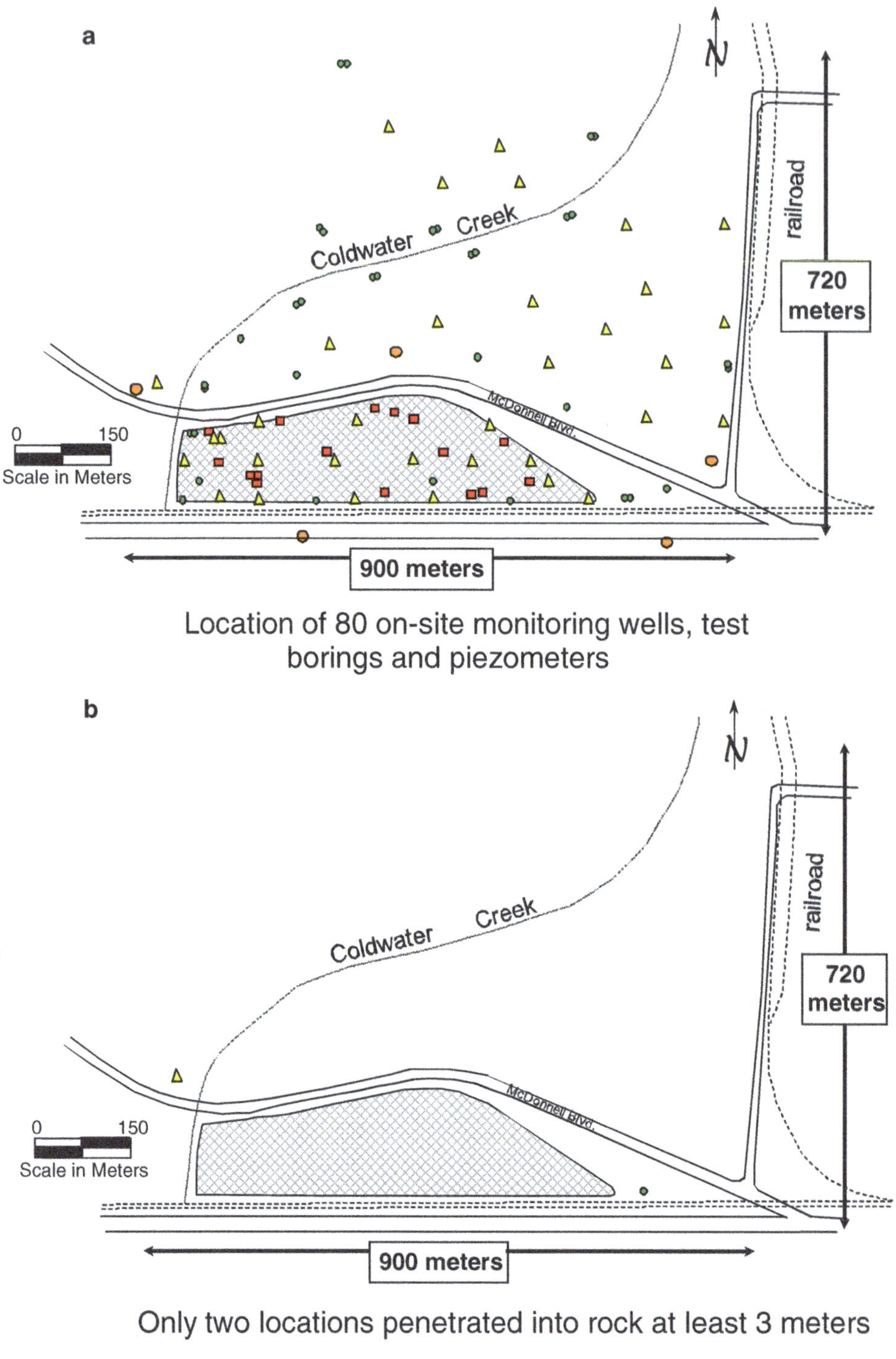

Fig. 12.11 At first glance the distribution of 80 borings over a 67 ha karst site appears to be reasonable (**a**). However, only 2 of the 80 borings penetrated bedrock by about 3 m (**b**)

to identify deficiencies errors and inadequacy of any of the data used in the site characterization process including all data from others along with data obtained from all databases (Yuhr 1998).

If errors, data gaps or inappropriate data are allowed to be entered into the project database and are not detected they can impact all future analysis. The results can significantly impact conceptual models, interpretations, calculations, risk assessment and modeling resulting in inaccurate site characterization, and engineering analysis. Therefore, it is critical that all data must be assessed and verified before being integrated into the project database.

Errors come from a variety of sources and may include but not be limited to:

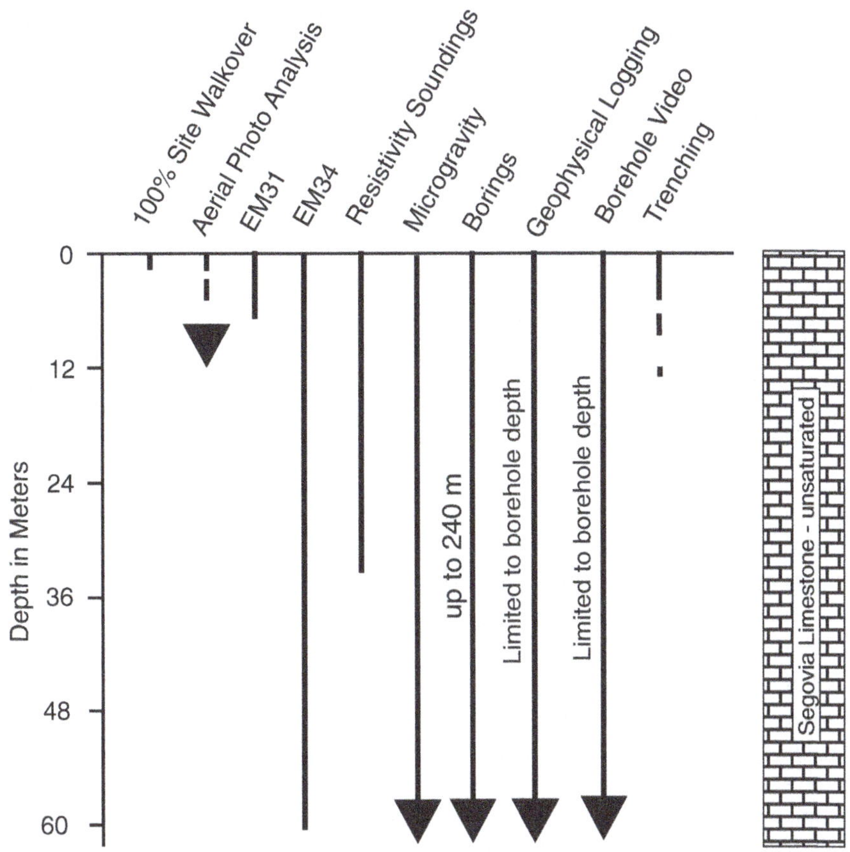

Fig. 12.12 An example of the range of methods and their depth of measurement used on a site characterization project in southwest Texas

Fig. 12.13 A creek bed running adjacent to the site appeared almost dry most of the time (**a**). The conditions are very different as a result of a rare storm event (**b**)

- spatial locations such as improper location (x, y, z) of well or measurement;
- reporting errors where work done in the field is not documented or incorrectly documented in a report (such as the construction of a well);
- data gaps from inadequate spatial sampling (laterally or vertically) or inadequate temporal sampling missing some significant features or change in conditions.
- impossible data can include data outside of the spatial boundaries (for example water level measurements below the total depth of the well or coordinates of a well outside the study area) or outside temporal boundaries (for example water quality measurements made before the well was installed).

Two types of data must be considered for evaluation, factual data (e.g. the elevation at a well head, or water level in a well) and interpreted data (e.g. a contour map of water levels). Errors in the factual data need to be identified and corrected. Errors in interpretation are more difficult to identify. If an interpretation is largely subjective, caution should be used until it is either supported by data or discounted by data.

In many cases, it will be found that considerable work has already been done at the site including; borings, installation of monitor wells, hydraulic testing, and laboratory work. This work is often done by a variety of consultants and contractors, sometimes over many years. In such cases, the results must be checked in detail to assure its quality and applicability to the site characterization.

12.2.3.1 Errors in Data and Databases

A variety of databases are available from local, state and federal agencies. Unfortunately, errors occur in virtually all databases (local, state and federal) as well as consultant reports. For example, in the western US data from water wells are often contained in two databases, one for the well itself and another for the water use. It is not uncommon to find differences in the data for the same well from these two databases such as well location, sometime listed kilometers apart. In addition, the nature of databases is that they are a compilation of data over time. Over time the people who gather the data or input the data have likely changed or sources of the data has changed. These changes may result in inadvertent errors. For example, how many ways are there to report a water level? Measurements could be reported in depth below ground surface as a positive or negative number or an elevation. We have experienced cases where all three were used in the same database.

An example of this was discovered at a landfill site where groundwater flow and contaminant transport modeling was to be completed (Benson and Sharma 1994, 1995). More than 1,000 monitoring wells were listed in the original database for the site. After assessing the well data accuracy, it was found that only about 100 of the wells could be used for the purpose of modeling. The others were omitted due to incomplete or erroneous data, which could not be reconciled. Many of the errors identified were simply impossible data such as a water level below the bottom depth of the well.

12.3 Key Steps in the Site Characterization Process

Site investigations should be carried out in a logical sequence and should focus upon developing a preliminary conceptual model of site conditions and verifying it or updating as the project proceeds. The proposed approach utilizes older existing approaches such as the "Observational Method" by Terzzagi (Goodman 1999), and strategies used by TVA (Hopkins 1977) as well as the later strategies of DOE's ESC philosophy discussed in ASTM D 6235-98a (ASTM 1998). These key steps are based upon the strategies developed over many years by others and have been modified by decades of field experience by the authors.

There are always compromises made when planning each project. Project objectives, desired site coverage, resolution of measurements and spatial sampling will always be balanced with time, budget, logistical and political (non-technical) constraints. However, it can only succeed if the owner is supportive and there is a small team of dedicated professionals who are in the field directly involved with the site characterization process from beginning to end.

The following sequence of tasks presented in Fig. 12.14 is an idealized one. There may be changes in the sequence, due to technical or non-technical reasons. There may also be iterations within a task and between tasks as data is obtained and perceived conditions are better defined. All aspects of the program must be flexible in order to change direction and activities as new data is obtained and our understanding of conditions is improved.

12.3.1 Project Preparation

12.3.1.1 Defining the Problem

The first step is to clearly define the objectives of the work by talking with the client and other interested parties. This may include county, state, and federal agency staff who may be reviewing and approving the project. If specific interests of others can be identified and incorporated into the site characterization effort early, it is more likely that their approval will be more forthcoming at the end of the project. All too often there will be a wide range of divergent ideas from different parties and it is necessary to get all parties in agreement so that clear and realistic objectives can be defined. Good communication is essential at this point.

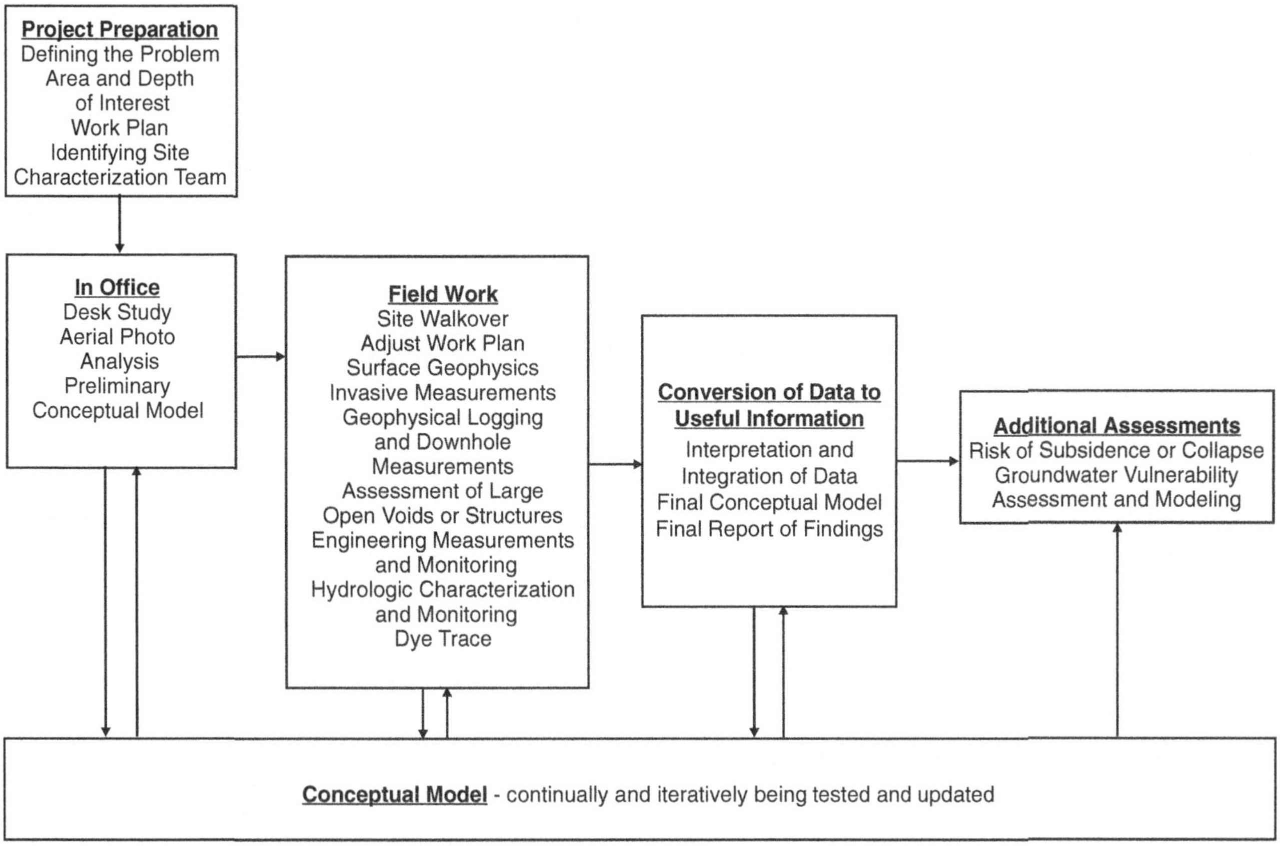

Fig. 12.14 An ideal sequence of activates in a site investigation

During the initial discussions we attempt to obtain the clients understanding and objectives for the project. It is also at this point where conditions at a site (whether actual or perceived) are presented. It is important to identify what assumptions limit our understanding of current site conditions. In many cases we find that the client does not have a good understanding of the importance of the geologic issues and their impact to the project, and the necessary steps to satisfactorily resolve them. Our role is to provide understanding of the nature and extent of potential karst problems and make recommendations and provide guidance for the project. It should be clear that the initial definition of the project objectives might change as the project proceeds and preliminary data may reveal unforeseen conditions. The client's initial expectations must be balanced with the time and budget as well as other possible external constraints such as regulatory issues. This can be accomplished by helping the client recognize what is feasible and realistic within the limits of technology, budget, time and other constraints. A practical level of reality must be defined prior to planning the site characterization effort. A balance of risk versus level of effort (both time and cost) should be discussed to assist the owner in assessing their goals and level of comfort.

12.3.1.2 Area and Depth of Interest

Part of defining the problem should include defining the area and depth of interest for the investigation. The area of interest needs to be established from the client's perspective. In many cases, artificial boundary limits are placed on the site characterization process. It is not uncommon for clients to suggest that work only be carried out within the immediate confines of the site. While the majority of the detailed investigation work will be done within the site boundaries, it should be made clear that the regional and local setting should not be ignored since it may provide significant information to achieve a complete and accurate site characterization.

There may be quite a bit of ambiguity in the early stages of a project regarding the area and depth of interest. The project objectives and initial estimates of area and depth of interest along with data coverage, will commonly change as questions are answered, new data is obtained, and as conceptual models are developed during early phases of the project.

12.3.1.3 The Work Plan

The initial work plan should be focused on meeting project objectives as well as challenging the opinions and assumptions of site conditions. The work plan identifies the type of data needed to confirm the conceptual model or change it depending upon findings. In the ESC process, the work plan is generic and flexible allowing the on-site core team to make the necessary decisions in the field on where and how to obtain the needed data based upon current findings and professional judgment. There will always be unforeseen conditions and new data will be used to guide the fieldwork.

In addition, a number of other issues need to be addressed

- A site survey grid needs to be established for all tasks so that spatial correlation between data can be easily accomplished.
- Any site clearance or access issues need to be addressed
- If there is not an existing land survey of the site, one needs to be obtained.
- If additional aerial photos or topography are needed they need to be ordered.
- Contractors need to be identified for any unique services not provided by the core team.
- If needed a test laboratory for the project needs to be identified.
- Any other unique services or supporting team members that may be needed should be identified.

Since the site characterization process is one of building and verifying the conceptual model, the results of one or more sets of observations or measurements may suggest additional data is needed in an area or that a different set of data is needed for correlation. If unforeseen issues arise, the field plan is modified so that appropriate and adequate data can be acquired to address these new issues. This iterative process continues until a sound conceptual model is developed and unknowns and uncertainties are sufficiently minimized.

Prior to beginning fieldwork a common survey grid must be established for the site. While there are numerous means to record both data and position digitally using GPS, a local survey grid is needed to provide a reasonable degree of quality control in locating measurements and to allow a wide range of concurrent activity at the site. The authors have always used a semi-permanent local grid system for any project work. This local survey grid allows:

- Data to be more easily and accurately collected along a series of straight parallel, survey lines,
- Preliminary field data to be reviewed and anomalous conditions to be readily re-located without the use of GPS,
- Different field teams to communicate and integrate data more clearly whether it is during the same field effort or months apart.

A site map should be developed that utilizes this local reference grid and includes all pertinent site features and topography. The survey grid can use paint marks on the ground (in open areas), surveyor flags or wooden stakes located at suitable intervals along a line or over a grid. The survey grid remains in place until all work has been completed at the site. This step may seem like common sense, however, it is often overlooked. Getting data to correlate or seeing trends in data sets is often based upon a having correct and accurate locations for all data.

12.3.2 The Conceptual Model

The conceptual model is an important building block in our site characterization effort. The term "conceptual model" or "conceptual framework" is a convenient designation for a means to visualize the sites expected geologic and hydrologic characteristics. It will typically consist of simple graphical presentations, usually cross-sections and plan views as needed to present the spatial perspective of geologic and hydrologic conditions at a site. The conceptual model provides a means to document the current understanding of the site geology, hydrology and cultural conditions from which we continually test our understanding of these conditions.

"The preliminary conceptual model is your first sophisticated guess about the subsurface and what you are likely encounter at the site. This sketch or model is used to stimulate an approach to geologic conditions and uncertainties with ordered thought. There is a significant amount of deductive thinking required to construct the preliminary conceptual geologic model" (Hatheway 1993). In addition, Hatheway (1993) suggests to "let the sketch of the preliminary conceptual model sit for a day or two (if you have that much time), then go through the sketch and explain to yourself or a co-worker why you have made your decisions". This provides you with some time to be away from the topic and then to come back to it with a fresh look. This process is recommended for all key issues including interpretation, integration of data and the final report. On the spot, rush interpretations without proper time for thinking and integration with other data will often lead to improper conclusions.

A good model of site conditions allows both engineers and geologists to understand the spatial relationship and interactions of the many geologic and hydrologic components that make up the site (Hoek 1999). The model is most effective during these early phases of the investigation by identifying what is known, what is not known, what is conjectured along with uncertainties and where significant gaps in knowledge may lie (Fookes and Shilston 2001). One of the benefits of the conceptual model is in its ability to help in anticipate conditions (Fookes 1997).

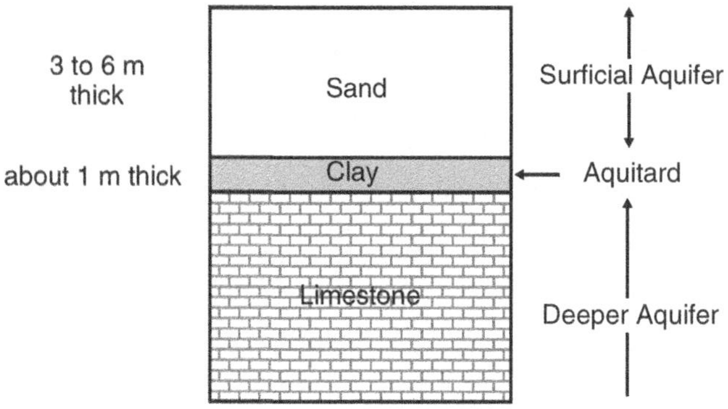

Fig. 12.15 The preliminary conceptual model may be very simple sketch with a brief narrative

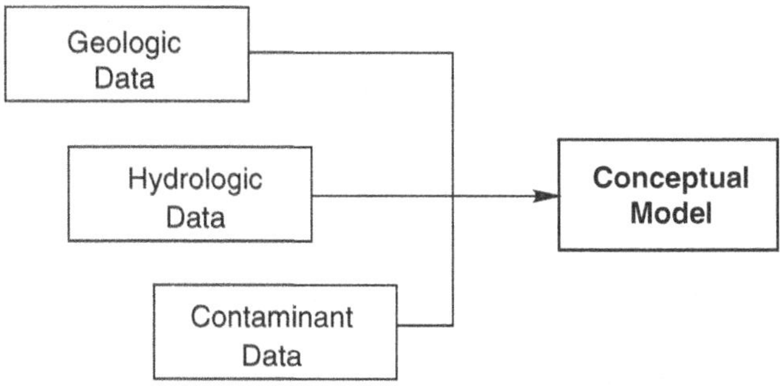

Fig. 12.16 Geologic, hydrologic and often contaminant data may be used to develop the conceptual model at a site

Initially the conceptual model may be very simple sketch (Fig. 12.15) with a brief narrative of our early understanding of site conditions. It may only be an approximation of reality since we are in the early stages of site characterization and may have only limited site-specific data available. At a minimum, the conceptual model will include both geology and hydrologic features. When dealing with contaminant sites the conceptual model will also include contaminant data (Fig. 12.16).

In developing the conceptual model of site conditions we need to keep in mind that the present geologic conditions at a site are the result of its total geological and geomorphological history. This includes the stratigraphy, the structure, the past and present climatic conditions (temperature, precipitation, and vegetation) along with the impact of weathering, glaciation, as well as changes in sea level and uplift or subsidence of the land mass (Fookes 1997; Fookes et al. 2005). By understanding the geologic history of the site we can develop a reasonable understanding of the events that may have controlled what we see today. While we don't need to make a detailed study of each of these issues we need to be aware of them and how they may have impacted the site. The conceptual model is continuously tested and developed through iterations during the site characterization process. It should be challenged and changed to incorporate new features and conditions as additional data is obtained in each step of the site characterization (Fig. 12.17).

The concept of a conceptual model for a site characterization is not new. LeGrand and Rosen (2000) discuss the development of a preliminary model of regional geologic conditions using LeGrand's 1983 "A Standard System for Evaluating Waste Disposal Sites" or DRASTIC by Aller et al. (1987). Motyka (1998) discusses conceptual models of hydrologic networks in carbonate rocks and presents various conceptual models for chalk, limestone and dolomite along with their hydrologic parameters.

12.3.3 In the Office

The early stage of a site characterization is where the learning curve is steep, with our understanding rapidly increasing. There may be a large amount of existing data and information available on both regional and local geologic and hydrologic conditions as well as cultural conditions that may constrain the investigation or impact the site. The steps for the site characterization at this stage includes:

- A desk study that covers the review of existing regional, local and available, site-specific literature and data
- Develop a preliminary conceptual model of site conditions based upon available information
- Acquisition of existing aerial photography and analysis of aerial photos

With these steps completed, the fieldwork can begin with some insight and logical direction. However, the team should remain flexible and anticipate changes to accommodate the unknowns.

12.3.4 The Field Effort

The field effort is all about obtaining appropriate, adequate and accurate data. The data should be designed to verify and build upon the preliminary conceptual model, meet project objectives and answer questions regarding the key unknowns at the site. This effort should make up at least half of the total site characterization effort. The sequence and number of steps will be site-specific to meet project needs. Ideally, these steps include:

- An initial site walkover and possibly a flyover
- The use of surface geophysical measurements to provide necessary data density

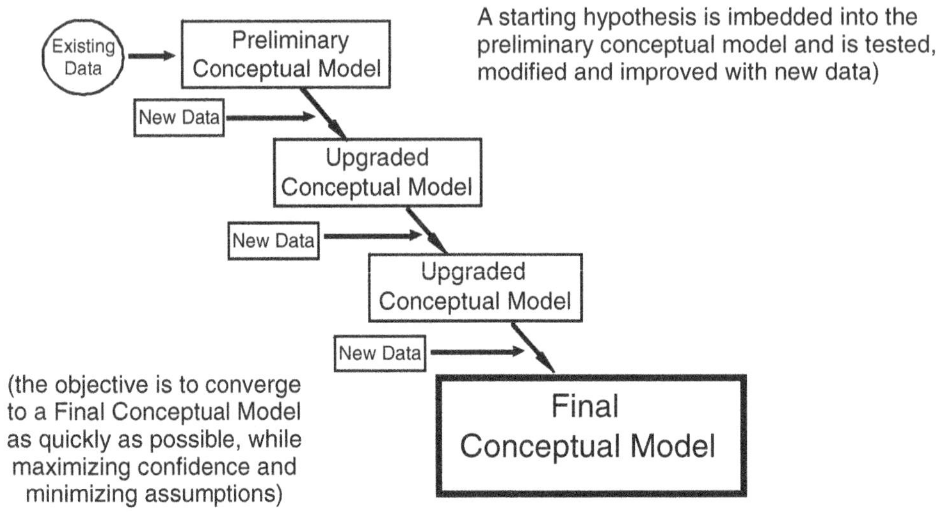

Fig. 12.17 The development of the conceptual model will go through many iterative cycles as new data is obtained and as the understanding of site conditions evolve

- Invasive investigations by direct push measurements, borings and trenching
- Downhole geophysical logging and measurements
- Assessment of large open voids or structures, as needed
- Engineering measurements and monitoring
- Hydrologic characterization and measurements (including chemistry)
- Dye tracing

Throughout the field effort as data is acquired and interpreted, the conceptual model is being continually verified, improved and expanded. Details are added as they become available and the conceptual model is changed if the data shows unexpected conditions. In addition, the initial work plan should be reviewed and modified as necessary to make sure it is meeting project objectives, as the conceptual model changes.

12.3.5 Conversion of Data to Useful Information

During the project, as appropriate, adequate and accurate data has been obtained it is checked for completeness and accuracy. Each data set is then interpreted independently and results added to the conceptual model as appropriate. However, there remains the process of converting the large amount of data that has been acquired into useful information. This is an intensive effort of data assessment, analysis, and integration. This part of the site characterization includes:

- Data assessment for completeness and accuracy.
- Data analysis and interpretation of each independent set of data.
- Integration of the data sets
- Final conceptual model with supporting data
- Reporting and presentation of findings

12.3.6 Additional Studies

Once the site characterization is completed and a final conceptual model has been developed, additional activities can be carried out, if needed. They may include a subsidence or sinkhole collapse risk assessment or groundwater risk assessment and modeling. Confidence levels in these activities are greatly improved by the solid foundation of data and understanding provided by the site characterization.

12.4 Summary

The strategies presented here are embedded throughout the remainder of Part II. The sequence of methodologies presented follows that of an ideal site characterization effort. In addition, more than half of the site characterization effort should be devoted to field activities. In practice however, it is recognized that technical or non-technical factors may modify the order in which the site characterization is carried out.

The strategies, knowledge and technology, to solve the problem of locating, mapping and characterizing geologic anomalies in karst and pseudokarst are presently available. The methods range from the traditional boring, sampling, and laboratory testing to dye tracing, mapping of caves, the many remote sensing, geophysical and minimally invasive methods. These "off the shelf" technologies provide an extremely diverse range of measurements and methods for our "tool box" which can be applied to site characterization of karst, pseudokarst or other complex geologic conditions.

It should be remembered that there is no single, universally applicable method, group of methods, or software that can be used to achieve a complete and accurate site characterization under all geologic site conditions. While a given method or group of methods may be successful in one situation, they may not be in another. In addition, most methods will provide some data, but it will usually take a combination of methods as well as critical thinking and subjective judgment to put the pieces of the geologic puzzle in place. There is no easy answer and there never will be. The strategy and methods must be selected to meet the specific project needs of each site.

The many methods available for site characterization are discussed in the following sections with examples and mini case histories to illustrate their applications. Each of the methods has its advantages and limitations. Every site characterization may use a slightly different set of technologies and measurements and may apply them in difference ways. The "how to do it details" have been deliberately avoided. These details are extensive and are covered by college courses and many other books, along with specialty short courses. Numerous references are provided to fill this gap.

References

Aller L et al (1987) DRASTIC: a standardized system for evaluating ground water pollution potential using hydrogeologic settings. U.S. Environmental Protection Agency, Robert S Kerr Environmental Research Laboratory, Office of Research and Development EPA/600/2-87/035

American Society for Testing and Materials (1998) Standard practice for expedited site characterization of vadose and ground water contamination at hazardous waste contaminated sites, ASTM D 6235-98a. ASTM International, West Conshohocken

Benson RC, Sharma D (1994) The pitfalls in site characterization and how to avoid them. Short course presented by Environmental Education Enterprises, Salt Lake City Utah

Benson RC, Sharma D (1995) Pitfalls of site characterization and how to avoid them. Short course presented by Environmental Education Enterprises, Charlotte, NC, November

Benson RC et al (1984) Geophysical techniques for sensing buried wastes and waste migration. US EPA Environmental Monitoring Systems Laboratory, Las Vegas, NV. Contract No. 68-03-3050

Benson RC et al (1996) Uncertainty in the geologic setting and its impact on site characterization. In: Shackelford CD et al (eds) Proceedings of uncertainty '96, uncertainty in the geologic environment: from theory to practice. ASCE Geotechnical Special Publication No. 58(1):91–103

Da Cunha AP (1990a) An introduction to scale effects. In: Da Cunha AP (ed) Proceedings of the 1st international workshop on scale effects in rock masses. Balkema, Rotterdam, pp 3–27

Da Cunha AP (ed) (1990b) Scale effects in rock masses, Proceedings of the 1st international workshop on scale effects in rock masses. Balkema, Rotterdam

Fookes PG (1997) Geology for engineers: the geological model, prediction and performance. The geological society of London. Q J Eng Geol Hydrogeol 30:293–431

Fookes PG, Shilston DT (2001) Building the geological model: case study of a rock tunnel in SW England. In: Griffiths JS (ed) Land surface evaluation for engineering practice, Geologic Society of England Special Publication 18:123–128

Fookes PG et al (2000) Total geological history: a model approach to the anticipation, observation and understanding of site conditions, Proceedings of the international conference on geotechnical and geological engineering, Melbourne, Australia. Technomic Publishing Co., Lancaster, pp 370–460

Fookes PG et al (eds) (2005) Geomorphology for engineers. Whittles Publishing, Caithness

Franklin AG et al (1981) Foundation considerations in sitting of nuclear facilities in karst terrains and other areas susceptible to ground collapse. Geotechnical Laboratory, U.S. Army Engineers Waterways Experiment Station, NUREG/CR- 2062, Vicksburg, MS

Goodman RE (1999) Karl Terzaghi, the engineer as artist. ASCE Press, Reston

Hatheway A (1993) Preliminary site geologic model: your first sophisticated guess about the subsurface, perspectives no. 17. AEG News 36(4):36–39

Hoek E (1999) Putting numbers to geology – an engineers viewpoint. The second glossop lecture. Q J Eng Geol Hydrogeol 32(1):1–19

Hoek E, Bray JW (1977) Rock slope engineering, 2nd edn. Institute Mining and Metallurgy, London

Hopkins R (1977) TVA philosophy and methods for site investigation. In: Butler D (ed) Symposium on cavity detection of subsurface cavities. US Army Waterways Experiment Station, Vicksburg, pp 25–26

Hunt RE (2005) Geotechnical engineering investigation handbook, 2nd edn. CRC Press, Boca Raton

LeGrand HE (1983) A standardized system for evaluating waste-disposal sites, 2nd edn. National Water Well Association, Worthington

LeGrand HE, Rosen L (2000) Systematic makings of early stage hydrogeologic conceptual models. Groundwater 38(6):887–893

Motyka J (1998) A conceptual model of hydraulic networks in carbonate rocks, illustrated by examples from Poland. Hydrogeol J 6(4):469–482

Nelson P (1984) Hydrologic properties of rock and rock masses. Short course notes, p 32

Olhoeft GR (1994) Geophysical observations of geological, hydrological and geochemical heterogeneity. In: Bell RS, Leper CM (eds) Proceedings of symposium on the application of geophysics to engineering and environmental problems, vol 1, 27–31 March 1994. EEGS, Boston, pp 129–141

Wilson WL (1995) Sinkhole and buried sinkhole densities and new sinkhole frequencies in karst of northwest Peninsular Florida. In: Beck BF (ed) Proceedings of the 5th multidisciplinary conference on sinkholes and the engineering and environmental impacts of karst: karst geohazards, Gatlinburg, Tennessee. Balkema, Rotterdam, pp 79–91

Worthington SRH (1999) A comprehensive strategy for understanding flow in carbonate aquifers. In: Palmer MV et al (eds) Karst modeling, Special publication 5. Karst Waters Institute, Leesburg, pp 30–37

Worthington SRH et al (2000) Matrix, fracture and channel components of storage and flow in a Paleozoic limestone aquifer. In: Sasowsky ID, Wicks CM (eds) Groundwater flow and contaminant transport in carbonate aquifers. Balkema, Rotterdam, pp 113–128

Yuhr L (1998) Managing data and its integrity in the ESC process, Symposium on the application of geophysics to environmental and engineering problems, EEGS, Chicago, 22–26 March 1998

Yuhr L et al (1996) Achieving a reasonable level of accuracy in site characterization in the presence of geologic uncertainty. In: Shackelford CD et al (eds) Proceedings of uncertainty '96, uncertainty in the geologic environment: from theory to practice. ASCE Geotechnical Special Publication No. 58(1):195–209

The Desk Study

13

Abstract

The desk study is where a basic understanding of the site is developed. We also begin to identify what we know and do not know about the site along with a preliminary conceptual model for the site. The desk study includes the acquisition of relevant existing data that covers the regional setting, its geology and geomorphology (the sites geologic history) as well as local and possibly site-specific conditions. During the desk study, reconnaissance aerial photos, topographic and geologic maps should be obtained to provide an overview of the site and surrounding area. Information should be gathered and mined from a variety of sources including government agencies, sinkhole, cave and well databases, site-specific reports and even anecdotal information. This information and data is then used to develop the preliminary conceptual model of site conditions.

13.1 What We Know and Don't Know

During this phase of work, the process of identifying what we know and what we don't know about the site should begin. Tables 13.1, 13.2, and 13.3 provide an overview of the many questions that may be asked regarding possible geologic, hydrologic and karst conditions along with cultural factors that may impact the site. These tables include generic topics, which are applicable to most site characterization. These tables are meant to promote a discussion of some of the many features that may impact a site characterization and are not intended to be inclusive of all possible topics. Not all of these issues will be included or are applicable on every project. Some of the questions may be answered during the desk study, aerial photo analysis and site walk over, while other questions will not be resolved until further site-specific work has been completed.

13.2 Sources of Existing Information and Data

A wide variety of sources for regional and sometimes site-specific geologic and hydrologic data are available from databases or published literature. Initial data can be obtained from:

- Government agencies (local, county, state or federal including USGS)
- Publications from geologic and engineering organizations and their field trips,
- In-house client libraries or files which may contain previous work at the site by others such as consulting firms and drilling contractors reports
- Research by local universities
- Newspaper reports
- Observations and experiences of the owner's, employees and neighbors.

The wealth of existing data that is easily attainable has increased greatly since data of all sorts have become available on the internet. Government agencies often have reports and maps available on-line and some even have digital maps ready for integration into geographic information systems (GIS) used for the project. Field trip reports organized by various professional organizations and local geologic groups are often quite useful since they tend to include unique and critical aspects of the geology for an area. All existing data should be used in context and any of its inherent limits of scale or detail should be recognized. In addition, all data should go through a review for errors, this is particularly true for data that will be used for site-specific purposes.

R.C. Benson, L.B. Yuhr, *Site Characterization in Karst and Pseudokarst Terraines*,
DOI 10.1007/978-94-017-9924-9_13

Table 13.1 Geologic and karst issues

General geologic and geomorphic issues	What is the regional and local geologic stratigraphy?
	What rock types will we be dealing with? Soluble rocks such as limestone, dolomite, gypsum, salt, or non-soluble rocks or possibly some combination
	Are there possible outcrops, quarries, and road cuts where rock may be observed?
	What is the characteristic of the rock matrix (age, density, porosity, degree of fracturing, etc.)?
	What is the nature of the un-weathered intact rock, is it massive rock or is it well jointed, is it thin or thickly bedded strata?
	Are there possible fissures or lineaments present?
	What is the fracture spacing, if present?
	Is there any structure, synclines or anticlines or faults present?
	What is the sites geomorphic history (i.e. changes climate, sea level, uplift, and subsidence and its potential impact)?
	What is the thickness and character of overburden soils, sediment, or fill material?
Karst specific issues	What type of karst features can be expected, their stage of evolution and maturity (see Karst classification Fig. 4.1)?
	Identify recent sinkholes, and paleocollapse features at and surrounding the site along with their proximity to the site using databases and aerial photography
	Are there any sinkhole or karst databases or sinkhole maps available?
	What type of sinkholes are present (Fig. 3.1)
	What is the expected cause of the sinkholes along with their depth of origin?
	Are there any spatial trends of sinkholes that are evident?
	Is there the potential for buried sinkholes associated with historic sea level stands?
	Identify culturally induced sinkholes (pseudokarst) vs natural ones?
	What is the nature of the epikarst, the degree of weathering at the top of rock or the top of rock profile (rockhead relief)?
	Obtain locations and maps of caves from databases, agencies, NSS or cave grottos
	Characterize caves proximity to the site, their depth, classification (fracture, bedding plane or structure) along with the geologic unit(s) in which they are formed

Table 13.2 Hydrologic and karst issues

General hydrologic issues	Identify the presence of rivers, lakes, springs, and recharge
	Establish the general direction of surface water flow
	What is the seasonal and annual precipitation?
	Assess the groundwater system including basin boundaries, depth, flow directions, and velocities
	Identify groundwater recharge and discharge points
	Determine aquifer characteristics; thickness, porosity, hydraulic conductivity, storativity, isotropy, homogeneity and flow
	Obtain water quality data including; turbidity, color, iron staining, oil sheen, noticeable odor, etc. as well as field parameters such as specific conductance, temperature, and pH
	Evaluate temporal changes; in flow, variation of hydraulic head, and water-quality parameters
	Identify the presence of an aquitard, layers of clay or shale
	Are there any zones of flooding?
	Dry or intermittent creek beds
	Identify the presence of dams, lakes, impoundments, and man-made recharge zones
	Identify areas of groundwater or other fluid withdrawal (possible subsidence) or injection or spray fields
	Are contaminants expected at the site, if so, the type, extent and any details already known?
Karst specific issues	What is the depth thickness and nature of the epikarst along with its lateral variability, is it saturated or unsaturated?
	Determine the presence and direction of flow in caves and springs
	Have any dye traces been carried out?
	Assess changes in water levels due to excessive rainfall or drought
	Assessment of past sea level changes at the site
	Evaluate the past or present fresh water lens and the salt water interface (where appropriate)
	Assess changes in water levels due to groundwater withdrawal or dewatering from near by quarries or mines, if present

Table 13.3 Cultural issues

Above ground	Develop an understanding of past and present land use. This is particularly relevant when working in an area that has been developed for many decades
	Identify quarries and their dewatering
	Identify areas of drainage and changes in surface water flow over time
	Identify areas of concentrated runoff from structures and paved areas
	Identify the presence of dams, lakes, impoundments, and recharge zones
	Identify areas of contaminants, leakage or spills
	Assess changes in surficial conditions (cut and fill, modification of surface drainage and recharge)
	Has there been building stress and settlement due to subsurface conditions?
	Identify changes in vegetation and trees
	Identify access issues including; roads and trails, along with limitations due to vegetation, wet areas, ponds, lakes and rivers, fences and property boundaries that may limit or provide access
	Identify overhead utilities and metal fences (which may impact access or affect some types of measurements)
Below ground	The location of active and abandoned mines as well as mine shafts, tunnels and their dewatering
	Identify areas of withdrawal of groundwater from community well fields or private wells, along with the injection of fluids
	Locate underground utilities, tanks and other infrastructure that may impact conditions or impact measurements
	Identify areas of possible fill, dumping, trash, etc.

13.3 Type of Data Available

The type of data that may be available from any one of the above sources is quite varied. A focus should be made on the following:

- Topographic and geologic maps
- Aerial photos (current and historical)
- Geologic and hydrologic reports
- Databases of sinkholes and caves or cave maps
- Databases of borings, water wells, and monitor wells
- Maps showing cultural features and changes over time
- Anecdotal information

13.3.1 Topographic and Geologic Maps

Topographic maps of the site are one of the most fundamental starting points of the desk study. Topographic maps of various scales are available from the USGS. The most useful are the 7.5 min quadrangle maps at a scale of 1:24,000. These maps provide topographic data as well as identify hydrologic and cultural features. Sinkholes and springs are often obvious along with cultural features such as roads, mines, quarries etc. These maps are now available in both printed or digital formats. These maps provide context for the regional setting and local conditions, which is essential during the site walkover.

Geologic maps tend to cover large areas rather than local details. However, it is that regional context that can be important when working on a complex site characterization. The presence of faults, synclines, anticlines or overturned beds while all large-scale features will all impact the interpretation of data on a local or site-specific basis. In addition, these type of features are likely to also have a hydrologic impact.

13.3.2 Aerial Photos

As part of the desk study, aerial photos should be obtained as early as possible to provide an overview of the site and surrounding area. Sources for aerials can range from easily accessed websites or images with GIS overlays of data from commercial or government agencies. In the past, a photo mosaic of the area could be constructed by assembling a sequence of 9″ by 9″ photos into a mosaic (Fig. 13.1). Recent aerial photos provide the team with further insight and typically more up to date information than the topographic maps.

Google Earth or Microsoft Virtual Earth provides worldwide coverage with user friendly GIS overlays with both vertical and oblique angles along with zoom capabilities. While not useful for a detailed photo analysis (to be discussed later) they provide an excellent overview of the site and surrounding area to familiarize the team with the site and its setting.

Using the topographic maps and the preliminary aerial photos we can begin to develop a feel for the level of karst maturity for the site (Table 4.1 and Fig. 4.1). We can also begin to identify:

- Geologic windows – Areas where rock outcrops can be observed should be identified. This includes mines, quarries, outcrops and road cuts. Time and effort should be allotted in the site walkover to view these outcrops and document conditions found. Geologic field trip guides often include these locations as well.

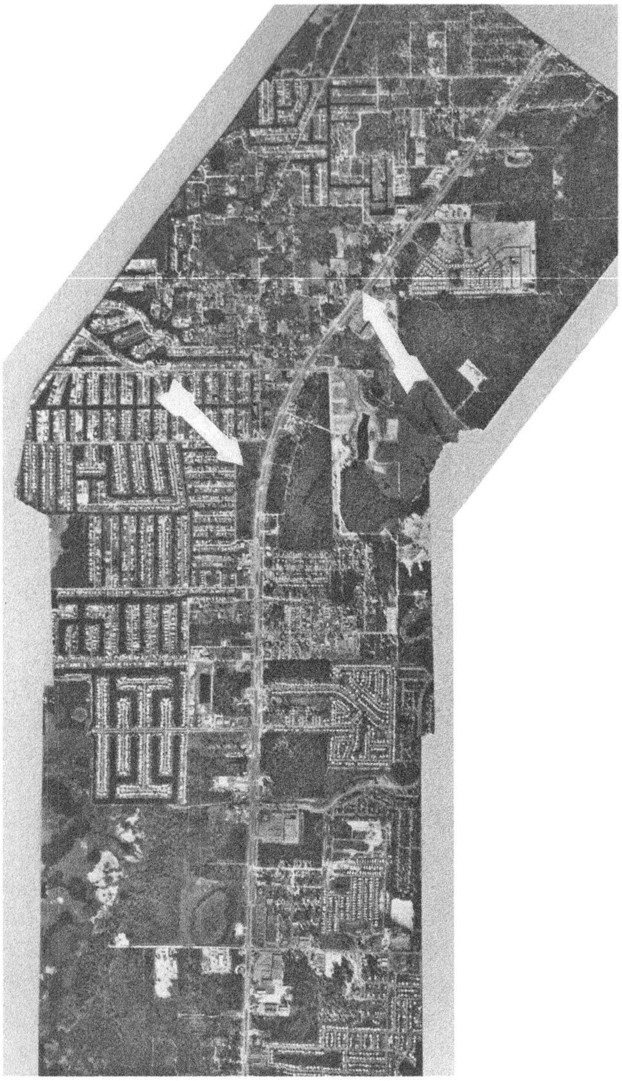

Fig. 13.1 An aerial photo of the site and surrounding area is essential early in the project. Such photos can be ordered or constructed from a group of (9×9) aerial as a mosaic

- Hydrologic conditions – Surface water bodies (lakes, rivers, creeks and ponds) should be identified. Some larger springs may be located on topographic maps and other locations are found in USGS literature and State Geologic Survey reports. Smaller discharge points along with most recharge points (sinks) will be found during the site walkover.

13.3.3 Geologic and Hydrologic Reports

Most areas will have some regional geologic data available in the scientific literature. Geologic and hydrologic reports are commonly available from the USGS and state geologic surveys. Much of the information available at the beginning of the site characterization process often lies in the regional literature. Field trips reports organized by various professional organizations and local geologic groups are also often quite useful since they tend to include local or site-specific aspects of the geology.

While access to the site itself for detailed investigation is usually not a problem, legal or physical access to some of the surrounding area may not be available. Should this be the case, having an understanding of the regional geologic setting from the desk study will help fill the gap between the site itself and the surrounding properties.

Sometimes we are lucky and considerable detailed information is available in the literature. The following two examples are from Southwest Texas and the Kansas City area.

13.3.3.1 Southwest Texas

In southwest Texas, Freeman (1968) had observed indications of elongated collapse ranging in width from a couple of meters wide and several meters long to as much as 450 m wide and almost 5 km long. These features were for the most part aligned in a northeast direction and many had angular branching suggesting a relation to joint sets. Faults along one or both sides of the long narrow areas were not uncommon. Subsidence of 12–15 m was typical and as much as 76 m was demonstrated.

While this report focused on an area a several kilometers south of the site we were working on, it alerted us to major paleokarst features in the area. These features were subsequently extrapolated to the site by detailed aerial photo analysis. This turned out to be a critical aspect of the karst investigation for the site of interest.

13.3.3.2 Kansas City Area

Within the Kansas City Area, Hasan et al. (1988) have provided the "Geology of Greater Kansas City" a series of documents published by the Association of Engineering Geologists covering many cities of the US. This document covers a broad range of geology and hydrology for the area including problems with limestone mines. Another document published by AEG was also very useful. Gentile (1984) discusses deep-seated paleocollapse structures in the Kansas City area.

These documents describe the Pennsylvanian cyclothemes, a sequence of limestone and shale about 152 m thick that outcrop in the Kansas City area. Each of the shallow limestone strata is generally thin and do not support the development of large cavity systems. In contrast, the underlying deeper and massive Mississippian limestones are more than 120 m thick. Large cavernous systems have developed within the massive Mississippian limestone throughout the Midwest. Some have collapsed in the Kansas City area resulting in visible subsidence at the surface (Gentile 1984). These two documents,

Hasan et al. (1988) and Gentile (1984) provided critical support in developing the preliminary conceptual model for the case history presented in Chap. 25 in Part III.

13.3.4 Sinkhole Databases

Sinkhole databases have been developed for some karst areas. For example, the Florida Geologic Survey maintains such a database. These databases can provide the location, date of occurrence and often contains some details such as size, shape and depth, along with the circumstances of collapse.

It is often possible to search the database to determine the number of sinkholes within a radius of a site or the number of new sinkholes per year in a given area (NSH) (Wilson 1995). The number of sinkholes within a given radius of a site will provide an initial indication of nearby sinkhole activity. The NSH can be an initial indication of sinkhole activity for an area. However, it should also be noted that very mature karst (Table 4.1 and Fig. 4.1) can have a small number of NSH. Such examples occur in Puerto Rico and the cockpit karst of Jamaica.

If sinkholes are present, at or near the site, they should be identified for further follow up as part of a site walkover. Information that may be relevant could include:

- Their estimated size (diameter) and depth.
- Are the sinkholes old, recently formed or reactivated
- Are the sinkholes dry, part of a river or lakes
- Sinkhole density (number per unit area) and or typical spacing between sinkholes
- Their geometry (a circular single sinkhole or an elongated composite of merged smaller sinkholes)
- What is the thickness of overburden
- What is there depth of origin (i.e. near top of rock, or deeper within the rock matrix)
- In what geologic formation do they originate
- Are there any spatial trends that are evident such as alignment between sinkholes forming linear trends

In addition, there may be sinkhole risk or susceptibility maps and groundwater vulnerability maps developed for some karst areas. An early sinkhole risk map of Florida (Fig. 3.10) by Sinclair and Stewart (1985) is based upon sediment cover thickness and type of sinkholes. Such regional risk maps are useful indicators of the type of sinkhole activity expected at a site.

These regional sinkhole susceptibility maps and sinkhole databases provide an excellent overview of sinkhole occurrence. Keep in mind that databases tend to list only recent sinkhole activity and tend to be biased towards population centers. Such data may be of interest depending upon its proximity to the site under investigation; however, it may not necessarily be an accurate indicator of site-specific sinkhole activity.

13.3.5 Cave Maps and Databases

A great deal of information about an area can be obtained from cave maps and databases (Sect. 3.4). Such data can be obtained from State agencies that maintain such records and by consultation with the National Speleological Society (NSS) and their caving members who have formed local groups called grottos (www.cave.org). If cave maps are available, the cross section of a cave will allow us to make an assessment of minimum and maximum depths of a cave and its relation to the geologic stratigraphy. It will also reveal the location of any large chambers, if they are present. This information can be useful in determining preferential depth or patterns of dissolution that can support the conceptual model for the site.

Cave databases are also available in some areas. This information can often be searched to provide number of caves near a site under investigation. Figure 13.2 is an example from West Virginia where a prison facility was being located. The cave locations were plotted at a 1.6, 3.2 and 4.8 km radius from the site. A total of 58 caves where located within about a 10 km radius of the site as per communication with the Virginia Speleological Survey (Lucas, 1997, personal communication). Although detailed cave maps were not provided for each cave, this data did provide a first approximation to cave density within the area of interest.

13.3.6 Databases of Borings, Water Wells, and Monitor Wells

Databases are often available for borings, supply wells and monitor wells in many counties and states. These databases may include the well location, geologic data, depth, and screen interval of the well, along with construction information. This information is often very useful in providing a regional perspective. In some case, site-specific data may exist. This may include simplified geologic data or depths of water production zones. This regional geologic and hydrologic data is often sufficient to develop a preliminary conceptual geologic model for the site itself although the data may be many kilometers from the site (Fig. 13.3).

13.3.7 Cultural Features and Changes

Development on and around the site may have caused changes over time. Changes could be significant such as topography, drainage or access. Other changes that may limit access and can impact some of our measurements may include:

- Current land use and changes in land use over time
- Roads, trails and fencing that can affect access

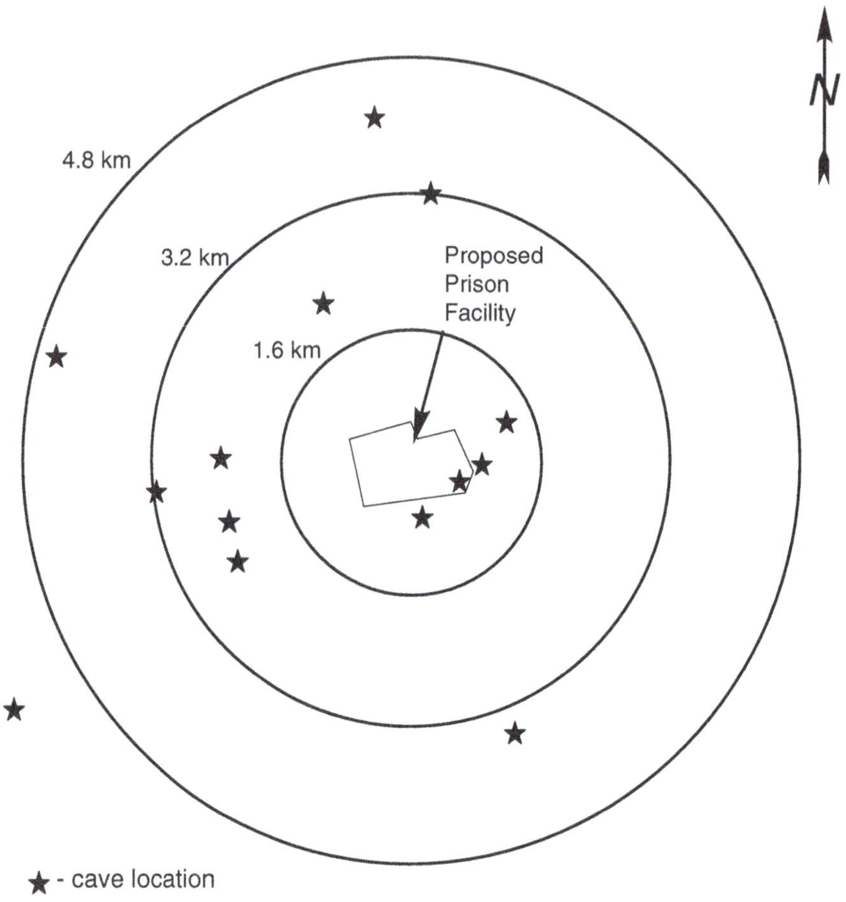

Fig. 13.2 This example shows the locations of known caves based upon a cave database. The cave locations are shown at distances of 1.6, 3.2 and 4.8 km (1, 2, and 3 mile) radius from a site of interest

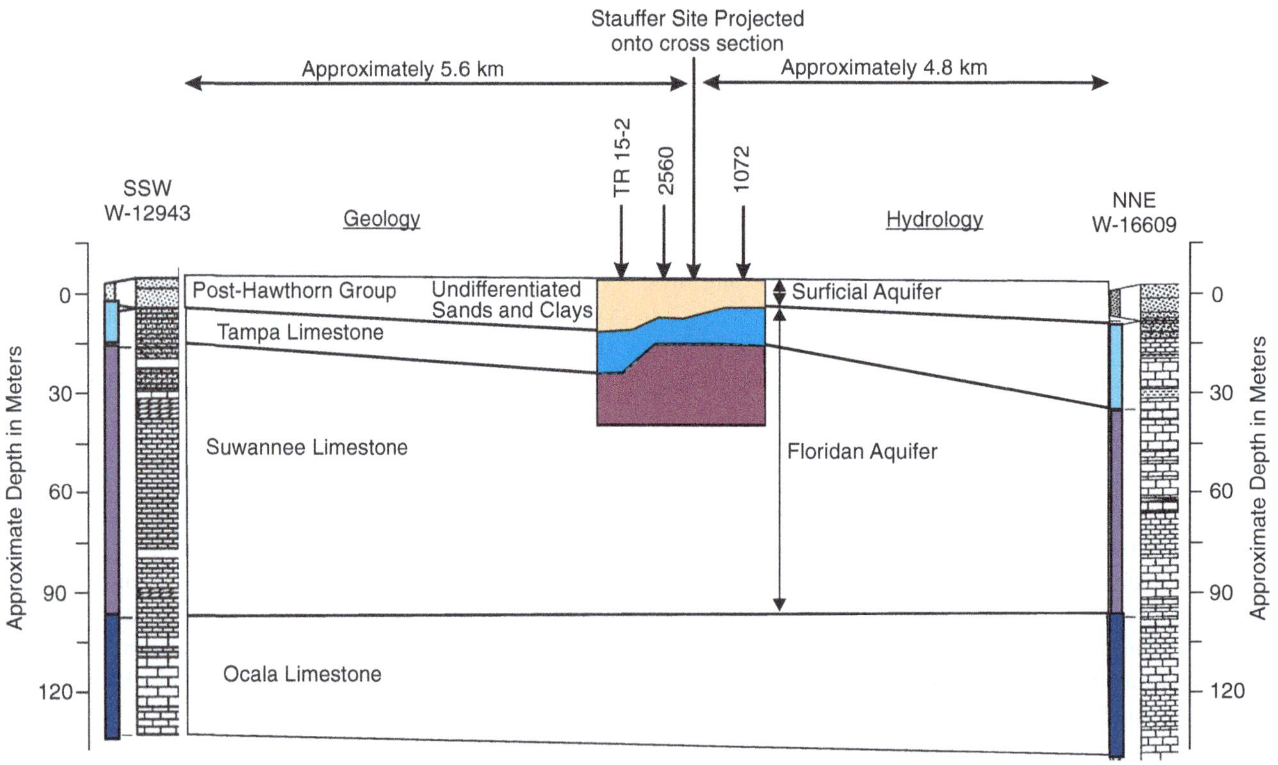

Fig. 13.3 The initial regional geologic and hydrologic data may be many kilometers from the site itself

- Overhead and underground pipelines, power lines, and other utilities should be identified since these may impact access and affect our field measurements
- Major supply wells, well fields and well injection sites near the site should be identified

It is just as important to note the lack of changes at a site. This could provide some evidence of stability at a site. How the site has been used over time may also be important. Stresses and loads at a site that did not cause any sinkhole activity. This information is all part of the site history and should be considered when developing a conceptual model.

13.3.8 Anecdotal Information

Stories of sinkhole activity are often obtained from local newspapers and persons living in the area. Caution is recommended in believing everything you read and hear. Newspaper stories are just that and are not necessarily reliable sources of technical data and quite often the reported sinkholes are due to water or sewer line breaks (pseudokarst) not natural sinkholes.

Observations or anecdotal stories by others must be verified because memories may not present a clear picture of the events of one or two decades ago or in some cases only a few weeks ago. However, these sources of information can often provide insight to local or even site-specific conditions that may not be available elsewhere.

13.4 Data Mining and Review

When a project is initiated at an existing facility or site, the owners will often have a sense of starting with a solid base of information. There may have been considerable work that has already been completed at the site by others. However, all too often when the factual data is reviewed, the amount of useful data and its accuracy is often found to be much less than originally expected and perceived conditions change as a result (Yuhr 1998).

As an example, one project included the compilation, review and physical inspection of a county-wide saltwater intrusion monitoring well network consisting of 167 wells. An initial visual inspection to confirm well locations and access revealed damage or loss of almost half the wells (destroyed, paved over, built over, etc.). The remaining half of the wells were inspected in more detail and included confirmation of well construction, its total depth and internal conditions using a downhole camera. Almost 50 % of the remaining wells were impacted due to collapse, obstruction or incorrect construction/depth from that noted in the database. The impact of reducing the number of reliable monitoring wells from 167 to about 45 is obvious when evaluating county-wide water quality conditions or developing a groundwater model. But consider the impact of utilizing the database information without realizing these errors (Yuhr 1998).

It is also possible to identify useful data from unexpected sources. This is often possible at a site that has been in operation for many years and therefore has a large amount of site-specific data available. Much of this data is often buried or scattered in old reports and requires a bit of persistence to find.

An example of this occurred during a site characterization at a landfill in Independence, Missouri. The landfill was built upon a shale layer that was to act as a barrier to the groundwater. As the existing data was reviewed, it was found that prior to construction of each new landfill cell a small number of geotechnical borings were installed and packer tests completed in the shale layers underlying the site (upper 3 m). This was part of the regulatory requirements for the facility. The data was submitted to the agency and filed away. During a site characterization where the objective was to develop a single document summarizing hydrogeologic conditions and review the groundwater monitoring plan for the site, these data were discovered. A map of all existing borings and monitoring wells was created. The greatly enhanced density of borings revealed a portion of a syncline underlying the site, which impacted the groundwater flow. In addition, packer tests completed within the shale layer actually showed the presence of water and a level of transmissivity that was not expected of a barrier layer of shale (aquitard). These under-used data provided a solid base of existing data that could be used at the site, and the initial understanding of site conditions was dramatically changed prior to beginning any field efforts.

At the end of the desk study there is usually a large amount of data available including notes, aerial photos, maps, topographic sheets, boring logs, portions of databases and GIS documents along with some site-specific reports. This data makes up the initial database for the site and will be added to as the project proceeds. Therefore, appropriate standard software and or GIS system should be selected for the project. This will facilitate retrieval, comparison, plotting and interpreting data throughout the project.

With the project objectives finalized and this initial data a work plan for field efforts can be developed along with the preliminary conceptual model of site conditions.

13.5 The Preliminary Conceptual Model

The data obtained during the Desk Study will provide the basis for our preliminary conceptual model of site conditions. While a basic understanding of the site has been used during contracting, discussions with the site owner's and developing a work plan, the initial data is now available to evaluate what we really know and don't know about the site. We should be able to include the regional setting and the geomorphology that may impact the site, as well as local and possibly even site-specific data.

The preliminary conceptual model will be developed and modified throughout the project, as data becomes available. Changes and modifications to the conceptual model should only be made if they are supported by solid data. When stripping out unsupported interpretations, it is not unusual to find that the initial understanding of site conditions was based largely on opinions and assumptions.

References

Freeman VL (1968) Geology of the Comstock-Indian Wells area Val Verde, Terrell, and Brewster Counties. Texas USGS professional Paper 594-K

Gentile RJ (1984) Paleocollapse structures: longview region, Kansas City, Missouri. Bull Assoc Eng Geol 21(2):229–247

Hasan S et al (1988) Geology of greater Kansas City, Missouri and Kansas, United States of America. Bull Assoc Eng Geol 25(3):281–341

Sinclair WC, Stewart JW (1985) Sinkhole type, development, and distribution in Florida, Florida Geologic Survey Map Series #110, FGS Tallahassee, Florida

Wilson WL (1995) Sinkhole and buried sinkhole densities and new sinkhole frequencies in karst of northwest Peninsular Florida. In: Beck BF (ed) Proceedings of the 5th multidisciplinary conference on sinkholes and the engineering and environmental impacts of karst: karst geohazards, Gatlinburg, Tennessee. Balkema, Rotterdam, pp 79–91

Yuhr L (1998) Managing data and its integrity in the ESC process, Symposium on the application of geophysics to environmental and engineering problems, EEGS, Chicago, 22–26 March 1998

14 Aerial Photography and Remote Sensing Data

Abstract

With the desk study completed, the objectives of the site characterization should be well defined, the area of investigation known and the regional geologic setting established. While the overview aerial photo or photo mosaic of the area of interest has provided an initial familiarization of the site, we are now concerned with additional details that can be obtained from a detailed analysis of the aerial photos.

Aerial photography provides a very cost-effective method of getting a regional and local overview. While topographic and geologic maps are one of the primary sources for this type of information, not all details can be presented on a single map. Furthermore, such maps are not frequently revised. It is here, where aerial photography can provide additional details on regional and local conditions as well as temporal data often over many decades.

In addition to aerial photography, selected imaging and remote sensing technology has been included in this section. This type of data is generally limited to more regional mapping projects due to the scale of information. However, some very useful applications of imaging and remote sensing data have been applied karst projects. For example, aerial photo and imaging technologies provide both spatial and temporal data for developing a karst inventory. Management of karst areas has utilized spatial and temporal analysis of terrains for planning, development, and zoning.

Lillesand and Kiefer (1994), Paine and Kiser (2012), and Johnson and Pettersson (1988) provide detailed discussions of aerial photography and remote sensing image interpretation.

14.1 Availability

The selection of photography or imagery will depend upon the specific-site characterization needs, the area involved, the availability of existing imagery and the cost of new imagery. There is an increasing array of imaging methods being used for special applications. Many types of aerial photos or satellite imagery are available. They all have advantages and disadvantages (cost, coverage, scale, range of wavelengths, etc.). Here, the focus is upon the most commonly used data for general geologic and karst assessment.

Aerial photography is available in the US from a number of federal agencies such as United States Geological Survey (USGS) and the Soil Conservation Service (SCS). Many state agencies, such as Department of Transportation (DOT) also may have aerial photography available as well as private sector companies.

In more recent years, internet-based access to aerial photography, imagery and satellite data has expanded and become quite useful for reconnaissance purposes. Sources such as Google Earth and Microsoft Virtual Earth provide world-wide coverage with user friendly GIS overlays with vertical and oblique angles as well as zoom capabilities. The imagery includes scale as well as elevation data and labeling of key roads and features. While not used for a quantified photo analysis they provide an excellent overview for location, planning purposes, and orientation of field team members. This is an area that is continually expanding availability of coverage and features. For example, older images are now available on Google Earth to look at changes over time.

R.C. Benson, L.B. Yuhr, *Site Characterization in Karst and Pseudokarst Terraines*,
DOI 10.1007/978-94-017-9924-9_14

14.2 Scale

As with all types of measurements, scale of the photos or imagery is important (Fig. 14.1). Landsat images of 1:500,000–1:1,000,000 provide a synoptic view of the regional geologic setting over 100's of km^2. High altitude stereoscopic aerial photographs at scales from 1:60,000 to 1:120:000 are used for intermediate scale analysis. Finally, stereo-paired aerial photo interpretation with 1:10,000 or 1:20,000 scale provides a high level of detail for more local photo analysis. Real estate/tax assessment aerial photographs at a scale of 1:300 are readily accessible for many developed areas. Such detailed photos provide local site-specific cultural details such as roads and buildings.

14.3 Coverage

Aerial photo coverage, at a minimum, would normally cover the site itself and the surrounding area. Regional coverage might extend for many kilometers from the site. For example, if the purpose of the site characterization is one of site selection or if there is complex geomorphology impacting the site. Aerial photography traditionally has been flown on a fairly regular basis dating back to about the 1930s. Maps showing photo coverage for most, if not all states, are available to order 9×9-in. aerials. These can be used individually, as stereoscopic pairs for a detailed analysis.

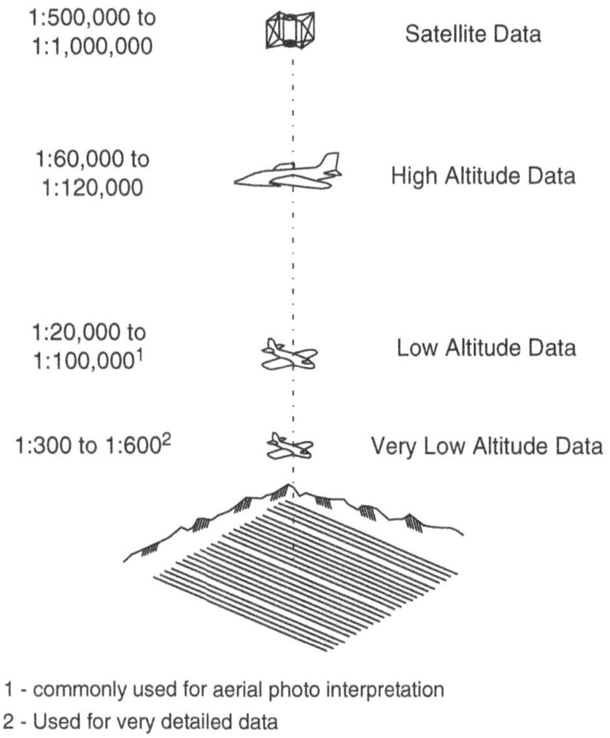

Fig. 14.1 Range of scales that are commonly used for acquiring aerial photography and satellite imagery

14.4 Aerial Photos

Aerial photos have been used for site characterization since the early 1900s. Their application can be qualitative or quantitative in nature. Black and white aerial photos have been the standard for analysis because of their availability and ease of qualitative and quantitative interpretation. But more importantly because of their continued acquisition over many decades which provides a temporal record of geologic, hydrologic, and cultural changes.

Qualitative interpretation consists of simply viewing the aerial photos and noting obvious features and providing initial familiarization with the site for the early phase of any site characterization project, this is typically done as part of the desk study. Initial familiarization may include assessment of site access and logistical planning which may be impacted by features such as vegetation, lakes, rivers, roads and trails as well as cultural changes due to development at the site. Current land use including property boundaries, fences, areas of fill, landfills, structures, major pipelines and transmission lines, along with both quarries and mines can be identified. In addition, a review of aerial photos over a period of time, such as a few decades, will provide insight into land use changes. One of the most common uses of aerial photos for karst studies is to locate sinkholes including their size and spacing. By comparing older aerial photos (some go back to the 1930s) with more recent aerials one can determine sinkhole activity over time.

Historic aerial photography can be used to identify old sinkholes where more recent land use has caused sinkholes to be completely covered, partially covered or filled-in. Panno and Luman (2013), working in a agriculturally-rich southwestern Illinois, digitized and orthorectified 1940s black and white aerials. Those older aerial photos allowed the near-surface geology to be detected through the mature summer crop canopy. Approximately 30 % more sinkholes are present than were delineated on the USGS 7.5 min topographic quadrangle maps of the study area.

Quantitative interpretation consists of a detailed analysis to make a "photogeologic assessment of conditions". Quantitative analysis is also often used later in a site characterization when there are very specific questions to be answered. Quantitative analysis requires the skills of a trained and experienced photo interpreter. This type of analysis can provide data such as topography and drainage, fracture patterns, geologic mapping and hydrologic conditions. This is often accomplished by looking at variations in location, size, shape, shadows, tone/color, texture, patterns, etc. of various features (Clayton et al. 1982, 1995). All of the

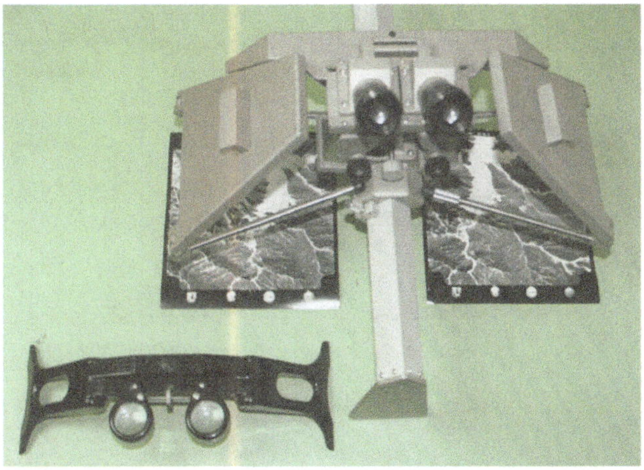

Fig. 14.2 Two standard stereoscopes for stereo-pair analysis (desk mirror stereoscope and portable stereoscope)

aerial photo and imaging methods are responding to the upper-most surface of the ground. From this we can often make interpretations of deeper geologic and geomorphic conditions. Traditionally, this was done manually using stereo-pairs (Fig. 14.2). Today, the aerials are digitally acquired and analyzed using computer software programs.

For example, an array of circular fissures on the ground surface were initially thought to be due to a large 7 ha collapse within an abandoned limestone mine at a site in the Kansas City area. The fissures lay directly over the area of the mine collapse. To confirm this cause and effect relationship, historical aerial photos, prior to mining, were reviewed. Evidence of the surface fissures was confirmed prior to the time of mining. It became clear that these fissures were not due to the mine collapse, but were instead associated with deep-seated paleocollapse, not uncommon in the area (Gentile 1984). Chapter 25 provides more detail on this case history.

In general, the type of information that can be obtained from an analysis of aerial photos might include, but are not limited to:

- Topography of the site and the surrounding area. This is particularly important when access to the surrounding area is restricted for any reason.
- Obvious sinkholes are generally easy to identify on aerial photos
- Fracture orientation and spacing along with lineament analysis (Fig. 14.3)
- The geology of a site can often be interpreted from landforms, drainage patterns, vegetation and land use (Fig. 14.4).
- Surface water and drainage conditions of the site and surrounding area can be assessed including rivers, lakes, ponds, and springs, along with poorly drained ground and flood zones.
- Development and use of the land over time can be determined by a sequence of photographs taken over decades.
- Special areas of concern can be identified such as areas of instability (sinkholes and landslides), sources of materials from quarries and mines, both active and abandoned.
- Site accessibility can be determined to help plan movement of drill rigs or other heavy equipment via roads or trails.

14.4.1 Fracture Trace and Lineament Analysis

Fracture traces and lineaments are natural linear features consisting of topographic, vegetation, or soil tonal alignments, visible primarily on aerial photographs and sometimes on satellite imagery and topographic maps. They are commonly straight in plan view and are considered surface manifestations of vertical or near-vertical zones of fractures. Fracture traces are defined as continuous features less than a kilometer or so and lineaments greater than a couple of kilometers in length (Parizek 1988). Joints and fracture traces can be recognized on 1:20,000 aerial photos and lineaments are best recognized on 1:500,000 Landsat images. It is best to use the terms such as "photo fracture trace" or "photolineaments" which indicate that the feature was interpreted from aerial photos as opposed to other means such as mapped in the field or interpreted from surface geophysical data.

Water resource development has utilized fracture trace and lineament analysis to locate wells in fracture zones to optimize water recovery. Sinkholes often tend to occur along linear patterns of fracture traces or lineaments (Fig. 14.5). Littlefield et al. (1984) have suggested that sinkhole probability increases at major lineament intersections and decreases as the location moves further from a lineament. Short-term sinkhole development is most probable along the largest photolinear features.

Parizek (1988) cites 21 features by which fracture traces and lineaments can be recognized, some of which include:

- Long straight river channels are indications of major lineaments while abrupt and angular changes in stream channels over short distances are indications of local fracture trends
- Sinkholes and depressions that form a linear pattern
- Alignment of vegetation
- Alignment of soil tonal patterns

Lattman (1958) and Lattman and Parizek (1964) have described vegetation and soil tonal pattern may also be subtle indications of lineaments. Greener areas or early growth may be an indication of increased moisture content. Sometimes a

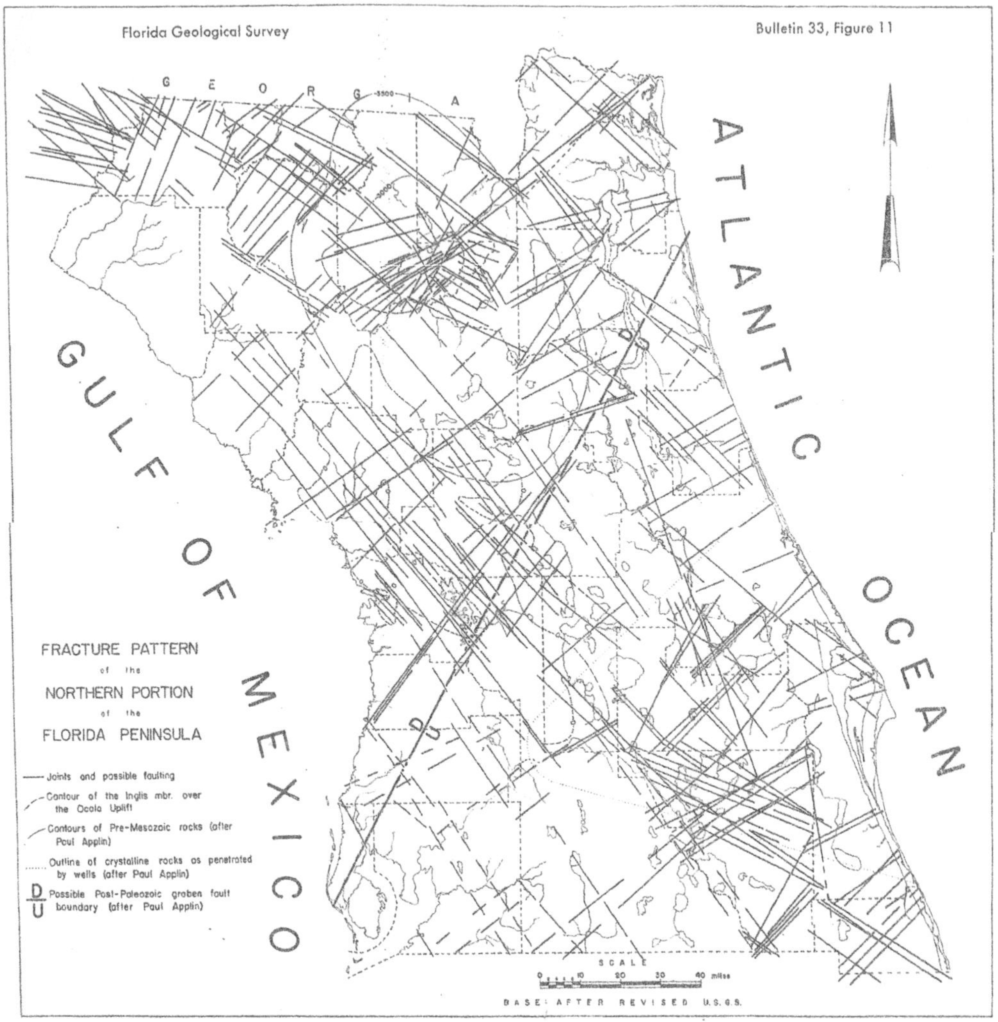

Fig. 14.3 Fracture pattern map from Florida (Vernon 1951)

linear trend of tree and vegetation growth may be an indicator of a fracture or fault due to the presence of sediment infilling, weathering and recharge (Norman 1968). A linear trend may also be caused by cultural features or man's activity such as utilities, roads, and cattle trails. Caution must be used in the interpretation of lineaments and they must be field checked to verify their presence and that they are not related to other non-geologic causes (Allum 1966).

Parizek (1988) lists features which could be mistaken for geologic conditions, some of which are cited below:

- Game and livestock trails
- Old fence lines
- Field boundaries
- Old abandoned railway right of ways
- Old roads and foot paths
- Property lines consisting of trees
- Old cultivated fields
- Old wagon and military trails
- Overhead telephone and power lines
- Buried pipelines

While some obvious fracture traces are easily identified, only experienced photo-interpreters can aid in characterizing more subtle ones. Detailed systematic examination of stereo-pairs is beyond the skill of most geotechnical engineers and engineering geologists (Clayton et al. 1982, 1995). While good use can be made of aerial photos by the inexperienced, the specialist aerial photo interpreter is necessary to carry out a detailed analysis.

Photo analysis is normally done on land, but it also has been used in shallow water covered areas often up to 10 m deep. Finkl has carried out aerial photo analysis of near shore conditions and reefs in southeast Florida using experimental water penetrating film and Lidar (Finkl and Warner 2005; Finkl et al. 2005).

Photo analysis was carried out between the Florida mainland and the upper-most Florida Keys (Key Largo)

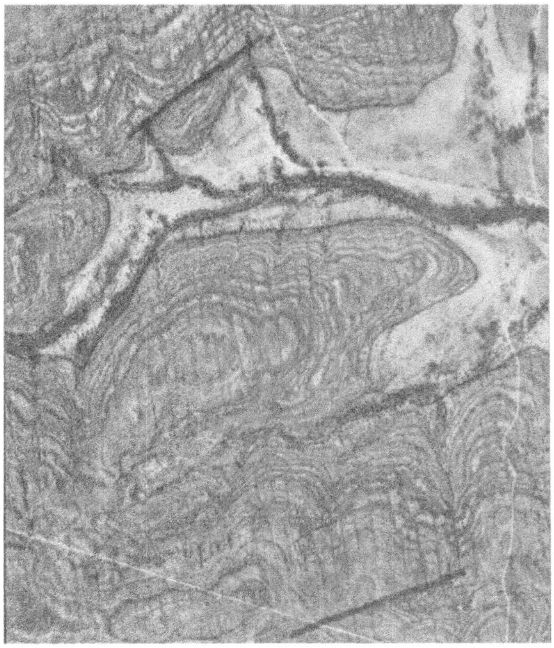

Fig. 14.4 Aerial showing fracture patterns and landforms in southwest Texas

where a new bridge was to be constructed in an area that was covered by mangroves and shallow water. This was an unusual area to attempt aerial photo interpretation. Two independent aerial photo analyses were carried out on the same set of black and white stereo photos. An experienced aerial photo interpreter (Finkl 1994) completed one analysis using a simple stereoscope. Another independent analysis was completed with the same photos scanned into a computer using pattern recognition and edge detection software. It is interesting to note that both independent interpretations came up with essentially the same results. A number of linear trends were identified including a linear pattern cutting through the mangroves and portions of the water-covered area roughly perpendicular to the highway and centered within the area of interest. By itself this feature may not have had much credibility, but as other independent data became available, all data indicated an anomalous paleokarst collapse centered on the linear trend identified by aerial photo interpretation. This is an excellent example of the correlation of multiple data sets. Chapter 26 provides more detail on this case study.

14.4.2 Oblique Aerial Photos

Oblique photos can provide additional perspective and information about site conditions and may be better understood by lay people during presentations. Paine and Kiser (2012) discuss oblique aerial photos and point out benefits:

- An oblique photo includes a greater area when taken from the same altitude as a vertical photo;
- The view is more natural making features more recognizable.
- Some objects not visible on vertical photos may be seen in oblique photos (for example caves or objects concealed by forest cover).

An oblique historical photo from a superfund site (Fig. 14.6) shows the location and type of operations at the site including the large plant facilities, the supporting railway facility, the nearby river and stockpiles of raw materials. When work started at this site, almost the entire facility has been decommissioned and removed with only a few remaining buildings. While old site drawings showed the footprint of the facilities, the oblique aerial photo effectively illustrates the magnitude of the operation. This photo also illustrated the large piles of materials and railroad track that were present during operation from the 1940s to 1980s. These features would have provided heavy loading and vibrations over a portion of what was later found to be a large zone of paleocollapse. The presence of the heavy loads and vibration for 40 years, over this area supported the conclusion that this karst feature would be stable if left untouched. Chapter 27 provides more detail on this case study.

Advancements in aerial photography over the years includes going from black and white to color and from analogue to digital. More recently a method of acquiring low altitude high-resolution views from multiple oblique angles has been developed and is called Pictometry®. This is a patented process (Pictometry International Corporation 2002), which includes overlapping images (12–20) collected from many directions for any given point on the ground. This "bird's eye view" allows a more intuitive interpretation when different angles of the same feature are reviewed (Alexander et al. 2013).

14.4.3 Aerial Photos and Video from Small Unmanned Aircraft

In the past decade, or so, interest has focused upon the use of small unmanned aircraft (model airplanes, helicopters and drones) to obtain both aerial photography and video. Photos and video are obtained digitally and, in some cases, are transmitted to the ground real-time as the Unmanned Aerial System (UAS) fly's its mission. As an example, the Colorado DOT has used small-model, fixed-wing aircraft and helicopters to provide a wide range of low altitude, high-resolution photographs for both planning construction and monitoring highway conditions (Hotchkiss 2007). CDOT's UAS fleet is comprised of three types of remote controlled aircraft includes a pusher-type airplane, helicopters and a "hexa-copter"

Fig. 14.5 Linear trends associated with in sinkholes along a portion US19 in west-central Florida

Fig. 14.6 An oblique photo shows many of the site features not easily understood in a vertical photo. This historical photo illustrates the magnitude of site activity and diverse uses. Most of these structures had been removed before the field investigation began

Fig. 14.7 Unmanned aerial systems (UAS), fixed-wing aircraft, a helicopter and hexa-copter used to obtain aerial photos and video (Photos courtesy of the Colorado Department of Transportation)

(Fig. 14.7). The UAS's have two cameras; one mounted for a pilots view and another one to take the photographs. Photos can be vertical to almost any oblique angle needed for the project.

CDOT's Aerial Reconnaissance Program has and can be used for project documentation, rockfall and slope stability analysis, landslide analysis, environmental review, archeological sites, and resource inventory. The general use of the UAS's are applied to inaccessible areas such as canyons, where manned aircraft are unable to fly or obtain photographs at the detail required.

These types of small aircraft were used as part of the inspection of a sinkhole collapse at the Corvette Museum located in Bowling Green, Kentucky. The sinkhole formed under a portion of the museum housing vintage corvettes. The collapse was recorded by building security cameras and eight of the corvettes ultimately fell into the sinkhole. The collapse feature was estimated to be 12 m across and about 10 m deep (The National Corvette Museum 2014). Video inspection using a drone helicopter was carried out within the collapse by the University of Western Kentucky engineering department (Wall Street Journal 2014).

These small aircraft provide high resolution, low altitude photographic documentation at a reasonable cost as compared to manned aircraft. This approach is also easily mobilized with reliable equipment, a quick access to final photos and flight access to populated areas or small partially enclosed areas where standard aircraft would have limited or no access. Disadvantages include weather restrictions, the same as with standard aircraft, limited visibility due to precipitation or fog and photos are not easily geo-rectified.

14.5 Beyond Black and White Aerial Photos (Other Formats and Methods)

While black and white aerial photos will often provide most of the visual information needed for both the regional and site-specific scale site characterization there will be occasions where other types of data may be needed. There is a wide range of other imaging and remote sensing measurements available (Lillesand et al. 2003; Paine and Kiser 2012). Those imaging techniques that are most commonly used for karst projects include:

- **Color photos** can provide additional information of particular use in geologic interpretation. The human eye is capable of resolving about 100 times more color ranges than gray scale values. While color photos are also limited by sun angle and cloud cover they provide additional details of vegetation and wet areas (Lillesand et al. 2003).
- **Infra-red (false color)** film is used in forestry, agriculture and vegetation studies. Since water totally absorbs infrared radiation, the use of false color infra-red photography is useful for drainage studies including springs and seeps (Clayton et al. 1995). While limited by sun angle and cloud cover, Infra-red provides additional details of vegetation and wet areas.
- **Thermal imagery** allows one to map variations in temperature with or without illumination and is usually flown at low altitudes (few 100 m). Applications range from military use to geologic mapping of vents, springs, seeps, cave openings, etc. on land or water as long as a temperature contrast exists.
- **Synthetic aperature radar interferometry (inSAR)** provides ground deformation data with centimeter precision. Newer methods of processing have enabled ground movements to be monitored down to 1 mm per year over wide areas providing ground vulnerability mapping (Browitt et al. 2007). Such measurements can be used night or day and penetrates cloud cover and vegetation.
- **LiDAR (light detection and ranging)** is a remote sensing (laser) technology that can be used to map topography at a very high resolution and accuracy, which has been applied to karst terrain for regional classifications, investigations and monitoring. LiDAR can be used night or day and penetrates cloud cover, tree cover and vegetation (Stennett and Grusky 2008; Haneberg 2008).

14.5.1 Aerial Thermography

When developing an inventory of site features for watershed studies in karst terrain locating the various swallets, springs, (recharge and discharge) areas can be tedious. Aerial thermography can detect variations in temperature as small as 0.05°, allows coverage over large areas and can minimize laborious field efforts to identify these features on the ground.

Aerial thermography surveys have been conducted to locate sources of springs discharging water on the bottom of lakes, creeks and rivers. Aerial thermography surveys provided a means of locating many unknown springs along the length of the river system in central Florida (Davis 2007). The locations of the larger springs were easily established, but many of the smaller features had not been located. Figure 14.8 shows a portion of a thermography survey completed at Silver Springs as part of the Florida Springs Initiative. Data was collected with a 2-m pixel resolution and entered into a GIS system. The thermal anomalies identified by aerial thermography surveys were ground-truthed in the field by divers. Three main discharge points were identified along this section. The thermography survey successfully identified known springs as well as several new discharges throughout Silver Springs area (Munch et al. 2006).

14.5.2 InSAR

Interferometric Synthetic Aperature Radar (InSAR) measurements are used to measure deformation of the earths surface from a satellite. By subtracting two scans of the same area at different times, it is possible to make precision measurements of changes in elevation. Under ideal conditions it is possible to resolve changes in elevation on the order of 10 mm or less. InSAR has been used to map land subsidence associated with geothermal fields, oil and gas fields and aquifer system compaction in California, Nevada and Texas (Bawden et al. 2003). InSAR is ideally suited for measurements of regional deformation and can be less expensive than obtaining sparse point measurements. The regional results can be used to optimize the location of point measurements or survey lines at a more local scale (Bawden et al. 2003).

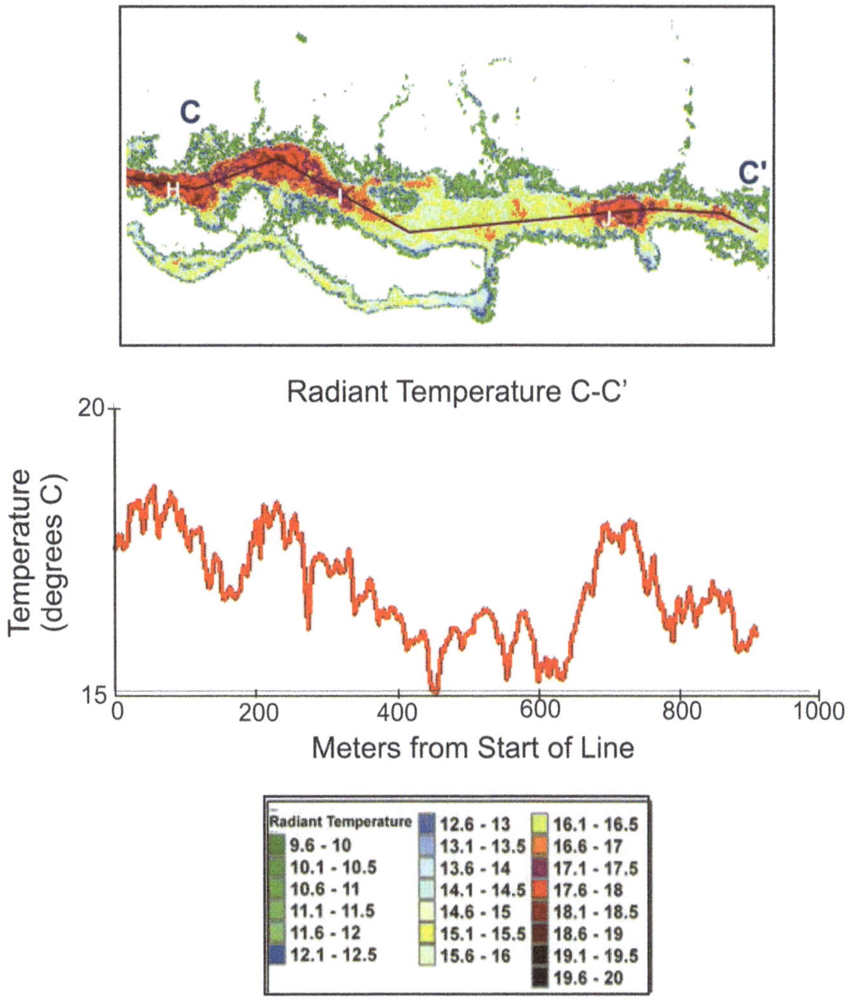

Fig. 14.8 Thermography data from Silver Springs, Florida showing three main points of discharge within the river (Munch et al. 2006)

14.5.3 LiDAR

LiDAR (light detection and ranging) is a remote sensing technology that can be used to map topography at a very high accuracy, covering large areas by aircraft. About 150 systems exist throughout the world, in 2008 with about half of these active in the private sector. Aircraft mounted lasers record elevations at a rate of 2,000–5,000 pulses per second and have a vertical precision of 15 cm (Stennett and Grusky 2008). These measurements can be made in difficult areas of access and highly vegetated areas, and have been used to develop detailed bathymetric data in shallow water depths to 10 m (Finkl et al. 2005). This technology can provide such detailed data that a national effort was initiated to establish total coverage of the US using LiDAR.

One of the myths is that Lidar can see through a canopy of tree cover. Reality is that LIDAR makes measurements only in the openings or gaps in the tree cover. If there is 80 % tree coverage then only about one in five shots will make it to the ground. But with repetition rates of 150,000 per second or more (around 10 shots/m^2 and sometimes triple that for helicopter work) reasonable surface coverage can be achieved in tree covered areas (Stennett and Grusky 2008).

LiDAR data can provide elevation measurements of the ground surface through most tree cover. As a result, detailed surface topography can be obtained which can reveal the presence of sinkholes and depressions within vegetated areas, which do not normally show up in aerial photography. One of the many applications of this method is to monitor areas and map changes in topography over time. It can also be used to locate lineaments and possibly some cave entrances. As an example, the use of LiDAR in this process has indicated a fourfold increase in the number of sinkholes mapped for Winona County, Minnesota (Rahimi and Alexander 2013).

LiDAR digital elevation models (DEM's) aren't restricted to sun position on a given day, and can be looked at with an infinite number of lighting combinations on the computer. Therefore, LiDAR–based topography maps and low sun angle shaded relief images have the potential to highlight features such as subtle areas of subsidence as well as obvious sinkholes. As with all methods there are inherent limitations. Haneberg (2008) cites some of the practical limitations and discusses the accuracies that can be obtained.

In the best cases, LiDAR can produce gridded DEM's with horizontal spacing on the order of meters (roughly an order of magnitude better than conventional USGS 10 m or 30 m DEM's) and a nominal vertical accuracy on the order of centimeters. Some LiDAR providers are claiming 7 and 8 cm, more typically only 15 cm is reasonable (Stennett and Grusky 2008).

However, LiDAR can produce DEM's and shaded relief images that are difficult to interpret, filled with ambiguous data, and processing artifacts. Depending upon the area of interest and type of karst terrain the non-karst features may include old building foundations, small quarry pits, cattle wallows, tree tip-up and in areas of past glaciations ice block melting processes (Alexander et al. 2013). Even with these limitations, this technology can provide an amazing amount of detail over large areas, in difficult terrain and with vegetated cover. So while LiDAR is widely being used to map sinkholes and other closed-depressions, standardization for its use need to be developed so that databases from different interpreters or regions can be compared and subjectivity can be minimized (Doctor and Young 2013).

A recent article in Arcnews, published by ESRI discussed efforts of the US Geological Survey who is taking the lead in the nation-wide 3D elevation program (Sugarbaker 2014). Both LiDAR and InSAR will be used to acquire three-dimensional elevation data. The USGS has developed the acquisition criteria to be met in order to maintain a uniform, quality database. These efforts are being met with a network of federal, state and tribal partnerships in order to create a database that is available for use by a variety of government and private industries. For Lidar and InSAR data availability go to www.coast.noaa.gov/inventory.

References

Alexander S et al (2013) Combining LiDAR, aerial photography and pictometry® tools for karst features database management. In: Land L, Doctor DH, Stephenson BJ (eds) Proceedings of the 13th multidisciplinary conference on sinkholes and the engineering and environmental impacts of karst, National Cave and Karst Research Institute, Symposium 2, 6–10 May, Carlsbad, pp 441–448

Allum JAE (1966) Photogeology and regional mapping. Pergamon Press Ltd, Oxford

Bawden GW et al (2003) Measuring human-induced land subsidence from space. U.S. Geological Survey Fact Sheet 069-03

Browitt C et al (2007) Terra not so firma. Geoscientist 17, 6(22):18–22

Clayton CRI et al (1982) Site investigation, a handbook for engineers. Halsted Press, New York

Clayton CRI et al (1995) Site investigations, a handbook for engineers, 2nd edn. Blackwell Science, Cambridge

Davis J (2007) Springs eternal. Prof Surveyor Mag 27(9):8–12

Doctor D, Young J (2013) An evaluation of automated GIS tools for delineating karst sinkholes and closed depressions from 1-m LiDAR-derived digital elevation data. In: Land L, Doctor DH, Stephenson BJ (eds) Proceedings of the 13th multidisciplinary conference on sinkholes and the engineering and environmental impacts of karst, National Cave and Karst Research Institute, Symposium 2, 6–10 May, Carlsbad, pp 449–458

Finkl C (1994) Report on FDOT Jewfish Creek Bridge. Consultants report, The Coastal Education & Research Foundation, CERF Remote Sensing Lab, Ft. Lauderdale

Finkl CW, Warner MT (2005) Morphologic features and morphodynamic zones along the inner continental shelf of southeastern Florida: an example of form and process controlled by lithology. J Coast Res Spec Issue 42:79–96

Finkl CW et al (2005) Interpretation of seabed geomorphology based on spatial analysis of high-density airborne laser bathymetry (ALB). J Coast Res 21(3):501–514

Gentile RJ (1984) Paleocollapse structures: longview region, Kansas City, Missouri. Bull Assoc Eng Geol 21(2):229–247

Haneberg W (2008) Elevation errors in a LiDAR digital elevation model of west Seattle and their effects on slope stability calculations. In: Baum RL et al (eds) Landslides and engineering geology of greater Seattle area, Washington, Geological Society of America, Reviews in Engineering Geology, 20, pp 55–65

Hotchkiss A (2007) Low altitude large scale reconnaissance or is this a great job or what? 58th annual highway geology conference, Pennsylvania 16–18 October 2007, pp 60–71

Johnson AI, Pettersson CB (1988) Geotechnical applications of remote data transmission. ATSM STP967-EB. American Society for Testing Materials, Philadelphia

Lattman LH (1958) Technique of mapping geologic fracture traces and lineaments on aerial photographs. Photogramm Eng 24(4):568–576

Lattman LH, Parizek RR (1964) Relationship between fracture traces and the occurrence of ground water in carbonate rocks. J Hydrol 2:73–91

Lillesand TM, Kiefer RW (1994) Remote sensing and image interpretation, 3rd edn. Wiley, New York

Lillesand TM et al (2003) Remote sensing and image interpretation, 5th edn. Wiley, New York

Littlefield JR et al (1984) Relationship of modern sinkhole development to large scale-photolinear features. In: Beck BF (ed) Proceedings of the 1st multidisciplinary conference on sinkholes: their geology, engineering and environmental impact, Orlando, Florida. Balkema, Rotterdam, pp 189–195

Munch D et al (2006) Fifty-year retrospective study of the ecology of Silver Springs, Florida. St. Johns River Water Management District; University of Florida – Department of Fisheries and Aquatic Sciences; and Wetland Solutions, Inc. prepared for Florida Department of Environmental Protection as part of the Florida Springs Initiative

Norman JW (1968) Photogeology of linear features in areas covered with superficial deposits. Trans Inst Min Metall Appl Earth Sci 77B:60–77

Paine DP, Kiser JD (2012) Aerial photography and image interpretation, 3rd edn. Wiley, New York

Panno SV, Luman DE (2013) Mapping palimpsest karst features on the Illinois sinkhole plane using historical aerial photography. Carbonates Evaporites 28(1–2):201–214

Parizek R (1988) Lineament analysis: how and why. Engineering & Geology of Karst Terranes, A short course sponsored by the Florida Sinkhole Research Institute, Orlando

Pictometry International Corporation (2002) US Patent Ser. No. 60/425,275 filed November 8, 2002

Rahimi M, Alexander EC Jr (2013) Locating sinkholes in LiDAR coverage of a glacio-fluvial karst, Winona County, MN. In: Land L, Doctor DH, Stephenson BJ (eds) Proceedings of the 13th multidisciplinary conference on sinkholes and the engineering and environmental impacts of karst, National Cave and Karst Research Institute, Symposium 2, pp 469–480

Stennett T, Grusky SW (2008) Lidar: fact or fiction. Professional Surveyor Magazine, April 28(4):16–20

Sugarbaker L (2014) US Geological Survey to lead ambitious 3D elevation program. ArcNews published by ESRI, spring p8 (www.csc.noaa.gov/inventory)

The National Corvette Museum (2014) National corvette museum media alert sinkhole collapse in skydome, www.corvettemuseum.org. Accessed 13 Mar 2014

Vernon R (1951) Geology of citrus and levy counties, Florida. Florida Geologic Survey, Bulletin 33, FGS, Tallahassee, Florida

Wall Street Journal (2014) Drone footage from inside Corvette Museum Sinkhole. Youtube, February 12, 2014. Accessed 26 Mar 2014

Site Walkover

15

Abstract

While a site visit may have taken place during the contracting phase of work, these are rapid tours of a site with little time to stop and observe or take detailed notes. The site walkover is when we get our first chance to make good observations and an initial assessment of general site conditions. It provides an opportunity to evaluate and expand our preliminary understanding of site conditions obtained from the desk study and the review of aerial photos. This initial effort should allow adequate time for the site to be covered. The observations made and information gathered from a site walkover should include geologic and hydrologic observations, inventory of karst features, indications of subsidence or sinkhole activity as well as an assessment of cultural features that may impact access, safety or a particular method of measurement used on site.

While the site walkover has many objectives, it should progress from a reconnaissance overview to obtain first impressions and then begin to gradually focus upon the site-specific details. On-site observations must not be limited to a single site walkover, but as the field effort proceeds, additional site walkovers and observations should be incorporated throughout the site characterization process.

15.1 The Initial Site Walkover

During the initial site walkover we begin to test the concepts in our preliminary conceptual model against actual site conditions. It is our first close look at the geologic, hydrologic and karst aspects of the site and the surrounding area. The walkover also provides an opportunity to obtain site-specific details not previously identified during the desk study, on the topographic maps, or the aerial photos. Finally we need to identify any site constraints that will limit access or impact safety of our future field operations. These may include simple access for personnel or heavy equipment. At the same time we can identify any obstacles for the field investigation, including topography, streams, rivers and lakes, as well as dense vegetation, fences, structures and both above ground and buried utilities. In addition, site security, along with health and safety aspects can be evaluated.

The objective of the initial site walkover is to leave the field with sufficient notes, sketches, and photos to reasonably document surface conditions, which may be of interest for further detailed investigation, or may impact the field efforts. While the focus is upon the site itself, the surrounding area should also be inspected to the extent needed and where access can be obtained.

15.2 Importance of Observations

Observations are one of the most important tools we have in our tool box and the primary tool used during the site walkover. It is here where experience and judgment aided by skilled observations can make such great contributions. However, some individuals are better observers than others. For this reason the site walkover field team should consist of at least two persons, that way, observations and ideas can be shared and discussed on the spot. "Quite often the most essential services I rendered to my clients had no relationship to the original assignment. They grew out of casual observations I made while inspecting the site" (Terzaghi

R.C. Benson, L.B. Yuhr, *Site Characterization in Karst and Pseudokarst Terraines*,
DOI 10.1007/978-94-017-9924-9_15

1957). The authors have also found that observations have been key to solving numerous engineering geologic puzzles over the years.

Unfortunately field observational skills are becoming a lost art and are not easily taught. An early lesson in observation was taught to the senior author in an undergraduate geologic field trip to northern Minnesota. As the group gathered to look at a fault, the professor asked Benson if he would like to comment upon the fault they were here to observe. After a pause and a blank look on Benson's face the professor suggested that the group should notice that Benson had one foot on either side of the fault. That was a good early lesson in observation and attention to details.

Many karst or pseudokarst features will not be recognizable on a topographic map or aerial photos and may only become apparent during the site walkover. This may be due to their size or being masked by vegetation or cultural features. In some cases, there is only subtle evidence of subsidence or sinkholes present at the surface. Therefore, it will be necessary to locate, map, and evaluate karst features with a site walkover. For example, of the 535 sinkholes in Winona County, Minnesota 85 % were not shown on the 7.5-min topographic maps and were only found by fieldwork, (Dangleish and Alexander 1984). Fewer than 5 % of the springs in the Mammoth Cave area of Kentucky are shown on 7.5-min topography maps. The remaining 95 % were found by fieldwork (Quinlan 1989, 1990). All of these details provided important clues as to the nature of the site conditions.

There are many subtle details that can be obtained from careful observations yet can easily be overlooked or simply dismissed as irrelevant. An example of subtle features was seen on our initial visit to an earthen water retention dam located near Phoenix, Arizona. Small desiccation-like cracks were noted perpendicular to the axis of the dam. They were subtle and would have been missed unless attention was focused upon surface details. Surveyor flags were then placed at each crack over a distance of approximately 120 m and then the spacing between them was measured. Of the 20 survey flags, there spacing was found to be 5.8 m ± 1 m. This suggested that the mass of the earthen material making up the dam was fracturing (due to settlement and or desiccation) at an interval related to the mass of the material properties.

The importance of observation cannot be over emphasized. "If you do not know what you should be looking for in a site investigation, you are not likely to find much of value" (Fookes 1997). "Only with experience and attention to detail can the relative importance and significant of these features be interpreted in the field" (Clayton et al. 1982, 1995). Observational skills may be the best tool we have available to us in the field.

15.3 Some Tools for the Field

There are obvious tools such as topographic maps, aerial photos, rock hammer, a field notebook, camera and a compass. Limited sampling of soil and rock may be obtained for initial analysis and reference. On a new site development, the initial area in which a project is to be located can be many km^2. While a site walkover is done mostly on foot, larger sites may require an ATV (Fig. 15.1a) or a 4×4 pickup truck as a means of transportation. A GPS unit and topographic

Fig. 15.1 An ATV is used to cover a large site (**a**) and field mapping has been aided by the use of GPS and digital field notebooks (**b**)

sheets are a must in wide ranging reconnaissance of 100's–1,000's of hectares particularly over areas with few landmark or reference points. With the wide-spread use of GPS one might think that pacing would be a lost art, however, pacing remains effective for field reconnaissance in wooded areas, underground or in structures. Distances by pacing can be better than 2 % on flat smooth surfaces, but will be less in rough, hilly, or vegetated terrain. Field mapping can be improved by the use of a GPS system, hand-held computers and digital topographic maps and aerial photos (Fig. 15.1b).

Sketches were once standard for professional fieldwork and were well done by many (See Fookes 1997 for a number of excellent examples). The art of field sketching has largely been academically "pushed off-screen" by the encroachment of other trendy and "more modern" technologies. Few engineers or geologists today are taught topographic surveying, and field sketching is largely neglected in class or camps teaching field geologic mapping. Field sketches can be an important part of the site walkover. The very act of sketching forces you to stop and observe what site conditions are while they are before your very eyes (Hatheway 1998) (Fig. 15.2). Sketches should be made showing features at the site, which are not shown on existing topographic sheets or aerial photos. While on-site photographs are a necessary component of the site characterization effort, the process of taking a picture

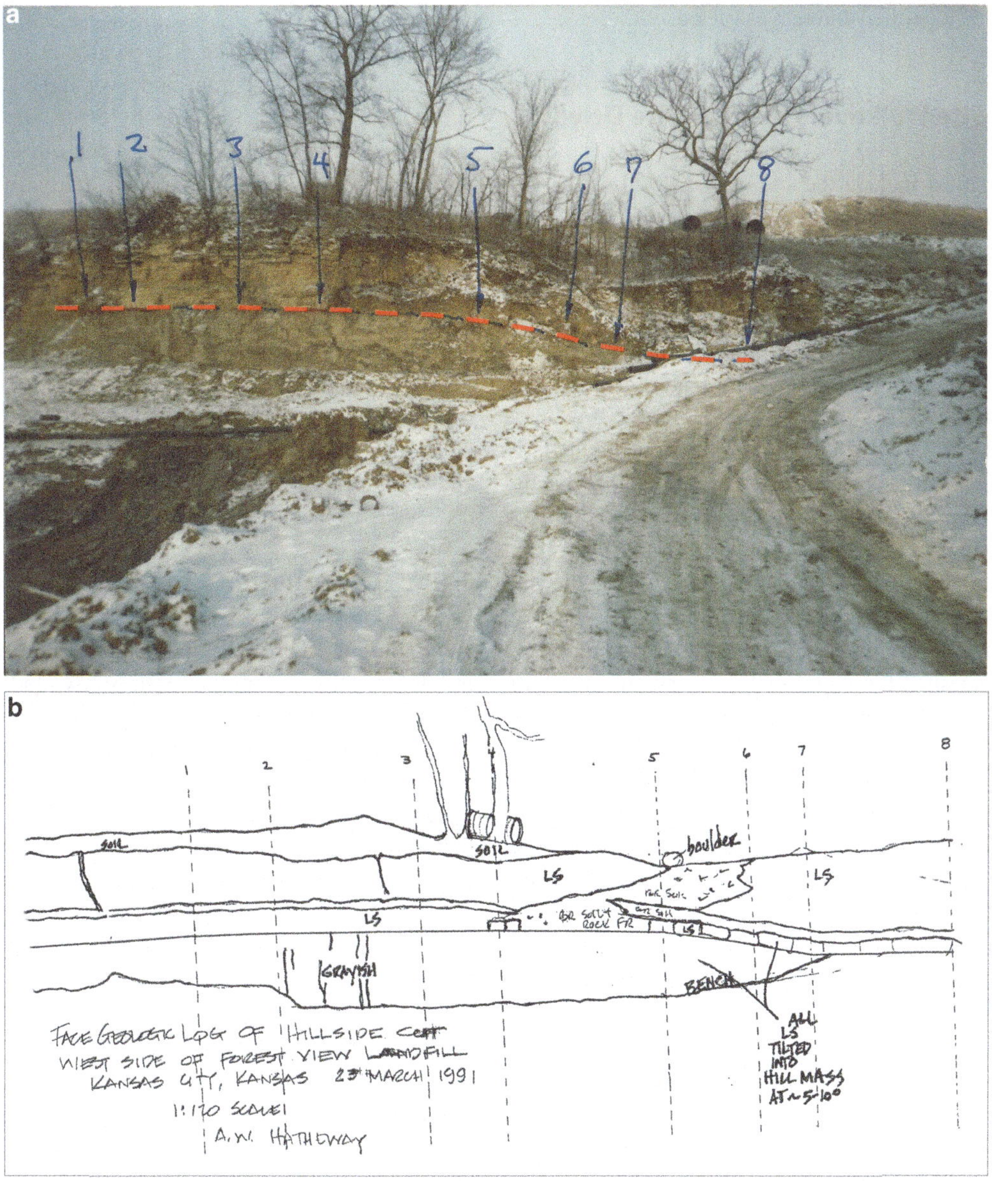

Fig. 15.2 An excavation has exposed a cross section of geology indicating the edge of a paleocollapse. It was photographed (**a**), sketched (**b**) and elevations were obtained along the top of a shale layer to document conditions

does not force one to take the time to see the details that sketching does. It is often the little details that are most important.

While today's field staff is probably less talented in terms of sketching ability, (that includes the senior author), a reasonable alternative to detailed sketching is the use of photo documentation. Almost anyone can take reasonable good photos in the field that can be checked immediately with today's automated digital cameras with a wide angle to telephoto lens. However, good observational skills are still necessary in order to acquire photo documentation that is useful. Field notes need to be taken for each photo, including the direction of the photo and a simple sketch to note the key elements of interest. That way, when the dozens or even hundreds of photos are being reviewed, at a later time, their location and the points of interest can easily be identified and reconstructed.

15.4 On-Site Walkovers and Off-Site Drives

The walkover must include the site itself as well as the surrounding area, which may require reconnaissance over many square kilometers. One should cover enough of the surrounding area to minimize future surprises, this will usually require driving. The area that needs to be covered and detail that is required will depend upon specific project objectives. In general, the further away from the site these features are located, the less important the feature would be. However, if a site is being evaluated from a hydrologic point of view, particularly if it involved contaminants, the karst inventory would typically be basin-wide. For these projects, even small features that are well connected may have an impact at great distances.

Artificial boundary limits are often placed on the site characterization process. It is not uncommon for owners to suggest that work only be carried out within the immediate confines of the site (this commonly occurs on sites with environmental contaminant issues). The issue of limiting off-site access is often driven by legal concerns. This can be an awkward situation since quite often; critical elements of geology are commonly discovered off-site. While the majority of the detailed investigation work will be done within the site boundaries, it is a serious error to blindly ignore the regional and local setting.

A classic example of this occurred during our work at Love Canal in the early 1980s. The health and safety officer had not yet shown up at the site and our team's access was denied prior to a health and safety briefing. The time was spent driving around the site to familiarize us with the area. A new sewer line was being installed by a local contractor, approximately 600 m to the east and parallel to the Love Canal site. What was of interest was the open trench which exposed a massive clay (a clean clay with the consistency of modeling clay) which the contractor said extended throughout the entire extent of trench from well north of the Love Canal site to the south of it. The exception according to the contractor was the "underground river" he had encountered approximately due east at the north end of the site with water flowing away from the Love Canal area. Upon returning to the project office, the team was chastised by the project manager for going beyond the site (actually into the next county just a few blocks away). To this manager, the only thing of interest was the Canal property itself and the regional geologic setting was of no interest. Unfortunately this type of thinking, or more correctly a lack of understanding of the site characterization process, is all too common.

15.5 Site Coverage

The initial site walkover is typically at a reconnaissance effort early in the program and over large areas. However, subsequent efforts can approach 100 % site coverage over key portions of the site, later in the site characterization effort. This may be accomplished gradually as work progresses at the site or as a single focused field effort. Either way it is essential that the observations and notes made get summarized and added to the project map or database. The focus and detail that is required will depend upon specific project objectives. For example, a site to be used for an engineered structure will be interested in site-specific conditions that would impact structural integrity. If a site is being evaluated from a hydrologic point of view, the focus would typically be basin-wide.

15.6 Observations and Mapping

Geologic and hydrologic mapping can be optimized using the information from the desk study and aerial photos. General conditions for a site should be established in a preliminary conceptual model at this point, what type of conditions might be expected and therefore what to look for. The initial site walkover would typically only confirm locations and document essential information. A separate and focused field effort may be required for specific geologic field mapping in areas with complex geomorphology or when a detailed inventory of karst features is necessary.

15.6.1 Geologic Observations

Field mapping would be focused upon finding and utilizing geologic windows or rock exposures in road cuts, rivers, creeks, quarries or possible local excavations. Inspection of rock exposures is a key aid in estimating the fabric and character of the rock including: joint spacing, and aperture as

Fig. 15.3 Three road cuts illustrate the valuable geologic information available from these geologic windows. (**a**) Southwest Texas. (**b**) Puerto Rico. (**c**) Kansas City, Kansas

well as degree of weathering and presence of karst features. In some cases, open fractures or cavities can be found in other cases they may be sediment filled. These windows into the shallow subsurface often provide excellent information.

Road cuts are great opportunities to observe rock characteristics that are otherwise not accessible. The road cuts may be adjacent to the site or represent the region surrounding the site. The information provided needs to be put into context with relation to the site under investigation so that information is not over extrapolated.

A road cut along Highway 90 in southwest Texas (Fig. 15.3a) shows two smaller areas of in-filled collapse.

Fig. 15.4 The floor of the Missouri Portland Cement Quarry in Independence, Missouri reveals the extent and spacing of fractures in the Bethaney Falls Limestone

These features are relatively close to one another and they extend from the surface to the roadway indicating that the feature extends deeper well below the road. A road cut in Puerto Rico (Fig. 15.3b) shows a highly weathered and irregular limestone profile. In the Kansas City area, the uppermost Argentine Limestone (Fig. 15.3c) is commonly exposed along road cuts showing extensive fractures and weathering.

Quarries are another window of opportunity to observe rock characteristics. But because quarries are selected to be in the best rock, they are not necessarily the best places to observe karst. However, when a quarry is available it must be included in the site walkover and offers an opportunity to observe a major cross section of geologic conditions. The quarry geologist or engineers are usually pleased to show you around and can usually provide a wealth of information.

A limestone quarry floor adjacent to a site under investigation in Missouri revealed the extent and spacing of vertical fractures in the Bethaney Falls Limestone (Fig. 15.4). This plan view of the rock fracturing was unique and would not have been as readily perceived from any other data or observations.

A large limestone quarry very near a project site in Fredrick, Maryland had a perimeter of approximately 2,925 m and a maximum depth of approximately 76 m. Figure 15.5a shows a portion of the western quarry wall. This quarry provided an excellent geologic window. However, only three small cavities were seen on the western side of the quarry at a depth of approximately 36 m below grade. The largest had a diameter of about 1 m, and the other two about 0.3 m in diameter (Fig. 15.5b). An area of spring flow was observed near the upgradient base of the quarry at its north end (Fig. 15.5c). The cross section of the springs was approximately 1 m^2. This is an example of the relatively low density of cavity systems within the 223,000 m^2 quarry wall. This illustrates the density of cave systems and the difficulty of finding them.

15.6.2 Hydrologic Observations

All significant surface hydrologic features should have been identified during the desk study and they should be verified during the site walkover. However, many smaller features, not readily identifiable on aerial photos or topographic maps such as small ponds, creeks, sinks, springs and seeps, may be found during the site walkover. The location, elevation and estimated flow rate need to be added to the site map and database along with the date of the observation and a photograph taken to document the feature. Further data may include the temperature and conductivity data for these waters. This is particularly important, if a dye trace study is part of the program since it would help to identify possible injection and sampling points.

Streams sometimes become sinking streams and disappear, recharging the groundwater. As you follow the streambed water may reappear, discharging to the surface as a spring further downstream. The location of a sink or spring may also move over time as water levels change and systems become plugged with sediments and debris. In addition, there are some hydrologically significant features that may only reveal themselves during heavy rainstorm events. Since fieldwork is typically avoided during a heavy rainstorm, a special walkover may be required during such events.

Temporal changes in hydrologic conditions may occur over a short period of time or a longer period of time. Figure 12.13a shows a small creek adjacent a site under investigation. A rainstorm during fieldwork rapidly changed conditions over a period of a few hours (Fig. 12.13b). Figure 15.6a is taken from the dry creek bed in the Peace River in west-central Florida showing the sign for canoeists in wetter seasons. This dry season occurred over a period of months. Open joints are now revealed as a result of the dry season (Fig. 15.6b). Knowing about such unique temporal conditions can provide insight that would otherwise be missed.

15.6.2.1 Flow of Air from Fractures, Caves, Mines, and Boreholes

Cool air flowing from the entrance of caves, mines and their ventilation shafts is common. In addition, air flowing from exposed fractures in rock as well as from well casings whose

Fig. 15.5 A section of the westernmost quarry wall (**a**), the three cavities identified high on the western wall (**b**) and the springs (**c**) located low on the north wall

screen or open hole has extended into the unsaturated zone. Air is about 1,000 times lighter than water and air currents require much less energy and have greater diffusion than water. As a result flowing air can be used as a very sensitive tracer (Milanovic 2004). Vertical permeability based upon air flow in the unsaturated soils has been described by Weeks (1978, 1987).

Cavers will often follow flowing air within a cave and have reported air flow from fissures on the surface which has lead to cave systems. Air flow from caves and mines may be constant in one direction or may reverse flow. Lewis (1991, 1992, 1996) proposes a number of factors that may cause and control air flow from caves including:

- Changes in barometric pressure
- Surface winds
- A water siphon effect from fast turbulent flow
- A water fall in a narrow shaft (water jet pump)
- Bernoulli effect at a tee shape junction between higher and lower entrances
- Humidity effect due to differences in air density between moist and dry cave air, and
- Venturi effect of surface winds drawing air out of a cave entrance

Some modeling of the fluid-thermal dynamics of air flow in caves has been done to simulate two-dimensional air flow in Carlsbad Cavern, NM. This work focuses on buoyancy and natural convection due to geothermal heating (Boston et al. 2005).

The flow of cool air from well casings and fractures in rock exposed at the surface are indicators of interconnection within fractured rock or possibly a cave in the subsurface.

- The authors noted a significant air flow from the casing of a windmill well in southwest Texas which was located in

dry creekbed usually active with canoeists

open fracture found in dry creekbed

Fig. 15.6 The Peace River in west-central Florida is normally a small quiet stream used by canoeists. However, during periods of low rainfall the river will dry up completely (**a**), note the sign on the cypress tree indicating direction for canoeists. When dry, open joints in the rock are exposed (**b**)

limestone where the water table was about 150–180 m below grade. Cool air flowed freely from the well casing indicating that the borehole had encountered open fractures in the rock, (or possibly a cave system). Other observations in the area revealed obvious air flow from fractures in the surface rock. These were good indicators of the interconnections of fractures within the rock mass and or the possibility of the presence of a cave. Measurements of temperature can be made to support flow measurements.

- At the Nevada Test Site "Faultless" an underground atomic bomb of intermediate size (200 k ton–1 M ton) was detonated some 900 m below grade in volcanic tuff. The water table was about 90 m below grade. A significant rim fracture with a displacement of about 3 m extended about 450 m south of the blast site. A monitor well about 90 m south of the rim fracture had an obvious and continuous air flow from it. Measurements of the permeability have been made at several nuclear chimneys at the Nevada Test Site (Rosza et al. 1975)
- Thermal imagery of borings used for backfilling of a mine in the Kansas City area would quickly indicate which boreholes had encountered an open mine since cool air would commonly be flowing from them.

These are examples where attention to detail can provide additional data regarding the interconnections of fractures within the rock mass and or the possible presence of a cave or mine.

15.6.3 Inventory of Karst Features

An inventory of karst features both on and off-site is standard in any site characterization in a karst terrane. This inventory includes possible features identified from Tables 13.1, 13.2, and 13.3 in the desk study. This effort usually starts at the regional setting and then moves to the site-specific setting. A list of existing or possible karst features would be developed from topographic maps, aerial photos, regional literature, databases and even site-specific reports. During the site walkover these features should be confirmed and additional information on them gathered. The purpose of the karst inventory may:

- Provide an idea of what type of karst features are present in and around a site;
- Reveal patterns of occurrence;
- Provide an indication of subsurface activity (or lack of activity) in an area; and
- Establish injection and sampling points for a dye trace study.

When sinkholes, sinks, springs or caves are encountered, their location should be identified on a topographic map, aerial photo, or site map. In some cases, where access is safe additional insight can be gained by entry. Small caves can often be found that may or may not have been previously identified or mapped.

At a proposed federal prison facility in Virginia, a small cave was encountered during the site walkover (Fig. 15.7a). This large property had not been previously developed or surveyed so there was no existing information on this cave. Documentation included basic dimensions and geologic formation along with evidence of collapse or presence of

Fig. 15.7 Caves can be encountered during a site walkover. Measurements of a cave were made to establish the cave's height and span in West Virginia (**a**). A small collapse was found at the edge of a house. A quick inspection indicated that this was a small cave (**b**)

flowing water. Although this cave was not within the footprint of development, it provided an indication of the type of karst feature that may be presence elsewhere on the property.

While the site walkover starts with an indication of what to expect in the field, an open mind needs to be maintained. A project along the southeast coast of Florida is not necessarily where you would expect to find a cave. However, a small cave was found extending under a house (Fig. 15.7b) after a collapse occurred as a result of a rain down spout concentrating water at the edge of the house. The small cave extended under portions of the house, the swimming pool and much of the yard.

An example of a unique karst inventory comes from a proposed hazardous waste site in southwest Texas. The initial site covered an area of 15 km^2, designated by the owner. The final area of proposed construction was ultimately narrowed down to 252 ha as the site characterization progressed. However, the reconnaissance walkover (and in this case drive over) was carried out over 165 km^2 and focused upon a known caves, paleokarst collapse features, springs, and road cuts in the area.

Later in the site characterization effort, Jim Quinlan, a member of the advisory committee for the site, requested a very detailed site walkover. Based upon the type of facility that was planned and the area of karst in which it was to be located, a 100 % visual site walkover was completed. This effort included the 252 ha as well as a zone of 76 m around the entire perimeter. At this point in the project, we knew that no obvious karst features were present over the 252 ha. The purpose of the walkover was to identify any subtle signs of subsidence, soil piping, fissures, or depressions at the surface.

Survey stakes were installed on a 76 by 76 m grid. A team of 11 people (local high school students) were used to form a line in an east-west direction over the 76 m distance between survey stakes. Prior to the walkover, the team was briefed and shown examples of surface features they may encounter such as localized depressions, soil cracks and fissures and animal borrows. Each person was responsible for observing the ground surface directly in front of them and about 4.5 m to each side. They were instructed to place a surveyor's flag at any surface feature they observed and if in doubt, they were instructed to place a flag. Several professional staff members walked behind the line to monitor the team's position, stop the forward movement when necessary, to answer questions and to provide experienced observations. At the end of each 76 m swath the team would stop and be debriefed before continuing on to the next 76 m section.

A total of 128 features were identified during the 2-day walkover. Each feature was mapped and assessed to determine its true nature. Many of the features were found to be old animal burrows or small areas where water would pond. These features were abandoned and their flags removed. Of the remaining 42 features, fissures dominated the type of feature found and they extended about a meter below grade based upon shallow excavations. Figure 15.8 illustrates some of the typical features identified.

When these features were plotted on a site map they were clustered in the southeast portion of the site and indicated a clear northeast-southwest orientation. The orientation of the features identified in the site walkover and the aerial photo analysis matched the orientation of the many large paleokarst collapse features identified by Freeman (1968) a few kilometers south of the site (Fig. 15.9).

One could have easily dismissed these features as local desiccation cracks. However, their correlation with the aerial photos features and alignment with the large paleocollapse features mapped by USGS some 4.8 km to the south indicated an increased significance, which were incorporated into the conceptual model of the site.

15.6.4 Indications of Subsidence and Sinkhole Activity

There are many indicators of subsidence and sinkhole activity that occur at the ground surface that can be obtained from careful observations. Many of these indicators can be quite obvious while other more subtle indicators can easily be overlooked or simply dismissed. Because of the scale of these features they will not be recognizable on topographic maps or aerial photos. Field observations are the only means to locate, map and evaluate these features.

There are a number of precursor indicators of subsidence and sinkhole development. These indicators can occur naturally or manifest themselves in man-made structures (Fig. 15.10). The following has been adapted from Newton (1987).

- Depressions in the soil or pavement commonly resulting in the ponding of water (Fig. 15.10a).
- Cracks, fissures in the soil, pavement or concrete (Fig. 15.10b, c).
- The tilt of trees and fence posts.
- Holes in the ground that have formed over short periods of time and which may drain surface water (Fig. 15.10d).
- Turbidity and mudding of well water or a sudden decline of water level in a well.
- Sudden drainage of a surface water impoundment or a new impoundment designed to hold water.
- The localized presence of vegetation in the middle of an otherwise open field (some plants will flourish with an excess of water pounding at the surface).
- Localized wilting vegetation due to lack of water, (some plants will diminish as moisture drains into a sinkhole).

Fig. 15.8 Three examples from a total of 42 localized features and fissure found during a 100 % site coverage walkover. Excavation of these fissures by shovel and backhoe revealed that they extended at least a meter below grade

The presence of one or more of these indicators does not necessarily confirm the presence of an impending sinkhole, but it does provide a basis for further investigation. Careful observations may provide clues as to subsurface activity and build the conceptual model for the site characterization.

When structures are impacted by subsidence or sinkhole activity, the initial results are often visible cracks within the structure walls, walkways or driveways (Fig. 15.10c). General indicators of foundation movement have been adapted from Freeman et al. (2002) and are summarized below:

- Isolated cracks at weak points within the structure
- Cracks taper from top to bottom
- Cracks occur both externally and internally at the same location
- Cracks consistent with a pattern of movement
- Changes in windows and doors sticking and not closing properly
- Walls and floors measurably out of level or plumb
- Settlement or tilting of a portion of the floor slab
- Drains and utility services interrupted
- Broken pipes and utilities

Besides a sinkhole, other causes of structural damage to buildings may include shrink swell clays, a foundation built on organics or trash, mine subsidence, slope instability, soil erosion or long-term groundwater withdrawal resulting in subsidence. These should be kept in mind as one assesses structures.

Freeman et al. (2002) and Audell (1996, 2004) provide an excellent description of the cracks in structures using a Crack Classification System (CCS). While theses three publications do not address subsidence due to sinkholes

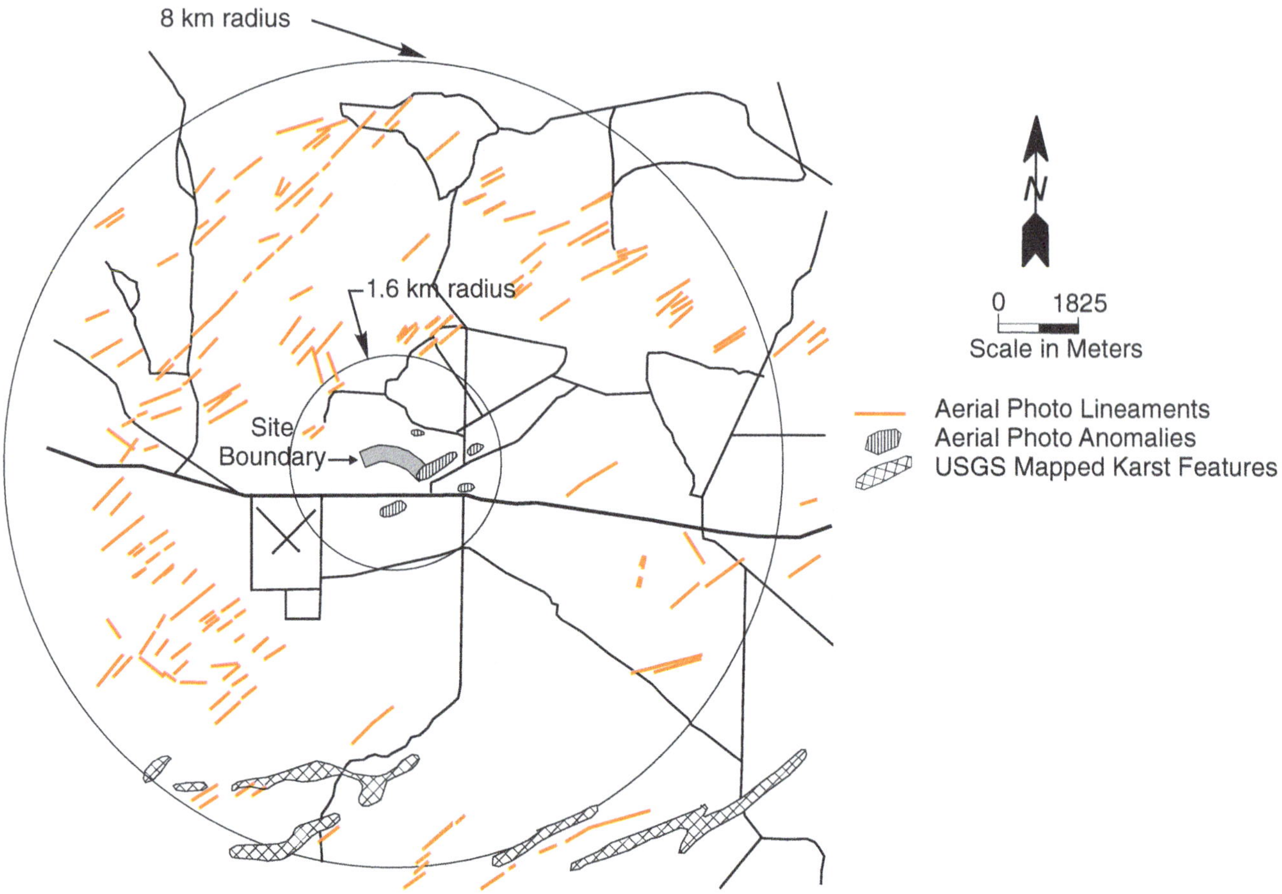

Fig. 15.9 The location of photo lineaments, the aerial photo anomalies and the larger paleocollapse features mapped by USGS just south of the site all show the same general orientation

they provide useful data for evaluating cracks in structures due to other causes and can be useful in resolving sinkhole claims.

15.6.5 Cultural Factors, Utilities and Other Site Limitations

During the initial site walkover observations should be made of possible cultural activity, which can impact the logistics, safety or possible measurements at a site. This might include remnants of previous building foundations, drainage channels, debris piles, vegetation, fences and ditches.

Above ground and below ground utilities are also an important feature that could impact site access and measurement locations. Their presence and locations must be identified. Utility locations on public right of ways will be identified by the responsible utility company or marked by a one call service. On private property utility locations must be obtained by utility location devices such as pipe and cable locators, or ground penetrating radar. In some cases, utility drawing as-built plans may be available. However, caution should be used since errors are common on such "as-built" drawings. As-built plans often turn out to be as designed plans and therefore may not contain deviations that occurred during construction or modifications after construction.

Surrounding cultural features may also impact a site. These features may include spray fields, water mains, major utility corridors, transmission lines, communication towers, private and local pumping wells, or well fields. Small community well fields can result in triggering sinkholes under the right conditions. A single small sinkhole developed in the gypsum beds along a roadway in east-central Michigan where open pit gypsum mining had been carried out. The nearest abandoned mine was some 460 m from the sinkhole had been inactive for decades. The only recent change to groundwater levels had been the new well field that had been installed between Lake Huron and the roadway (a distance of about 210 m) to supply a new housing development. This was likely the change in water level that triggered the sinkhole.

Fig. 15.10 Early subsidence is commonly indicated by the presence of circular depressions that retain water (**a**), cracks, fractures or local settlement in the soil (**b**), cracks in asphalt or concrete driveways (**c**) and localized depressions or drop-outs (**d**)

15.6.6 Eye Witness and Anecdotal Information

Discussions with local workers or residences can often be quite effective to discover unique site conditions that are not noted or reported elseware including temporal events. Even if no obvious occurrence of sinkhole collapse has taken place, subtle indications may have been witnessed. For example, while working along a bridge alignment, we would stop at a local shop for lunch. The owner was curious about what we were doing. Our conversation revealed a spot in the road as you approached the existing bridge that repeatedly had been filled and patched. This subtle piece of information ended up correlating well with other data that indicated a paleocollapse at this location (for more details see Chap. 26).

Verbal information from persons who have worked at the site or who live nearby can be helpful. These sources of information can often provide insight to local or even site-specific conditions that may not be available elsewhere. However, caution is called for since people's memory may be inaccurate and they may report issues in error or with bias.

15.7 Fly Over

As part of the site walkover, it is sometimes useful to make a site fly over which can provide a rapid means of familiarization with the site itself and its relation to the overall setting. A fly over is especially useful over large areas and those with difficult or limited access on the ground. The fly over may be done before or after the site walkover depending upon access and project needs.

A fly over is commonly completed using a small fixed wing aircraft or helicopter. A slow speed aircraft with an overhead wing is the best choice for visibility and obtaining oblique aerial photos. The senior author has used a two seat Piper Cub with an overhead wing to provide the best visibility and with the window open to improve photographic quality. Helicopters are quite maneuverable but are also noisy and have a fair but of vibration. The senior author has utilized the Goodyear blimp a number of times in the 1970s to monitor dye traces from sewerage outfalls in the ocean and to track dye traces of tidal flow in coastal waters and inlets. Today however, we will be seeing more unmanned aerial systems (UAS) (Fig. 14.7) being used for such reconnaissance.

Fly over's often proceed at a rapid pace with many distractions including communication with the pilot. During the flight looking down at your maps for a few seconds and then looking out the aircraft window one can easily lose track of your location and orientation. Recording audio for notes or video and audio for more complete documentation is recommended and good communication and planning with the pilot prior to the flight is essential.

A DOE, Expedited Site Characterization (ESC) program was being carried out by the Ames Laboratory with the assistance of the authors and others at the Central Nevada Test Area (CNTA). The area of interest extended over hundreds of hectares and included various disposal areas and a large mud pit. The site was mostly an open area with few reference points. While maps and vertical aerial photos provide an accurate overview of the site, oblique aerial photos often will provide a better prospective and a feel for access and logistics including areas that may be inaccessible. A site fly over was completed by a key member of the ESC team which provided perspective of site conditions which were difficult to obtain from maps and observations on the ground alone.

15.8 Updating the Conceptual Model

The site walkover will commonly provide an opportunity to resolve a number of the issues identified in Tables 13.1, 13.2, and 13.3 discussed in the desk study. In addition there will have been other features that will have been discovered during the site walkover. These will provide the first on-site geologic, hydrologic and karst data to improve the conceptual model.

The following is an example of the early evolution of a preliminary conceptual model at a mine collapse site. After completing a few days of site walkover and a few days inside of a mine, we were starting to form some ideas and come to initial conclusions about a number of issues at the Tobin mine site in Kansas City where a 7 ha mine-roof collapse had occurred. The following morning at breakfast Allen Hatheway presented me with a hand written note in which he raised a number of questions about our early thoughts and opinions we had started to form about site conditions. The note was titled: "Dicks Moment of Truth Exercise" and contained the following:

1. We can't really show there is paleokarst involved. Yet we want to employ such to help explain the mine failure phenomenon.
2. The literature on subsidence is complex – no clear answers
3. We don't know where the mine water is coming from, or going to, and how much is coming and going.
4. Why can't the mine collapse propagate all the way to the ground surface through 51 m of limestone and shale, next week?
5. What makes you think the surface fissures in the loess are older than the mine?
6. Why can't the loess develop arcuate-trace surface fissures, or such features oriented perpendicular to the ground surface?
7. Our concepts and present knowledge of mine elevations as tied to major features is lousy.

Such discussions between team members are necessary during the site characterization process to keep challenging preconceived ideas. After much more time and data were collected at this site these issues were all resolved. See Chap. 25 for more details.

The revised conceptual model may include improved geologic cross sections and maps, block diagrams, and photos along with the text necessary to explain them. Possible conditions that may not have been identified, but might be reasonably expected should be mentioned. The basis for any assumptions should be noted along with levels of uncertainty. This model (which is based upon knowledge of the site conditions, not guess work) will then be used to guide further on-site work.

15.9 Updating the Work Plan

Now that the site walkover and a possible fly over have been completed, we have a reasonable first approximation of the nature of the geology and hydrologic conditions to be expected at the site. We also may have a first approximation of the karst conditions at the site.

For example if the focus is on sinkholes, topographic maps and aerials will have provided data on the type, density and patterns of sinkholes and fractures. A review of sinkhole databases along with comparison of old and recent aerial photos provides an indication of sinkhole activity. The karst maturity of the site should be able to be determined based upon the five classes (Fig. 4.1) developed by Waltham and Fookes (2003). Sinkholes on-site should be able to classified (Fig. 3.1). If cover collapse sinkholes are present, we might can insight into the nature of the void spacing with the rock based upon their size (volume of material lost). Similar data can be developed for caves, springs, fracture patterns, etc. While this is based upon existing, readily available data, it should provide a level of focus to the subsequent field efforts.

In addition, a reasonable understanding has been established for the various natural and cultural conditions that may limit access and control our approach to field work. We have filled in some of the answers to questions about the site and identified other new issues that need resolution. Based upon this information we can update our work plan for the subsequent field investigation and confirm that the methods and coverage are all appropriate and adequate.

References

Audell HS (1996) Geotechnical nomenclature and classification system for crack patterns in buildings. Environ Eng Geosci 2(2):225–248

Audell HS (2004) Field guide to crack patterns in buildings, a guide to residential building cracks caused by geologic hazards. Association of Engineering Geologists, Southern California Section, Special Publication No. 16

Boston PJ et al (2005) Micrometeorological modeling of idealized caves: predictions and applications to Carlsbad Cavern, NM, USA. NSS Convention Abstracts. J Cave Karst Stud

Clayton CRI et al (1982) Site investigation, a handbook for engineers. Halsted Press, New York

Clayton CRI et al (1995) Site investigations, a handbook for engineers, 2nd edn. Blackwell Science, Cambridge

Dangleish J, Alexander EC Jr (1984) Sinkhole distribution Winona County Minnesota. In: Beck BF (ed) Proceedings of the 1st multi-disciplinary conference on sinkholes: their geology, engineering and environmental impact, Orlando, FL. Balkema, Rotterdam, pp 79–85

Fookes PG (1997) Geology for engineers: the geological model, prediction and performance. Geol Soc Lond Q J Eng Geol Hydrogeol 30(4):293–431

Freeman VL (1968) Geology of the Comstock-Indian Wells area Val Verde, Terrell and Brewster Counties. Texas Geological Survey, Professional Paper 594-K, USGS Washington, DC

Freeman TJ et al (2002) Has your house got cracks? A homeowner's guide to subsidence and heave damage. Institution of Civil Engineers and Building Research Establishment, 2nd edn. Telford Publishing, London

Hatheway A (1998) "Site sketches" the essential picture, perspectives no. 32. AEG News 41(1):36–39

Lewis WC (1991) Atmospheric pressure changes and cave airflow: a review. NSS Bull 53:1–12

Lewis WC (1992) Natural blowholes and blowing wells. NSS News, October, pp 258–263

Lewis WC (1996) Constant airflow anyone? NSS News, July, pp 181–182

Milanovic T (2004) Water resources engineering in karst. CRC Press, Boca Raton

Newton JG (1987) Development of sinkholes resulting from man's activities in the eastern United States. USGS Circular 968

Quinlan JF (1989) Ground-water in karst terranes, recommended protocol and implicit assumptions. USEPA Las Vegas Laboratory, EPA/600/X-89/050

Quinlan JF (1990) Special problems of ground-water monitoring in karst terranes. In: Neilson DM, Johnson A (eds) Ground water and vadose zone monitoring, ASTM special publication 1053, West Conshohocken

Rosza R et al (1975) Permeability of a nuclear chimney and surface alluvium, Area 2, ERDA NTS, UCID-16722 Lawrence Livermore Laboratory, CA

Terzaghi K (1957) Consultants, clients and contractors. Presented at the Boston Society of Civil Engineers Section of ASCE, Reprinted by Risk Analysis & Research Corporation for the Association of Soil and Foundation Engineers

Waltham AC, Fookes PG (2003) Engineering classification of karst ground conditions. Geol Soc Lond Q J Eng Geol Hydrol 36:101–118

Weeks EP (1978) Field determination of vertical permeability to air in the unsaturated zone. Geologic Survey Professional Paper 1051, Washington, DC

Weeks EP (1987) Effect of topography on gas flow in unsaturated fractured rock, concepts and observations. Am Geophys Union Geophys Monogr 42:165–170

Surface Geophysical Methods 16

Abstract

We now have a work plan that is based upon regional and local existing data and possibly some site-specific data. A conceptual model has been developed and focuses our attention on what we know, what we need to know and what we might expect on site. As we begin to acquire on-site data using surface geophysical methods, we will begin to test the conceptual model. Their primary benefit is that of greatly increasing the data density over the site, which improves the definition of background and anomalous conditions. These measurements allow us to more accurately locate invasive measurements based upon data rather than guesses thereby improving the chances of accurately defining geologic conditions.

This section provides an introduction to the many surface geophysical methods. We have focused on considerations for their use and how to select methods rather than how to make the measurements. In addition, a large number of examples showing their application are provided to show how these methods can be used in a site characterization effort.

16.1 Introduction

The geophysical methods encompass a wide range of airborne, surface, marine and downhole measurements (Fig. 16.1). Some geophysical measurements can be made from an aircraft such as electromagnetic (EM), magnetic, thermal imagery, and radiometric measurements. Airborne geophysical surveys are cost effective when coverage of large areas is required, especially when the area is not easily accessible for ground surveys. Here we will focus upon the surface geophysical methods, which represent an extremely powerful array of tools to help characterize complex subsurface geologic and hydrologic conditions. While geophysical methods applied to exploration for oil are looking down 1,000's of meters, our focus is typically much shallower, from 1 to 30 m or so. Surface geophysics have a wide range of applications including: engineering, environmental, water and mineral resources, and even archeological applications. Its use is referred to as applied engineering or environmental geophysics but it is all the same thing. Applied geophysics can be defined as the measurement and mapping of subsurface hydrogeologic conditions by remote sensing one or more of its properties.

The focus of this section is on applied surface geophysical methods used on land. Although many of the surface geophysical methods can also be used on water (rivers, lakes and estuaries) or over frozen bodies of water or even underground in tunnels, mines and caves. The various downhole geophysical logging and borehole methods are discussed later in the site characterization process after the borings have been located and drilled.

This section presents an overview of surface geophysics. It is not intended to provide detailed knowledge of individual methods, but rather to introduce the broad array of measurement techniques available and their applications. Numerous examples are provided to illustrate the wide range of methods and applications to characterize karst or for that matter any other complex geologic condition. Further discussion of geophysical methods and their applications can be found in Ward (1990), Butler (2005), Reynolds (2011), Burger (1992), Milsom and Eriksen (2013), Sharma (1997), McCann et al. (1997), Nazarian and Diehl (2000), and Wightman et al. (2003).

R.C. Benson, L.B. Yuhr, *Site Characterization in Karst and Pseudokarst Terraines*,
DOI 10.1007/978-94-017-9924-9_16

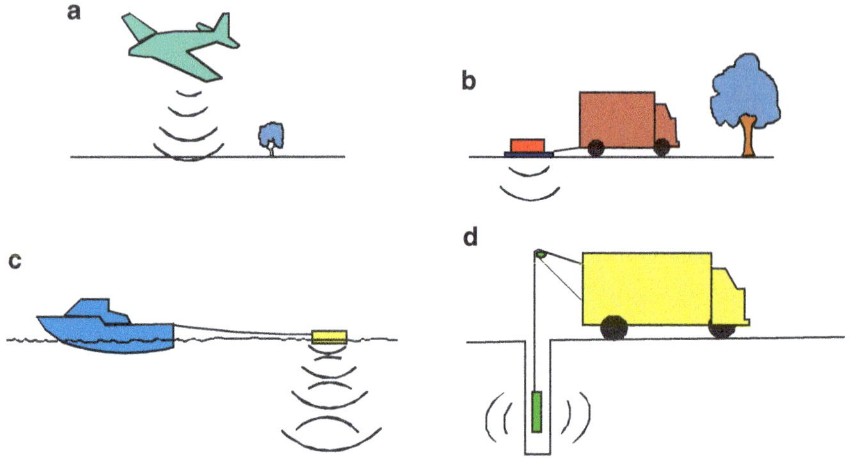

Fig. 16.1 Geophysical measurements can be made from the air, on the surface, over water and from within boreholes. (**a**) Airborne or satellite remote sensing. (**b**) Surface geophysics. (**c**) Marine geophysics. (**d**) Downhole geophysics

Recall our early discussion of the limited spatial extent of caves (something on the order of less than 5 % of the surface area) and the low probabilities of their detection by borings alone even when a large number of borings are used (Fig. 12.2). The great advantage of the surface geophysical methods is that of greatly increasing the spatial density of data. Thereby improving our ability to detect and map spatial changes in complex geologic conditions and providing a means to accurately locate invasive measurements based upon data rather than guesses or rules of thumb. This increases accuracy and confidence levels in site characterization process.

The authors have played a significant role in pioneering and promoting the application of surface geophysical techniques over the years. This includes:

- Developing the application of geophysical methods to the site characterization process including application of the methods to a wide variety of site characterization problems (Benson et al. 1984; Benson and Yuhr 1996, 1997)
- The development of many of the ASTM guidelines for surface geophysical methods,
- The development of DOE's Expedited Site Characterization program (ESC) in the 1990s as described in Beam et al. (1997), Benson et al. (1998), Benson (1997), and Yuhr (1998).
- The development of the ASTM guideline for Expedited Site Characterization (ESC) (ASTM 2010a).

Others have also emphasized the need for surface geophysics in the site characterization process including:

- The National Research Council's "Seeing into the Earth", (2000)
- DiMaggio (2007) Commentary: Advancements and Disappointments in Geo-Engineering and Geo-Construction – A 30–Year Reflection. "More tools for rapid, if not real time monitoring, readout and digital data reduction".
- Mitchell and Kavazanjian (2007) Geoengineering Engineering for the twenty-first Century, "An improved ability to see into the earth perhaps the most important need in geotechnical engineering".

As a result this section is longer and contains more details in order to emphasize and illustrate the importance of the geophysical methods.

16.2 A Brief History of the Surface Geophysical Methods

Geophysical methods are not new; they have been used for many decades in the exploration for oil and gas as well as mineral and water resource exploration dating back to the early 1900s. In the 1950s and 1960s early geotechnical applications used relatively simple seismic refraction measurements and analysis to determine top of rock profile and to measure the P-wave velocity of rock to determine its rippability. Early engineering seismographs begin to appear around the 1960s.

The nuclear power plant siting in the 1960s utilized geophysics to aid in site characterization by adapting seismic reflection methodology to assess the presence of faults. In the early 1970s, ground penetrating radar (GPR) systems first became available from Geophysical Survey Systems, Inc. and the shallow electromagnetic induction systems became available from Geonics, Ltd. With the GPR and EM systems available, we now had two extremely powerful and versatile tools for characterizing shallow geologic conditions. Both of these methods were capable of providing continuous high-resolution data along a profile line. By the early 1970s, sub-bottom profiling (high resolution seismic reflection on water) and side scan sonar (acoustic imaging of the sea bed) were being applied to marine projects.

Although many of the surface geophysical methods are capable of making deep measurements, it is the near surface environment, within the upper 30 m or so, that is usually of interest since it is the region most susceptible to human activity and has the greatest impact on geotechnical and environmental projects. Engineering and environmental applications of the geophysical methods are relatively new (roughly 1970s) and are somewhat different in their application because they are typically used to obtain relatively shallow, higher resolution data than applications for oil and gas exploration and mining. The application of the shallow high-resolution geophysical methods to hazardous waste sites (Benson et al. 1984) and critical structure siting of nuclear facilities (Franklin et al. 1981) investigations began in the mid to late 1970s and early 1980s. With expanding coastal populations and the need for additional fresh water in the 1980s and 1990s, there has been an effort to better characterize the extent of saltwater intrusion using geophysical measurements.

Today geophysical methods are commonly applied to geotechnical, environmental and karst problems worldwide. In the US, the semi-annual Sinkhole Conferences (www.sinkholeconference.com) along with the Environmental and Engineering Geophysical Society (EEGS) and their annual Symposium on the Application of Geophysics to Engineering and Environmental Problems (SAGEEP) conferences have been major factors in promoting the use of the geophysical methods. In 1977, the US Army Engineer Waterways Experiment Station in Vicksburg, Mississippi held a symposium on Detection of Subsurface Cavities (Butler 1977). During the 1980s and early 1990s, the Colorado School of Mines sponsored Tunnel Detection Meetings. The Engineering Waterway Experiment Station in Vicksburg, Mississippi and the US Army also carried out extensive programs in tunnel and cavity detection. In the mid-1990s the Federal Highway Administration (FHWA) promoted the applications of geophysics to karst related problems along roadways and infrastructure. An application guide on the use of geophysical methods for highway related problems, was published by the FWHA (Wightman et al. 2003). In 2000, the Federal Highway Administration and the Missouri Department of Transportation sponsored the First International Conference on the Application of Geophysical and Non-Destructive Testing (NDT) Methodologies to Transportation Facilities and Infrastructure. Some of these conferences continue to provide information and examples on the application of geophysical methods.

16.3 An Overview of Surface Geophysics

Table 16.1 lists the common surface geophysical methods used for site characterization along with the parameter measured and their typical applications. This table provides a general understanding of what can be done with the wide array of methods that are readily available. We recognize that there are other techniques or adaptations of techniques that may be applicable and could be considered but have limited our discussion to those methods commonly used by the authors and many others. Figure 16.2 illustrates some of the methods and ways in which geophysical data can be acquired.

One of the most critical factors affecting the assessment of geologic conditions at any site is the uncertainty of the data from limited spatial sampling (Fig. 12.2). This is often the fatal flaw of many site characterization efforts. Achieving a reasonable statistical sampling of geologic site conditions would require borings or other methods of limited spatial sampling to be placed in a close-order grid, which would greatly increase time, effort and cost, as well as reduce the site to "swiss-cheese". Yet relying on data from a limited number of borings remains a weak link in many site characterization efforts (Benson et al. 1984; Franklin et al. 1981).

Unlike direct sampling and analysis, such as obtaining a soil, rock or water sample and sending it to a laboratory, the geophysical methods provide non-invasive, in-situ measurements of the subsurface geologic conditions with little or no disruption of the surface. The geophysical measurements integrate a larger volume of the subsurface compared to borings. As such they are averaging much more material into their measurement, which has both benefits and disadvantages. Sampling larger volumes increases our chance of detecting changes in geologic conditions. On the other hand if we sample too great of a volume we can lose the sensitivity and resolution required to detect smaller changes in geologic conditions. Selecting the appropriate method(s) whose scale of measurement will optimize our chances of detecting and resolving the geologic variability of interest at a specific site is critical.

The geophysical methods can be used at two different stages of the site characterization process.

- Early in the site characterization process when our objective is simply to locate the anomalous geologic conditions by increasing sampling density. Here, the use of geophysics is often qualitative and highly intuitive (Greenhouse et al. 1995).
- Later on, in the site characterization process, geophysical data may be used more quantitatively to determine specific engineering or hydrologic parameters such as: P-wave velocity for rippability, calculating elastic constants from P-wave and S-wave velocities, locating the fresh-seawater interface, and characterizing the size and

Table 16.1 Surface geophysical methods their measurement and application

Method and common terminology	Typical depth of measurement (m)	Parameter measured	Typical applications	Used over water
Ground penetrating radar (GPR) (radar)	3–10 m typical, much less in clays, conductive soil, rock or pore fluid	Dielectric constant, electrical conductivity and magnetic susceptibility	High resolution cross sections of soil and rock showing strata, fractures, voids, cavities and also buried utilities, drums and tanks	Fresh water only
Electromagnetic (EM) methods				
Frequency domain (FDEM)	1–60 m, depending upon instrumentation used	Electrical conductivity and magnetic susceptibility	Detect lateral variations in soil and rock, also used to locate underground metal utilities, drums and tanks	Fresh water or waters of low electrical conductivity
Time domain (TDEM)	5–1,000 m	Electrical resistivity	Sounding measurements. Top of bedrock. Salt water intrusion	Not normally
Very low frequency (VLF)	About 20 m typical	Tilt and phase angle of electric field or field strength, vertical and horizontal In-phase and quadrature components	Tabular conductors such as faults and fracture zones. Used to locate water and mineral zones	Not normally
Electrical methods				
DC resistivity	Up to 100 m	Electrical resistivity	Mapping lateral and vertical resistivity changes in soil and rock	Fresh and salt water
2D imaging	Up to 100 m	Electrical resistivity	2D profiles mapping lateral and vertical soil and rock changes	Fresh and salt water
Capacitance measurements	Up to 12 m	Electrical resistivity	2D profiles mapping lateral and vertical soil and rock changes	Not normally
IP/complex resistivity	Up to 100 m	Electrical resistivity and chargeability	2D profiles mapping lateral and vertical soil and rock changes	Not normally

Self potential (SP) also called natural potential or spontaneous potential	Typically shallow <20 m	Surface voltages resulting from flowing water or electrochemical effects in soil or rock	Seepage of water at dams, impoundments, flow in conduits or caves, electrochemical effects due to recharge zones	Fresh and salt water
Seismic methods				
Seismic refraction	100 m or less	Compressional V_p and or shear V_s wave velocity	2D profiles of top of rock profile, weathered zone, rippability	Fresh and salt water
Seismic reflection	300 m or more	Compressional V_p and or shear V_s wave velocity	2D profiles of stratigraphy, and structure	Fresh and salt water
Multiple analysis of surface waves (MASW)	15–45 m	Shear wave V_s velocity	2D profiles of variations in stratigraphy and for engineering calculations	Fresh and salt water
Microgravity	100 m or more	Changes in the local acceleration of gravity due to changes in density of subsurface materials	Variations in depth to rock, location of buried valleys, fractures, faults, cavities, cave systems and mines	Not typically used over water
Magnetic	100 m or more	Magnetic susceptibility of soil and rock	Lateral variations of geology related to magnetic variations in soil and rock or location of underground ferrous metal utilities, drums and tanks	Fresh and salt water
Metal detectors and pipe cable locators	1–10 m	Changes in electrical conductivity or resistivity of metals, or to induced EM signal from detector	Location of buried utilities and location and mapping of buried drums, tanks and trash	Fresh and salt water
Thermal imagery	At surface	Temperature of the surface	Location of springs, caves and mine entrances and air flow from wells and fractures	Fresh and salt water
Radiometric	At surface	Natural gamma radiation	Location of fractures, faults and caves along with higher permeability zones within the soil and rock	Fresh and salt water

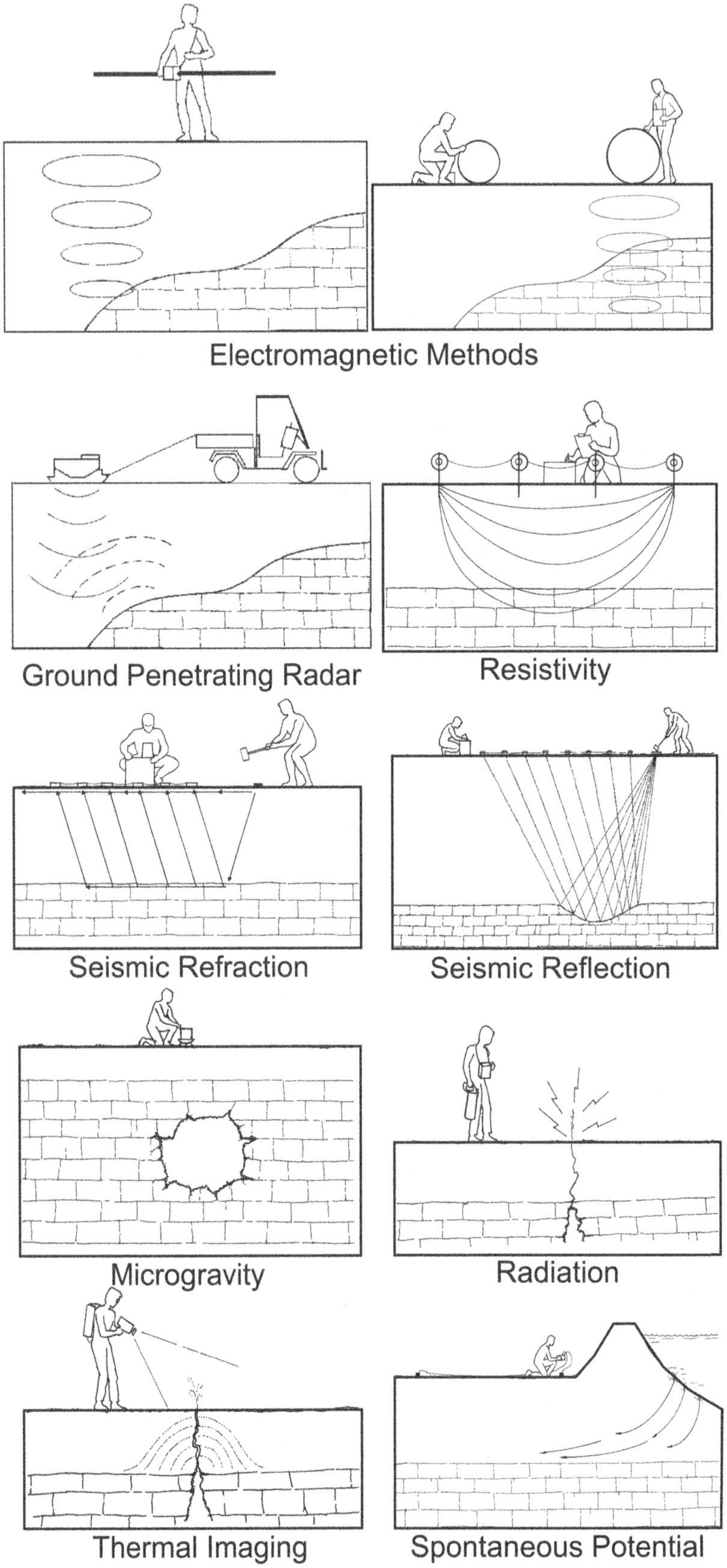

Fig. 16.2 A wide range of surface geophysical instrumentation is available to acquire subsurface data

dimensions of anomalous conditions such as fractures faults and cave systems.

16.3.1 Parameter Measured

Each of the surface geophysical methods responds to the changes of some physical, chemical, electrical, or radiological property of the subsurface (Table 16.1). The geophysical methods respond to specific frequencies over a wide range of the seismic/acoustic and electromagnetic spectrum (Fig. 16.3). Changes in geologic and hydrologic conditions will commonly be associated with a change in one or more of the measured parameters. The surface geophysical methods are able to detect lateral and or vertical changes in a given parameter, which can then be interpreted to provide an assessment of subsurface geologic or hydrologic conditions.

The geophysical methods do not typically detect the soil/rock interface itself, but the change in electrical resistivity, density, P-wave velocity, etc., that occurs at this interface. When applied to karst, these methods do not typically detect a cavity itself, but the change in density or electrical resistivity of the cavity compared to the surrounding materials. The methods do not detect the flow of water but the resulting streaming potential (voltage) on the surface created by flowing water in a fracture or conduit. Regardless of geophysical method selected, there has to be a change or contrast in the measured parameter related to the target of interest for the results to be successful.

For example, Caterpillar Tractor Company uses seismic refraction measurements to obtain compressional wave seismic velocity to establish rock rippability. The combination of seismic velocity along with a visual determination of the rock character (ie degree of fractures and bedding thickness) will determine if the rock can be ripped and which size tractor will be required (Caterpillar 1988). Whiteley (1983) provides a discussion of the relationships between some of the geologic, hydrologic and engineering parameters and the geophysical measurements.

There are also correlations that exist between water quality and geophysical measurements. Inorganic contaminants such as landfill leachate or natural contaminants such as saltwater intrusion, can be mapped using electrical/electromagnetic techniques due to their elevated pore fluid conductivity levels (Benson et al. 1985). Where degradation of organic contaminants has occurred an increased electrical conductivity values have also been noted (Sauck et al. 1998; Cassidy et al. 2001; Atekwana et al. 2002). These correlations allow electrical and electromagnetic measurements to be use to map some organic contaminant plumes, identify their boundaries and sometimes identify flow pathways.

16.3.2 Anomalies

The term anomaly is used to describe any unexpected deviation of the data from expected or background conditions. For example, an anomaly can represent a change in depth to rock, a change in clay content or the presence of a cavity. An anomaly can also represent a cultural change due to a pipeline, buried tanks or contaminants. When a geologic anomaly is identified by geophysical measurements, there are a variety of rules of thumb or calculations that can be used to provide an approximate depth and size of a subsurface target that resulted in the measured anomaly.

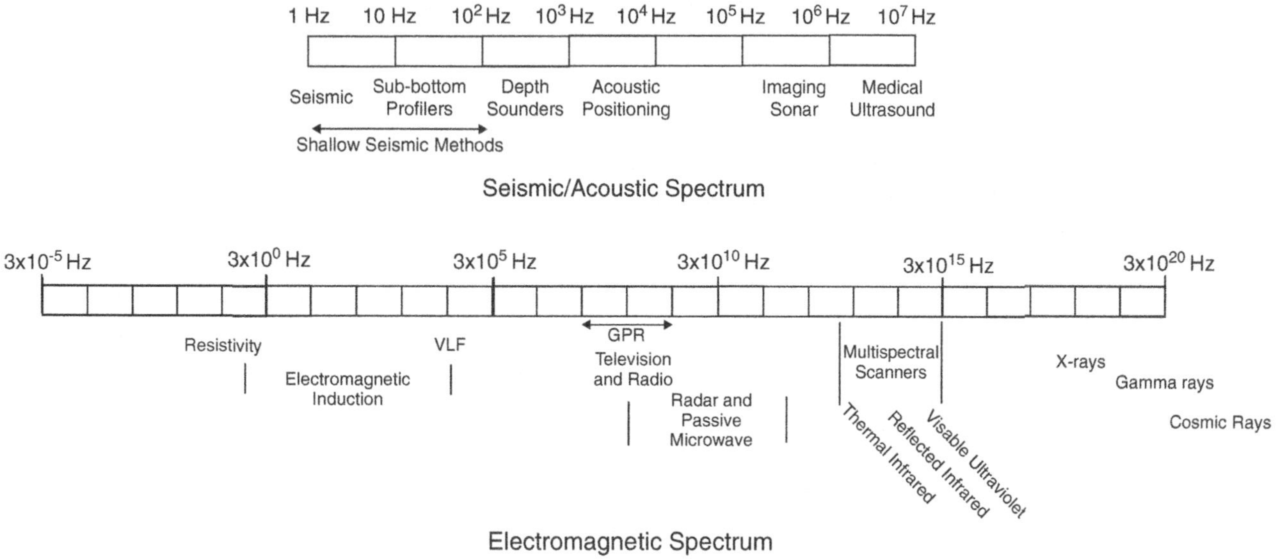

Fig. 16.3 The geophysical measurements encompass a wide range of the seismic/acoustic and electromagnetic spectrum

A geophysical anomaly can be due to many different geologic or hydrologic conditions. For example, an increase in electrical conductivity could be due to inorganic contaminants, increase clay content, increase moisture content, thicker soil layer, etc. It is our job to have enough data and information about the site to know what conditions to expect. With a developing conceptual model in place, based upon existing data, we should be able to narrow down the possible cause of the anomaly. Knowing the location and possible depth of the anomaly we can then locate additional measurements such as borings for further characterization.

16.3.3 Direct and Indirect Detection of Anomalous Conditions

Direct and indirect detection are two approaches that can be used to detect anomalous subsurface conditions using the surface geophysical methods.

- Direct detection can be accomplished by some of the surface geophysical methods when the feature of interest such as a fracture or cave system is large enough, has a sufficient contrast to its surrounding materials, and is within the depth range of the specific measurements (Table 16.1). This approach provides a very high level of confidence in subsurface conditions.
- Indirect detection can be accomplished by the use of near surface indicators (NSI) or halo effects, which refers to detecting the presence of an anomaly such as a major fracture or cave system without direct detection of the feature itself. Very often a cave system at depth (may be beyond the range of our measurement) but will show signs of its presence and possible activity by NSI or halos within the shallow sediment or rock. These NSI or halo effects may include:
  - Soil piping, raveling or dipping strata that may be occurring in the shallow sediment as a result of a deep collapse, which has not yet reached the surface (Benson and Yuhr 1992).
  - A fracture system over a cave may have caused a local increase in permeability and provide an area of groundwater recharge resulting in geochemical changes over the cave. As a result, a local change in electrical conductivity, resistivity, or SP voltage may provide an indication of the presence of an underlying cavity (Benson and Yuhr 1992).
  - Thermal gradients or flowing air may exist at the surface, associated with permeable zones, fractures or caves within the unsaturated zone (Thompson and Marvin 2006; Straley 2008; Bogle and Loy 1995; Campbell and Singer 2001).
  - Radiation halos (gamma or radon) may exist at the surface over areas of increased permeability, fractures or a cave system (Banwell and Parizek 1986; Upchurch et al. 1987; Armstrong and Heemstra 1973; Hansen 1975).

Another halo effect associated with cave systems may include the stress fractures surrounding the cave and the many tertiary fractures associated with cave development that are too small for exploration and mapping by cavers. These features surrounding the main cave create "halo effects" and may increase a gravity anomaly by a factor of 1.5–2.5 (Butler 1994, 2008; Llopis et al. 2005), making the cave system more easily detectable.

The use of NSI or halo effects provides a means of indirectly detecting the presence of deeper cave systems.This is not unlike the use of vegetation stress in aerial photography to indicate the presence of subsurface lineaments (ie fractures, faults and recharge). Depending upon the nature of the project, both direct and indirect methods may be utilized.

16.3.4 Penetration of Measurements

Table 16.1 provides a typical maximum depth of measurement or penetration of the surface geophysical measurements. Again, these are generalized and the selection of instrumentation, set-up parameters and site-specific conditions will all impact the actual results. For example, both frequency domain electromagnetic (FDEM) and GPR measurements are limited by high conductivity values of the soil, rock, and their pore fluids. If the surface sediments are highly resistive (i.e. quartz sands or dry clay-free limestone) GPR depths can be greatly extended. The depth of resistivity measurements will be reduced when there is a highly resistive layer such as clean quartz sand, which would limit adequate current getting into the ground. These are just a couple of examples showing the impact of site-specific conditions on the depth of measurements.

16.3.5 Resolution

Resolution is the ability to resolve two geologic conditions or cultural features, which are close together. Both the lateral and vertical resolution of the geophysical method can impact the data and its interpretation. There is a difference between detecting a subsurface feature and resolving that feature. Detection does not necessarily include determining the size, thickness or shape of a subsurface feature. As a result the requirements for detection are less stringent than for resolving a feature (National Research Council 2000).

The resolution is an inherent function of the method selected, how the method is used, along with the depth, nature and size of the target. As an example, radar can provide resolution of centimeters when mapping individual steel reinforcement in concrete (Fig. 16.4) while resolution may be reduced to a meter or so when mapping the top of rock at a depth of 10–20 m.

16.3.5.1 Lateral Resolution

The lateral resolution of surface geophysical measurements is an inherent function of the method used, the volume sampled by the specific measurement, and the spacing between measurements. The spacing of measurements along a survey line and the distance between survey lines or grids need to be close enough to laterally resolve the feature of interest. Selecting the station and line spacing for measurements requires insight regarding the likely size, depth and nature of the feature of interest (Fig. 12.9), the general geologic variability and noise conditions along with the physics of the sensing method. This insight only comes from years of experience.

Data from some methods may be obtained "nearly continuously" along a survey line, resulting in very high lateral resolution. Such measurements include GPR, FDEM, towed capacitively-coupled resistivity, thermal imaging, radioactive measurements, and most of the marine methods.

By "nearly continuously" we mean that measurements made by most systems today have a fairly high digital sampling rate of 10–30 or more measurements per second. If we are walking at 4.8 km/h and have a low sample rate of 10 samples per second then we will have acquired data at 7.2 samples/m. Even at speeds of 16 km/h using a vehicle-towed system, the sample rate would be almost 3 samples/m, which for all practical geologic purposes is continuous.

Figure 16.5 shows two sets of frequency domain electromagnetic (EM34) data along the same profile line. The data in Fig. 16.5a was collected at a station spacing of about 60 m, while the data in Fig. 16.5b was collected nearly continuously. The continuously sampled data provides greater lateral resolution and has clearly defined the fracture patterns within the gypsum rock buried under about 2 m of sediment.

16.3.5.2 Vertical Resolution

Vertical resolution is the ability to separate the response from two distinct layers or targets at different depths. Regardless

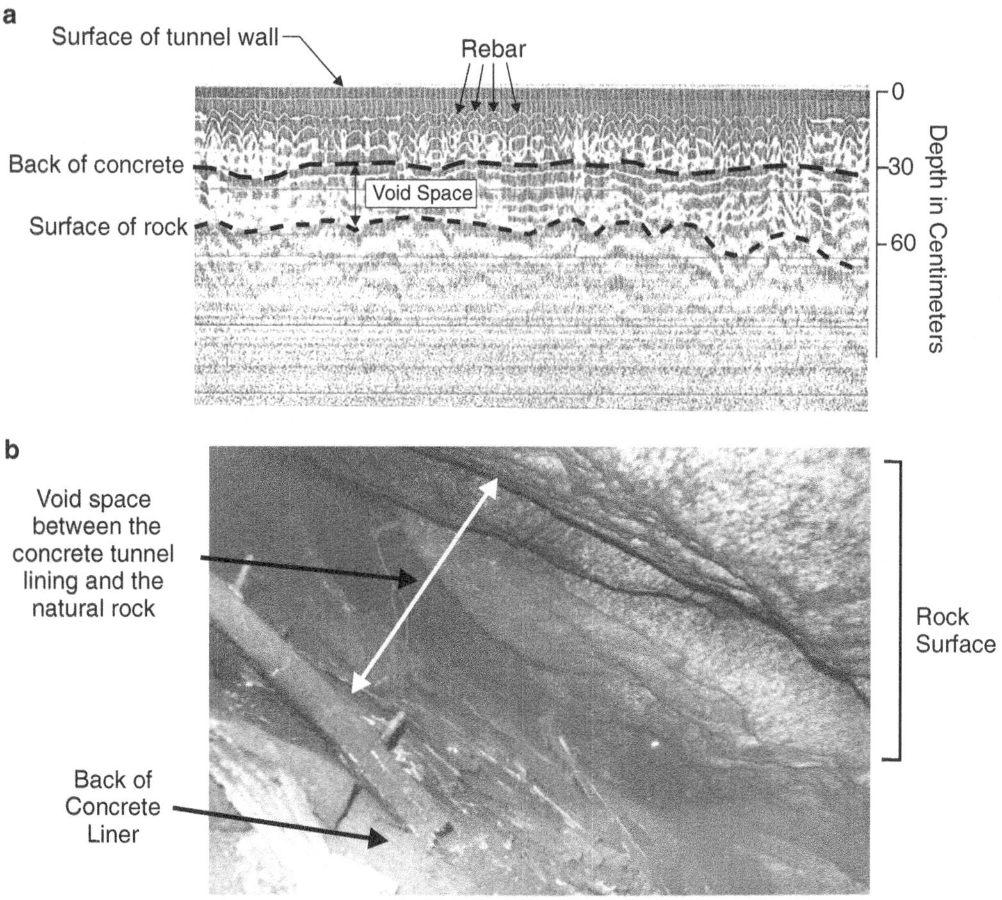

Fig. 16.4 An example of shallow radar data illustrating the level of detail that can be achieved shows the thickness of the concrete tunnel lining, the reinforcing steel bars within the concrete and the void space behind the concrete (**a**). A photo behind the concrete tunnel lining (**b**) shows the variable void space

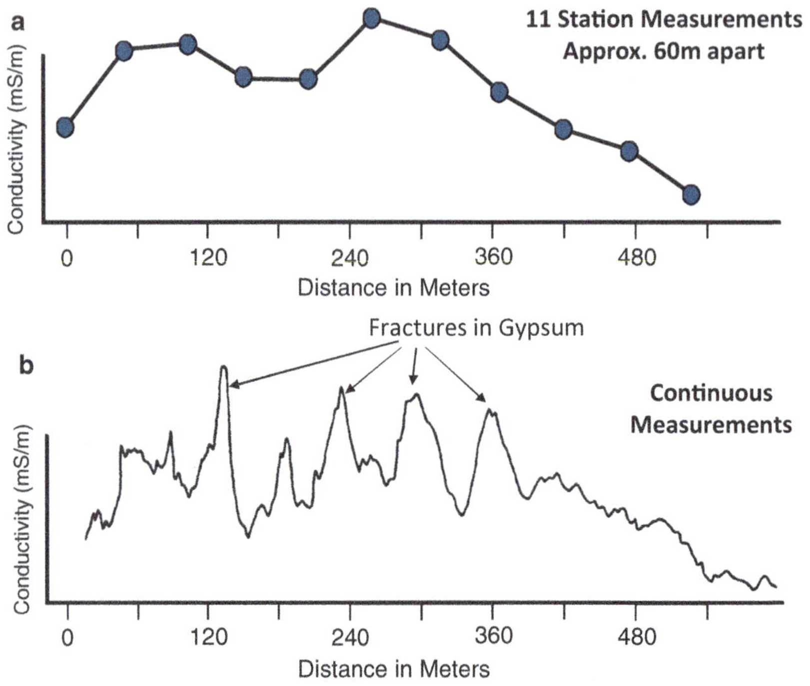

Fig. 16.5 A comparison of station measurements (**a**) versus continuous measurements (**b**) over the same survey line. The continuous electromagnetic measurements detect a pattern of fractures in gypsum that is undetected in the station measurements

of the method of measurement, the resolution of all surface geophysical measurements decrease with increased depth. As a particular layer of interest or target becomes deeper, a greater contrast is necessary between the target and surrounding materials and or its size must be greater in order for it to be detected and its size and shape to be resolved.

16.3.6 Processing and Presentation of Data

The results of many surface geophysical methods can be interpreted directly from the field data (Fig. 16.6a) or after the data have been plotted as a profile or have been contoured. This allows preliminary in-field assessments to be made. For example, GPR, frequency domain EM, resistivity profile data, as well as thermal imaging and radioactive measurements can often be interpreted without any processing other than for presenting the data. In addition, some marine data (bathymetry, side-scan, and subbottom profiling) can also be interpreted without additional processing.

In some cases, simple steps may be carried out to improve the data and its presentation. For example, radar data collected along a survey line can have topographic corrections applied (Fig. 16.6b). When topographic correction is applied, the data is presented in a corrected visual format, which can improve understanding and interpretation of the data.

Many of the methods (such as; resistivity soundings, 2D resistivity, seismic refraction and reflection, MASW, SP, and gravity) require varying degrees of processing and modeling to convert the raw field data into useable information. These models may include:

- 1D models consisting of a sounding or vertical log that shows variations in the measured parameter with depth
- 2D models consisting of a cross section or contour map
- 3D models consisting of isometric figures combining 1D or 2D models
- 4D models consisting of time-series plots of 1D, 2D or 3D data from the same location

Geologic conditions exist in a three-dimensional (3D) world. Hydrologic conditions can change with time and may be four-dimensional (4D). In many cases, we must simplify the models by holding some of the variables constant. In addition, care should be taken not to over interpret data or create more detailed models than the data actually supports.

Nowadays, data is often presented in 3D. This is often accomplished by acquiring data in 2D and processing the data to create a 3D presentation. For example, ground penetrating radar data is gathered as a series of parallel or perpendicular lines and can then be processed to yield 3D block of data. When data is acquired with the intent of processing a 3D image, extremely high density data is required (both laterally and vertically). Because of the need for such high density data, three-dimensional data are typically limited to small areas. This is due to equipment limitations as well as

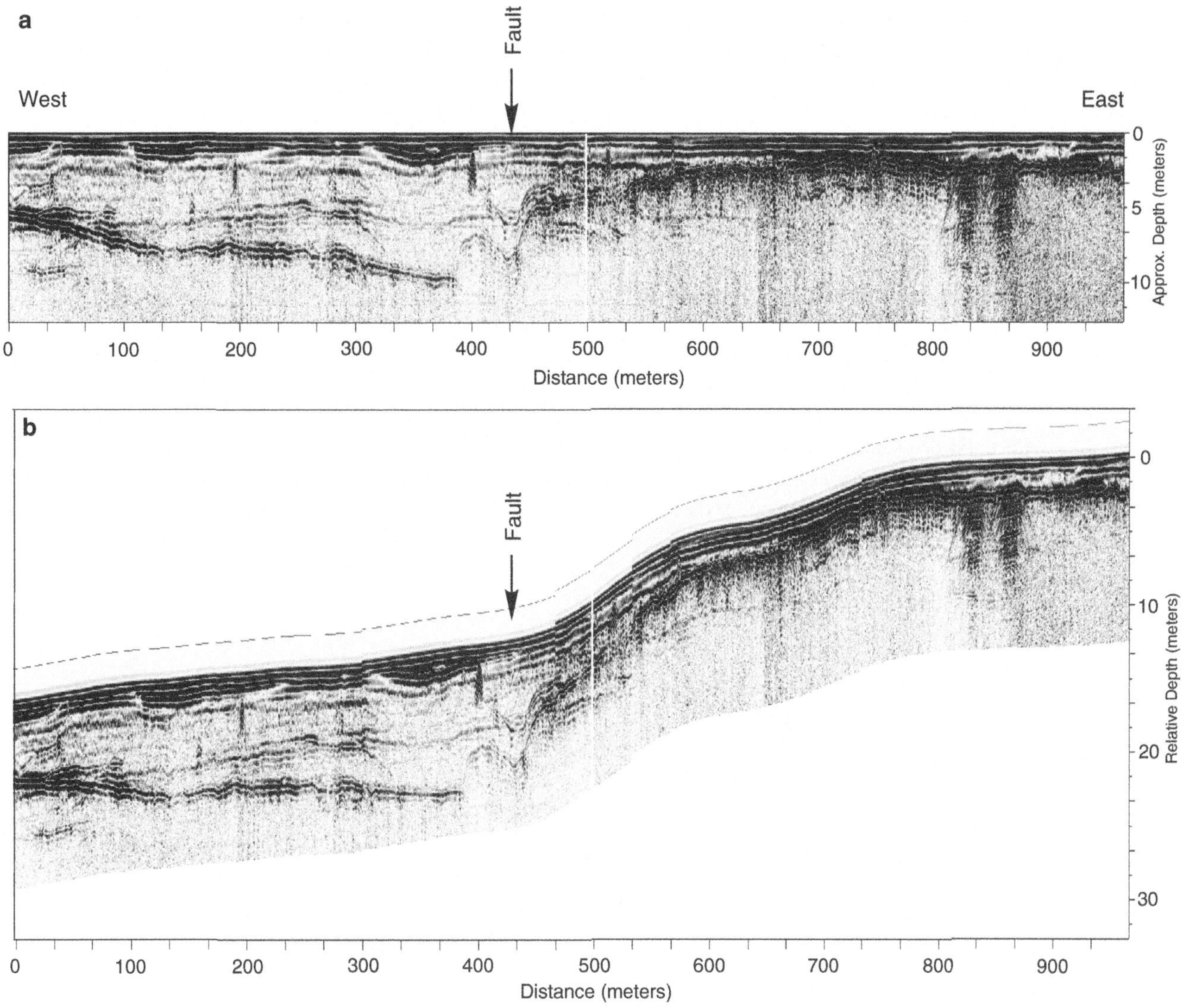

Fig. 16.6 An example of radar data using a 100 MHz antenna without topographic corrections (**a**) and with topographic corrections (**b**)

time and cost limitations. Therefore 3D data should be strategically used to characterize known anomalous conditions which have already been located but require additional detailed data or confirmation and not for reconnaissance purposes.

16.3.7 Use of a Survey Grid

Surface geophysical measurements are commonly made along a survey line, as a series of parallel survey lines or over a grid. An important aspect of any geophysical survey is to establish a survey line, parallel lines or grid along with a standard coordinate system. The best data in the world are useless if no one knows where it came from or cannot easily get back to the specific location. In addition, the geophysical data must be able to be correlated to other data. Therefore, the use of a standard coordinate system for the entire site characterization process is necessary. See discussion of survey grid in Sect. 12.3.1 Project Preparation.

Global Positioning Systems (GPS) can now be integrated with geophysical measurements and used to record locations in real-time. However, the sole dependency on GPS systems can be problematic and does not avoid the need for an established local survey grid system for all data collected on-site. The survey grid allows spatial correlation of all data, allows for corrections in positioning errors and aids in orienting positions on-site. Some of the geophysical methods, such as radar, utilize a time based sampling or distance established by a counter wheel. In some cases, position will be manually noted on the data record as one passes a fixed station mark. When errors in positioning occur, and they do, they can be corrected prior to data processing and interpretation. Most survey work on water will utilize GPS, often incorporating a steering system to guide the helmsman. Range markers and or reference buoys are also used to guide the helmsman. These visible references provide quick visual orientation on the water.

16.3.8 Location of Utilities, Buried Drums, Tanks and Trash

Many of the surface geophysical methods can be utilized to detect and map such man-made features such as utilities, buried debris, underground storage tanks (UST's), drums, etc (Benson et al. 1984). While this is not the focus of this book, there are many cases in which these conditions exist at a site, along with karst or pseudokarst conditions, and can interfere or impact site investigation efforts. The various surface geophysical methods used for detecting these man-made targets include:

- GPR to locate buried utilities and (UST's);
- Frequency domain EM systems, can be used to locate either ferrous or non-ferrous metal utilities as well as trash, conductive contaminants (soils or pore fluids), and UST's;
- Magnetometers are use to locate ferrous metal drums, UST's and utilities;
- A variety of metal detectors and pipe and cable locators can be used to locate utilities as well as ferrous or non-ferrous metal trash.

Figure 16.7 shows the results of selected measurements over drums, buried debris, utilities and UST's. Figure 16.7a is the response from a metal detector over a trench that contained 55-gal drums. Figure 16.7b is a contour map of electromagnetic (EM31) data (0–6 m) revealing a conductive contaminant plume, buried debris and various utilities. Figure 16.7c is a set of data that includes electromagnetic (EM31), magnetometer and GPR data over four abandoned UST's, each 10,000 gal.

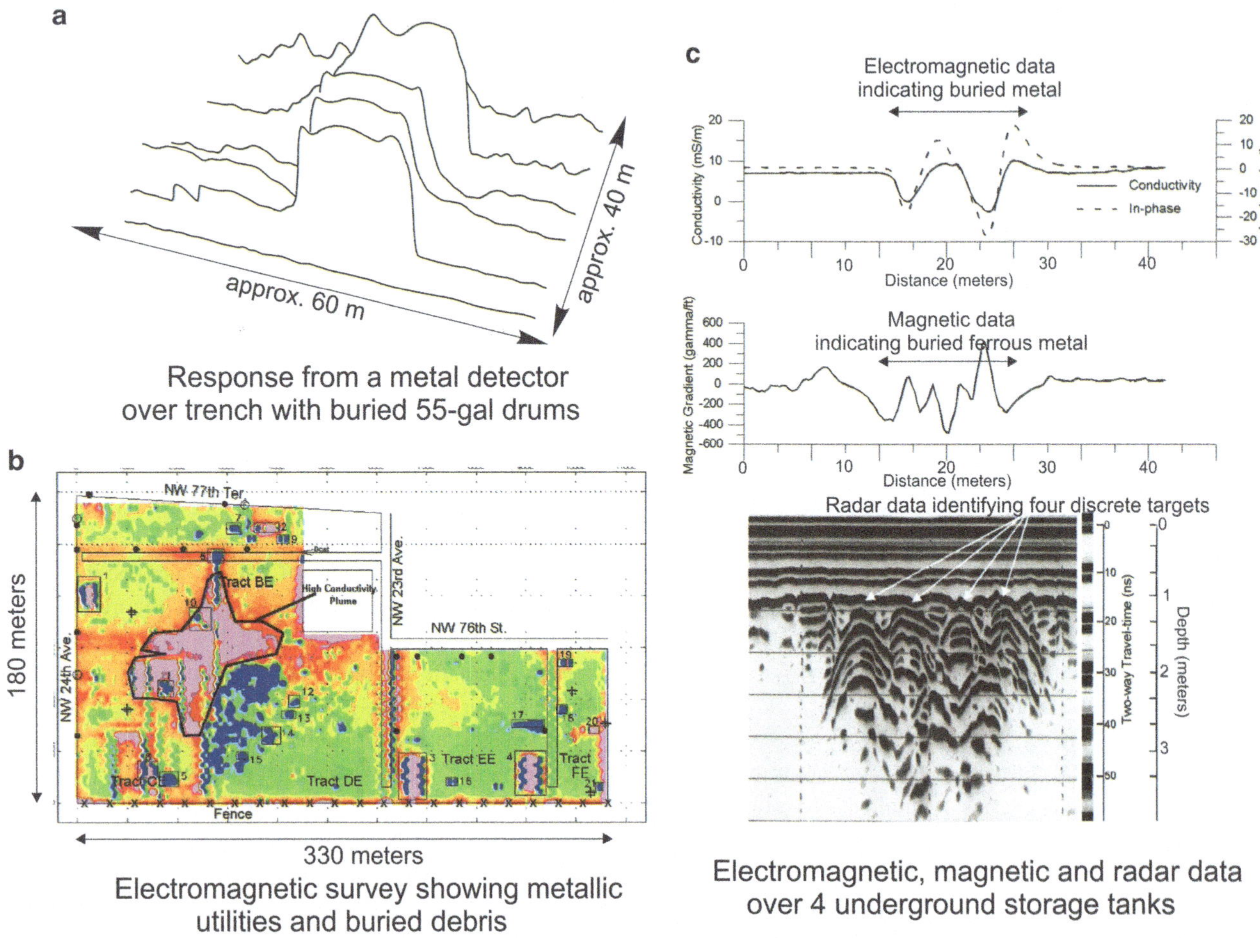

Fig. 16.7 Some of the geophysical methods can be used to map man-made features such as buried steel drums (**a**), utilities and buried debris (**b**) and underground storage tanks (**c**)

16.3.9 Ease of Use

Ease of use refers to the level of effort necessary to make the geophysical measurements. The level of effort includes whether the measurements are acquired on a station-by-station basis or continuously, the size of the field crew required for the measurements, or whether the measurements require an intimate physical contact with the ground surface.

- Some measurements such as, GPR, frequency domain EM, capacitively coupled resistivity, magnetic, thermal, and radioactive measurements are relatively easy to make since they can be made by station measurements or by continuously walking or driving along a survey line.
- Gravity measurements are made on a station-by-station basis without probes driven into the ground. However, microgravity measurements do require highly accurate (first order) relative elevation data for each gravity station.
- Others measurements such as resistivity, time domain EM, self-potential (SP), and seismic measurements require that coils, probes or geophones be placed in contact with the ground along a profile line and cables are deployed and connected. Both resistivity and seismic measurements typically require intimate contact with the ground. In some cases a towed cable array may be used to speed up data acquisition for some seismic and resistivity measurements using a weak ground contact.

These factors need to be considered when addressing site conditions and logistics. Site conditions such as heavy vegetation, steep slopes, and rocky surface will impact the selection of methods to be used as well as sources of natural noise such as wind and storms, cultural noise such as buried utilities, vibrations from machinery and transmission lines.

16.3.10 Surface Geophysical Data Can Be Acquired Over Water

Most of the surface geophysical methods are designed for use on land but many have been adapted for use over bodies of water such as inland rivers, streams, canals, lagoons, reservoirs, water-filled quarries, lakes and even ice-covered bodies of water. These water-covered areas often provide additional "geologic windows" into the subsurface, which are free from many cultural effects. When water covered areas are adjacent to a site they often provide unique opportunities to obtain continuous, high resolution geologic data.

Traditional land-based measurements have been adapted to be used over water. Both radar and EM can be used over water, but over fresh water only (Fig. 16.8a). Resistivity measurements (Fig. 16.8b), spontaneous potential, thermal imaging and radioactive measurements can be used over fresh or salt water. All of these measurements are made "continuously" as the survey boat moves through the water using a high sampling rate.

There is an array of marine seismic methods commonly used in fresh or salt water that includes:

- Seismic refraction and seismic reflection, which provide subsurface data in cross section similar to those methods on land (Fig. 16.8c).
- Bathymetry or echo sounding measurements are made to determine the depth of water.
- Sidescan sonar provides an acoustic image of the bottom.

16.3.11 Limitations

All geophysical methods have advantages as well as limitations. Some of the surface geophysical methods are limited by site-specific conditions. Some of the methods may be impacted by cultural features, such as nearby metal fences, utilities, or structures, and some are sensitive to ground vibrations from vehicle traffic, wind, or even remote earthquakes. Site characterization will often occur in areas that have been previously developed, and have been used for a variety of purposes over long periods of time. These sites may include the presence of buried structures, fill materials, abandoned infrastructure, utilities, buried debris and contaminants which can complicate the investigation. These conditions can often eliminate a particular geophysical method from consideration. The selection and effectiveness of a particular surface geophysical method requires not only identifying the most appropriate method to meet a project objectives but also identifying the method, which will be successful under site-specific constraints.

16.4 Guidelines for the Selection of the Surface Geophysical Methods

American Society for Testing and Materials (ASTM) guidelines have been written for many of the surface geophysical methods. The terms guidelines and standards are often used interchangeably to describe these various documents. These documents were not intended to be standards but only to be used as guidelines. These guidelines were developed to provide a common approach for the use of these methods and general information on the selection and application of a particular method for non-geophysical experts. The ASTM standard guides for surface geophysical methods include:

- Ground Penetrating Radar Method (ASTM 2011b)
- Frequency Domain Electromagnetic Method (ASTM 2008)

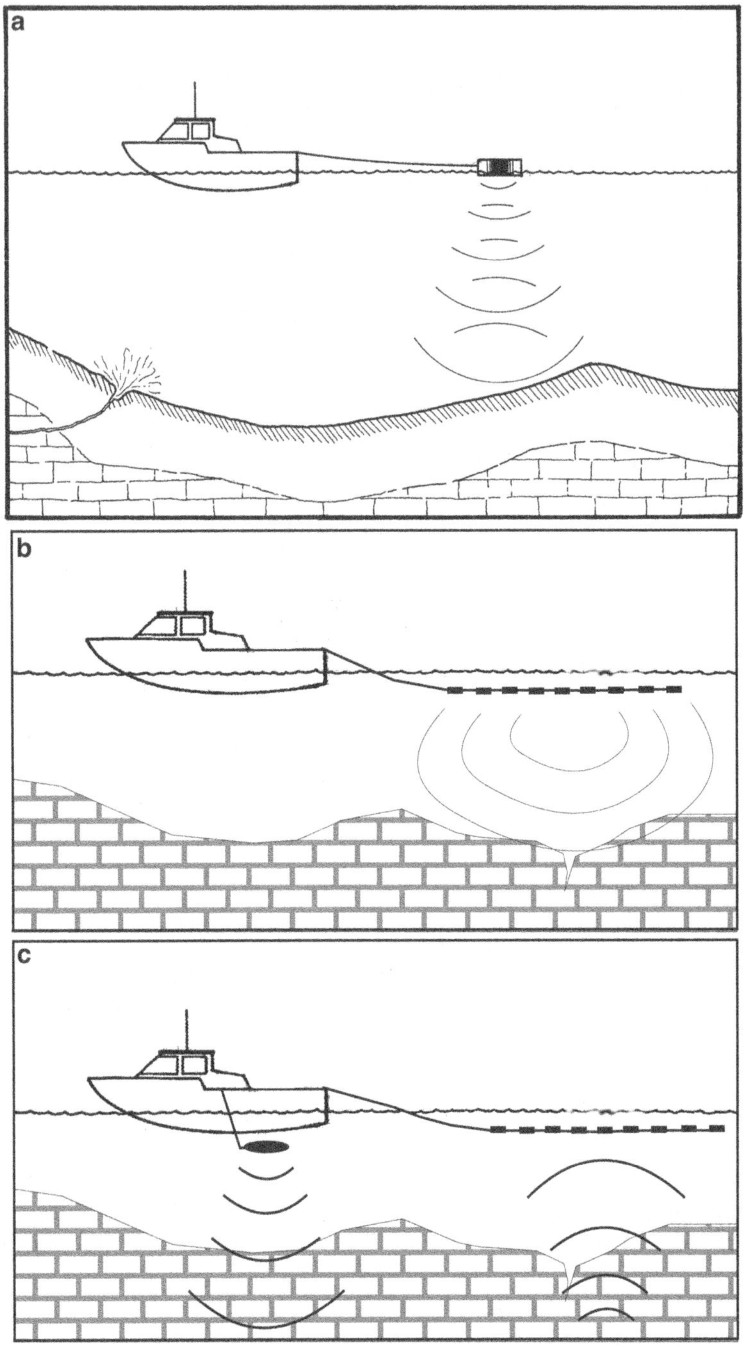

Fig. 16.8 A wide range of surface geophysical measurements can be employed over water. (**a**) Ground penetrating radar, electromagnetics (fresh water only), and magnetics. (**b**) Marine resistivity. (**c**) Seismic reflection (subbottom) and seismic refraction

- Time Domain Electromagnetic Method (ASTM 2007)
- Direct Current Resistivity Method (ASTM 2010c)
- Seismic Refraction Method (ASTM 2011a)
- Seismic Reflection Method (ASTM 2010d)
- Gravity Method (ASTM 2010b)
- Metal Detection Method (ASTM 2011d)

In addition, a guideline was developed that covers the selection of surface geophysical methods (ASTM 2011c). This document includes a table recommending the use of certain geophysical methods for certain types of problems. All such guidelines are inherently simplified and generalized and cannot possibly consider all of the variables involved in planning a site-specific investigation program. These guidelines do not discuss the wide range of options in which each method can be applied, their limitations, or their possible applications over water. Without an endless array of footnotes, qualifications, limitations and exceptions, such guidelines must be used with caution and only as a first approximation guide. However, for someone without any background in geophysics it is a good place to start, if only for introducing the wide array of possible techniques available for use. Further information on the wide array of surface geophysical techniques has been described by Saunders et al. (1999), Benson and Yuhr 1996, Benson et al. (2003), and McCann et al. (1997).

Anderson et al. (2003), Anderson and Ismail (2003), Olhoeft (2003), and Benson et al. (2003) have outlined some of the key steps and considerations for the selection and use of the geophysical methods. The following is a composite list from these papers and includes some of the key questions and issues to be considered in selecting geophysical methods and preparing for a site characterization:

- The first step is to clearly understand and define the problem or project objective.
- What existing geologic and hydrologic data is available for the site and or surrounding area
- What is the area of interest? And which methods can provide the necessary coverage in both a technical and cost effective manner
- What is the target or targets – isolated and discrete, linear or planer?
- What is the target's depth or range of depths?
- What is the required site coverage, spatial sampling and resolution both laterally and with depth to detect the target and which techniques can provide the desired results?
- What are the relevant physical properties of interest to define the geology and the target and which geophysical methods will respond to?
- Is there a possibility for NSI or Halo effects from deep-seated karst?
- What are the site-specific constraints (access, topography, structures and noise) and which geophysical methods can perform best under these conditions?
- Which techniques (geophysical and non-geophysical) can provide complementary data?
- What data are already available and what other data are required to interpret and or constrain the interpretation of geophysical data.
- What are the expected results?
- Who will make the decision regarding the methods to be used, plan the survey, make the measurements, analyze the data, integrate the geophysical data with other data to arrive at a realistic conceptual geologic model and write the report?
- Are there any cultural, access, environmental, or safety issues that may impact the measurements?

Note that some of these questions deal with non-technical issues, which are as critical to the overall success of the program as are the selection and proper application of the method(s). No guideline can cover all of these many interwoven issues involved in properly selecting and applying geophysical methods. It is not simply a question of selecting a method or methods from an ASTM guideline it is also a question of how it will be applied, processed, interpreted and integrated with other data along with considerations of the methods inherent advantages and limitations. The key is to select those methods and their adaptations, which are most appropriate for the project needs and have the best chance of providing useful results.

In practice, someone who has had first-hand experience with most of the methods should be making the final selection of appropriate methods. This experience should include acquisition, processing, interpreting and integration of data. Then the site-specific conditions and the project objectives must be considered in order to recommend an optimum approach.

16.5 Application of Surface Geophysical Methods

The following section provides examples of a variety of geophysical methods and illustrates their broad applications that have been used by the authors. These categories include:

- Soil piping and collapse within the sediments
- Epikarst along with the top of rock profile
- Fractures and cavities systems within the rock,
- Buried sinkholes and paleokarst, and
- Pseudokarst conditions.

The examples provided cover a few of the commonly used geophysical techniques listed in Table 16.1. The site locations are mostly within the United States. Each of these examples shows how a surface geophysical method was used to acquire a set of data that was critical to the site characterization. These examples illustrate the unique insight into subsurface conditions that surface geophysics can provide and how they are integrated into the site characterization process.

16.5.1 Soil Piping and Collapse Within the Sediments

The movement of soils into void spaces within the underlying rock (fractures, voids, and cavities) will typically result in surface depressions, small circular soil piping features and ultimately a surface collapse. These are often small, nuisance type of features that are triggered by flow of water or changes in water levels. The materials entering the void space are either being flushed through the system or in some cases the void space is large enough to accommodate the overlying material and eventually result in a collapse feature. The examples include:

- Active cover collapse sinkholes along a military railroad in North Carolina
- An active cover collapse sinkhole within a water-filled sand quarry in central Florida and
- Seepage at an earthen dam in New Mexico.

16.5.1.1 Active Collapse Along a Military Railroad (North Carolina)

Sinkhole activity began occurring in 1976 along a critical military railroad that provided supplies to the Sunny Point Military Ocean Terminal, 27 km southwest of Wilmington, North Carolina. There had been older sinkhole activity in the area for some time as indicated on topographic maps. Recent sinkhole activity began after construction of a nearby dam to create an artificial lake for a housing development. This resulted in a significant increase in hydraulic head near the railroad track triggering the appearance of springs and sinkholes. The sinkholes presented a high risk for the trains that provided critical supplies to navy ships in the nearby port. Titcomb and Keeton (1984) provide early details of sinkhole development at this site.

The area is covered by quartz sand with increasing clay content below a depth of 6 m and limestone occurring at a depth of 12 m or more. A conceptual model of site geology is shown in Fig. 16.9. The Corps of Engineers had drilled more than 100 borings along 1,200 m of railway to evaluate the problem. The sinkholes develop as small diameter piping features (<1 m diameter) caused by raveling of the unconsolidated sediments (Fig. 8.3a). Subsequently the diameter of the surface collapse could enlarge over time (Fig. 8.3b). The initial soil piping was of such small dimensions that it was almost impossible to detect prior to breakthrough at the surface by a drilling program alone.

Initial surveys with ground penetrating radar using an 80 MHz antenna resulted in excellent quality data and indicated that the total depth of the radar penetration was about 10 m. A 1,200 m reconnaissance radar survey was run along on both sides of the railroad track and identified 14 anomalous conditions. Each anomaly was then re-surveyed in detail using a series of parallel radar traverses 1 m apart to provide the exact center location and a detailed cross section of each anomaly.

Figure 16.10 shows two examples of radar data over anomalous conditions. Figure 16.10a shows the development of a small diameter piping feature breaking through the silty clay layer and migrating upward through the quartz sand. The white area in the radar record is caused by loose material that provided fewer reflections and indicates the zone of piping. This piping feature is estimated to be about 1 m in

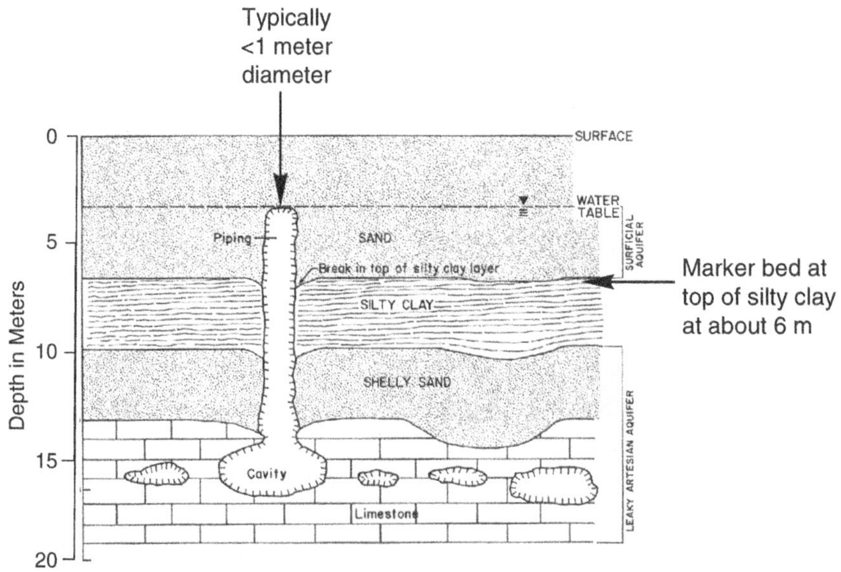

Fig. 16.9 The geologic conceptual model of conditions based upon borings and reconnaissance radar data. The cavity zone within the limestone was below the penetration depth of the radar, which was limited to about 10 m

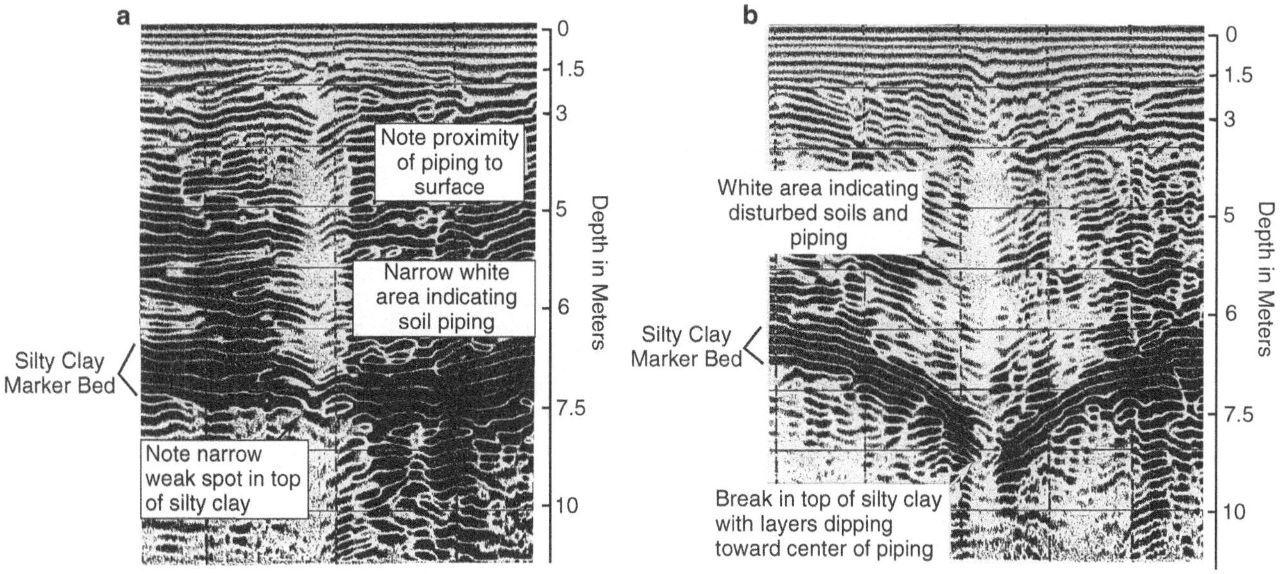

Fig. 16.10 The radar data from two of the anomalies showing the details of soil piping before it reached the surface, initial soil piping, a narrow feature <1 m in diameter (**a**), and more advanced development of soil piping where the silty clay and the sand layers have begun to dip downward (**b**)

diameter and is seen to be within 1.5 m from the surface. Figure 16.10b shows a more developed piping feature. Here the top of the silty clay layer has been deflected downward and the piping is much more developed than that seen in Fig. 16.10a. This piping feature has also progressed to within a meter or so of the surface. The strong reflector at a depth of about 6 m is a silty clay horizon, which was found consistently throughout the area of investigation and provided an excellent marker bed.

The cavities that were causing the sinkholes were located within the limestone at depths greater than 12 m (beyond the site-specific depth of penetration of the radar). Yet, the data clearly show evidence of soil piping and a dipping clay layer in response to the deeper cavity within the limestone. These shallow near-surface indicators (NSI) clearly identify the location and activity of the deeper cavity system without detecting the cavity itself.

The radar data clearly identified the piping features before they reached the surface. These anomalies could then be accurately located and a boring could be placed over the center of the small diameter anomaly that was then remediated by grouting. This case history shows very high quality data and the level of detail that can sometimes be obtained by radar surveys. Such data is relatively easy to interpreted from unprocessed records with little independent supporting data. The radar data at this site were of excellent quality due to the presence of a clean quartz sand at the surface. The radar survey was repeated every 6 months over a two-year period to monitor conditions for any collapse activity during construction of a major railway by-pass system (Benson and Yuhr 1987).

16.5.1.2 Sinkhole in a Sand Quarry (Central Florida)

A hydraulic dredge was used to excavate sand at a quarry in central Florida. During operations water levels began to drop rapidly due to a sinkhole that developed in the bottom of the quarry. Water levels had already dropped by more than 3 m (Fig. 16.11a) before an investigation was initiated to identify the location of the sinkhole throat and remediate the conditions. Meanwhile dredging operations continued.

A review of the recent dredging operations records was used to focus in on the most likely area to start the search. An echo sounder was used to map the bottom profile of the quarry along a series of profile lines. Figure 16.11b shows the bottom profile over the sinkhole. A weighted line was used to determine the depth and exact location of the sinkhole and a float was used to mark the location of the sinkhole. Once the sinkhole was located, it was sealed using sediments with a high clay content and cement. This simple, yet effective survey approach located the throat of the sinkhole within about an hour.

Since we had a reliable record of the last areas in which the dredge had worked, the search was able to quickly focus in on the mostly likely problem area. However, if a large area was to be searched or there was great uncertainty in location, other methods would have been more appropriate such as side scan sonar. Side scan sonar can cover a large area of the bottom in a single pass and provide a high resolution image of the bottom.

a

b

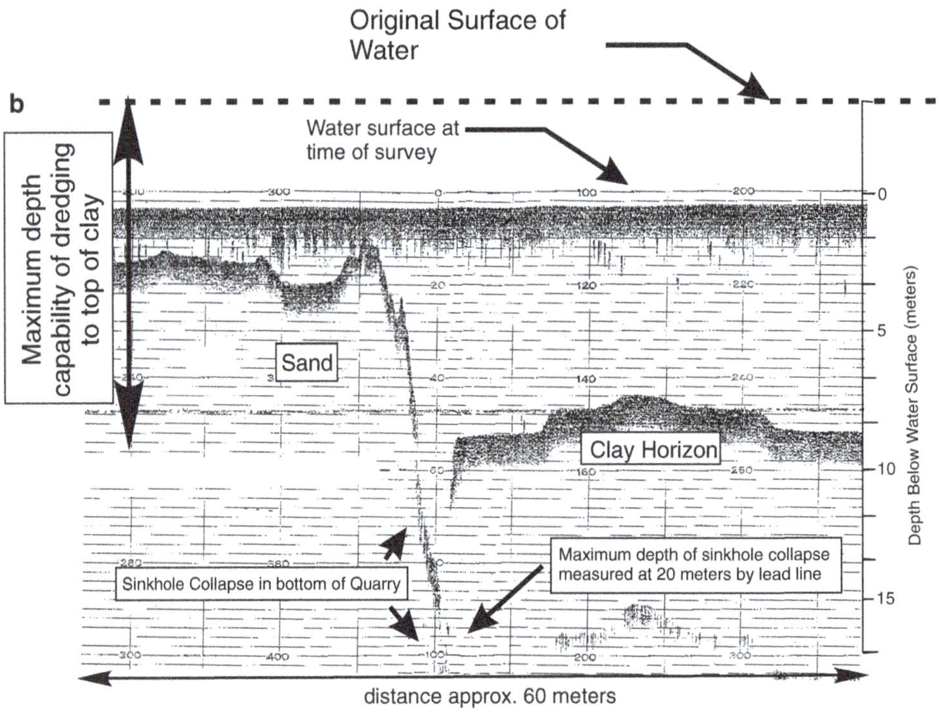

Fig. 16.11 The quarry shows the original water level along the quarry wall (**a**) and the bathymetry data that clearly shows the location of the recent sinkhole in the bottom of the quarry (**b**)

16.5.1.3 Seepage at an Earthen Dam (New Mexico)

An investigation was initiated due to seepage and small collapse features at an earthen flood-water retention dam near Carlsbad, New Mexico. There had been four different boring investigations at the site, but none had identified the problem. The owner was at the point where a decision had to be made on whether to repair or breach the dam. At the time of this work the dam was dry both up and downstream.

A visual inspection of the site revealed conditions that had not been previously identified, including small periodic fractures perpendicular to the dam's axis along with small localized soil piping features. The client had requested the use of ground penetrating radar, which was a new popular method at the time and the initial effort focused upon acquiring radar data. While radar had limited penetration due to the silty clayey sediments, it did provide indications of shallow cavities within the alluvial soil, associated with soil piping.

The radar data (Fig. 16.12a) shows an anomaly about 0.3 m below the surface. Several of the radar anomalies were excavated and found to be cavities due to soil piping (Fig. 16.12b). While the radar penetration was limited to about 2 m it provided a means of identifying near surface indicators (shallow soil piping), which were related to deeper problem areas, but their cause remained unresolved.

An extended site walkover revealed a dry streambed upstream from the dam that terminated into a small sinkhole. Discussion with the local geologist indicated that the area was underlain by gypsum. Because radar penetration was limited to 2 m we now considered an alternative method and chose electromagnetic (EM34) measurements using a 10-m coil spacing providing data to a depth of about 15 m. A series of station measurements were made along the upstream toe of the dam, which revealed significant variation in conductivity (Fig. 16.5a). Then a series of continuous EM34 measurements were made using a truck mounted system. The continuous measurements revealed the presence of localized areas of high conductivity (Fig. 16.5b). These areas of high conductivity were thought to be associated with fractures in the underlying gypsum, which contained more moisture. Trenches at the upstream toe of the dam uncovered fractures of 0.15–0.3 m wide in the gypsum that contained some water confirming our interpretation. Not encountering these very narrow vertical fractures by four boring programs was understandable.

Finally a series of survey lines were run parallel to the dam both up and downstream. The data (Fig. 16.13) identified localized increases in conductivity on consecutive parallel survey lines showing linear trends upstream, and downstream of the dam. Note the correlation of some fractures over distances of 60 m or more and the spatial changes in fracture patterns of 30 m or less. The results clearly identified a group of fractures within the gypsum running under the dam. The cause of the seepage had been identified and sufficient data was provided to allow a decision to be made on whether to repair or breach the dam.

This example illustrates the importance of flexibility to switch methods in the field, selecting the appropriate method to solve the problem. It also illustrates the benefit of the use of "continuous" measurements whenever possible.

Fig. 16.12 A shallow radar anomaly was seen within the alluvium (**a**), a trench was cut exposing a small cavity over a soil fissure (**b**)

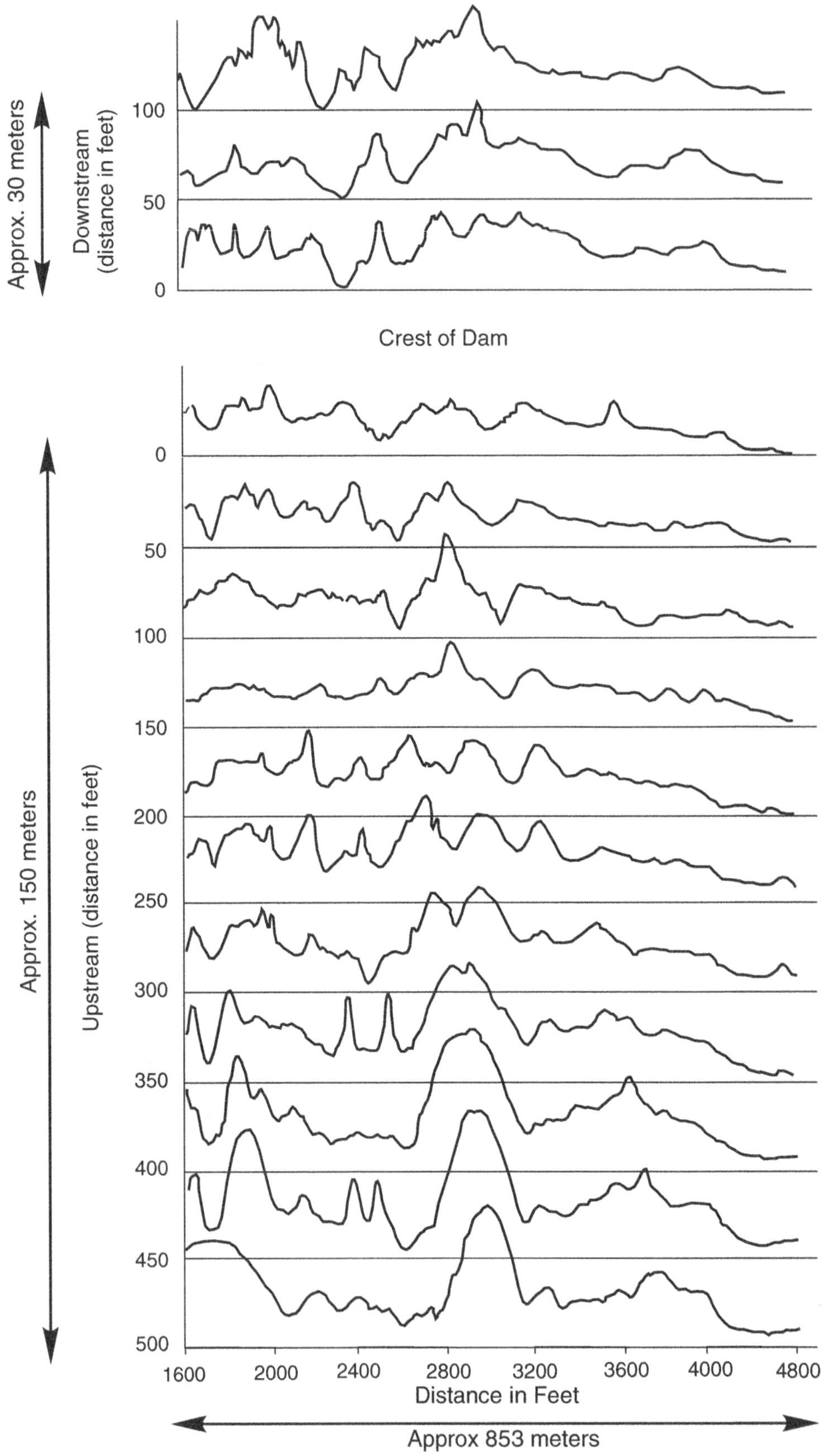

Fig. 16.13 Continuous electromagnetic data was plotted as parallel survey lines both up and down stream of a flood retention dam. The linear trends in the EM data identified fracture zones within the gypsum, which continued under the dam and were verified by trenching

16.5.2 Conditions Within the Epikarst, Top of Rock or Rockhead

The epikarst or the "skin of the karst" is the dissolutionally weathered upper portion of the bedrock (Figs. 3.12 and 3.13). This zone is characterized by its high degree of variability in the depth of dissolution (i.e. its thickness) and the degree and type of sediment fill within the zones of dissolution. The top of rock profile (the unweathered rock) or rockhead is a common concern to engineers for excavations, piles and foundations. This zone is also of concern when dealing with contaminants because of its extreme geologic and hydrologic variability. If the top of unweathered rock is relatively shallow, less than 15 m or so below grade, there is a good chance at defining the epikarst and the top of rock profile using the surface geophysical methods. The following are examples of characterization of this zone and include:

- A diesel fuel spill along a highway in Kentucky,
- A landfill expansion in Florida,
- A site of proposed power plant in Florida, and
- An expansion of a power plant in Alabama.

16.5.2.1 Diesel Fuel Spill Along Highway (Kentucky)

A diesel fuel spill of 14,000 l occurred along a section of a highway in Kentucky and there was obvious concern about cleanup. Three 2D resistivity survey lines of data were obtained, one on each side of the divided highway and one in the median (each about 23 m apart). An electrode spacing of 3 m was used to provide a reasonable degree of resolution.

Figure 16.14 shows the three parallel lines of 2D resistivity cross sections obtained. Unweathered rock with high resistivity values on the order of 600 ohm-m or more is seen at depth. Weathered rock had resistivity values between 100 and 600 ohm-m and extends from near the surface to depths of more than 12 m. Zones of clay and increased moisture with low resistivity values, less than 100 ohm-m, are scattered near the surface and in pockets within the weathered rock. The resistivity data was successful at this site because of the significant electrical resistivity contrast between the overlaying sediments (with low resistivity) and the limestone rock (with higher resistivity). The top of unweathered rock varied between 3 to more than 14 m below grade, providing a thick and highly variable epikarst zone (Stephenson et al. 2003).

This example illustrates the highly complex nature of the epikarst zone and the difficulty of sampling and remediation within such geologic conditions. With insight from the 2D resistivity survey, boring locations and sampling were optimized.

16.5.2.2 Landfill Expansion (Florida)

A landfill in west central Florida was undergoing expansion when a small sinkhole developed in an excavation. Electromagnetic measurements (EM) were used to map variations in the top of rock. The EM data was acquired using an EM31 instrument that measures bulk electrical conductivity to a depth of about 6 m (Fig. 16.15) and indicated a highly variable bedrock topography. This was similar to that observed at a nearby excavation which had exposed the top of rock. Where the conductive clayey soil is thicker the EM values are higher. The closely spaced changes in conductivity values are due to clay-filled fractures that occur at an approximately spacing of 9–15 m. Another major trend in the conductivity occurs at approximately 120 m interval and suggests deeper weathered zones or deeper rock where there was more conductive clayey soil present.

The original top of rock contour was developed from a limited number of borings (Fig. 16.16a). An EM survey was completed using a series of parallel survey lines over a portion of the landfill. The resulting conductivity contour map showed variability across the site. Fourteen boring locations were then selected based upon the EM contour map to provide depths to rock for correlation purposes. This data was then used to develop a revised top of rock contour map (Fig. 16.16b). A remedial plan was designed that focused on the fractures and deep areas of rock which were grouted to stabilize conditions and eliminate future subsidence or collapse beneath the landfill.

16.5.2.3 Site of a Proposed Power Plant (Florida)

A site was being investigated for a potential power plant in west-central Florida. Seismic refraction survey lines were located through the site based upon aerial photos and physical access. The purpose of the investigation was to define the top of rock in areas of potential site development. Seismic refraction tomography data was acquired using a geophone spacing of 3 m. The data shows a high degree of variability in the depth to the top of unweathered rock that was interpreted as the 7,000 ft/s contour. Three significant lows in the top of unweathered rock are located along the line shown in Fig. 16.17. A 30 m deep boring was located just north of the survey line near station 410 and did not encounter any rock. These data were used in part to evaluate the suitability of the site and plan locations for the large structures associated with a power plant.

16.5.2.4 Expansion of a Power Plant (Alabama)

At a proposed power plant site, seismic refraction data were used to characterize the top of rock. The site is underlain with relatively hard limestone containing linear weathered zones based upon photo-lineaments. A high-resolution seismic refraction tomography survey was used to define the top

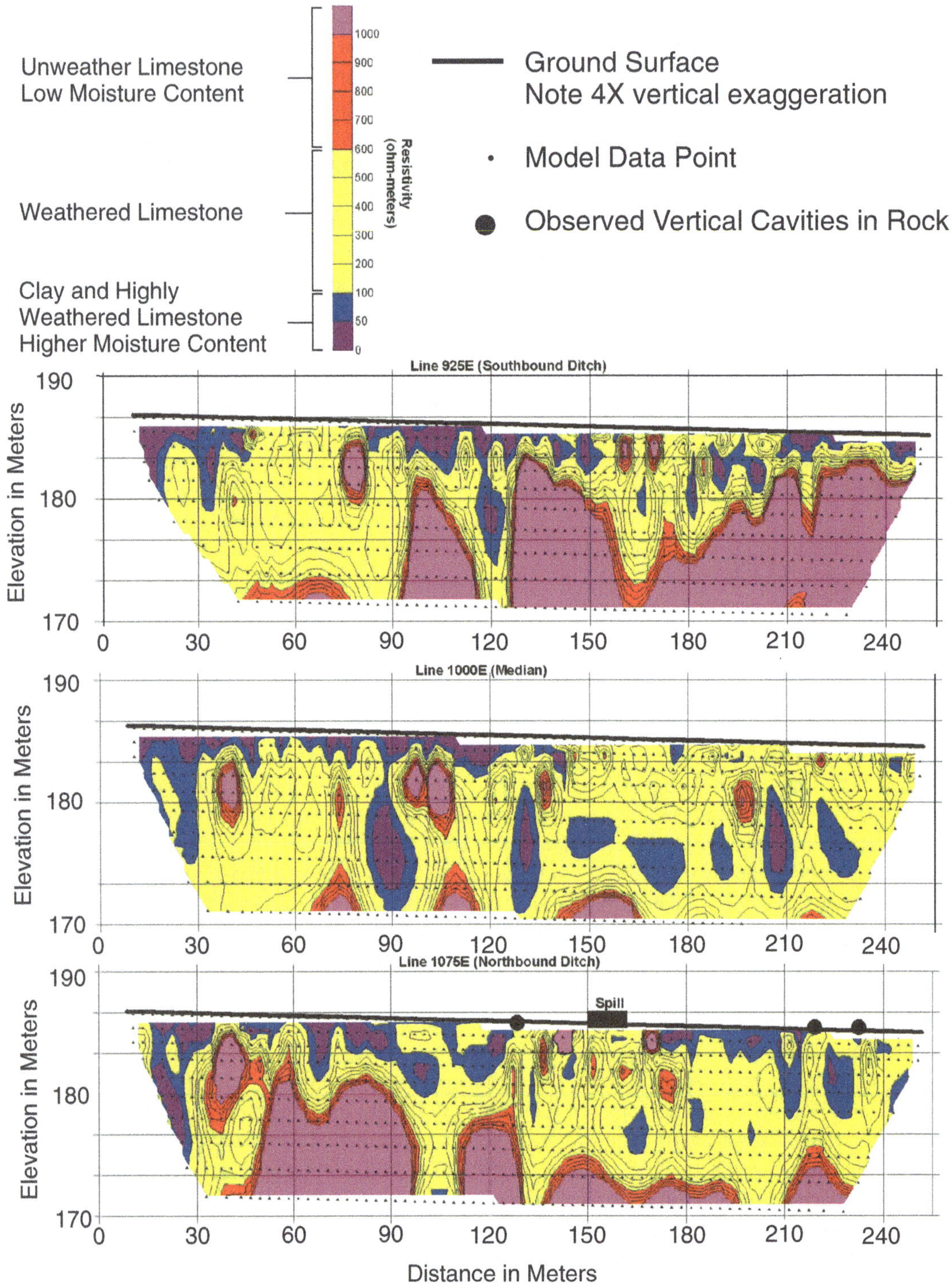

Fig. 16.14 Three parallel 2D resistivity cross sections illustrate both the lateral and vertical complexity of the epikarst. The extreme geologic variability is seen along each survey line as well as between the three survey lines, which are about 23 m apart

of rock and weathered zones within the limestone. Almost 5 km of data were acquired over the area of interest, with a geophone spacing of 1.2 m. The resulting data were contoured to show the depth to top of unweathered rock (Fig. 16.18), which clearly identified two erosional lineaments. The resulting data allowed borings to be positioned in anomalous and background locations and were used to aid in planning development at the site.

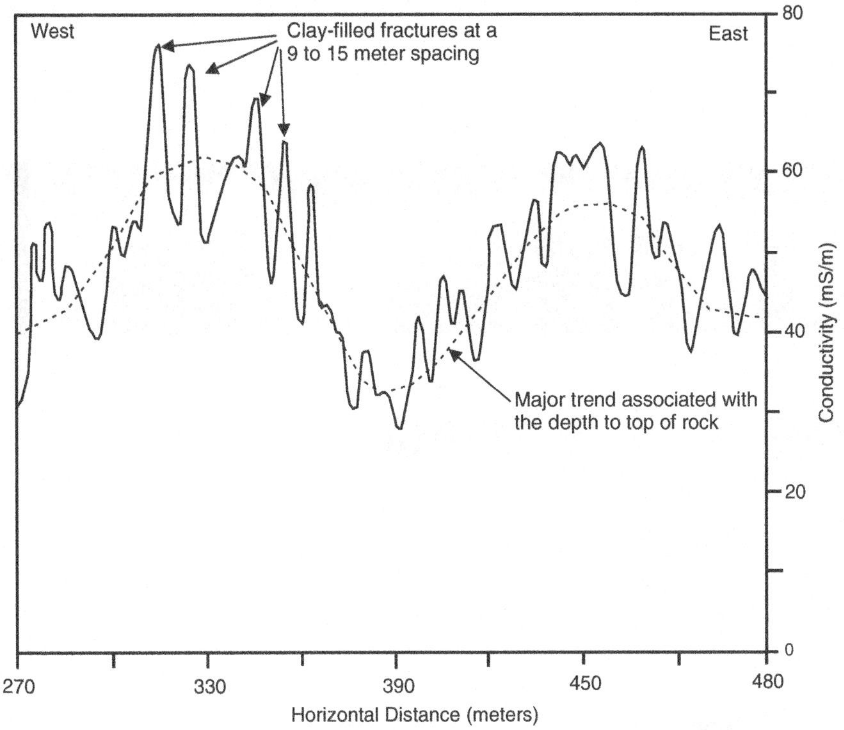

Fig. 16.15 Electromagnetic data along one of the profile lines at the base of a landfill excavation shows highly variable conditions associated with clay-filled fractures within the rock as well as variations in the depth to top of rock

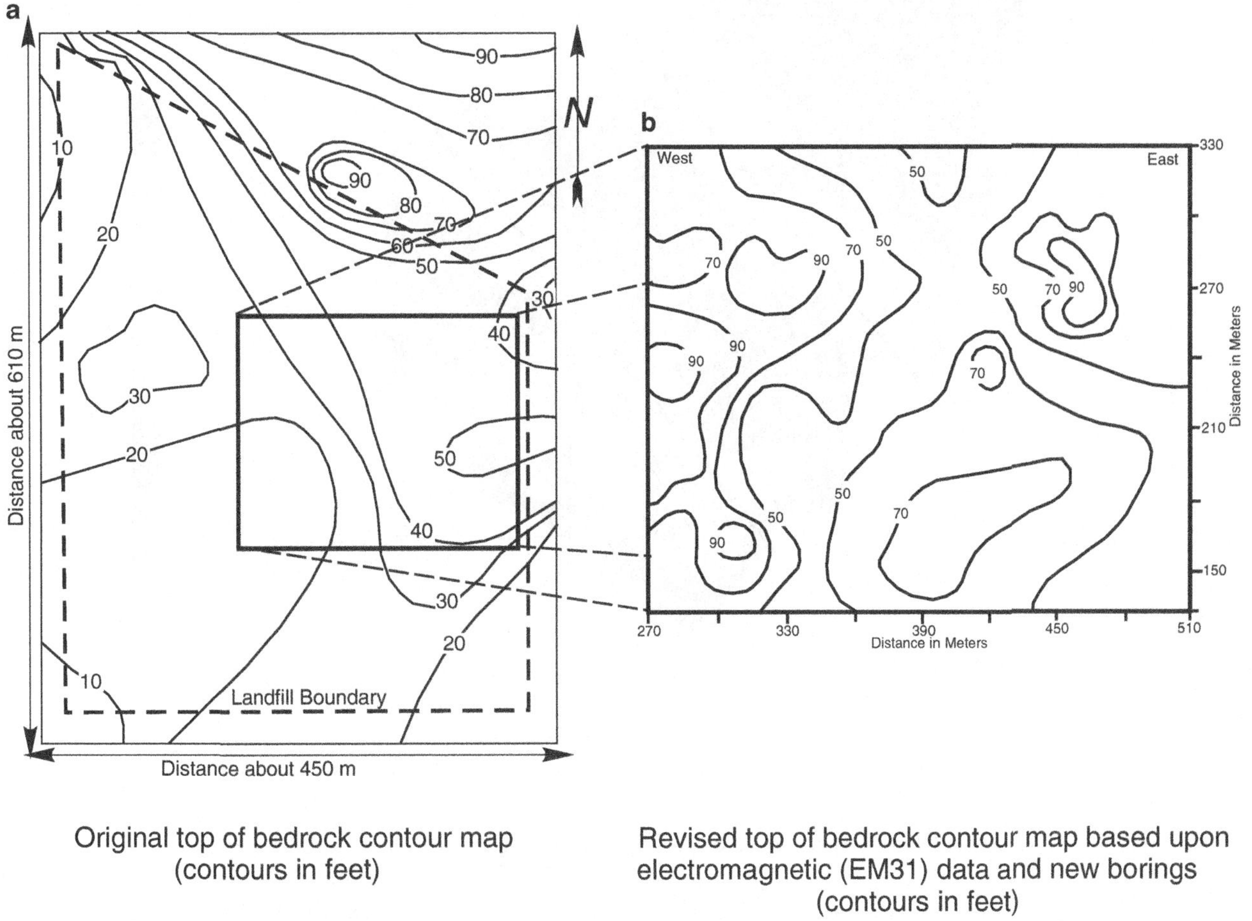

Fig. 16.16 A top of rock *contour* map was based upon widely spaced borings (**a**), a more detailed top of rock contour map was (**b**) based upon electromagnetic data (Fig. 16.15) and 14 new borings (*contours* in feet)

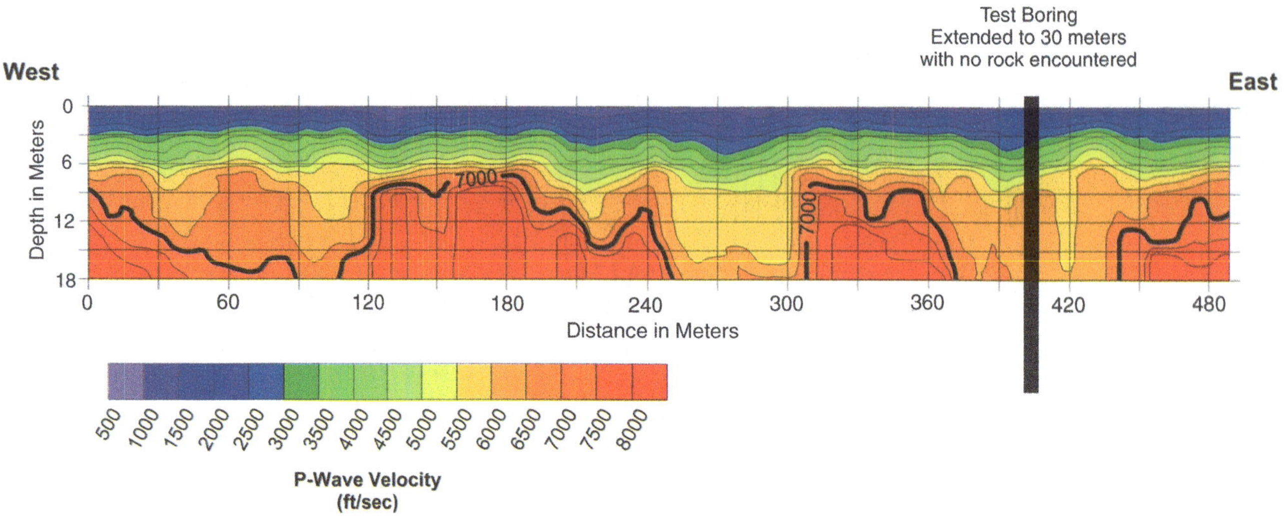

Fig. 16.17 The seismic refraction data indicates extreme lateral changes in the depth to top of unweathered rock based upon the 7,000 ft/s P-wave velocity *contour line*

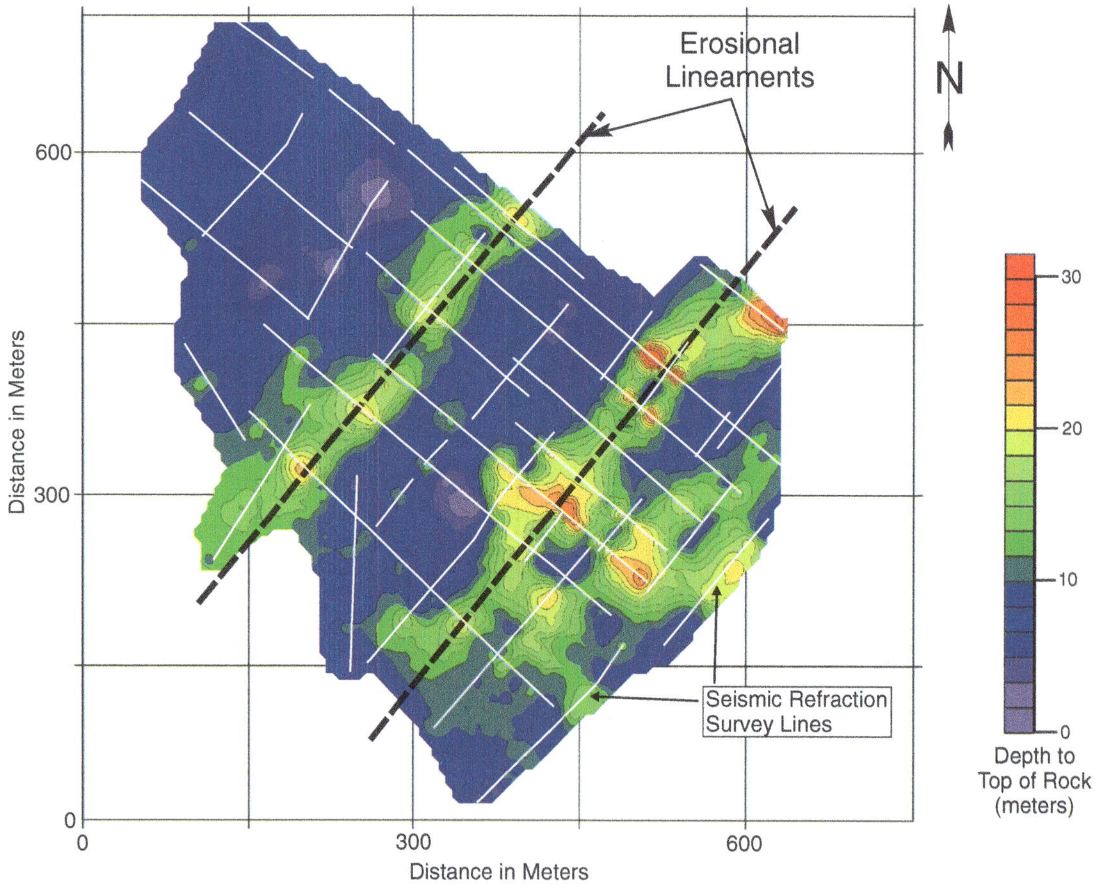

Fig. 16.18 The depth to unweathered bedrock based upon seismic refraction data reveals two major southwest to northeast erosional lineaments

16.5.3 Fractures and Cavities Within the Rock

Below the epikarst zone within the unweathered rock, our focus shifts to fractures, cavities, conduits and cave systems. These features are more difficult to detect yet may have a significant impact to a site. These features can be air-filled, water-filled or soil-filled, can be closely or widely spaced, or even random. Larger caves and conduit systems will typically occupy less than 5 % of the surface area and are difficult to locate by borings alone (Quinlan J 1999 personal communication). The following examples illustrate how the surface geophysical methods can be utilized to locate and characterize these features. These examples include:

- Identifying fractures along a new dam alignment in Alabama,
- Identifying a fracture zone at a low-level radioactive site in Missouri,
- Sinkhole collapse at a military site in Guam,
- Deep conduits in the Woodville Karst Plain in Northern Florida, and
- Detecting a large cave room under a road widening in Alabama.

16.5.3.1 New Dam Alignment (Alabama)

Anomalous seepage at a dam had been on-going since its construction in the late 1960s. A new dam was being built downstream. A detailed site characterization was being completed for the new dam that included identifying and remediating of karst features. An initial resistivity survey identified anomalous areas within the rock and had been excavated exposing the bare limestone. More detailed resistivity data was acquired using two parallel 2D resistivity survey lines run over the anomalous area with an electrode spacing of 1.5 m. The resistivity data (Fig. 16.19) illustrates the location of a major fracture zone of about 6 m wide dipping to the south. Both 2D resistivity lines indicated similar results. Lower resistivity values (<400 ohm-m) are probably associated with the presence of weathered rock and increased moisture within the fracture zone.

16.5.3.2 Low Level Radioactive Site (Missouri)

The St. Louis Airport Site (SLAPS) is a 9 ha area located on the north side of the St. Louis airport. The SLAPS site was used from 1946 through 1969 for storage of residues from uranium ore processing. Most of these residues were removed and building facilities were demolished and buried on-site. The site was then covered with up to 1 m of clean backfill. The state of Missouri is known to have well developed karst features including many large sinkholes, caves and spring systems. The Missouri Department of Natural Resources (DNR) had raised the issue of potential karst underlying the site.

The area of investigation encompasses the original 9 ha plus surrounding areas for a total of about 67 ha. Previous work at the site had not addressed any aspects of karst. Of the 80 existing borings only 2 extended into limestone a short distance, less than 3 m (Fig. 12.11). As part of the DOE Expedited Site Characterization Technology Program, a site characterization to assess karst conditions was carried out but without the benefit of additional borings. This included a review of available data, aerial photo analysis, a site walkover and observations of outcrops, microgravity data and time domain EM soundings (TDEM), geophysical and gamma spectrometer logging of monitor wells as well as soil sampling and analysis.

One component of the site characterization was a cross section created from a single survey line of 50 TDEM soundings. The TDEM sounding measurements were spaced 15 m apart. Each 1D sounding was interpreted using a four-layer resistivity model. The 50-1D models were then combined to create a 2D profile across the site (Fig. 16.20). This resistivity profile identified the presence of a major fracture zone. The TDEM measurements were responding to the contrast between the thicker clay and moisture within the fractures (low resistivity) verses the surrounding unweathered limestone (higher resistivity). This data spatially correlates with the lineament analysis from aerial photos and logging data from existing wells and helped to improve the understanding of site conditions.

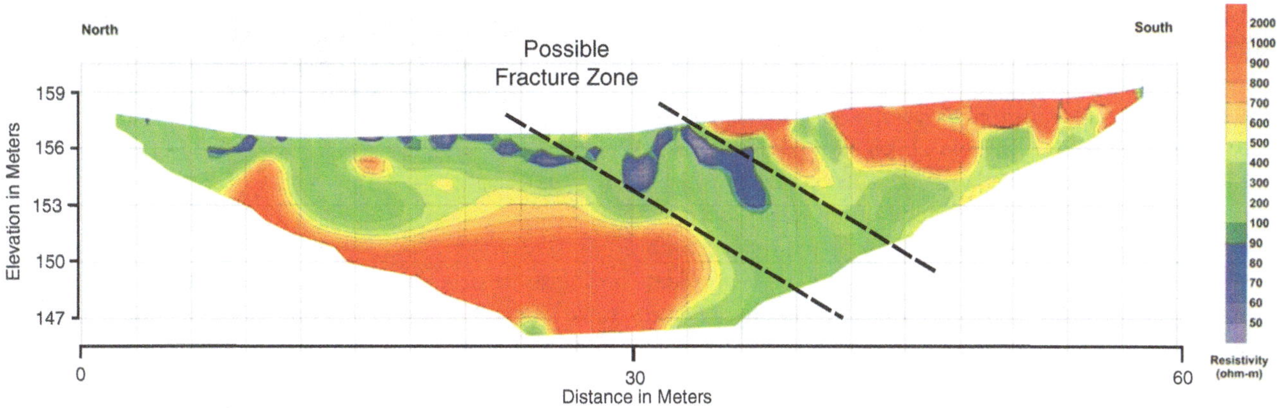

Fig. 16.19 2D resistivity data identified a dipping fracture zone along a proposed dam alignment

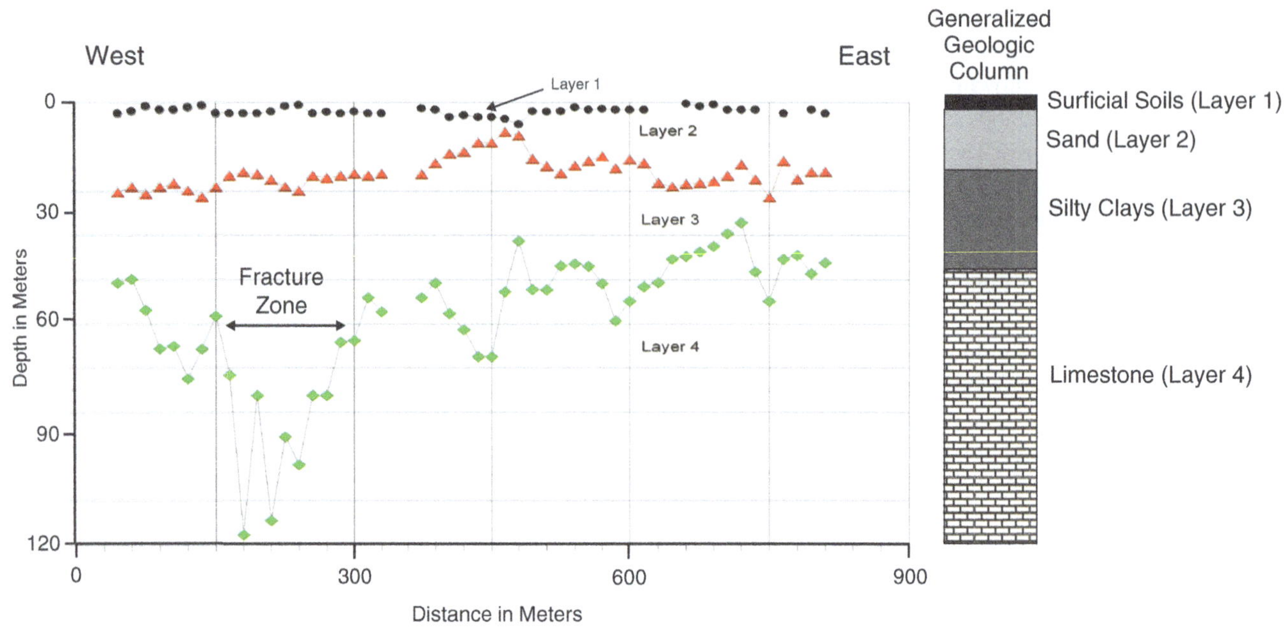

Fig. 16.20 Cross section showing 50 four-layer TDEM sounding models defining the top of limestone and a fracture zone within the bedrock

16.5.3.3 Sinkhole Collapse at a Military Site (Guam)

A high security military facility was located on a cliff (almost 180 m high) on the northwest coast of Guam. The facility was experiencing localized sinkhole activity (Fig. 16.21). Areas of surface collapse were typically 1–2 m in diameter opening into small cavities within the limestone rubble fill used as soil cover. Concern arose regarding possible subsidence or collapse that could impact the critical military structures at the site. Radar was selected to survey the site using an 80 MHz antenna. The survey objectives were to identify cavities, joints and other areas of concern, which might impact the structures. The radar data quality was excellent and provided a depth of penetration to 24 m due to the dry, clean limestone conditions (Fig. 16.22a). However, the range was reduced to 12 m to provide higher resolution data within the shallower zone of concern (Fig. 16.22b). A series of parallel radar survey lines 7.5 m apart were run oriented perpendicular to the cliff.

Borings were then located based upon the cavities identified in the GPR data (hyperbola). Each air rotary boring was geologically logged then a borehole video system was used to investigate the vertical and lateral extent of any fractures or cavities encountered. Cavities encountered ranged from 1.8 to 3.6 m in diameter. Small joints (0.3–0.6 m wide) within the rock appeared to be allowing soil to migrate into the cavities, causing localized collapse features.

While large, deep vertical shafts in the limestone are seen on Guam, this problem was not a deep-seated karst problem and there was no eminent hazard of major collapse to structures at the facility. The combination of radar data, borings and borehole video data provided increased confidence levels in the interpretation of subsurface conditions. A remedial program was designed and undertaken to stabilize the surface conditions by grouting within the upper 12 m.

Fig. 16.21 One of many sinkholes at a military site on the northwest coast of Guam

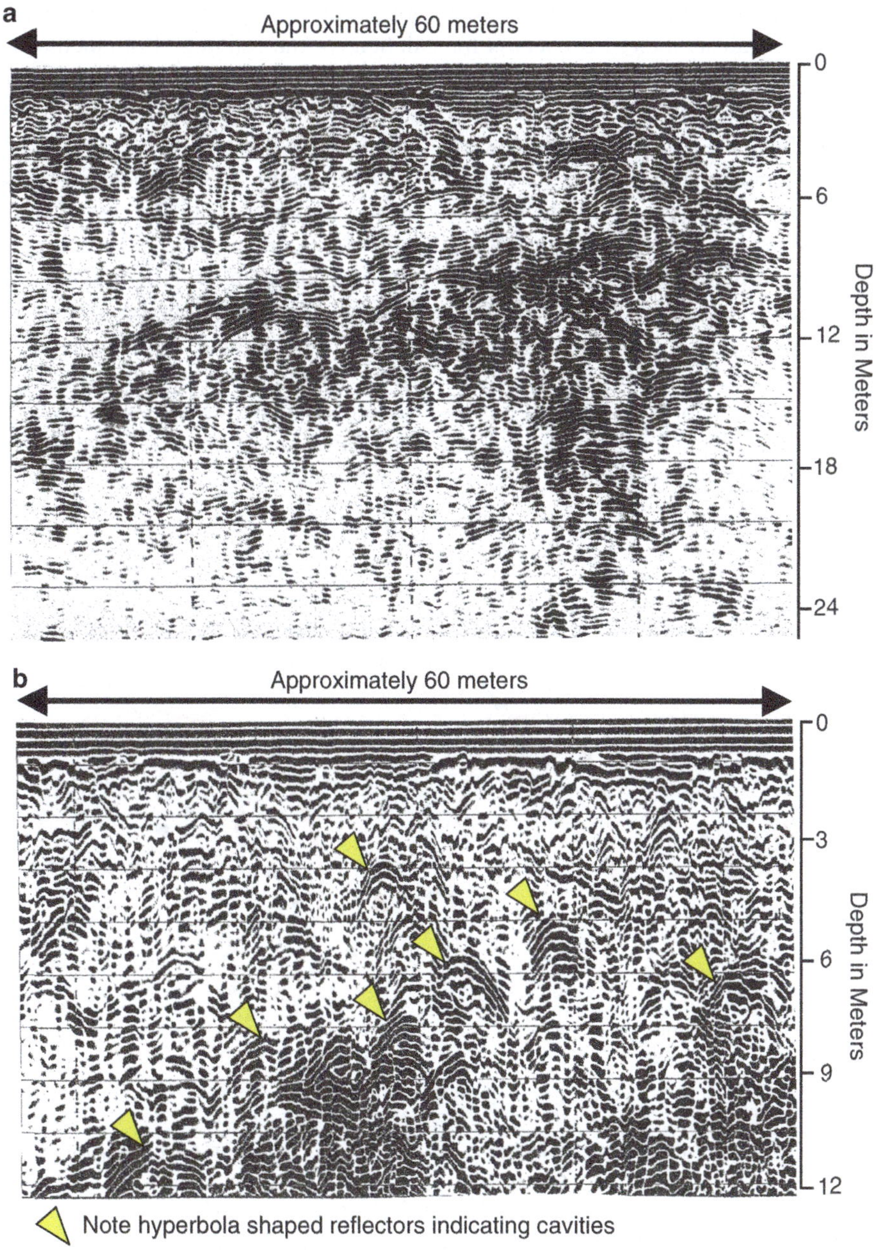

Fig. 16.22 Radar data was initially acquired to a depth of approximately 24 m (**a**) while it was of excellent quality only the upper 12 m were of interest for engineering purposes (**b**). Note the numerous hyperbola (possible cavities) were identified in the upper 12 m of limestone

16.5.3.4 Woodville Karst Plain (Florida)

The Woodville Karst Plain (WKP), previously introduced in Sect. 7.2.2, covers approximately 1,200 km^2 and is located in northern Florida between Tallahassee and the Gulf of Mexico. The abundance of sinkholes provides direct connections between the surface and groundwater making the groundwater vulnerable to contamination. Groundwater contamination due to nitrates has been blamed for the declining water quality conditions in Wakulla Springs (Kincaid et al. 2005).

One of the many features being mapped within the WKP is the extensive underwater cave and conduit system. Over 63 km of caves and conduits have been mapped by cave divers (Wisenbaker et al. 2007) but it may only represent a small portion of the total conduits and caves within the WKP. There was a need to better characterize the conduits that feed the springs in order to develop more accurate hydrologic understanding and, eventually, numerical models to predict groundwater flow and contaminant transport within the Woodville Karst Plain.

A variety of geophysical measurements were made at the Leon Sinks area in order to evaluate their effectiveness for mapping large conduits. These measurements included microgravity and spontaneous potential (SP) measurements as well as several other techniques. Two large conduits were known to run perpendicular under the road at this site. Their depth and size were known from divers and their location had been established by cave radio. This site provided an excellent location to test the response from various geophysical measurements over known conditions (Yuhr et al. 2008).

Figure 16.23 shows both the SP and microgravity data collected at this site. Both of these techniques were used to acquire measurements at a spacing of 6 m and both sets of data were successful at this site. The SP data shows a clear response to the flow within the conduits and the microgravity data shows a clear response due to the mass deficits of the conduits. Both methods also show a broader response from the deeper, discrete conduit at station 155, indicating a larger area of weathering, fracturing, etc. extending to the east.

Microgravity measurements were selected as the most technically appropriate and cost-effective method to identify large deep conduits over a large portion of the WKP area. Over 44 km of microgravity data with station spacing of 30 m were acquired along survey lines upgradient of Wakulla Springs. The purpose of the study was to identify potential large conduits not previously identified or mapped. Significant microgravity anomalies were identified that may be related to large karst conduits 60 m deep or more (Kaufmann and Dehan 2007).

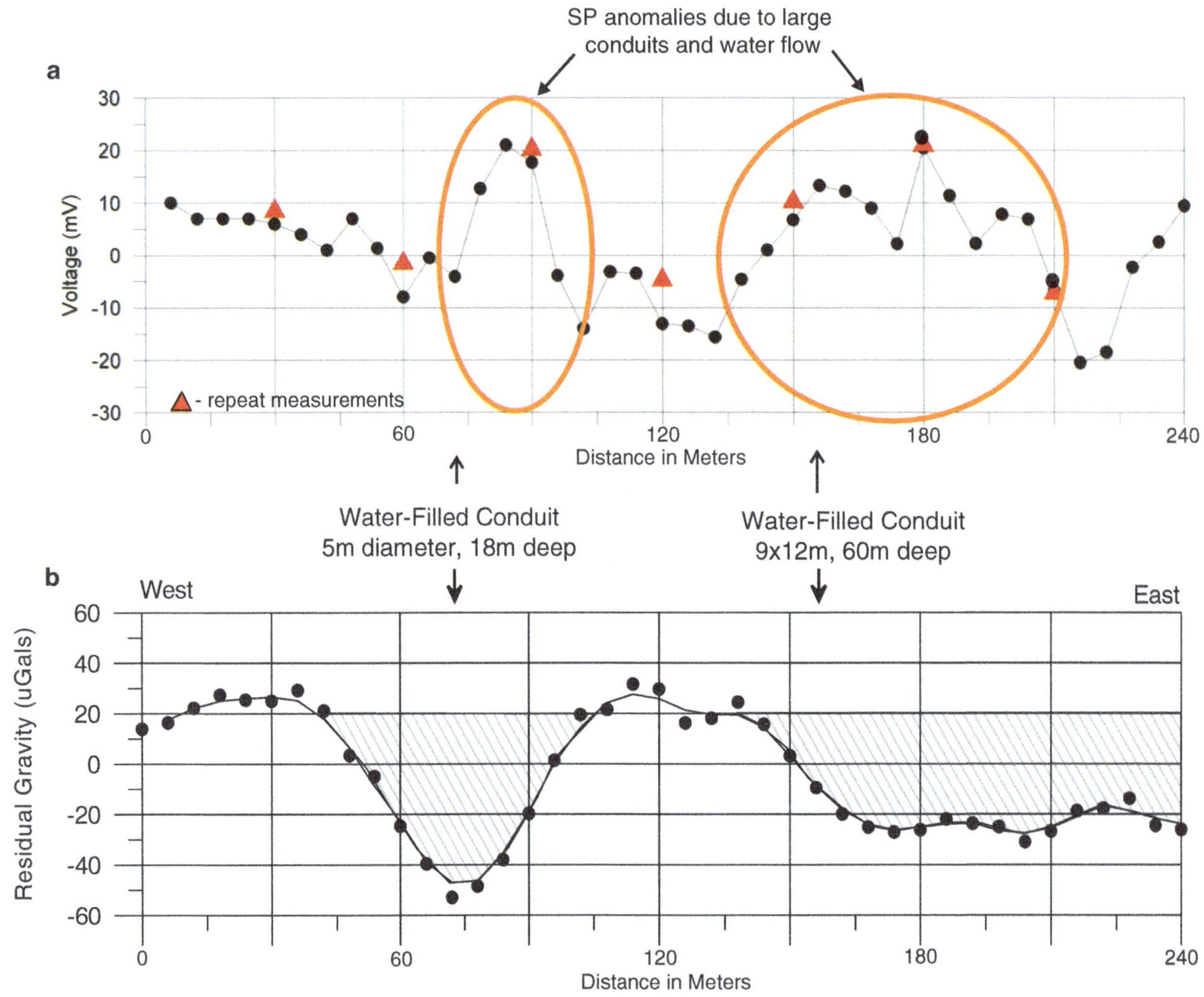

Fig. 16.23 SP (**a**) and microgravity measurements (**b**) over known water-filled conduits at the Leon Sinks site. The *triangles* show repeat SP measurements as a means to check noise level and repeatability of the data. Note that the deeper cave system to the right has a broader SP and gravity anomaly

An example of the production gravity data (Fig. 16.24a) clearly shows the presence of the three known conduits along with another deeper conduit, which had not been previously identified. A cross section (Fig. 16.24b) shows the mapped conduits and their dimensions along with an additional large conduit modeled from the gravity data. The results of these surveys aided in the placement of new borings (Yuhr et al. 2008) and advance the on-going hydrogeologic characterization of the karst system to refine the hydrogeologic models of the area (Kincaid et al. 2012).

In addition, as part of the Woodville Karst Plain (WKP) study, marine geophysical measurements were made in the Spring Creek area along the coastline to locate springs. The Spring Creek area is a tidal stream in Apalachee Bay and contains up to 14 submarine springs within its lower reaches. Figure 16.25 shows a portion of the marine 2D resistivity and echo sounder data over two of the many springs. Both sets of data show a response from the springs. The echo sounder data shows a physical low in the bottom conditions at each spring. The marine 2D resistivity data shows an area of lower resistivity around each spring indicating a zone of weathered, more permeable rock.

16.5.3.5 Roadway Widening (Alabama)

A road was being widened in Madison County, northern Alabama. A portion of the expanded roadway crosses over Burwell Cave, which had been mapped in the 1960s. The cave contains a large room 14 m high and 5.5–6 m wide. The ceiling of the cave was expected to lie about 3 m below the proposed road expansion. The exact location of this room was not known and access to the cave was no longer available. Therefore, surface geophysical measurements were acquired to determine the location of this large room in the cave. The microgravity data (Fig. 16.26) and the 2D resistivity data (Fig. 16.27) show the response from the large cave adjacent to the road. The combination of the microgravity

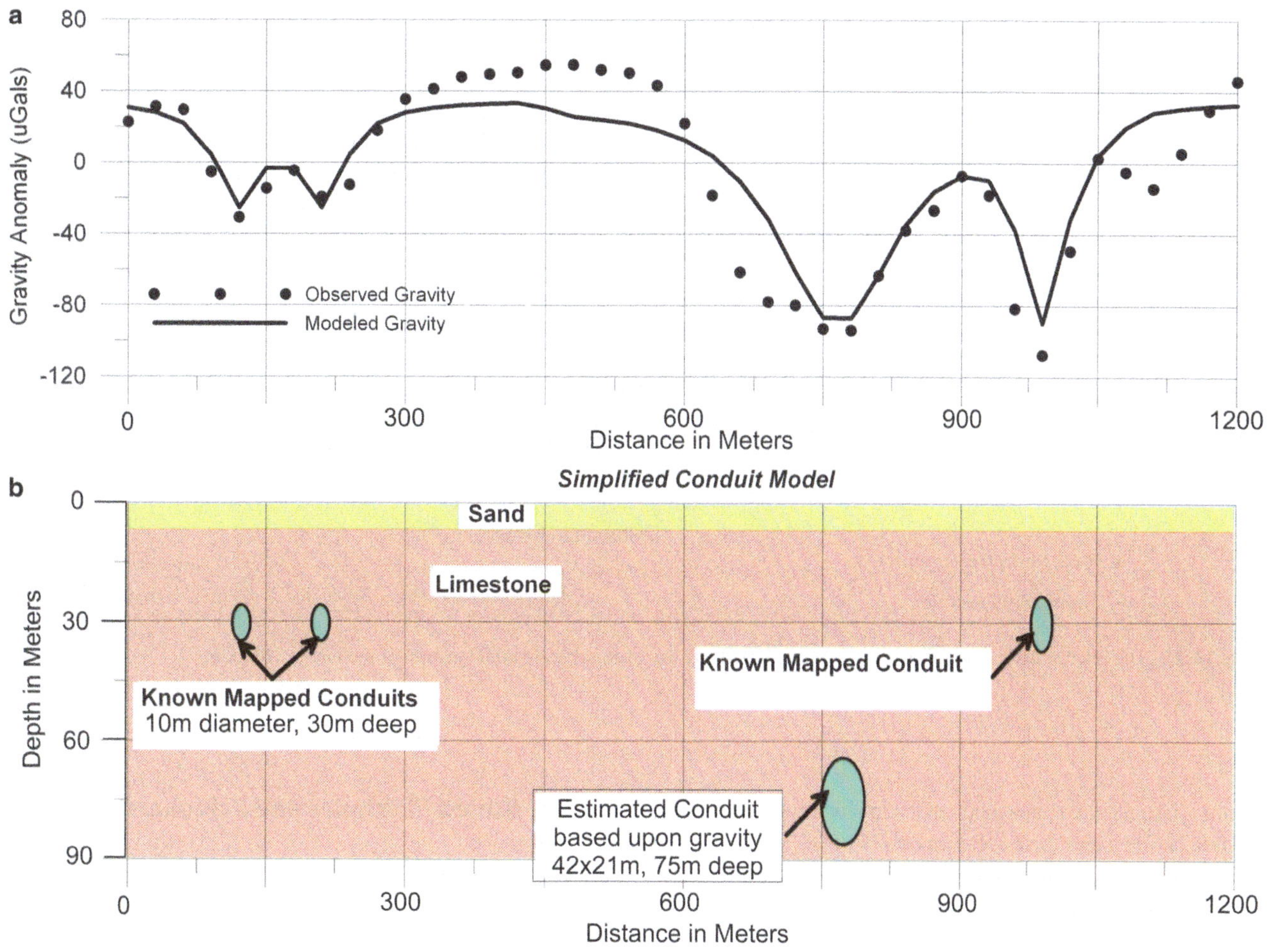

Fig. 16.24 Microgravity measurements (*dots*) were made over known conduits. An inverse model using the gravity data and known geologic conditions (*solid line*) are shown for comparison (**a**). In addition to the known cave systems an additional low gravity anomaly at about station 760 was detected and modeled. (**b**) shows the depths and dimensions of the known and estimated conduits

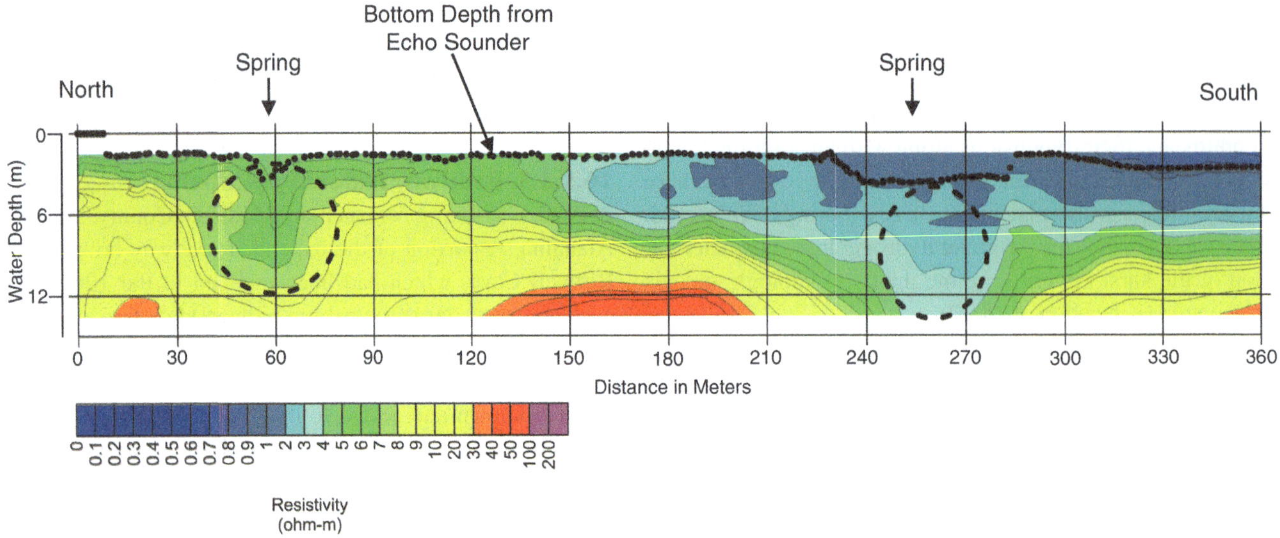

Fig. 16.25 The two of the many springs within Spring Creek were located both by echo sounder data (see *dots* in the *upper edge* of the image) and 2D marine resistivity data. The resistivity data suggests a zone of weathered, more permeable rock around each spring

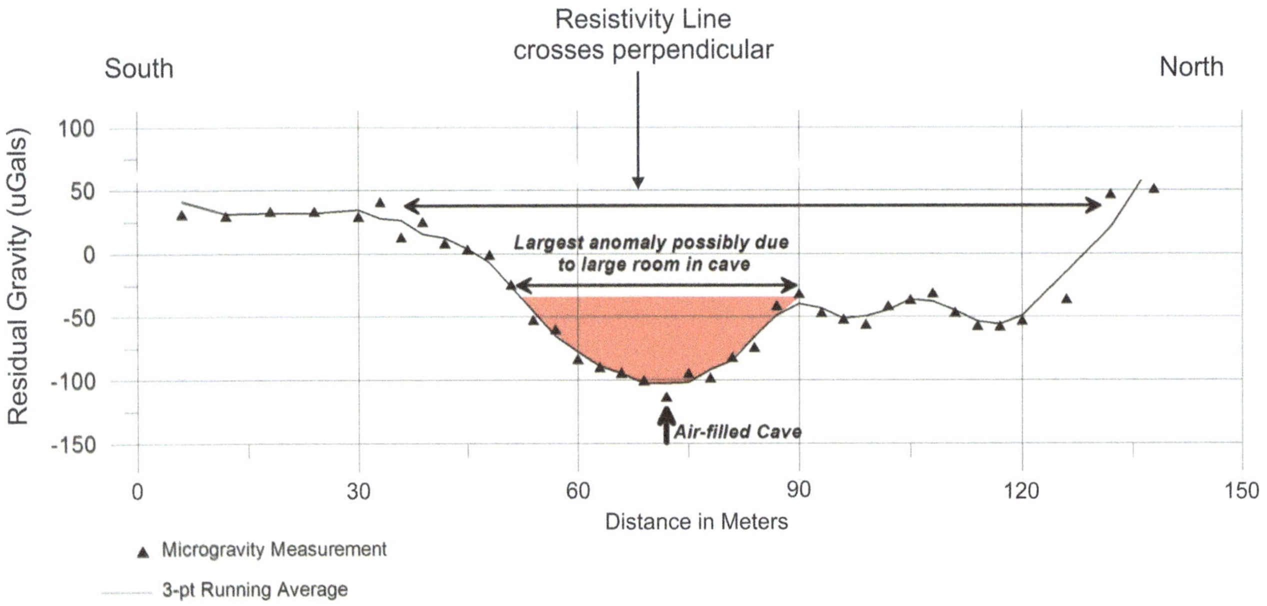

Fig. 16.26 The microgravity data parallel to a proposed roadway expansion detected the location of the large cave room

and resistivity data provided optimum locations for borings to determine exact cave location and depths.

Both microgravity and resistivity measurements were successful because the cave was large and very shallow. The site, while wooded, was relatively flat and required no major topographic corrections for the microgravity data. In addition, no major utilities, fencing or guardrails were present to provide interference for the resistivity data.

16.5.4 Buried Sinkholes and Paleokarst

We commonly find paleokarst features on topographic maps or in historical aerial photos as sinkhole lakes, circular lows in topography or old collapse features. However, paleokarst features can also include buried sinkholes or collapse features that have no surface expression. These features may be stable from a collapse point of view, but they can impact a

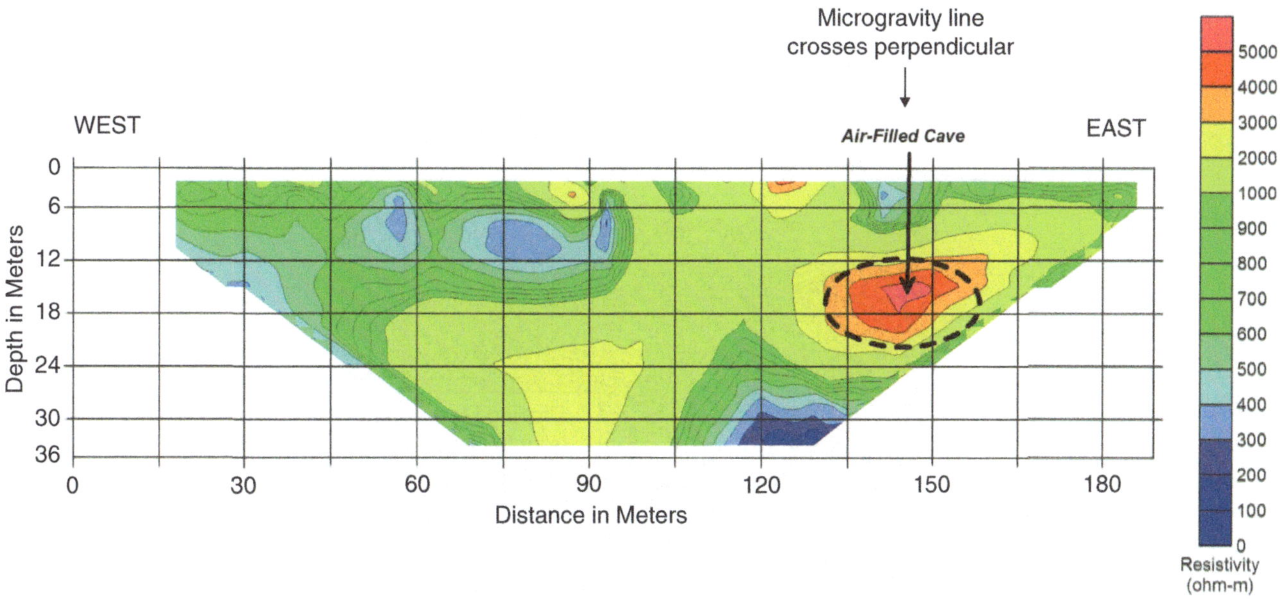

Fig. 16.27 To further assist in the location of the cave a 2D resistivity survey line was run perpendicular to the road and the gravity survey line. The resistivity data shows the location, lateral extent and the approximate depth of the large cave room about 9 m below grade

site in other ways. These features may present areas of weaker materials impacting construction or increased permeability that could impact groundwater flow and future settlement. In addition, there is also a small risk for reactivation of paleocollapse features. All three of our case histories in Part III include the detection and characterization of a buried paleocollapse feature with little or no surface expression. The examples for this section include:

- A residence in central Florida,
- A proposed hazardous waste site in southwest Texas,
- A tunnel boring under a shipping channel in south Florida, and
- Radiation halo over paleo sinkholes in central Florida.

16.5.4.1 Residential Subsidence (Florida)

One of the big concerns in Florida is establishing whether or not a sinkhole is the cause of settlement in order to resolve insurance claims. When a sinkhole is obvious at the surface (Fig. 3.2a) the answer is clear. However, without the presence of an obvious sinkhole, the cause of subsidence often becomes a debate between the owner, their insurance company, their lawyers and the various consultants. The State of Florida has attempted to address these complex issues in regulations (Florida Senate 2010).

In some cases, the problem can be resolved with the aid of surface geophysical methods. Figure 16.28 shows the two parallel radar profile lines run along opposite sides of a house in which the cause of settlement was in question. Excellent quality data was obtained to a depth of 9 m due to the clean quartz sands. A marker bed of sand with green-gray silt/clay is seen at a depth of about 6 m. The radar data clearly defines the marker bed dipping downward on both sides of the house identifying the presence of a paleosinkhole. Radar data is one of the recommended methods of investigation in the State of Florida (Schmidt 2005).

In this case, the house happened to be built directly over a paleosinkhole filled with sand that had no surface expression prior to building. The presence of a cavern within the deeper limestone that caused the paleosinkhole is at a depth, beyond the radar data penetration and could be at depths of 15–30 m or more. The cause of the recent settlement is associated with a pumping well located alongside of the house resulting in fluctuation of the local water table, possibly aided by concentration of surface runoff from the house roof, patios or driveways. The settlement is clearly related to the paleosinkhole under the house, however the degree of settlement is uncertain. It may be a small amount of settlement that would eventually stabilize, or may be a precursor to a reactivation of this paleocollapse feature. In either case, the radar data has clearly identified the basic issue of whether it is sinkhole related or not. Further assessment and monitoring was required in order to determine the potential impact of the paleosinkhole and its possible remediation.

16.5.4.2 Proposed Hazardous Waste Site (Texas)

An investigation was being carried out to evaluate a property regarding its suitability for a hazardous waste site in southwest Texas. A previous site about 8 km to the east had been abandoned due to the presence of paleokarst and a large cave. It had been proposed that this alternate site would not

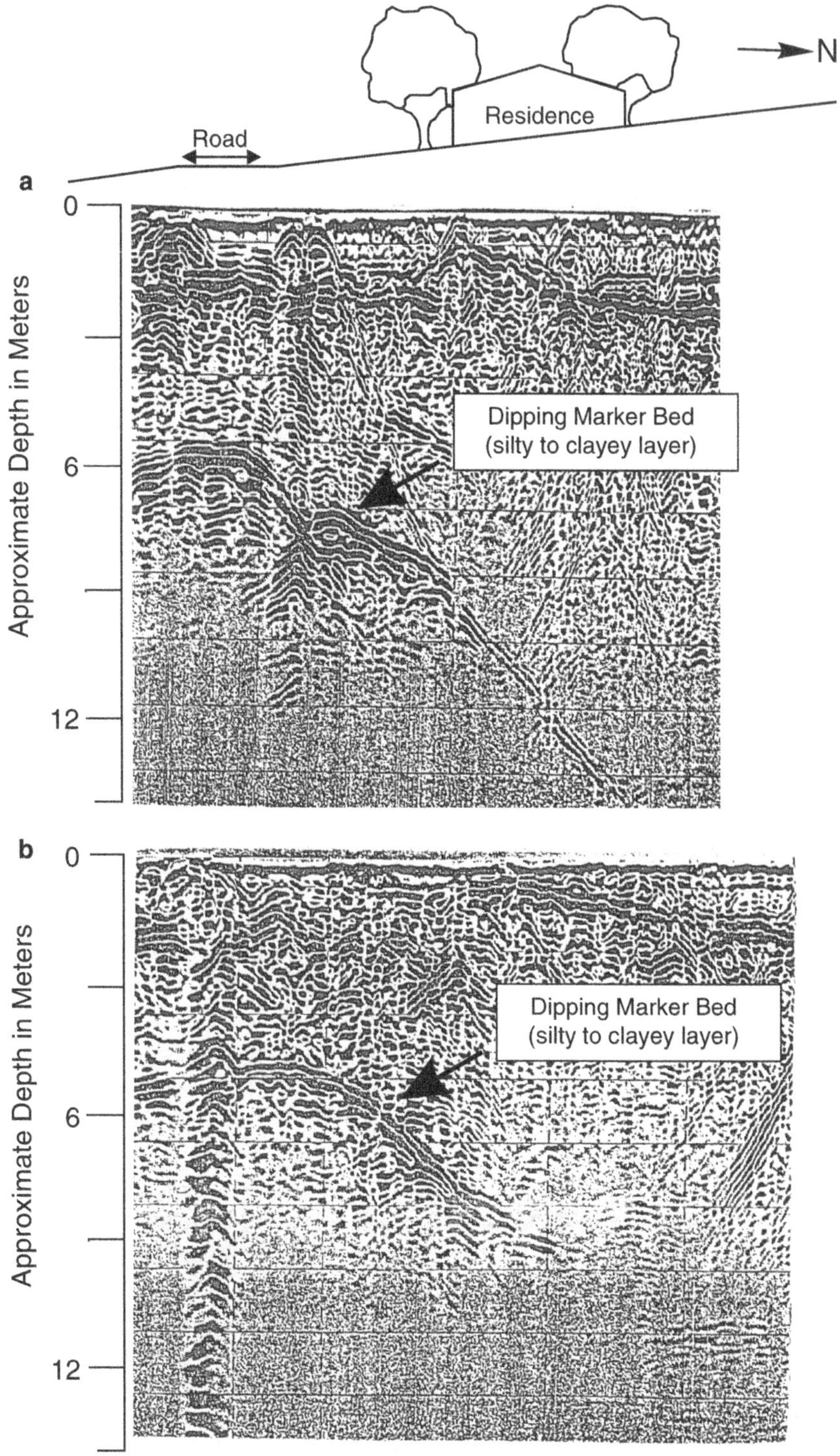

Fig. 16.28 Radar data acquired on either side of a house show clear evidence that the house is located over a paleosinkhole. (**a**) Radar data run along west side of residence. (**b**) Radar data run along east side of residence

have any karst, and the objective at this alternate site was to locate areas where the alluvium was thickest to provide ample material for excavation and fill. The initial area investigated for the site selection was 3,300 by 4,500 m. Electromagnetic (EM34) sounding measurements were selected as a rapid and economical means to characterize the alluvium thickness over this large area. Six lines 3,300 m long were run north-south, spaced 900 m apart. These widely spaced lines and measurements provided initial reconnaissance data over this large area.

EM34 measurements were made to depths of 15, 30 and 60 m along each of the survey lines at intervals of 150 m. Figure 16.29a illustrates the EM data from one of the survey lines. The interpretation is shown in Fig. 16.29b identifying an area of thicker alluvial sediments. The alluvial sediments were more electrically conductive than the massive limestone. Where the alluvial sediments become thicker, a higher electrical conductivity is measured by the EM34. Ultimately, it was determined that the area mapped on-site with thicker alluvial sediments was a buried stream channel.

The final site selected was 300 by 750 m. A detailed site characterization was completed and included desk study, site walkover, EM34 measurements, microgravity, extensive borings, geophysical logging, borehole video logs and trenching (Yuhr et al. 1993).

The desk study had identified a USGS report where extensive paleokarst features had been identified some 5–11 km to the southeast of the site (Freeman 1968). These features were clearly identified by upturned rock around their boundaries (Fig. 16.30) and were as large as 450 m wide and 5 km long (Fig. 15.9).

One of these paleokarst features had been identified in the site walkover to the east of the site. The presence of upturned rock at the surface was observed along a railroad cut and marked the boundary of the paleokarst feature. There was no other indication of subsidence or collapse at the surface. Surface geophysical measurements (microgravity and electromagnetic EM34 measurements) were acquired along a single survey line over the feature and a single boring had been placed in the center of the collapse. The four sets of independent data (observations, a boring log, and two set of surface geophysical data) provided both the lateral extent and depth of the paleocollapse (Fig. 16.31) and were used to develop a reliable conceptual model of this paleocollapse (Yuhr et al. 1993).

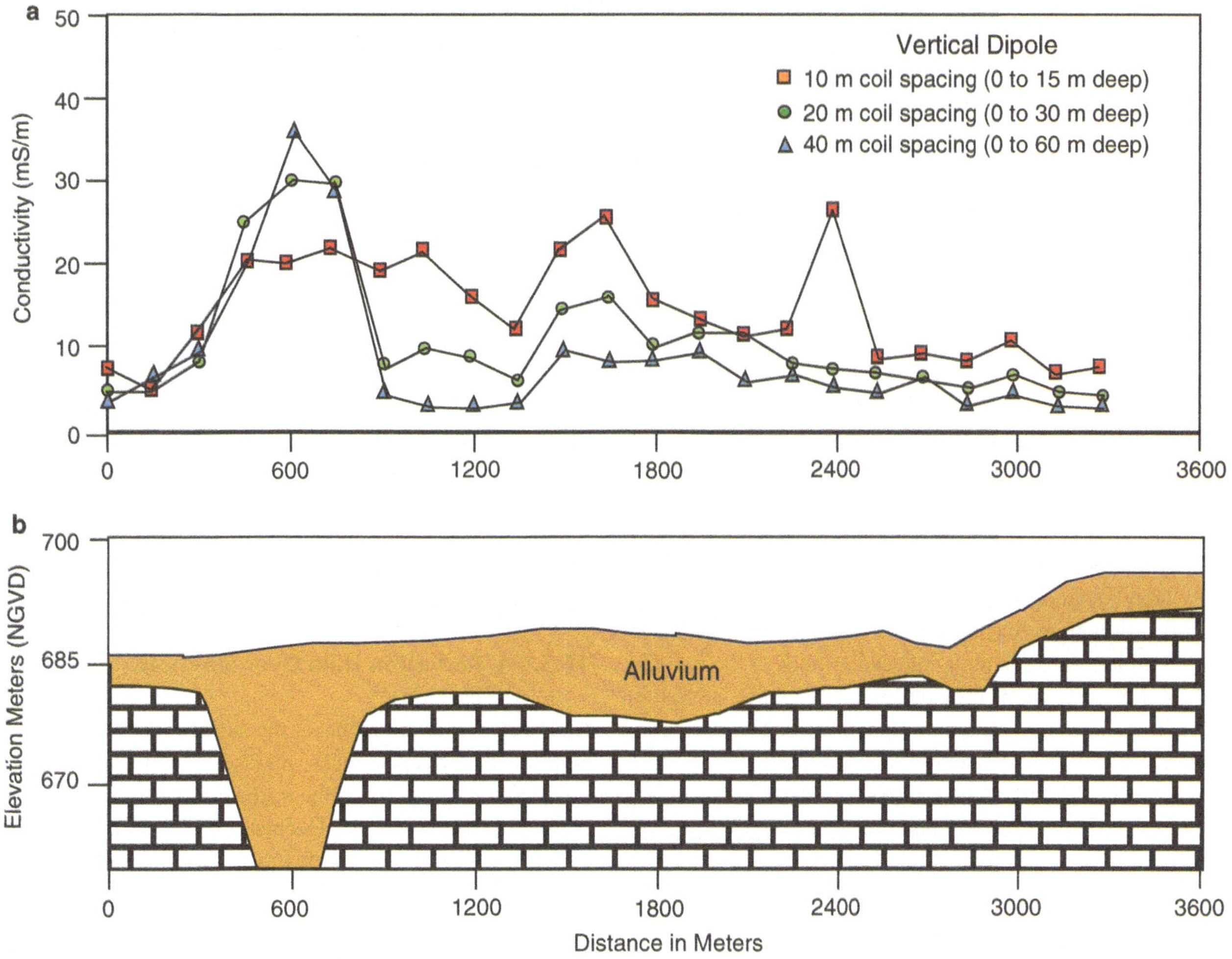

Fig. 16.29 Electromagnetic (EM34) station measurements (**a**) detect an area of increased conductivity. The interpreted model (**b**) indicates thicker alluvium centered at station 600, possibly a buried channel or paleokarst

Fig. 16.30 Frank Woodward (**a**) observes edges of a paleocollapse feature defined by upturned rock (**b**)

16.5.4.3 Tunnel Boring Under a Ship Channel (Florida)

An underground tunnel linking Interstate 395 on Watson Island and the Port of Miami on Dodge Island in Miami-Dade County, Florida had been proposed to alleviate traffic in downtown Miami. The proposed link would consist of two bored tunnels crossing beneath Government Cut, which accommodates both passenger cruise ships and cargo ships to the Port of Miami.

A suite of marine geophysical methods were used to investigate the conditions and determine engineering properties within Government Cut. The surveys included bathymetry (echo sounder), sub-bottom seismic reflection, and seismic refraction. Six lines of survey data were obtained, two of which were run along the center lines of the two tunnels. All three methods provide spatial insight to subbottom variability. Seismic refraction data identified variations in compression (P-wave) velocity. Areas of Low P-wave velocities measured by the seismic refraction data correlated well with low N-values (blow counts) obtained from existing SPT borings. Figure 16.32 shows the seismic refraction data from one of the survey lines. The top of "hard" rock is defined as 8,000 ft/s. The localized low velocity areas could represent possible zones of sand and sediments in areas between massive coral reefs, or paleosinkholes filled with sediments. In either case, these areas represent areas of loose or less cemented materials that may present problems for the tunneling machine.

This project was completed in May of 2014. Traffic is now flowing through the tunnels which are 1,280 m long consisting of two parallel tubes 11 m in diameter and dipping to about 30 m below sea level. This is one of the most expensive and elaborate transportation projects in south Florida history. Asked about technical problems during tunneling, the project manager said that it was the "softer zones" in the sea bed. They had to be frozen to 27° below zero so that the tunnel boring machine could work effectively (Chardy 2014). The "soft zones" were probably the areas identified by the geophysical measurements (Fig. 16.32).

16.5.4.4 Radiation Halo Over Sinkhole (Central Florida)

Radioactive measurements at the surface are based upon the concept that deeper rocks release uranium and it's by-products that migrate to the surface via various permeable pathways Armstrong and Heemstra (1973). Upchurch et al. (1987) found that gross-alpha radiation is correlated within 0.3 km of the axis of fracture traces in west-central Florida. Banwell and Parizek (1986) found similar results for radon emanating from fracture traces in Pennsylvania. There has been limited use of radioactive measurements for geotechnical applications and karst. Possible applications include the

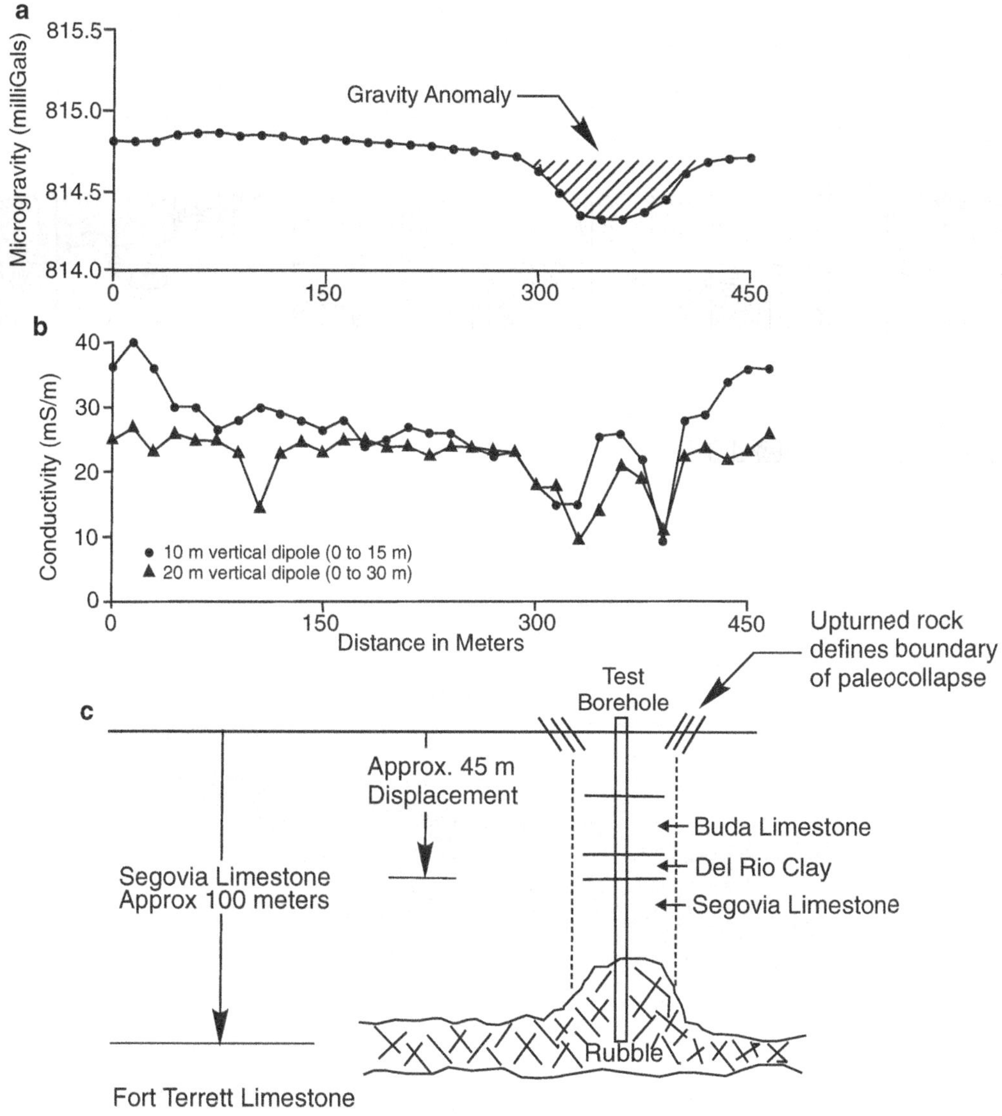

Fig. 16.31 The boundaries of a paleocollapse feature could be seen on the surface adjacent to the railway cut by upturned rock similar to that seen in Fig. 16.30. Microgravity measurements (**a**) and electromagnetic (EM34) measurements (**b**) were made over the collapse and a boring was drilled in the center, all resulting in the conceptual model (**c**)

location of cave openings, and buried sinkholes as well as man-made tunnels and mine shafts which provide a more permeable pathway to the surface.

Radiometric measurements were made with a gamma ray spectrometer along a road passing near the edge of two sinkhole depressions in central Florida. Radar data was acquired along the side of the road shows the edges of two sinkholes (Fig. 16.33a). Florida sinkholes are commonly sand filled providing a more permeable pathway for upward migration. Total count measurements were made along the side of the road (off of the asphalt) to determine if radioactive halos were detectable over the edges of the two sinkholes. The radioactive measurements were made at 15 m intervals and the resulting data showed two areas of elevated values adjacent to the two sinkholes (Fig. 16.33b). Radioactive halos have been found to be present at many other sinkhole sites, buried paleokarst and fracture zones in Florida due to areas of increased permeability, which allow upward migration of radioactive materials.

16.5.5 Pseudokarst Conditions

Pseudokarst can be caused by a wide range of conditions, typically man-induced, such as subsidence due to withdrawal of fluids, collapse and subsidence associated with mining, or cavities due to leaking acids. Regardless of the cause, the site characterization commonly utilizes the same strategy as

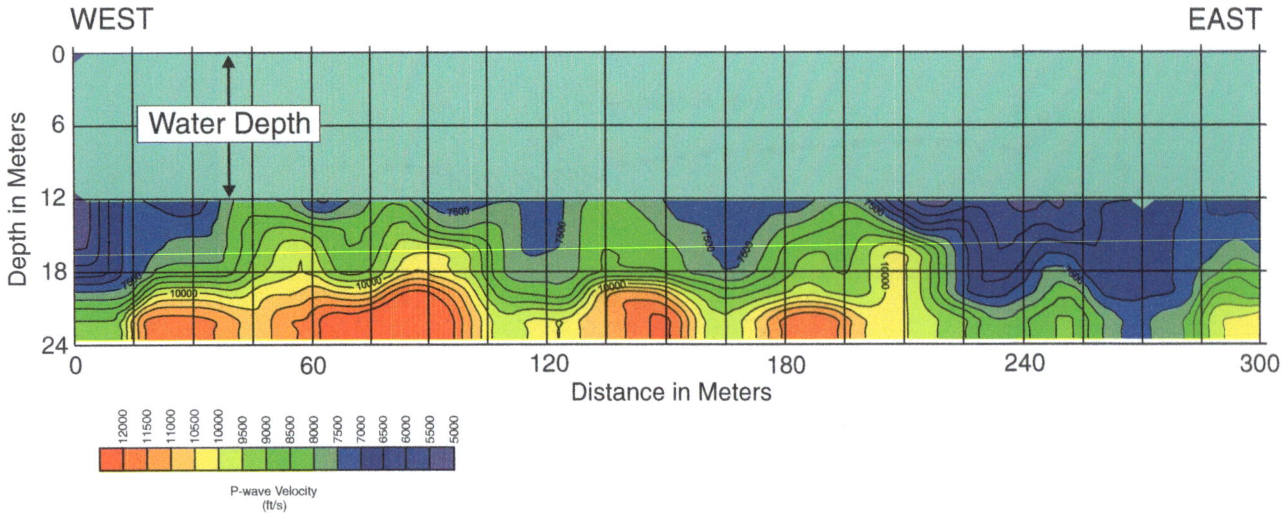

Fig. 16.32 One of the six seismic refraction survey lines illustrates the variation in seismic velocity and depth of harder rock based on (8,000 ft/s velocity)

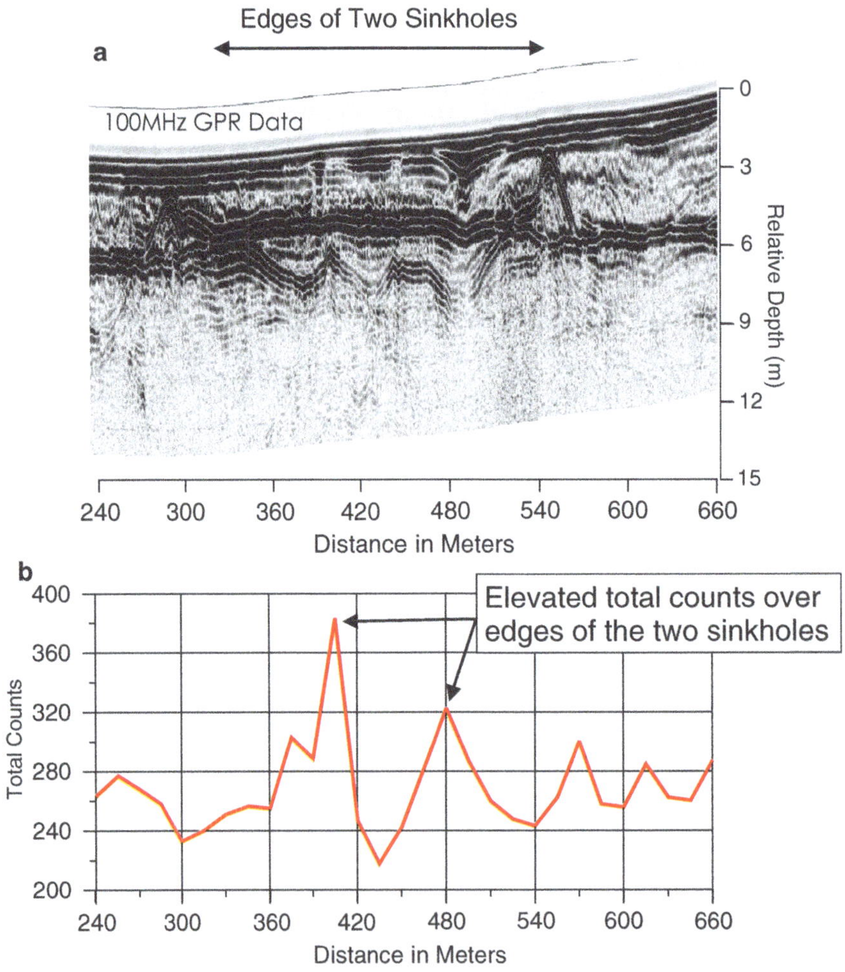

Fig. 16.33 Ground penetrating radar data using a 100 MHz antenna was acquired along the road, parallel to two sinkholes, the edges of the two sinkholes can be seen as dipping strata in the radar data (**a**). The total count gamma radiation measurements show elevated values when passing the two sinkholes (**b**)

natural karst conditions. The following examples illustrate some of the applications of surface geophysics used to characterize pseudokarst conditions and include:

- Fissures on an Air Force runway in California, and
- Mapping of an abandoned mine in Ohio.

16.5.5.1 Fissures on Air Force Runway (California)

Edwards Air Force Base is located north of Los Angles, California, and has long earthen runways, (one as long as 11 km) for test fights of advanced aircraft and shuttle landings. Groundwater withdrawal from a deep aquifer has lead to the development of extensive surface fissures along the runways (Ward et al. 1992). When the fissures break through the surface (Fig. 16.34a) they are repaired. However, when they lay hidden from view just below the surface, they present a hazard for aircraft whose wheels may encounter them, causing breakthrough and possible damage to the aircraft.

After their mission, the Space Shuttle would sometimes land at Edwards Air Force Base. The shuttle's 86,184 kg hits the bare earth desert floor at 340 km/h. The surface had been breached on three separate occasions due to the weight of aircraft; one of these occurred when the space shuttle landed (Kratochvil et al. 1992).

Colonel Kratochvil, who had just returned from supporting the USA tunnel detection program at the demilitarized zone in Korea, was assigned to NASA to evaluate the conditions. He had tried available airborne methods without success (Kratochvil 1989) and finally resorted to trying surface geophysical methods.

In 1988, Kratochvil requested ground penetrating radar data to detect the fissures. However, radar was not expected to be effective because of the clay content of the runway materials. But because the sediments were very uniform along the runway and the fissures were relatively shallow (within 1 m or so), radar proved to be quite effective. Figure 16.34b is an example of the radar data using an 80 MHz antenna that for the most part provided an extremely uniform radar response over most of the runways. When crossing a fissure the radar responds to the subtle change in conditions of void space and an increased moisture level. Radar was used as a reconnaissance tool and data was acquired along ten parallel survey lines that ran the length of each runway (Technos 1988).

The radar survey provided the approximate location of anomalous conditions along the long runways (within a meter based upon a distance counter on the survey vehicle). After being located by radar, a special cone penetrometer was used to confirm the presence, width and nature of the fissure (more details are provided in Sect. 17.2.2). A remediation team then carried out repairs of the fissures.

16.5.5.2 Mapping Abandoned Mines (Ohio)

Abandoned underground mines are quite common throughout the United States. As infrastructure and development continue, the location of these abandoned mines becomes a concern. Maps of abandoned mines are commonly not

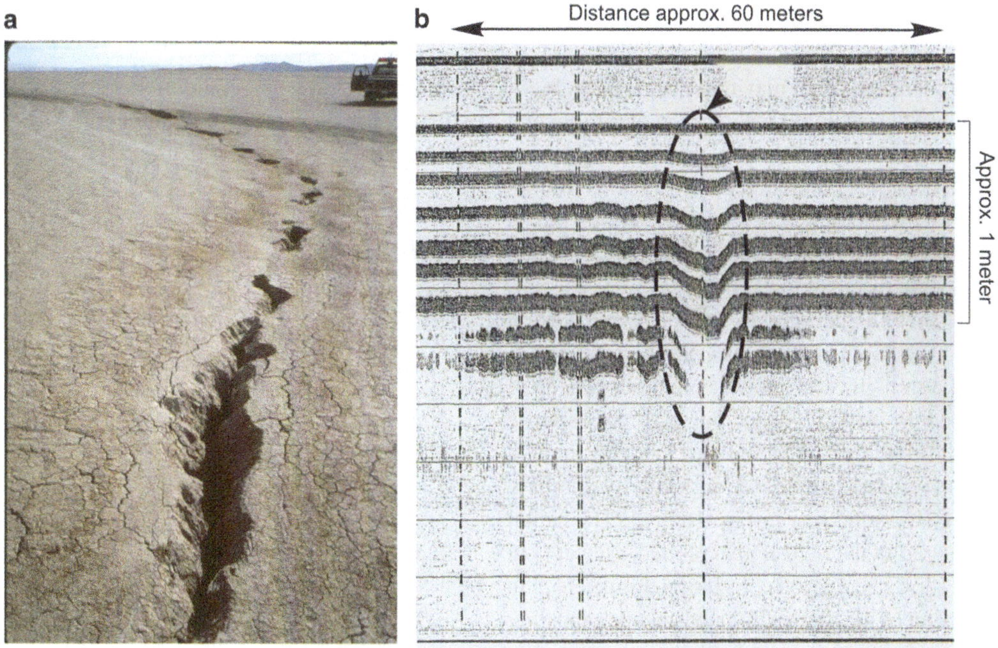

Fig. 16.34 A typical surface fissure along the earthen runways at Edwards Air Force Base California (**a**). Radar data obtained along the earthen runways at Edward's Air Force Base shows the location of a typical fissure before it has broken through the surface (**b**)

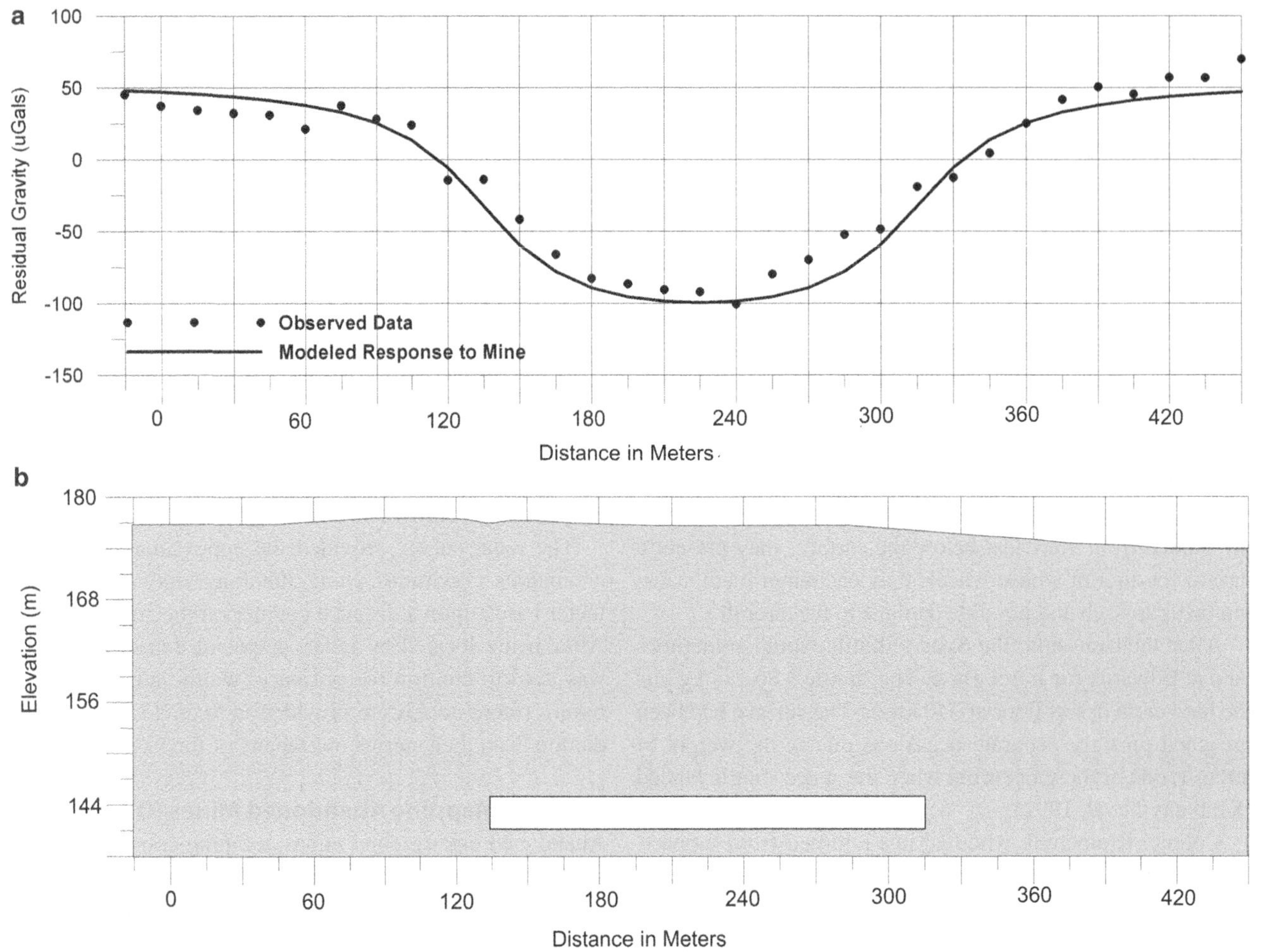

Fig. 16.35 The measured microgravity data (*dots*) and the modeled response (*solid line*) over a known abandoned mine (**a**). The location and depth of the known mine is about 30 m deep and 3.4 m high based upon existing mine maps (**b**)

accurate or incomplete and often may not even exist. A highway realignment in Ohio required the location of a room and pillar gypsum mine to be defined. While old mine maps were available and had been pieced together from various mining companies, the exact location and height of the mines was in question. Microgravity was used as one of the methods tested for its ability to locate the abandoned mines.

Figure 16.35a shows a microgravity profile over a portion of a known mine. The depth and thickness of the mine (30 m deep and about 3.3 m high) was known with a fair degree of certainty and used to estimate what the gravity response would be over such conditions. The forward model estimated that the gravity survey should detect a large gravity anomaly (>100 microgals), (see solid line in Fig. 16.35a). The actual gravity data measured in the field is also plotted, (see dots in Fig. 16.35a), and shows very good correlation. Figure 16.35b shows a cross section of the known mine based upon existing mine maps.

This test phase was quite successful in determining that microgravity measurements could effectively determine the location of a mine and could model both its depth and height with reasonable accuracy (Butler 1980, 1994; Yuhr et al. 1993).

References

Anderson N, Ismail A (2003) A generalized protocol for selecting appropriate geophysical techniques. Geophysical Technologies for detecting Underground Coal Mine Voids Forum, Lexington, KY

Anderson N et al (2003) Selection of appropriate techniques: a generalized protocol based on engineering objectives and site characteristics. Presented at the 3rd international conference on applied geophysics, Orlando, FL, 8–11 Dec 2003

Armstrong FE, Heemstra RJ (1973) Radiation halos and hydrocarbon reservoirs: a review. US Department of the Interior, Bureau of Mines, Information Circular 8579

ASTM (2007) Standard guide for the use of the time domain electromagnetic method for subsurface investigation. Designation: D6820-02. American Society for Testing Materials, Philadelphia

ASTM (2008) Standard guide for using the frequency domain electromagnetic method for subsurface investigation. Designation: D6639-01. American Society for Testing Materials, Philadelphia

ASTM (2010a) Standard practice for expedited site characterization of vadose zone and ground water contamination at hazardous waste contaminated sites. Designation: D6235-04. American Society for Testing Materials, Philadelphia

ASTM (2010b) Standard guide for using the gravity method for subsurface investigation. Designation: D6430-99. American Society for Testing Materials, Philadelphia

ASTM (2010c) Standard guide for using the direct current resistivity method for subsurface investigation. Designation: D6431-99. American Society for Testing Materials, Philadelphia

ASTM (2010d) Standard guide for using the seismic reflection method for shallow subsurface investigation. Designation: D7128-05. American Society for Testing Materials, Philadelphia

ASTM (2011a) Standard guide for using the seismic refraction method for subsurface investigation. Designation: D5777-00. American Society for Testing Materials, Philadelphia

ASTM (2011b) Standard guide for using the surface ground penetrating radar method for subsurface investigation. Designation: D6432-11. American Society for Testing Materials, Philadelphia

ASTM (2011c) Standard guide for selecting surface geophysical methods. Designation: D6429-99. American Society for Testing Materials, Philadelphia

ASTM (2011d) Standard guide for using the metal detection method for subsurface investigation. Designation: D7046-11. American Society for Testing Materials, Philadelphia

Atekwana EA et al (2002) Geophysical investigation of vadose zone conductivity anomalies at a hydrocarbon contaminated site: implications for the assessment of intrinsic bioremediation. JEEG 7(3):103–110

Banwell GM, Parizek RR (1986) Potential for permeability characterization of crustal fracture zones using helium-4 and radon-222 concentrations in groundwater and soil gas. Geologic Society of America program with abstracts, Boulder, CO, 18(6)534

Beam P et al (1997) Lessons learned, a brief history of site characterization. HazWaste World/Superfund XVIII, Washington, DC, 2–4 Dec 1997

Benson RC (1997) ESC: just another new acronym or maybe something meaningful. Environ Eng Geosci 3(3):453–456

Benson RC, Yuhr LB (1987) Assessment and long term monitoring of localized subsidence using ground penetrating radar. In: Beck BF, Wilson WL (eds) Proceedings of the 2nd multidisciplinary conference on sinkholes and the environmental impacts of karst, karst hydrogeology: engineering and environmental applications. Orlando, FL. Balkema, Rotterdam, pp 161–169

Benson RC, Yuhr L (1992) A summary of methods for locating and mapping fractures and cavities with emphasis on the geophysical methods. In: Proceedings of the symposium on the application of geophysics to engineering and environmental problems, EEGS, Chicago, IL, 26–29 Apr 1992, vol 2, pp 471–486

Benson RC, Yuhr L (1996) An introduction to geophysical techniques and their applications for engineers and project managers, short course. Presented at SAGEEP, 28 Apr 1996, Keystone, CO

Benson RC, Yuhr L (1997) An overview of the geophysical methods for geotechnical and environmental site characterization. Presented at ASCE GeoLogan conference, Logan, UT, July

Benson RC et al (1984) Geophysical techniques for sensing buried wastes and waste migration. US EPA Environmental Monitoring Systems Laboratory, Las Vegas. Contract No. 68-03-3050

Benson R et al (1985) Correlation between field geophysical measurements and laboratory water sample analysis. Presented at National Water Well Association and Environmental Protections Agency 2nd national conference on surface and borehole geophysical methods in groundwater investigations, San Antonio

Benson RC et al (1998) ESC: how it differs from current state of the practice. In: Symposium on the application of geophysics to environmental and engineering problems, EEGS, Wheatridge, CO, pp 531–540

Benson RC et al (2003) Some considerations for selection and successful application of surface geophysical methods. In: 3rd international conference on applied geophysics, geophysics 2003, Orlando, FL, pp 8–12

Bogle FR, Loy K (1995) The application of thermal infrared thermography in the identification of submerged springs in Chickamaunga Reservoir, Hamilton County Tennessee. In: Beck BF (ed) Proceedings of the 5th multidisciplinary conference on sinkholes and the engineering and environmental impacts of karst: karst geohazards, Gatlinburg, TN. Balkema, Rotterdam, pp 415–424

Burger HR (1992) Exploration geophysics of the shallow subsurface. Prentice Hall, Englewood Cliffs

Butler DK (1977) Proceedings of symposium on detection on subsurface cavities. US Army Engineer Waterways Experiment Station, Vicksburg

Butler DK (1980) Microgravity techniques for geotechnical applications. US Army Engineer Waterways Experiment Station, Miscellaneous Paper GL-80-13

Butler DK (1994) Detection and delineation of subsurface cavities, tunnels and mines. Am Soc Civil Eng Ill Sect Geotech Lect Ser 3:1–29

Butler DK (ed) (2005) Near-surface geophysics. Society of Exploration Geophysics, Investigations in geophysics No. 13, Tulsa, OK

Butler DK (2008) Detection and characterization of subsurface cavities, tunnels and abandoned mines. In: Xia J (ed) Near-surface geophysics and human activity. Proceedings of the 3rd International Conference on Environmental and Engineering Geophysics (ICEEG). Science Press USA, Monmouth Junction, New Jersey, pp 578–584

Campbell CW, Singer M (2001) Application of thermography to groundwater monitoring at Arnold Air Force Base, Tennessee. In: Barry BF, Herring JG (eds) Proceedings of the 8th multidisciplinary conference on sinkholes and the engineering and environmental impacts of karst: geotechnical and environmental applications of karst geology and hydrology, Louisville, KY, pp 353–358

Cassidy DP et al (2001) The effects of LNAPL biodegradation products on electrical conductivity measurements. JEEG 6(1):47–52

Caterpillar (1988) Handbook of ripping, 8th edn. Caterpillar Inc., Peoria

Chardy A (2014) Port tunnel dream finally becomes reality. Miami Herald, Sunday 18th May 2014

DiMaggio (2007) Commentary: advancements and disappointments in geo-engineering and geo-construction – a 30–year reflection. Geo-Strata July-August, p 12, 13, 26

Florida Senate (2010) Interim report 2011-104 December. Committee on banking and insurance, issues relating to sinkhole insurance

Franklin AG et al (1981) Foundation considerations in sitting of nuclear facilities in karst terrains and other areas susceptible to ground collapse. U.S. Army Engineers Waterways Experiment Station Vicksburg Miss. Prepared for U S Nuclear Regulatory Commission NUREG/CR-2062

Freeman VK (1968) Geology of the Comstock-Indian Wells Area, Val Verde, Terrell and Brewster Counties, Texas. USGS Professional Paper 594-k

Greenhouse J et al (1995) An introduction to application of surface geophysics in environmental investigations. A short course presented a the symposium on the application of geophysics to environmental and engineering problems, environmental and engineering geophysical society, Orlando, FL

Hansen D (1975) Geological applications manual for portable gamma ray spectrometers. Exploranium, Toronto

Kaufmann RD, DeHan RS (2007) Microgravity mapping of karst conduits within the Woodville Karst Plain of North Florida. In:

Proceedings of the symposium on the application of geophysics to environmental and engineering problems, environmental and engineering geophysical society, April, Denver, CO, pp 1517–1526
Kincaid T et al (2005) Quantitative groundwater tracing and effective numerical modeling in karst: an example from the Woodville Karst Plain of North Florida. In: Beck B (ed) Proceedings of the 10th multidisciplinary conference on sinkholes and the engineering and environmental impacts of karst, vol 144. ASCE Geo-Institute Geotechnical Special Publication, San Antonio, pp 114–121
Kincaid T et al (2012) Demonstrating interconnection between a wastewater application facility and a first magnitude spring in a karstic watershed: tracer study of the southeast Farm Wastewater Reuse Facility, Tallahassee, FL. Florida Department of Environmental Protection, Florida Geological Survey Report of Investigation No. 111
Kratochvil GL (1989) Geotechnical evaluation of potential hazards to shuttle landing operations on Rogers Dry Lakebed Edwards Air Force Base, California. Astronaut Office NASA Lyndon B Johnson Space Center, Houston
Kratochvil G et al (1992) Fishing for fissures. The Military Engineer Jan–Feb, pp 66–68
Llopis JL et al (2005) Tunnel detection along the southwest US Border. In: Proceedings of the symposium on the application of geophysics to environmental and engineering problems, environmental and engineering geophysical society, pp 430–443
McCann DM et al (1997) Setting the standard for geophysical surveys in site investigation. In: McCann DM et al (eds) Modern geophysics in engineering geology, Geologic Society London, Engineering Geology Special Publication No. 12:3–34, London, England
Milsom J, Eriksen A (2013) Field geophysics, 4th edn, The geological field guide series. Wiley-Blackwell, West Sussex
Mitchell JK, Kavazanjian E Jr (2007) Geoengineering engineering for the 21st century. ASCE Geo-Institute, Geo-Strata July/August, pp 14–18
National Research Council (2000) Seeing into the earth, non-invasive characterization of the shallow subsurface for environmental and engineering application. National Academy Press, Washington, DC
Nazarian S, Diehl J (ed) (2000) The use of geophysical methods in construction. ASCE GeoInstitute Special Publication Number 108, Reston, Virginia
Olhoeft GR (2003) Quality control in geophysics. Geophysical technologies for detecting underground coal mine voids form, 28–30 July 2003, Lexington, KY
Reynolds JM (2011) An introduction to applied and environmental geophysics. Wiley, New York
Sauck WA et al (1998) High conductivities associated with an LNAPL plume imaged by integrated geophysical techniques. JEEG 2(3):203–212
Saunders W et al (1999) Selecting surface geophysical methods for geological, hydrological, geotechnical, and environmental investigations: the rational for the ASTM provisional guide, ASTM D 6429-99(2011). In: Durham GM, Marr WA (eds) Proceedings for field instrumentation for soil and rock, ASTM STP 1358, 18–19 June 1998, Atlanta, GA, pp 392–403
Schmidt W (2005) Geological and geotechnical investigation procedures for evaluation of the causes of subsidence damage in Florida. Florida Geologic Survey Special Publication No. 57, FGS Tallahassee, Florida
Sharma PV (1997) Enviromental and engineering geophysics. Cambridge University Press, Cambridge
Stephenson JB et al (2003) Environmental investigation of a diesel spill in a karst area of Kentucky. In: Barry BF (ed) Proceedings of the 9th multidisciplinary conference on sinkholes and the engineering and environmental impacts of karst, ASCE Geo-Institute Geotechnical Special Publication No. 122, pp 415–425
Straley D (2008) How thermal imaging of caves got its start. NSS News 66(5):16–17
Technos, Inc. (1988) Final report, ground penetrating radar investigation of the shuttle landing zone, Rogers Dry Lakebed, Edwards Air Force Base, California. Prepared for NASA's National space transportation system, June, Consultants report
Thompson J, Marvin M (2006) Experimental research using thermography to locate heat signatures from caves. In: Proceedings of 17th national cave and karst management symposium, Albany, NY, November, pp 102–115
Titcomb EP Jr, Keeton JM (1984) Sinkhole activity in the vicinity of the Sunny Point, NC Military Ocean Terminal. In: Beck BF (ed) Proceedings of the 1st multidisciplinary conference on sinkholes: their geology, engineering and environmental impact, Orlando, FL. Balkema, Rotterdam, pp 344–352
Upchurch SB et al (1987) Spatial and temporal controls on gross-alpha radiation, central Florida phosphate district, natural radiation in Florida. In: Proceedings of Natural Radiation and Technologically Enhanced Natural Radiation in Florida, Clearwater, Florida, pp 50–61
Ward SW (ed) (1990) Geotechnical and environmental geophysics. Investigations in Geophysics No. 5, Society of Exploration Geophysics Tulsa, Oklahoma
Ward AW et al (1992) Geologic setting of east Antelope Basin with emphasis on fissuring on Rogers Lake, Edwards Air Force Base, Mojave Desert, California. In: Prince KR et al (eds) US Geological Survey Subsidence Interest Group Conference, Edwards Air Force Base, Antelope Valley, California, USGS Open File Report 94-532, pp 35–37
Whiteley RJ (1983) Recent developments in the application of geophysics to geotechnical investigations. In: Ervin M (ed) In-situ testing for geotechnical investigations. Balkema, Rotterdam, pp 87–110
Wightman WE et al (2003) Applications of geophysical methods to highway related problems. Federal Highway Administration, Publication No. FHWA-IF-04-021, Lakewood, Colorado
Wisenbaker M et al (2007) Explorers connect America's longest underwater cave to Wakulla Springs. NSS News 65(12):4–11
Yuhr L (1998) Managing data and its integrity in the ESC process, Symposium on the application of geophysics to environmental and engineering problems, EEGS, 22–26 Mar 1998, Chicago
Yuhr L et al (1993) Characterization of karst features using electromagnetics and microgravity: a strategic approach. In: Symposium on the applications of geophysics to engineering and environmental problems, Environmental and Engineering Geophysical Society, Englewood, CO, pp 209–228
Yuhr L et al (2008) Mapping and drilling of conduits within the Woodville Karst Plain of North Florida. In: Yuhr LB, Alexander EC Jr, Beck BF (eds) Proceedings of the 11th multidisciplinary conference on sinkholes and the engineering and environmental impacts of karst. ASCE Geotechnical Special Publication No. 183, pp 340–350

Invasive Methods

17

Abstract

By now the conceptual model has become site-specific with increased details provided by the results from the surface geophysical data. Both background and anomalous conditions have been defined along with their lateral extent. Our focus now shifts to obtaining site-specific geologic details along with engineering, hydrologic and contaminant data. We can now, with a reasonable degree of certainty, develop an effective invasive program and locate direct push measurements, borings, sampling points, trenches and excavations in a logical way so that they provide representative measurements of background and anomalous geologic conditions. Furthermore, by increasing the spatial representativeness of this data, we can obtain sufficient data using fewer invasive measurements.

17.1 Introduction

There are a wide variety of invasive methods to choose from. They include minimally invasive methods such as direct push methods and cone penetrometer testing (CPT) to invasive drilling methods, along with excavations and trenches (Fig. 17.1). These methods can provide a wealth of information and data whether for geotechnical, groundwater resources, or environmental contaminant projects. Recent decades have seen advancements in direct push technology and drilling along with improved instrumentation for in-situ testing. These invasive methods and resulting samples provide a high level of detailed data for the site characterization process. Nielsen (2006) provides more details on these topics.

Traditionally, the location of invasive measurements and the spacing of measurements along a profile line or spacing over a grid is always an issue. Many rules of thumb provided in the literature depend upon whether the scope of the investigation is an initial exploration or a detailed investigation. Regularly spaced borings along an alignment or a fixed number of boreholes for a given area are often used in exploration. The spacing is then decreased when erratic conditions are encountered. In any case, all invasive measurements should be located in a manner to provide the best possible data needed for the investigation (Hvorselv 1949). Another common practice has been to drill one boring in each corner and one in the middle of a small site (Fig. 12.10). Regulatory issues will sometimes tend to dominate the location of monitoring wells with some states requiring layouts on a grid pattern while others impose minimum numbers of wells based upon area (Sara 2003). These are not unreasonable approaches when no other information is available or likely to be obtained.

Without prior geologic knowledge the question of how many, what spacing, and how deep is not obvious, particularly in karst. In addition, the interpolation or extrapolation of boring data can only be trusted, if the spacing between borings is less than the lateral changes in geologic conditions (which is most often an unknown). However, based upon the surface geophysical measurements, we should have general understanding of the geologic spatial variability as well as expected geologic features at this point. It is now possible to locate invasive measurements more effectively.

17.2 Direct Push Methods

The direct push methods include both percussion (using a rod driven into the ground with blows from a hammer device) and the cone penetrometer (using a rod pushed into the ground

R.C. Benson, L.B. Yuhr, *Site Characterization in Karst and Pseudokarst Terraines*,
DOI 10.1007/978-94-017-9924-9_17

with hydraulic ram using the weight of the truck). Both of these systems are in common use today and provide minimally invasive characterization of subsurface conditions. These two techniques are applicable in areas that have an unconsolidated soil cover in which the probes can penetrate.

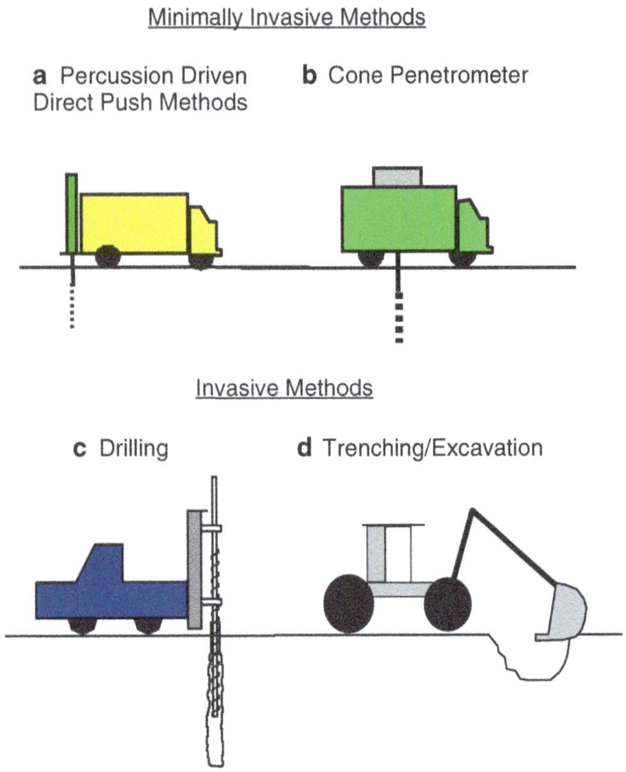

Fig. 17.1 Methods of minimally invasive (**a** and **b**) and invasive investigation (**c** and **d**)

Several ASTM standards have been developed which cover cone penetrometer testing techniques. These standards were not developed for investigating karst sites specifically, but they do provide general information on the technology. Sara (2003), Hunt (2005), McCall et al. (2006), and Ruda and Farrar (2006) all provide further details of these methods and their use.

17.2.1 Percussion Driven Direct Push Methods

Percussion driven direct push technology can be used to rapidly characterize subsurface conditions in unconsolidated materials. Direct push measurements are made by driving a small diameter probe into the sediment using a hammer device. The direct push percussion equipment is commonly mounted on the back of a small truck (Fig. 17.2a) and can be adapted in a variety of ways for access to unique site conditions (Fig. 17.2c).

The direct push methods eliminate any drilling fluids or soil waste from the hole. This can be advantageous when working at hazardous waste sites. The small diameter holes can be easily grouted when the probe is extracted. Depths of 9–15 m are commonly obtained and in ideal conditions maximum depths of up to 30 m or so have been reached. Depth of penetration is limited by the presence of dense sands, gravel and cobbles. Penetration rates of 0.5–7 m/min are typical.

The type of data acquired includes the rate of advancement into the sediment (blows/m or speed of penetration). This is an indicator of the materials resistance to penetration by the probe and can be used to help characterize the materials encountered and can often be tied to stratigraphy.

Fig. 17.2 Typical direct push equipment mounted on a truck (**a**), sediment samples obtained using direct push (**b**) and a modified portable direct push system (**c**)

Soil samples can be obtained (Fig. 17.2b) and an electrical tip can be used at the end of the probe can measure the electrical conductivity of the material to identify clays or pore fluids (Christy et al. 1994; Schulmeister et al. 2003). These push technologies are now commonly used for environmental site characterization, providing sampling of soils, water and contaminants. See ASTM D6001 a guide for direct push technology (ASTM 2012c).

The direct push method using an electrical conductivity tip was used at the Central Nevada Test Site (CNTA) to characterize sediments in old drilling mud pits. This area was used in 1968 to test atomic bombs by detonation deep underground. A large mud pit (about 1.2 ha) was a result of drilling a 2.4 m diameter borehole, 975 m deep for a test detonation. The mud pit materials were semi-consolidated and would flex as one jumped upon the surface. The pit would have likely liquefied with the load of a truck and the vibration of a percussion driven sampler. As a result, sampling was carried out using a modified direct push system that only required two people using a portable percussion hammer to obtain samples (Fig. 17.2c).

In west central Florida the direct push percussion method using an electrical conductivity tip was used extensively at a 52 ha EPA superfund site. A thin layer of clay over weathered rock formed a semi-confining layer between the sandy surficial aquifer and the underlying limestone Floridan aquifer at a depth of 3–6 m. The presence of this layer, its thickness and continuity were critical to the sinkhole risk assessment and contaminant migration. Where this layer was missing or dipped deeper was potentially an indication of sinkhole activity or paleokarst conditions.

Direct push technology was used in conjunction with ground penetrating radar to characterize the semi-confining layer. Radar was used to acquire shallow data over the entire site mapping the top of the semi-confining layer identifying lows or missing sections. While the radar data provided good lateral coverage of the site showing the top of this layer, it did not provide details through the semi-confining layer due to its clay content. Direct push technology with an electrical conductivity tip was then used to provide data through the semi-confining layer. The electrical conductivity tip was very effective at characterizing the semi-confining layer due to its clay content. The direct push measurements provided a confirmation of the depth to the top of semi confining layer along with its thickness, relative clay content, and lateral continuity. Figure 17.3 shows the correlation between the radar data and the direct push electrical conductivity data. The radar data shows a localized low area in the semi-confining layer. Direct push electrical conductivity data indicates the presence of clay on both sides and within the low area. Higher clay content is indicated by increasing electrical conductivity values (to the right). This case history is provided in detail in Chap. 27 of this book.

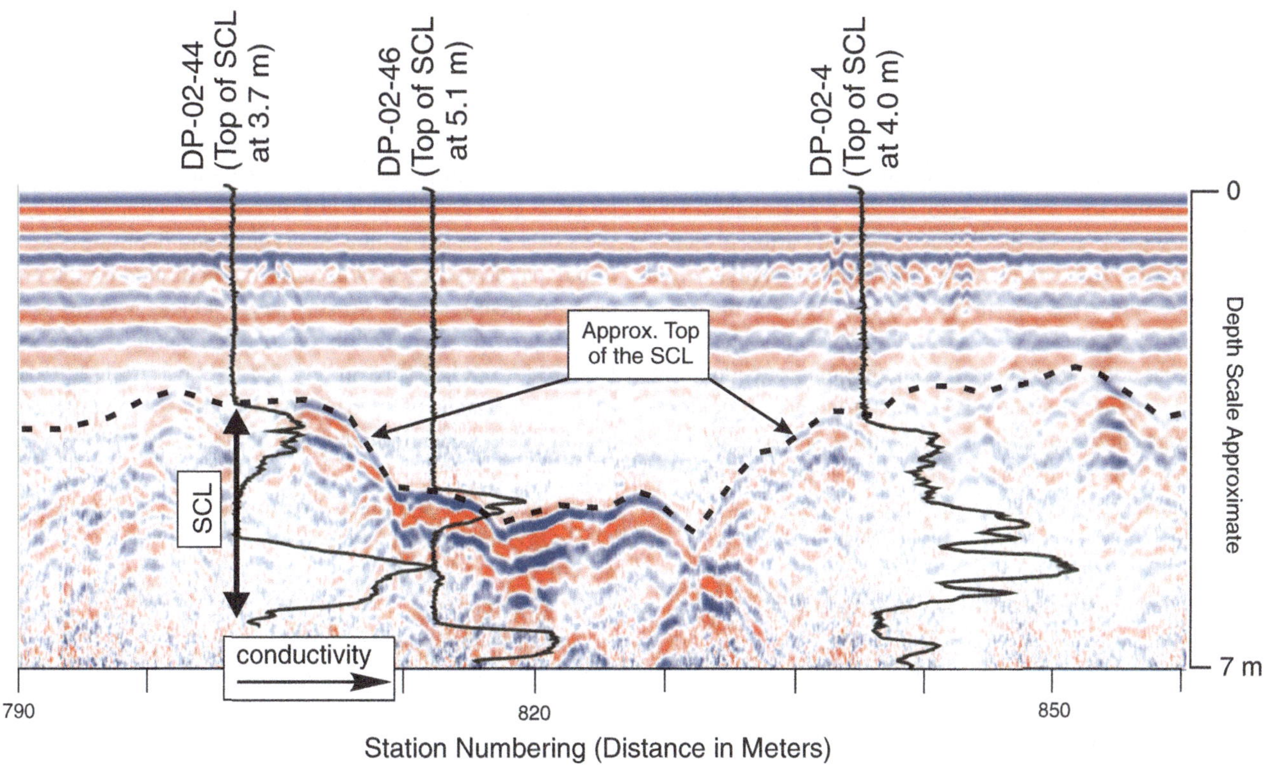

Fig. 17.3 Direct push electrical conductivity data plotted over radar data used to assess the presence, thickness and clay content of the epikarst zone which included a semi-confining layer

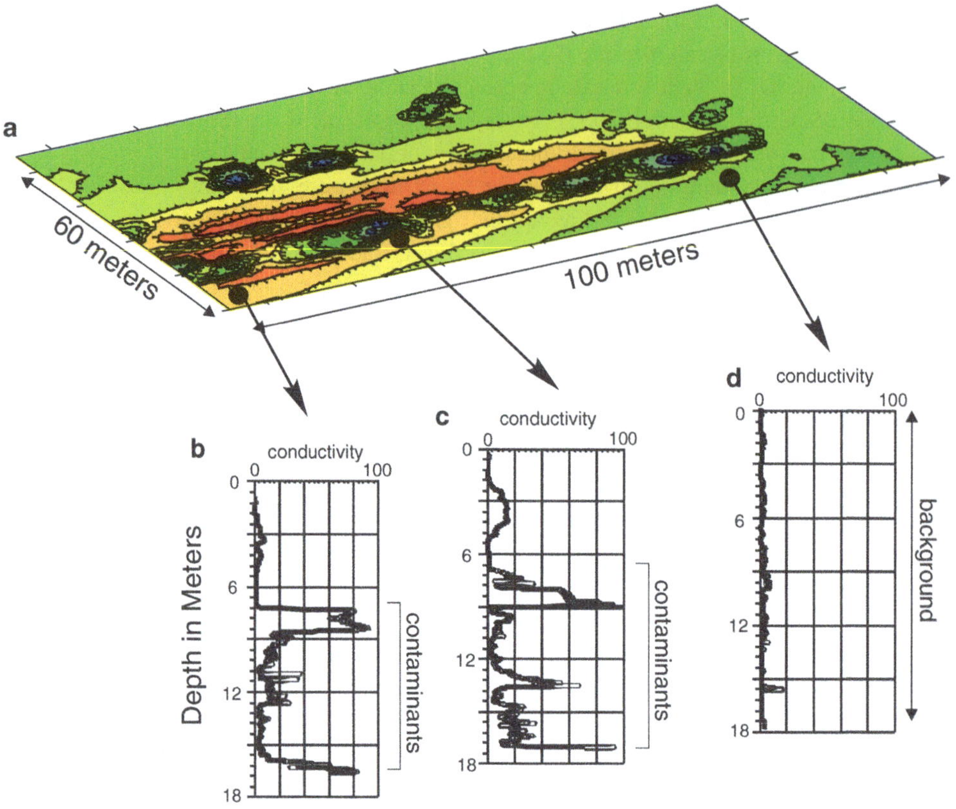

Fig. 17.4 Contour of electrical conductivity (EM31) data over a buried waste site and direct push electrical conductivity plots showing the vertical extent of contaminants (**b** and **c**) as well as background conditions (**d**)

At the Savanna River Site (SRS), in southeast South Carolina, a low-level radioactive waste site, an electromagnetic survey using an EM31 instrument mapped the lateral extent of buried contaminants over a small area that had a high conductivity signature (Fig. 17.4a). The direct push method with an electrical conductivity tip was then utilized to characterize the vertical extent of the contaminants. Two of the pushes (Fig. 17.4b, c) clearly identified the presence and depth of a high conductivity zone (contaminants) within the plume identified by surface electromagnetic measurements. One of the pushes (Fig. 17.4d) clearly identified background conditions (lack of contaminants) outside of the plume identified by surface EM measurements. A field gas chromatography unit was used to analyze soil samples and laboratory analysis of water samples was used to identify and characterize the contaminants.

17.2.2 Cone Penetrometer Testing (CPT)

Cone penetrometer testing is carried out by pushing a small diameter probe into the sediment using the weight of the truck and a hydraulic ram. The hydraulic pressure and the rate of advance into the sediment can be used to characterize the materials encountered. Figure 17.5 shows a typical CPT rig being used to evaluate soil conditions at a contaminated site in a karst setting.

The traditional cone penetrometer data (Fig. 17.6) includes sleeve friction, tip pressure, a ratio between the sleeve friction and tip pressure, and pore pressure as well as depth, time, and pushing force, which are measured as the probe is pushed hydraulically into the ground. Penetration rates of a <1 m per minute are typical and depth of penetration depends upon soil conditions but is limited by the presence of dense sands, gravel, and cobbles. McCall et al. (2006) provide more details on this topic along with ASTM D-5778 (ASTM 2012b) and ASTM D-6067 (ASTM 2010a).

A wide range of measurement options supplement the regular CPT measurements such as:

- Seismic probe to obtain P wave velocity.
- Hydrocarbon detection
- Video
- VOC detection
- Soil moisture/resistivity profiles
- Soil, gas, and water sampling
- Gamma radiation profiles

At a site in west central Florida, CPT data was an integrated part of a site characterization. The site had a thick

Fig. 17.5 A CPT rig is being used to evaluate soil conditions at a contaminated site in a karst setting (**a**) and a close-up of the probe beneath the truck (**b**)

cover of clean quartz sand that provided an ideal application for ground penetrating radar and CPT methods. The radar data provided a means of rapid cost effective site coverage to identify potential problem areas that showed signs of sinkhole activity or the presence paleokarst. A cone penetrometer was then used to further evaluate the potential problem areas by identifying loose materials, voids and changes in conditions.

As a means of data correlation, a test site was selected in background conditions without any active or paleokarst based upon the radar data. Data obtained included a boring with standard penetration tests (SPT) values, a geologic log and geophysical logs along with CPT measurements made immediately adjacent to the borehole. Figure 17.7 shows the results of this test data.

This example clearly illustrates the impact of scale on a variety of measurements. The drilling data sampled every 1.5 m and describes a 0.6 m thick clay layer at a depth of 14 m based upon a visual inspection of the sample from the hollow stem auger. The SPT acquired during drilling and sampling indicated loose materials immediately above this clay layer. The natural gamma geophysical log provides essentially continuous data throughout the depth of the borehole and a measured volume of about 30-cm diameter around the sensor. This method detects variable clay content from 11 to 17 m based upon natural radiation associated with clays. The CPT friction ratio data indicates a variable clay content starting at a depth of 9 m and continuing to 14 m where refusal was encountered. Note that each set of data was acquired correctly and is accurate, but tells a slightly different story.

A specialized cone penetrometer was used for a very shallow investigation of surface fissures along the earthen runways at Edwards Air Force Base (EAFB) in California. The fissures were a result of groundwater withdrawal from a deep aquifer (Fig. 7.10). Radar surveys had been carried out along the long runways to locate possible fissures before they broke

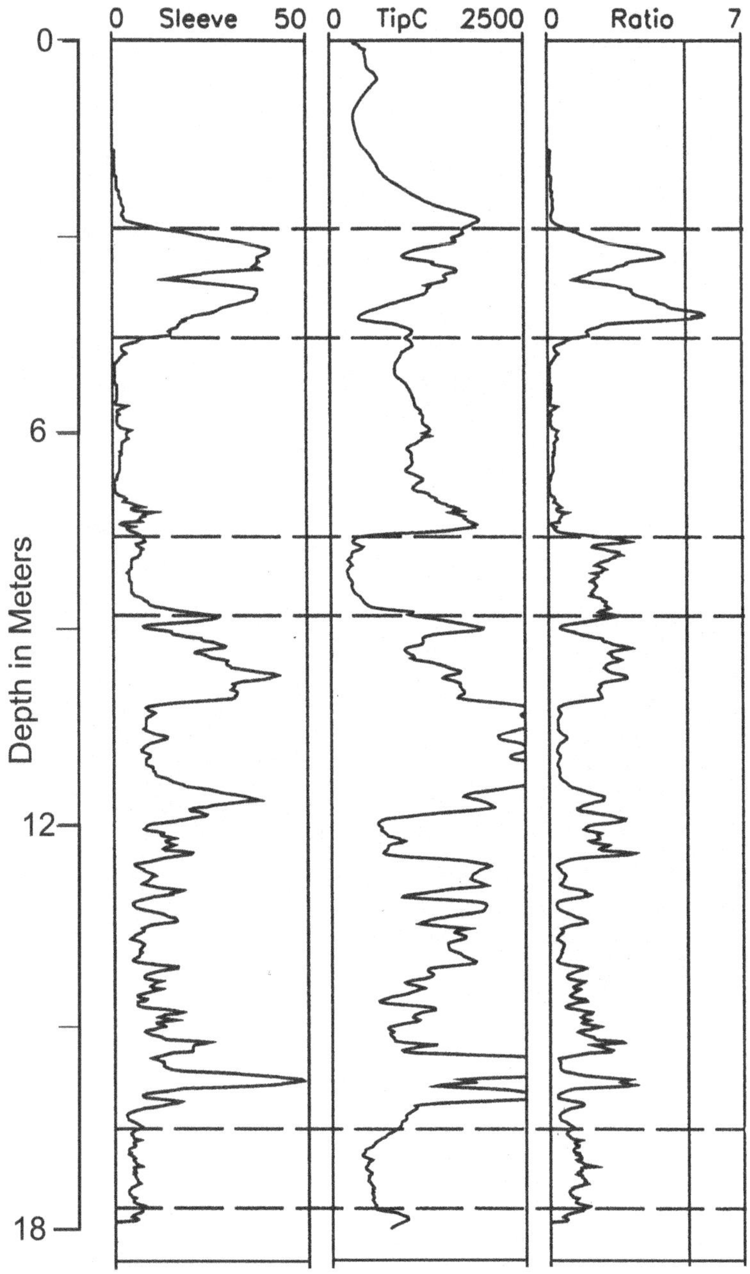

Fig. 17.6 An example of CPT data showing sleeve friction, tip resistance, and friction ratio

through to the surface (Fig. 16.34). Ten parallel survey lines were run along each of four runways for a total of 362 km of radar data. Radar data was used to provide an approximate location of the fissures (within <1 m). Additional data was acquired at these locations using a specialized cone penetrometer (Kratochvil 1989; Kratochvil et al. 1992).

An automated cone penetrometer, mounted on the front of a truck (Fig. 17.8) would push the probe to depth (about 30–60 cm) at a station, then the probe assembly would move a short distance and repeat the process. Up to seven measurements could be made over a distance of about 2 m without moving the truck. This CPT data provided a detailed assessment of subsurface conditions by using the recorded tip pressure in PSI (pounds per square inch). This specialized CPT rig could rapidly identify the lateral boundary of the fissure (a loose zone of material). The data in Fig. 17.9a shows one of the seven measurements encountering a localized loose zone. This indicated the fissure's exact location and that the fissure was narrow, probably in its early stage of development. Figure 17.9b shows a second set of data with a wide range of soil conditions from three of the six pushes, indicating that the fissure was wider and more mature. Once the location and character of a fissure was identified, repairs were then carried out. Each site was trenched, backfilled, and

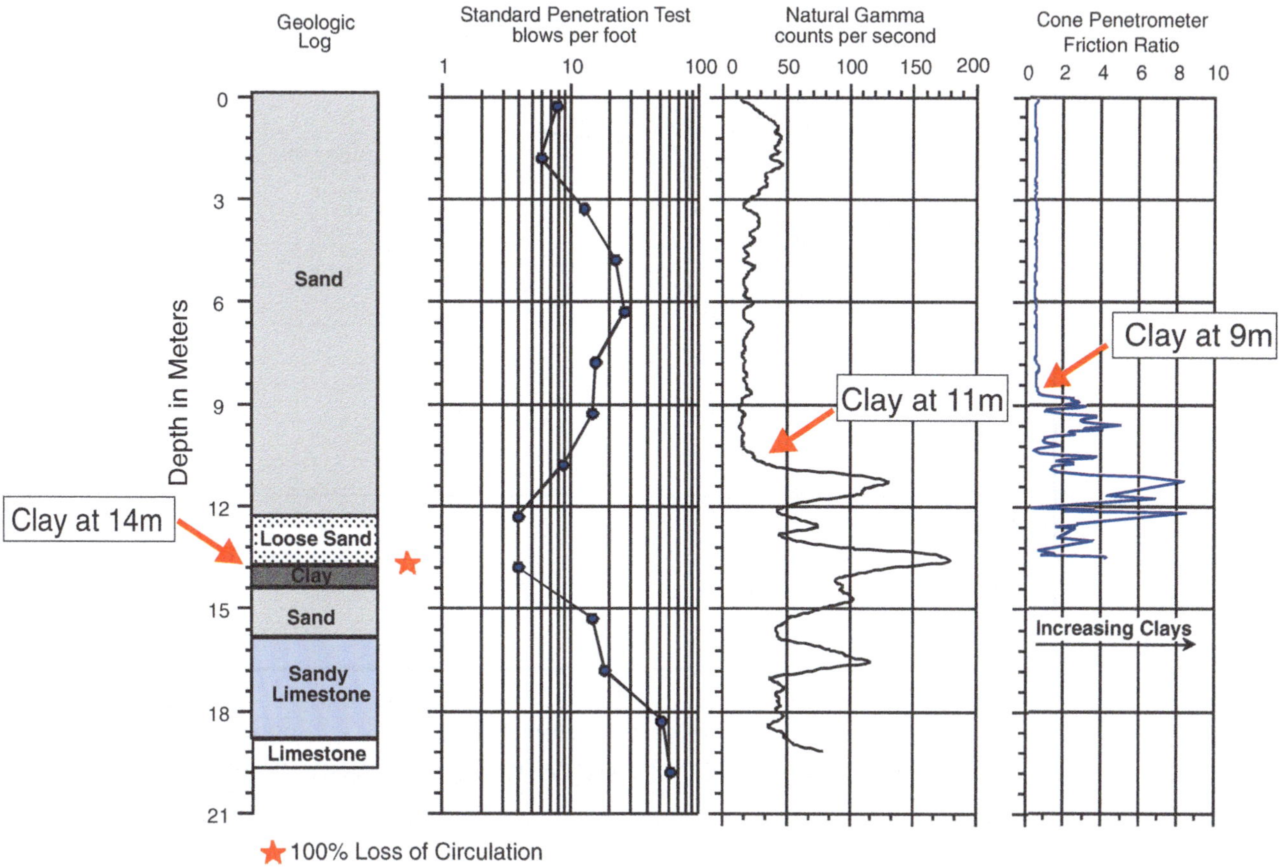

Fig. 17.7 Test site showing geologic log, SPT data and natural gamma log from the sample borehole. The cone penetrometer data is adjacent borehole about 1 m away. Note 100 % loss of fluid in zone of loose sand

Fig. 17.8 Photo of the specialized shallow CPT system used to evaluate the fissures at Edwards Air Force Base (*EAFB*)

compacted. Khedr et al. (1985) provides details on this automated cone penetrometer system.

17.3 Borings

Borings have been and continue to be one of the primary methods for site characterization in geotechnical investigations and environmental studies. Boring logs remain the mainstay and are often a critical piece of data in solving the geologic puzzle for the site characterization process. Borings provide soil samples and standard penetration tests (SPT) "N" values, or rock core (RQD) values as well as water levels and samples. Standard driller's or geologic logs provided a description of geologic conditions, depth of contacts or stratigraphic changes. Depending upon the type of drilling method used, the logs may include rate of drilling, rod drops or loss of drilling fluid, which can all be indications of karst conditions. In addition borings also provide access for running a variety of geophysical logs, making in-situ geotechnical and hydrologic measurements of the soil and rock, and may eventually be used to install monitoring wells or piezometers. Borings also provide samples of the soil or rock for our geologic descriptions and for laboratory testing,

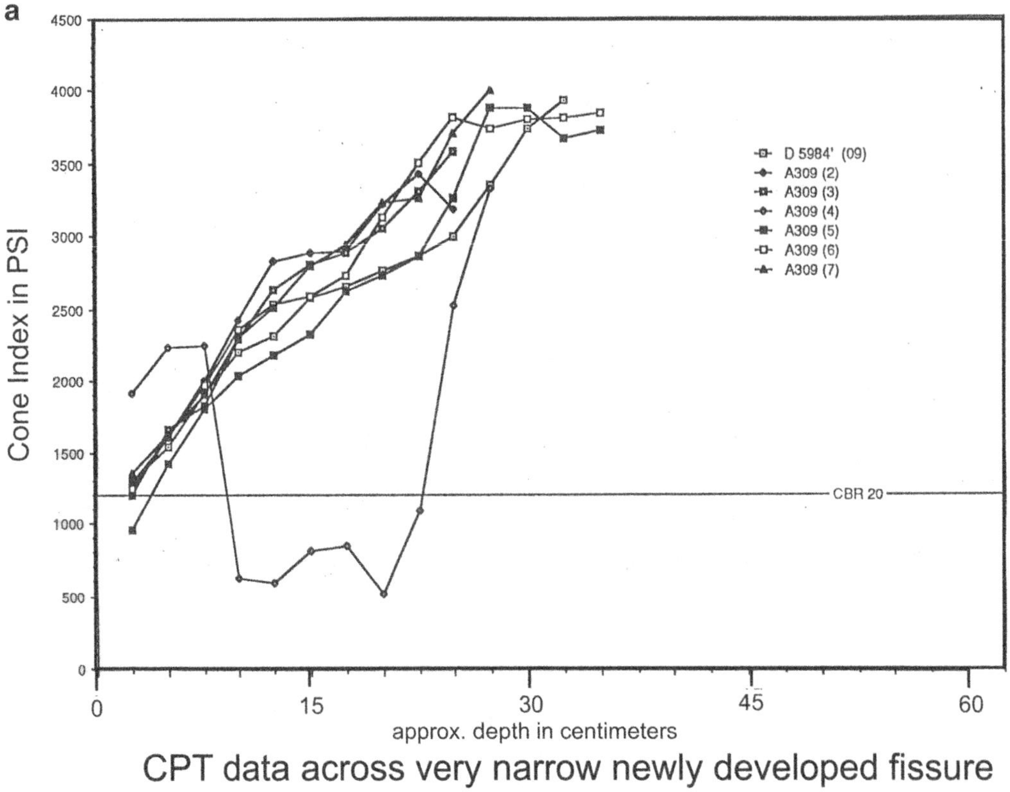

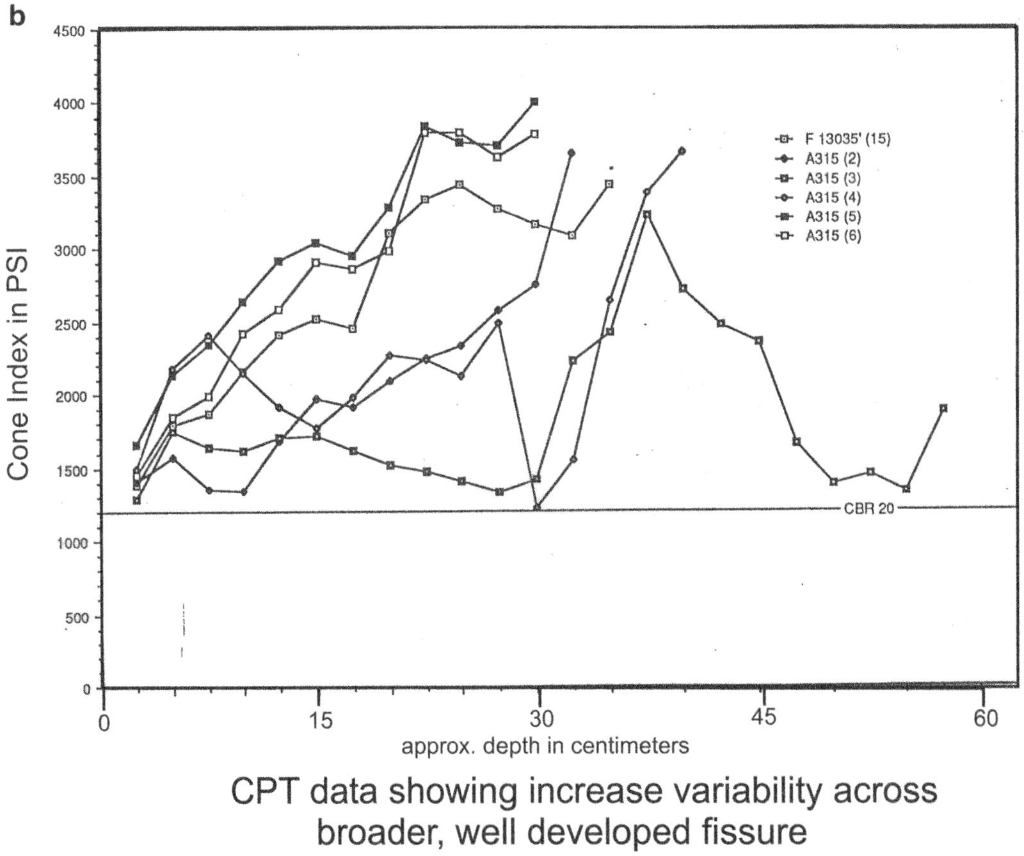

Fig. 17.9 Two examples of CPT data from EAFB. One set of data shows a narrow fissure (**a**) while another set of data shows a wider fissure (**b**) with considerable variation

providing the engineer with much of the data necessary for design.

Borings are often the first method to be used to determine top of rock profile or rockhead. Hencher (2012) discusses the problem of determining the rockhead. This approach is problematic in karst due to the highly variable conditions often encountered (Fookes 1997). It is here that we find unpredictable variations in the depth of rock, its degree of weathering, frequency of fissuring, buried pinnacles, the extent of isolated loose blocks of rock and undercut pinnacles supported only by the surrounding soil. It is often difficult to determine the difference between a large boulder, a pinnacle, and the upper surface of continuous in-situ massive rock (Waltham and Fookes 2003).

17.3.1 A Drilling Plan

Ideally drilling would not be done until the desk study and site walkover had been completed, a preliminary conceptual model developed and the surface geophysical investigation carried out. There are always exceptions. In an area where there is little or no geologic data available (a most unusual situation since there will be at least regional data in most areas) it may be necessary to drill a few borings early in the field program to establish the general geologic conditions. In such cases, locations are based upon the best available data at the time.

Drilling can be an expensive procedure and the information obtained from each borehole must be maximized. The objective of an effective geotechnical or environmental site characterization should be to minimize the number of boreholes, piezometers and wells while maximizing their representation of background and anomalous geologic conditions. Always avoid drilling a hole or placing a well unless you are reasonably sure of its location and purpose. A field plan for drilling should be based upon all prior data and the resulting conceptual model. The drilling plan should identify not only locations and depths of borings but the nature of the data required and possible anomalous conditions expected along with other long-term use of the boring and subsequent data.

Selection of the most appropriate drilling method should be made by the engineer or geologist along with a knowledgeable drilling contractor and must take into consideration site conditions, objectives, availability of equipment, personnel, and costs. Often, the best approach when selecting a drilling method is to talk to drilling firms who are experienced with the local geologic conditions. Sometimes choosing a lesser level of drilling technology is better if one can find a driller who is familiar with the local geology and has proven track record. Drilling is, in part, an art and its success is dependent upon good practice and skill of the driller. An experienced driller is an invaluable asset to the site investigation effort.

In all cases, flexibility must be maintained in the drilling program to accommodate changes in conditions that are discovered as the site characterization effort proceeds. This includes changing locations of borings and adding additional borings where unusual conditions have been identified. These decisions must be the responsibility of the site characterization team who are in the field.

17.3.2 Drilling Methods

There are a wide variety of drilling methods (Table 17.1). They range from simple hand probes or soil augers to a wide variety of jetting, percussion and rotary drilling methods.

Table 17.1 Common direct push and drilling methods

Method	Use in soil or rock	Use of casing	Use of drilling fluids	Geologic samples
Direct push methods				
Direct push (percussion)	Soils, soft rock	Possible	N	Undisturbed
Cone penetrometer testing (CPT)	Soils, soft rock	Possible	N	Undisturbed
Jetting or percussion methods				
Wash boring	Soils	N	Water	None
Cable tool	Soils and rock	Y	Water	Chips
Downhole hammer	Rock	Y	Air, water, foam	Chips
Auger methods				
Hand probes and augers	Soils	N	No	Disturbed
Solid stem augers	Soils, soft rock	N	No	Disturbed
Hollow stem augers	Soils, soft rock	Y-the auger itself	No	Disturbed
Rotary methods				
Direct rotary	Soils and rock	Optional	Air, water, mud, foam	Chips
Reverse circulation	Soils and rock	Optional	Air, water mud foam	Chips
Dual tube reverse circulation	Soils and rock	Y	Air, water mud foam	Chips
Rotosonic	Soils and rock	Y	Air, water, or mud optional	Undisturbed

With any group of methods, they all have their advantages and limitations. Some drilling methods work well in unconsolidated materials and others in solid rock. Some of the drilling methods provide rapid penetration, but are destructive, disturbing the sample as a result of the drilling process. Other drilling methods are relatively slow but can recover relatively undisturbed samples of soil or rock. Drilling fluids (air or water and sometimes additives) are commonly used to carry cuttings to the surface and maintain a clean, open borehole.

In some cases, the borehole can be advanced without concern of collapse within the borehole. In other cases borehole collapse is a common problem (such as with thick unconsolidated sands) and must be considered when selecting the drilling method. An unstable borehole is maintained by the use of drilling mud or by advancing casing as the hole is drilled. Casing effectively stabilizes the hole from collapse, but limits access to the formation for direct tests (ie packer tests and certain geophysical logs), and observations (by video, acoustic or optical televiewer), or sampling of fluids and flow.

Environmental aspects of installing piezometers and monitor wells, geotechnical measurements and instrumentation as well sampling of water and contaminants will often tend to dominate drilling requirements. In some cases the method of drilling will be dictated by what equipment is available. These factors and more need to be considered when selecting the appropriate method for a given geologic conditions and project objective.

Poor drilling procedures, casing installation, installation of instrumentation, can lead to subsequent problems with the data acquisition, and monitoring. While specifications are easily written on paper, it is another thing to carry out the necessary details in the field and to verify them once in place.

Sara (2003), Hunt (2005), and Ruda and Farrar (2006) provide further details on drilling methods and their use. Several ASTM standards have been developed which cover hollow stem augering, SPT and split barrel sampling as well as rock core drilling and sampling techniques (ASTM 2006, 2011, 2014).

17.3.2.1 Jetting Methods

Wash boring or probing with a hand probe and water under pressure is a simple means of probing shallow soil conditions without a drill rig and is readily adapted for work in shallow water. The method has been used to define top of clay or rock, and probe mud thickness of sinkhole ponds and lakes. A small centrifugal pump provides a jet of water from a 2 cm steel pipe, which is pushed into the sediments until it encounters resistance or rock.

A simple wash boring probe was used to probe the bottom of a circular pond that appeared to be a sinkhole. Ground penetrating radar data obtained up to the edge of the pond indicated strata dipping toward the center of the pond. A profile line of wash borings was completed from the pond edge to the pond center. The probes could extend to a maximum depth of 6 m. Refusal was encountered at the edge of the pond and progressively got deeper toward the center. The combination of the wash borings and radar data confirmed our interpretation that this was a sinkhole. See case history in Chap. 27.

17.3.2.2 Percussion Methods

A variety of percussion methods of drilling are available including repeatedly dropping a cutting tool (cable tool) or by the use of a surface or a downhole hammer. These are destructive drilling methods where only chips are recovered. However, they are generally a fast and efficient means of drilling.

For example, small sinkholes were occurring at a military site on the cliffs of Guam near critical radar structures. The site was located at an elevation of about 180 m above sea level with a water table near sea level. A detailed radar survey was done to identify the lateral locations and distribution of cavities at the site. See Sect. 16.5.3.3 for further discussion along with Figs. 16.21 and 16.22. A percussion drill rig was used to investigate selected radar anomalies. The drilling method was quick and effective since no samples were necessary. Voids were indicated due to loss of air circulation and rod drops. Each boring was then logged with a borehole video camera to visually document the cavities encountered. These cavities were typically 0.5–2 m in diameter with a few larger ones of 2.5–3.5 m in diameter. Once identified the cavities were grouted up.

17.3.2.3 Auger Methods

Auger drilling rigs are commonly available and are used in many site investigation efforts. Augers can range from hand augers (Fig. 17.10a) to small portable augers (Fig. 17.10b) to large truck mounted augers with SPT capabilities (Fig. 17.10c).

Solid stem augers provide a rapid means of drilling and sampling of unconsolidated materials. Sampling of disturbed soil is done as the auger spins-up the sample or by pulling the auger out of the ground to extract the sample on the auger. Hollow stem augers are one of the most commonly used and readily available methods for drilling and sampling in unconsolidated material. Standard penetrating tests (SPT) "N"-values are often obtained as part of the drilling and sampling process. Augers are unable to penetrate very dense compacted materials and boulders. No fluids are used and the auger remains in place to maintain the open borehole. More details on auger drilling are provided in various ASTM guidelines (ASTM 2006, 2009, 2011).

Large diameter augers provide a means of developing holes up to 1 m in diameter. Besides their common use for poured in-place piles, these larger holes are sometimes used to provide direct access for a person to observe and log geologic conditions.

Fig. 17.10 A variety of auger devices are used. A hand auger (**a**), a small portable auger rig (**b**) and a common hollow stem auger (HSA) drill rig (**c**) which are commonly used for drilling in unconsolidated material

17.3.2.4 Rotary Methods

Rotary drilling using air or water as a drilling fluid can be used in all types of unconsolidated materials and rock. Air rotary drilling can function in areas in which a supply of water is not available. Rotary percussion drilling uses both a rotary and percussion of the drill bit to fracture the material. Geologic sampling is by cuttings, which are carried upward by return air or water, which flows between the borehole wall and the drill pipe and aids in maintaining a stable borehole wall. Caving in unconsolidated materials can be avoided by the use of drilling mud or by advancing casing as the hole is drilled.

Dual Tube Reverse-Circulation

Dual-tube reverse circulation rotary provides very rapid drilling in both unconsolidated materials and rock formations (Fig. 17.11a). This method of drilling is useful in difficult conditions and reducing lost circulation by providing a casing as part of the drilling process. The inner tube rotates the drill bit and the outer tube functions as the casing. Drilling fluid flows down between the two tubes, out around the drill bit and up the inner tube. As a result the method is useful in fractured rock, boulders, and cavernous conditions because it minimizes problems with lost circulation and borehole stability. Sampling is by chips returned to the surface by air, water or mud (Fig. 17.11b, c).

Dual tube reverse circulation drilling is commonly used where thick sands or loose sediments are a problem. In many areas of Florida, thick loose sand and sediments are known to collapse into the borehole when casing is removed. Where such problems are known to occur, the borehole can be drilled by the dual tube method and with the casing still in the ground selected geophysical logs can acquire data through the steel casing. In this manner, we have both the geologic log based upon chips from drilling and the continuous geophysical logs (natural gamma, gamma-gamma density, and neutron porosity) that can be run through steel casing providing additional geologic data from the boring for correlation.

Rotosonic (Vibratory) Methods

Rotosonic drilling uses vibration of 50–50 Hz of the drill stem in addition to downward pressure and rotation of the bit drill to rapidly drill through most materials. The sonic method has proven fast and effective in a wide variety of geologic environments. Drilling can be done in most unconsolidated materials and softer bedrock (sandstone, limestone, shale, and slate) and is very rapid. Harder rock can be drilled but at increased cost.

Long continuous cores (typically 3 m) with core size of 7.5–25 cm can be obtained. The core sample is removed from the core barrel into a plastic sleeve or stainless steel tray with a minimum of disturbance. Samples may also be collected with clear plastic liners or stainless steel split liners inside of the core barrel. The unconsolidated sample may be somewhat disturbed (by vibration and handling) but are essentially continuous. Figure 17.12 shows a soil core being removed from the drill rig at the Savanna River Site (SRS), screened for radiation using a Geiger counter and then boxed for storage.

An outer casing provides a means to keep to borehole from collapsing while the core is retrieved and minimizes cross contamination during drilling, if fluids are used. It also makes the installation of monitor wells or geotechnical instrumentation very efficient. Drilling can be done with or without fluids, (sometimes just water is used) and there are no drilling wastes. For more details see ASTM D6914 on sonic drilling (ASTM 2010c).

Fig. 17.11 A typical dual tube air rotary rig (**a**) commonly used when drilling in loose sediment. The outer tube acts as casing to maintain an open hole in loose sediments that are commonly found in Florida. A pan to catch the cuttings (**b**) is changed at regular intervals in order to identify cuttings from a certain depth range (**c**)

Fig. 17.12 A sediment sample from a rotosonic drill rig is being placed into a plastic sleeve (**a**), which is then monitored for radioactivity (**b**). The core is then labeled and stored (**c**)

17.3.3 Indications of Karst When Drilling

17.3.3.1 Drilling in Unconsolidated Materials

There are several indicators of subsidence or sinkhole activity when drilling in unconsolidated materials. These include low SPT or blow counts along with rod drop and fluid losses. SPT results are commonly found in driller's logs and provide a means of measuring the resistance of the unconsolidated materials to the penetration of the sampler using the blow count of a standard hammer. The "N" values (blows per foot) from SPT are useful for identifying loose or raveling zones in areas of suspected subsidence. More details on SPT measurements are provided in an ASTM guideline D1586 (ASTM 2011).

A slurry wall was being designed along the north side of a landfill in Florida. Drilling was completed along the proposed alignment with a spacing interval of about 15 m and went to depth of refusal. SPT data was collected and used along with soil samples to identify the top of a hard clay layer in which the bottom of the slurry wall was to be placed. This was a data set that appeared to have excellent data density. However, due to concerns with potential geologic variability in this karst setting, a surface geophysical survey was completed along the alignment using MASW method (mapping shear wave velocities). Figure 17.13 shows a contour of shear-wave velocities from the MASW data along with the corresponding SPT data from the drilling. The SPT data and the MASW data showed remarkable correlation. Sharp increases in blow counts (from less than 10 to >20) corresponded to an increase in shear wave velocity, roughly at 900 ft/s contour. While both sets of data are valid, the MASW data shows some of the variability not detected by the SPT data. For example, see stations 125 and 157 in Fig. 17.13.

17.3.3.2 Drilling in Rock

When drilling in rock, poor core recovery, rod drops and fluid loss are a good indication of the presence of highly weathered zones, open fractures, voids or cavities. These types of conditions are not uncommon and if they are small and isolated may not be of concern. However, a spatial concentration of such indications laterally or at a particular depth may be significant and should not be disregarded. It is then necessary to verify the significance of this data by other independent data such as geophysical logs or borehole imaging methods. See ASTM guideline D2113 for details on rock core drilling and sampling (ASTM 2014).

For example, initial drilling had been carried out for a new bridge into the Florida Keys. Borings were spaced at approximately 75 m intervals along 2.6 km to depths of 20–47 m with the majority extending to 30 m. The 34 boring logs had been plotted on a large sheet and indicated relatively uniform geologic conditions. Upon close review of details, 27 % of all rod drop and fluid loss had been noted in two adjacent boreholes. Two additional nearby borings accounted for an additional 34 % of all fluid loss, but this was not mentioned in the geotechnical report. This turned out to be one of the many indicators for an area of a large paleokarst collapse zone. This was a case of boring encountering indications of potential problems yet the data was buried within a large data set and went unnoticed. If these anomalies had occurred in widely spaced borings they probably would have little significance, but the fact that the data clustered spatially is what caught our attention. This case history is presented Chap. 26.

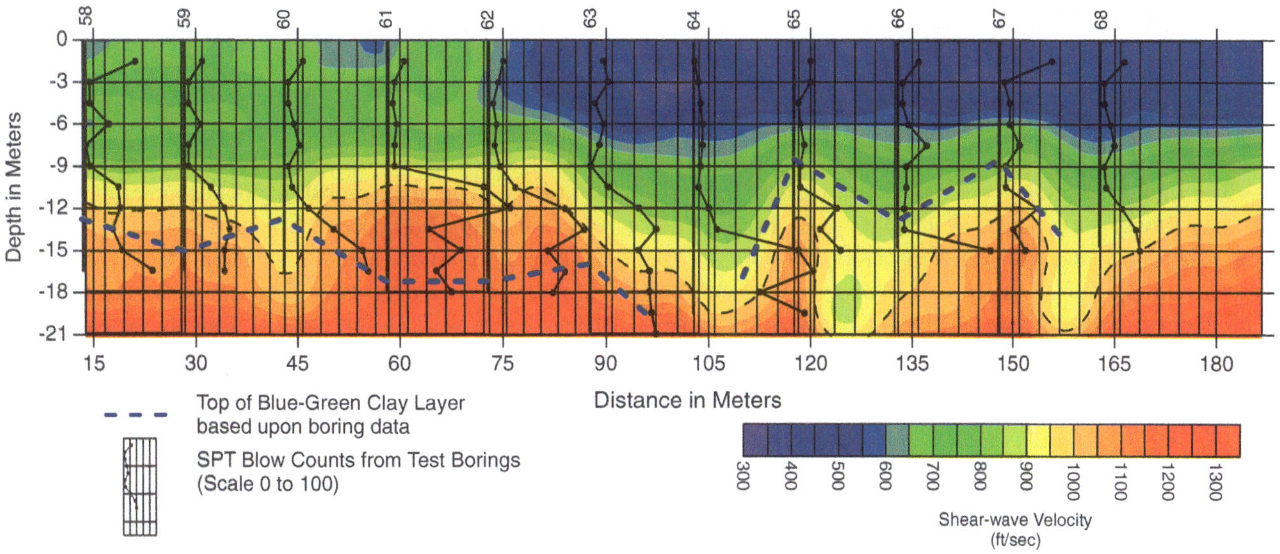

Fig. 17.13 SPT data and shear wave velocity data (MASW data) is used to define top of clay layer for a slurry wall

17.3.3.3 Geologic Sampling

Most of the engineering and hydrologic properties of soils, and rock have been determined from measurements made within a borehole along with laboratory measurements from borehole samples. There is a long history of laboratory measurements with highly quantified results. The results of localized field tests and laboratory analysis on small samples (with dimensions of <0.3 m) are commonly used to represent conditions of a large mass of soil and rock (with dimensions of ten's to hundreds of meters). Unfortunately, it is the better quality intact samples that are usually sent to the laboratory for testing and the worse samples are never retrieved or tested. This leads to biasing of the data and assumes that the intact samples recovered are representative of the actual site conditions. This practice is not acceptable in the heterogeneous conditions found in fractured rock and karst. See ASTM D5434 for details on logging of soil and rock (ASTM 2012a).

Soil Samples

Split spoon samples provide a majority of the data in unconsolidated materials. While somewhat disturbed, they provide reasonably good samples for geologic and engineering purpose. It is not unusual to see samples obtained at intervals of 1.5–3 m. However, continuous sampling should be used when dealing with any complex geologic conditions, such as karst. Once soil samples are collected there are a long list of laboratory tests that can be performed such as grain size analysis, permeability testing and moisture content. See ASTM guide D1586 for STP and split-barrel sampling of soils (ASTM 2011).

Rock Samples

Rock core is the standard for assessing conditions in rock. Rock core samples may be obtained by a variety of rotary drilling and core sampling methods in a variety of diameters. A variety of tests on the rock core can be run such as compression tests. ASTM D2113 provides information on the practice of rock core drilling and sampling of rock (ASTM 2014).

The amount of core or percent recovery is an indication of rock quality or the presence of fractures/voids. Figure 17.14 shows core collected at two different karst sites. Figure 17.14a was acquired in the Kansas City area and shows fairly good core recovery of limestone and shale with lots of near horizontal fracturing. Figure 17.14b was acquired in the south Florida area and shows poor core recovery and a highly porous relatively soft limestone.

Core recovery or percent recovery compares the total volume recovered to the total volume of the core run. Rock Quality Designation (RQD) values are a more rigorous evaluation for rock core than simply a core recovery. This classification uses a percentage based upon total length of core pieces longer than 10 cm versus the total length of core run, to provide an evaluation of the rock quality. These RQD values range from 0 to 100 with 0 being poor quality and 100 being excellent quality rock. ASTM D5878 describes rock mass classification systems (ASTM 2008). Bell (2004) and Hunt (2005) provide further discussion of the RQD designations along with the properties of soil and rock for engineering purposes and methods of testing. Hencher (2012) describes various methods of rock characterization.

Many rotary methods are destructive when drilling and result in chips of rock brought to the surface by drilling fluids. How the chips are collected, how often, who is describing them and if they are saved in chip trays will all affect the quality of data retrieved. At a Superfund site in west central Florida, a rotary drilling technique was used for monitoring well installation. Samples of the chips were acquired every 0.75–1.5 m, described and used to created geologic logs by a geologist and saved in chip trays (Fig. 17.15a) for further documentation. The chip trays were used to create a visual cross section (Fig. 17.15b) through an anomalous area (a buried paleocollapse) at the site. The change in materials from background conditions consisting of clean, light-colored sands, gravels and limestone to anomalous conditions consisting of variable, dark organic and peat materials can clearly be seen in the chip trays. This visual cross section provided a very effective way to illustrate conditions at a public meeting. See the full case history in Chap. 27.

17.3.4 Special Considerations When Drilling or Using Drilling Data

17.3.4.1 The Need for Angle Borings

While the great majority of borings are vertical, inclined or angle borings often become necessary to characterize fractured rock, fault zones and to characterize complex geologic conditions. When strata are close to horizontal, joints will generally be close to vertical. Dissolution will often occur along these joints in soluble rock. Encountering these vertical joints with vertical borings becomes an almost impossible task based upon the probability of encountering the joint. As a result angle borings are necessary.

Sowers (1996) discusses an example where TVA carried out a site characterization at a nuclear plant site in central Tennessee. A very closely spaced grid of vertical borings was used to investigate the horizontal bedded strata. The investigation included several hundred vertical borings followed by geophysical logging. No angle borings were made because the drilling equipment was limited to vertical holes. Later excavations for a cooling water canal encountered several dissolution enlarged joints 0.6–1.2 m wide and 1.2–2 m high at a depth of 12 m in the limestone. These features had been completely missed by the vertical drilling.

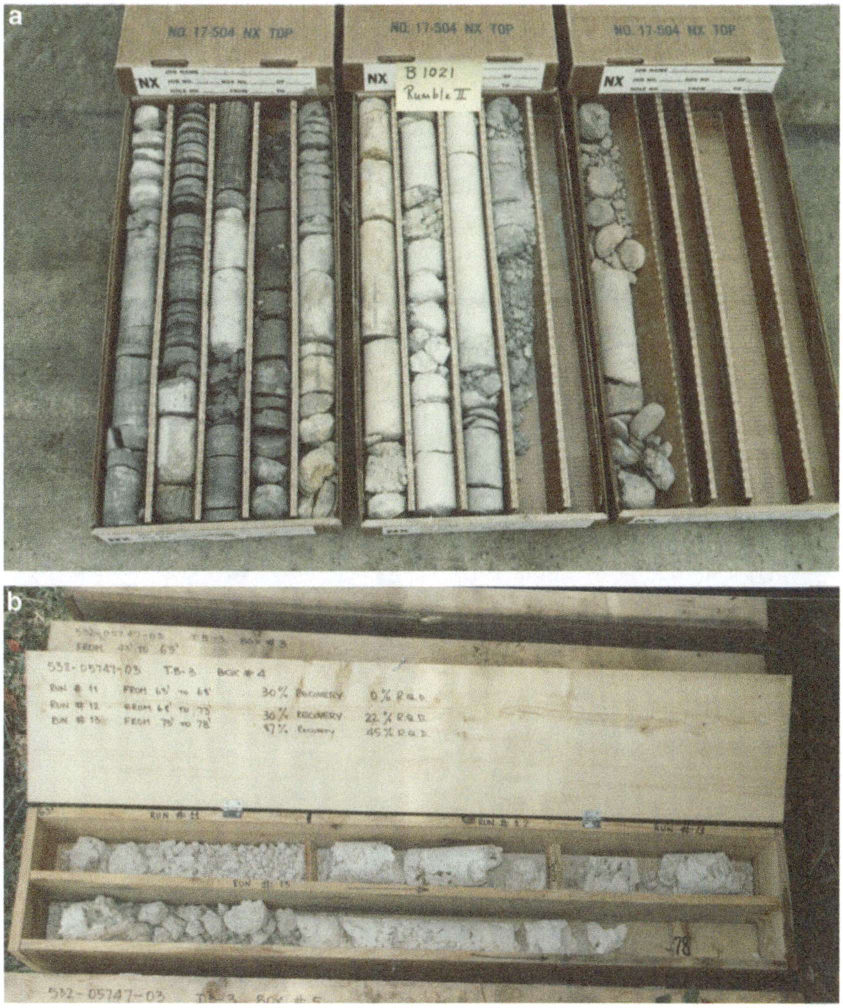

Fig. 17.14 Core samples from a Kansas site (**a**) indicate almost total core recovery while core samples from south Florida (**b**) indicate poor recovery and weather core

Angle borings were successfully used to confirm the presence and evaluate a fracture zone at a site in Kansas City, Kansas. A small soil piping feature was observed within a surface fissure in the loess soils at the site. The loess soils were removed revealing an obvious fracture zone in the rock (Fig. 17.16a). To evaluate the fracture zone, angle borings were made to intersect the fracture. Borehole video located fractures in the shale and large quantities of water (1,000 gal at 10 gpm) were poured into these two angle holes without being rejected (Fig. 17.16b). The exposed fracture was then used as an injection point for a dye trace to evaluate the connection of the fracture zone into the underlying mine. See the complete case history in Chap. 25.

Inclined borings are also used to provide access under existing structures. Figure 17.17a shows a drill rig squeezed between two buildings at a resort in the Florida Keys drilling at an angle to obtain data under an existing building. At another site, microgravity measurements indicated a large cavity zone had developed under a large structure due to an acid leak. Angle borings were then used to confirm the interpretation of the microgravity data and to provide further data to define the exact boundaries of the cavity. These same borings were then used for grouting of the cavity. Figure 17.17b shows the drilling plan to assess conditions under a large structure within a well-developed industrial setting.

17.3.4.2 Drilling May Sometimes Trigger a Sinkhole

In some cases, drilling can be done without concern of collapse. In other cases drilling can trigger a collapse. The hammer blows, drilling pressure and drilling fluids (both water and air along with excessive pressure) can be enough to trigger already unstable conditions. The process of drilling by itself does not create a cavity, but it can be the final event to trigger a collapse by simply encountering an existing geologic weak zone. This is discussed with examples in Part I, Sect. 8.1.2.

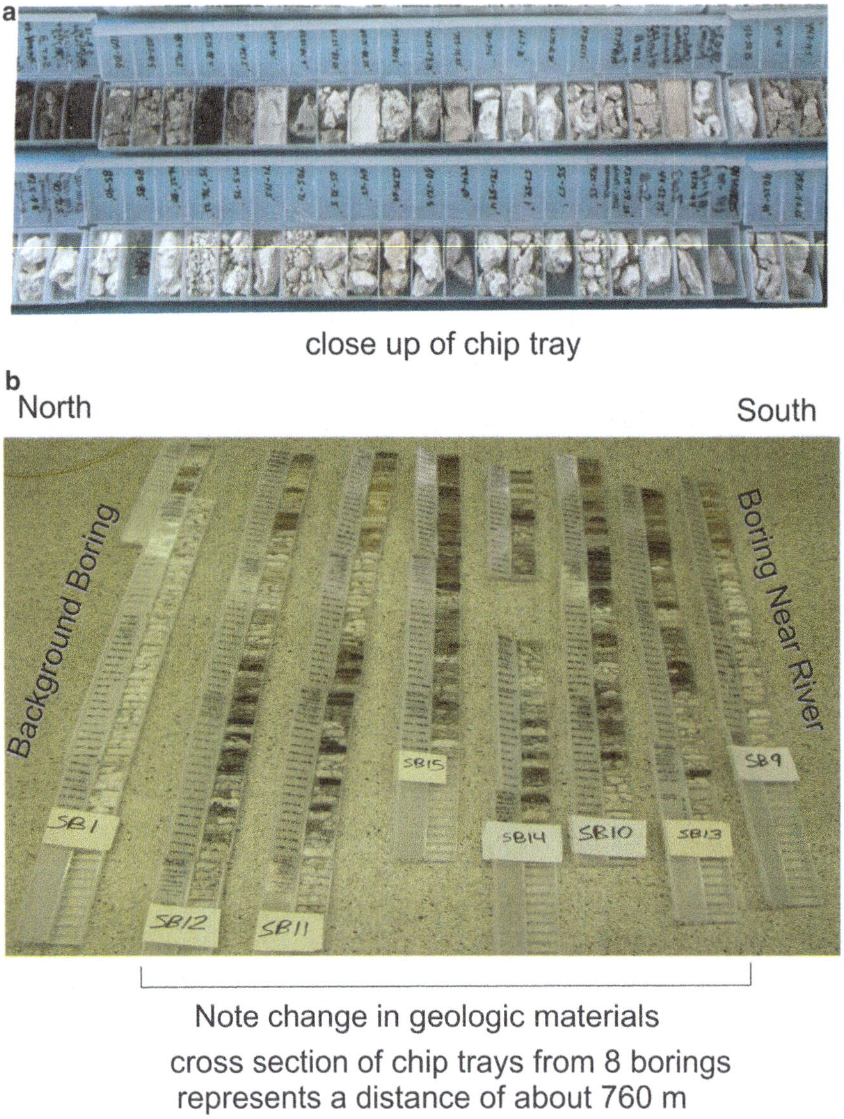

Fig. 17.15 Chip samples are placed in chip trays for storage (**a**). Chip trays from eight borings were laid out from north to south to illustrate the change in geologic conditions at the site (**b**)

17.3.4.3 Flexibility When Encountering Anomalous Conditions

All voids in limestone were interconnected at one time because they were formed by the flow of water (Waltham and Fookes 2003). When a void is encountered within the rock mass it should not be assumed to be an isolated feature, but it should be considered part of a conduit system developed by groundwater flow. An effort should be made to determine its lateral dimensions and its orientation, assuming it is a part of a fracture or cavity system. For example, an 2.4 m thick cavity was encountered at a depth of 33 m while drilling a borehole in limestone at a proposed nuclear power plant site. This was a bit of fracture or surprise to the on-site staff and they promptly grouted it, which was their standard procedure. The better approach would have been to take advantage of this opportunity to determine the nature and orientation of the cavity. This could have been done using a video inspection or optical televiewer log (above the water table) or an acoustic televiewer log (below the water table). This information would have helped to determine a possible zone of cavity development so that further investigation efforts could have been focused upon likely zones of cavity development. This is similar to the strategy used by TVA (Hopkins 1977).

17.3.4.4 Drill Rig Access

In many cases, borings have been located where the drill rig had easy access rather than where the problem lies. Rex Morey, the developer of ground penetrating radar, was demonstrating the use of radar to the Florida Department of Transportation (FDOT). The top of rock had been identified

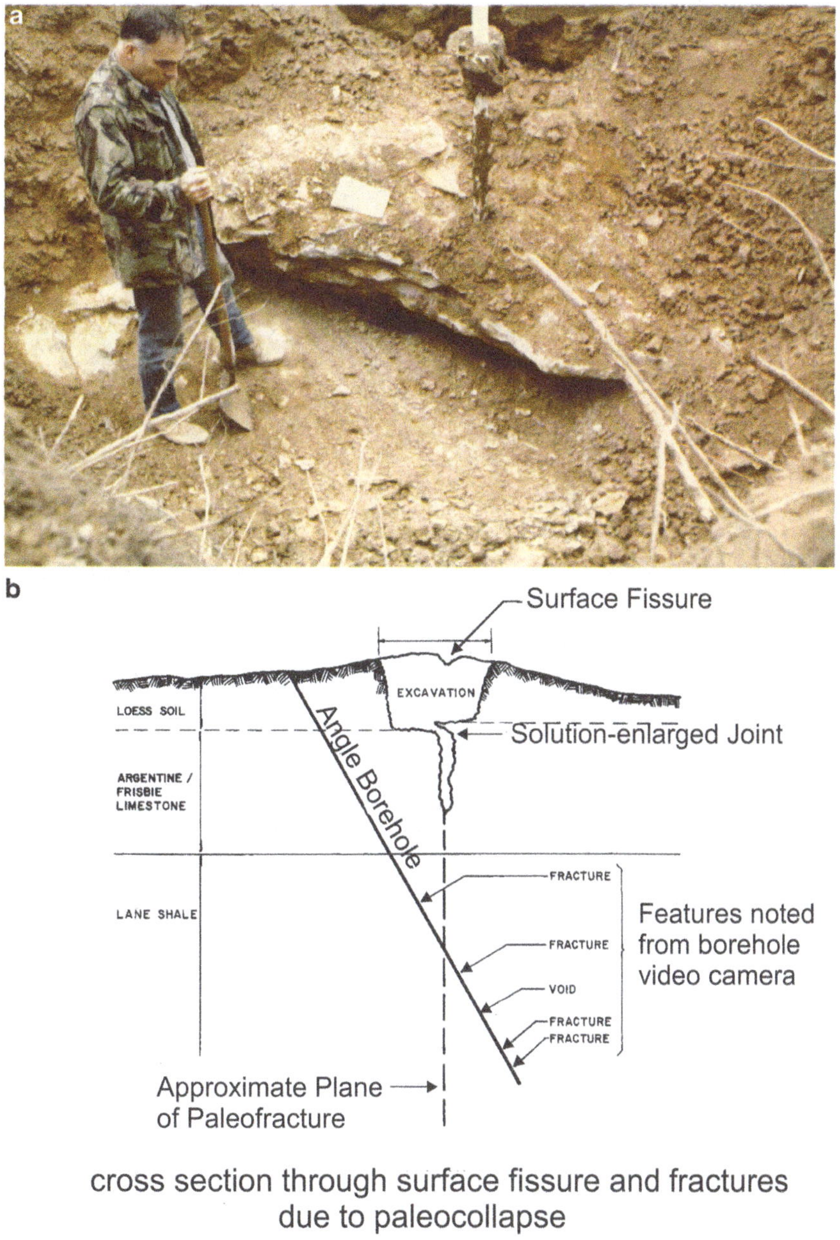

Fig. 17.16 A fissure in the loess soil cover was excavated and exposed a major opening into top of rock (**a**). Two angle borings indicated fractures that would not retain water (**b**)

as well as a number of possible isolated cavities near the top of rock within a ditch parallel to a road. The FDOT drillers were sent to the field to verify conditions, and reported finding no cavities. This was most unusual since the radar data was quite clear showing the presence of many cavities. Much later, a discussion with the senior FDOT geologist revealed that the drill rig was actually located at the edge of the road because of easy drill rig access. This was 6 m or more from the line of radar data in the ditch where recharge was occurring. In this case, a one to one correlation between the radar data and the boring data was impossible because the drill rig was not located along the line of radar data.

17.3.4.5 Borehole Deviation

The driller will not usually be aware of possible deviation of the borehole from vertical (Fig. 17.18), yet it can be critical when making subsequent measurements from that boring. Deep water level measurements where there is little gradient can be significantly affected by borings that are not vertical. Some geotechnical and hole-to-hole geophysical

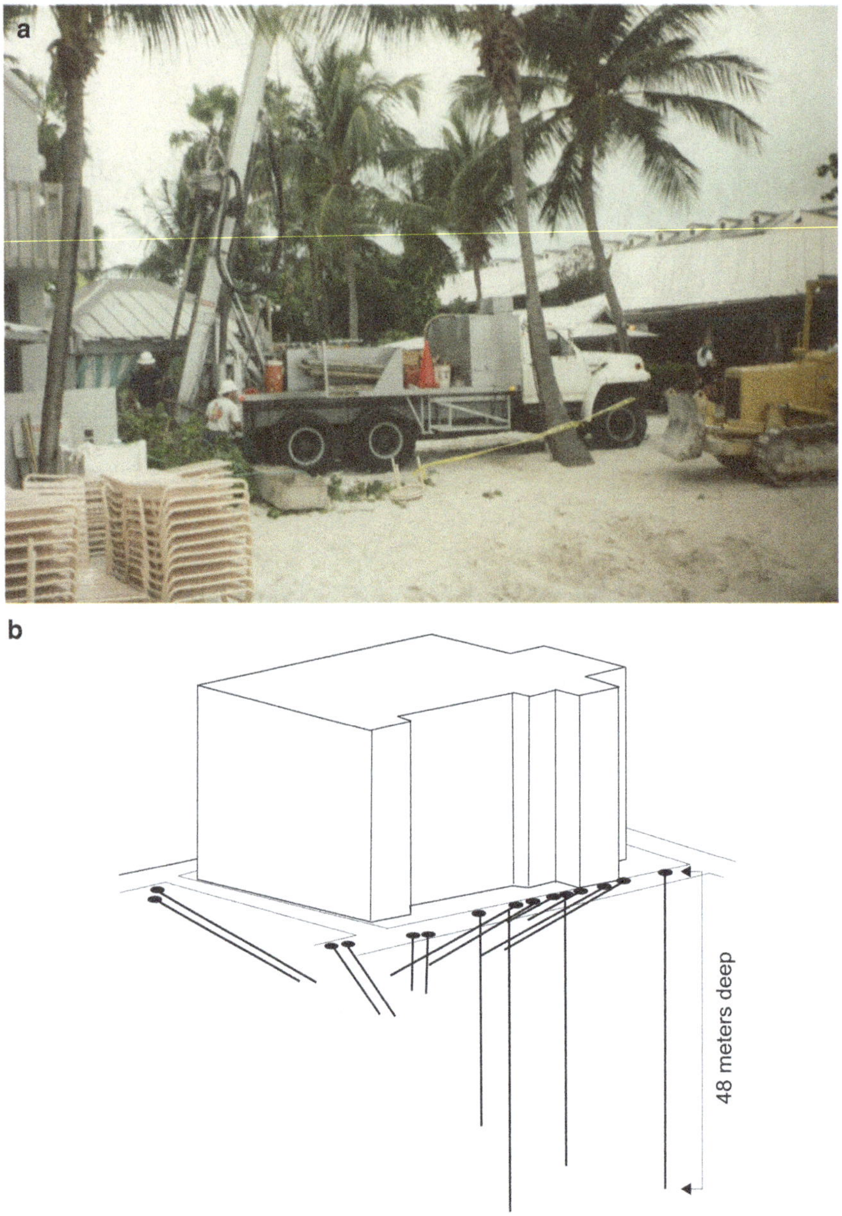

Fig. 17.17 Drilling of angle borings allows access to conditions under a structure (**a**) and may require a combination of vertical and angle boreholes may be required to sufficiently assess conditions under a structure (**b**)

measurements are also impacted by errors introduced by borehole deviation. See the example of borehole deviation (Fig. 18.13) in Chap. 18 geophysical logging section.

17.3.4.6 Drilling Inside Structures

It is not uncommon to find industrial facilities, in karst settings, that have been impacted by the presence of voids or cavities. While deeper features may be investigated through angle borings from outside of the building, smaller, shallower features may require coring through the concrete floor.

Acid leaks had occurred at an electroplating plant in southwestern Missouri, and voids were known to exist under the concrete floor. Settlement and cracking of the concrete floor was observed at a number of locations. New equipment was to be installed and there was concern that the failure of the floor could cause damage to the production line.

A detailed assessment of existing conditions of the building floor was made including the time history of equipment installation dates, and subsequent leaks as well as areas of visible subsidence and cracks in the floor. Then a ground penetrating radar survey was carried out to identify voids under the concrete floor and their lateral extent.

These data were used to determine specific locations for direct sampling. Core holes were drilled through the concrete and the depth of the void could be directly measured. An auger was used to remove soil samples to test for the presence of acid (Fig. 17.19a). A side looking borehole video camera (Fig. 17.19b) was used to evaluate the lateral extent of voids. All of this data was used as the basis for providing recommendations for remediation.

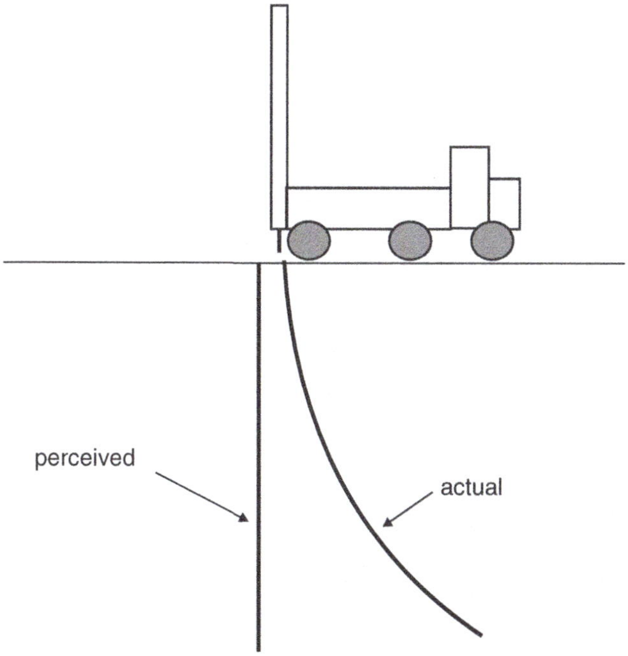

Fig. 17.18 Borehole deviations from vertical can impact certain measurements yet the driller will not be aware of the deviation

17.3.4.7 Misinterpretation of Data

While the boring data itself may be accurate, the interpretation may not be Sara (2003). The example in Fig. 17.20 illustrates how drilling data can be used to create a cross section for a site. In this case, the drillers or geologic log may be totally correct by itself but the data may be wrongly interpreted. Widely spaced borings and or interpreting pinnacles and a floating boulder as the top of rock or low areas between pinnacles as the top of rock will result in an inaccurate top of unweathered rock profile. Sometimes the true details are only discovered when the site has been excavated. In addition, cross sections developed from boring data will often have boring data projected over large distances onto the cross section. In areas of karst, where there is a high degree of spatial variability, this practice can be misleading. The location of all borings on the cross section should be shown on a map, so that the data used in the development of the cross section can be clearly evaluated.

At a site in Kansas, both borings and seismic refraction data were used to determine the depth to top of rock. The seismic measurements resulted in depths much deeper than that determined by the boring data for the same locations. It was only after persistent questioning that the reason for this discrepancy was identified. The top of rock had been defined by lawyers as the "first trace of rock" in the boring logs and included isolated boulders well above the top of massive

Fig. 17.19 A cored borehole was cut through the concrete floor of an electroplating plant to assess suspected voids under the floor and acquire samples (**a**). A borehole video camera was used to assess the extent of the void under the floor (**b**)

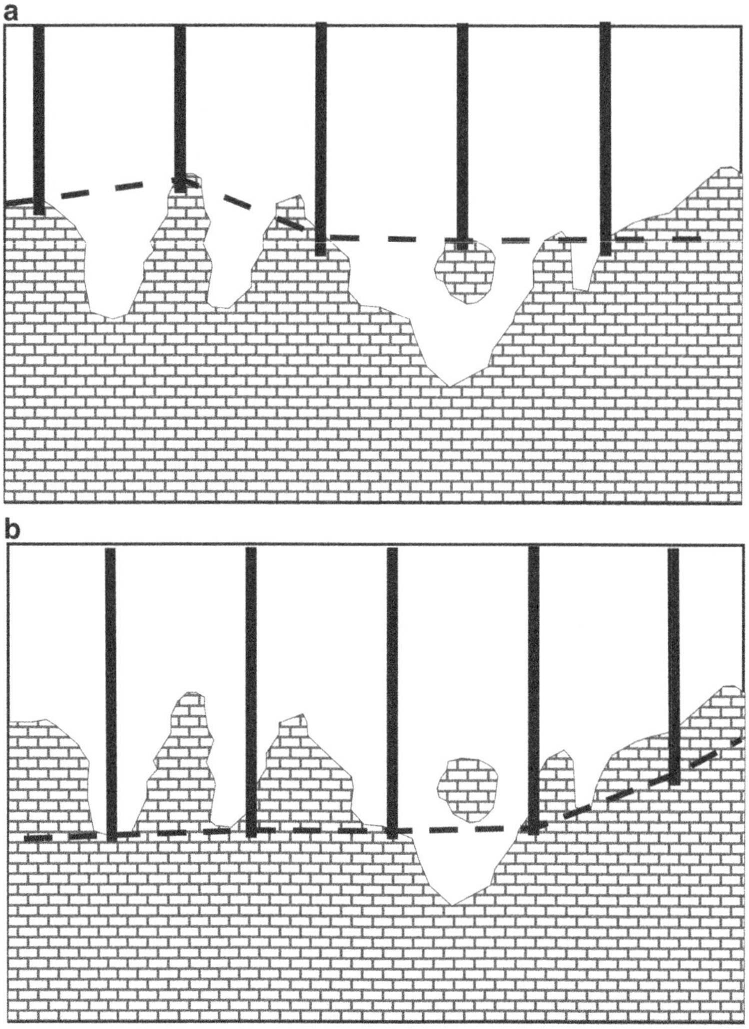

Fig. 17.20 A cross section can be easily misinterpreted from boring data alone. In both cases (**a** and **b**) the driller's or geologic log may be totally correct by itself but the overall interpretation may be wrong

in-place rock. These results, although technically incorrect, were now accepted as the legal definition of top of rock and were not allowed to be changed. The seismic refraction data was measuring the top of massive in-situ unweathered rock. While both sets of data were acquired accurately, the misinterpretation of the driller's logs by the lawyers forced the seismic data to be in question rather than resolve the erroneous decision of the legal definition.

17.3.4.8 Ground Truth and Hard Data

For decades we have seen borings referred to as the ultimate means of "ground truth" or "hard data" and are the standard from which other data and interpretations must be referenced. However, boring logs whether done by a driller, geologist, or engineer, (like any other source of data) are subjective and are also subject to a variety of errors, even with experienced personnel.

Drillers logs (made by the driller) or geologic logs (made by an engineer or geologist) include descriptions of the soil and rock samples, as well as water levels, the rate of drilling, rod drops, fluid loss and other related data. It is interesting to consider the manner in which boring logs are developed. While we have many ASTM guidelines and standards for the field aspects of drilling, and recording of geologic data, these guidelines or standards alone do not assure quality results. It's the experience and dedication of people that provide quality results and we don't have a standard for that.

It is almost too much to ask of the driller and his helper to manage the drilling operation, keeping an accurate drillers log, recording all of the details and changes in drilling conditions. There is considerable multitasking (recording drilling rate, fluid loss, rod drops, SPT measurements, and sampling and packaging of sediments or labeling of core) taking place while operating the drill rig (including safety issues) all of which can easily lead to gaps in the data. Considering that most drilling is contracted on a unit basis (and often at low bid) under such conditions the best drillers and the drilling companies are forced to compromise. Furthermore the work is often carried out under difficult weather conditions and the field logs may become wet and muddy and some portions of a page may not even be legible. Then back in the office the logs are often handed to someone (often other than the person in the field) to be typed, and details get lost. It is here where the potential for transcription errors, misinterpretations and omissions to occur.

Unfortunately, it has become common to think of the task of drilling, logging of the borehole and acquiring samples as a trivial one, which can be done by almost anyone if given some guidelines. In fact, it is as difficult a task as any of those in the site characterization process.

There can be significant differences due to training and experience of the persons doing the interpretation of geologic conditions and record keeping. Unfortunately all too often the geologist or engineer sent to the field are the most junior staff with little experience and often no formal training in drilling or logging of a borehole in soil or rock. Most universities do not teach the fundamentals associated with drilling and geologic logging. In many situations today we find the most junior professional staff assigned to drilling operations. We have seen cases when a young geologist with a master's degree was assigned to a drill rig for the first time. He had no training in college on the aspects of geologic logging and was placed in the position of logging borings with little more than the companies reference document in hand and "good luck, the driller will help you".

Another example was from a project that specified the requirement for a geologist to log the core (but did not define the level of experience required). The drilling company did not have a geologist on staff and hired a recent graduate part time for the job in order to comply with the project specifications. Unfortunately the geologist had no training or experience in such work. Luckily the driller was an old hand who had spent many years drilling and logging in the specific formations, including extensive work on the specific site. The driller provided the descriptions to the new geologist who then made the field notes. The project succeeded because of the experienced driller and the inexperienced geologist had a great learning experience. In some cases a driller who has considerable experience in a particular geologic setting will produce better quality drillers logs than an inexperienced geologist or engineer.

17.3.4.9 Maximize the Data from Boreholes

Some of the borings completed on-site will be used to acquire additional data such as pump tests, engineering properties, water levels or water quality samples over time. Some of these tests or measurements require open boreholes, some require monitoring wells or piezometers to be installed. The field plan should maximize the data from each of the borings. For example, installing small diameter piezometers will limit the amount of data that can be obtained, simply due to the diameter of the casing. The plan should also consider the longevity of each boring. Existing boreholes can provide the opportunity for additional data, if needed, avoid grouting the borings too quickly.

17.3.5 An Optimum Approach for Drilling and Sampling

The following is a proposed strategy to optimize borings, geophysically logging and sampling. It is based upon the site characterization strategies presented and assumes that a solid set of data are used to properly select boring locations. This approach maximizes the amount of data from each boring and at the same time minimizes the number of borings. This strategy also minimizes the number of samples that need to be obtained and stored. It provides independent sets of data (boring logs and geophysical logs) to improve confidence levels. It follows the same philosophy that TVA had used for the extensive work in the construction of dams and nuclear power plants, (Hopkins 1977), and further defined by DOE ESC recommendations in ASTM D6235 (2010b).

- Initially two or three borings are drilled to establish on-site geologic conditions. These would include detailed drillers logs and continuous sediment and core samples. The continuous soil samples and rock cores can be examined in detail and tested for various geotechnical parameters.
- Then each of these initial borings would be geophysically logged with an appropriate suite of measurements to provide an independent set of geologic data for correlation with the driller's logs. This combination of two independent sets of data (the geologic log from drilling and the geophysical logs) provides a geologic reference for the site.
- Subsequent drilling can be done destructively and all borings logged with the same suite of geophysical tools. This set of data can then be compared to the reference borings previously completed to aid in their interpretation.
- If and when geologic conditions are seen to change a few additional reference borings may be required (continuously cored, geophysically logged, and samples sent for testing, as needed).

All too often core samples are missing in zones of weathered rock and voids, which is of particular concern when working in karst. In addition, the need to acquire and safely store large amounts of core (and oriented core) can be logistically difficult. Both of these issues can be resolved by obtaining a high resolution 360° image of the borehole using an acoustic televiewer (ATV) in water-filled boreholes or an optical televiewer (OTV) in dry boreholes. Both the acoustic and optical televiewer provide an excellent means of obtaining a continuous and detailed in-situ image of geologic conditions within an open borehole even in areas where samples cannot be recovered. The ATV and OTV data is also digital and can be presented as a virtual core sample, which can be rotated for viewing. A discussion of geophysical logging is provided in the following section.

17.4 Excavations and Trenches

Outcrops and core from borings usually provide only a fragmented picture of soil and rock conditions. Excavations and trenching offer one of the most definitive of all invasive exploratory methods since it provides a means of direct observation of geologic conditions over a larger area. Trenches are necessary when attempting to assess details of geologic conditions, assessing fractures and fissures, settlement, subsidence, or the dating of the movement induced by an earthquake.

Trenches can range from small shallow shovel cut trench by hand to backhoe (Fig. 17.21) and large dozer cut trenches. Safety becomes an issue if the trench is narrow and deep, then bracing to prevent cave-in is required. A properly braced trench will allow safe entry. Federal (OSHA) and State regulations cover the safety aspects of trenching.

The success of trenching depends upon getting the trench in the right location, sufficiently deep to expose the area of interest, and the skills necessary to provide detailed documentation of the conditions using cross section or face maps. A thorough cleaning of the trench walls and a detailed grid or string lines must be established over the surface of the trench wall. The trench wall must be carefully cleaned (much as with an archeological excavation) and many more features and detail should be mapped on the face map, then should be little left to conjecture (Hatheway 1982). This requires a high level of observational and documentation skills. Age dating methods may include petrographic analysis, clay mineralogic analysis and radiometric techniques. An objective trench log attempts to portray equally all physical features in an impartial manner without regard to relative importance and subjective interpretation is minimized during the recording process.

Although photographs and video may be obtained, a detailed sketch along with a detailed descriptive analysis of the geology is generally the primary means of documentation. Wide excavations might consider the use of electronic distance meter (EDM), total station, photographic or video documentation, (Keaton 2009; Haneberg 2009).

Excavation for landfill expansion provided a large exposed geologic cut through an area being studied in Kansas City, Kansas. The excavation was first photographed (Fig. 17.22a)

Fig. 17.21 Shallow trenches whether by backhoe (**a** and **c**) or dug by hand (**b**) can provide critical observations in the assessment of subsurface conditions

Fig. 17.22 An exposed section of limestone and shale at the edge of a paleocollapse. The top of the shale was marked and elevations surveyed (**a**). Cleaning the face of the dozer cut further exposed fractures in the shale (**b**)

and then sketched in detail. Then elevations were established across the entire face of the excavation along a distinct geologic unit. This revealed a low area of 3 m in the center of the 167 m wide excavation. This was the northernmost edge of a large 7 ha paleocollapse at the site. Localized areas were further excavated using shovels and hand picks and exposed open fractures within the underlying shale (Fig. 17.22b). This geologic window was unexpected and invaluable since it provided cross section through the edge of an obvious paleocollapse. Furthermore, it exposed open fractures in the uppermost shale at the site, which impacted on-going landfill design. See Part III, Chap. 25 for a more detailed description of this project.

At another site in Pennsylvania, excavations were utilized to assess the investigation of geologic pathways that might be controlling movements of organic contaminants from a military base. Excavation revealed the complex top of rock conditions (Fig. 17.23). The layers of limestone can be seen dipping which played an important part in controlling the movement of contaminants at this site.

An example of a very large trench is shown in Fig. 17.24. This is from a site being characterized for a proposed

Fig. 17.23 An excavation clearly shows the complex nature of the top of limestone

Fig. 17.24 A large trench is being excavated to expose the top of rock and assess possible buried channel or paleokarst conditions

hazardous waste facility in southwest Texas. The site had a thick alluvium over limestone. Some anomalous areas (possible buried channels or paleosinkholes) had been identified by surface geophysical measurements. This site was a large open area, and heavy excavating equipment was readily available on-site so trenching was an easy and effective option to provide more detailed data. This large trench was more than 15 m wide and about 11 m deep, which exposed the top of limestone. The large trench allowed detailed mapping to be completed and evaluated possible karst conditions at the top of rock.

Further details of trench mapping are provided in Hatheway and Leighton (1979), Hatheway and McClure (1979), and McCalpin and Shlemon (1996).

References

ASTM (2006) Standard guide for use of hollow-stem augers for geoenvironmental exploration and the installation of subsurface water-quality monitoring devices. D5784-95(2006). American Society for Testing and Materials, West Conshohocken

ASTM (2008) Standard guide for using rock-mass classification systems for engineering purposes. D5878-08. American Society for Testing and Materials, West Conshohocken

ASTM (2009) Standard practice for soil exploration and sampling by auger borings. D1452-09. American Society for Testing and Materials, West Conshohocken

ASTM (2010a) Standard practice for using the electronic piezocone penetrometer tests for environmental site characterization. D6067-10. American Society for Testing and Materials, West Conshohocken

ASTM (2010b) Standard practice for expedited site characterization of vadose and ground water contamination at hazardous waste contaminated sites. D6235-04(2010). American Society for Testing and Materials, West Conshohocken

ASTM (2010c) Standard practice for sonic drilling for site characterization and the installation of subsurface monitoring devices. D6914-04(2010). American Society for Testing and Materials, West Conshohocken

ASTM (2011) Standard test method for standard penetration test (STP) and split-barrel sampling of soils. D1586-11. American Society for Testing and Materials, West Conshohocken

ASTM (2012a) Standard guide for field logging of subsurface explorations of soil and rock. D5434-12. American Society for Testing and Materials, West Conshohocken

ASTM (2012b) Standard test method for electronic friction cone and piezocone penetration testing of soils. D5778-12. American Society for Testing and Materials, West Conshohocken

ASTM (2012c) Standard guide for direct-push groundwater sampling for environmental site characterization. D6001-05(2012). American Society for Testing and Materials, West Conshohocken

ASTM (2014) Standard practice for rock core drilling and sampling of rock for site investigation. D2113-14. American Society for Testing and Materials, West Conshohocken

Bell FG (2004) Engineering geology and construction. Spoon Press, London

Christy CD et al (1994) A percussion probing tool for the direct sensing of soil conductivity, Geoprobe technical paper no. 94-100, Geoprobe Salinas, Kansas

Fookes PG (1997) Geology for engineers: the geological model, prediction and performance. Geol Soc Q J Eng Geol 30(4): 293–424

Haneberg WC (2009) Rapid prototyping of computer models to characterize discontinuous rock masses. Geo-Strata, March/April, pp 22, 24–26

Hatheway AW (1982) Trench, shaft, and tunnel mapping. Bull Assoc Eng Geol 19(2):173–180

Hatheway AW, Leighton FB (1979) Trenching as an exploratory method. In: Hatheway AW, McClure CR Jr (eds) Geologic society of America, geology in the siting of nuclear power plants. Rev Eng Geol 4:169–195

Hatheway AW, McClure CR Jr (eds) (1979) Geology in the siting of nuclear power plants. Geol Soc Am Rev Eng Geol 4:169–195

Hencher S (2012) Practical engineering geology, applied geotechnics, vol 4. Spoon Press, London

Hopkins R (1977) TVA philosophy and methods for site investigation. In: Butler D (ed) Symposium on cavity detection of subsurface cavities. U.S. Army Engineers Waterways Experiment Station Vicksburg, Vicksburg, pp 25–26

Hunt RE (2005) Geotechnical engineering investigation handbook, 2nd edn. CRC Press/Taylor and Francis, Boca Raton

Hvorselv MJ (1949) Surface exploration and sampling of soils for civil engineering purposes. The Waterways Experiment Station, Vicksburg

Keaton JR (2009) Close-range terrestrial photogrammetry for geotechnical characterization and visualization. Geo-Strata, March/April, pp 18–20

Khedr SA et al (1985) Automated cone penetrometer a non-destructive field test for subgrade evaluation. Annual transportation research board meeting, National Academy of Sciences, Washington, DC

Kratochvil G (1989) Geological evaluation of potential hazards to shuttle landing operations on Rodgers Dry Lakebed. Astronaut Office NASA Lyndon B Johnson Space Center Houston Texas, Edwards Air Force Base

Kratochvil G et al (1992) Fishing for fissures. The Military Engineer, Jan–Feb, pp 66–68

McCall W et al (2006) Use of direct push technologies in environmental site characterization and ground-water monitoring. In: Nielsen DM (ed) Practical handbook of environmental site characterization and ground-water monitoring, 2nd edn. CRC Taylor & Francis, Boca Raton, pp 345–471

McCalpin JP, Shlemon RJ (1996) Trenching as a tool in geologic investigations. Prof Geol 33(12):19–22

Nielsen D (ed) (2006) Practical handbook of environmental site characterization and ground water monitoring, 2nd edn. CRC Taylor & Francis, Boca Raton

Ruda T, Farrar J (2006) Environmental drilling for soil sampling, rock coring, borehole logging, and monitoring well installation. In: Nielsen DM (ed) Practical handbook of environmental site characterization and ground water monitoring, 2nd edn. CRC Taylor & Francis, Boca Raton, pp 297–344

Sara MN (2003) Site assessment and remediation handbook, 2nd edn. Lewis Publishers, New York

Schulmeister MK et al (2003) Direct-push electrical conductivity logging for high-resolution hydrostratigraphic characterization. Ground Water Monit Remediat Summer 23(3):52–62

Sowers GF (1996) Building on sinkholes, design and construction of foundations in karst terrain. ASCE Press, Reston (out of print)

Waltham AC, Fookes PG (2003) Engineering classification of karst ground conditions. Geol Soc Lond Q J Eng Geol 36(2):101–118

Geophysical Logging

18

Abstract

At this stage in a site characterization, we have refined the conceptual model based upon the surface geophysics and located new borings or monitor wells with insight. While drilling these boreholes has provided geologic logs and samples at some interval, there is additional information that can be obtained. This is where geophysical logging techniques are applied. This suite of measurements provides detailed, often continuous, vertical data within open boreholes and cased wells.

In addition, there may be existing borings, monitoring wells, supply wells on-site that provide opportunities for gathering additional detailed geologic and hydrologic data. This chapter provides a brief introduction to geophysical logging along with many considerations for using this type of measurement. Examples are also provided showing a variety of the geophysical logging measurements and their application.

18.1 Introduction

Geophysical logging provides measurements of a wide variety of parameters down the length of an open borehole or cased monitor well, similar to the surface geophysical methods (Fig. 18.1). Geophysical logging is commonly used in the oil and gas industry as part of exploration and development. This logging is done by specialty logging contractors with large dedicated trucks and sophisticated logging tools, which may be many cm in diameter and several meters long. Here, we are discussing applied geotechnical and hydrologic logging in small diameter boreholes and wells.

Figure 18.2 is a typical small logging system, from Mount Sopris, used for characterizing conditions within open boreholes and cased monitoring wells. The systems consist of a laptop computer, system control unit, winch (Fig. 18.2a), tripod over the borehole or well (Fig. 18.2b) and a variety of logging probes that are lowered into a borehole to acquire data (Fig. 18.2c). This system has been installed in a vehicle, however, some systems are small enough to be carried to remote locations. Most logs can be run in holes as small as 5 cm diameter and therefore can be run in most cased monitor wells. Logging is carried out by lowering or raising a logging tool in an open or cased borehole. The digital data is displayed on the computer screen and recorded on a computer. The raw data can also be printed as a hard copy in the field.

The logging tools provide in-situ measurements of one or more parameters of borehole fluids and undisturbed soil and rock conditions surrounding the borehole. Most logging tools provide essentially continuous measurements at a sample rate of 0.3 cm as the probe is moved up or down the hole. Some logging tools use station measurements that are acquired as the probe is held at a specific depth over some time interval. The depth or penetration of the measurement varies from the fluid within the borehole to the borehole wall up to a meter into the soil or rock depending upon the particular log being used.

While drillers and geologic logs are subjective based upon personal experience, quality of samples retrieved and visual interpretation, the geophysical logging measurements are very non-subjective. The geophysical logging measurements provide in situ measurements and are virtually 100 % repeatable. It is the interpretation of the logging data that can be subjective.

The objective of geophysical logging is to obtain more detailed and less subjective measurements from a borehole

R.C. Benson, L.B. Yuhr, *Site Characterization in Karst and Pseudokarst Terraines*,
DOI 10.1007/978-94-017-9924-9_18

or well than can be obtained from drilling and sampling alone. Drilling any kind of a borehole or well is an added cost to the project and the borehole provides data at just one point over the site. Therefore, the data from each borehole or well should be maximized with the addition of geophysical logging. The combination of a driller's log, or geologic log, with geophysical logs provides an increased confidence level in our interpretations of soil and rock conditions.

While the resolution of all surface geophysical measurements decreases with depth, geophysical logging measurements maintain their resolution independent of depth providing high resolution, nearly continuous, in-situ information along the length of a hole. These data can be used to characterize geologic and hydrologic and engineering properties of the unsaturated and saturated zones within open or cased wells and help to constrain the vertical geohydrologic details in our conceptual model of the site.

Further details about logging can be found in Hearst et al. (2000), Chapellien (1992), and Killeen (1986).

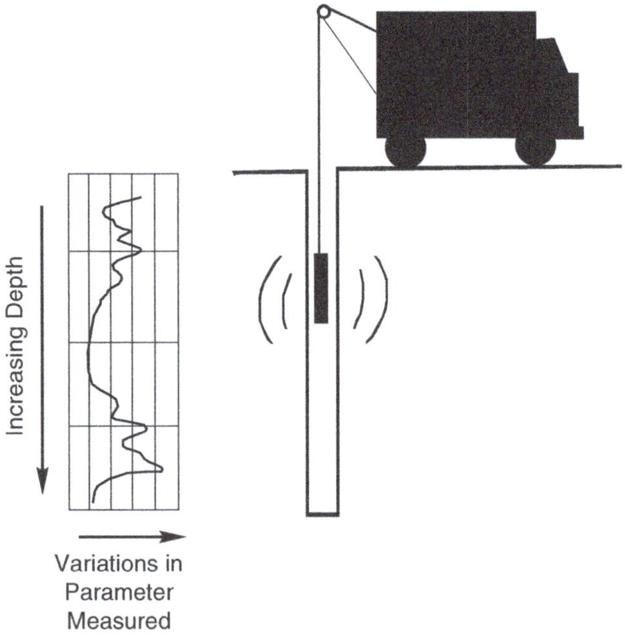

Fig. 18.1 Sketch of typical downhole logging measurements

18.2 Geophysical Logging Measurements

A wide range of geophysical logs are available that can be used to gather different types of data for determining the geologic and hydrologic conditions. The purposes for logging may include:

- acquiring detailed geohydrologic data after new borings have been properly located,
- recovering geohydrologic data from existing borings and
- evaluating the condition of existing wells.

Geophysical logging data provides an independent means of obtaining geologic and hydrologic data from old and new borings. This provides a consistent set of data that allows the geologic log and the geophysical logs to be compared within a single borehole as well as between boreholes. The geophysical logs can also be used as a means of quality control, provide multiple data sets for correlation and to aid to interpretation.

Table 18.1 lists the more commonly available logs along with the parameter that is measured or calculated, and the applications in which the logs may be used. The most generic name for the particular log is used avoiding names applied by specific manufacturers and service companies. The logs

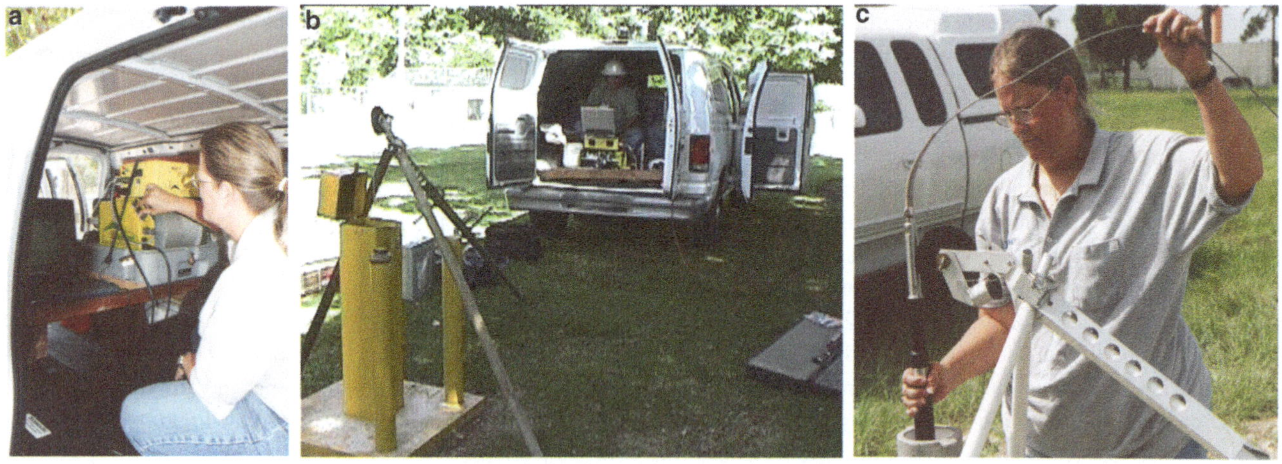

Fig. 18.2 Photos of typical logging system in operation. (**a**) Control unit, laptop and winch. (**b**) Tripod over well. (**c**) Probe being lowered into well

Table 18.1 Summary of common geophysical logs

	Log	Parameter measured or calculated	Open hole, PVC or steel	Saturated or unsaturated borehole	Radius of measurement (cm and m)
Electrical/EM logs	Induction	Bulk electrical conductivity of soil, rock, and pore fluids and magnetic susceptibility of soils and rock	Open hole or PVC	Both	About 75 cm
	Resistivity	Electrical resistivity of soil, rock and pore fluids	Open hole	Saturated	Up to a meter or so
	Single point resistance	Electrical resistance of soil, rock and pore fluids	Open hole	Saturated	Small/variable
	Spontaneous potential (SP)	Voltage – responds to electrochemical effects of differences in borehole fluids, oxygen reduction of minerals and streaming potential due to movement of pore fluids	Open hole	Saturated	Small/variable
Fluid logs	Conductivity	Electrical conductivity of borehole fluids	Open hole	Saturated	Within the borehole
	Spinner flowmeter	Flow rate of fluid within the borehole	Open hole	Saturated	Within the borehole
	Heat pulse flowmeter	Flow rate and direction of fluid within the borehole fluid	Open hole	Saturated	Within the borehole
	Temperature	Temperature of borehole fluid	Open hole	Saturated	Within the borehole
	Chemical sensors	Selected chemical parameters of the borehole fluid	Open hole	Saturated	Within the borehole
Mechanical logs	Caliper	Diameter of borehole or casing	Typically open hole	Both	Borehole wall or casing
	Deviation	Inclination and direction of borehole (deviation from vertical)	Open hole or PVC Steel using gyro system	Both	Borehole
Visual/imaging logs	Television	Visual image of borehole or casing	Typically open hole	Both	View range a few cm to about a meter
	Optical televiewer	A 360° optical image of the borehole wall (or casing)	Typically open hole	Both	View range up to about 30 cm
	Acoustic televiewer	A 360° acoustic image of the borehole wall (or casing)	Typically open hole	Saturated	View range 10 cm
Nuclear logs	Natural gamma	Total count rate of natural gamma radiation	Any	Both	15–30 cm
	Natural gamma-spectrometry	Identification and quantitative analysis of the radioisotopes that contribute to the total natural gamma count rate	Any	Both	15–30 cm
	Gamma-gamma	Relative density or true density, if calibrated	Any	Both	15 cm
	Neutron	Relative moisture/fluid content in unsaturated zone, porosity in saturated zone	Any	Both	15 cm

(continued)

Table 18.1 (continued)

	Log	Parameter measured or calculated	Open hole, PVC or steel	Saturated or unsaturated borehole	Radius of measurement (cm and m)
Other logs	Radar	Provides EM reflection data at greater distance from the borehole	Open hole or PVC	Both	About 3–15 m
	Gravity	Provides changes in density at greater distance from the borehole	Any	Both	Up to 30 m or more
	Full waveform sonic	Travel time and magnitude of seismic signal in soil and rock	Open hole	Saturated	Up to a meter

have been divided into six groups based upon the way they make measurements.

Electrical/Electromagnetic Logs – These logs measure electrical properties of the subsurface [conductivity, resistivity, resistance or spontaneous potential (voltage)]. They aid in characterizing lithology/stratigraphy, identifying fractures and cavities, detecting inorganic pore fluids (*ie*. salt water intrusion, inorganic contaminants, etc.) and in some cases aid in detecting flow. These logs typically require an open borehole with the exception of the electromagnetic log, which can be used in a PVC-cased borehole.

Fluid Logs – The fluid logs measure a property of the borehole fluid such as fluid conductivity, temperature, flow of fluid (spinner flowmeter or heat pulse logs), and tracers within the fluid or specific chemistry of the fluid. These measurements require an open borehole in which the fluids are in equilibrium in order to be representative and are made by slowly lowering the probe down the borehole, while most other logs are made coming up the hole. If there is fluid flow through fractures in the borehole wall this flow can be detected by a flowmeter or changes in temperature or fluid conductivity to measure the presence and rate of flow.

Mechanical Logs – the mechanical logs include caliper and deviation logs. The caliper log is mechanically measuring the borehole diameter allowing identification of variations caused by fractures, cavities, wash-outs, etc. It can also be used in cased boreholes to confirm casing diameter, where casing ends and open borehole begins, or where damage to a casing has occurred. The deviation tool measures the deviation of the borehole from vertical, providing corrections for other measurements made within the same borehole.

Video and Imaging Logs – video logs provide an image down the axis of the borehole or the borehole wall. This can be used to identify lithology as well as fractures, cavities in an open hole or conditions within a cased well. The imaging logs (acoustic televiewer and optical televiewer) provide a digital 360° image of the borehole wall that can be used for quantitative analysis of strike, dip and aperture of fractures. They can also be used to inspect cased wells for damage. The acoustic log requires a water-filled borehole while the optical log can operate in either a dry or water-filled borehole.

Nuclear Logs – nuclear logs are measuring either natural radiation or back scattered radiation from a radioactive source used in the logging tool. The natural gamma log measures natural radiation, and is a more commonly used log to map lithologies such as clays, shales and weathered zones. A spectral gamma log also measures natural radiation and can differentiate between radioactive isotopes. The logs utilizing a radioactive source are much less commonly available but are very useful when mapping variations in density (gamma-gamma log) and porosity (neutron log). All of these nuclear logs can be used in PVC or steel-cased wells.

Other Specialty Logs – there are a few specialty logs such as sonic waveform, ground penetrating radar and gravity. Sonic logs can provide an in situ measurement of P and S wave seismic velocity within a borehole. Borehole radar measurements can be made from a single borehole with a transmitter and receiver antenna located in the same borehole or between holes with the transmitter and receiver in different boreholes. The radar range is dependent upon soil or rock type. In resistive rocks ranges of 10–40 m can be achieved. However, in conductive, clay-rich or silty rock ranges can be limited to 5 m or less. Borehole radar has been used for a variety of applications including bedrock fractures, voids, and tracing tests. Singha et al. (2000) provides a brief summary of borehole radar theory and its application.

Unlike conventional nuclear density logs, borehole gravity measurements provide density measurement for large volumes of rock surrounding the borehole and are not affected by borehole conditions. The effective radius of investigation is about 5 times the spacing between measurements and can be 10's of meters or more. Detection of structure, faults, caverns and mines can be detected at significant distance from the borehole. More information can be obtained from the manufacturer, Scintrex Ltd. (www.scintrexltd.com).

While most logging tools are measuring one parameter at a time, there are a few that measure two parameters at a time such as resistance/spontaneous potential or temperature/fluid

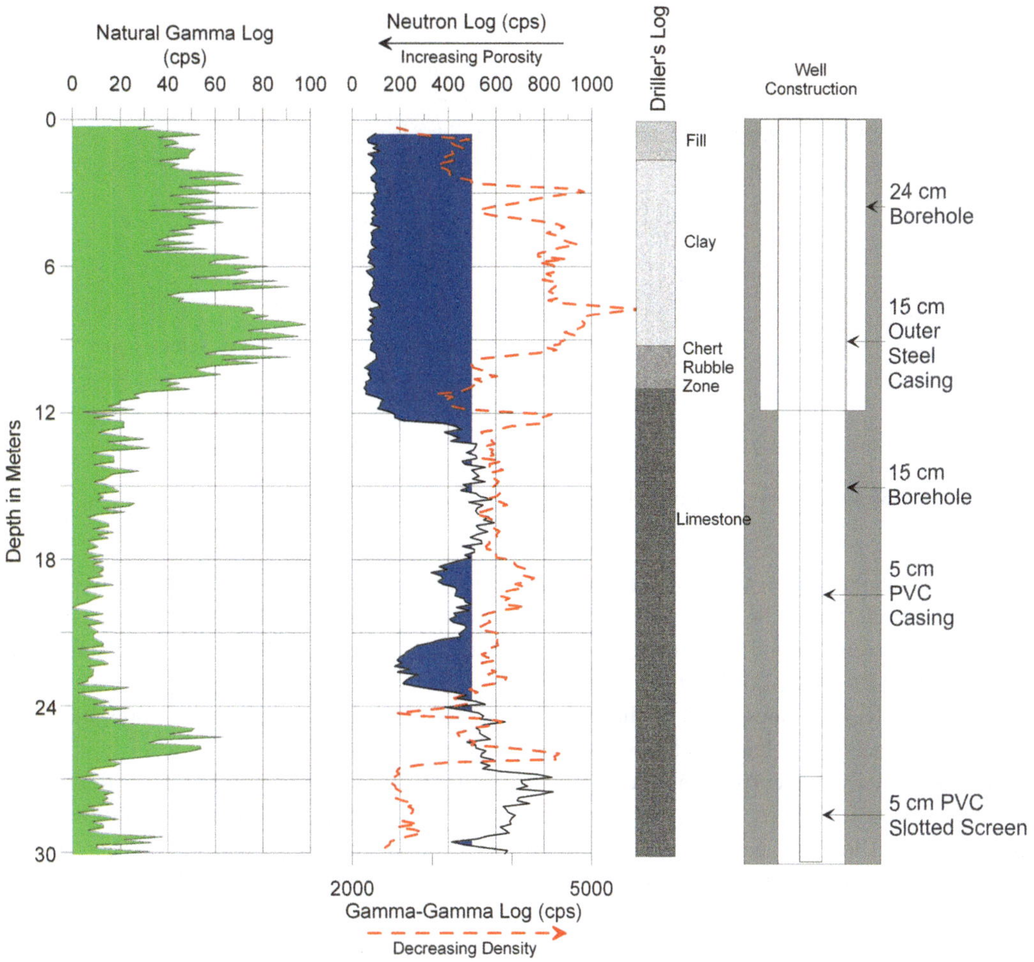

Fig. 18.3 Typical presentation of geophysical logs including the drillers log and well construction

conductivity. In addition, some manufacturers sell probes that can be combined so that one pass in the borehole can measure multiple parameters. A common combination includes natural gamma, resistance and spontaneous potential.

Figure 18.3 is a typical presentation of geophysical logging data along with the corresponding geologic log and well construction information. The site is in northern Alabama where there is a clay residuum overlying limestone and is unsaturated. The project included the logging of all existing wells. The purpose was to provide a uniform set of data in which to assess geologic conditions and to identify possible karst conditions. Three geophysical logs were used that all provide data within steel cased wells and included natural gamma, neutron (porosity), and gamma-gamma (density). In this example, the existing well had a steel surface casing to about 12 m deep. The three logs are all responding to relative variations in three different properties:

- natural gamma log – variation in clay content,
- neutron log – variations in porosity, and
- gamma-gamma log – variations in density.

No significant geologic variations were noted in the driller's log. However, the geophysical logs show a distinct change in conditions between 24 and 27 m deep. This area is within the limestone and shows the presence of clay and has a lower density than the rock above and below indicating a possible filled cavity, fracture zone, etc. It is this ability to detect these subtle conditions that makes an array of geophysical logging tools such an asset to the site characterization process.

18.2.1 Key Aspects of Geophysical Logging

The selection of logs to be used for a project will primarily be based upon the objectives of the project. However, logistics and site-specific conditions will also impact the selection. For example, if you are working in newly drilled boreholes, will the open borehole be stable or is a temporary casing needed, and are the boreholes saturated or dry. If you are logging existing monitoring wells, what are their diameter and casing materials? There are a wide variety of logging

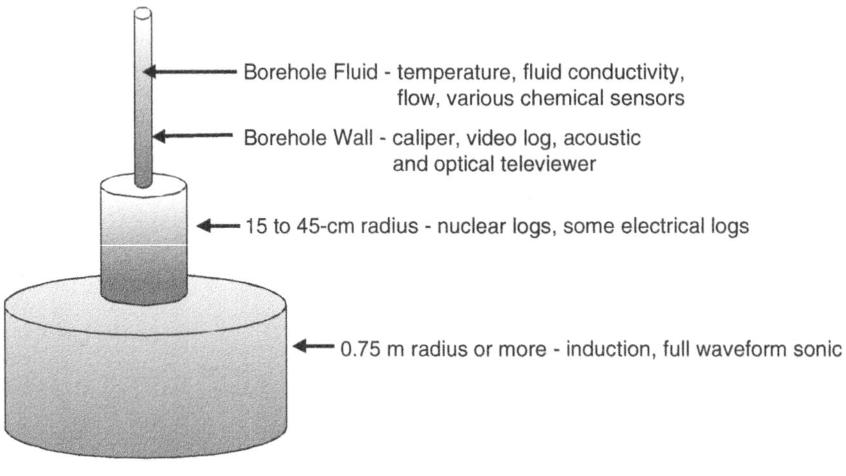

Fig. 18.4 The radius of measurement of some geophysical logs

measurements that can be made in a variety of borehole and well conditions. Although many logging measurements require an open borehole, some limited measurements can be made from within steel or PVC cased wells with a minimum diameter of 5 cm. Table 18.1 includes some of the generic conditions under which each of the logs can be used. There are other considerations when incorporating geophysical logging into a site characterization. Some of these key aspects of the geophysical logging methods that should be considered when selecting appropriate logs include:

- Parameter measured,
- Radius of measurement,
- Borehole conditions,
- Vertical resolution and speed of logging,
- Sequence of logging, and
- Processing and presentation of data.

18.2.1.1 Parameter Measured

Each of the borehole logging measurements are responding to changes of some physical, electrical, or chemical property of the borehole, its fluids or geologic conditions (Table 18.1). Changes in geologic and hydrologic conditions will be associated with a change in one or more of the measured parameters. For example, if detecting fractures is the objective you could possibly use one or a combination of logs depending upon site-specific conditions. If an open borehole is available, then caliper logs may be your first choice as a simple straightforward method. If more information on the fracture is necessary, then you might use downhole video or imaging using acoustic or optical televiewer logs. If determining flow along the fracture is of interest then the options include spontaneous potential, temperature and fluid conductivity which all may provide an indication of flow. If you want to quantify the flow then the options are a spinner flowmeter or heat pulse logs.

If logging open boreholes is not feasible, and all measurements must be made through casing then you are limited to the possible logs to choose from which includes:

- Natural gamma log, if the fractures are clay filled (PVC or steel casing),
- Induction conductivity log if the fractures are clay filled or have slightly conductive pore fluids (PVC casing only),
- Gamma-gamma log to detect an decrease in density (PVC or steel casing), or
- Neutron log to detect an increase in porosity (PVC or steel casing).

18.2.1.2 Radius of Measurement

Most geophysical logs measure 360° from the logging tool in all directions over some distance. The radius of sampling (or depth of penetration) ranges from measuring the fluid in the borehole, to the borehole wall, to a few centimeters into the borehole wall to a meter or more into the formation (Fig. 18.4). The measurements of in situ soil and rock properties become more representative as the volume of material sampled by geophysical logging increases beyond the borehole wall.

While not as commonly used, full waveform sonic can provide data up to a meter from the borehole, but typically much less. Ground penetrating radar can provide data up to several meters from the borehole and gravity measurements can provide data to tens of meters or more from the borehole.

18.2.1.3 Borehole Conditions

Variations in borehole conditions can affect or even limit the use of some logs. Some logs will allow measurements to be made in the unsaturated zone as well as the saturated zone, while others can only be run in the saturated zone. Most logs require an open hole while some logs will provide measurements from inside plastic or steel casing (Table 18.1).

The borehole or well diameter has to be large enough to allow the logging tool to pass smoothly up and down the borehole or well. The longer the logging tool, the more susceptible it is to bends in the casing or deviations in the borehole. Therefore, you need a borehole diameter larger than the logging tool and large enough to accommodate those bends or deviation without causing the tool to get stuck. However, it is also possible to have a borehole diameter that is too large. If a logging tool only measures out a few cm in radius then a larger borehole would reduce the effectiveness of the measurements. In a very large diameter borehole, the logging tool can be constrained by a spring up against one side of the borehole (using a decentralizer spring) to obtain more effective measurements.

18.2.1.4 Resolution and Speed of Logging

Vertical resolution will depend upon the digital sample rate and the speed at which the logging tool is moved through the borehole. Typical sampling for most logs is on the order of every 0.3 cm with a typical logging speed of 5 m/min. This provides essentially continuous data, for geologic purposes, over the length of the borehole. The radial resolution of the geophysical logs is constant with depth and is dictated by the specific logging tools penetration of measurement.

While a typical logging speed for many logs is about 5 m/min, some logs require a much slower speed of logging. Video logs, which are recording images, utilize a much slower speed of logging so that the images can be viewed effectively. Temperature and fluid conductivity logs are also run much slower (about 1 m/min) to allow the sensors to respond to changing conditions and minimize the disturbance of the borehole fluids.

18.2.1.5 The Sequence of Logging

The sequence of logging or the order the tools are used may affect the quality of the data depending upon the logs selected for a project.

Measurements within the water column require the borehole fluids to be in equilibrium with ambient conditions. After drilling, and possibly installing a well, sufficient time must be allowed so that the borehole can return to equilibrium. This may require from 10 to 100 times the duration of the drilling. Even the logging operation mixes the borehole fluid. Therefore, logs measuring fluid properties (temperature, fluid resistivity, and fluid sampling) should be run prior to other logs to minimize mixing of borehole fluids.

Consideration should also be given to when video logs are run because some logging tools may degrade borehole conditions and borehole fluid clarity. For example, caliper logs or those using bow-springs that contact the borehole wall should be run late in the logging sequence. The contact with the borehole wall creates a greater possibility of material falling into the borehole, degrading visibility and creating conditions, which may cause tools to be stuck in the borehole.

In general, all logs except fluid properties and video should be run with the probe moving up the borehole to reduce depth errors. Most logs are run continuously through the length of the borehole. Some measurements such as (P and S waves and gravity) are made at discrete intervals along the borehole.

Nuclear logs with radioactive sources such as gamma-gamma (density) logs and neutron (moisture/porosity) logs present special issues. The terms of licensing requires the operators to minimize the chances of a radioactive source being stuck down a borehole. Because of the consequences of loosing a tool with a radioactive source, unstable boreholes should not be logged with radioactive probes. An approach used by the authors is to only run nuclear logs through the drill stem, after final installation of the well or to have the borehole temporally cased with a PVC casing in which to run the nuclear logs. If using a temporary casing, it can then be removed and other logs can be acquired in an open borehole. If a nuclear log is run in an open hole, the hole must be run with a video log first to assure a clear path and stability of the borehole.

18.2.1.6 Processing and Presentation of Data

The typical geophysical logs are printed as raw data versus depth (Fig. 18.3). Many of the logs are used to identify relative changes and can be interpreted directly from the field data once the data have been plotted. A few of the logs such as full waveform sonic, ground penetrating radar and gravity will require different levels of processing and presentation.

Geophysical logs can be used in a qualitative or quantitative manner, depending upon project objectives. For example, a gamma-gamma (density) log can be used to indicate that one zone of rock is more or less dense than another zone. This same data can be expressed in density units if the probe has been calibrated to known conditions or samples from the borehole have been tested in the laboratory for comparison. While quantitative results can be obtained from most geophysical logs, caution must be used in over-interpreting and quantifying of data, if proper calibration to site-specific conditions or correlation to laboratory analysis has not been properly made.

ASTM has developed a guide for planning and conducting borehole geophysical logging measurements D5753-95 (ASTM 2010a) along with a few guidelines for specific geophysical logging measurements. They include:

- ASTM D6274-98 for gamma logging (ASTM 2010b),
- ASTM D6726-01 for electromagnetic induction logging (ASTM 2007a),
- ASTM D6727-01 for neutron logging (ASTM 2007b), and
- ASTM D6167-97 for mechanical caliper logging (ASTM 2011).

18.3 Various Applications for Geophysical Logs

There are a wide variety of applications for the geophysical logging methods. The application requires a borehole or existing well and a contrast in the parameter being measured by the specific logging tool. The applications can generally be categorized into six areas and include:

- Assessment of natural geologic and hydrogeologic conditions;
- Measurement of soil and rock properties for engineering applications;
- Detection and mapping of contaminant plumes and spills;
- Evaluation of construction and conditions within existing monitoring wells and piezometers;
- Evaluation of man-made soil, rock or concrete structures; and
- Groundwater and mineral exploration and evaluation.

The following examples illustrate some of the wide range of applications for the geophysical logs as applied to geology, hydrology and karst problems. These are but a few examples to highlight some of the logging methods. There are a wide variety of logging tools and variations on their application that are available.

18.3.1 Mapping Stratigraphy

Geophysical logging was used effectively to map the limestone and shale layers at a site in Kansas City, Kansas. The site was a landfill that was expanding over an abandoned limestone mine (see case history in Part III, Chap. 25 for more details). The natural gamma log provided an extremely repeatable response to the limestone and shale layers as well as strong responses to three key marker beds of shale (Fig. 18.5). The limestone is clearly shown by the lower gamma values and certain shale beds by higher gamma values.

The three obvious shale beds in the natural gamma log were used as marker beds for correlation between boreholes. In this case, the consistency of the gamma logs between boreholes (Fig. 18.5) is a measure of uniform geologic conditions over the site. Thirty-nine natural gamma logs were acquired across the site using existing mine fillholes and old mine vent holes. Surface elevation for each borehole was established and the elevation of the uppermost shale layer seen in the natural gamma log was used to develop the contour map of this shale layer. The contour map clearly shows a distinct depression in the shale layer, as much as 5 m (Fig. 18.6). This data helped define a paleocollapse zone, which originated 180–240 m below the mine.

18.3.2 Low Density Zones, Fractures and Cavities

Development on Grand Cayman, Cayman Islands in the Caribbean, required an injection borehole. A suite of geophysical logs was run in a borehole in order to identify an optimum depth for injection. The objective was to identify a more permeable zone with fractures or cavities. The limestone of Grand Cayman is known for its dissolution features, so cavities were expected as well as low density weathered rock and small to large fractures. Figure 18.7 shows a portion of the geophysical logs acquired within this borehole in the saturated zone. The suite of logs all indicated a broad zone of cavities extending from 63 to 69 m deep at the contact between two formations. Each of the logs responded to this cavity zone:

- The caliper log indicated an increase in borehole diameter;
- The resistance log indicated a decrease in resistance associated with water-filled cavities;
- The spontaneous potential indicated an increase in voltage (due to possible flowing water);
- The gamma-gamma (density) log indicated a decrease in density (an increase in cps); and
- The neutron (porosity) log indicated an increase in porosity (a decrease in cps).

This is an excellent example in which multiple logs all indicated a zone of increased fractures, porosity and cavities. The use of multiple logs also improved confidence levels in the interpretation.

Another example of logging data within a complex hydrogeologic system is from the Woodville Karst Plain (WKP) in northern Florida. After a test phase utilizing several geophysical techniques, microgravity was selected as the most effective means for regional surveying within the WKP in order to detect large mass deficits associated with conduits and caves (see Part II, Sect. 16.5.3 for more details on this project). A total of 43 km of gravity data was obtained. Four of the large regional microgravity anomalies (70–90 uGals) indicating mass deficits within the subsurface were drilled and geophysically logged (Yuhr et al. 2008a).

The field plan for the geophysical logging required two phases. The first phase was a suite of logs that were acquired through a temporary 5 cm PVC casing because radioactive source logs were being used. This included the nuclear logs natural gamma, gamma-gamma, neutron-neutron and an induction log. The temporary casing was then removed and fluids within the boreholes were allowed to stabilize for 24 h. The second round of logging included temperature, fluid conductivity, and caliper. Downhole video was added in all boreholes after the caliper log got stuck in the first borehole at a depth of 75 m.

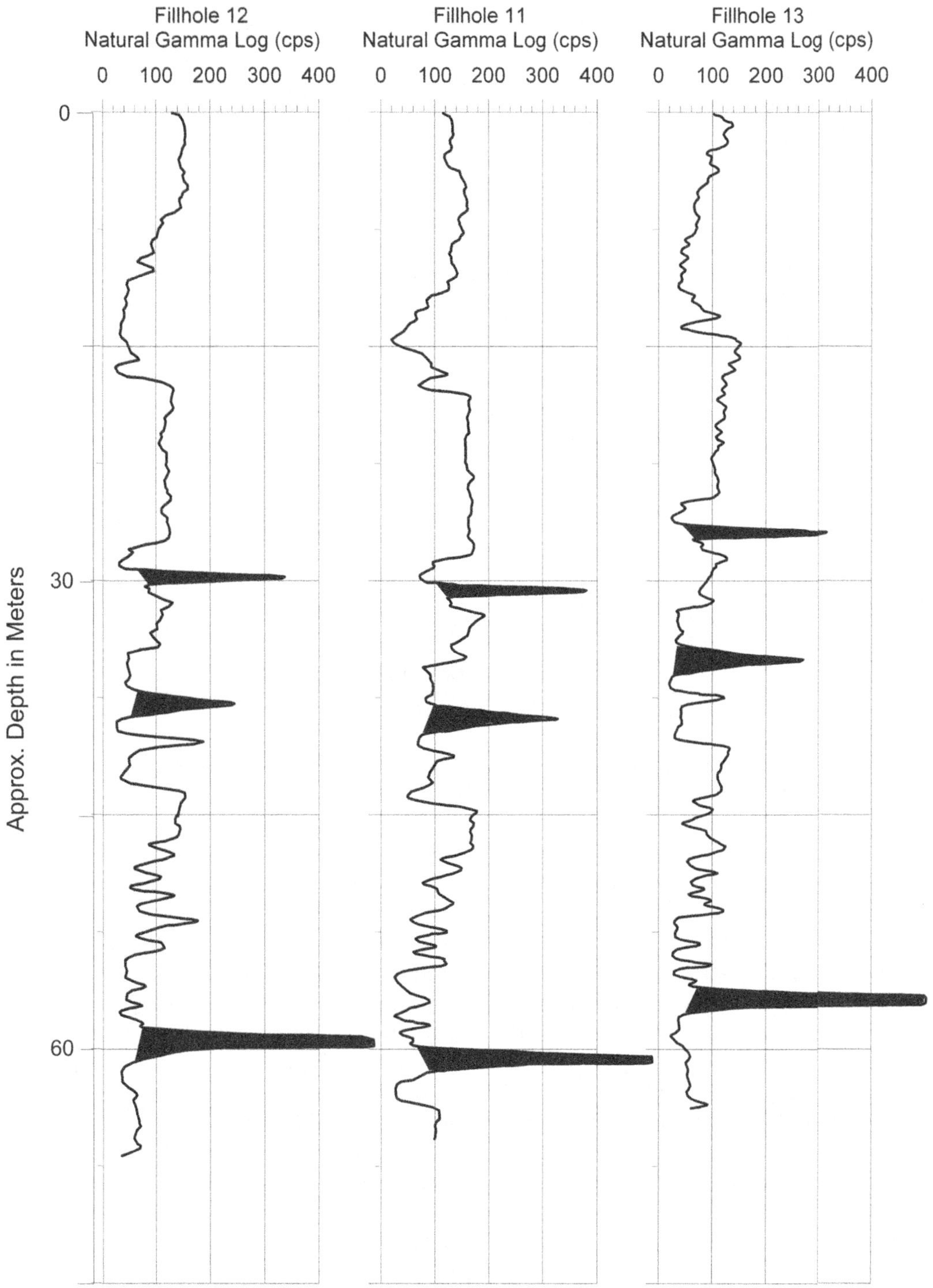

Fig. 18.5 Correlation of stratigraphy using geophysical logs. Note the good correlation between boreholes using the shale marker bed

Figure 18.8 shows the complete suite of logs acquired in one of the four boreholes (B6-1). An open borehole was difficult to maintain during drilling due to the numerous fractures and cavities encountered. A steel surface casing extended to about 70 m and a temporary PVC casing was used inside the steel casing extending to the bottom of the borehole at a depth of 90 m. The first phase of logging was completed through the PVC casing, then it was removed. The fluid logs (temperature and fluid conductivity) were run at the bottom of the open borehole (70–90 m). The steel casing was then removed and the fluid logs were then run in the upper 70 m. The caliper log was the last log to be run and could not be lowered to the

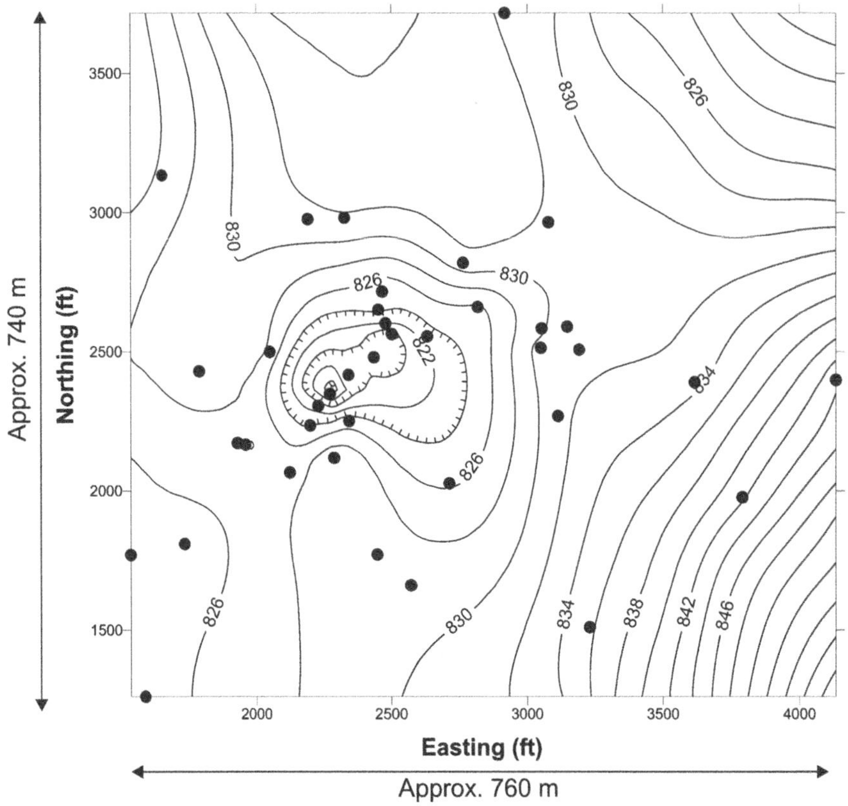

Fig. 18.6 A contour of the uppermost shale bed identified in the geophysical logs (Fig. 18.5) from 39 boreholes shows a depression of about 5 m that corresponds to a paleocollapse. Note *contour lines* are elevation in feet

bottom of the borehole due to an obstruction at 70 m. Therefore data is only from the upper 70 m. This borehole is an example of some of the difficulties that are often encountered when drilling and logging in karst conditions.

Figure 18.9 shows the gamma-gamma (density) and caliper logs from all four boreholes. This data clearly show the broad low-density zones within each of the boreholes. There were also discrete cavity zones indicated in three of the borings. One of the boreholes B2-1 was not run with a caliper due to the tool getting stuck and not being able to maintain an open uncased hole.

The downhole video revealed turbulent flow at discrete cavities and upward flow in two of the borings (B2-1 and B6-1). This was based upon movement of particulate matter in the water column. The other two borings (B5-1 and B5-2) showed little to no particulate movement, however observations of a small crawfish-like creature in one of these borings (B5-1) indicate some level of connectivity to a more open conduit system.

The logging data confirmed the presence of low-density zones mapped by microgravity data. This then supported the use of the microgravity data to estimate the locations and frequency of higher porosity zones within the WKP beyond those mapped by cave divers and dye tracing. This data could also be used to support a groundwater model of the WKP.

Another example is from an investigation of the proposed new Detroit River Crossing Bridge that utilized a wide range of borehole geophysical measurements to characterize the geology and possible brine cavities from the dissolution mining of salt. Data included 13 borings, core analysis, geophysical logs (natural gamma, density, ATV, and deviation logs), vertical seismic profiling (VSP), crosshole and borehole gravity measurements. The borehole gravity measurements were effective in identifying depth and thickness of possible brine cavities (Michigan Department of Transportation 2008).

Further details of the wide range of borehole based geophysical techniques intended for investigation of deep waste depositories along with world-wide examples including the Detroit River Crossing Bridge are found in Williams et al. (2009). These same methods are also applicable to site characterization of karst and pseudokarst conditions.

18.3.3 An Alternate to Core Samples or Oriented Core

Core sampling is a standard approach for detailed geologic assessment of rock and necessary in order to provide samples for engineering tests and laboratory analysis. Oriented core has been used to determine not only the presence of fractures, but their orientation, strike and dip. However, if detailed mineralogical diagnostics or geotechnical laboratory analysis are not required from the core then an alternative approach is

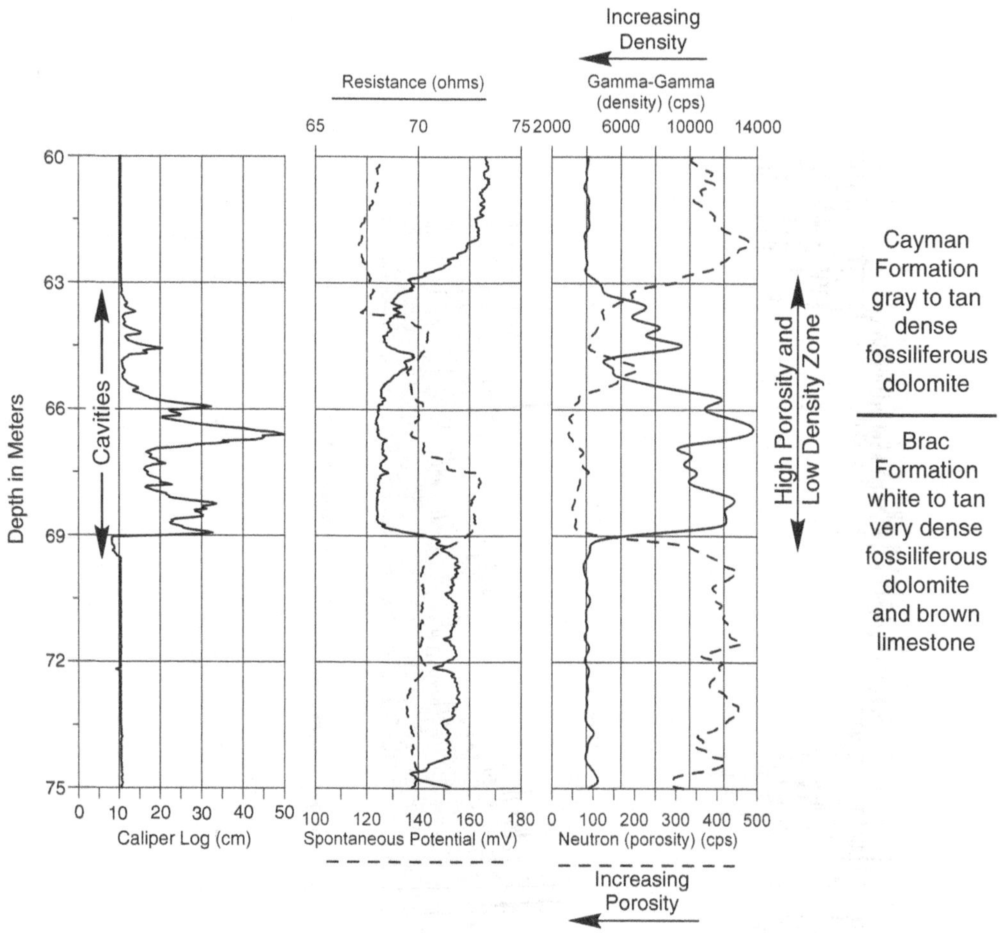

Fig. 18.7 A suite of geophysical logs from the Cayman Islands identifying a porous/cavity zone for injection of fluids

available. Both the acoustic televiewer (ATV) and optical televiewer (OTV) provide an excellent means of obtaining a continuous and detailed in-situ image of geologic conditions within an open borehole similar to a core (Fig. 18.10). This data was obtained by Colog, who was part of the site characterization team at Marshall Space Flight Center in Huntsville, Alabama. The level of detail is as good or better than that obtained from oriented core. In addition, the log does not induce fractures that might occur during coring and the log provides data in weaker zones where core may not retrieved. Furthermore, the 360° data allows a pseudo-image of the core to be developed. Quantitative structural analysis can be done such as fracture aperture, strike, dip, rose diagrams and stereo-net plots. Since the logs are digitally recorded, the storage and maintenance of core is avoided.

18.3.4 Groundwater Flow and Contaminants

Mapping of geologic features such as cavities, fractures, void, etc. is only one part of the investigation. The connectivity and flow associated with these features is required when developing groundwater models, establishing vulnerability mapping or assessing contaminant transport and distribution. Some of the geophysical logging tools can indicate flow by changes in a temperature or fluid conductivity log. Other geophysical logging tools such as impeller or heat pulse flow logs can measure flow directly.

Another method that has been developed is called HydroPhysical logging (Colog 2005). This technique uses an open borehole filled with deionized water. Time series fluid electrical conductivity logs are then run. Changes in the fluid conductivity are measured as the borehole returns to ambient conditions and natural groundwater moves both into the borehole, displacing the deionized water. Figure 18.11a is a plot of the time series fluid conductivity logs in a single borehole. The data has been annotated showing the flow into and out of the borehole. The changes measured in the fluid electrical conductivity logs over time can be used to calculate various hydrologic parameters for specific intervals or a discrete feature. These calculations include flow rates, hydraulic conductivity and transmissivity. These tests can be run under ambient or pumping conditions. This is particularly useful when hydrologic

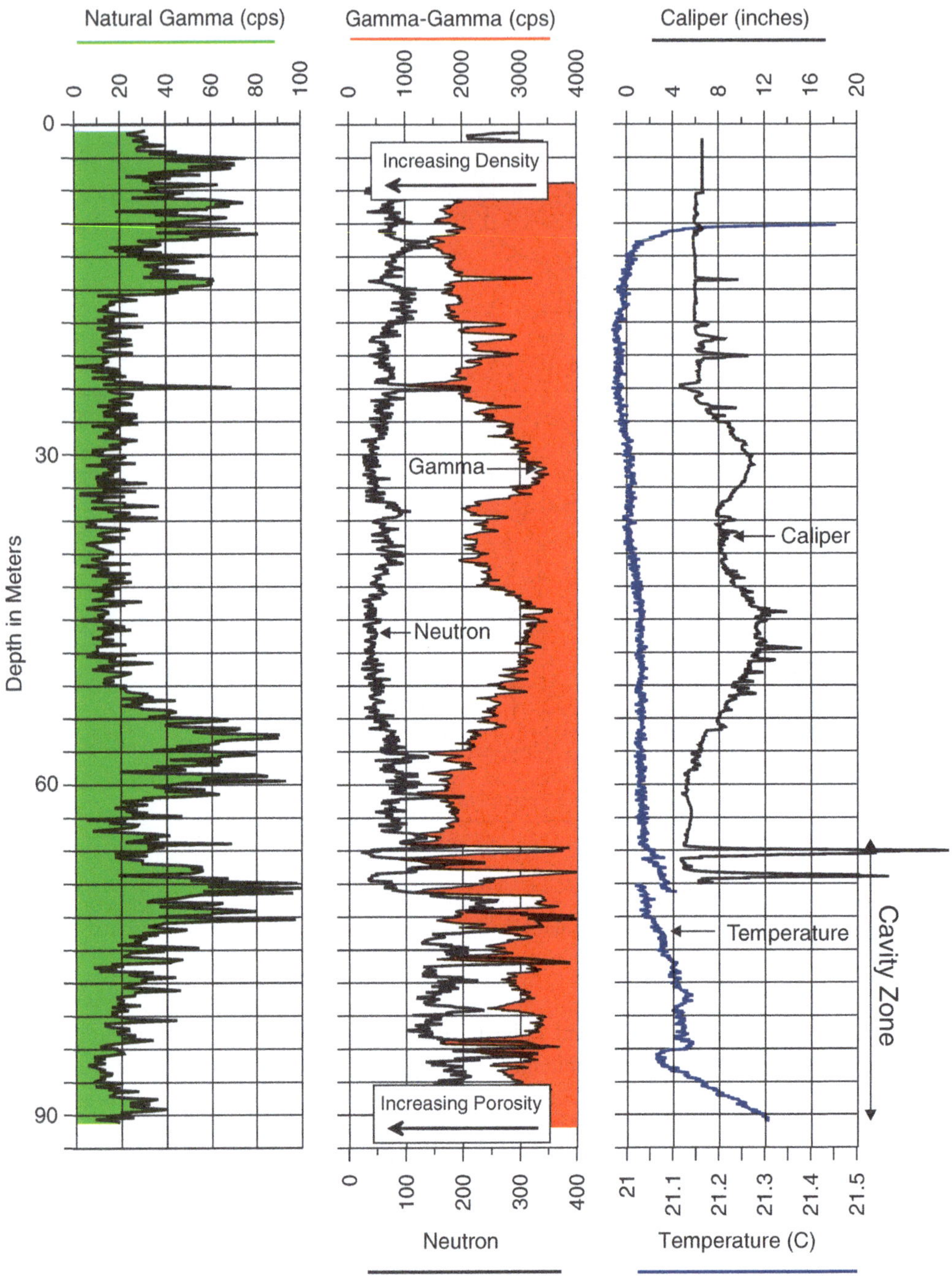

Fig. 18.8 A suite of geophysical logs from the Woodville Karst Plain in northern Florida showing a cavity zone

conditions or connectivity of karst conditions are of concern.

A geologic and karst characterization was completed at the Marshall Space Flight Center (MSFC) in northern Alabama. It was a multi-phase project that incorporated multiple surface geophysical methods, drilling and borehole geophysical logging methods. The primary objective of the surface geophysical investigation was to evaluate the extent, occurrence and hydrogeologic significance of the transition (epikarst) zone and deeper karst features that may provide preferential groundwater flow paths (Yuhr et al. 2005). This information was integrated with a diverse suite of existing data and then used to guide the location of twenty-eight (28) deep bedrock borings to assess groundwater flow and quality.

The integration of surface geophysical data with existing data proved to be highly beneficial when selecting the locations for bedrock wells to intercept groundwater flow. However, the drilling and logging program indicated that, even in the most significant anomalous areas, no major karst features were encountered (e.g. old collapse features, large dissolutioned zones, caves, etc.). Instead, small discrete

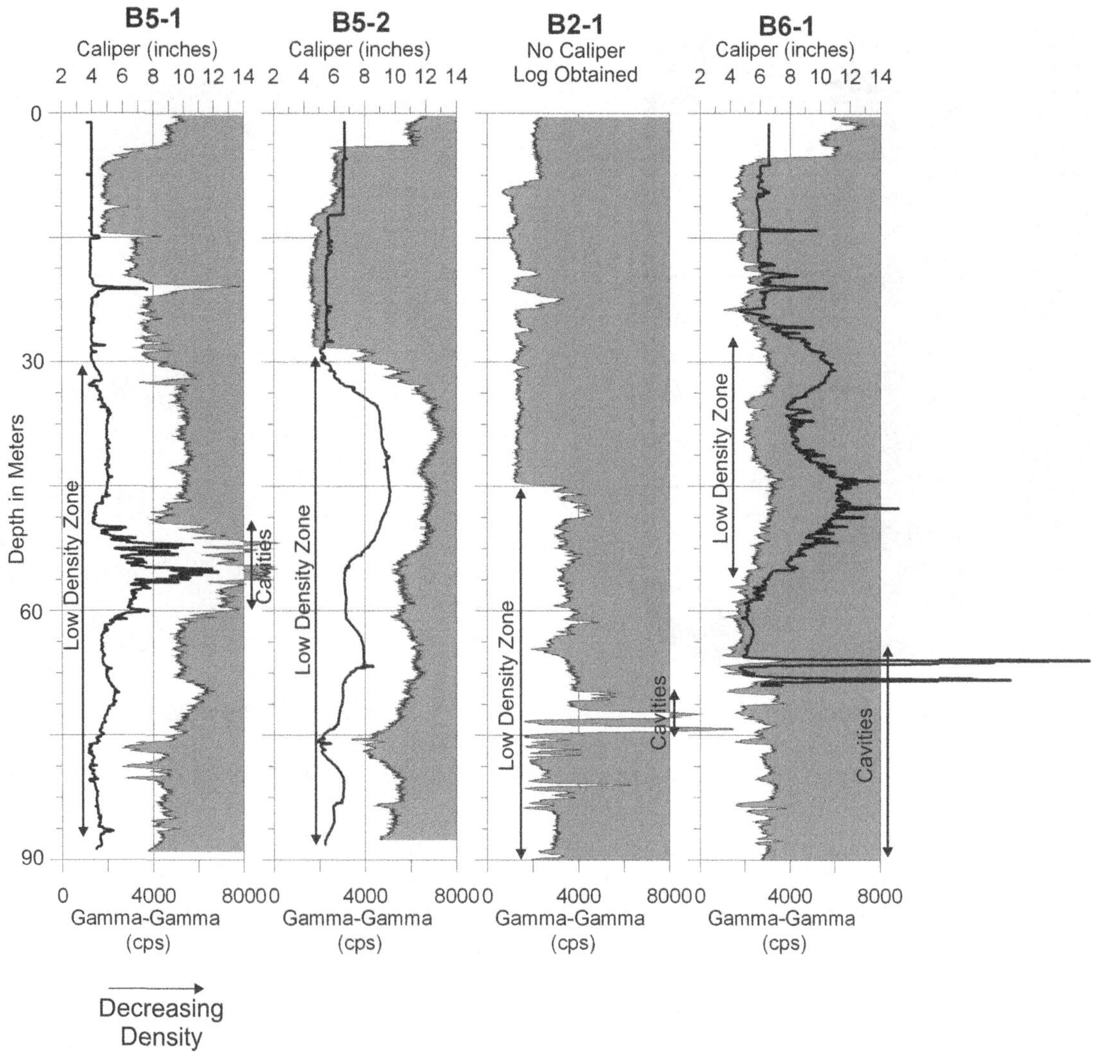

Fig. 18.9 The caliper and gamma-gamma logs from the four boreholes in Woodville Karst Plain. The caliper is plotted as a *single line* and the density log is *shaded* with density decreasing as values increase to the right

fractures or fracture zones were detected and were pervasive throughout the depths drilled (75–90 m deep). Figure 18.11b plots the hydraulic conductivity of the flow zones (calculated from the HydroPhysical™ logging) versus their depth and thickness to illustrate the spatial distribution of the flow zones. The highest flow was focused within the upper 22 m. This upper 22 m high flow zone corresponds to the weathered limestone in our conceptual model (Yuhr et al. 2008b). The small discrete fractures and limited fracture zones that were detected appear to be the key factor in controlling groundwater flow.

18.3.5 Pseudokarst Due to Acid Leaks

Logging can be very effective whether evaluating naturally occurring karst or pseudokarst. This example of pseudokarst was formed by acid leaks into the underlying limestone at an industrial facility. This is type of pseudokarst can rapidly dissolve the limestone creating problems over short periods of time (tens of years).

Two small adjacent sinkholes had developed along side of a large manufacturing facility in Puerto Rico. They were each less than a meter in diameter and one was about 3 m deep while the other was about 11 m deep. An initial assessment was made with a borehole video camera lowered into the deeper sinkhole. There appeared to be a problem with the camera as it shut down when it was lowered into the deeper sinkhole. While examining the camera, it became apparent that it was hot to the touch and likely overheating. After examining the camera, which appeared to be operational, the problem was found to be an increase in temperature with depth in the subsurface that was overheating the camera.

While the site is located in a karst terrain, it was becoming obvious that these sinkholes were not naturally occurring.

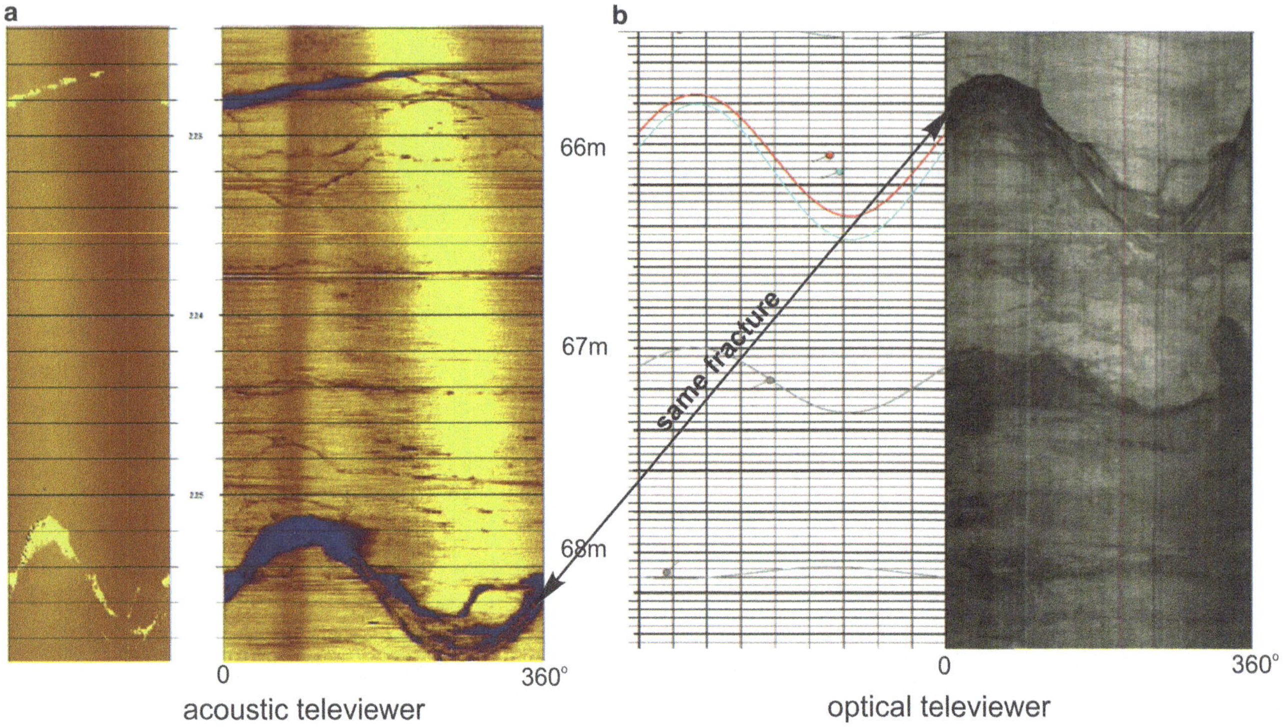

Fig. 18.10 Examples of ATV log (**a**) and OTV log (**b**) illustrating their ability to characterize fractures (Colog 2005)

The elevated temperatures and unusual odor within the sinkholes indicated that there might be acid leaking from the plant and actively dissolving the limestone. These two small sinkholes were an early indication of possibly larger voids, which could critically impact the structure.

Microgravity measurements were made in and around the building to assess the potential mass deficit in the subsurface due to dissolution. The gravity survey indentified a 100-microGal anomaly centered at the two sinkholes. The mass deficit calculated from the gravity data was approximately 650 m^3, much larger than observed at the two sinkholes. A drilling program was then initiated using both vertical and angled boreholes along with geophysical logging, which confirmed significant dissolution and voids to a depth of 45 m.

Two geophysical logs were used in the borings, a natural gamma log to indicate variations in clay content and an electromagnetic induction log to indicate variation in conductivity. Figure 18.12 shows an example of these logs from three of the vertical borings. The consistent low values of 10–20 cps in the natural gamma log suggest that there was no clay present. Background conductivity values for limestone are typically 10–20 mS/m. The induction logs measured conductivity of 50 mS/m with values increasing to greater than 200 mS/m. The combination of no clay indicated by the natural gamma logs, high conductivities measured in the induction log, and high temperatures in the subsurface all indicated the presence of extensive acids.

A conceptual model was developed for the void space and a grouting plan was designed based upon this investigation. A total of 400 m^3 of grout were injected in the upper 23–30 m to stabilize the structure.

18.3.6 Corrections Due to Borehole Deviation

Most borings are drilled vertical and we assume that the borehole is vertical throughout its total depth. However, it is not uncommon for a borehole to deviate significantly from the vertical. The impact of borehole deviation can be negligible in many applications, but can induce significant errors in certain measurements, such as deep water level measurements with a relatively flat water table, or hole-to-hole seismic measurements and other geotechnical measurements. Therefore, for those critical measurements, we need to know if the borehole deviates from vertical and the extent of the deviation.

A commonly used deviation log (or inclinometer probe) utilizes magnetic and accelerometers sensors, and provide continuous data. But these probes are not effective within steel casing. For steel cased holes or areas with strong magnetic soil or rock a gyro-based system can be used. The older Fotobar deviation probes, which use bubble levels and a 16 mm camera are still in use and provide data at discrete depths.

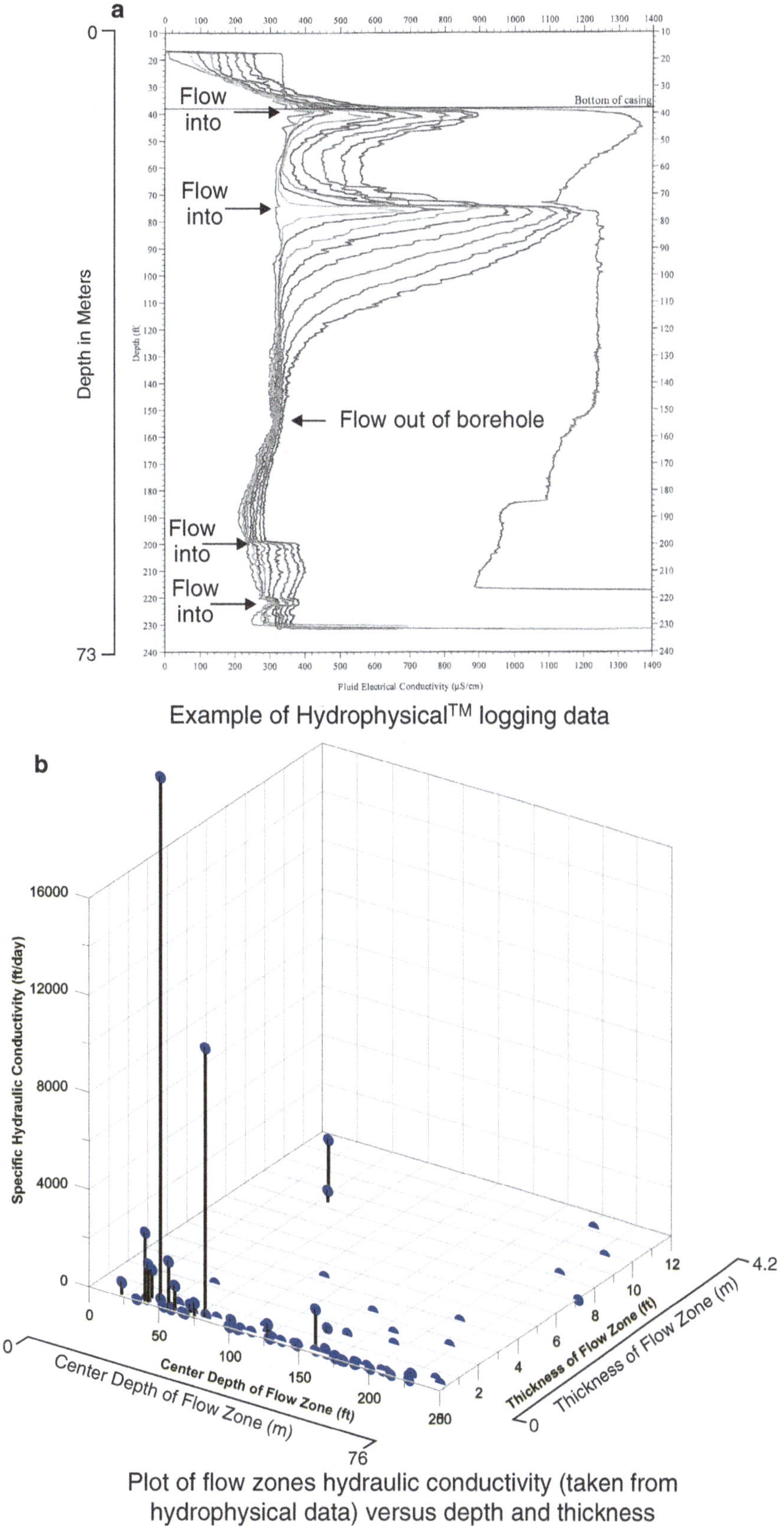

Fig. 18.11 An example of HydroPhysical logging (Colog 2005) (**a**) and the spatial distribution of flow zones identified by the HydroPhysical data (**b**)

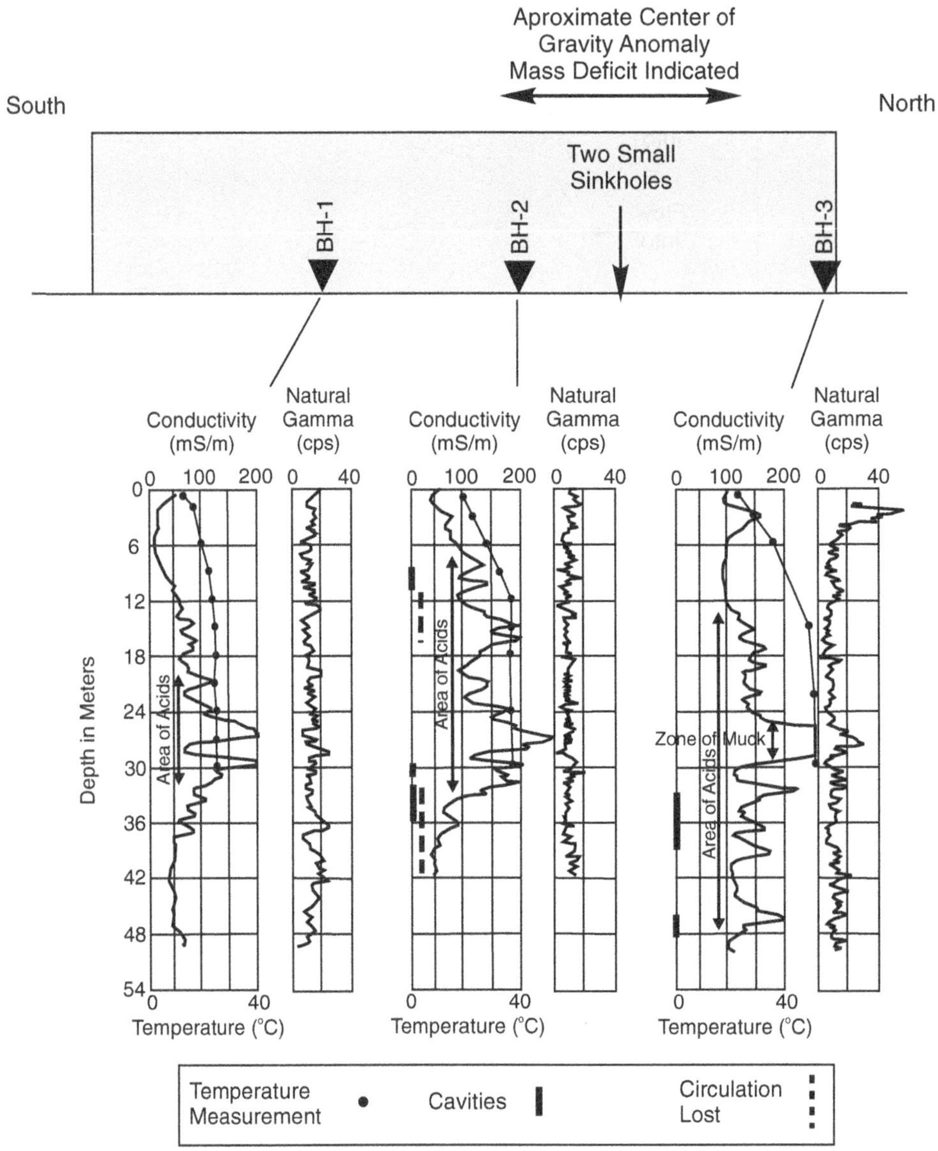

Fig. 18.12 Natural gamma, electromagnetic and temperature data indicating the presence and distribution of acids in three borings

A situation in which borehole deviation may need to be considered is that of determining water levels. If the water table is relatively flat and deep (more than 60 m) borehole deviation alone can result in significant errors in the measurement of water level. For example, water table measurements at a depth of 76–82 m in northwest Puerto Rico were affected by borehole deviation. Deviation logs were acquired in 16 deep wells that all had elevation errors in the range of 0.25–1.44 m due to deviation. Figure 18.13 shows the borehole deviation in one deeper borehole in Puerto Rico.

An extreme example of deviation was noted when hole to hole geophysical measurements were being made by Tennessee Valley Authority (TVA). The two probes became stuck at a depth of about 30 m in two boreholes 10 m apart. It was then discovered that they were entangled with each other indicating that the boreholes were not parallel as assumed, but deviated significantly intersecting one another (Hopkins, 2013, personal communication).

18.3.7 Reconstructing Geohydrologic and Well Construction Data

There are many sites where existing borings or wells are available. These borings are often installed over time, by different contractors, using different criteria for construction. In these situations, there are a number of reasons to acquire geophysical logs in these wells (Turner and Benson 1986). For example:

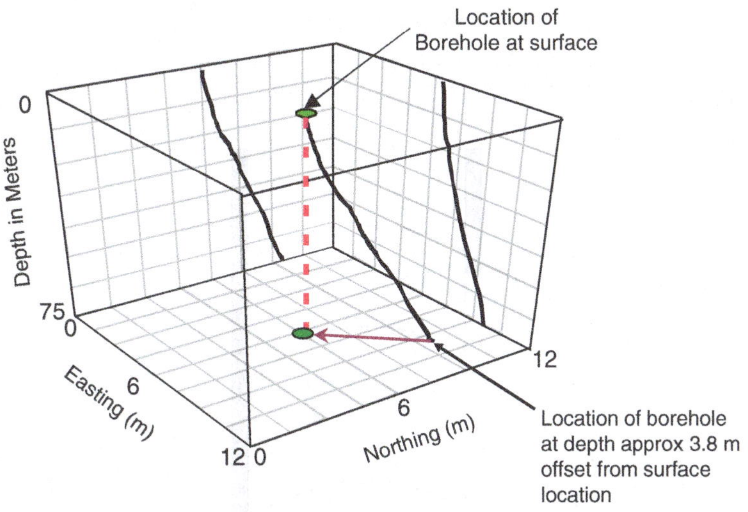

Fig. 18.13 A deviation log showing the distance from vertical a deep borehole was actually drilled

- If there is no geologic log of a borehole or well, or where there are questions about such data, geophysical logging can often provide the necessary geologic information. Even if a boring log is available, subtle variations in the geology or hydrologic conditions may be of interest and the geophysical logs may be used to verify the original geologic logs and provide additional data.
- If correlation between new and existing wells needs to be established, using a consistent suite of geophysical logs in all wells can be useful.
- If well construction details need to be determined or verified, such as, casing material, depth of casing, diameter, screen interval and type of screen.
- If an evaluation of existing conditions is required to assess damage or deterioration or other changes in cased or open hole conditions geophysical logs can provide the needed data.

18.3.7.1 Consistent Data Set for Geologic Evaluation

A Superfund site in west-central Florida, had been in operation for about 40 years as a phosphate processing plant. They had numerous existing wells on-site that included piezometers, monitoring wells, and deeper supply wells. As part of the site characterization, geophysical logging was completed in all existing wells and new borings in order to provide a consistent data set that was used to aid in the evaluation of the geology at the site. This data set consisted of natural gamma (clay content), gamma-gamma (density), neutron (porosity) and induction (conductivity) logs. By logging the existing wells, we took advantage of having these vertical geologic windows and added 35 spatial data points to support the conceptual model. This information was specifically used to develop a contour map of the semi-confining layer at the site, evaluate the epikarst and identify the presence of karst features. For more information see the case study in Part III, Chap. 27.

18.3.7.2 Reconstructing Geologic Data

Brine water associated with oil exploration wells was collected in above ground storage tanks in southwest Florida. During operations at one site, brine was spilled on the surface creating a significant contaminant plume. A suite of monitoring wells had been installed over the site, but chemical sampling did not reveal elevated values or a clear plume boundary based upon chloride data. The conductive brine contaminants provided an excellent target in which to use both surface and downhole electrical measurements. An investigation using surface geophysical electromagnetic methods to map the lateral extent of the plume and downhole electromagnetic induction logging to map the vertical extent of the plume as well as a thorough review of regional literature and all chemical data was completed.

The review of existing literature identified an aquitard, which separated the surficial and deeper aquifers. The review of on-site drilling data showed that the majority of the monitoring wells were shallow and completed well above this aquitard. The few wells that were drilled below the aquitard were sampled at a 3 m interval when drilling and had missed identifying the aquitard. The deeper wells had their screens installed just below the aquitard.

The results of the surface electromagnetic conductivity data showed a very clear high conductivity plume surrounding the above ground tank. Spatially the wells were located within the high conductivity plume and should have defined it based upon chloride data. However, the chloride values

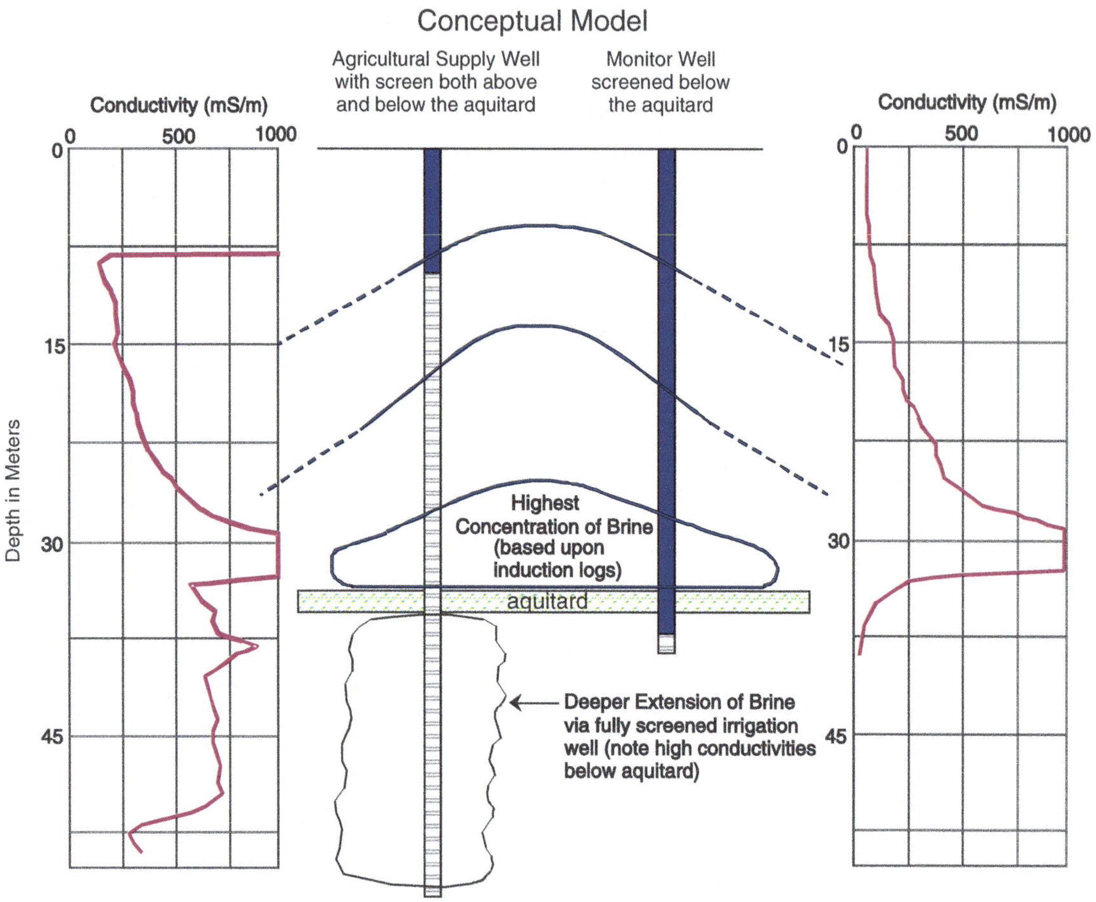

Fig. 18.14 Electromagnetic induction logs identified brine contaminant distribution

were all relatively low and did not indicate a coherent plume of contaminants.

Induction logging was completed in all monitoring wells on-site. While in the field we found another well associated with supplying water for agricultural purposes. This well was not part of the monitoring system and extended much deeper. This well was also logged with the induction logging tool.

Figure 18.14 shows two of the electromagnetic induction logs from the site that were located near the center of the high conductivity plume. The induction logs indicated very high conductivity (greater than 1,000 mS/m) centered at a depth of about 30 m just above the aquitard. These very high conductivity values confirm the presence of the brine and show that the dense brine fluid settled down on top of the aquitard.

The reason the chemical sampling data from the wells did not define the plume is because the shallow wells were screened above the plume and the deeper wells were sampling just below the aquitard, also missing the plume. Once the vertical extent of the brine was identified, screened intervals and chemical data from the existing monitoring wells could be put into proper perspective.

18.3.7.3 Monitoring Well Audit

Assessment of existing borings, wells and monitor wells can provide basic geologic and hydrologic data. This type of assessment can also indicate whether well conditions area as designed or even have deteriorated causing any data acquired from them to be invalid (Turner and Benson 1986). A landfill located in northern Indiana initially began in 1979 and was in the process of permit expansion in 1990. The 55 ha site is located in an area of glacial till (22 m thick), overlying a sand aquifer. Depth to bedrock is approximately 50 m.

As part of the landfill expansion, a comprehensive hydrogeologic report was being put together. Much of the data and information came from existing site-specific reports, borings and monitoring wells. Five (5) monitoring wells had been installed and sampled quarterly since 1979. The landfill was well designed with an engineered liner consisting of on-site glacial till and leachate sumps, which were pumped on a regular basis. No problems had been detected in any of the wells.

A monitoring well audit was completed at the site that included a surface inspection of the well and geophysical logging using a natural gamma and induction logs as well as

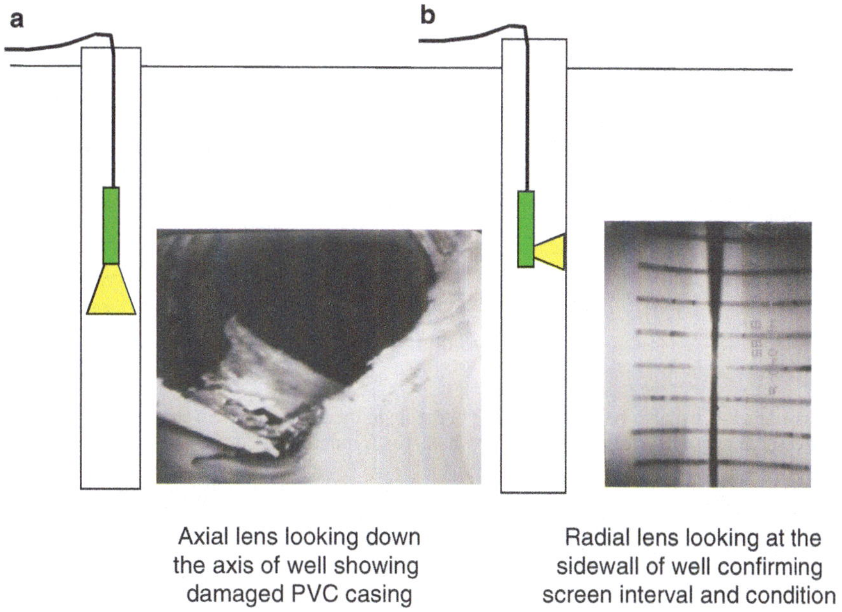

Fig. 18.15 Downhole video camera logging provides for assessment of damaged casing (**a**) and the presence of well screen (**b**)

a video camera. A simple check of the well elevations indicated a 1.5-m error in one of the five wells. This had a significant impact on the original water table map. Natural gamma and induction logs of existing and new wells indicated the presence of two aquifers, not one, based upon the presence of an aquitard. A downhole video camera was used with both axial (Fig. 18.15a) and radial (Fig. 18.15b) lens. The downhole camera in the wells showed that the formal construction diagrams for the wells were completely in error. The wells had broken casing (Fig. 18.15a) and screened intervals in some wells were created using a few hacksaw cuts. The screened intervals were also found to be interconnecting the two aquifers.

The monitor well audit revealed a number of issues that made the data from the original monitoring network useless. All five monitoring wells were replaced with a cluster of two wells (one in the shallow aquifer and one in the deep aquifer) and new water table and potentiometric maps were developed along with a new site-specific hydrogeologic report, which was all submitted to the state.

18.4 Downhole, Crosshole and Tomographic Measurements

Besides traditional geophysical logging in a single borehole there are a range of other measurements that can be made between the ground surface and a borehole (downhole) or between boreholes (crosshole). These methods significantly expand the volume of material that is sampled from the borehole. These measurements would normally be made after site-specific geologic conditions had been defined so that measurement locations can be selected to meet very specific objectives. While surface geophysical measurements are very effective in locating anomalous conditions, they often lack resolution or penetration required for deeper foundation design. It is here where downhole and crosshole measurements are necessary to obtain the detailed engineering data often needed for foundation design.

Figure 18.16 illustrates these types of borehole measurements. Measurements of P and S wave velocities using seismic methods can be made between the surface and a borehole (downhole measurements – Fig. 18.16a). These provide in-situ measurements of undisturbed soil and rock conditions from about 3 m or so away from a borehole. Measurements can also be made between two or more boreholes (crosshole measurements – Fig. 18.16b). The distance between boreholes is typically 3–6 m. These measurements of P and S wave velocities made at discrete depths within the borehole are used to calculate elastic modulus. A variety of measurements such as seismic, resistivity and radar can also be made between two or more boreholes (Fig. 18.16c) to provide topographic imaging of conditions between boreholes.

18.4.1 Downhole and Uphole Measurements

Many foundation and soil-structure interaction problems require values of elastic moduli. Determinations of elastic moduli can be made using laboratory tests on cores or in-situ measurements of seismic compressional and shear wave

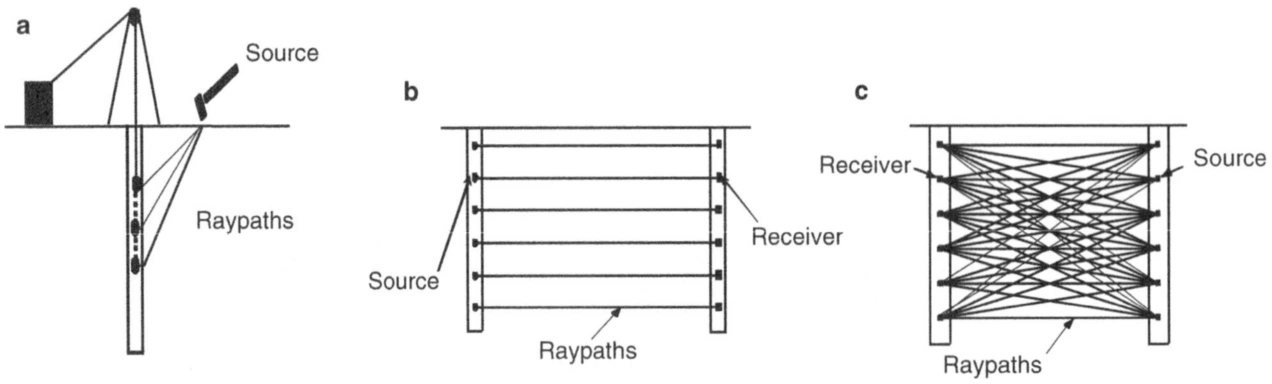

Fig. 18.16 The types of measurements that can be made from the borehole to the surface or between boreholes. (**a**) Surface to borehole or borehole to surface (downhole measurements). (**b**) Hole-to-hole or crosshole. (**c**) Tomography

velocities (P- and S-waves). Measurements from surface to hole, hole-to-hole, or crosshole methods not only provide data on in situ conditions but also sample a large volume of material. Values of elastic parameters such as shear modulus, Young's modulus and Poisson's ratio can then be determined by the measurement of P- and S-wave velocities, provided the density of the soil is known or can be reliably estimated (Anderson and Woods 1975; Ballard and McLean 1975; Stokoe and Abdel-Razzak 1975; Wair et al. 2012).

Seismic measurements made in a surface to borehole configuration are commonly referred to as Vertical Seismic Profiling (VSP). The velocity data are also used to model and interpret seismic refraction and reflection data. These measurements have been commonly made using a sledge hammer and plank at the surface as an energy source. By striking a vertical blow (Fig. 18.17a) compressional waves (P-waves) are developed. By striking the opposite ends of the plank (held in place by the wheels of a vehicle) opposing horizontal shear waves (S-waves) are developed (Fig. 18.17b). The seismic energy is recorded as a function of depth by geophones clamped into place in a dry borehole or hydrophones placed within a water-filled borehole at various depth intervals. These measurements will provide seismic velocities of both P and S-waves versus depth at a borehole location. See ASTM D7400-14 standard test methods for downhole seismic testing for more information (ASTM 2014b).

Hole to surface measurements simply reverses the procedures with the hammer or source in the borehole and the geophones at the surface. Because the energy source is downhole, there is an increase in frequency of the signal and improved resolution.

An example of surface to borehole seismic measurements comes from the karst investigation at the Superfund site on the west coast of central Florida. The location of a paleocollapse feature had already been identified and its stability and strength was part of the assessment. Boreholes were already installed both within the paleocollapse as well as in background conditions. Surface to borehole seismic measurements were completed in a number of test borings across the site. Figure 18.7c, d show plots of both the P-wave and the S-wave velocity data. It is interesting to note that there was very little variation in P-wave velocities across the site regardless of location. However, the shear-wave velocities show a slight but clear decrease in velocities within the paleocollapse feature due to the type of infill material. The results suggested that the paleocollapse was stable with reasonable strength characteristics. For more detailed on this case history see Part III, Chap. 27.

18.4.2 Crosshole (Hole to Hole) Measurements

Crosshole or hole-to-hole measurements (Fig. 18.16b) have been commonplace for years, making its entrance into the engineering profession in the early 1970s. Hole-to-hole measurements of seismic velocity will generally provide better vertical resolution of both P and S-wave measurements, than that obtained from either down or uphole measurements.

Measurements are commonly made between two or three boreholes in a line. Measurements are made by placing a source in one borehole and receivers in the other borehole or boreholes (Fig. 18.18a) with a spacing of 6–7 m or less between boreholes. Measurements are made at some interval moving down the borehole and both P and S-wave velocities can be calculated from the each measurement. The data in Fig. 18.18b was acquired every 1.5 m and shows much more detail than that presented in Fig. 18.17c, d. Receivers can be geophones in dry boreholes or hydrophones in water-filled boreholes. A source that can generate both P and S-waves is used. ASTM D4428 M-14 provides guidance for the spacing and construction of boreholes used for P and S-wave measurements for use in hole-to-hole seismic testing (ASTM 2014a). These same concepts are appli-

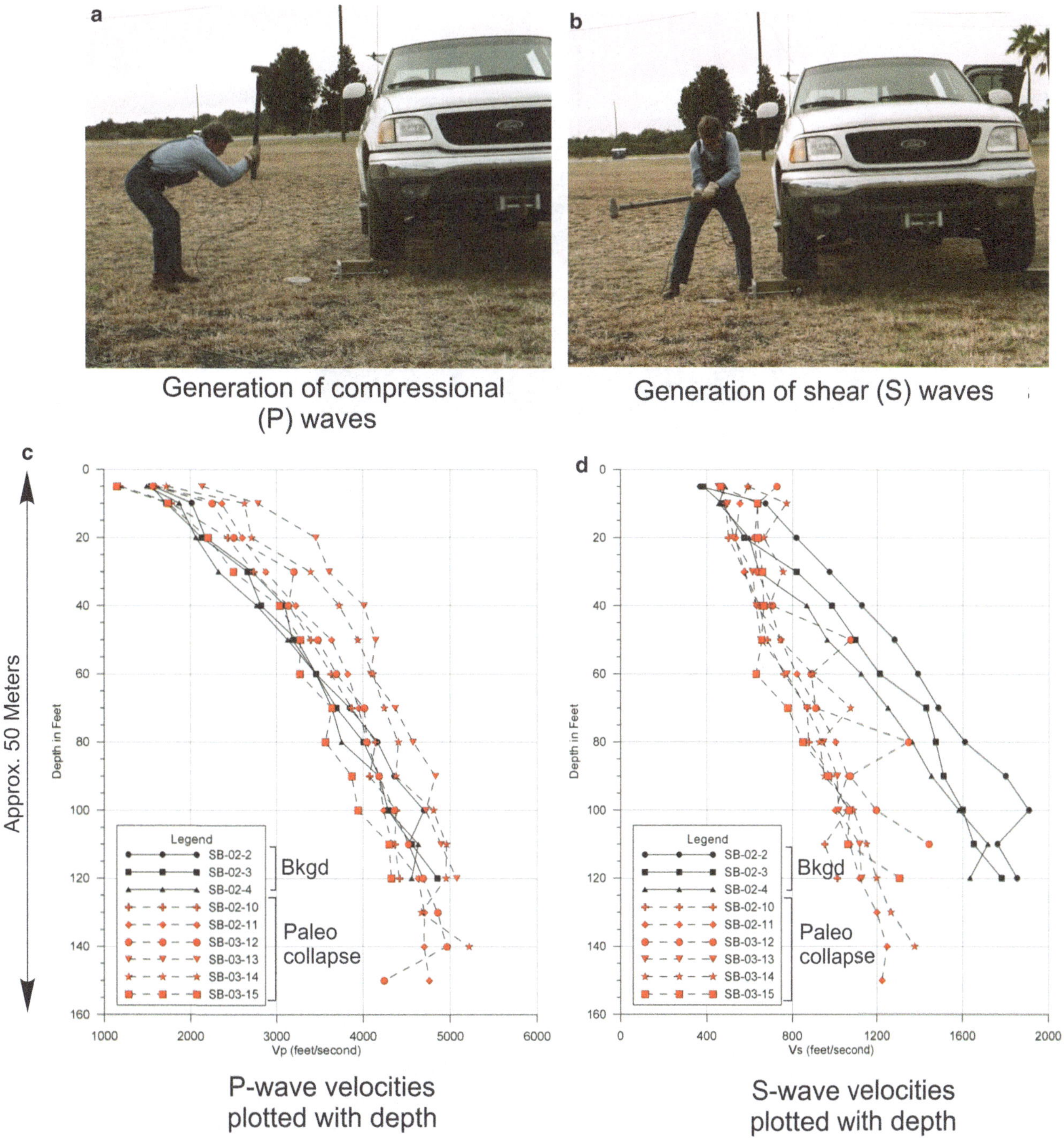

Fig. 18.17 Downhole seismic velocities were obtained using a hammer to generate compression P-waves (**a**) and S-waves (**b**) using a plank on the surface and geophones in the borehole. P-wave velocity (**c**) and S-wave velocity (**d**) are plotted versus depth for nine boreholes

cable to casing installations for other geotechnical measurement applications.

Other measurements can be made hole-to-hole including borehole ground penetrating radar (Haeni et al. 2002) and borehole resistivity. Borehole radar has been used in the reflection mode where the transmitter and receiver antennas are located in the same borehole. Radar is also used in the cross-hole mode where the transmitter antenna is located in one borehole and the receiver antenna is located in another hole some distance away (Fig. 18.19a). These systems can determine conditions between two boreholes and provide detection of fractures (Fig. 18.19b), cavities and other discontinuities such as pipelines and tunnels, which are not intersected by the borehole (Singha et al. 2000; Williams et al. 2009).

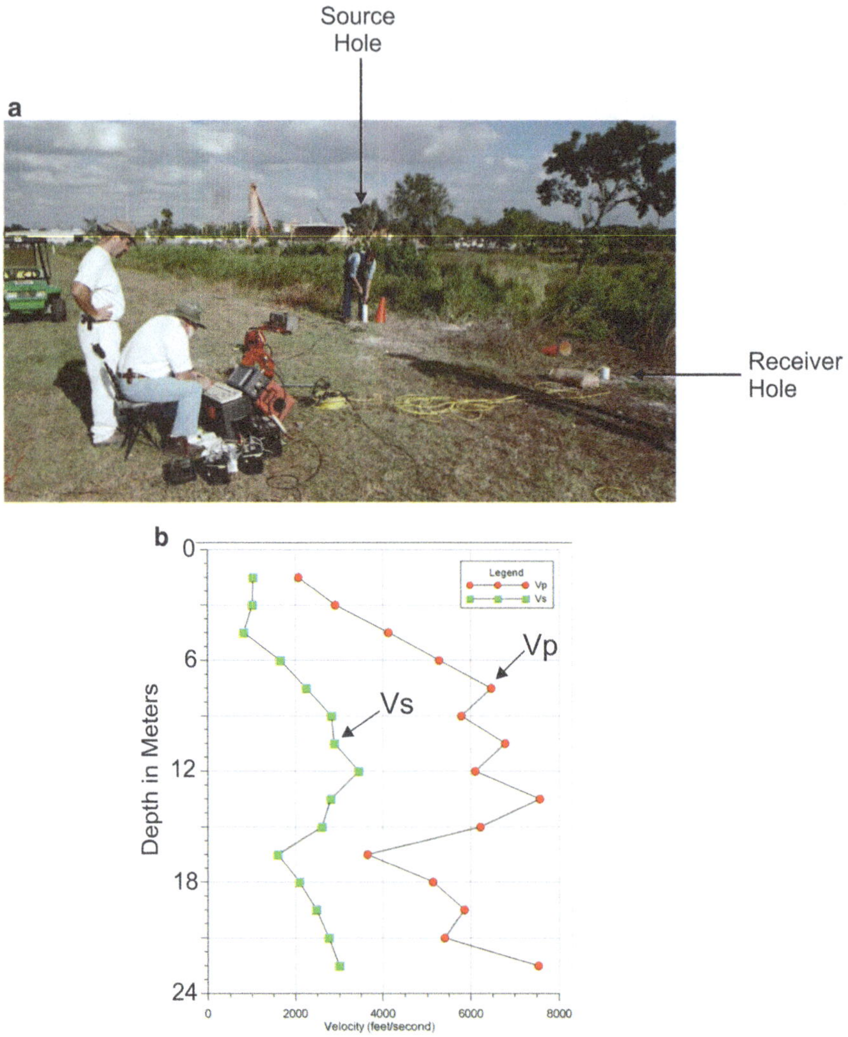

Fig. 18.18 Hole-to-hole seismic measurements between two boreholes and resulting velocity data. (**a**) Hole-to-hole seismic measurements. (**b**) Crosshole seismic data

18.4.3 Tomographic (Imaging) Measurements

Tomographic measurements can be used to image conditions between two or more boreholes (Fig. 18.16c). When measurements between two boreholes are made at a number of angles the data can be used to develop detailed 2D cross sections between two boreholes. Three-dimensional cross sections can be developed by making measurements between three or more boreholes. When the borings are installed around the perimeter of an anomalous area, the resulting data can provide an image of anomalous conditions between the boreholes. Seismic measurements are the most commonly used tomographic measurements, however, a number of other methods, including radar, EM and resistivity can also be used.

For example, seismic crosshole tomography can be used to obtain a 2D cross section of seismic P-wave velocities between two boreholes by measuring the seismic travel-times between a source (in one hole) and a receiver (in the other hole) at different combinations of source-receiver depths. In theory, a single source and single receiver can be used to make tomographic measurements by moving the transmitter and receiver to provide a combination of measurements. However, in practice, a series of receivers in one borehole are used with a transmitter in another borehole. This approach saves considerable time.

Both seismic and radar have been used in crosshole and tomographic measurements. Corin et al. (1997) describe the use of radar tomography to assess karst conditions between pairs of piers along a viaduct. The boreholes were 20–40 m deep and the distance between boreholes ranged from 12 to 20 m. They were able to resolve areas of massive limestone, weathered limestone and voids filled with sand based upon changes in the radar velocity due to water content.

a

Borehole radar being used in angled boreholes extending beneath building

b

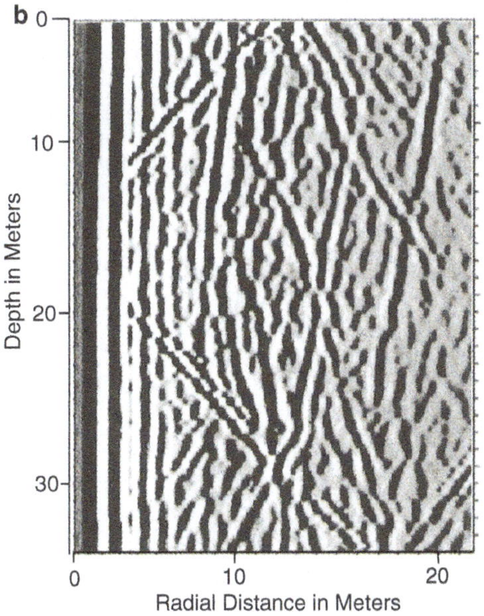

Example of 60 MHz borehole radar data showing planar fractures (Singha et al. 2000)

Fig. 18.19 Borehole radar data being acquired between two boreholes (**a**) and an example of radar data showing fractures (**b**) (Singha et al. 2000)

References

Anderson DG, Woods RD (1975) Comparison of field and laboratory shear moduli. In: ASCE proceedings of the conference on in situ measurement of soil properties, volume 1 in situ measurement of soil properties. North Carolina State University, June 1975. ASCE, New York, pp 69–92

ASTM (2007a) Standard guide for conducting borehole geophysical logging-electromagnetic induction, D6726-01(2007). American Society for Testing and Materials, West Conshohocken

ASTM (2007b) Standard guide for conducting borehole geophysical logging-Neutron, D6727-01(2007). American Society for Testing and Materials, West Conshohocken

ASTM (2010a) Standard guide for planning and conducting borehole geophysical logging D5753-05(2010). American Society for Testing and Materials, West Conshohocken

ASTM (2010b) Standard guide for conducting borehole geophysical logging-gamma, D6274-10. American Society for Testing and Materials, West Conshohocken

ASTM (2011) Standard guide for conducting borehole geophysical logging-mechanical caliper, D6167-11. American Society for Testing and Materials, West Conshohocken

ASTM (2014a) Standard test methods for crosshole seismic testing, D4428M-14. American Society for Testing and Materials, West Conshohocken

ASTM (2014b) Standard test methods for downhole seismic testing, D7400-14. American Society for Testing and Materials, West Conshohocken

Ballard RF Jr, McLean FG (1975) Seismic field methods for in situ moduli. In: ASCE proceedings of the conference on in situ measurement of soil properties, volume 1 in situ measurement of soil properties. North Carolina State University, June 1975. ASCE, New York, pp 121–150

Chapellien D (1992) Well logging in hydrogeology. Balkema, Rotterdam

Colog (2005) Hydrogeophysical™ and geophysical logging results Marshall Space Flight Center, Huntsville, AL. Consultants report June 20

Corin L et al (1997) Radar tomography applied to foundation design in a karstic environment. In: McCann DM et al (eds) Modern geophysics in engineering geology. Geologic Society engineering geology special publication no. 12. The Geological Society, London, pp 167–173

Haeni FP et al (2002) Detection and mapping of fractures and cavities using borehole radar. In: Proceedings of fractured rock, National Ground Water Association, Denver, CO, p 4

Hearst JR et al (2000) Well logging for physical properties, a handbook for geophysicists, geologists and engineers, 2nd edn. Wiley, West Sussex

Killeen PG (ed) (1986) Borehole geophysics for mining and geotechnical applications. Geological survey of Canada paper 85–27

Michigan Department of Transportation (2008) Detroit river international crossing study. Brine well cavity investigation program technical report, February, Detroit, Michigan

Singha K et al (2000) Borehole-radar methods: tools for characterization of fractured rock. USGS Fact Sheet 054-00 May

Stokoe KH, Abdel-razzak KG (1975) Shear moduli of two compacted fills. In: ASCE proceedings of the conference on in situ measurement of soil properties, volume 1 in situ measurement of soil properties. North Carolina State University, June 1975. ASCE, New York, pp 422–449

Turner MS, Benson RC (1986) Auditing existing monitor wells. In: Focus on southeastern ground water issues conference. National Water Well Association, Westerville, Ohio

Wair BR et al (2012) Guidelines for estimation of shear wave velocity profiles, PEER report 2012/08. Pacific Earthquake Engineering Research Center, Berkley, 68p

Williams MME et al (2009) Review of borehole based geophysical site evaluation tools and techniques, Nuclear Waste Management Organization NWMO TR-2009-25

Yuhr L et al (2005) An integrated geophysical approach for a karst characterization of the Marshall Space Flight Center. In: Beck BF (ed) Proceedings of the 10th multidisciplinary conference on sinkholes and the engineering and environmental impacts of karst. ASCE Geo-Institute. Geotechnical Special Publication No. 144, pp 572–579

Yuhr L et al (2008a) Mapping and drilling of conduits within the Woodville Karst Plain of north Florida. In: Yuhr LB et al (eds) Proceedings of the 11th multidisciplinary conference, sinkholes and the engineering and environmental impacts of karst. ASCE Geotechnical Special Publication No. 183, pp 340–350

Yuhr L et al (2008b) Karst characterization of the Marshall Space Flight Center: two years later. In: Beck BF (ed) Proceedings of the 11th multidisciplinary conference on sinkholes and the engineering and environmental impacts of karst. ASCE Geo-Institute. Geotechnical Special Publication No. 183, pp 98–109

19 Assessment of Larger Open Voids and Structures

Abstract

On occasion, larger open voids, caves, conduits and structures such as large pipes, tunnels, shafts and mines will be a part of the site characterization. These features may be dry or water-filled. While direct visual inspection and photographic or video documentation would be the preferred approach to these openings or structures, it is not always possible. In addition, there may be a need to obtain more quantified measurements and data. There are however, a range of tools that can be used whether access is easily obtained or limited through a borehole or crevice.

19.1 A Variety of Methods

When faced with a situation where a space (natural or man-made) requires inspection as part of the site characterization a unique approach may be necessary. There are a variety of methods that can be employed to image and evaluate such conditions, all of which are fairly common and available. The range of methods for assessment include:

- Direct visual inspection – when conditions permit direct access safely, visual inspection, supplemented with sketches and photographic or video documentation provides the most comprehensive assessment of such conditions.
- Photographs and video – can provide conditions of water-filled and air-filled open spaces.
- Cave mapping systems – can provide coordinates and dimensions of air-filled caves as well as mines and other open spaces.
- Laser measurements – can provide size, shape and conditions of large air-filled open spaces.
- Sonar measurements – can provide size, shape and conditions in large water-filled open spaces.
- Robotic and autonomous vehicles equipped with video, laser, or sonar – can provide a means to access air-filled or water-filled spaces and can carry a variety of sensors and tools.

Table 19.1 summarizes the typical conditions and limitations in which these methods can be used to assess dimensions and conditions of large voids or openings.

19.2 Visual Inspection

The most powerful tool we have available is one of observations. Whenever possible, visual inspections should be made along with sketches, photo or video documentation, but particularly when unusual conditions present themselves. Peck (1972) warns that observations made by an intelligent person are often overlooked as the instrument of choice. Only when we cannot make direct observations should more specialized instruments be employed.

Visual inspections and observational skills have been stressed throughout the site characterization process. Here we are addressing more unique physical areas or conditions that require "thinking outside the box". Figure 19.1 shows some of the unique conditions in which direct observations can be made in a borehole, water intake pipe to a nuclear power plant or underwater cave.

R.C. Benson, L.B. Yuhr, *Site Characterization in Karst and Pseudokarst Terraines*,
DOI 10.1007/978-94-017-9924-9_19

Table 19.1 Conditions and limitations of methods for assessing large voids or openings

Method	Conditions	Minimum diameter opening for access	Useful distance	Limitations	Comments
Visual	Dry	About 60 cm for human access	Typically 15–30 m or so without special lighting	Size of opening. Safe physical access. Limited visibility by dust or fog in air	Visual observations supported by sketches, photos, or video
	Water filled (by divers)	About 60 cm or more for human access	0–10 m	Size of opening. Safe physical access. Limited visibility by particulate matter in water	Visual observations supported by sketches, photos, or video
Photos or video documentation	Dry	5 cm diameter for borehole systems	Up to 30 m depending upon lighting used	Limited by dripping water and debris in a borehole, or by dust and fog in large open voids or mines	Visibility can sometimes be extended by using lighting in separate boreholes further from camera
	Water filled	5 cm diameter for borehole systems	0–10 m	Limited visibility due to particulate matter	Visibility can sometimes be improved by using flocculants
Laser	Dry	7–45 cm, depending upon equipment	From 1 to >50 m	Type of surface may limit reflections	Laser scanning system can provide 3D models
Sonar	Water filled	7–30 cm depending upon equipment	From 1 to 100 m	Sonar image quality is function of frequency	At higher frequency and short range images can approach photographic quality
Robotic vehicle (ROV)	Air filled or water filled depending upon vehicle	Typical diameter opening for a sewer inspection system 20 cm. A small ROV about 40 cm opening	Limited by cable length (up to 300 m)	Limited visibility without sonar. Cable drag and snags. Knowledge of accurate position	Can be equipped with camera, video, sonar, or laser
Autonomous vehicles (AUV)	Air filled or water filled depending upon vehicle	Most AUV's will require an opening of about 1 m or more	16–160 km or more	Limited by obstructions and battery life	Can be equipped with camera, video, and or sonar

Fig. 19.1 First-hand visual observations under unusual conditions. (**a**) Large borehole. (**b**) Intake pipe. (**c**) Underwater cave (Photo **a** courtesy of Dick Woods)

19.2.1 Concerns for Deep Foundation Piles

The use of 60 cm bucket auger borings and downhole visual geologic inspection has been widely used in California to characterize landslide problems. Critical to this approach are the regulatory and safety issues in a borehole, which are discussed by Scullin (1994). The same concept can be adapted to inspection of geologic conditions within vertical mine shafts, vertical pits into caves, etc.

A new manufacturing facility was being constructed which was to contain vibration sensitive equipment. The design required 2.5 m diameter piers to be founded on unweathered shale to provide adequate stiffness. Surface wave seismic measurements indicated about 6–8 m of clay with nearly constant shear wave velocity over a thin stratum of weathered shale over intact shale. The precise depth and thickness of the weathered zone was not uniform over the site, so at each caisson, the quality of the rock needed to be determined. If the shale was not un-weathered or stiff enough, the caisson would be excavated deeper and the rock retested. Visual inspection and testing of the rock quality using a hand-held Housel Penetrometer was seen as the best option. Dr. Richard Woods was the vibration consultant to the owner and performed both the surface wave measurements and caisson rock testing. He is seen in the photo descending into a caisson for inspection of rock quality at the bottom of the borehole (Fig. 19.1a).

19.2.2 Power Plant Ocean Water Intake System

Sometimes site characterization is not a geotechnical issue. The 2.5 m diameter ocean water intake line at the St Lucie Florida nuclear power plant was being restricted by rapid barnacle growth. The ocean intake line was inspected using divers and photographs to monitor the rate of growth on different antifouling materials. The intake canal was inspected using an echo sounder and bottom sediment sampling, and the power plant intake was visually inspected and photographed including the grizzly intakes and piping up to the condenser (Fig. 19.1b) to evaluate both biofouling and corrosion.

19.2.3 Mapping by Cavers and Cave Divers

Cavers have provided much of the data on cave locations and maps. Most caves will have an opening to access the cave, however, one caver has a more aggressive approach. John Ackerman has excavated sinkholes to uncover the entrance to a caves (Faulkner 2008). When caves are underwater direct observations becomes more complicated and cave divers are necessary to make observations. In the late 1960s cave divers begin to explore the many sinkholes and caves in Florida, first as a hobby, then slowly shifting to serious diving technology, cave mapping and scientific measurements. Their original equipment was pretty much standard SCUBA gear using compressed air along with a few homemade pieces of specialized equipment such as dive lights. As time passed equipment became more advanced including: rebreathers, special gas mixtures and scooters enabling them to make longer and deeper penetrations into the cave systems.

While the authors were never directly involved with cave diving, many of their friends were, including Tom Mount, Steve Cawthon, Rick Freeze and Bob Friedmann. These early cave diving pioneers from the Miami area provided the authors with maps of many of the caves they explored in the late 1960s. Here Cawthon is making a reconnaissance inspection using an underwater scooter (Fig. 19.1c).

More recently, extensive studies have been completed to characterize the Woodville Karst Plain (WKP) in northern Florida. There are two major springs in the area, the Wakulla Spring and the Spring Creek system further south at the edge of the Gulf of Mexico (Kincaid et al. 2012). Cave divers have made tremendous contributions to the descriptions and connections of the underwater network of caves within the WKP. Cave divers began exploring the sinkholes and cave systems in the Leon Sinks area and Wakulla area in 1970s. After years of exploration it was thought that the two areas were connected. That connection became the focus of two divers and their support team. The connection was finally made between Turner Sink (the southern most end of the Leon sinks area) and Wakulla Springs by Jarrod Jablonski and Casey McKinlay in July 2007. It was a 6.5 h dive followed by 14 h of decompression. Their support team consisted of approximately 50 members.

This is the longest underwater cave system in the United States (a total of 45 km of mapped surveyed passages) and the fourth largest in the world. The depth ranged from 15 to 85 m, the five largest mapped caves range from >2 km to >23 km in length and conduit diameters ranged from less than 2 m to greater than 30 m averaging approximately 10–15 m. The diver's amazing story is told by Wisenbaker (2006a, b), Wisenbaker and McKinlay (2006) and Wisenbaker et al. (2007).

19.2.4 Tarpon Springs Bridge Failure

Several sections of the northbound US19 bridge over the Anclote River in west central Florida collapsed in the late afternoon in 1968. One person was killed and five were

Fig. 19.2 A geologist prepares to make a dive to inspect conditions of a bridge collapse (Photo courtesy of Paul Beam)

injured in the incident. Rescue divers at the site reported that the bridge pilings of the collapsed bridge sections went straight down into the river bottom. Officials from the Florida Department of Transportation (FDOT) stated that the cause of collapse was believed to be a large sinkhole, which are common in the area. The bridge was built in 1946 and no structural problems had been previously noted.

Paul Beam, a young geologist, from the FDOT Research and Testing Laboratory supervised the investigation of the collapse. Three drill rigs (one on a barge) were used to sample subsurface conditions. After 4 weeks of subsurface core sampling by the three drill rigs, Beam reported that there was no evidence of sinkhole activity. Being professionally curious, Beam used SCUBA gear and an underwater camera to further examine the area (Fig. 19.2). The bridge pilings were steel H-beams with a concrete jacket around them to protect them from sea water corrosion. However, tidal action and river flow had scoured the river bottom beneath the concrete jacket allowing corrosion to occur on the now exposed H-beam piles. Mr. Beam discovered softball-sized holes in some of the pilings of the still standing sections of the bridge. Based upon drilling data indicating no sinkholes and his personal observations he concluded that the bridge collapse was not caused by sinkhole activity but by corrosion of the steel piles (Beam, 1984, personal communication).

This is another case where curiosity and observations provided the answer to a problem. The combination of extensive drilling data under the supervision of a professional geologist as well as first hand underwater observations by the same geologist provided a solid basis for the interpretation of the cause of the bridge failure.

19.2.5 Road Widening Adjacent Sinkhole

Another example of observations is taken from a water-filled sinkhole only 30 m away from US19 in the town of Hudson in west central Florida near the Gulf of Mexico. In 1980, plans to widen the highway had created concern about the stability of the road widening near the edge of the sinkhole. An investigation of the sinkhole was made by divers. They measured the depth and shape of the sinkhole using tape and compass measurements (Figs. 19.3 and 19.4). They also obtained vertical profile measurements of conductivity and temperature and photographed key geologic conditions within the sinkhole. Ground penetrating radar measurements were made on the surface parallel to both sides of US19.

The Suwannee Limestone formation outcrops at the surface and is approximately 45 m thick. The Suwannee Limestone is underlain unconformably by the Crystal River formation. Much of the cave development in the area appears

Fig. 19.3 A diver enters water-filled sinkhole to map, photograph conditions and measure conductivity/temperature (**a**). Note block fall photographed at a depth of 42 m (**b**)

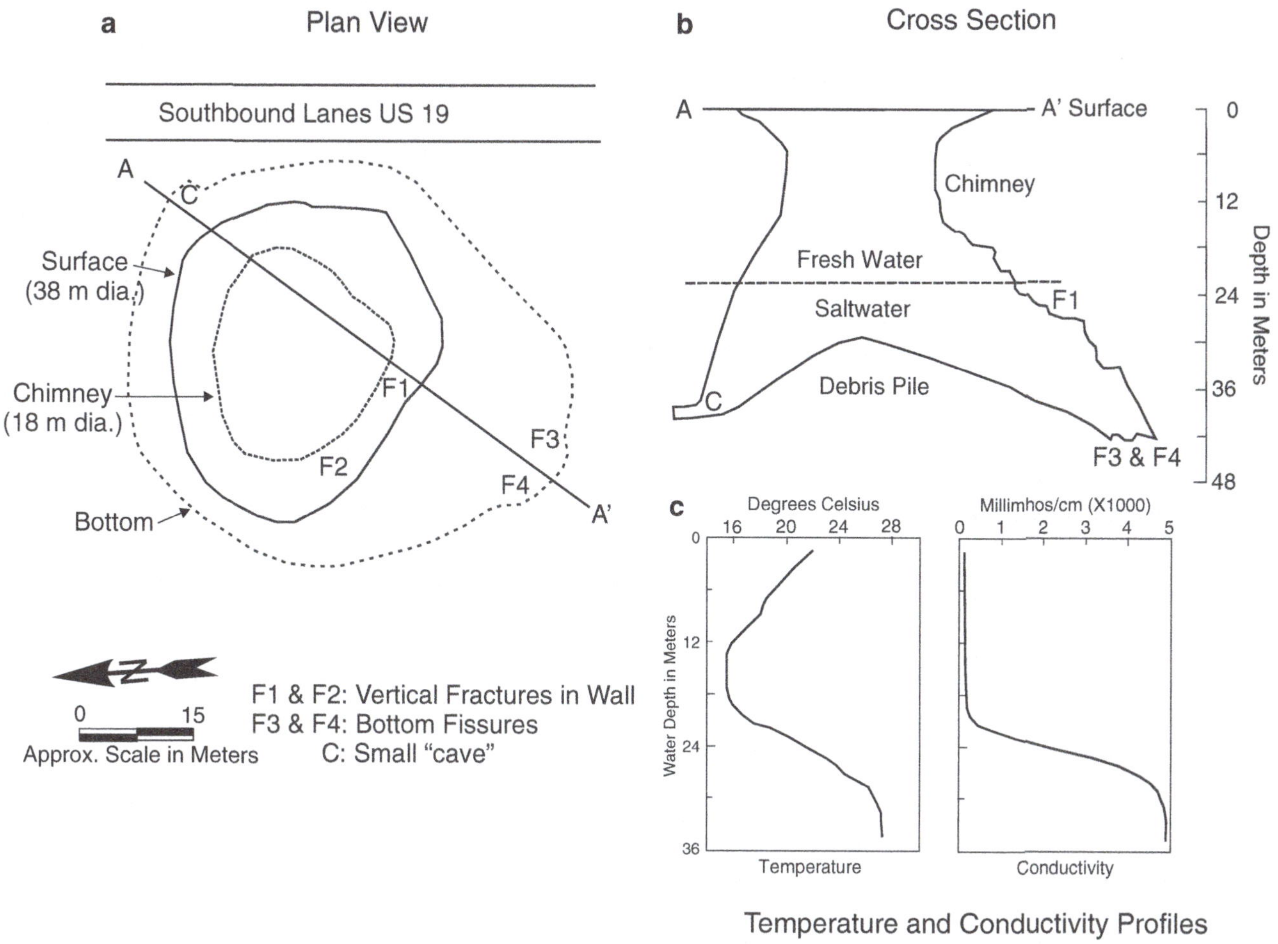

Fig. 19.4 Plan view (**a**) and cross section (**b**) of mapped sinkhole along with temperature and conductivity profiles (**c**) measured within the sinkhole

to have developed near this contact and sinkholes are common in the area. Cave divers have noted both fracture and bedding control within many of the local sinkholes (Freeman, 1980, personal communication).

The sinkhole has a classic bell shape typical of many sinkholes in Florida (Fig. 19.4b). It is 38 m in diameter at the surface, about 18 m in diameter at the chimney. The major conduit from which the sinkhole developed lies at a depth of about 45 m and is oriented in the NE to SW direction coincident with the reported fracture patterns in the area. A 1 m wide fissure, 5 m high was seen in the southwest wall of the sink at a depth of 22–27 m. An undercut of the wall and a fallen block was noted at the bottom of the sinkhole on the southwest side of the sink (Fig. 19.3b). No recent signs of collapse activity were noted by the two divers, one of whom was an experienced cave diver, the other a geologist.

A ground penetrating radar survey was carried out between the sinkhole and US19. There were no anomalous conditions seen in the radar data. However, further east of US19, about 1.2 km, sinkholes were detected by radar and confirmed by drilling.

Vertical profile measurements of conductivity and temperature were made. This data indicates that the fresh-salt water interface is at a depth of about 21 m below grade (Fig. 19.4c). The fresh-salt water interface (the zone of dissolution) now occurs well above the original depth of the cave, which was developed at a time of lower sea level.

While the eastern toe of the sink extends almost to the edge of the southbound lane of US19 (Fig. 19.4a), there is more than 30 m of massive limestone above the easternmost extent of the sinkhole. There was also no indication of sinkhole activity in the radar data between the sinkhole and highway. The conclusions were that this paleosinkhole appears to be dormant and the risk of collapse is virtually zero, if no other factors impact this area.

19.3 Photographic and Video Documentation

Both cameras and video systems provide one of the simplest and most available means of documenting conditions. A distinct advantage of the digital systems is that the results can be viewed immediately so that one can retake the photo, if needed. Some systems are water-proof or at least splash resistant which is necessary in certain environments.

There are inherent distance limitations when using cameras or video to determine conditions within a large, dark cave or mine. In an air-filled system, limitations include dirt and dripping water on the lenses, as well as dust and fog, and the inherent distance of the lighting system. In a water-filled space visibility may be as little as a 30 cm or so due to particulate matter. In clear water visibility can be 3 m or more.

During the monitoring of collapse conditions in a limestone mine (1989–1993), which was partially flooded, a Nikonos V underwater camera using ASA 400 film and a SunPak 355AF flash were used to photograph collapse conditions (see Figs. 25.18 through 25.21 in Part III). These photos provided adequate documentation of individual collapse areas where distances between pillars was approximately 18 m. Details beyond a distance of about 15–22 m were not clearly illuminated due to limited lighting from the single flash system. However, the use of a million candle power 12 V handheld spotlight provided localized viewing to more than 30 m, well beyond the range of the flash system, but the field of view is limited by the width of the narrow spotlight beam (Benson and Hatheway 1994). Photographic documentation was effectively used to document both stable collapse areas as well as new collapse areas as they occurred. The description of conditions to site owners and regulators was easily conveyed using this method.

Inexpensive chemical light sticks (cyalume) were used as an effective means to mark the boundaries of areas of interest prior to taking photographs. They were also effective as safety markers to mark our trail so that we could quickly orient ourselves and as a safety means to find our way out of a mine or cave should all lighting fail.

Many cavers have obtained extraordinarily well-illuminated photos of large caverns over distances of much greater than 30 m (Fig. 9.1). This is accomplished by the use of multiple lighting locations, special lighting sources, tripods and long exposures. This is something not normally available for on-site inspections and would require an experienced photographic team. For more details on how this was accomplished see *Shooting the Big Room of Camp's Gulf* (Anderson 2000).

Many of the cavities or caves requiring assessment are not accessible and must be accessed using equipment through boreholes, small crevices or collapse areas.

Borehole video camera systems are typically intended for viewing borehole conditions within a few centimeters to a meter or so from the borehole. The authors have used a downhole video camera to assess conditions within dry mines and have found that effective viewing distances range from 15 to 20 m using 500 to 2,000 W of omni directional lighting from the same hole as the video camera (Fig. 19.5a). Where additional boreholes are available, lighting can be lowered into boreholes further from the video camera to increase the distance of viewing (Fig. 19.5b).

Besides those video camera systems that are used in boreholes, there are a wide range of other off-the-shelf video systems available which can be adapted to unique conditions for inspection and documentation. One system referred to as a snake, and is used by plumbers to inspect sewer lines. The video camera is located at the end of a long flexible rod, which can be used for inspection in small diameter pipes, as

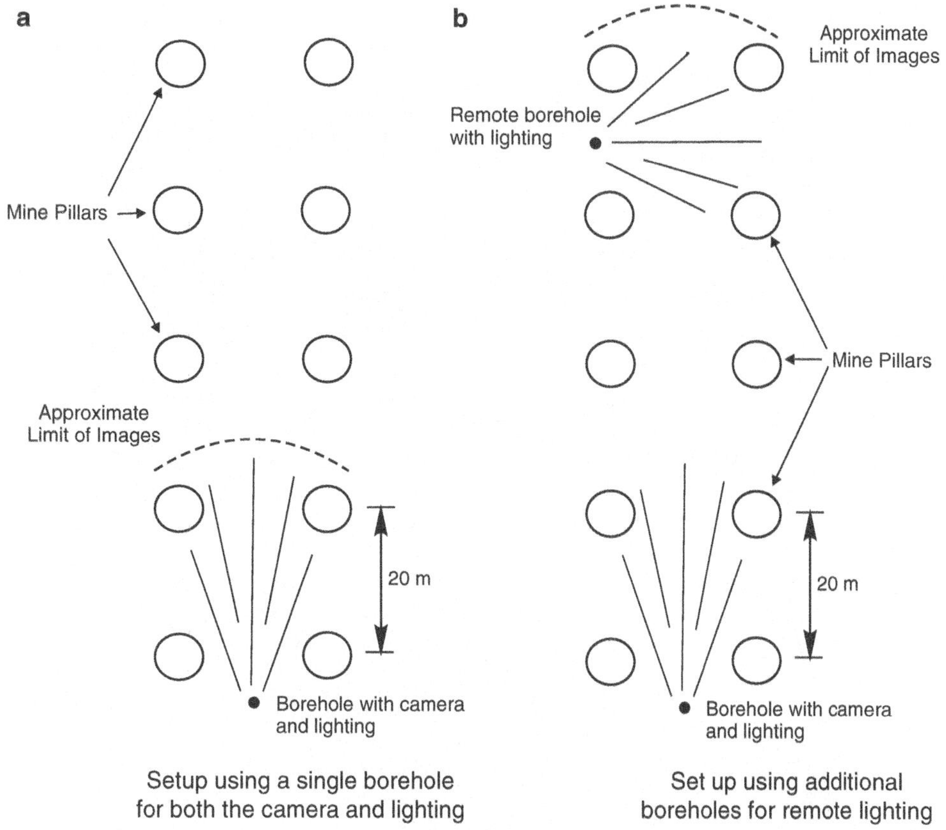

Fig. 19.5 The range of visibility for borehole video camera and lighting in the same borehole is limited to about 23 m (**a**). The distance can be extended considerably by locating additional lighting in a boring further from the camera system (**b**)

small as 3 cm in diameter. These can also be used to probe into small openings within a cave or mine. Another system mounts the video camera at the end of a ridged rod, which can be used to probe into inaccessible or dangerous areas. These systems are used for inspecting shallow confined areas and sewer lines and have been used to aid in search and rescue missions.

Sewer inspection video systems are in common use by most municipalities. These systems are typically available as truck mounted systems but are also configured as portable systems. The video cameras are mounted on wheels or tracked platforms, which are steerable systems to provide directional control of the platform for inspection of sewer lines and tunnels. The cameras commonly have pan and tilt capabilities and offer better lighting for viewing longer distances than most borehole video cameras. While intended for inspection in horizontal sewer lines, the authors have lowered such systems down a borehole or shaft to inspect caves, tunnels, and mines.

A small collapse had occurred at the corner of a residence located in southeast Florida (Fig. 19.6a). The house was located on the coastal ridge facing the Gulf Stream. Six shallow borings and a ground penetrating radar survey were completed at the site to evaluate the extent of any voids around the building and over the property. Two of the borings and a few radar survey lines indicated a void from 3 to 4 m below grade in the area of the collapse. A borehole video camera was lowered down two of the boreholes, which had encountered a void and revealed the possibility of a small cave system extending away from the house. A local resident reported that that he and his friends used to play in this cave when they were kids many years ago. The entrance was in the neighbor's yard but had been filled with debris. The cave had gone unnoticed by the builder and the problem did not appear until a downspout at one corner of the house had focused enough run-off to erode sufficient material into the cave so that a local collapse occurred (Fig. 19.6a).

A light was lowered into one of the boreholes and it exposed the larger portion of the cave (Fig. 19.6b) in which kids had played. Then a video camera was then attached to the end of a long pole and from within the collapse area adjacent to the house (Fig. 19.6a) it was used to probe into areas that were physically inaccessible. The void extended under a portion of the house 0.3–1 m in height (estimated at about 45 m^3) and under the swimming pool, but also expanded into a larger cave system away from the house where kids had played. This larger area was estimated to be about 3 m wide and 1–2 m in height and its volume was estimated to be (approximately 141 m^3). A grouting program was then

a

Downhole video camera mounted on pole being extended into collapse beneath home

b

View from access point in (a) with additional lighting from adjacent borehole (circled) one could detect a larger cave in the distance

Fig. 19.6 The video camera was attached to the end of a long pole in order to provide visual inspection of inaccessible areas (**a**). A light lowered into one of the boreholes provided a partial illumination of the larger portion of the cave (**b**)

designed to stabilize the portion of the void space under the house and the larger cave under the swimming pool.

Although the boring and GPR data provided initial insight to the problem, the details only became clear as we acquired direct observations, photos and video data which enabled us to piece together bits of data into a final sketch of the cave along with estimates of its dimensions.

19.3.1 Its Not Always Easy

Detailed maps of a mine in the Kansas City area along with extensive visual and photo documentation had been acquired. At one location near the mine portal entrance a sump had been installed along with a large pump about 1 m tall, which was painted red. Years later we were required to relocate this specific area of the mine from the surface, to verify its location for backfilling. Given the mine maps and surveyors coordinate system we were able to locate the area on the surface. Once a borehole had been drilled into the open mine space, a standard sewer inspection camera was lowered down the borehole to verify our location. Observing the pump with the video camera would provide confirmation that we were in the right location. It took about 4 h of tedious field work (tweaking with horizontal scanning, along with the pan and tilt system and adjusting video system variables) to obtain a fuzzy image of the pump which was approximately 15 m from the borehole. Problems with dripping water and small particles and a slight fog as well as insufficient lighting prevented a clear image of the pump. Only by repeated panning of the area with a variety of optical changes were we finally able to identify the pump and verify its location with a vague video image.

19.4 Cave Mapping Systems

Cavers and cave divers have made great contributions to locating and mapping both dry and underwater cave systems. Cave maps have been traditionally developed using tape, compass and clinometer. The care taken by cavers has generally produced reasonably accurate maps. Digital advancements include laser distance measuring devices instead of the tape measure along with digital compass and clinometer. Data is now entered directly to a hand-held computer instead of written in a notebook. Special cave survey software and data management software including GIS is now available and being used. This equipment and software results in faster measurements, accurate recording and processing providing nearly real-time digital cave maps to the field team similar technology can be utilized in mapping abandoned mines.

19.4.1 Inertial Navigation Systems (INS)

GPS is now the choice for most positioning in open areas not restricted by tree cover or tall buildings. There are alternatives for use inside of structures and underground. An Inertial Navigation Systems (INS) utilizes the data from accelerometers and gyros feed into a computer system. After establishing known control points to reference the system, one can walk about with the INS in a back pack to map 3D space and the INS will keep track of your position relative to the control points. By stopping at each survey point for a few seconds the position of that point is recorded. At the end of the survey the data is referenced to the control points. A professor from Missouri State University has used Inertial Navigation Systems (INS) to map caves (Vickery 2010).

19.4.2 Cave Radio

If a cave is accessible by cavers or cave divers then a "cave radio" (more correctly a "cave location" system) can be used. A cave radio can be used to locate and determine an approximate depth of a point within the cave by measurements on the surface. Radio location is used to provide a means to verify the locations of long traverses made with compass and tape and to provide a location on the surface to drill into the cave system for sampling and monitoring.

A cave radio consists of a transmitter and a receiver coil. The coils are less than a meter in diameter. The system operates at very low frequency in the kilohertz range. The transmitter coil is leveled on the cave floor so that the center axis of the coil is vertical. Then the receiver coil is moved over the ground surface until an electromagnetic null is achieved. By establishing a null at two or more locations on the surface, the intersections of the lines will be directly over the transmitter coil in the cave (Fig. 19.7a). By measuring the tilt angle and off-set distance from the location, the depth of the cave can be estimated (Fig. 19.7b). Typical location accuracy of better than 0.5 m is obtained under ideal conditions (away

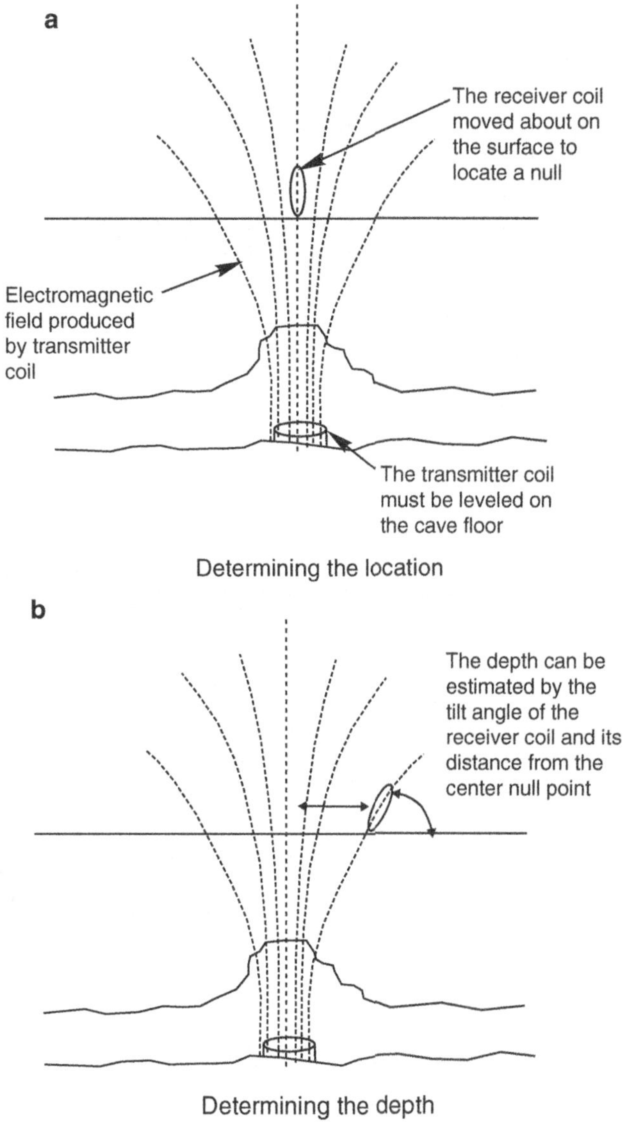

Fig. 19.7 The cave radio location system set up to provide a single location (**a**) as well as to estimate the depth of the cave (**b**)

from buildings, metal fences, and power lines). Measurements to a few hundred meters depth can be made and measurements have been made in both air caves and under water caves.

Under optimum conditions it should be possible to determine depth within 3 %. As with most technology, one should seek out those with field experience, who have mastered the skill of radio location. Further details on cave radios and communication can be obtained from the US based NSS section for cave electronics (caves.org/section/commelect) or from the UK based cave radio and electronics group (CREG) (www.bcra.org.uk/creg/).

19.5 Laser and Sonar Systems

It is often difficult to accurately determine the distance and size of an object such as the wall of a cave, a mine wall or pillars, or the aperature of fissures from photos or video alone. Accurately determining distances and shape can be accomplished by the use of lasers (in dry environments) or by sonar (in water-filled environments). They can be used to map and define the size and shape of larger open spaces such as caves, mines, tunnels and shafts. However, some means of access is required such as a shaft or a borehole. Figure 19.8 illustrated how a laser or sonar device can be used to make such detailed measurements.

19.5.1 Lasers

There a few different type of laser distance measurements. A simple single laser distance measurements can provide the distance to an object. Small hand-held laser range finders have working ranges from 1 to a few 1,000 m with accuracies of ±1 mm to 1 m. A pair of laser beams from different boreholes can be used to triangulate location. Collimated laser beams can be projected onto an object or fracture so that an accurate dimension of a feature can be obtained. More sophisticated laser scanning systems can be used to provide a 3D digital image of the shape of caves or mines.

A multi-electrode 2D resistivity survey had identified a significant anomaly in an area where new sinkholes had occurred each year. One borehole was drilled into the approximate center of a suspected cavity based upon a 2D resistivity survey made at the surface. A 2 m high void was encountered below approximately 7 m of bedrock. The cave interior was roughly estimated based upon video observations from the one borehole. Then four other boreholes were drilled into the cave around its perimeter based upon the 2D resistivity survey and the results of a downhole video camera in the first hole.

A simple off-the-shelf laser distance measurement device was modified and attached to the end of a ridged rod, which was lowered down a borehole and was supported by a tripod. The laser was placed at a given depth and could be rotated

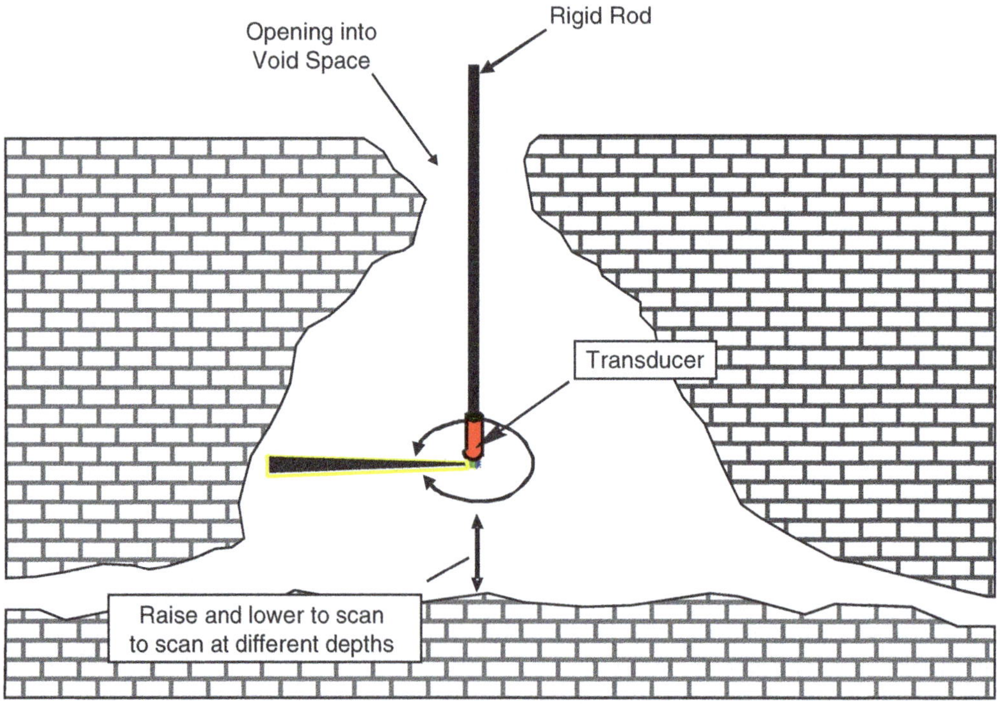

Fig. 19.8 A laser or sonar system can be used to develop a 360° scan of a void space at a given depth. Repeating this process at different depths will develop a complete 3D image

360°. The rotation angle was determined at the surface by dial plate on top of the tripod. By making measurements around 360° the perimeter of the cave could be accurately determined. The lasers were then lowered a small distance (about 0.5 m) and the 360° scan is repeated. The process was repeated through the entire depth of the cavity. The approximate shape of the cavity could then be reconstructed from this set of data. The cave was found to be 2 m high and approximately 4 m long and 1 and 0.5 m wide (Roth et al. 2004).

This is another example of some fairly simple measurements with off-the-shelf hardware being applied to define the size and shape of a cave with a reasonable level of detail and accuracy.

The use of 3D mapping by LiDAR (light detection and ranging) is now commonly used to create detailed cloud maps of buildings and complex structures. The use of 3D LiDAR and digital photography was used to map the interior of Devil's Sinkhole in Texas. The combination of the two methods provides a 3D photo realistic model with very high resolution (Neubert et al. 2008).

A laser system called the Dry Ferret was developed by Workhorse Technology, (a spin off from Carnegie Mellon University). The system utilizes a laser device to produce three dimensional views of dry underground spaces and can be deployed in a 20 cm borehole. The system is gyro and compass stabilized and also includes a low-light color video system (www.workhorsetech.com).

The device is lowered down a borehole and scans 360° then is lowered some distance and makes another scan. This process is continued in order to acquire data from the ceiling to the floor of the mine, cave or other open area. The system also has pan and tilt capabilities, which can image an area from one position. This extensive set of point cloud data is then processed to create the 3D Mesh Model of the shape and boundaries of the void space.

This system was used to aid in assessing open areas remaining during a mine backfilling project. Once the backfill limited access into the mine, the only way to confirm the extent of fill, was to drill boreholes and view conditions using a downhole video camera. The Dry Ferret laser system was also used to evaluate these areas.

19.5.2 Sonar

Extending the distance of measurements beyond that of cameras and video in water-filled environments can be accomplished by the use of sonar. Depending upon the application, sonar can provide data from a few tens of meters to a few hundred meters or more with resolution from a few centimeters to a few meters. The frequency determines the range and resolution of the sonar system. A lower frequency of 100 KHz provide longer ranges of 600 m with lower resolution. Higher frequencies of 1 MHz have less range of 50 m but result in an image with much greater resolution.

Sonar is commonly used as an echosounder on a boat, which is used to determine the depth of water as well as the presence of fish. Simple adaptations have been made using the small sonar transducer from a simple fisherman's echosounder or fish finder mounted on the end of a pole and lowered down a borehole or shaft to develop a vertical profile of a void. An alternative is to mount the transducer horizontally and rotate 360° to produce a cross section of the opening or void. The sensor is then lowered some distance and the 360° scan is repeated. The process can be repeated through the depth of the opening to provide a number of 2D cross section of the opening or void (Fig. 19.8).

Special systems have the motorized capability to provide 360° scanning in a horizontal plain by electronic control from the surface. Travel time of the acoustic signal is measured and converted to distance to create an image of the area being scanned. A variation of the standard scanning sonar is a pan-and-tilt version. In addition to the standard 360° horizontal scan, the sonar head also tilts to provide a profile from horizontal to vertical.

Scanning sonar is commonly used in the ocean industry for inspection of oil rigs and pipelines and hazards such as wrecks. The oil and gas industry has used scanning sonar to determine the size of large cavities around a borehole. The navy and research organizations make extensive use of sonar.

The Wet Ferret, a scanning sonar, was also developed by Workhorse Technology. The system utilizes sonar to produce two-dimensional views of water-filled underground spaces and can be deployed in a 15 cm borehole. The system is gyro and compass stabilized and also includes a low-light color video system. The data is acquired digitally and software is available to develop 2D or 3D views with a resolution of 10 mm at distances up to 100 m (www.workhorsetech.com).

Figure 19.9 is a single 360° sonar scan of an abandoned mine showing the location and spacing of pillars.

19.6 Remotely Operated and Autonomous Vehicles for Inspection

There are a number of remotely operated and autonomous vehicles that can be used for inspection and mapping in both dry and water filled environments. These systems can be used as platforms for a variety of instrumentation.

19.6.1 Remotely Operated Vehicles (ROVs)

There are a number of small and inexpensive mini-ROV systems available, which are referred to as observational class

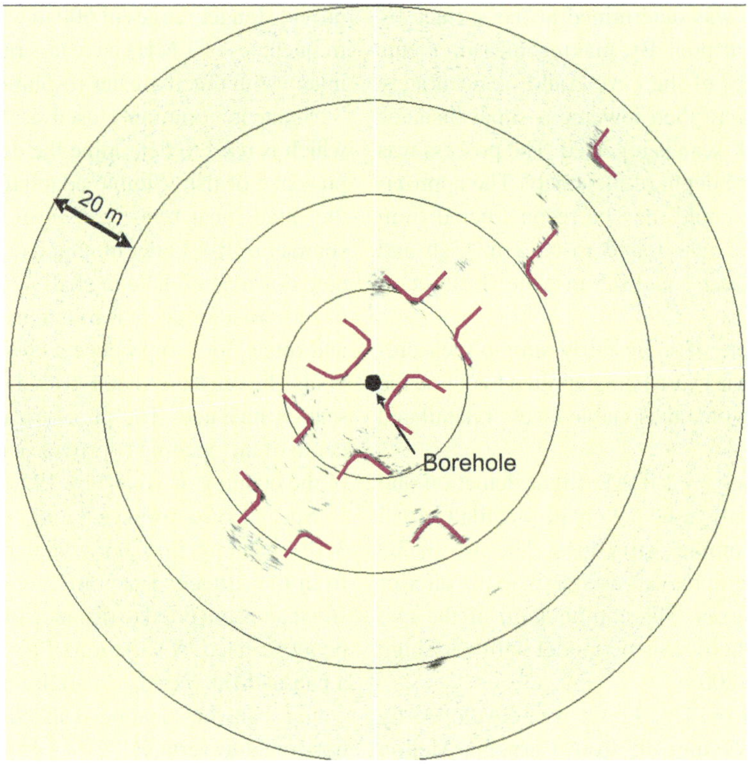

Fig. 19.9 A single sonar scan made with the Wet Ferret in a water-filled mine. The sonar probe is located between two pillars. Line of sight reflections are occurring from pillars up to 67 m away. Pillar spacing is known to be about 20 m

Fig. 19.10 Two small mini-ROVs, one dry mini-ROV was used in a mine to inspect an area of active roof-collapse (**a**) and the other was used to inspect conditions in a water-filled sinkhole (**b**)

ROV systems. Figure 19.10 shows two such systems used by the authors, one is a land based system (Fig. 19.10a) and one is underwater system (Fig. 19.10b). Many of the features found on the larger commercial ROV systems used in the offshore energy business have been adapted to these mini-ROV systems such as powerful thrusters and maneuverability, high quality video and a large field of view, as well as a robotic arm and the ability to retrieve samples. They include intuitive controls along with information displays and auto functions. These small ROV's can be deployed from a small boat by hand without major rigging or large winches. They are used to inspect pipelines and structures in power plants and dams and to enter and inspect remote, inaccessible, and dangerous areas in both wet and dry conditions. They are also being used in security, and search and recovery operations. One marine ROV manufacturer has added the capabilities for crawler mode by mounting the ROV on a four-wheel drive platform.

Besides the common problem of visibility, particularly underwater, knowing where you are is one of the major problems in using a small tethered robotic vehicle. Most ROV's are equipped with depth sensors, compass and a number of turns indicator. However, after changing direction a number of times it can become difficult to keep track of your position within a large sinkhole, cave or mine. In some cases, acoustic pingers can be used to add direction, range or position data by triangulation. For further details on small observational ROV's and their use see Christ and Wernli (2007).

19.6.1.1 Using an ROV to Inspect Conditions in a Flooded Mine

In 2006, a uranium ore mining operation in Canada suffered a roof collapse approximately 450 m below grade resulting in flooding of the mine at rates of 15,000–20,000 gal per minute. By the time flooding had stabilized, water had completely flooded all of the mine and water level had risen to within 17 m below grade.

Snyder and Cook (2008) summarize the approach used in the inspection of this flooded mine. They cite the operation of an ROV system within the low visibility and complex mine tunnels with unexpected obstructions to provide an inspection of conditions prior to remediation. Access was only available from the vertical shaft where the ROV was lowered to the 450 m level to explore the lateral tunnel and to inspect conditions of a jammed high-pressure bulkhead door that had failed to close.

Accurate maps of the mine were available and a 3D map and model of the mine along with known conditions such as audits, location of heavy equipment and utilities were incorporated into the model. As a result there was a good deal of data and maps available prior to launching the ROV. The ROV was equipped with a variety of high resolution sonar systems along with pan and tilt cameras and lighting. The issue of real time navigation required an inertial navigation system supported by a doppler-velocity log, a gyroscope, and accelerometers in three dimensions. High resolution sonar on the ROV provided avoidance of unknown obstacles (Snyder and Cook 2008).

The combination of an accurate 3D map and model of the mine along with the sophisticated navigation system enabled the ROV operator at the surface to maneuver through the mine passages. It is interesting to note that this work overcame limitations and was carried out in this very complex environment by incorporating off-the-shelf hardware and software including the ROV.

19.6.2 Autonomous Vehicles

Untethered autonomous underwater vehicles (AUV's) have become more common in the oceanographic industry over the past decade (Merrill 2007). Autonomous vehicles are now routinely used in oceanographic surveys and have also been used for inspection of flooded tunnels. An AUV equipped with video cameras was used to inspect a 24 km water supply tunnel 2.25 m in diameter for damage. The concrete tunnel, in Germany, had not been inspected visually in more than 40 years (Kalwa 2012).

Workhorse Technology has also developed an AUV called the Cave Crawler that provides a means of inspecting and documenting conditions in abandoned mines and areas to dangerous for human entry. The Cave Crawler can be operated from a joystick control or may be completely autonomous within air-filled voids with walk-in access. The vehicle weighs 180 kg and is about the size of a small "Bobcat" construction vehicle. Its positional accuracy is 5 cm and it is equipped with an inertial positioning system with gyro and compass. Other mapping sensors may include a 3D laser, low-light video, high resolution still camera, radar and thermal image system depending upon mission objectives (www.workhorsetech.com).

Deep Phreatic Thermal Explorer (DEPTHX) is an unmanned autonomous robotic vehicle that was a NASA funded effort using an autonomous vehicle to explore deep cenotes in the state of Tamaulipas, Mexico. The project was a 4-year effort to map and obtain chemical data along with water and rock samples from four large cenotes.

The DEPTHX vehicle is approximately 1.5 m in height by 1.9 m in length and width with a dry weight of 1,500 kg. The vehicle can move at speed of about 0.2 m/s. It is equipped with sonar, depth sensors, and a navigation system for mapping, as well as multi-parameter water quality measurements, water samplers rock samplers video and still cameras. The final test of DEPTHX was carried out in Zacaton, the deepest underwater vertical shaft and the second deepest underwater cave in the world (319 m) (Gary et al. 2008).

References

Anderson C (2000) Shooting the big room of Camps Gulf. NSS News 58(7):201

Benson RC, Hatheway AW (1994) Underground lighting. NSS News 52(7):214–215

Christ RD, Wernli RL Sr (2007) The ROV manual: a user guide for observation-class remotely operated vehicles. Elsevier, New York

Faulkner S (2008) In the media. NSS News 66(5):18–20

Gary M et al (2008) 3D mapping and characterization of Sistema Zacaton from DEPTHX (Deep Phreatic Thermal eXplorer). In: Yuhr L et al (eds) Proceedings of the 11th multidisciplinary conference on sinkholes and the engineering and environmental impacts of karst, ASCE Geo-Institute, Geotechnical Special Publication No. 183

Kalwa J (2012) Sea cat AUV inspects water supply tunnel. Sea Technol 53:43–44

Kincaid T et al (2012) Demonstrating interconnection between a wastewater application facility and a first magnitude spring in a karstic watershed: tracer study of the southeast Farm Wastewater Reuse Facility, Tallahassee, Florida. Florida Department of Environmental Protection, Florida Geological Survey Report of Investigation No. 111

Merrill ML (2007) Unmanned untethered submersible technology. Marine Technology Reporter, October, pp 38–41

Neubert B et al (2008) 3D photo real modeling of Devil's Sinkhole-Rocksprings, Texas. In: Yuhr L et al (eds) Proceedings of the 11th multidisciplinary conference on sinkholes and the engineering and environmental impacts of karst, ASCE Geo-Institute, Geotechnical Special Publication No. 183

Peck RB (1972) Observation and instrumentation: some elementary considerations. US Dep Transp Fed Highw Adm Highw Focus 4(2):1–5

Roth et al (2004) Measuring cave dimension remotely using laser pointers and a downhole camera. In: Proceedings of the symposium on the Application of Geophysics to Engineering and Environmental Problems, Environmental and Engineering Geophysical Society, pp 1307–1314

Scullin CM (1994) Subsurface exploration using bucket auger borings and down-hole geologic inspection. Bull Assoc Eng Geol 31(1):91–105

Snyder JZ, Cook M (2008) ROV ops in low visibility, 3-D models and inertial navigation. Marine Technology Reporter, May, 51(4):24–31

Vickery K (2010) Inertial navigation systems (INS) not just for land anymore. Professional Surveyor Magazine, May, pp 43–45

Wisenbaker M (2006a) A history of exploration in America's longest underwater cave, Part 1. NSS News, June, pp 4–7, 30

Wisenbaker M (2006b) A history of exploration in America's longest underwater cave, Part 2. NSS News, July, pp 4–7, 30

Wisenbaker M, McKinlay C (2006) A history of exploration in America's longest underwater cave, Part 3. NSS News, August, p 13

Wisenbaker M et al (2007) Explores connect America's longest underwater cave to Wakulla Springs. NSS News 65(12):2–11

20 Engineering Measurements and Monitoring

Abstract

At this stage of the site characterization some geotechnical data has already been obtain which may include soil and rock samples, STP blow counts, and RQD values. Geophysical logging data has provided additional geotechnical details from borings. Some laboratory analysis may have been made on selected samples from borings. With a refined conceptual model of conditions in hand, the engineering measurements and monitoring is now focused upon obtaining very site-specific measurements and data to address specific engineering questions and quantify additional engineering parameters. This is a broad range of measurement types and data. Within this chapter, we will focus specifically on the engineering measurements and monitoring of subsidence and collapse. We summarize some of the common methods for measuring and monitoring land subsidence and potential collapse, cover the general applications, some considerations and provide a few examples. As always, each technique or measurement has an associated scale of measurement, advantages and limitations.

While the many details for making these geotechnical measurements are beyond the scope of this book, further details can be found in Dunicliff (1988), Hunt (2005) and by specialty conferences (DiMaggio and Osborn 2007) who all provide a solid foundation for these types of measurements. Due to the advances that continue to occur in geotechnical instrumentation this information should be supplemented with recent manufacturers literature and recommendations.

20.1 In-Situ Geotechnical Measurements and Monitoring

In their book, *Geotechnical Instrumentation for Monitoring Field Performance* Dunnicliff (1988), with the assistance of Gordon Green, provide an excellent coverage of the topics of measurement and monitoring soil and rock conditions for engineering projects. While their book is more than three decades old it remains one of the best references for such measurements and provides numerous guidelines for achieving quality results. The details of the instrumentation will have changed due to rapidly changing technology but the basic principles of measurement remain the same. Both Dunnicliff and Green are practicing experts in the field and have also conducted excellent short courses on the topics. We refer readers to their book for the many details of geotechnical instrumentation, measurements and monitoring which are beyond the scope of this book. The topics covered in their book include measurement of groundwater pressure, total stress in soil, stress changes in rock, deformation, load and strain in structural members and temperature. They also emphasize the many pitfalls in acquiring such data and refer to the "25 links in the chain" required to achieve reliable results.

Selection of measurement type and instrumentation requires consideration of scale as well as sensitivity. Some of the measurements can be quite local while others are regional in nature. Figure 20.1 and Table 20.1 summarize the range of these type of measurements. In addition, sensitivity of the instrumentation needs to fit the expected results, which may require some guess work. Instrumentation for such measurements may include:

R.C. Benson, L.B. Yuhr, *Site Characterization in Karst and Pseudokarst Terraines*,
DOI 10.1007/978-94-017-9924-9_20

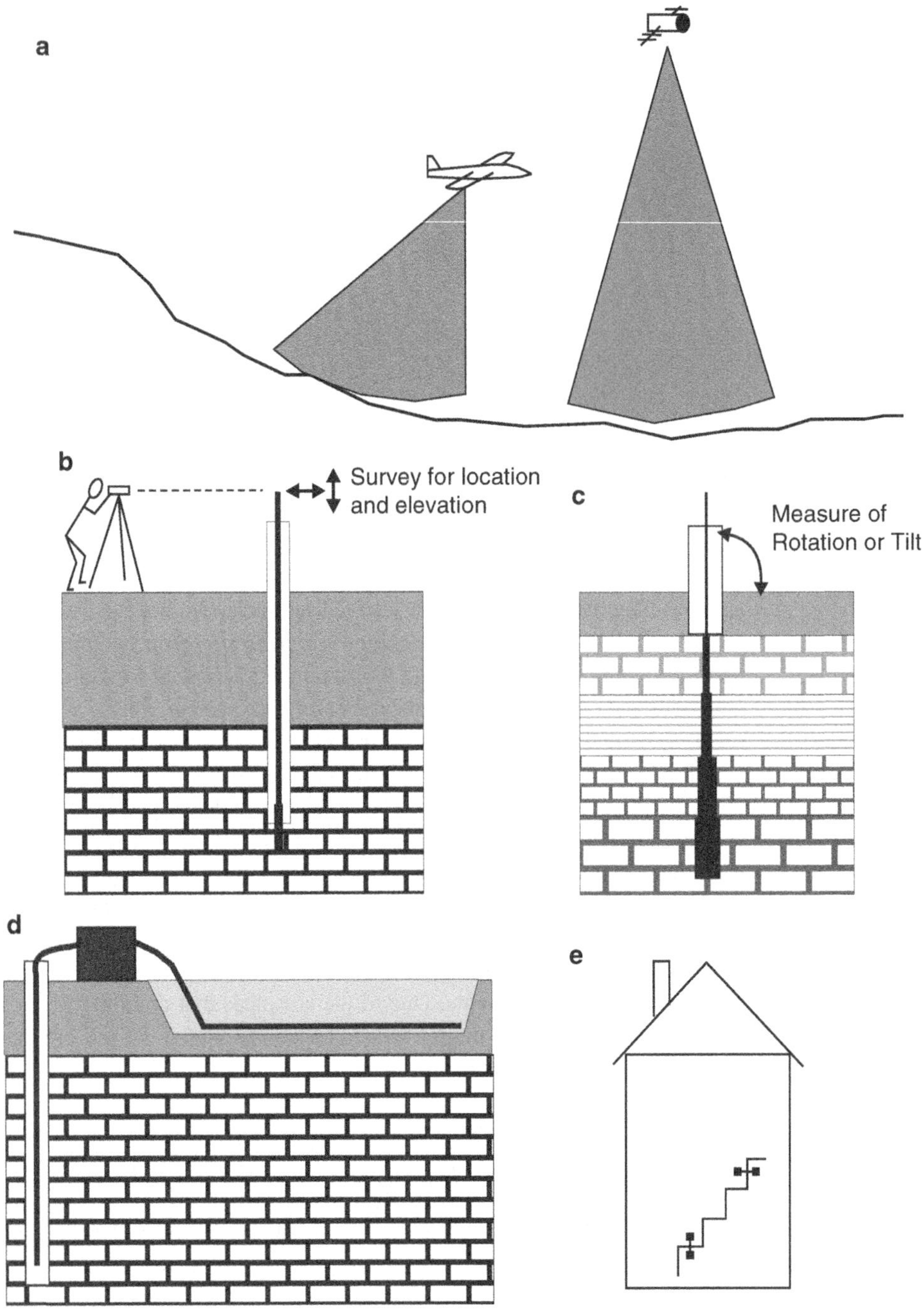

Fig. 20.1 Some of the many measurements used to measure and monitor subsidence. (**a**) LiDAR (from airplane) or InSAR (from satellite) for mapping surface deformation over large areas. (**b**) Benchmarks for monitoring horizontal or vertical movements by optical survey or GPS. (**c**) Tilt meters or Inclinometers for monitoring rotation or movement on multiple planes. (**d**) TDR or OTDR for monitoring horizontal or vertical movement. (**e**) Strain gage type of measurements, very localized horizontal or vertical movement

- Level survey (by first order land survey procedures)
- Piezometers (a variety of devices for measurement of pore-water pressure and water levels).
- Inclinometers (to measure lateral earth movements or deflection of retaining walls and piles under load and to provide a means of measuring settlement of embankments foundations and other structures)
- Tilt meters (used to monitor changes in tilt of a structure).
- Borehole extensometers (to monitor settlement, heave, and lateral deformation in soil and rock).
- Time domain reflectometry (TDR) and optical domain reflectometry (OTDR) measurements (provide a means of

Table 20.1 Comparison of land movement and subsidence measuring techniques

Method	Displacement	Resolution (millimeters)	Spatial scale of measurement
InSAR	Range (distance to ground)	5–10	Regional
Lidar	Range (distance to ground)	70–150	Regional
Land survey (sprit level) to benchmark	Vertical and horizontal	0.1–1	Local line or network
Land survey (geodimeter)	Horizontal	1	Regional line or network
Borehole extensometer	Vertical or any angle	0.01–0.1	Point
TDR and OTDR	Vertical or horizontal or any angle	Varies with distance, 75 mm at 60 m distance	Line
GPS	Vertical and horizontal	20 vertical 5 horizontal	Network, usually local
Tilit meter	Rotation about a fixed point	10 arc sec	Single point
Inclinometer	Horizontal movement from the vertical	±6 mm per 25 m	Length of the borehole
Strain gauges	Any orientation	±5 μ strain	Localized on structure

monitoring changes in moisture or water level and is also used to monitor subsidence).

- Acoustic emissions used to detect abrupt movement of soil and rock due to failure.
- Strain gauges and load cells (provide a means of measuring strain on steel and concrete structures).

Again, we must be aware of scale effects in our measurements and how they can impact the results. DaCunha (1990) discusses the scale effects and the concept of representative elementary volume (REV) associated with determination of deformability, strength, internal stresses in rock masses as well as hydraulic properties.

Some of the applicable engineering measurements are also discussed in the chapters on surface geophysics (Chap. 16) and geophysical logging (Chap. 18). For example, seismic measurements of compression (p) wave and shear (s) wave velocities are used in calculation of elastic modulus. The measurement of velocity can be accomplished by surface seismic refraction measurements, MASW measurements, downhole or by hole-to-hole measurements (ASTM 2014a, b and Wair et al. 2012).

20.1.1 Drilling and the Installation of Instrumentation

Many of the engineering measurements are made from within a borehole. Measurements made from within a borehole are affected by both the location and the volume sampled (REV). An essential prerequisite to the efficient design and monitoring of a valid instrumentation program is having an understanding of the geology and the proposed measurement. This improves the chances that the data are representative of the overall geologic system including both best and worst case conditions.

Whenever instrumentation is to be installed and measurements made in boreholes the method of drilling, casing installation along with other factors need to be of special concern.

- A clean, stable, and open borehole without the use of drilling mud is necessary for some geotechnical instrumentation and measurements.
- Borehole deviation may be of concern with certain measurements.

Detailed specifications for drilling and installation need to be developed and executed in the field in order to assure quality measurements. In his forward to Dunnicliff's (1988) book, Ralph Peck states that even with the utmost care and diligence following the manuals, equipment and their resulting measurements can be unusable if a critical step is overlooked. There are also unique conditions where the manual should be purposely set aside. In both of these cases, a solid geotechnical background and experience with the particular equipment are necessary for success.

20.2 Monitoring Subsidence

Monitoring of subsidence may be required as a warning of potential collapse where known cavities, caves or mines are located. Areas with signs of active subsidence may be monitored to confirm subsidence and assess subsidence rates. There are a number of different ways to measure and monitor land subsidence ranging from regional surveys to local site-specific monitoring (Fig. 20.1). The complexity of the measurements can range from

- simple observations and documentation (photographic/spray paint/etc.)

- repetitive surveying measurements – horizontal and vertical changes
- In place instrumentation for monitoring, some instrumentation can also have data transferred directly to the office.

The strategy, as well as the technology to measure and monitor subsidence and collapse in karst and over mines, is currently available (Benson 2001). Table 20.1 summarizes some of the common methods for measuring and monitoring land subsidence. The methods are relatively simple in concept, but can be difficult in practice and as always each technique or measurement has advantages and limitations. All of the many details for making these geotechnical measurements are beyond the scope of this book. Advances continue to occur in geotechnical instrumentation including measurement technology, fiber optics, along with digital data acquisition and transfer. For further details, see Dunnicliff (1988) supplemented by recent manufacture's literature and recommendations.

20.2.1 Regional Subsidence

Regional subsidence monitoring requires very detailed measurements over a large area. This type of measurement can only be effectively obtained from airborne or satellite data such as InSAR or LiDAR (Fig. 20.1a). These techniques are summarized in Bawden et al. (2003) and have been discussed in Sects. 14.5.2 and 14.5.3.

These are valuable tools for developing regional subsidence maps or showing frequency of new sinkhole development. Many papers are being written about the use of these methods along with their inherent limitations. It has been shown that field checking results, even on a random basis, is necessary for accuracy (Doctor and Doctor 2012; Doctor et al. 2008).

20.2.2 Site-Specific Subsidence

Monitoring for site-specific subsidence or potential collapse may be required in population areas to minimize loss of life and property. There are a number of options available that can be used for developing a monitoring network over a large area or more limited monitoring along a single survey line over a very specific area of concern.

20.2.2.1 Reference Datum and Benchmarks

For a site-wide assessment of subsidence, benchmarks or monuments can be installed and surveyed at intervals over time. Measurements can be relative or absolute when tied to a stable reference datum. Benchmarks are the reference datums used to detect vertical deformation while horizontal control stations are used for horizontal deformation (Dunnicliff 1988). A benchmark typically consists of a rod or pipe anchored into rock at depth. The rod is isolated from contact with the surrounding soil and rock by being surrounded by a larger diameter pipe. Since shallow datum points are subject to a variety of factors that can render them unstable, deep benchmarks should be considered (Figs. 20.1b and 20.2). Elevation at the top of rod is established and monitored over time in order to detect subsidence.

Benchmarks are commonly used to monitor subsidence related to sinkholes, mines, ground subsidence due to fluid withdrawal and monitoring dam stability and other critical structures. An optical deformation survey provides a direct measurement of the movement of a surface monument in both the vertical direction and sometimes in the horizontal direction. By repeating measurements at regular intervals, a time series model of structural movements can be developed providing accurate data for failure prediction. See Dunnicliff (1988) for categories of instruments for measuring deformation and the surveying methods that can be used.

A subsidence or deformation analysis requires the design, observation and adjustment of a high accuracy survey network. Surveyors who have performed first order geodetic surveys are familiar with the basic principles. However, a deformation survey usually involves even stricter accuracy requirements and requires virtually all aspects of the survey to be rigidly controlled. A vertical network can be expected to detect vertical displacements on the order of 0.00021 m using first order measurements (Boston Survey Consultants 1990).

A system of telescoping benchmarks (TBM) was utilized in South Africa as a result of a devastating loss of lives and structures in the gold mining area (Brink 1984). This system of TBMs was utilized to act as a warning device to monitor the development of a sinkhole beneath existing structures. Each TBM consisted of a number of pipes, one within another, with their bottoms cemented into the rock at different depths. If a sinkhole or subsidence began to occur, the deepest rod would detect movement before the shallower rods providing an early warning of subsidence.

In another example, benchmark monuments were installed over an abandoned limestone mine in the Kansas City area to determine if any surface subsidence was occurring associated with a 7 ha collapse within the mine. The possibility of surface subsidence due to mine collapse was determined to be highly unlikely by all experts who had inspected the mine conditions. However, surface fissures that coincided with the area of mine collapse caused the question of mine stability to remain a concern by the state.

The monument design was modified from that in Dunnicliff (1988). Each monument consisted of a 2 cm steel rebar with bolt on coupling anchored in a minimum of 1.4 m into the uppermost rock (Argentine Limestone) with concrete.

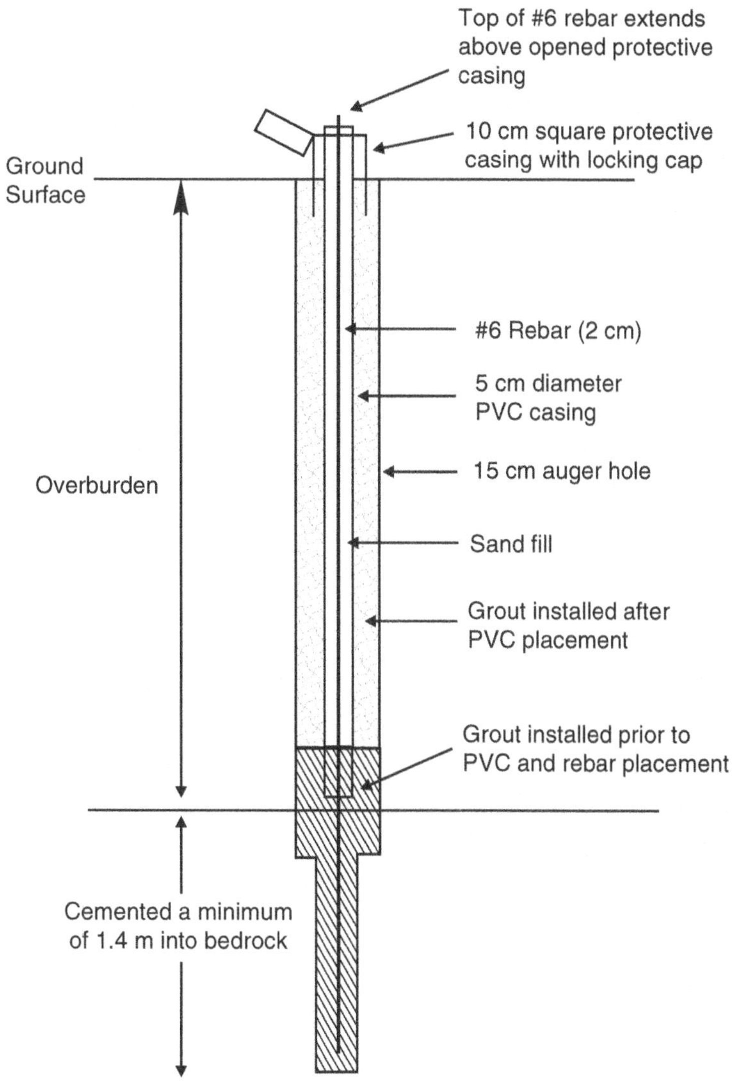

Fig. 20.2 A benchmark consists of a rod or pipe anchored at depth

The rebar was encased within a 5 cm PVC casing, which was filled with loose sand. The outside of the PVC casing was then grouted in place (Fig. 20.2). A protective steel casing was installed over the PVC casing. Three of the monuments were equipped with thermocouples at 30 cm intervals so that thermal expansion/contraction of the subsidence monuments could be accounted for over seasonal measurements. The vertical network could be expected to detect vertical displacements of the order of 0.00021 m using first order vertical measurements.

A total of 11 shallow monuments were initially installed embedded in the uppermost rock (Fig. 20.3). Nine of the monuments were over the mine and two were off the mine and provided stable reference points. As a result of early measurements that showed subsidence activity two additional shallow and two deeper monuments were installed. The deeper monuments were more than 30 m deep anchored in a massive limestone.

Figure 20.4 shows examples of the subsidence data obtained from this site. Three of the shallow monuments both on and off of the paleocollapse area are presented in Fig. 20.4a that showed little or no movement. Figure 20.4b shows three of the monuments that did show subsidence. Monuments 6 and 10 are both shallow, monument 10 showed a very gradual subsidence while monument 6 had an abrupt increase in subsidence. Monument 13 was a deep monument installed into the massive limestone after subsidence was detected in the shallow monuments. It is very close to monument 6 and has a similar rate of subsidence indicating that both the shallow and deep rock is showing subsidence. Further details are discussed in Part III, Chap. 25 case histories.

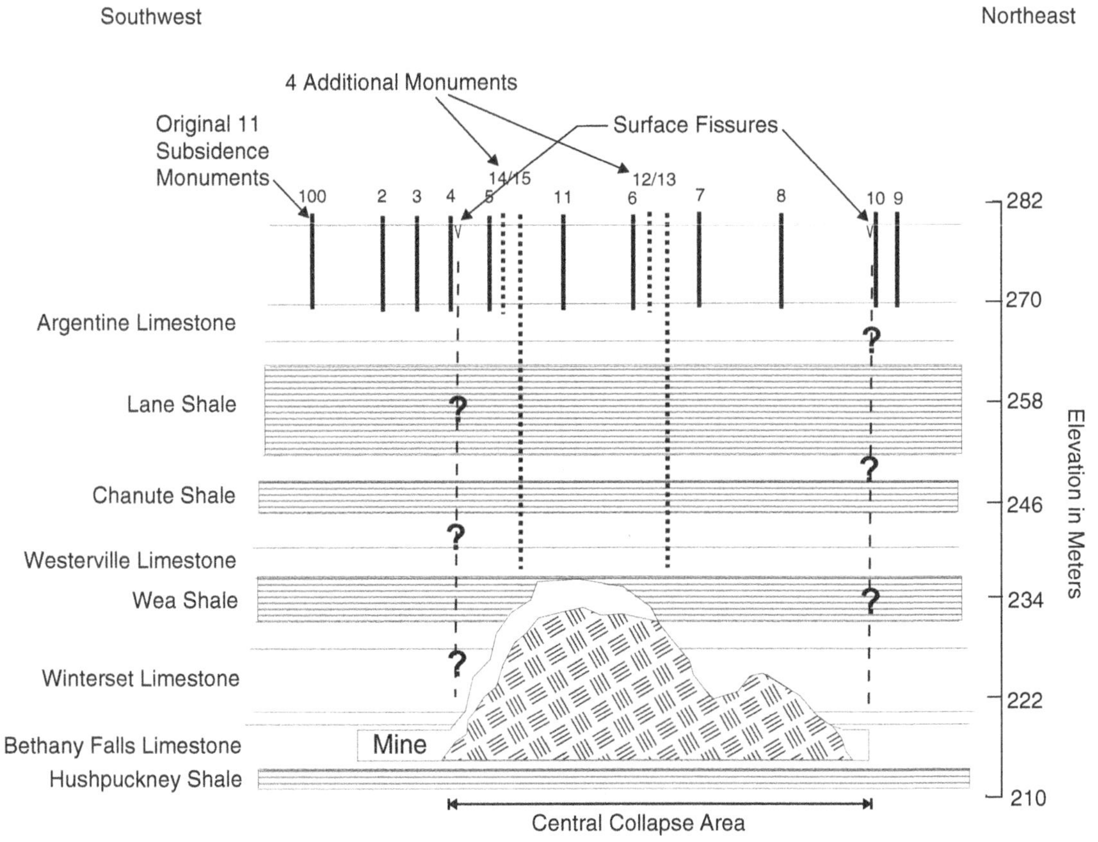

Fig. 20.3 The locations of subsidence monuments are shown over an area of mine collapse

This project required extreme quality control measures in order to assure the accuracy of the data required. The team consisted of the authors who designed and installed the monuments, Boston Survey Consultants who checked our monument design and layout, provided checks of data during the survey sequence and carried out an independent survey as a means of quality control. A local survey firm who had experience with highly accurate survey measurements made periodic measurements and provided the data. This entire group worked closely throughout the program to minimize pitfalls cited by Dunnicliff (1988) when carrying out such work.

20.2.2.2 Inclinometers and Tilt Meters

Inclinometers and tilt meters are another means of measuring changes in the surface or subsurface movement (Fig. 20.1c). Inclinometers typically monitor horizontal movement from the vertical. Instrumentation consists a probe that runs along the length of a specialized designed track that is installed in a borehole. Time series measurements can detect changes in any lateral movement in multiple planes through the length of the borehole. Tilt meters monitor the change in rotation of a fixed point on the surface, within a borehole or on a structure.

An example of this type of instrumentation comes from a site in southeastern New Mexico where large surface collapse due to man-made cavities is not uncommon. The injection of fresh water into salt formations creates brine, which is pumped out and used as drilling fluids for the oil industry. In the process of creating this brine fluid, the natural salt formation is dissolved leaving cavities of substantial size. The resulting surface collapse is often on the order of 100 m in diameter or more (Fig. 7.11). These features are typically in remote areas with little direct impact to communities and infrastructure. However, two brine wells were located in the city of Carlsbad, New Mexico within 50 m of each other. This is a populated area at the intersection of two major highways, a railroad track, a drainage canal and a trailer park immediately adjacent. Collapse in this area would have a major impact on life and property (Land and Veni 2012).

Brine production has ceased and an investigation to estimate cavity size and extent was completed. One of the investigations utilized a 2D seismic reflection survey (Goodman et al. 2009) and another utilized an electrical resistivity survey (Land and Veni 2012). High-resolution tilt meters were installed in the two borings as a monitoring system over the cavity. This early-warning system should provide indications

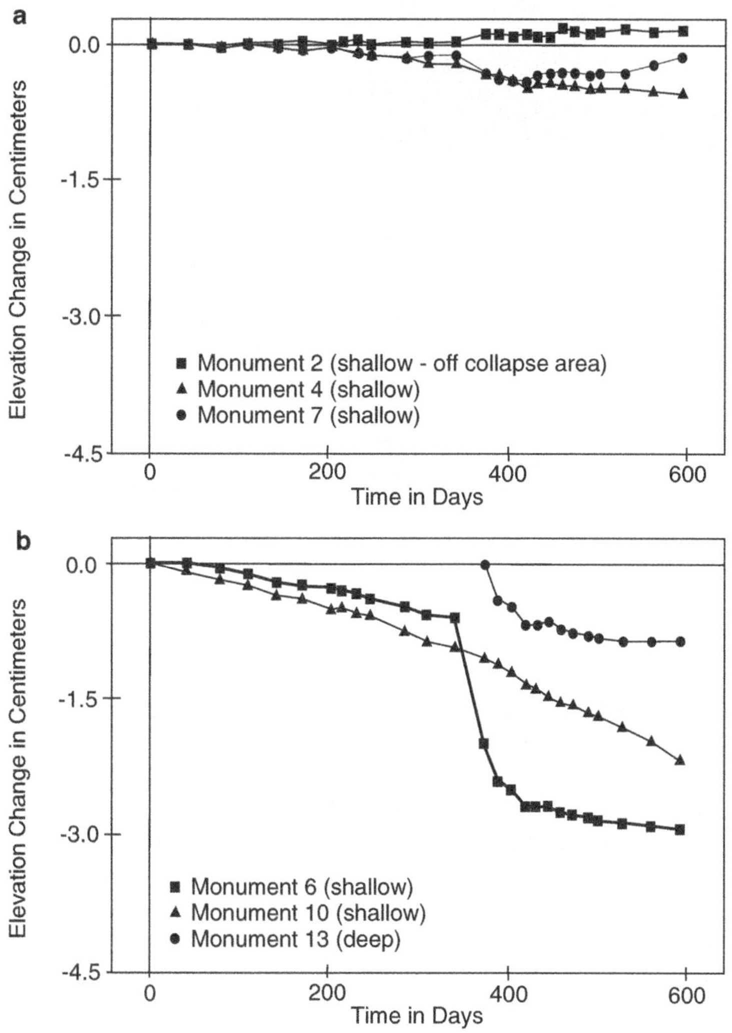

Fig. 20.4 Example of subsidence data over a paleocollapse area with monuments showing little to no movement (**a**) and small but abrupt movement (**b**)

of subsurface subsidence that may precede a collapse. Surface subsidence is also being monitored directly over the cavern by optical survey measurements. In addition, two wells were instrumented with pressure transducers and data loggers to detect sudden changes in water levels (Goodman et al. 2009).

20.2.2.3 Time Domain Reflectometry (TDR) and Optical Domain Reflectometry (OTDR) Measurements

Time domain reflectometry (TDR) measurements use an electrical impulse sent down the length of an electrical cable. When a significant distortion or a break in the cable is encountered a reflection occurs and is detected by the instrument. By measuring the travel time of the impulse (reflection), the location of the break or distortion in the cable can be obtained. Optical domain reflectometry (OTDR) is similar to TDR but uses a fiber optic cable and light rather than an electric pulse (O'Conner and Dowding 1999).

These methods are being used in geotechnical and environmental applications to monitor changes in deformation of soil and rock as well as changes in moisture content, water level and density of soil (ASTM 2012). The cable may be installed vertically in a borehole or horizontally in a trench (Fig. 20.1d) or at any angle from the vertical to the horizontal. TDR measurements can also be used within reinforced concrete to monitor strain and cracking. Prior to installation the TDR cable can be deformed (by crimping the cable) at intervals to provide a reflection from a known distance for calibrating measurements. A significant benefit of these

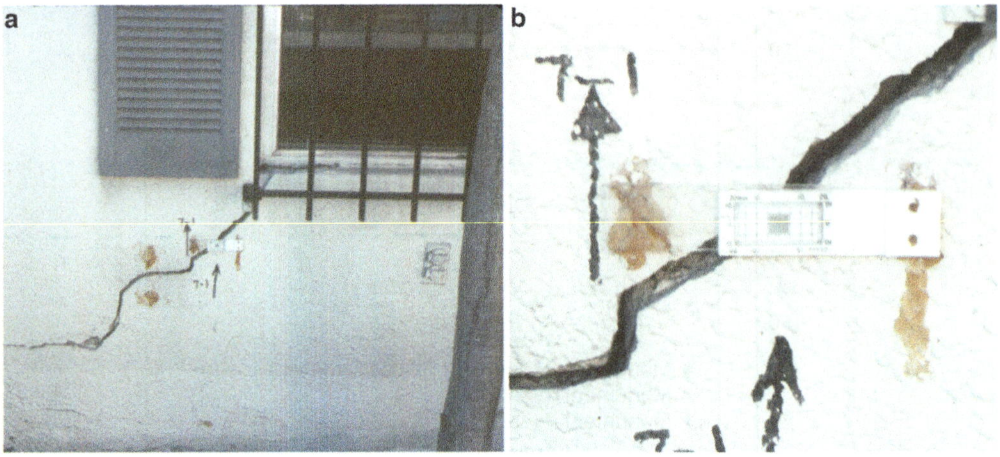

Fig. 20.5 An example of a strain gauge mounted across a crack in a structure (**a**) to monitor any change in movement, (**b**) shows a close up of the strain gauge

measurements is their ability to be continuously monitored having a telemetry network transfer data to an office. This enables TDR to be readily adapted to monitoring of karst or mine related subsidence. Examples of TDR measurements to monitor soil and rock slopes are given by Anderson and Welch (2000).

The details of the field installation become critically important. The borehole diameter should be small to maximize coupling between the soil or rock and the TDR cable. The grout strength should be less than the bearing capacity of the soil so that any movement of the soil or rock is coupled to the TDR cable.

An example of TDR measurements is from a longwall coal mine that was advancing at a depth of approximately 150 m beneath I-70 in Pennsylvania. TDR measurements were made in seven boreholes along with 32 tilt meters to detect the subsidence as the mine face advanced (O'Conner et al. 2001). Precursor shear deformation was detected by TDR when the face of the mine was more than 760 m from the TDR cable. The rate of deformation increased when the mine face was within 60 m of a TDR cable. Tilt meter response began as the mine face moved underneath the highway and reached a peak value as the mine face moved past.

Dowding and O'Conner (2000) have compared TDR and inclinometer measurements for slope monitoring. As with all measurements there are advantages and disadvantages, both methods provide useful information. The TDR measurements respond best to localized shear and inclinometers are especially sensitive to gradual changes in inclination or soils undergoing general shear.

20.2.3 Localized Settlement or Subsidence

Local site-specific subsidence is typically focused on a specific structure to address specific concerns often after signs of subsidence are noted such as cracks in a concrete structure, warping of a sidewalk or localized subsidence in the ground surface. A variety of strain gauges and transducers are available for monitoring these very localized conditions. They are discussed in detail along with their application in Dunnicliff (1988).

Strain gauges essentially measure the changes in the width of a fracture or crack. They can be employed to monitor movement of cracks on structures (Fig. 20.1e). This type of measurement is quite simple, very localized and can be used in the horizontal or vertical direction (Fig. 20.5). This type of measurement is typically used to monitor existing cracks to assess if they are old and stable or new and active.

References

Anderson N, Welch D (2000) Practical applications of time domain reflectometry (TDR) to monitor and analyze soil and rock slopes. In: Marr WA (ed) Geotechnical measurements lab and field, ASCE GeoDenver, Reston, Virginia, pp 65–79

ASTM (2012) Standard test method for water content and density of soil in situ by time domain reflectometry (TDR) D6780/D6780M-12. American Society for Testing and Materials, West Conshohocken

ASTM (2014a) Standard test methods for crosshole seismic testing, D4428M-14. American Society for Testing and Materials, West Conshohocken

ASTM (2014b) Standard test methods for downhole seismic testing, D7400-14. American Society for Testing and Materials, West Conshohocken

Bawden GW et al (2003) Measuring human-induced land subsidence from space. US Geological Survey Fact Sheet 069-03, December

Benson RC (2001) Strategies of site characterization and risk management. A presentation to National Research Council, committee on coal waste impoundment failures subcommittee on impoundment site characterization, National Research Council, Washington, DC

Boston Survey Consultants (BSC) (1990) Design and analysis of the subsidence monitoring network at forest view landfill. Consultants report, Boston, Massachusetts

Brink ABA (1984) A brief review of the South African sinkhole problem. In: Beck BF (ed) Proceedings of the 1st multidisciplinary conference on sinkholes: their geology, engineering and environmental impact, Orlando, FL. Balkema, Rotterdam, pp 123–127

Da Cunha AP (ed) (1990) Scale effects in rock masses. In: Proceedings of the 1st international workshop on scale effects in rock masses. Balkema, Rotterdam, pp 3–27

DiMaggio J, Osborn P (eds) (2007) Field measurements in geomechanics. In: Proceedings of the 7th international symposium, Boston, MA, 24–27 September 2007, ASCE Geo-Institute, Geotechnical Special Publication No. 175

Doctor DH, Doctor KZ (2012) Spatial analysis of geologic and hydrologic features relating to sinkhole occurrence in Jefferson County, West Virginia. Carbonates and evaporites, Special Issue on Integrating Science and Engineering to Solve Karst Problems Part I June 27(2):143–152

Doctor DH et al (2008) Bedrock structural controls on the occurrence of sinkholes and springs in the northern Great Valley Karst, Virginia and West Virginia. In: Yuhr L, Alexander CE Jr, Beck BF (eds) Proceedings of the 11th multidisciplinary conference on sinkholes and the engineering and environmental impacts of karst, ASCE Geo-Institute, Geotechnical Special Publication No. 183, pp 12–22

Dowding CH, O'Connor KM (2000) Comparison of TDR and inclinometers for slope monitoring. ASCE GeoDenver, Colorado

Dunnicliff J (1988) Geotechnical instrumentation for monitoring field performance. Wiley, New York

Goodman WM et al (2009) Two-dimensional seismic evaluation of the I&W brine cavern, Carlsbad, New Mexico. RESPEC Consulting and Services, Topical Report RSI-2083. http://www.emnrd.state.nm.us/OCD/documents/RSI2083.pdf

Hunt RE (2005) Geotechnical engineering investigation handbook, 2nd edn. CRC Press, Boca Raton

Land L, Veni G (2012) Electrical resistivity surveys of anthropogenic karst phenomena, southeastern New Mexico. Geology 34(4):117–125

O'Connor KM, Dowding CH (1999) Geo measurements by pulsing TDR cables and probes. CRC Press, Boca Raton

O'Connor KM et al (2001) Real time monitoring of subsidence along I-70 in Washington and Pennsylvania. Transportation Research Record Journal of the Transportation Research Board No. 1772, pp 32–39

Wair BR et al (2012) Guidelines for estimation of shear wave velocity profiles, PEER report 2012/08. Pacific Earthquake Engineering Research Center, Berkley, 68p

Hydrologic Characterization and Measurements

21

Abstract

This chapter focuses attention upon the unique hydrologic characteristics of fractured and karst aquifers and the wide range of measurements that can be utilized to characterize them. At this stage of the characterization we should have a well-developed geologic conceptual model of conditions with many of the questions about the site already answered (Tables 13.1, 13.2 and 13.3). We should already know the type of surface hydrologic features at the site; the surface water and groundwater boundaries, in general if not in detail; the depth, thickness and level of complexity of the epikarst; whether the aquifer is confined or unconfined, and whether there is more than one. In addition, we will have preliminary data from existing monitoring wells or initial borings on water levels, potentiometric surfaces and flow directions. The objectives for this phase of work will focus upon the details to quantify and support our conceptual model. This is obviously important for projects focused on groundwater resources or contaminant remediation. But surface and groundwater in a fractured or karst setting will also impact most engineered structures, if not managed well they can result in unstable conditions.

21.1 A Complex System

The hydrologic cycle is a large integrated system. Yet, the topics of surface water and groundwater have migrated into separate topics in text books, courses and practice. In a karst setting the rapid transmission of surface water into the aquifer through sinkholes or sinking streams, results in surface and groundwater being an intimately coupled system and must be treated as such by practicing professionals.

Even though it is an integrated system we must consider the hydrologic parameters that may be associated with each zone of the complex karst system from the surface down (Fig. 21.1). An individual feature such as a spring may be dry unless heavy rainfall occurs. The epikarst layer may act as a unit that both stores fluid and yet transmits it quickly at discrete points. Each of the saturated zones may have a flow component that is different from those units (layers) above and below or even adjacent features. This wide range of hydrologic conditions require a variety of measurements at a range of scales and that can be sampled over time in order to accurately characterize them.

In a karst setting, there are a number of potential impacts on the hydrology (water levels, flow, basin boundaries, etc.) that can come from a number of sources. A site-specific characterization must include consideration of these variables that may include:

- Karst heterogeneities themselves including fractures, conduits, sinkholes, and springs
- Structural features that may modify groundwater flow include anticlines, synclines, and faults. These features can concentrate groundwater flow or provide barriers to groundwater flow
- Temporal variations such as recharge from rainfall, drought conditions, tidal effects, etc.
- Man-made changes can affect water levels or recharge such as dams/reservoirs, supply wells, concentrated runoff, etc.

R.C. Benson, L.B. Yuhr, *Site Characterization in Karst and Pseudokarst Terraines*,
DOI 10.1007/978-94-017-9924-9_21

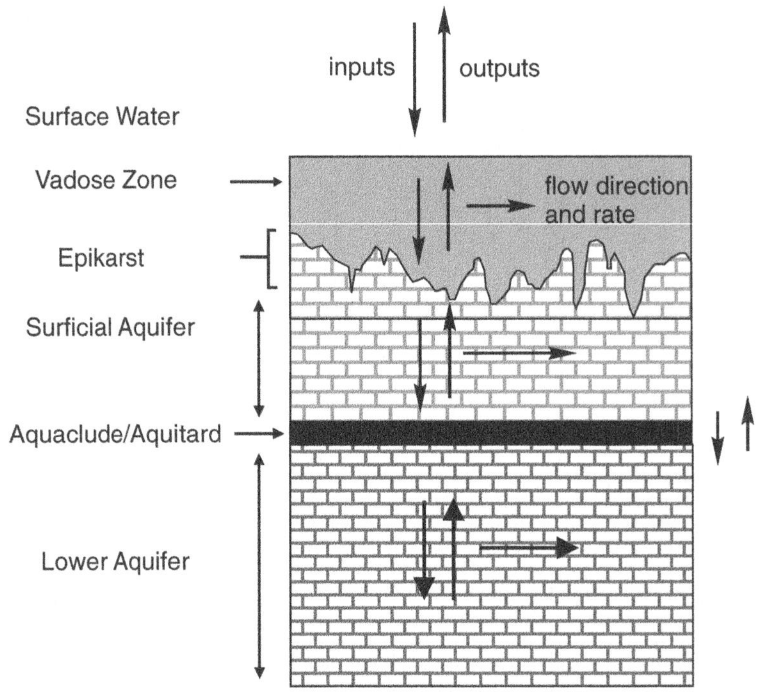

Fig. 21.1 A conceptual cross section with water flow options that may need to be considered or measured within the vadose zone, the epikarst, any aquifers, and any aquitards. Within each of these areas will be specific features such as fractures, conduits, weathered zones that will also need to be considered and measured

Therefore, there are a multitude of methods for field and laboratory measurement of hydrologic properties and sampling of groundwater and contaminants. For details, the reader is referred to Freeze and Cheery (1979), Sara (2003), Nielsen (2006), and Hunt (2005). Sasowky (2000), and Palmer (2006) also provide a practical discussion of groundwater hydraulics and its measurement specific to karst. Also see ASTM D5717, a standard guide for developing groundwater monitoring systems specifically for karst and fractured-rock aquifers (ASTM 1998).

The next and often necessary step in hydrologic characterization is that of groundwater modeling. However, the details of groundwater modeling in karst are well beyond the scope of this book. Palmer et al. (1999) is a publication specifically devoted to modeling of groundwater in karst. It provides an excellent overview of the many aspects of modeling karst flow. Topics include the various models that are employed, along with papers on the acquisition and application of field data. Palmer suggests that the acquisition of valid and sufficient field data is by far the most limiting factor in successful modeling of groundwater flow in karst. In addition, they point out fatal flaws that prevent successful modeling of groundwater flow in karst include:

- The lack of approaching the karst system in its entirety, and
- The limited acquisition of valid and sufficient field data.

Others support these ideas. The devil is in the details including the issues of scale and obtaining sufficient appropriate and adequate data for this immensely complex hydrologic system (Palmer 2006; Kovacs and Sauter 2007). Efforts for groundwater modeling would benefit greatly from a more complete and accurate site characterization.

21.2 Karst Is a Multiple Porosity System

Porosity is a primary factor when describing karst hydrology:

- Primary porosity is that associated with the original rock as it was deposited and lithified.
- Secondary porosity is associated with post deposition activity caused by strain and stress resulting in fractures.
- Tertiary porosity is caused by enlargement of secondary porosity features by dissolution. These tertiary porosity zones are commonly found as localized zones along fractures or bedding planes. The size of these openings can range from a few centimeters to many tens of meters. Here again the issue of scale becomes important.

Although the rock mass can provide most of the storage of groundwater, flow in the rock mass is only a small percentage of the volume of water discharging from a karst aquifer. Flow is primarily through fractures and conduits. A major factor controlling the exchange of water between the rock matrix and a conduit is the hydraulic gradient (Fig. 21.2). During low flow conditions, the water can flow from the rock matrix to the conduit. During high flow conditions, the flow between the rock matrix and the conduit may reverse with water flowing from the conduit to the rock matrix.

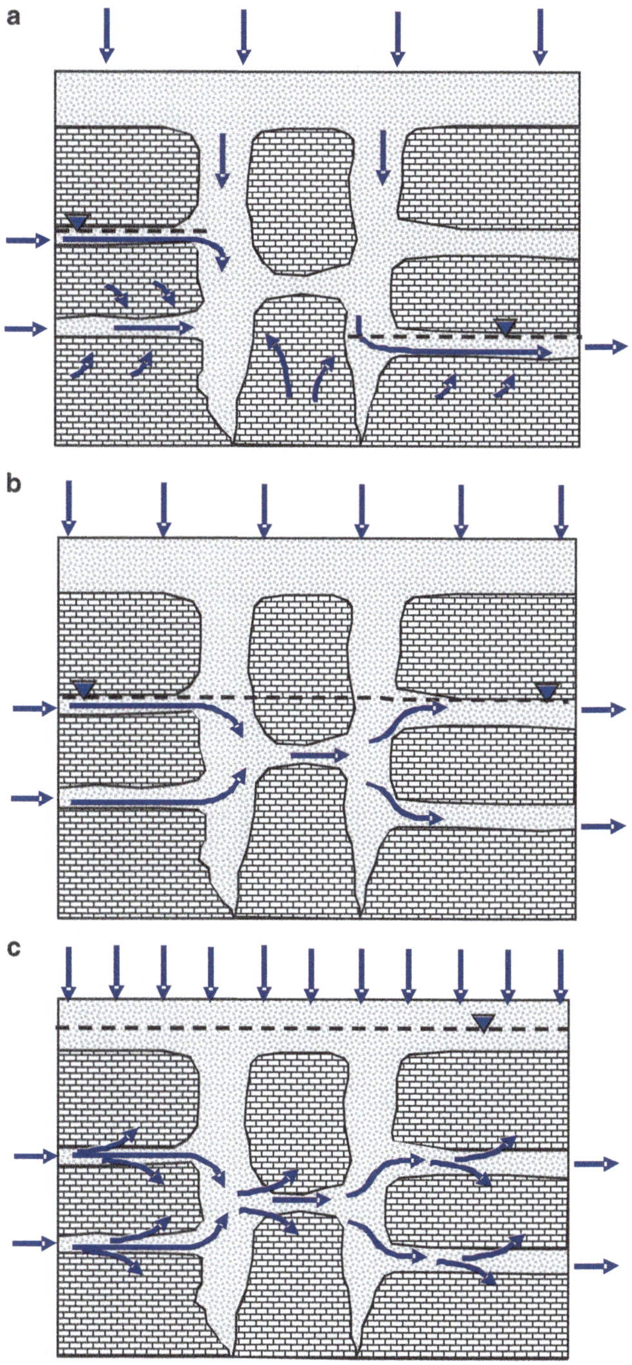

Fig. 21.2 Flow directions and water levels respond differently with low, moderate and high recharge (**a**) Low rainfall and limited inflow from upgradient (**b**) Moderate rainfall and flow from upgradient (**c**) High rainfall and large flow from upgradient (Adapted from Sowers 1996)

In addition, the flow of groundwater through a well-integrated network of fractures or conduits commonly transports water at rates several orders of magnitude greater than those encountered in non-karst groundwater systems. Aley (2008) summarizes the mean straight-line groundwater velocities from ten tracer tests, which ranged from 234 to 5,840 m/day. These rates are four to six orders of magnitude greater than groundwater flow rates commonly reported in the literature for most aquifers. Such data demonstrates that karst groundwater flow can rapidly transport contaminants for considerable distances. However, flow may need to be considered in all the hydrologic units, from the surface, through the unsaturated zone and the epikarst into the saturated zones (Fig. 21.1). Hydraulic conductivity may vary by many orders of magnitude when going from a relatively tight rock matrix to essentially pipe flow in cavernous limestone. In most cases tertiary porosity will dominate the flow.

21.3 Lets Revisit the Issue of Scale

The issue of scale is important to our hydrologic measurements and their interpretation. The concept of REV, presented in Chap. 12, can be applied to measurements of both hydrologic and engineering properties (Da Cunha 1990). Benson and Yuhr (1993) present a discussion of spatial sampling considerations in site characterization and its importance in characterizing fractured rock and karst systems. For example, measurements from the same location (i.e., one borehole) but taken at different scales (volumes measured) will often yield very different values of hydraulic conductivities of fractured rock. A laboratory permeability test run on an intact piece of core will generally yield lower hydraulic conductivities than an in-situ slug test. A slug test will typically yield lower values than a packer test and a packer test will usually yield lower values than a pump test. Each of these measurements is integrating a larger volume of rock. If the values from these measurements are all similar, we can then assume that the system is uniformly fractured at a scale similar to the smallest volume measured. On the other hand, if the measurements disagree then we have identified scale differences in the fracture system.

Rules of thumb suggest that to obtain representative hydraulic conductivity values from fractured rock requires that measurements be made on a volume of rock whose dimensions are ten times the fracture spacing (Nelson 1984). Da Cunha (1990) shows a variation of seven orders of magnitude variation in hydraulic conductivity as the scale of tests range from laboratory to basin scale.

The scale of measurements can be divided into three levels, regional, intermediate (or site-specific) and local near-field details. A regional scale of measurement can cover many square kilometers and encompass the entire site and its surroundings. These measurements would include basin-wide observations of hydrologic features along with measuring discharge and recharge areas and dye trace testing. Intermediate scale (or site-specific scale) of measurements would include pump tests, local dye tracing and surface geophysical surveys. Local scale within the site would include boreholes, sampling and measurements such as slug, packer and pump tests, geophysical logging along with laboratory testing of core samples (Fig. 21.3).

21.4 Temporal Aspects

In addition to the spatial aspects of measurements, the temporal aspects of measurements must also be considered. Temporal changes can include both natural (rainfall, tides, drought) and man induced activities (pumping supply wells, discharges from reservoirs). These changes can also occur rapidly or gradually. Dramatic changes in conditions over very short periods of time are common in karst hydrology. Adequate temporal measurements must be made to avoid gaps and aliasing of the data. Since observations and many measurements are usually intermittent in nature they can easily miss anomalous temporal conditions. To detect and record anomalous temporal conditions, continuous recording instruments are necessary so that the rare events are detected and can be quantified.

Two examples illustrating the need for temporal measurements have been previously presented and represent changes in surface water conditions. One example was the small creek that one could step across without getting your feet wet (Fig. 12.13a), which turned into a raging torrent (Fig. 12.13b) that would sweep you away as a result of heavy rainfall in the nearby mountains. Had we not been on-site during this event, the speed and magnitude of this change in conditions would have been missed. As it was, our staff gauge measuring water levels in the creek was insufficient for documenting this event. The second example comes from the Peace River in southwest Florida. This area is normally a small quiet stream used by canoeists, but on occasion will become dry (Fig. 15.6). While this is not a rapid change in conditions,

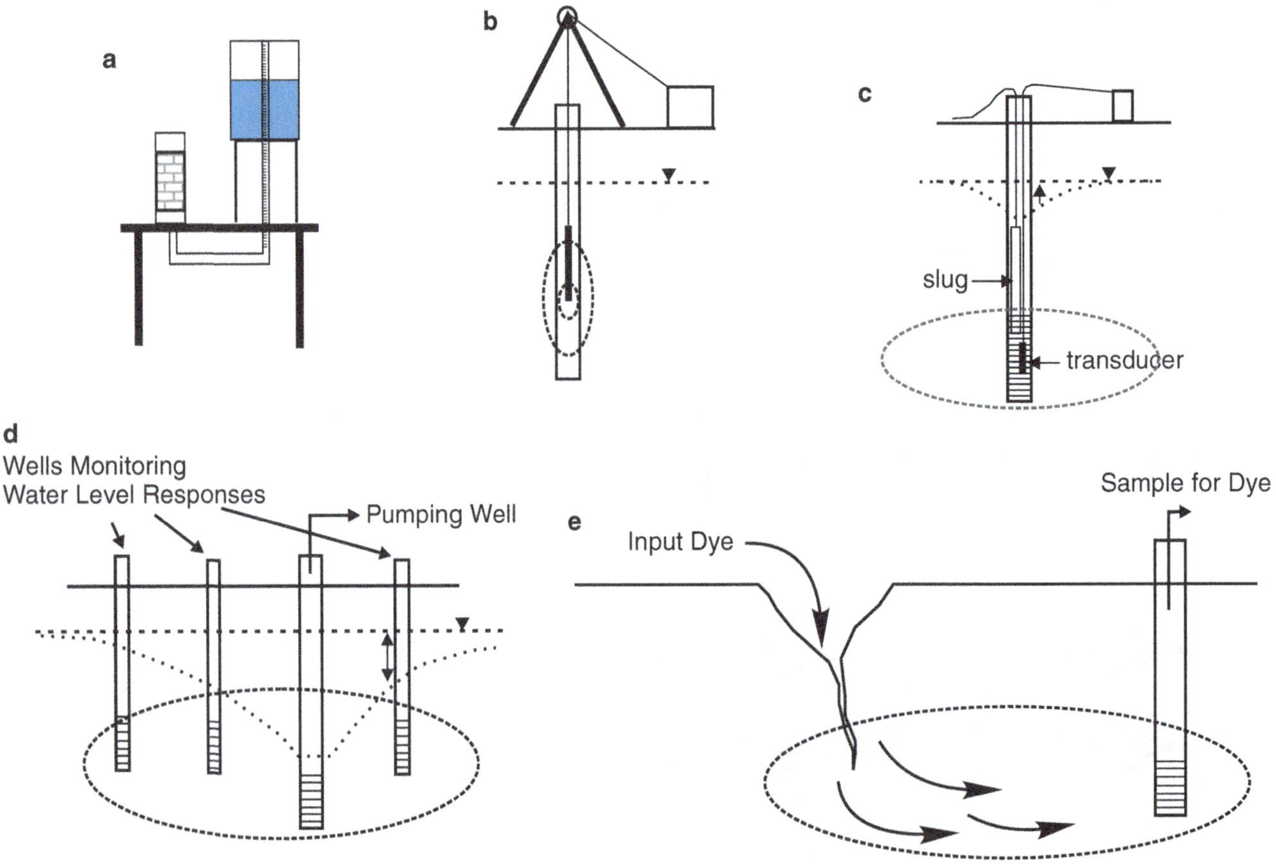

Fig. 21.3 Some of the wide range of measurements that can be used to characterize hydrologic conditions, each is associated with an increasing volume of measurement from detail to regional (**a**) Laboratory tests on core sample (few tens of cm) (**b**) Logging tools measuring conditions within the borehole or well (immediately surrounding the borehole) (**c**) Slug test or Packer test measuring conditions surrounding the screened or packed interval (up to a few meters) (**d**) Pump test measuring conditions over network of wells (to tens of m^2) (**e**) Dye trace tests measuring conditions over large distances (to hundreds of meters)

depending upon the timing of work in this area, the hydrologic impact and assessment of this river in its dry condition could have easily been missed.

Temporal changes in groundwater conditions also occur and need to be anticipated. Fractures and conduits respond quickly to recharge events and may spill over into higher laying conduits or fractures (Fig. 21.2). In fractured and karst rock it is not uncommon to see large variations in water levels in response to recharge events. The degree of response in a given well is dependent upon the size and interconnection between fractures and conduits encountered by the well and their connections to the surface. Monitor wells in fractured rock and karst conditions may see large variations in water levels in immediate response to recharge events.

Changes in water level data due to recharge from rainfall can add information regarding the connectivity between the surface and groundwater to the conceptual model. In fractured-rock and karst aquifers with minimal overburden, recharge tends to be rapid. Milanovic (2004) cites changes in water level of 90 m in 10 h and a maximum change of 312 m in 183 days in eastern Herzegovina. These changes in water level elevations can cause flow to move in different directions on both a short term and seasonal basis resulting in significant temporal changes. A fractured rock or karst aquifer with a thick overburden may have a long temporal lag similar to that of a granular aquifer. Recharge may be distributed through an aerially extensive network of fractures or through soil (dispersed recharge), or may be concentrated at points that connect directly to the aquifer (point recharge via a sinkhole). The percentage of point recharge strongly influences the character and variability of its discharge and water quality. These factors become critical when dealing with contaminants.

21.5 Hydrologic Measurements

The need for identifying and quantifying certain hydrologic conditions will vary depending upon project objectives. As with all types of measurements, hydrologic measurements should start with the simple and move to the more complex. These measurements should address the project needs such as:

- Water levels and potentiometric surfaces – spatial and temporal impacts or changes
- Flow – directions (laterally and vertically), rates (under various rainfall events or seasonal conditions)
- Interconnections between surface, epikarst and groundwater.
- Aquitards, barriers and impermeable layers or zone
- Water quality – natural conditions and contaminants, if needed

In order to optimize any hydrologic measurements, particularly those using borings or wells, the locations need to be put into context at the site. Are they in the right location, drilled and screened at the correct depth to be representative of conditions at the site? Conditions of existing wells should have been checked to confirm that they were constructed properly and have not deteriorated which could impact any data from them. An excellent example was provided in Sect. 18.2.7 where a monitoring well audit revealed that all five existing monitoring wells at a landfill were invalid. There were a number of construction errors along with deterioration and they were installed assuming a single aquifer when there were actually two. These problems invalidated all data from these wells (water levels and water quality) that had been collected over many years.

Figure 21.3 illustrates a few common hydrologic measurements and is presented from the smallest volume of measurement to the largest. The types of measurements can provide specific hydrologic parameters at various scales to evaluate and characterize various hydrologic conditions. Some of these measurements include:

- Laboratory testing (Fig. 21.3a) includes tests of water quality as well as soil or rock samples.
- Staff gauges and water level meters are very commonly used to monitor water levels at the surface or in a well.
- Flumes and weirs are structures placed in surface water bodies to direct a portion of the flow through an opening that can be instrumented to measure flow.
- Geophysical logging measurements (Fig. 21.3b) can measure flow within a borehole directly (impeller flow meters) or indirectly (temperature logs). See Chap. 18 for more details on geophysical logging.
- Single well tests such as slug test or packer tests (Fig. 21.3c) utilize man-induced changes in water levels from wells or borings which can allow us to measure recovery times and calculate various aquifer parameters.
- Pump tests (Fig. 21.3d) utilize a single, often centrally located well to be pumped to create a drawdown of the water level, surrounding wells are then monitored for changes in water levels.
- Dye tracing (Fig. 21.3e) uses special chemical dyes introduced into surface or groundwater at one point and sampled at other points to confirm connections and travel times. The topic of dye tracing is often a critical component for a site characterization in karst, therefore, we have included a separate chapter (Chap. 22) addressing dye trace measurements.
- Thermal infrared and optical time domain reflectometry have been mentioned for application in the sections on surface geophysics (Chap. 16) and engineering measurements (Chap. 20). These techniques can also provide

effective hydrologic measurement options in unusual conditions.

- Surface geophysical measurements can sometimes be used to measure the presence of moisture/water or indicate flow by measuring the conductive component of the groundwater associated with inorganic contaminants.

21.6 Surface Water

Surface water flow is often the easiest to measure since it can be seen and is accessible. Direct measurements of surface flow can be made by weirs, flumes, flowmeters or by monitoring a staff gauge (Palmer 2007). Surface water is often affected the most by temporal changes (rainfall, evaporation, runoff, discharges). Within the surface water body, localized flow can be coming from a spring or going into a sinkhole. These submerged features are often difficult to detect but can be a major contributor of flow to or from an aquifer.

Simple observations of surface water flow can often reveal low areas or locations and patterns of groundwater recharge that were previously overlooked. Identifying areas of concentrated surface water that may be flowing into the subsurface can often trigger soil raveling and ultimately collapse features. This is particularly important in developed areas that have changed the natural conditions (parking lots, stormwater drainage systems or stormwater ponds) (Fig. 8.1).

Surface water is also often where contaminants are introduced into the system. The sources can range from street run-off to wastewater spray fields to agricultural use of fertilizer/pesticides to accidental contaminants spills. These can be point sources or broader areas of infiltration.

21.6.1 Submerged Spring Flow Within Rivers, Lakes and Off-Shore

Springs with active flow on the surface are relatively easy to locate and measure (Fig. 3.11b). To locate those hidden sources of spring flow within rivers and lakes or off-shore are more problematic. An early method was to tow a temperature and conductivity sensor (or array of sensors) through the water to detect changes in temperature or conductivity associated with flow from springs. However this required that the sensor pass within the area of flow, which was problematic. The use of airborne thermal imaging (infrared camera) to located submerged spring flow has been very successful and provides a means to cover larger areas improving the chance of encountering these features (see Sect. 14.5.1 and Fig. 14.8). Anderson (2005) discusses the topic of assessing flow, using heat as a tracer.

The use of optical time domain (OTDR) has been used to detect the flow of spring waters into rivers and lakes. Using fiber optic cable, Lane et al. (2008) have demonstrated the application of this method for stream-aquifer interaction and estuary–aquifer interaction. They suggest that the method provides a means to observe temperature over large areas with fiber optic cables as long as 30 km, temperature resolution from 0.1 to 0.01 °C, a spatial resolution of 1 m and temporal resolution on the order of seconds to minutes. This type of data provides both the location as well as change in temperature and data can be sent to a central monitoring station.

This method of measurement provides extremely good lateral resolution along the length of the fiber optic cable. It can also provide a response off to the side of the cable if the cooler spring water is forming a plume of some size. However, these measurements are inherently limited by the specific location of the cable since the extent of the temperature plume may be off to one side of the cable, beyond the range of the cables sensitivity.

Off-shore spring flow is quite common in Florida and many of their general locations are known by fishermen and divers. An example of detecting a near-shore spring using marine resistivity has been presented in Chap. 16, Fig. 16.25. This spring provided an electrical contrast in water quality at the point of discharge, which was detected in the marine resistivity data.

In southeast Florida, Finkl and Krupa (2003) have measured the flow and water quality from off-shore seeps and springs using seepage meters. Flow was measured and water quality samples acquired. These localized data were used to estimate total flows and nutrients levels being discharged off-shore via submerged seeps and springs.

21.7 The Unsaturated Zone

Seepage of water and flow within the unsaturated zone can be complex and extremely difficult to determine with any level of confidence. The unsaturated zone may include soil overburden, the epikarst or bedrock. Subtleties such as slightly cemented layers or variations in clay/silt content and root structures can greatly affect flow direction and speed within the shallow unsaturated soil zones. The epikarst can act as a trap for groundwater or provide localized pathways to the saturated zone. Unsaturated bedrock containing fractures, bedding planes, localized cavities and conduits can be open or soil-filled impacting their connectivity. These are very local details, which are difficult to measure and are easily overlooked.

Flow within the unsaturated soil zone and the epikarst are usually of little concern unless there is a contaminant spill or raveling of soils into openings creating unstable surface conditions. An example of the complexity of the epikarst zone is illustrated by the 2D resistivity data acquired along a highway in Kentucky (Fig. 16.14). The importance of the epikarst is emphasized by the detailed study of the semi-confining layer (SCL) at the EPA Superfund case history presented in Chap. 27. In this case history the continuity of the SCL was critical to the prevention of shallow sinkholes as well as contaminant migration. The details of measurements and sampling within the unsaturated zone can be found in Wilson et al. (1994) and Nielsen and Johnson (1990).

There may be indications of connection and potential pathways for flow in the unsaturated zone indicated by air flow. Cool air flow is commonly noticed at the entrance of caves, mines and their ventilation shafts. In addition, obvious air flow has been noted at many sites coming from fractures in the rock as well as open boreholes and well casings (see Sect. 15.6.2 for further discussion). This can be an indication that the borehole had encountered open fractures in the rock or possibly a cave system.

21.7.1 Groundwater Monitoring for a Landfill in Karst

A groundwater monitoring plan was required for the expansion of a landfill over an abandoned underground mine in the Kansas City area known as the Tobin mine (see Chap. 25 for more details of this case history). A paleokarst collapse structure due to deep-seated cavities about 182 m below the mine floor had induced fractures in the overlying rock and resulted in fissures within the loess soil. When the mine was cut through the area of the paleocollapse feature, a major mine roof collapse of 7 ha occurred and was referred to as the central collapse area (CCA) (Benson et al. 1994).

The initial hydrologic conceptual model for the site included:

- Unsaturated bedrock between the mine and the surface, 52 m or more.
- The abandoned limestone mine was filling with water.

These issues impacted the development of the groundwater monitoring plan for the site.

21.7.1.1 Unsaturated Bedrock

Based upon existing monitor wells and extensive drilling into the mine it was determined that the bedrock above the mine (52 m or more) was unsaturated except for limited quantities of isolated perched water (Fig. 21.4).

Existing monitoring wells were mostly dry, only a few had limited amounts of water. There were many borings and mine backfilling holes available. Geophysical logs and downhole video camera had been run in many of these open borings and indicated that the majority of the strata overlaying the mine, (limestone and shale) were dry. While many fractures were identified, only one fracture with flowing water was observed.

Eight new wells drilled into the Raytown-Paola Limestone encountered small quantities of water, while none of the nine new wells drilled into the Drum Limestone encountered water. Downhole video camera inspection of the borings

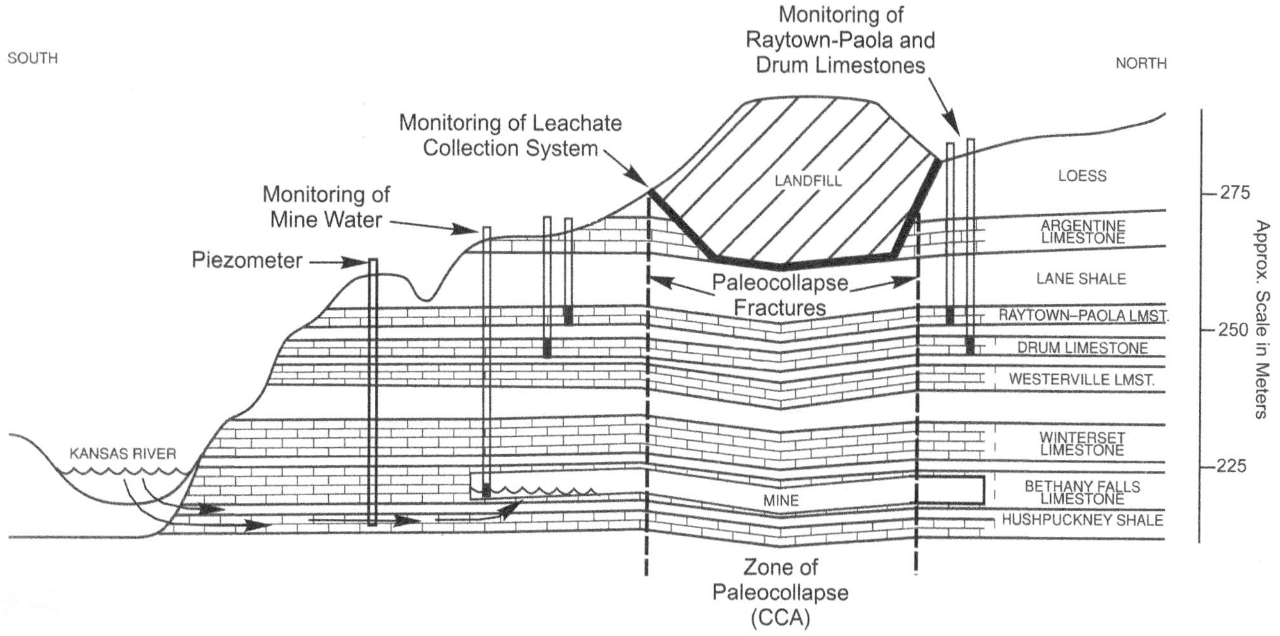

Fig. 21.4 Conceptual model for groundwater monitoring plan at a landfill overlying unsaturated bedrock and an abandoned limestone mine

showed isolated fracture flow within the Raytown-Paola Limestone. Additional drilling, hydrologic tests (core analysis and packer tests) and even hydrofracturing was completed to assess optimum locations for monitoring. Permeability testing in both limestones indicated values of 10^{-6} cm/s. Hydrofracturing of this zone improved the connection by two orders of magnitude. The purpose of the hydrofracturing was to improve the chance of the monitoring wells capturing any groundwater entering the bedrock.

21.7.1.2 The Source of Mine Water

The abandoned limestone mine was gradually filling with water, but the source of the water was unclear and little, if any, water was found in the bedrock units between the surface and the mine. The former mine superintendent, Nicholson (1988 personal communication), indicated that the mine was generally dry until mining extended to the southwest. Then a truck of water about 3,700 l was removed every few days. As the mine extended further westward, a number of sumps were installed below the mine floor and pumped as needed to control water. After mining operations ended in 1980 and without continued pumping, the mine began to slowly fill with water from the southwest.

Monitoring of water level within the mine was started in late 1987 using a digital water level recorder. A staff gauge within the mine provided back-up data in case the electronic equipment failed and was read on each trip into the mine. Measurements indicated that water was increasing at about 3.3 cm/month or about 40 cm/year. At that time there was an estimated 227×10^6 l of water in the mine increasing at an estimated rate of approximately 1.3×10^6 l/month.

Possible sources of water entering the mine had been suggested by others including:

- Seepage from the portal wall (about 45 m of exposed bedrock),
- Surface runoff into the mine from the portal road, and
- Seepage from the mine-roof.

None of these could account for the large amount of water entering the mine.

Considerable work had been carried out at two other landfills and a large mine and quarry some 29 km east in Jackson County, Missouri. The geology there was identical to that of the Tobin mine (Fig. 21.4). The bedrock units immediately below the mined Bethany Falls Limestone, were thought to be impermeable. These units have a total thickness of up to 3 m and include the Hushpuckney Shale, Middle Creek Limestone and Ladore Shale. However, lateral flow within these units had been observed at a number of locations at the mine and quarry due to the secondary porosity related to weathering and swelling of the shale (Camp 1988 personal communication). This suggests that this zone may have enough lateral hydraulic conductivity to serve as a pathway for significant water flow into the Tobin mine.

In addition, the Kansas River is located approximately 213 m to the south of the Tobin mine. The elevation of the mine floor in the southwest corner of the mine is about 7.6 m below that of the Kansas River. A piezometer was installed between the Kansas River and the southwest corner of the mine. Water level differences of about 7.6 m were measured between the river and the mine. River water would flow through the alluvium underlying the river into the permeable zone underlying the mine. This was a relatively simple solution to determine the source of the water filling the mine. The source for the mine water also supported the concept that the rock overlying the mine was essentially unsaturated.

21.7.1.3 Results of the Dye Trace

Although the rock overlying the mine was essentially unsaturated, a dye trace test was carried out to evaluate the possible connection from surface to the mine by the fractures associated with the paleocollapse. Section 17.2.4 discusses the excavation of a surface fracture and its evaluation using angle borings (Fig. 17.16). This fracture then was used as the dye injection point. A total of 106,000 l of water was used to pre-wet and post flush the dye.

This water along with dye was found to rapidly migrate latterly in the Raytown-Paola and Drum limestone formations up to 152 m from the injection site. This identified two zones within the unsaturated strata where monitor wells could be located. These two strata were also the uppermost limestone strata that would be continuous under the landfill.

21.7.1.4 The Proposed Groundwater Monitoring Plan

The proposed groundwater monitoring plan (Fig. 21.4) was based upon monitoring for possible landfill leachate at four points:

- The leachate collection system at the base of the landfill,
- Monitoring wells within the Raytown-Paola and Drum Limestones which are the first continuous limestone units below the landfill, and
- A monitoring well within the mine.

21.8 The Saturated Zone

Measurements within the saturated zone measure flow directly. The more traditional measurements include using either a single borehole or multiple boreholes for slug, packer or pump tests. Such measurements assume that the screen or open interval of the piezometer, well or borehole is representative of hydrologic conditions. Several geophysical logging techniques may also be used within a borehole or well, they can provide an indication of flow or are able to quantify the flow (see Chap. 18).

Traditional testing methods, such as slug tests, are limited to the immediate vicinity of the test well and its screen interval (Fig. 21.3c). Packer tests can be utilized in an open hole to measure hydraulic conductivity over specific intervals of the borehole. Pump tests are a way to obtain values of hydraulic conductivity over a much larger volume of the site and are therefore more representative of conditions further from the borehole (Fig. 21.3d). These measurements are usually limited to site-specific or detailed measurements. When data over larger areas are needed dye tracing is generally used. See Chap. 22 for more details on dye tracing.

21.8.1 Example of Fracture Flow in Southeastern Minnesota

A hydrogeologic investigation was completed at a landfill located within a karst area of southeast Minnesota. The investigation included data from a number of different measurements at a range of scales including:

- Topographic maps (regional scale)
- Aerial Photos (regional scale)
- Regional geology and on-site borings (regional and site-specific detail)
- Field observations of outcrops and measurements along adjacent river and in a nearby quarry (regional and site-specific scale)
- A cave map of Mystery Cave (site-specific scale off-site)
- Groundwater flow from monitor wells on-site (site-specific scale)
- Water quality measurements from monitor wells on-site (site-specific scale)
- The results of pump tests and (site-specific scale)
- Electrical resistivity soundings (site-specific scale)

21.8.1.1 Strong Linear Trends and Fracture Patterns

The topographic maps and aerial photo analysis (Fig. 21.5) indicated major linear trends and fracture patterns over the regional setting with linear trends up to 300 m long. Abrupt changes in the direction of the Root River bed including numerous 90° bends and two 180° bends are a strong indication

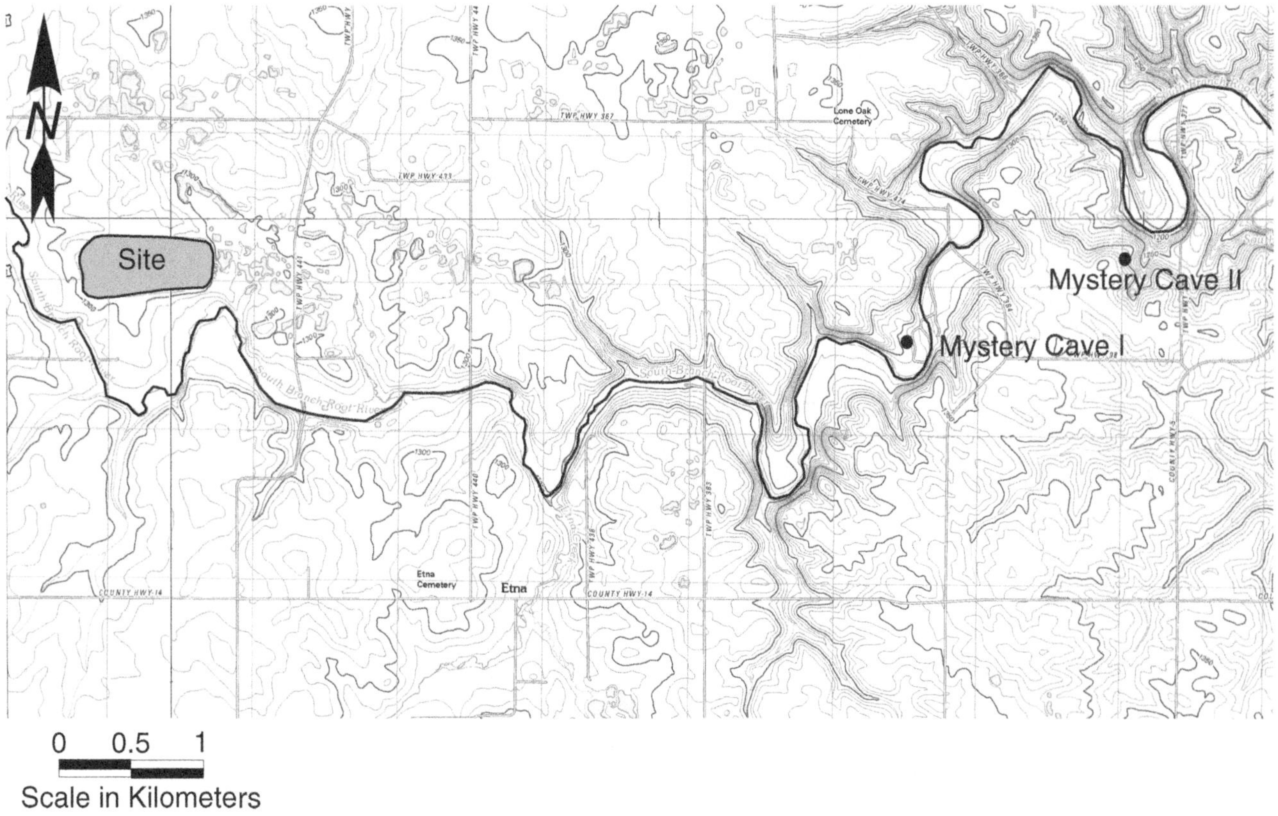

Fig. 21.5 Topographic map indicates a number of linear trends (lineaments) over distances of hundreds of meters along the river with numerous fracture trends (major valleys). A number of 90° bends are seen over distances of tens of meters within the river along with a 180° turn just south of the site (USGS 7.5 min Quadrangles Spring Valley, Wykoff, Ostrander and Cherry Grove, Minnesota, 1965)

of the fracture pattern at a scale of 30 m or so. Many deep, linear drainage valleys are seen adjacent to the Root River.

A cross section of the site geology, within the Cedar Valley, Maquoketa, Dubuque, and the Gelena Formations is shown in Fig. 21.6. Observations of outcrops in the Cedar Valley Formation were made along 3 km of the Root River downgradient of the site. Observations were also made of the Maquoketa, Dubuque, and the Gelena Formations in a nearby quarry. Both the outcrops along the river and those in the quarry indicated closely spaced fracture patterns with spacing of a meter to 10 m.

Rose diagrams were developed using the linear trends and fracture orientation observed from aerials and topographic maps as well as those fractures observed along the Root River and in the nearby quarry. The features observed in the aerials and topographic maps are at a regional or site-specific scale. The rose diagrams were plotted for features observed in the Cedar Valley and the Maquoketa Formations (Fig. 21.7a). Both of these formations indicate that these features have very distinct, well defined orthogonal orientations (NW-SE and NE-SW). The data from along the river and in a nearby quarry are more representative of site-specific (30 m or more) and local scales (3 m or so). These rose diagrams clearly show highly fractured formations with a more diverse and wide ranging orientation than those observed from the aerials and topographic maps Fig. 21.7b).

The general location of Mystery Cave (the largest in Minnesota) is located about 6 km to the east of the site within the Dubuque and Galena Formations (Figs. 21.5 and 21.6). Its passageways are highly joint and fracture controlled. A main east-west fracture system is superimposed with a NE to SW and NW to SE fracture system. Many short passageways are on the order of 30 m or so (Alexander et al. 2006).

21.8.1.2 Indications of Flow

Water level data from monitor wells indicated the direction of groundwater flow to the southwest of the site (Fig. 21.8). Specific conductance values from these monitor wells also indicated the conductive plume from the landfill flowing in a direction southwest of the site (Barr Engineering 1984).

Early results of a pump test initially indicated an elongated drawdown after one hour of pumping suggesting fracture flow in a roughly north to south orientation (Fig. 21.9a), after 7.5 h the drawdown became elliptical (Fig. 21.9b), and after 20 h became circular (Fig. 21.9c). This change in drawdown over time suggests that localized fracture flow initially dominates but becomes more uniform "matrix-like" flow as the time of the test increases (Barr Engineering 1984).

Nine resistivity soundings (to depths of more than 30 m) were made around the perimeter of the landfill including upgradient and downgradient in order to detect the inorganic contaminant plume, which has a distinct lower electrical resistivity signature. Figure 21.10 shows the locations of the resistivity soundings and an averaged electrical resistivity value at depth. Multiple soundings were completed at each location, varying the orientation in order to detect any impact or control due to the fractures/joints. The resistivity data at each location was repeatable, with no evidence of fracture control and clearly indicated a contrast in the groundwater quality within the limestone with relatively high background values upgradient of the site and uniform lower values under and downgradient of the landfill due to the presence of the leachate plume. The results of these resistivity soundings indicate that the plume from the landfill tends to be uniform throughout the underlying limestone based upon resistivity values and has probably moved below the nearby Root River to the south.

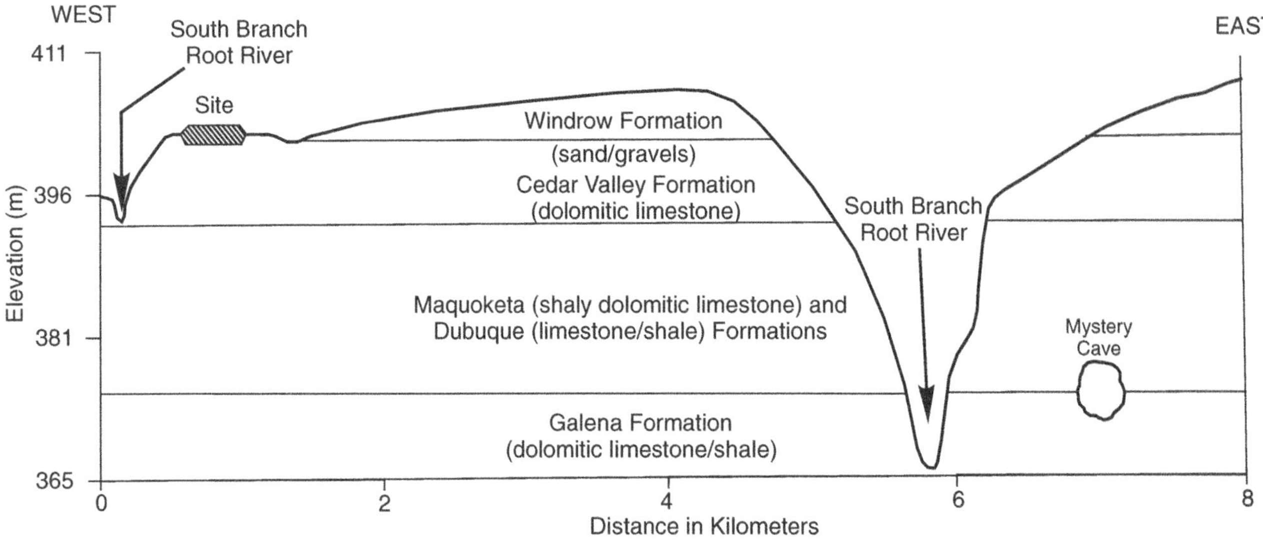

Fig. 21.6 Cross section shows the site location, the geologic section, the Root River valley and the nearby mystery Cave some 5.5–7 km east of the site

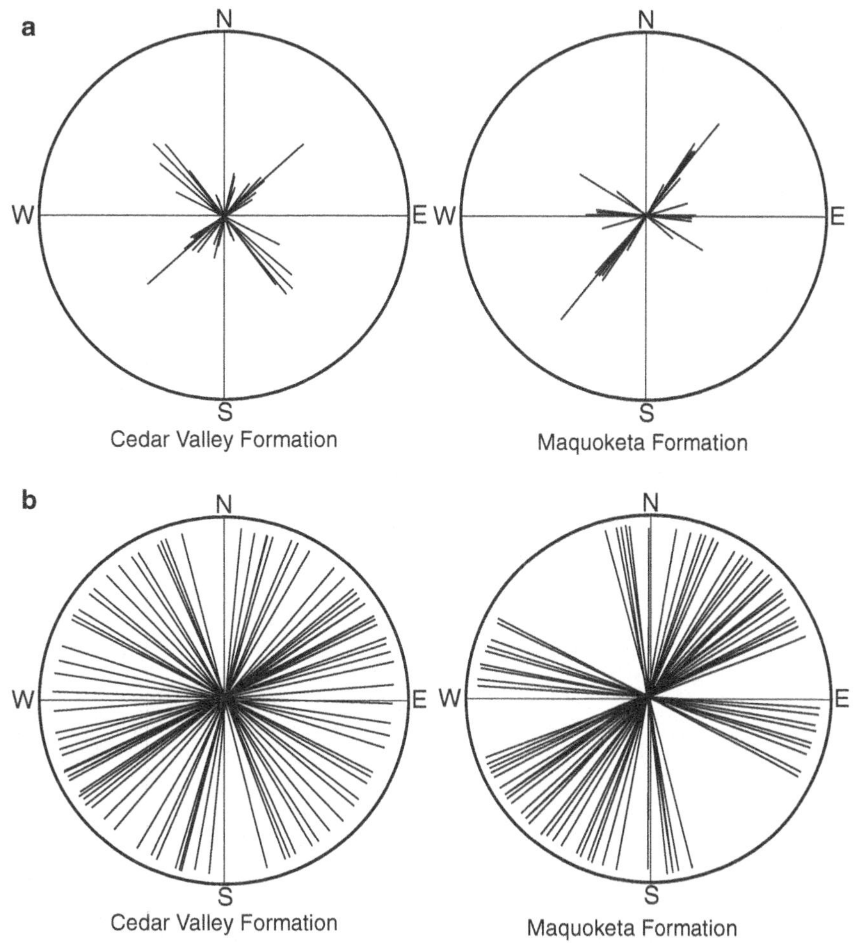

Fig. 21.7 Rose diagrams shows the fracture orientations obtained from aerials and topographic maps as compared to those obtained from the outcrops along the Root River and from the nearby quarry (**a**) Data from aerials and topographic maps (**b**) Data from outcrops and nearby quarry

21.8.1.3 Scale Effects on Flow

Strong major fracture patterns with a distinct and limited orientation were identified in both the topographic maps and aerials photos. These strong fracture patterns were also seen in the map of Mystery Cave. However, the more localized fracture patterns measured at outcrops along the river and in a nearby quarry in both the Cedar Valley and Maquoketa Formations show many more fractures in a more diverse range of orientations.

Additional data from localized measurements suggest a more uniform groundwater flow:

- The water levels from the monitor wells show a gradual flow to the southwest
- Pump test indicated that the fractures were well interconnected and flow was uniform over a larger scale (longer pumping time).
- The specific conductance values from the monitor wells indicate leachate migrating to the southwest.
- The resistivity soundings to a depth of 30 m suggested uniform flow though a highly fractured system of limestone with a coherent plume.

While each set of independent data by itself may not be totally convincing, the correlation of these independent sets of data over a wide range of scale provided a possible basis for interpreting the fracture flow at this location as equivalent porous media (EPM). In this example the interpretation of more uniform flow can be applied to the scale of the site itself, but may not be applicable to small areas within a site depending upon the degree of local fractures and their interconnections.

21.8.2 An Example of Flow from an Artesian Well

A 30 cm oil exploration well had been drilled to a depth of about 380 m in 1944 in southeast Florida. The well was

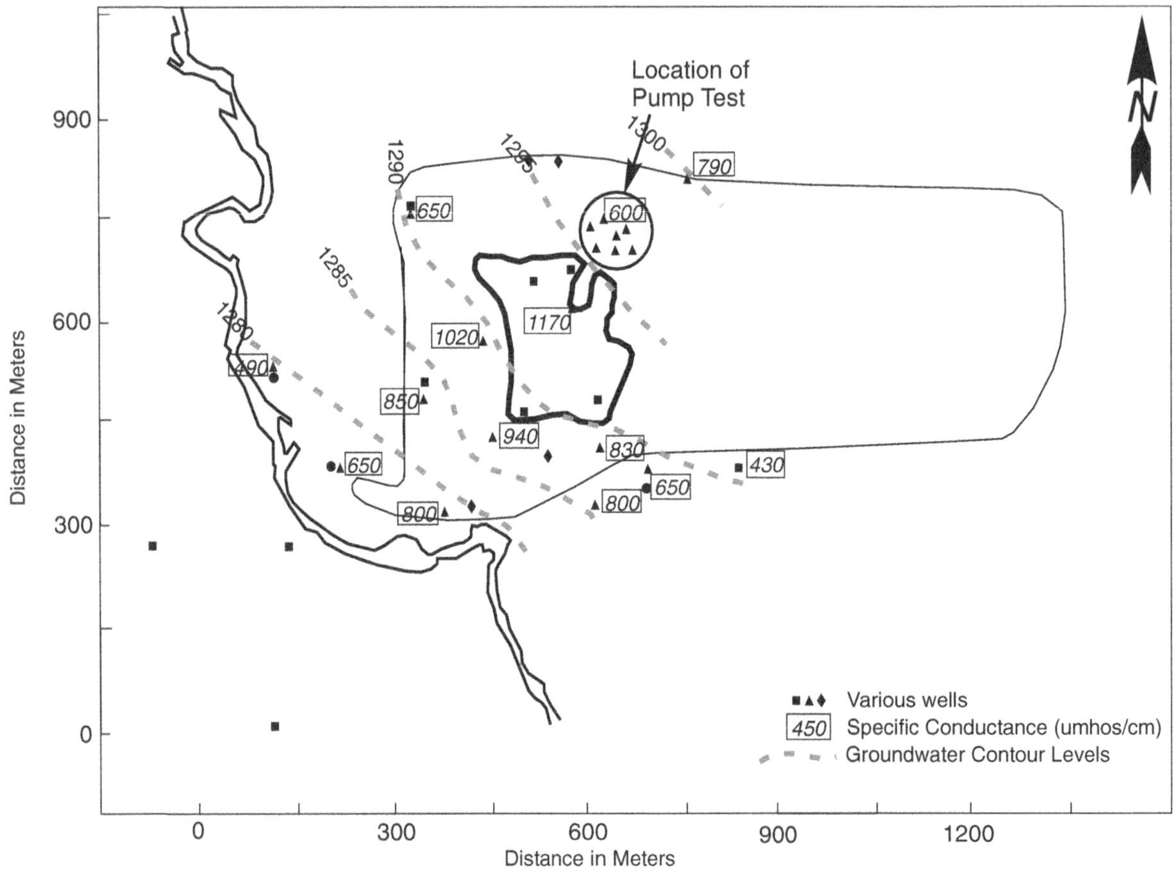

Fig. 21.8 The direction of groundwater flow was determined from water levels in monitor wells along with specific conductance values (**a**) After 1 h of pumping (**b**) After 8 h of pumping (**c**) After 20 h of pumping (Note the sharp turns in the Root River west and south of the site)

artesian from the deeper Floridan Aquifer and flowed with brackish water for 35 years before it was recognized as a groundwater problem for the surficial Biscayne Aquifer. The Grossman Hammock well is located within the Everglades National Park surrounded by a hardwood hammock currently referred to as Chekika Day Use Area. Initial artesian flow was 7,500 l/min when the well was drilled in 1944 and 3,750 l/min when the well was capped in 1995. That would be roughly 15×10^{10} l for the 50 years of flow. The Biscayne Aquifer lies within the upper 15 m, consists of young porous limestone, (Miami Oolite and the Fort Thomson Formation) and provides a sole-source of potable water for the southeastern Florida communities (Fig. 21.11) (Waller 1982).

The specific conductance of the artesian water (4,500–5,000 mg/l) was an order of magnitude greater than specific conductance levels of the surficial water (300–400 mg/l). This electrical contrast was used to map the plume using surface electrical geophysical techniques (electromagnetic and resistivity). EM34 electrical conductivity station measurements were made at the surface to map the lateral extent of the plume. A total of 133 station measurements were made over 30 km^2. Two measurements were made at each station to provide shallow data (0–7.5 m) and deeper data (0–15 m).

The shallow conductivity contours (0–7.5 m deep) indicate an 11.5 km long plume extending from the well, 1.5 to 3 km wide headed southeast toward Homestead Florida's wellfield (Fig. 21.12). Conductivity values of 8 milliseimens/m (mS/m) represent the outermost measurable boundary of the plume using the EM34 measurements. The deeper conductivity data (0–15 m deep) shows a very similar plume shape, although slightly smaller in lateral extent than the shallower conductivity data. The resistivity sounding data indicated that the plume is very uniform with depth. This was confirmed by wells at various depths, showing specific conductance was also very uniform with depth. Both the shallow and deep data indicate an off-shoot of the main plume extending to the east. However, the boundaries in this area are based upon more limited data due to access and therefore are not well defined. The cause of this portion of the plume is unknown but may be due to a porous zone developed at lower sea level stand.

Figure 21.13 is a plot of conductivity data crossing the plume. This cross section also runs along a canal where specific conductance measurements were made and are also plotted. There is a strong correlation between the conductivity measured by the surface geophysical methods and the water quality data of the canal. Both sets of data show sharp boundary conditions.

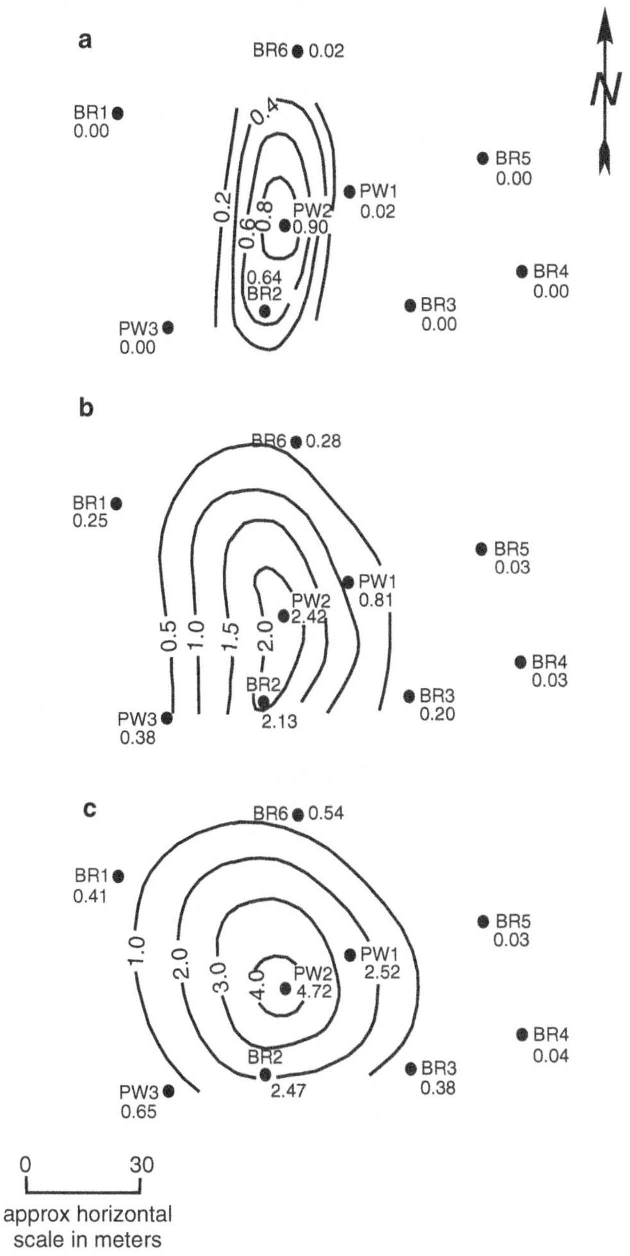

Fig. 21.9 The results of a pump test over time showing the water table drawdown at 1, 8, and 20 h of pumping

Once the plume boundaries were established, resistivity soundings were made to establish the thickness of the plume and the base of the Biscayne aquifer (upper Tamiami Formation). Data from existing test wells provided further information on water quality for correlation with the geophysical measurements.

The limestone of the Biscayne Aquifer, a couple of meters below the water table has enlarged porosity as high as 75 % (Sowers 1996). The coherent nature of the plume, its sharp boundaries and uniform conditions with depth indicate that the movement of the brackish water plume within the Biscayne Aquifer can be described as equivalent porous media as it flows towards the wellfield to the south.

In this case, the limestone is a relatively young, highly porous matrix, the water table is very shallow (<3 m bls) and brackish water flowed from the well for almost 40 years. The scale of the brackish water plume is about 12 km long and 1.5–3 km wide, minimizing the impact of localized heterogeneities. In this case, flow can be characterized as equivalent porous media (EPM).

21.8.3 Equivalent Porous Media

These previous two examples illustrate the application of EPM. In some limited cases the concept of Equivalent Porous Media (EPM) can be used to evaluate flow in fractured rock and limestone. When flow occurs through a young porous limestone such as the Miami Limestone or a well fracture media in which the fractures are closely spaced and uniformly distributed the system may be considered analogous to porous media EPM (at an appropriate scale). For EPM conditions to apply the observed vertical and horizontal fractures should be numerous, the distance between the fractures should be orders of magnitude smaller than the size of the site under investigation, and the fractures should show appreciable interconnection. When these conditions exist, then equivalent porous media (EPM) concepts can sometimes be applied to the overall setting.

While the EPM approach is not generally appropriate for modeling flow in most karst aquifers, it may be applied with certain restrictions Kovacs and Sauter (2007). Alexander (2005) cautions us about the use of Darcey's Law and the use of EPM in characterization of groundwater flow in karst.

21.9 Groundwater Contaminants

There is often the need to determine the flow or map the extent of various contaminants as part of a site characterization. The type of contaminants needs to be known so that their interaction with the geologic and cultural conditions can be anticipated. Are the contaminants inorganic or organic? If they are organics, are they floaters, mixers or sinkers? Would they be bound up in the soils or trapped in sand pockets or perched on clay lenses? Are they coming from a point source or being dispersed over large areas? The answers to these questions are necessary in order for the geologic conceptual model to be developed for the site and appropriate sample points to be determined.

21.9.1 Inorganic Contaminants

Inorganic contaminants typically have a higher electrical conductivity than groundwater. When there is an electrical

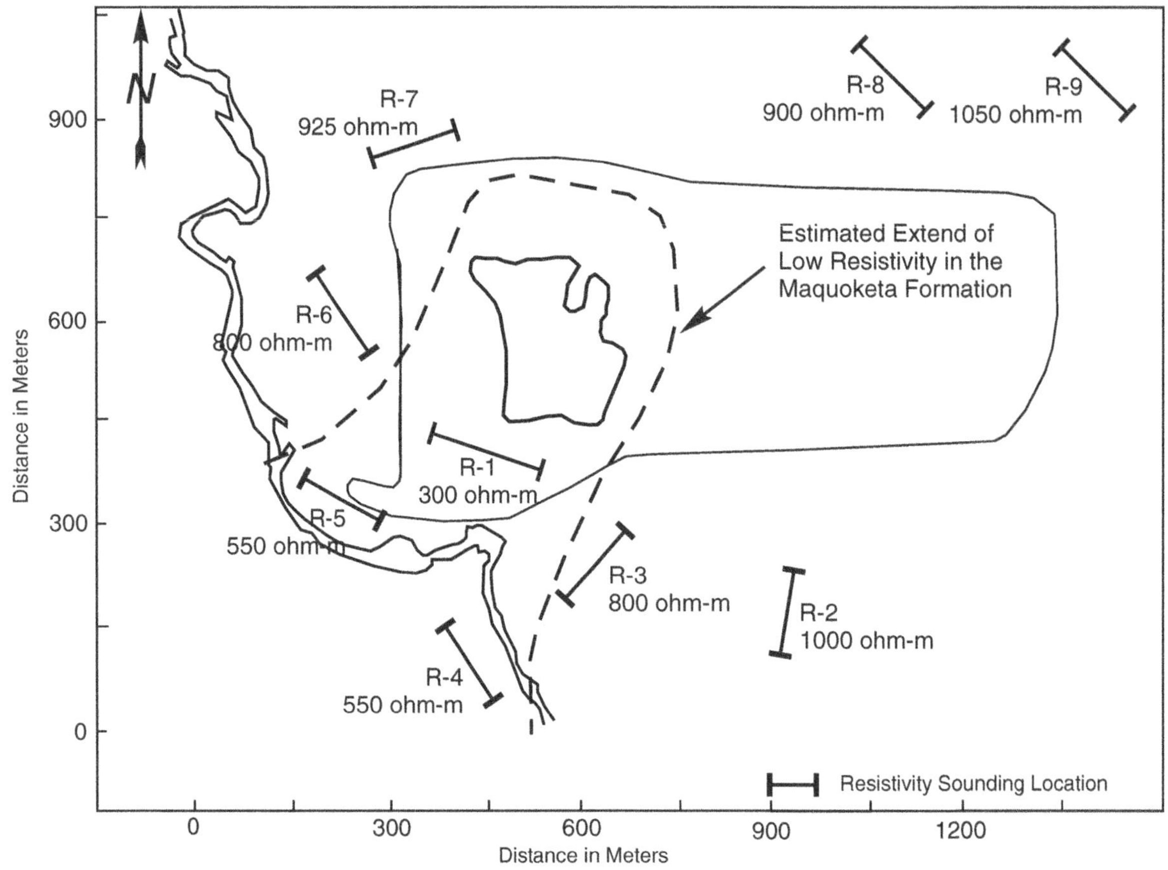

Fig. 21.10 Resistivity soundings R-1, R-4 and R-5 measured lower resistivity values (<600 ohm-m) indicating the presence of landfill leachate. Background values range from 800 to 1050 ohm-m

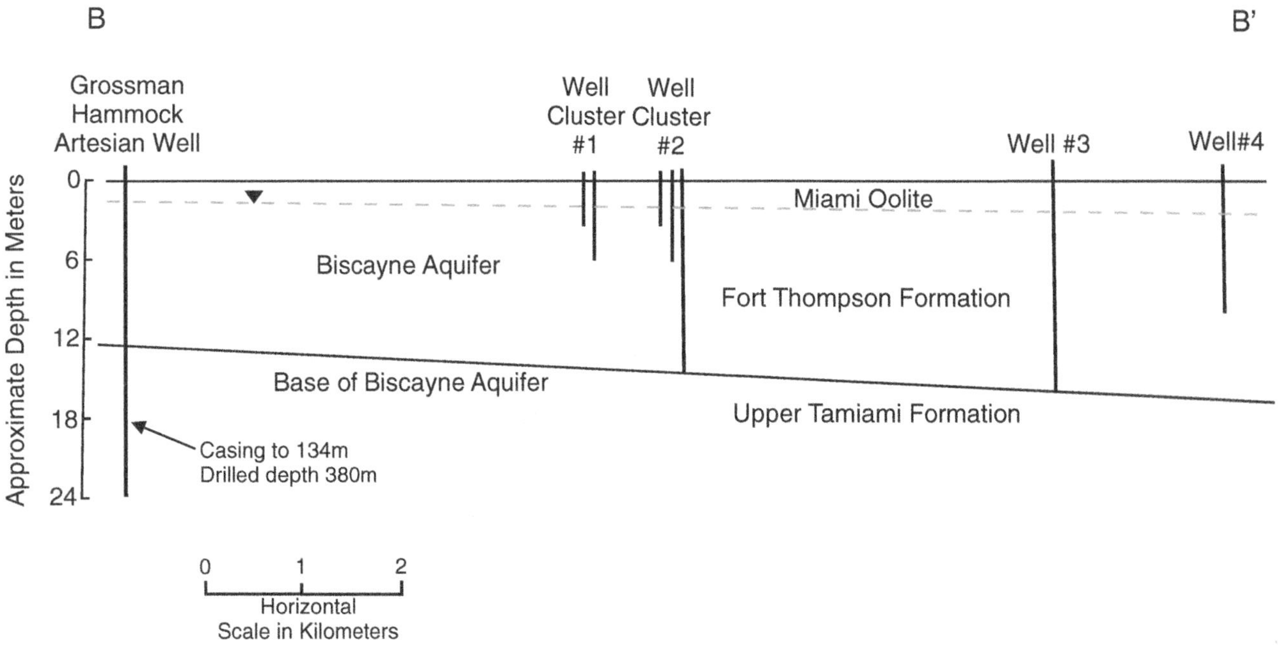

Fig. 21.11 The geologic section and monitor wells used to characterize the plume from the Grossman Hammock artesian well

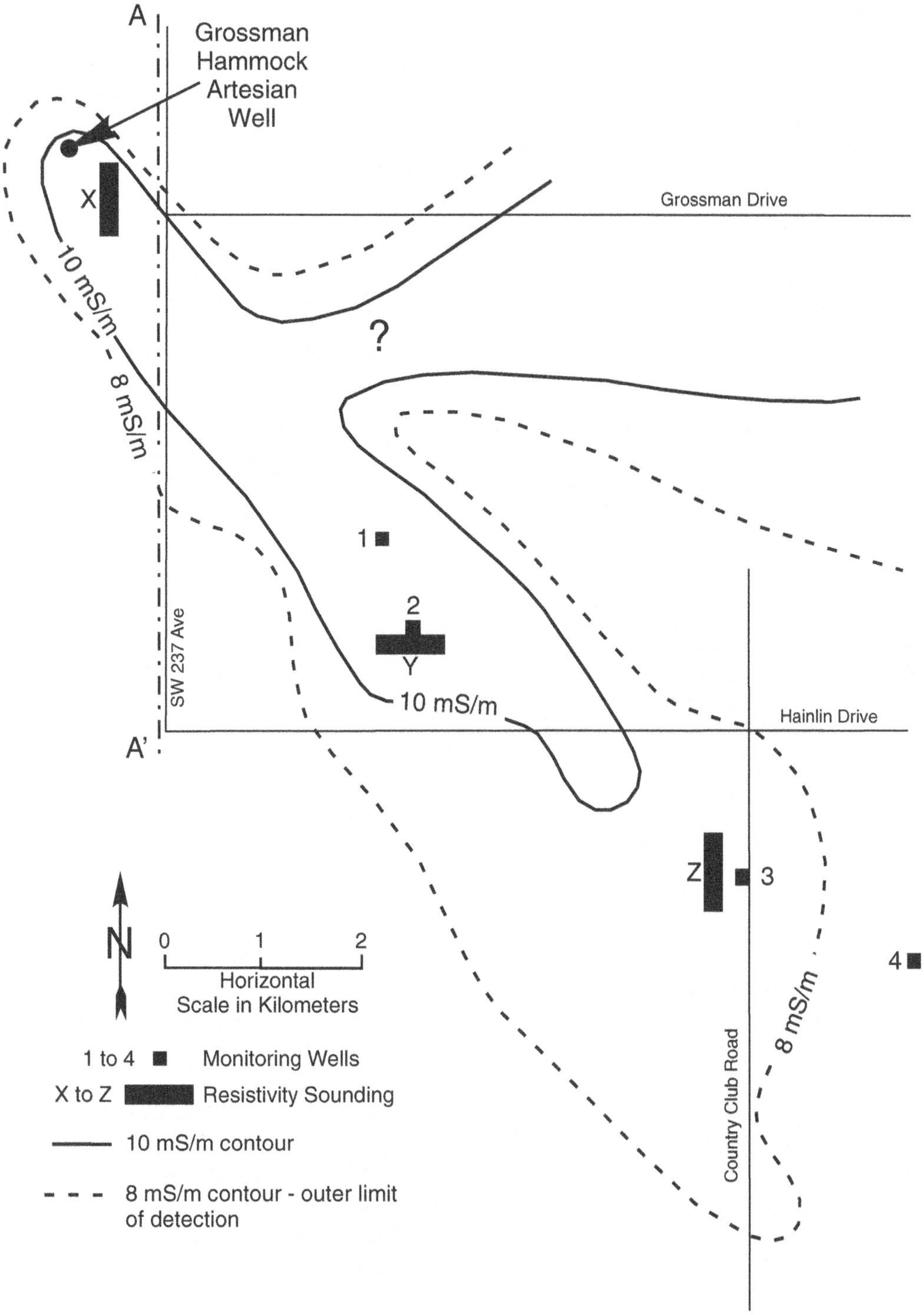

Fig. 21.12 A 11.5 km long plume headed southeast toward a wellfield based upon electromagnetic measurements (0–7.5 m deep)

conductivity contrast between the contaminant and surrounding pore fluids, soil and rock, surface geophysics can often be used to map the lateral extent and flow of the contaminants while borehole geophysics can be used to characterize the vertical extent of contaminants. The surface geophysical methods traditionally include resistivity, electromagnetic (EM31 and EM34) and radar measurements. The borehole geophysical methods traditionally include resistivity/resistance and electromagnetic measurements.

The survey results often provide insight into the plume behavior whether it is extending in the direction of the general groundwater flow or flowing along dominant fracture trends. Benson et al. (1985) have compared the results of

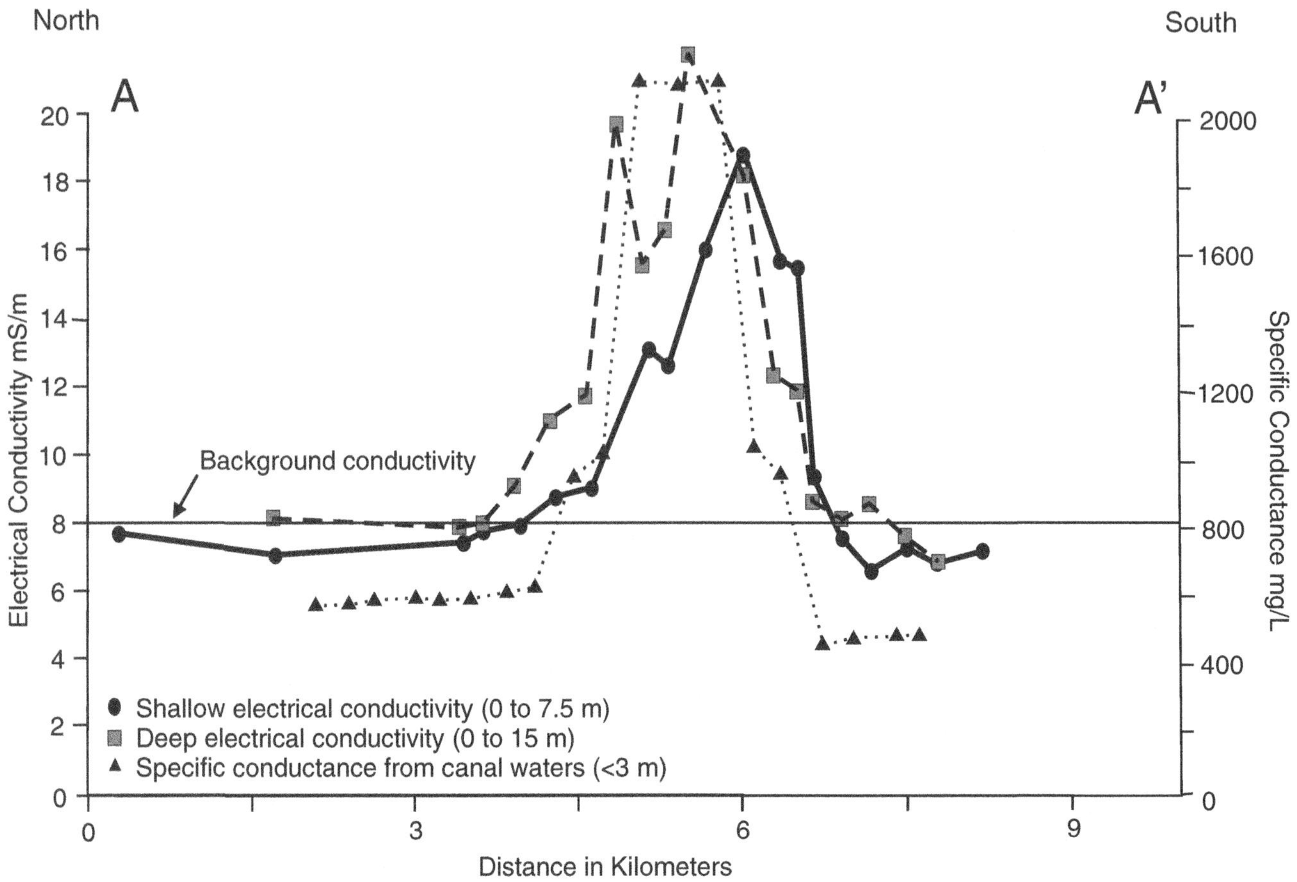

Fig. 21.13 The plume had very sharp boundaries defined by surface geophysical electromagnetic measurements and specific conductance measurements within canal waters

electrical geophysical survey methods and the water quality data from numerous landfills, using specific conductance as a leachate indicator parameter. A correlation of 90 % at the 95 % confidence level was found with both electrical resistivity and electromagnetic geophysical techniques.

For example, acid leaks had occurred at an electroplating plant in southwestern Missouri. Surface geophysical electromagnetic measurements were acquired at two different depths (EM31: 0–6 m and EM34: 0–15 m). The electrical contrast of the acids to background conditions provided a parameter for mapping their extent. Figure 21.14 shows both the shallow and deep contoured electromagnetic data. The shallower data (Fig. 21.14a) shows a broader plume within the soils (yellow to red). The deeper data (Fig. 21.14b) shows the contaminants (yellow to red) within the bedrock, largely being controlled by fractures.

21.9.2 Organic Contaminants

The detection and mapping of organic contaminants is a more difficult task. In very general terms organic contaminants can be broken into three categories:

- Floaters are low density hydrocarbons which tend to float at the top of the water table,
- Mixers are water soluble organic compounds which tend to migrate through an aquifer along with the groundwater flow, and
- Sinkers are high-density halogenated hydrocarbons whose movement is strongly influenced by gravity.

Some organic contaminants such as petroleum products will degrade resulting in an increase in electrical conductivity (Sauck et al. 1998; Cassidy et al. 2001), which can then be mapped using electrical geophysical methods. Knowing how the contaminants were introduced into the subsurface and how they may behave within the groundwater needs to be considered along with the geologic and cultural conditions present at a site.

21.10 Aquitards and Barriers

In the 1950s and 1960s clay soils were thought to be impermeable (Daniel 1993a). Our understanding of clays has changed significantly over the past few decades. Subtle vari-

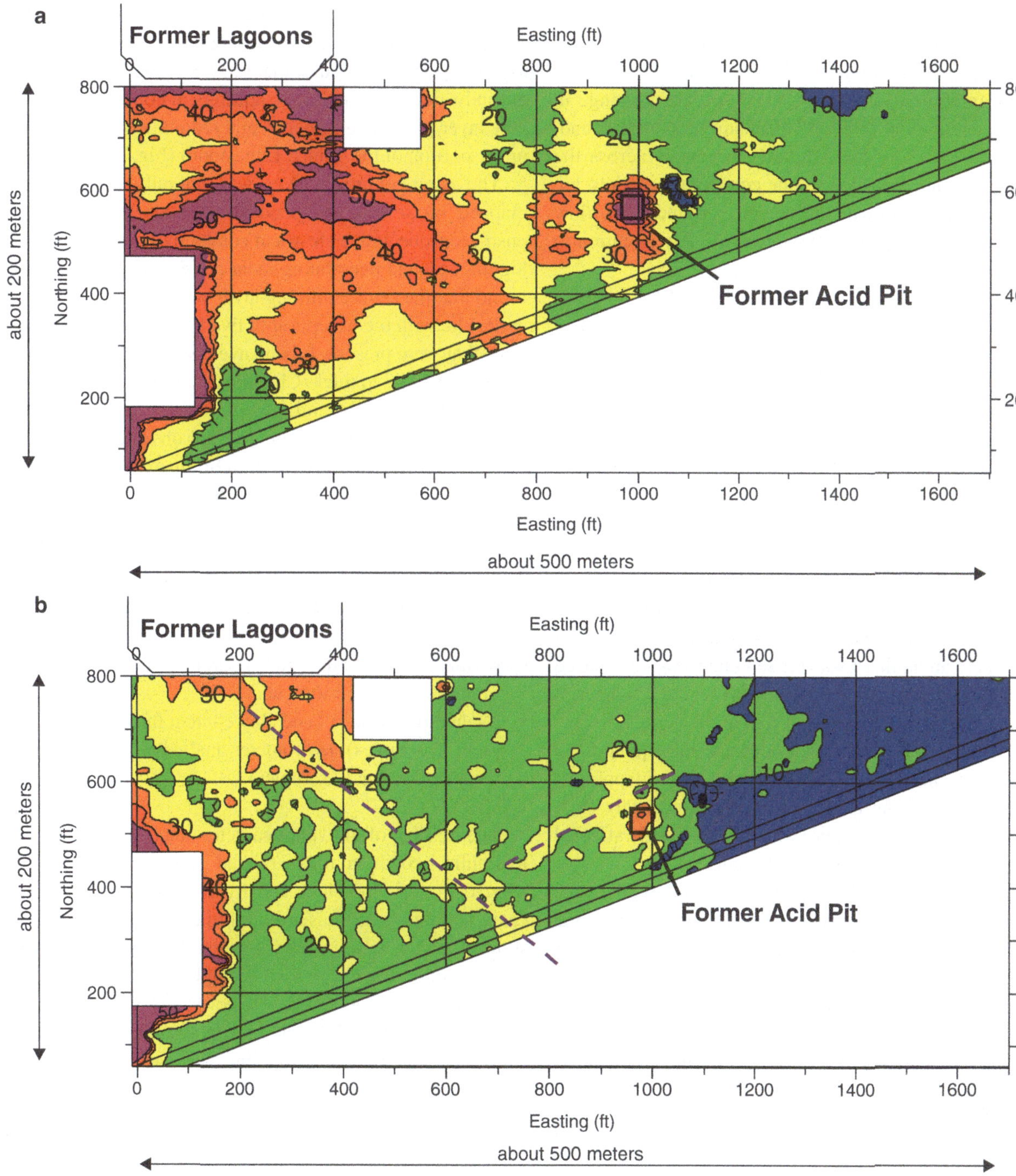

Fig. 21.14 Electromagnetic surveys were used to map inorganic contaminants (acids), which were being controlled by fractures (**a**) Electromagnetic survey (EM31: 0–6 m) (**b**) Electromagnetic survey (EM34: 0–15 m). Note: warmer colors indicate higher conductivity areas

ations in the clay/silt content or weakly cemented layers within soils can greatly affect unsaturated groundwater flow as well as contaminant transport. Clay liners that are used for landfills have leakage rates 10–100 times higher than predicted by laboratory tests (Daniel 1984, 1993b). Furthermore, laboratory tests have shown that the permeability of clay liners to various organic compounds may be 10–1,000 times greater than water (Daniel 1985).

Natural aquitards are often described as being present under an entire site and become the basis for design of landfills, groundwater modeling, risk analysis, and remediation plans. In subsequent work it is not unusual to find that the aquitard is sometimes found to be discontinuous or without a significant decrease in hydraulic conductivity to be classified as an aquitard. Two issues need to be addressed in characterizing an aquitard. First, is there a sufficient change in hydraulic conductivity to establish an aquitard, and second, is the lateral extent of the aquitard continuous across the site. Without such knowledge our conceptual models are often flawed and subsequent design, groundwater flow modeling and remediation may be in error. Pay attention to the details.

For example, at the Savannah River Site (SRS), a "green clay" which underlies a disposal facility had been described as an impermeable aquitard. The continuity and impermeability of this aquitard was the basis for limiting the depth of contaminant investigations, risk analysis and subsequent remediation. However, a sieve analysis of samples taken from this aquitard indicated it to be mostly coarse to very fine sand with only 10 % silts. Two electrical conductivity logs obtained with a geoprobe within 6 m of one another indicated that the properties of this aquitard are both vertically and laterally variable (Technos 1995).

In some cases, the presence of an aquitard may not have been identified. The example presented in Sect. 18.2.7 showed where an aquitard was missed due to widely spaced samples during drilling and no review of regional literature. This aquitard played an important role in the vertical migration of a brine contaminant plume. Because the aquitard and its impact were unknown, monitoring wells were inappropriately screened, missing the contaminants completely (Fig. 18.14). Not knowing an aquitard was present, clearly affected the understanding of conditions at this site.

Our case history from the Superfund site in Part III, Chap. 27 included an aquitard across the site referred to as the semi-confining layer (SCL) separating the surficial aquifer and the Floridan Aquifer in Tarpon Springs, Florida. The SCL played a critical role in limiting contaminant migration, indicating the presence of paleocollapse and developing a subsidence risk assessment for the site (Yuhr et al. 2003). The SCL had been identified in regional literature and during on-site investigations. This semi-confining layer is described as a low permeability unit that retards vertical movement of groundwater between the two aquifers. It consists of a sandy-clay layer that grade into a weathered clayey limestone (the upper portion of the Tampa Limestone) (Fig. 21.15).

The semi-confining layer was laterally and vertically mapped by several independent methods (drilling, direct push geoprobe with a electrical conductivity tip, geophysical logging, ground penetrating radar and 2D resistivity). It was found that the semi-confining layer was continuous over the entire site with the exception of two large paleocollapse features that were present. It is on average 2.3 m thick, but varies from 0.3 to 7 m thick. Chemical sampling in paired wells showed no chemicals migrating downward from the surficial to the deeper Floridan aquifer. Visual observations of samples showed that the lower portion of this layer was typically hard and dry. However, its ability to confine or impede flow between the aquifers was in question. Therefore, additional laboratory testing on samples of this layer were completed. This analysis was completed on both the upper and lower portions of the semi-confining layer and included:

- Permeameter testing for hydraulic conductivity (15 samples across the site), and
- Evaluation of the silt and clay content through sieve analysis (33 samples), hydrometer analysis (15 samples) and x-ray diffraction (7 samples).

The results of the permeameter analysis showed that hydraulic conductivities ranged from 2.84×10^{-4} to $<1 \times 10^{-8}$ cm/s. The sieve and hydrometer testing all indicated variability in the semi-confining layer, but in general showed a high silt and clay content. The lower hydraulic conductivities correlated well to high percentages of particles passing the #200 sieve. However, the x-ray diffraction showed that some of the clay-sized particles were actually quartz and calcite. While this unit was variable, it did have

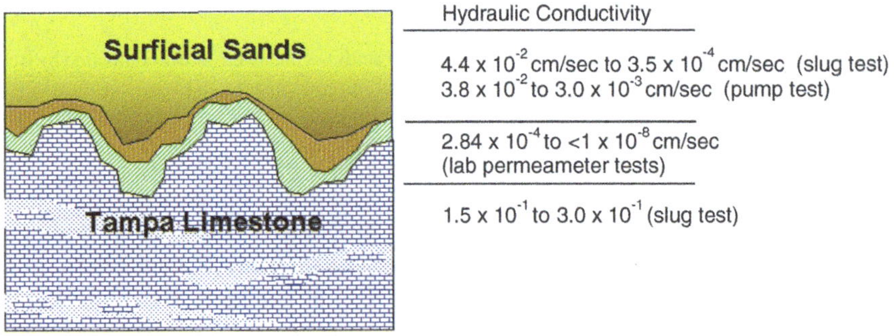

Fig. 21.15 The conceptual model of the semi confining layer (*SCL*) and the surficial and Florida aquifers along with measured hydraulic conductivities for each layer

hydraulic conductivities that were one to six orders of magnitude lower than the sands above or limestone below. This successfully limited contaminant migration into the underlying Floridan aquifer. Its lateral continuity across the site provided an indication of stability with regard to contaminant migration and sinkhole or subsidence activity. Where the SCL was absent correlated to two paleocollapse areas.

References

Alexander EC Jr (2005) Karst hydrogeology and the nature of reality revisited: philosophical musings of a less frustrated curmudgeon. In: Beck B (ed) Proceedings of the 10th multidisciplinary conference on sinkholes and the engineering and environmental impacts of karst, San Antonio, Texas September 2005. ASCE Geo-Institute Geotechnical Special Publication 144

Alexander EC Jr et al (2006) The karst geology of southeastern Minnesota. AIPG National Conference Field Trip, Sept 26

Aley T (2008) Chapter 15 tracer tests—dye. In: Weight WD (ed) Hydrogeology field manual, 2nd edn. McGraw-Hill, New York, pp 675–698

Anderson MP (2005) Heat as a ground water tracer. Ground Water 43(6):951–968

ASTM (1998) Standard guide for design of ground-water monitoring systems in Karst and fractured-rock aquifers D5717–95e1 (withdrawn 2005), ASTM International, West Conshohocken, Pennsylvania

Barr Engineering (1984) Remediation of the Ironwood Landfill. Consultants report

Benson RC, Yuhr L (1993) Spatial sampling considerations and their applications to characterizing fractured rock and karst systems. In: Beck BF (ed) Proceedings of the 4th multidisciplinary conference on sinkholes and the engineering and environmental impacts of karst: applied karst geology, Panama City, Florida. Balkema, Rotterdam

Benson RC et al (1985) Correlation between field geophysical measurements and laboratory water sample analysis. In: Proceedings of the conference on surface and borehole geophysical methods in ground water investigations, National Water Well Association, San Antonio, TX, NWWA, Westerville, Ohio, pp 178–197

Benson RC et al (1994) Ground water monitoring in unsaturated and saturated zones at a site with a paleokarst collapse structure. Am Inst Hydrol Hydrol Sci Technol 10(1–4):31–46

Cassidy DP et al (2001) The effects of LNAPL biodegradation products on electrical conductivity measurements. JEEG 6(1):47–52

Da Cunha AP (ed) (1990) Scale effects in rock masses. In: Proceedings of the 1st international workshop on scale effects in rock masses. Balkema, Rotterdam, pp 3–27

Daniel DE (1984) Predicting hydraulic conductivity of clay liners. J Geotech Eng 110(2):285–300

Daniel DE (1985) Effects of organic chemicals on clay liners. In: Proceedings of the 7th symposium on management of uranium mill tailings, low-level waste and hazardous waste, Ft. Collins CO, pp 373–390

Daniel DE (1993a) Introduction. In: Daniel DE (ed) Geotechnical practice for waste disposal. Chapman & Hall, London, pp 3–14

Daniel DE (1993b) Clay liners. In: Daniel DE (ed) Geotechnical practice for waste disposal. Chapman & Hall, London, pp 137–163

Finkl CW, Krupa S (2003) Environmental impacts of coastal-plain activities on sandy beach systems: hazards, perception and mitigation. J Coast Res SI 35 (Proceedings of the Brazilian Symposium on Sandy Beaches: Morphodynamics, Ecology, Uses, Hazards and Management):132–150. Itajai, SC–Brazil, ISSN 0749-0208

Freeze RA, Cherry JA (1979) Groundwater. Prentice Hall, Englewood Cliffs

Hunt RE (2005) Geotechnical engineering investigation handbook, 2nd edn. CRC Press, New York

Kovacs A, Sauter M (2007) Modeling karst hydrodynamics. In: Goldscheider N, Drew D (eds) Methods in karst hydrogeology, International contributions to hydrogeology 26, International Assoc. of Hydrogeologists. Taylor & Francis, New York, pp 201–222

Lane J et al (2008) Fiber-optic distributed temperature sensing: a new tool for assessment and monitoring of hydrologic processes. In: Proceeding of the symposium on the application of geophysics to engineering and environmental problems, Environmental and Engineering Geophysical Society, Apr 6–10, 2008, Philadelphia, PA

Milanovic PT (2004) Water resources engineering in karst. CRC Press, Boca Raton

Neilsen D (2006) Practical handbook of environmental site characterization and ground-water monitoring. CRC Press, Taylor & Francis Group, New York

Nelson P (1984) Short course notes for hydrologic properties of rock and rock masses. University of Texas, Austin, Texas

Nielsen DM, Johnson A (eds) (1990) Groundwater and vadose zone monitoring. ASTM STP 1053, ASTM, West Conshohocken, Pennsylvania

Palmer AN (2006) Digital modeling of karst aquifers – successes, failures, and promises. In: Harmon RS, Wicks CM (eds) Perspectives on karst geomorphology, hydrology and geochemistry – a tribute volume to Derek C. Ford and William B White: Geological Society of America, Special Paper 404, GSA, Boulder, Colorado, pp 243–250

Palmer AN (2007) Cave geology. Cave books, an affiliate of the cave research foundation. Cave Books, Dayton

Palmer AN et al (eds) (1999) Karst modeling. Special publication 5. Karst Waters Institute, Charles Town, West Virginia

Sara MN (2003) Site assessment and remediation handbook, 2nd edn. Lewis Publisher, New York

Sasowsky ID (2000) Carbonate aquifers: a review of thoughts and methods. In: Sasowsky W (ed) Groundwater flow and contaminant transport in carbonate aquifers. Balkema, Rotterdam, pp 1–14

Sauck WA et al (1998) High conductivities associated with an LNAPL plume imaged by integrated geophysical techniques. JEEG 2(3):203–212

Sowers GF (1996) Building on sinkholes, design and construction of foundations in karst terrain. ASCE Press, Reston (out of print)

Technos Inc. (1995) Final report, geologic, hydrologic and geophysical work at the D-area seepage basin, Savannah River Site Plant, Savannah Georgia for Ames Laboratory Technology Integration Program, Ames, Iowa. Consultant Report. Dec 7

Waller BG (1982) Aerial extent of a plume of mineralized water from a flowing artisian well in Dade County. USGS Water Resources Investigations Report 82-20, Tallahassee, Florida

Wilson LG et al (ed) (1994) Handbook of vadose zone characterization and monitoring. Lewis Publishers, CRC Press, New York

Yuhr L et al (2003) A case history of a large karst investigation presented at the 3rd international conference on applied Geophysics, Orlando

22 Dye Tracing

Abstract

Dye tracing is a valuable method to characterize flow in fractured-rock and karst aquifers. It is the method of choice for regional and intermediate flow measurements but can also be applied to some local measurements including the epikarst zone. In this chapter, we review some considerations for dye tracing, including types of dyes, along with sampling and analysis methods. A few examples are provided showing how dye tracing has been applied to answer some very specific questions including flow under a dam site, vulnerability of a cave habitat, confirmation of contaminant source and evaluation of capture zones for a major spring. A special thanks to Tom Aley of the Ozark Underground Laboratory, who provided assistance with this chapter as well as project examples based upon his lengthy experience with dye tracing.

22.1 Introduction

A dye trace study is a way to track water flow using man-made or natural tracers. Dye trace studies can be used to:

- Prove a direct connection between two points,
- Determine groundwater flow directions and rates,
- Identify basin boundaries, and
- Determine optimum monitoring frequency for groundwater quality.

Our good friend, Jim Quinlan, was fond of repeating the following phrase. "One well-designed tracer test, properly done and correctly interpreted, is worth 1,000 expert opinions or 100 computer models of groundwater flow". He had it printed on his business card. The truth in this statement lies in the fact that all appropriate field measurements, which are properly carried out and interpreted are always better than opinions and models. This is true not only of dye tracing but of all methods of measurement and monitoring.

A well designed dye trace test needs to consider a variety of conditions and variables prior to injecting any dye or tracer and should involve someone experienced in this field. This should include mapping karst features, assessing dye introduction and sampling points, developing a sampling program (intervals, methods of sampling and methods of analysis) as well as a clear objective and realistic expectations from the data. A dye trace could take a few hours for a short distance or be a basin-wide study that requires hundreds of samples and many months or years of time. In any case, the planning is critical to assure optimum results.

While dye trace tests are useful for determining flow connections and velocities in a system, they do not characterize the size of the conduits or provide the locations of a cave system, which are necessary for geotechnical investigations. Further details for dye tracing can be found in Aley (2002) *Groundwater Tracing Handbook* which can be downloaded from his website and Alexander and Quinlan (1992) *Practical Tracing of Groundwater with Emphasis on Karst Terranes*, a short course manual. In addition Smart (2005) discusses further details, strategy and errors of dye tracing.

22.2 Considerations for Dye Tracing

A tracer is a substance intentionally added to groundwater to give it a distinctive signature that makes the groundwater recognizable elsewhere. Numerous types of tracers are used,

R.C. Benson, L.B. Yuhr, *Site Characterization in Karst and Pseudokarst Terraines*,
DOI 10.1007/978-94-017-9924-9_22

Fig. 22.1 Fluorescein dye being introduced by Tom Aley into a sinking stream in Nevada (Photo courtesy of T. Aley and G. Baker)

both man-made and natural. The most common man-made dyes include fluorescein, eosine, sulforhodamine B, and rhodamine WT (Aley 2002, 2008). Natural tracers might include salts or temperature.

Dye tracing is a point-to-point method that is dependent upon the location of the dye introduction and sampling points. This would typically require an inventory of karst features as well as other natural and man-made windows into the groundwater system. There are many different ways in which dyes can be introduced into the groundwater system. Sites for the introduction of dyes include; sinkholes, surface streams that sink into the subsurface (Fig. 22.1), and directly into cave streams. In some cases wells (Fig. 22.2), borings, trenches, or ditches along a road may be used.

Selection of dye introduction and sampling points should be optimized to assure a successful dye trace. A saturated karst feature such as a sinking stream or water-filled sinkhole would provide a greater probability of getting dye into the subsurface karst system, than a dry, topographic depression that may lie well above groundwater. Sometimes dyes are placed into dry locations where they will be taken into solution by the first flow of water. This is very useful for stormwater management studies and in remote areas. However, a saturated karst feature will usually result in the shortest travel time and will also have the greatest chance to yield a positive result. The exact location of dye introduction will depend upon the objective of the dye trace. These selections must also take into consideration the logistical aspects of both dye introduction and sampling.

Fig. 22.2 Barry Beck introducing dye into a monitor well at a site in West Central Florida to determine its possible connection to a recent large sinkhole

In the case of a dry introduction site such as a dry well, dry fracture, or dry sinkhole the introduction area should be flooded with water prior to the introduction of the dye to

Fig. 22.3 A large water truck used to pre-flush and post-flush the dye introduction site in an unsaturated system

saturate the system. After the introduction of the dye, the system should be flushed again with large quantities of water to assist in flushing the dye through the fracture or karst system. This usually requires a large quantity of water. Figure 22.3 shows a water truck used for flushing dry systems.

The entire process of dye tracing in the epikarst is considerably different than in the karst groundwater system and requires additional knowledge and experience. Ground water tracing in the epikarst should be considered at waste sites or spills where the majority of contaminants are localized within the epikarst zone with the potential to be dispersed laterally and vertically through this zone. See Aley (1997) for more details regarding dye tracing within the epikarst.

22.2.1 Water Sampling Prior to Introducing Dye

There are numerous sources of dyes and related chemicals, which are used in a variety of manufactured products that find their way into the surface and groundwater system. In addition, it is not uncommon to find that dye trace testing has already been done in a karst area, which may impact the selection of dye.

An adequate quantitative sampling of all sample locations prior to dye introduction is essential to establish the natural background water quality and possible sources of interference. Ideally, in order to assess background conditions at a monitoring point, it should be monitored before and for several days or weeks, after a major recharge events.

22.2.2 Estimating Quantity of Dye

Many factors affect the quantity of dye to be used. These include:

- Analytical protocol. Laboratory analysis with a spectrofluorophotometer operated under a synchronous scan protocol is the most sensitive and reliable method. This is especially true at waste sites and in industrialized areas where multiple fluorescent compounds may be present. Other analytical approaches may require the use of more dye for credible results.
- Factors along the flow route that may degrade the dye or absorb it onto earth materials. Different dyes have different properties and are not all equally suited to all site-specific conditions.
- The amount of dye (called the dye equivalent) in the dye mixture being used can vary. As an example, Aley (2002) found dye equivalents in commercially available fluorescein dye mixtures varied from 2 to 80 %. If you do not know the dye equivalent you have no knowledge of how much dye was introduced.
- The intensity of fluorescence for the particular dye. As an example, after adjusting for dye equivalent percentages, fluorescein emission fluorescence peaks in water are about 13 times larger than rhodamine WT peaks (Aley 2002).
- Travel distances and volumes of water to be dyed.
- Nature of the flow system. Traces from a sinkhole to a spring routinely require much less dye than traces from most other features in the landscape. Traces to wells routinely require more dye than traces to springs.

Aley (2013, personal communication) indicates that less than 0.5 kg of dye is seldom used. In most karst areas 0.5–2.5 kg of dye is used. However, experience is the most effective method for determining appropriate dye quantities.

22.2.2.1 A Case of Too Much Dye

Rock quarries in western Dade County, Florida have operated since the 1950s providing rock and cement to Florida's construction industry. The quarries encompass an area of about 200 km^2 called the lake belt. After the mining, the quarries are left as open fresh water lakes. These quarries allow the surface water to be intimately connected to the groundwater over large areas.

The quarries lie just west of the northwest wellfield that provides drinking water to Dade County, Florida and draws water from the sole-source Biscayne Aquifer. The wellfield consists of 15 large wells providing water for residential and commercial use.

Concerns arose about possible contamination by Cryptosporidium, a parasite living in human and animal intestines, or from terrorist acts contaminating a quarry. A 800 m no mining protection buffer zone had been established around the wellfield based upon chemical and bacteria contamination, but more recent USGS studies have indicated rapid flow through the limestone of the Biscayne Aquifer. The present buffer zone is now thought to be insufficient to protect against Cryptosporidium. No Cryptosporidium or giardina have ever been detected in the lakes or the wells or the drinking water at the plant. Unfortunately Cryptosporidium is not controlled by chlorine, which is commonly used to treat groundwater.

USGS carried out a dye tracing test by introducing approximately 50 kg of rhodamine dye in a well about 100 m upgradient from one of the large (7,900 gpm) pumping wells. The dye was expected to reach the well in 2–3 days, instead it arrived in 4–6 h after dye was introduced (Renken et al. 2005). The operator of the water treatment plant noticed a bright red color to the water and called for the well to be shut down. The following newspaper accounts of the event reported it as "faucets flowed pink within hours after a dye was introduced into a nearby test well". Another reported "Beneath the Pink Underwear" because clothes in the wash machines turned red that day (Dudley 2003).

This is a good example in which the planning phase of the dye trace work was missing some of the geologic and hydrologic pieces of the puzzle (i.e. groundwater flow rate due to the large pumping wells 100 m away) and the quantity of dye had been grossly over-estimated.

22.2.3 Handling of Dye and Avoiding Cross Contamination

Small amounts of dye powder are easily lost to the air. Both powder and liquids can easily contaminate personnel, clothing, vehicles, hotel rooms, along with eating facilities and may not be detected. Minimizing all sources of cross contamination between transporting and handling of dye as well as sampling procedures are critical to maintaining a reliable analysis of all dye tracing efforts, yet little is written on this critical subject.

Fig. 22.4 Mixing of fluorescein dye prior to its introduction into a major fracture

However discussions with Jim Quinlan on this subject revealed these guidelines. The individual responsible for transporting, mixing, handling, and injecting dye should be isolated from other personnel and sampling points to avoid inadvertently cross contamination. If powder dyes are used, small amounts can be windborne leading to wide spread undetected cross contamination. Those handling and introducing the dye should use separate vehicles, hotel rooms and eating facilities. These precautionary procedures were used in a dye trace assessing flow into the abandoned Tobin limestone mine (see Part III Chap. 25). Figure 22.4 shows the mixing of the dye prior to its introduction. The person handling the dye used proper protective clothing and after introduction of the dye, they were removed from further activities at the site to minimize possible cross contamination.

Those collecting samples must use extreme care and be trained in proper sampling techniques to avoid contamination. If such practice is not followed the results of dye traces might be questioned due to possible cross contamination with the dye used.

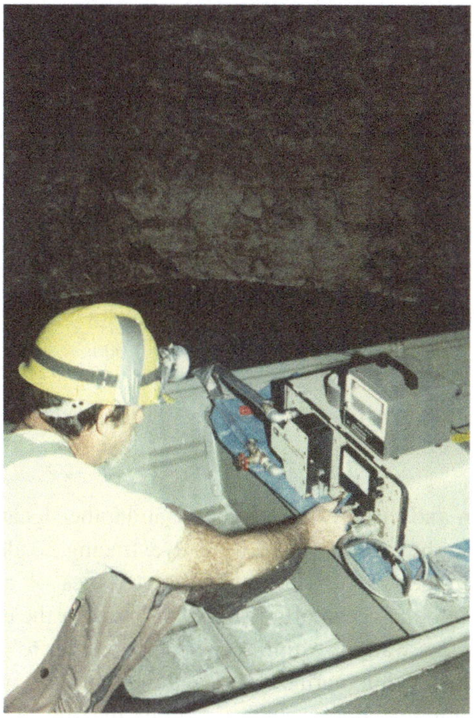

Fig. 22.5 A Turner fluorometer with water being continuously pumped through it to sample for the presence of any dye as the boat moves through the water within the Tobin mine. The results were recorded on a strip chart recorder

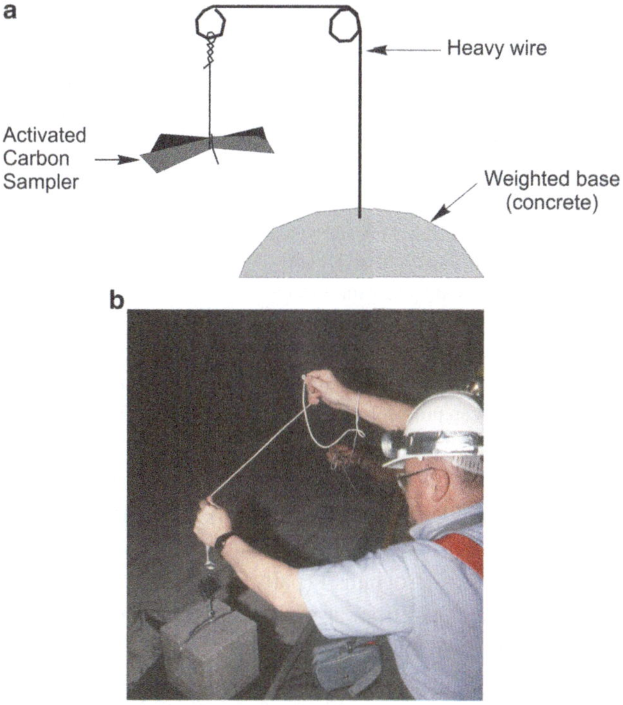

Fig. 22.6 A gumdrop sampler (**a**) made famous by the late Dr. Jim Quinlan. The gumdrop sampler suspends an activated charcoal sampler above a streambed by a few centimeters out of the bottom sediments. Here Jim installs a gumdrop sampler in the abandoned Tobin mine using a cement block as a base (**b**)

22.2.4 Sampling and Analysis Methods

Sampling for dye can be accomplished in a number of ways depending upon the specific needs.

- In some cases the presence of dye can be determined visually and no sample is taken.
- Discrete water samples may be obtained at selected time intervals.
- Sampling and analysis can be done using a field fluorometer. This provides a means of obtaining a continuous sample and analysis in real time (Fig. 22.5).
- The use of charcoal packets, called 'bugs', are commonly employed to acquire an integrated water sample over a period of time (Fig. 22.6a). The bugs are removed and replaced at selected sample intervals.
- Automated water sampling can be used at selected intervals using an automated pump and sampler system (Fig. 22.7).

Fig. 22.7 An automated water sampler can be used to obtain discrete samples at programmable intervals

Sample intervals would be selected to meet project objectives. However, typical sampling is more frequent during and immediately after the introduction of the dye and becomes less frequent with the increase in distance of travel and time after dye introduction. Use of a programmable automated water sampler (Fig. 22.7) such as manufactured by ISCO can be used to obtain discrete samples at periodic intervals. This provides a means of obtaining water samples without having to have personal on-site thereby reducing the cost of sampling.

Dye detection techniques have a range of over eight orders of magnitude from visual detection to the use of a spectrofluorophotometer. Dye detection limits for four commonly used dyes are shown in Table 22.1. Values are based upon fluores-

Table 22.1 Dye detection methods (Aley 2013 personal communication)

Analysis method	Fluorescein (75 % dye equivalent) ug/L	Eosine (75 % dye equivalent) ug/L	Rhodamine WT (75 % dye equivalent) ug/L	Sulforhodamine B (20 % dye equivalent) ug/L
Water sample analysis by spectrofluoro-photometer[a]	0.002	0.015	0.015	0.008
Elutant from carbon sampler and analysis by spectrofluorophotometer[b]	0.025	0.050	0.017	0.080
Dark room with flashlight beam; visually inspection by an experienced person	2	10	50	5
Visually in the field by an experienced person	7	135	125	50
Visually in the field by the general public	140	13,500	2,500	1,000

[a]Synchronous scan protocol with excitation slit of 3.0 nm and emission slit of 1.5 nm
[b]Synchronous scan protocol with excitation slit of 5.0 nm and emission slit of 3.0 nm

cein, eosine, and sulforhodamine B mixtures with 75 % dye equivalents and rhodamine mixtures with 20 % dye equivalent. All values micrograms/l (parts per billion). The concentration of dye eluted from an activated carbon sampler left in place for a week will be at least two orders of magnitude greater than the mean dye concentration in the water.

22.2.4.1 Charcoal Bugs

Activated carbon samplers, referred to as charcoal bugs, are commonly used in groundwater tracing studies. Activated charcoal is enclosed in a plastic mesh and placed at the sample location (Fig. 22.6a). These samplers adsorb and accumulate most dyes providing an integrated sample over time at a sample point.

An activated carbon sampler is left submerged for a week where there is at least a slightly visible current. The concentration of dye eluted from the sampler is routinely two orders of magnitude greater or more than the mean concentration in the water. The concentration factor in samplers placed in wells with minimal amounts of water circulation is typically less, but is routinely at least an order of magnitude greater than in water samples (Aley 2013 personal communication). The charcoal bugs are particularly applicable when working at very low concentrations of dye or when sampling at large time intervals.

A water sample should be retrieved along with the charcoal sampler and a new charcoal sampler would be put in its place. The water samples provide independent data on dye concentrations in the water at the time of sampling as a means of quality control. While the charcoal sampler is integrating the dye over the interval between sampling, the water sample is representing a single point in time.

22.3 Results and Analysis of Dye Trace Studies

Dye trace studies can be qualitative or quantitative. Simple dye trace studies to assess near field connections are often used. These can be a simple as introducing the dye in one location and visually identifying it at another location. For example the authors have used dye tracing to determine localized flow and dispersion characteristics of a sewage outfall into the ocean off of Miami, to monitor the exchange of surface waters within ocean inlets and to evaluate exchanges within marina basins and the Florida Bay.

Quantitative tracer studies require the introduction of known quantities of tracer and measuring tracer concentrations downgradient as they increase above background levels and then return to background levels. This data is used to develop a breakthrough curve. Figure 22.8 is a very simplified breakthrough curve where dye concentrations are plotted against time relative to dye introduction.

Breakthrough curves can be much more detailed and allow for complex analysis (Fig. 22.9). This more detailed data provides an increase level of confidence and accuracy for determining:

- Identification of introduced tracer
- Measurement of groundwater velocities
- Estimating hydraulic properties, and
- Assessment of hydraulic conditions.

22.3.1 Dye Traces Confirming Connections at a Dam Site

Crawford et al. (2005) describe the use of dye to determine the connection between upstream sinkholes and downstream springs and boils at a dam site. A hydrographic survey using sonar and video had identified five sinkholes in a lake upstream of the dam. There was a boil in the river about 550 m downstream and a spring 240 m downstream on land. Three dyes were introduced into three of the upstream sinkholes and samples collected downstream with an ISCO automatic sampler and activated charcoal bugs.

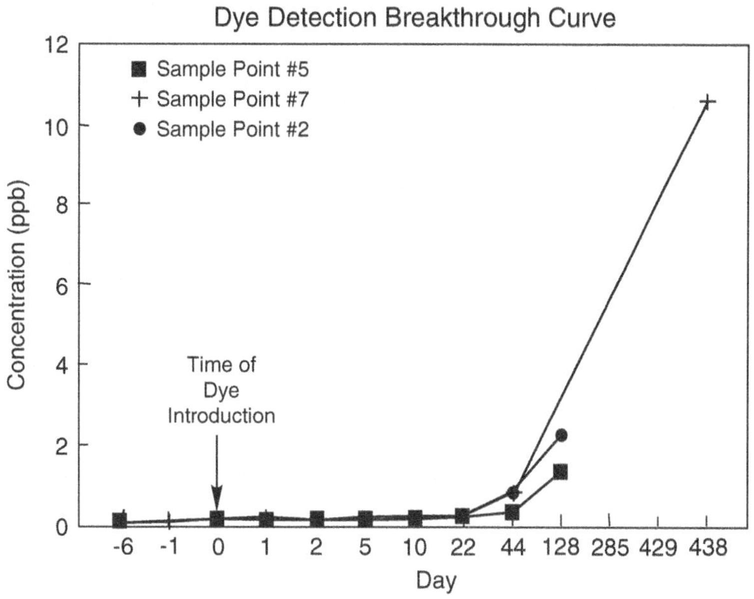

Fig. 22.8 A simple breakthrough curve from the Tobin mine showing time and concentration of dye at three sample points within the mine (Note: the time scale is not linear)

The dye trace clearly showed the connection between the upstream sinkholes and the downstream spring and the boil. The actual route taken by the water was unknown. Two conceptual models were proposed to account for the flow pathways. The details of the geologic setting and the dye test are provided in Crawford et al. (2005).

22.3.2 Assessing Vulnerability of a Cave Habitat

The following example was provided by Thomas Aley at the Ozark Underground Laboratory. A segment of federal highway in Missouri was to be repaved by chip-sealing. The highway segment was known to contribute water to Tumbling Creek Cave, which provides habitat for an aquatic snail that is on the federal endangered species list. There was concern that the process of chip sealing might yield detectable petroleum hydrocarbons in the first stormwater runoff event after the chip seal application, and that this might adversely impact the snail in the cave stream. Missouri Department of Transportation wanted to analyze water samples from the cave stream for Total Purgeable Hydrocarbons (TPH) that might be derived from the first flush of stormwater runoff after the chip sealing.

About 0.5 kg of fluorescein dye mixture containing 75 % dye equivalent was placed in each of two road culverts in the repaved area. The culverts were dry at the time of dye placement, but the dye was placed where it would be taken into solution by the first storm flow. An automatic water sampler was installed on the cave stream and programmed to pull water samples every 2 h and make composites of these samples every 8 h. Figure 22.10 shows the dye breakthrough curve resulting from this groundwater trace.

Water first flowed through the culvert at about 0500 h on November 15, 2006 (7 days after completion of the chip sealing and 35 h after the dye was introduced). Precipitation on November 15 and 16 equaled 4.6 cm. Dye first reached the sampler in the cave stream about 57 or 58 h after the first flow through the culvert. This water sample, and the subsequent water sample that had the highest detected dye concentration, were both analyzed for TPH. Both TPH concentrations were <100 ug/L, which was the detection limit. The straight-line travel distance through the karst groundwater system was about 5,100 m from one culvert and about 5,800 m from the other. Using an average straight-line travel distance of 5,450 m and a travel time of 57 h equals a mean straight-line groundwater velocity of 95 m/h. The resulting data were useful for evaluating potential TPH contamination of waters in the cave stream but also valuable for planning emergency actions in the event of a highway spill in the studied area.

22.3.3 Dye Traces in the Woodville Karst Plain

Wakulla Spring is a first magnitude spring that lies within the Woodville Karst Plain (WKP) in north Florida. The WKP was introduced in Part I, Sect. 7.2.2 and discussed in both the surface geophysical Sect. 16.5.3 and the geophysical logging Sect. 18.3.2.

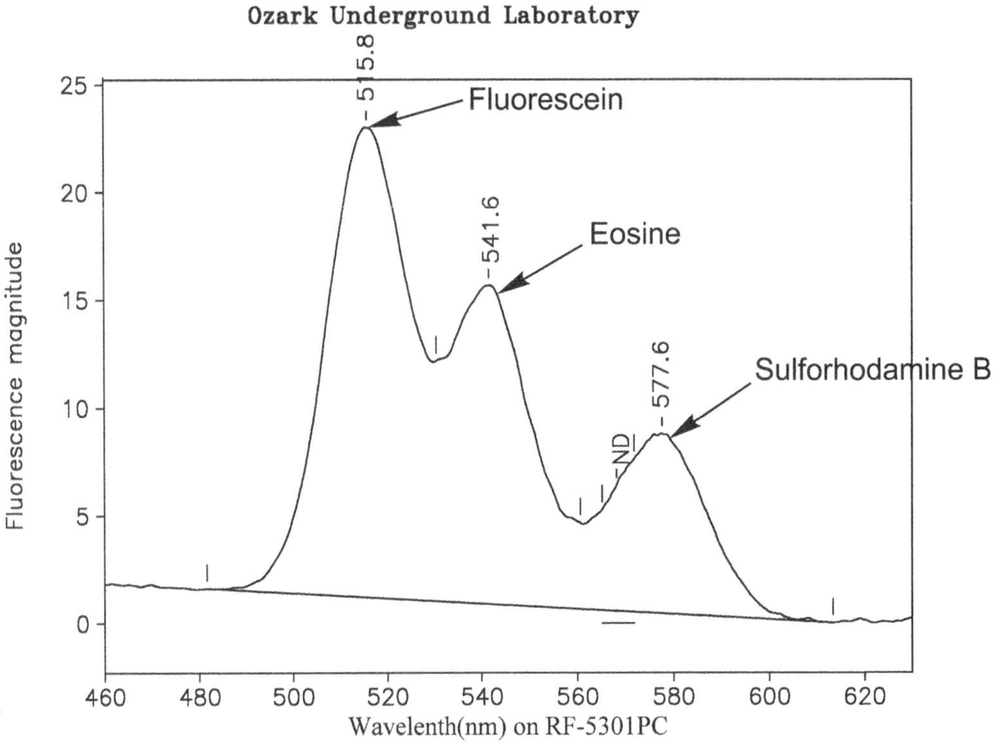

Station 33: Kaintuck Creek U/S Kaintuck Hollow Spring
OUL Number: W6758 Analyzed: 6/6/13
Matrix: Elutant Duration: 14.1 days
Date Placed: 5/14/13 1010 Collected: 5/28/13 1230

Peaks within the normal Range of tracer dyes:

Peak nm	Left X	Right X	Height	Area	Conc/Day	Conc.
515.8	481.6	530.4	21.82	463.54	0.681	9.60 Fl
541.6	530.4	560.7	14.79	310.23	0.752	10.6 Eo
568.2	565.2	571.8	0.00	0.00		ND
577.6	560.7	613.4	8.34	198.60	1.00	14.1 SRB

Peaks close to the normal range of tracer dyes:

Fig. 22.9 The results from the analysis from a carbon sampler elutant analyzed with a spectrofluorophotometer. The curve shows three dyes present in elutant from an activated carbon sampler in place for 14 days. The peak on the left represents fluorescein with a concentration of 9.60 ug/L; the center peak is eosine at a concentration of 10.6 ug/L, and the peak on the right is sulforhodamine B at a concentration of 14.1 ug/L (Figure courtesy of Aley)

The geology of the Woodville Karst Plain (WKP) consists of a thin veneer of unconsolidated Pleistocene quartz sand and shell beds overlaying a thick sequence of relatively horizontal carbonate rocks that comprise the upper Floridan aquifer. More than 1,000 sinkholes have been estimated to exist across the entire WKP, which is about 1,200 km^2 (Kincaid et al. 2012). The regional recharge for the area has been estimated to cover 2,500 km^2, (Gerami 1994; Davis 1996). Recharge to the Floridan aquifer in the Woodville Karst Plain occurs by:

1. Sinking streams,
2. Direct infiltration of precipitation through sinkholes,
3. Infiltration through the variable thick sands and soils overlaying the aquifer

Discharge from the Floridan aquifer under the WKP is through springs in the southern part of the region including Wakulla Spring, and submarine springs in Spring Creek Group at the edge of the Gulf of Mexico. Wakulla Spring is the largest inland spring in the WKP with an average discharge of 11 m^3/s, and the third largest spring in Florida. The Spring Creek group, which includes at least 14 underwater vents along the Apalachee Bay in the Gulf of Mexico is listed as the largest spring in Florida (Scott et al. 2002).

Nitrate has been blamed for the decreasing water clarity in Wakulla Spring by fueling growth of hydrilla and algae. The City of Tallahassee's Southeast Farm Wastewater Reuse Facility lies within the recharge area of the WKP. This facility sprays secondary treated wastewater onto the land surface

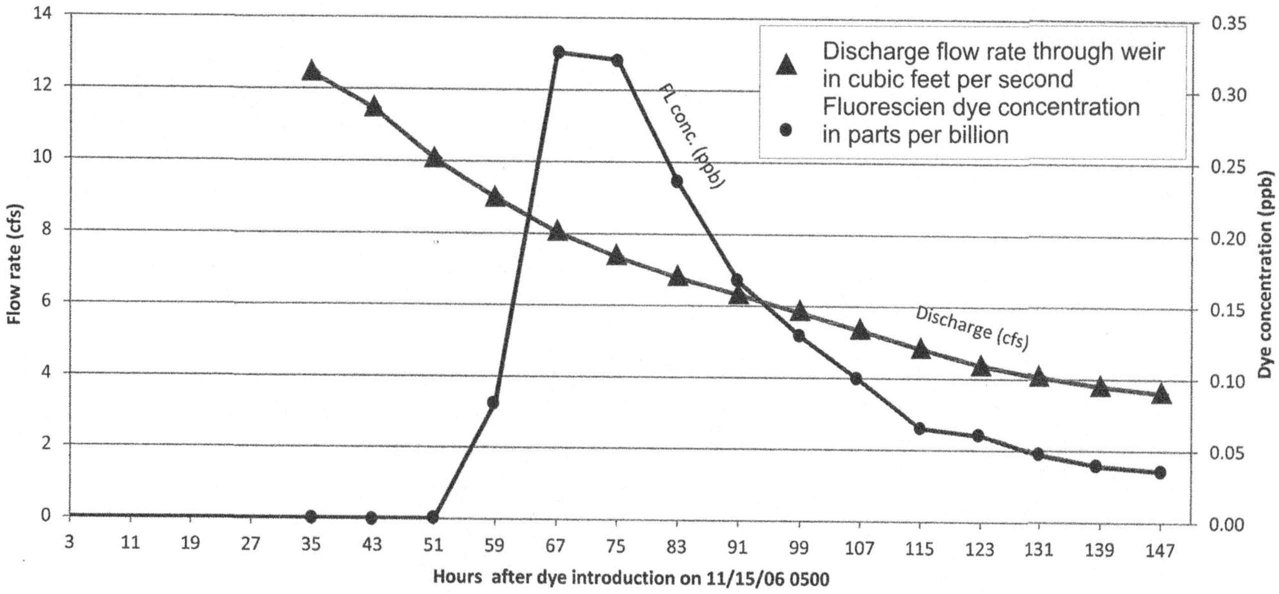

Fig. 22.10 A Fluorescien dye breakthrough curve at a weir in Tumbling Creek Cave. A rainfall event did not occur and create a flow rate until 35 h after the dye was introduced (Figure courtesy of Aley)

and was suspected as being a contributing source of nitrates at Wakulla Spring (Chelette et al. 2002).

A number of dye traces have been carried out within the WKP (Kincaid et al. 2005). The primary objectives of the most recent dye trace study was to show the connection between the wastewater facility south of Tallahassee and the springs and watersheds to the south, including:

- Developing tracer recovery curves,
- Evaluating hydraulic properties and mechanisms to be used in subsequent numeric modeling,
- Estimate mass recoveries for specific pathways, and
- Evaluate the effectiveness of in situ tracer detection devices for longer term studies.

Dyes were injected in and around the wastewater sprayfield. Automatic water sampling was used at the sampling points with high probability of detection, while periodic grab samples were used at lower probability of detection points. Samples were analyzed using a spectrofluorophotometer.

The results of the dye trace test indicated a clear connection from the wastewater facility to Wakulla Spring, a distance of about 16 km, at a rate of between 204 and 297 m/day (Fig. 22.11). The details of the dyes used, injection locations, sample locations and time, along with the results of analysis are described in Kincaid et al. (2012).

22.3.4 Evaluation of 2, 10 and 100 Year Capture Zone for Silver Springs, Florida

The following example was provided by T Aley at the Ozark Underground Laboratory. Silver Springs spring group is one of the largest first magnitude springs in Florida with discharges from 30 named springs and spring vent clusters located along the upper 1.2 km of the Silver River in Ocala, Florida. In 2004, the St. Johns River Water Management District delineated 2, 10, 100, and 1,000-year capture zones for Silver Springs based on a particle track simulation (MODPATH). One objective of the dye trace program was to evaluate the credibility of the delineated capture zones.

An extensive dye trace program that lasted almost 17 months and was a major component of the study (McGurk et al. 2012; URS et al. 2011). The traces were designed to identify dominant groundwater pathways and travel times between specific locations and the springs. Dye sampling placed primary reliance on activated carbon samplers since almost all dye concentrations at sampling stations were expected to be below the detection limits in water samples. These detection limits were 0.002 ppb for the fluorescein mixture, 0.015 ppb for the eosine mixture, 0.015 for the rhodamine WT mixture, and 0.008 ppb for the sulforhodamine B mixture.

Four separate dyes were used, one in each of four injection sites (numbered 1–4). Fig. 22.12 shows the locations of the dye introduction points with respect to the springs as well as the dye detection points.

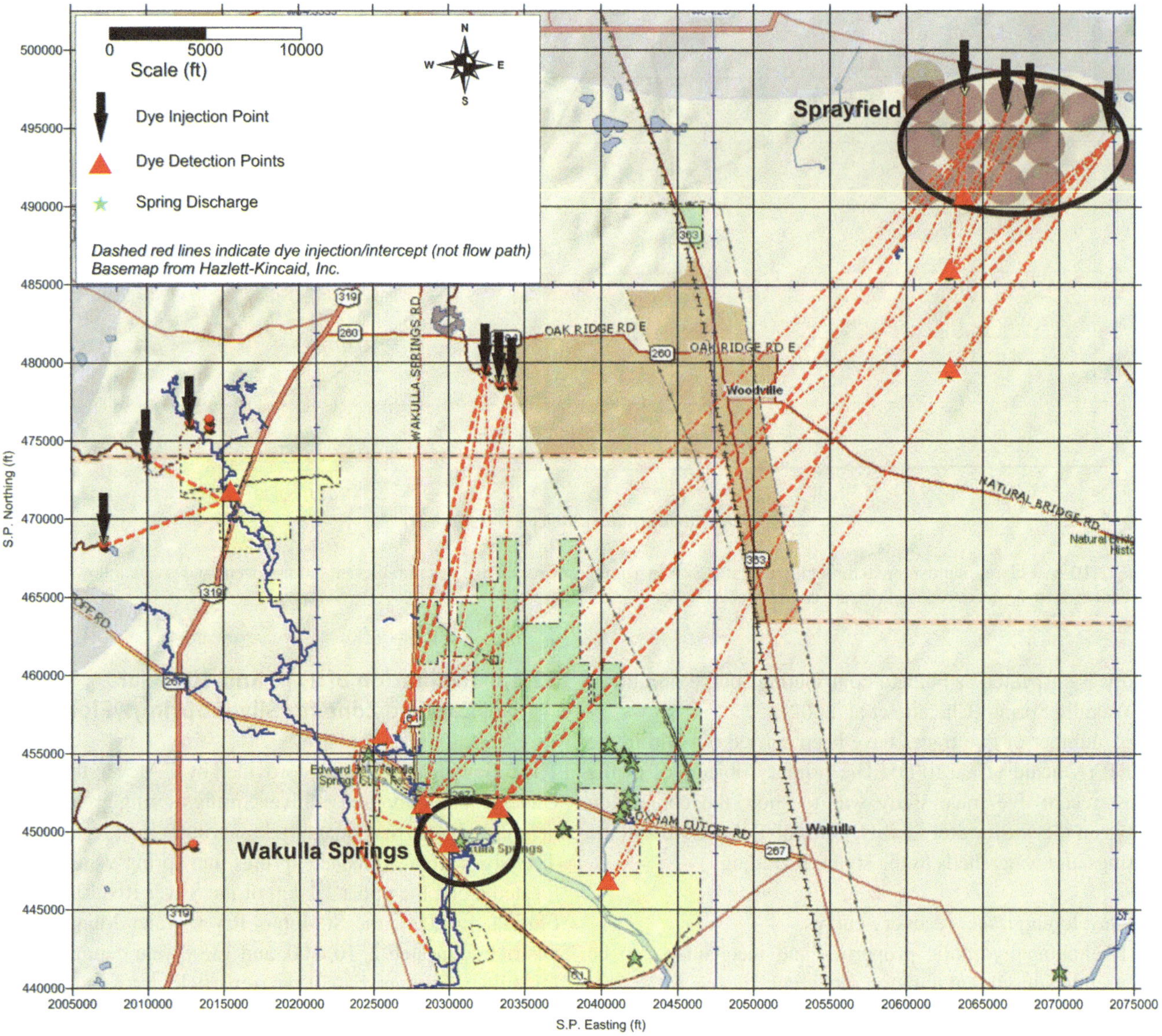

Fig. 22.11 The results of numerous dye traces within the Woodville Karst Plain from sinkhole to sinkhole as well as from the City of Tallahassee's waste water disposal spray field to Wakulla Spring (Kincaid et al. 2012)

1. Ocala Civic Theatre: 9 kg of rhodamine WT dye mixture with a 20 % dye equivalent was introduced into a recently formed sinkhole in a drainage retention area (at the Ocala Civic Theatre) 2.2 km from the nearest outlet for Silver Springs. This dye was detected at a number of the springs in the Silver Springs group. First arrival times for the dye ranged from 5 to 10 days at two of the closer springs and 39–45 days at one of the more distant springs. Based upon first dye arrival times mean straight-line groundwater velocities averaged 321 m/day. The dye introduction point was within the estimated 2-year capture zone for Silver Springs.
2. Tuscawilla Park: 13.6 kg of eosine dye mixture with a 75 % dye equivalent was introduced into a drainage well 8.2 km from the nearest spring in the Silver Springs Group at Tuscawilla Park. The first arrival time for the dye at this spring was between 295 and 312 days after dye introduction. This represented a mean straight-line averaged groundwater velocity of 27 m/day. The dye introduction point was within the estimated 10-year capture zone for Silver Springs and about 1,036 m from the outer edge of the 2-year capture zone.
3. Pontiac Sink: 22.6 kg of sulforhodamine B dye mixture with a 75 % dye equivalent was introduced into a large

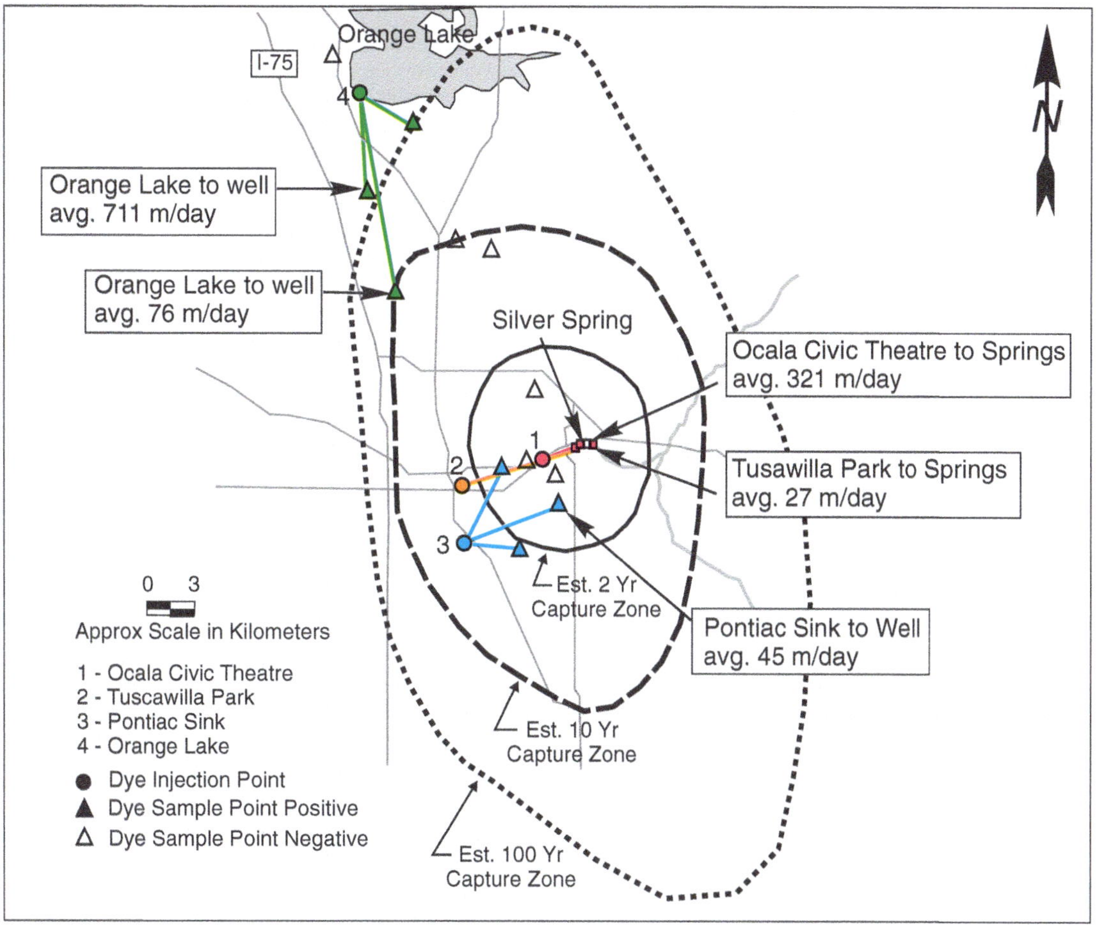

Fig. 22.12 Map of the Silver Springs spring group showing the 2, 10, and 100 year capture zones along with the points of dye introduction, points of detection, their travel times and flow velocity (McGurk et al. 2012)

sinkhole adjacent to a constructed wetland for treating stormwater runoff (Pontiac Sink). The sinkhole is 10 km from the nearest outlet for Silver Springs. This dye was detected at three public water supply wells roughly located between the sinkhole and Silver Springs. Distances from the sinkhole to the wells ranged from 4 to 7 km. First arrival times for the dye ranged from 50 to 57 days at the nearest well to 147–164 days at the furthest well. Based upon first dye arrival times at the furthest distance mean straight-line groundwater velocities averaging 45 m/day. The dye introduction point is about in the middle of the estimated 10-year capture zone for Silver Springs.

4. Orange Lake: 13.6 kg of fluorescein dye mixture with a 75 % dye equivalent was introduced into the Heagy-Burry sinkhole in the edge of Orange Lake. Water routinely drains from the lake into this sinkhole. Fluorescein dye from this introduction was detected at four public water supply wells between the sinkhole and Silver Springs. Distances from the sinkhole to the wells ranged from about 3.4 to 14 km. First arrival times for the dye ranged from 6 to 13 days at wells 6.75 km or less from the sinkhole to 181 to 194 days at a well 14 km from the sinkhole. Based upon first dye arrival times mean straight-line groundwater velocities ranged from 76 to 711 m/day. The dye introduction point is about 4 km outside of the boundary of the estimated 100-year capture zone for Silver Springs.

The groundwater travel rates determined by the dye trace study indicate that Silver Springs is fed by a multiple porosity system that includes both conduits and large pore flow. Groundwater travel rates were routinely faster than model estimates that were based upon porous media assumptions. The most notable example of this was provided by the fluorescein introduction at Orange Lake. This dye introduction point was 4 km outside of the boundary of the 100-year capture zone for

Silver Springs, yet it was first detected about half a year later at a well almost midway between the dye introduction point and Silver Springs. Aley (2008) reported a similar orders of magnitude differences between modeled travel times and those demonstrated by dye tracing in a karst aquifer in Maryland.

22.4 Limitations of Dye Traces

Even the best planned dye tract tests can encounter problems with varying levels of impact. Some of these problems may include:

- Background fluorescence can mask (in part) the tracer signature
- Very dark tannic waters may impact measurement accuracy of tracer concentrations
- Tracers discharge to unsampled points
- Tracer travel times are longer than the sampling period

Some of these problems can be avoided or minimized with a well-planned and executed dye trace program. Other problems are simply those conditions that are unexpected and cannot be anticipated.

References

Alexander EC, Quinlan JF (1992) Practical tracing of groundwater with emphasis on karst terranes, 2nd edition. A short course presented at the annual meeting of the Geologic Society of America, Boulder, CO

Aley T (1997) Groundwater tracing in the epikarst. In: Beck B, Stephenson JB (eds) Proceedings of the 6th Multidisciplinary Conference on Sinkholes and the Engineering and Environmental Impacts of Karst: the engineering geology and hydrogeology of Karst Terranes. Springfield, Missouri Balkema, Rotterdam, pp 207–211

Aley T (2002) Groundwater tracing handbook. Ozark Underground Laboratory available on-line from www.ozarkundergroundlab.com

Aley T (2008) Chapter 15 tracer tests—dye. In: Weight WD (ed) Hydrogeology field manual, 2nd edn. McGraw-Hill, New York, pp 675–698

Chelette A et al (2002) Nitrate loading as an indicator of nonpoint source pollution in the lower S. Marks-Wakulla Rivers watershed: Water Resources Special Report 02-1, Northwest Florida Water management District, Havana, Florida

Crawford NC et al (2005) Hydrologic investigation of leakage through sinkholes in the bed of Lake Seminole to springs located downstream from Jim Woodruff Dam. In: Beck B (ed) Proceedings of the 10th multidisciplinary conference on sinkholes and the engineering and environmental impacts of karst, San Antonio, Texas Sept 2005. ASCE Geo-Institute Geotechnical Special Publication 144, pp 489–499

Davis H (1996) Hydrologic investigation and simulation of groundwater flow in the upper Floridan Aquifer of north-central Florida and delination of contributing areas for selected city of Tallahassee, Florida water supply wells. USGS Water-Resources Investigation Report 95-4296, USGS, Denver, Colorado

Dudley S (2003) Beneath the pink underwear. The Miami News Times, 5 June 2003

Gerarmi A (1994) Hydrology of the St. Marks river basin, northwest Florida. MS Thesis Florida State University, Tallahassee Florida

Kincaid T et al (2005) Quantitative groundwater tracing and effective numerical modeling in karst: an example from the Woodville Karst Plain of North Florida: In: Beck B (ed) Proceedings of the 10th multidisciplinary conference on sinkholes and the engineering and environmental impacts of karst, San Antonio, Texas Sept 2005. ASCE Geo-Institute Geotechnical Special Publication 144, p 114–121

Kincaid T et al (2012) Demonstrating interconnection between a wastewater application facility and a first magnitude spring in a karstic watershed: tracer study of the southeast Farm Wastewater Reuse Facility, Tallahassee, Florida. Florida Department of Environmental Protection, Florida Geological Survey Report of Investigation No. 111

McGurk BE et al (2012) Silver Springs nutrient pathway characterization project, final report. St. Johns River Water Management District Special Publication SJ 2012-SP3

Renken RA et al (2005) Assessing the vulnerability of a municipal well field to contamination in a karst aquifer. GSA Environ Eng Geosci 11(4):319–331

Scott TM et al (2002) First magnitude springs of Florida. Florida Geological Survey Open File Report 85

Smart C (2005) Error and techniques in fluorescent dye tracing. In: Beck B (ed) Proceedings of the 10th multidisciplinary conference on sinkholes and the engineering and environmental impacts of karst, San Antonio, Texas Sept 2005. ASCE Geo-Institute Geotechnical Special Publication 144, p 500–509

URS et al (2011) Hydrogeologic evaluation to support nutrient groundwater pathway delineation near Silver Springs, Marion County, Florida. Final report with extended sampling prepared for St. Johns River Water Management District, Consultants Report

The Conversion of Data to Useful Information

23

Abstract

As the site characterization has progressed, many of the steps for processing, plotting, checking and interpreting the individual data sets have already been carried out. In addition, the conceptual model has been iteratively improved based upon these various sets of data. At this point we will have a wide range of data from different sources (our own as well as others). Much of this data is in different formats and scales. Now is the time for a final review and fresh look at all of the data when we have it all before us and the rush of field-work is over. The large amounts of data, gathered by the site characterization process, by themselves are not that useful. A formidable part of the site characterization process is having engineers, geologists and hydrologists assemble, conduct interpretations and integrate this diverse set of data. The conversion of raw data to useful information is a value-added process, which is achieved by careful professional analyses (Sharma 1994 personal communication). As the raw data is converted to useful information, a conceptual model or models are developed. The ultimate objective of the site characterization is to develop the most accurate conceptual models of site conditions possible within the time and budget constraints of the project. This is done by progressively minimizing the envelope of uncertainties in the interpretation of the data.

23.1 An Assessment of All Data

All of the data, both existing and newly acquired should already have been assessed in order to minimize the uncertainties in the data. The assessment should include steps to:

- Make sure the data acquired is appropriate for the project needs and specific to the site itself.
- Determine that adequate data has been obtained both spatially (laterally and with depth) and temporally to provide sufficient coverage over the site and at different scales of measurements.
- Confirm the accuracy of the data with thorough reviews and analysis. This includes identifying any errors, gaps, or impossible data. Correct them, if possible, or remove them from the active data set so that they cannot impact further interpretations or calculations.

After the data is acquired, a re-assessment of the data should be made. It is possible to meet all three criteria in a data set, yet the data is not useable, in part, due to unexpected conditions that caused noise in the data, interference that masks useable results, or impossible results. Situations like these should be identified.

Once these three criteria are met, then and only then, can we begin the assembly, interpretation and integration of disparate sources of data. This allows us to proceed with confidence in using the data to develop an accurate final conceptual model of the geologic and hydrologic site conditions (Yuhr 1998; Yuhr et al. 1996).

R.C. Benson, L.B. Yuhr, *Site Characterization in Karst and Pseudokarst Terraines*,
DOI 10.1007/978-94-017-9924-9_23

23.2 Managing the Data

A proper site characterization will encompass a wide range of professional disciplines and generate an enormous amount of data from a wide range of sources. Each data set may have a different scale of measurement, inherent bias, limitations and possible errors. Therefore, it should be recognized that the data are never totally complete or accurate. The line between factual data and interpretations and opinions is certainly a gray one. In addition it should be recognized that we cannot and will not know everything about a site (Fig. 11.1). However, the site characterization strategy presented in Chap. 12 will have optimized the data acquired and recognizes the different levels of uncertainty that arise in both data values as assembled and in their interpretations.

There are no magic tricks involved to improve confidence levels and minimize uncertainties, just common sense, perseverance and attention to detail. These steps began before data was even acquired, by selecting the appropriate methods that would meet objectives and overcome site-specific constraints. Adequate spatial or temporal sampling for each method or measurement was optimized and the data was checked for its accuracy. These three criteria (appropriate, adequate and accurate) were also used to evaluate any existing data and databases.

The next steps include a final assessment, processing, interpretation and integration of all the data as a complete group. However, many of these steps are overlooked or thought of as busy work for lower-level personnel. Yet, it is here, where the professionals with experience are able to fit the puzzle pieces together, identify the spatial trends and make correlations between diverse data sets, which is critically important. The task of data assessment and interpretation is not at all an easy one and there are a number of steps in the process that require considerable skill, experience and persistence.

23.3 Assembly of Data

The site characterization process is one of understanding spatial and sometimes temporal variations in geologic and hydrologic conditions. Being able to put data or results of data into spatial and/or chronological perspective is essential. However, integrating data from a wide variety of sources can be daunting.

23.3.1 The Use of Graphics

Almost all of the processes used for interpretation and analysis of data can be summarized as comparing, contrasting, and correlating data spatially or with time. It is here where graphics is an important tool since it is much quicker and easier to review graphic plots of data to the same scale and format than to review the same data in tabular form or other formats. The eye is a very powerful tool and can see spatial variations in data if it is put in graphic form.

Looking at a single piece of data at a time is not tremendously useful or effective. Reviewing massive amounts of printed data from different sources and in different formats is difficult and is often avoided. For example, looking at page after page of tabulated laboratory geotechnical measurements or chemical data can be confusing and may not readily identify subtle changes or trends, errors or gaps in the data. Another example is the process of reviewing 50 geologic boring logs done by different firms and individuals with different formats and level of detail. Both of these examples are difficult and confusing tasks. On the other hand graphic plots of the printed laboratory results or 50 boring logs plotted to the same scale in the same format can provide a more rapid and accurate review of such data.

The use of graphics allows subtle trends, errors, and gaps in the data or conflicts between disparate sets of data to be identified and resolved. When appropriate graphic formats are utilized, data can be much more readily checked, integrated and interpreted with confidence (Benson and Sharma 1994, 1995). The issue of visualization of data is an important one, both from the data interpreter's point of view and presentation of information to others. Presentation of the data with consistent scales and complete labeling is one of the objectives of software graphics (Benson and Sharma 1994).

23.3.2 Selecting Scales

A key step is to assemble all of the data into a common graphics format, with drawings to a common scale, whenever possible. As a rule, maps and cross sections should be shown to a consistent scale, when possible, so that they may be directly and easily compared. However, a comprehensive site characterization will encompass data that represents a wide range of scales including regional data (kilometers), site-specific data (hundreds of meters) and detailed data (meters). As a result three different scales will commonly be required for a site characterization. In some cases, more than three scales may be necessary. Selecting the wrong scale may hide subtle variations in some data that could have significance or may magnify variations in some data that are unimportant. This is where our professional judgment is utilized.

23.3.3 Developing Graphics

Contour maps, cross sections, and fence diagrams are widely used formats for presentation and interpretation of geologic and hydrologic conditions. Regardless of the graphics being

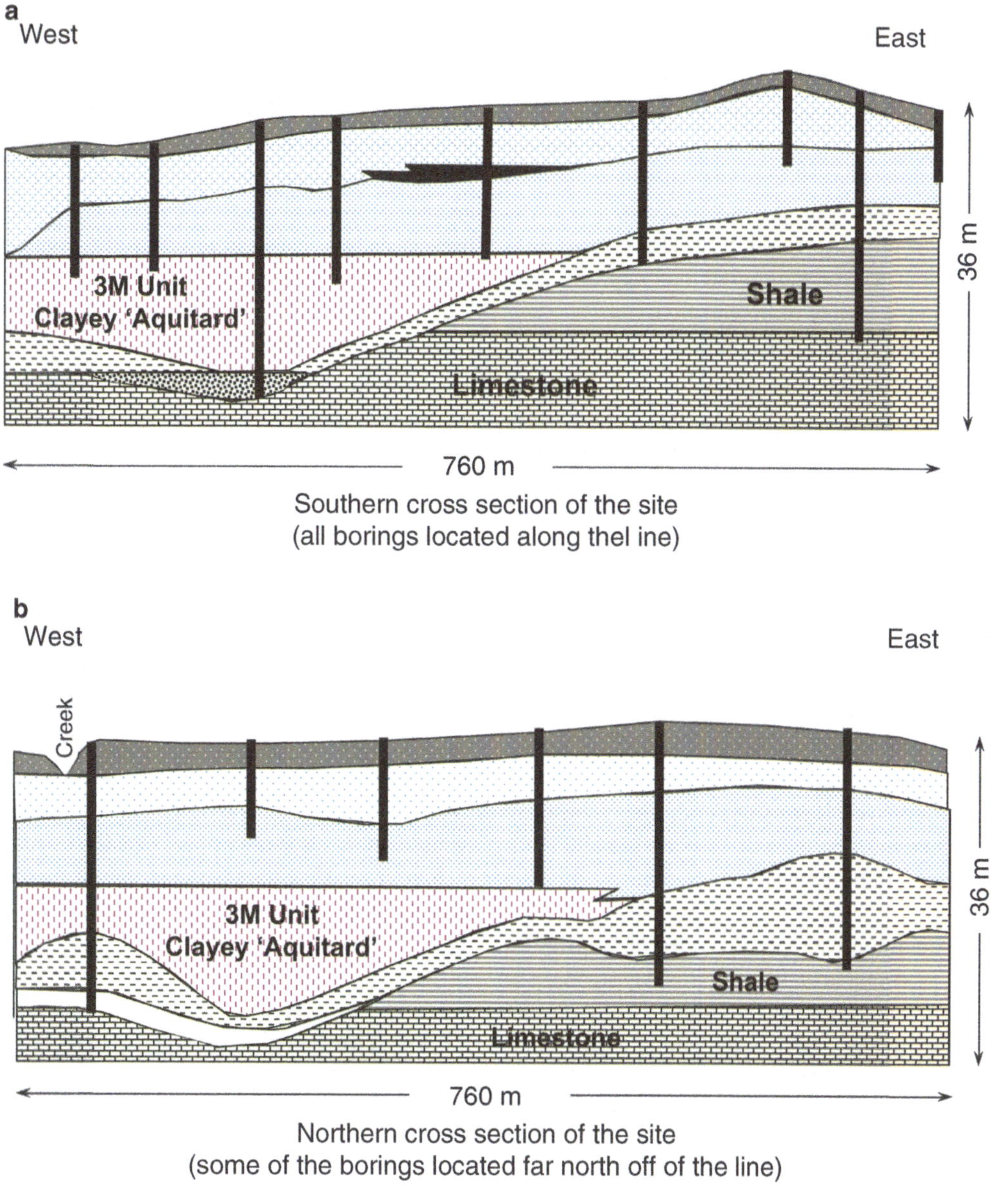

Fig. 23.1 Two cross sections based upon boring data. The southern cross section (**a**) is based upon a line of borings all on the cross section while the northern cross section (**b**) has borings projected over distances of 60–180 m. In addition, the presence, and thickness of the clayey aquitard along with the low in the top of rock underlying it, had been extrapolated onto the northern cross section

used, each figure must be supported by the data upon which it is based. For example, when a geologic cross section is based upon borings data the locations of borings should be shown on a plan view map. This will provide a means of assessing the reliability of the cross section. If all borings lie on or reasonably close to the cross section, there will be a higher level of confidence in the cross section compared with a cross section in which the borings are projected from long distances away. In addition to the boring locations, their depths and possibly a summary of the actual geologic log may be needed. This will allow the cross section interpretation to be readily evaluated.

An example is shown in Fig. 23.1 where existing geologic data was obtained from two lines of borings at a low-level radioactive disposal site north of the St Louis airport. The southern cross section utilized a line of borings that were all located along the line of the cross section (Fig. 23.1a). The northern cross section (Fig. 23.1b) utilized some borings that were off-line by 60–200 m. Only one boring along the south line was used to define the low in the top of rock seen in both cross sections. Based upon the geologic logs alone, it was not clear what other data, if any, were used to interpret a bedrock valley across the site. Later in the project, gravity data confirmed the presence of the buried valley seen in the cross section.

23.4 Processing of Data

Some data requires little, if any, processing such as our notes, sketches, and photographs. Other data will simply require transfer from our field worksheets to a standard format (water levels or geologic logs). Some data (surface geophysical data, geophysical logs, pump tests, dye tests etc.) will require some level of processing and result in a graphic and/or a quantitative value. The processing of data requires close attention to detail and a variety of diagnostic tools.

The proper software can greatly facilitate the processing, managing and integrating the data. There is a wide range of software available, each with its own advantages and limitations. It is also important to be aware of what is being done to the data by processing, the inherent assumptions in the software and the variables that are being selected. When data is processed strange things can sometimes happen such as errors in data entry, errors in data transfer from program to program, choice of inappropriate processing software or algorithms, inappropriate variables selected, over or under filtering of the data, etc. As such, processing of the data should be an integral part of the site characterization team's effort with proper control and assessment of the results by the team members to assure the output is consistent with reasonable geologic and hydrologic site conditions. The team members themselves must be the ones processing and managing the data thereby minimizing any opportunities for such problems.

A simple and common example is one of contouring of data. Contouring is a practical way for displaying data to show the third dimension on a map. Contours are easy to interpret, and a skill widely used in the geologic and engineering communities. Drawing of contour maps and interpreting contour maps used to be universally taught to all students of geology and engineering. In recent years, computer programs for contour data have become readily available and are sometimes used without any understanding of the impact to the data and its presentation.

There are a wide variety of gridding and contouring algorithms available in the different computer programs. They are each designed to effectively handle different types of spatial data. Needless to say that by the time we see the final contour, the spatial aspects of the raw data can be changed in many ways. The basic concepts and limitations of contouring and spatial interpretation and analysis of point data are presented in two papers by (Schreuder 1997, 1999) and are well worth reading by all those using contouring programs.

In addition, one must have an understanding of the type of data being contoured. Is the contour to represent a smoothly varying surface topography or highly variable top of rock? An understanding of the data will help determine the algorithm used for gridding and contouring and how much one needs to honor the raw data.

Figure 23.2 shows two contours of the same data. The data being contoured is the elevation of a shale layer from 39 borings over about 40 ha. The site has fairly flat-lying horizontal strata based upon a desk study, on-site observations and measurements. The contour shown in Fig. 23.2a used an inverse distance algorithm while the contour in Fig. 23.2b used a minimum curvature algorithm. Both of the contours in Fig. 23.2 used default variables and are mathematically correct but provide a very different presentation of the data. The contour using inverse distance algorithm is honoring the original data more effectively, but as a result creates a highly variable contours. The contour using the minimum curvature algorithm is honoring the original data less effectively and is smoothing out the variations in the data. The contour using the minimum curvature algorithm (Fig. 23.2b) is the most appropriate since it illustrates the general trend of localized subsidence over a paleocollapse feature, which was the objective of this data.

As shown in Fig. 23.2, contouring can greatly affect the presentation of data and its ultimate interpretation. This includes how closely the raw data is honored by the contours, the contour interval selected, whether the interval is linear or logarithmic and how color or shading is used. When using contours, the data point locations should be plotted on a contour to show the spatial extent of the data that is being contoured. When possible, raw data values should be plotted on the contour as well.

23.5 The Final Interpretation and Conceptual Model

23.5.1 Integration of Independent Data Sets

Throughout Part II of this book, we have repeatedly emphasized the need to make multiple independent observations and measurements of geologic and hydrologic conditions at different scales so that we can compare the results of one independent set of data to another. Once each individual set of data has been interpreted by itself, the resulting interpretation should then be checked against other sets of data and their interpretation to see if there is correlation or discrepancies. The correlation of two or more independent sets of data is extremely powerful to reduce uncertainty. When the results from different independent measurements agree, a higher level of confidence in our interpretation is achieved (Fig. 23.3). In addition, the resulting conceptual model is improved and uncertainties assumptions and opinions are minimized. At the same time, others can easily and confidently arrive at the same conclusions based upon the data.

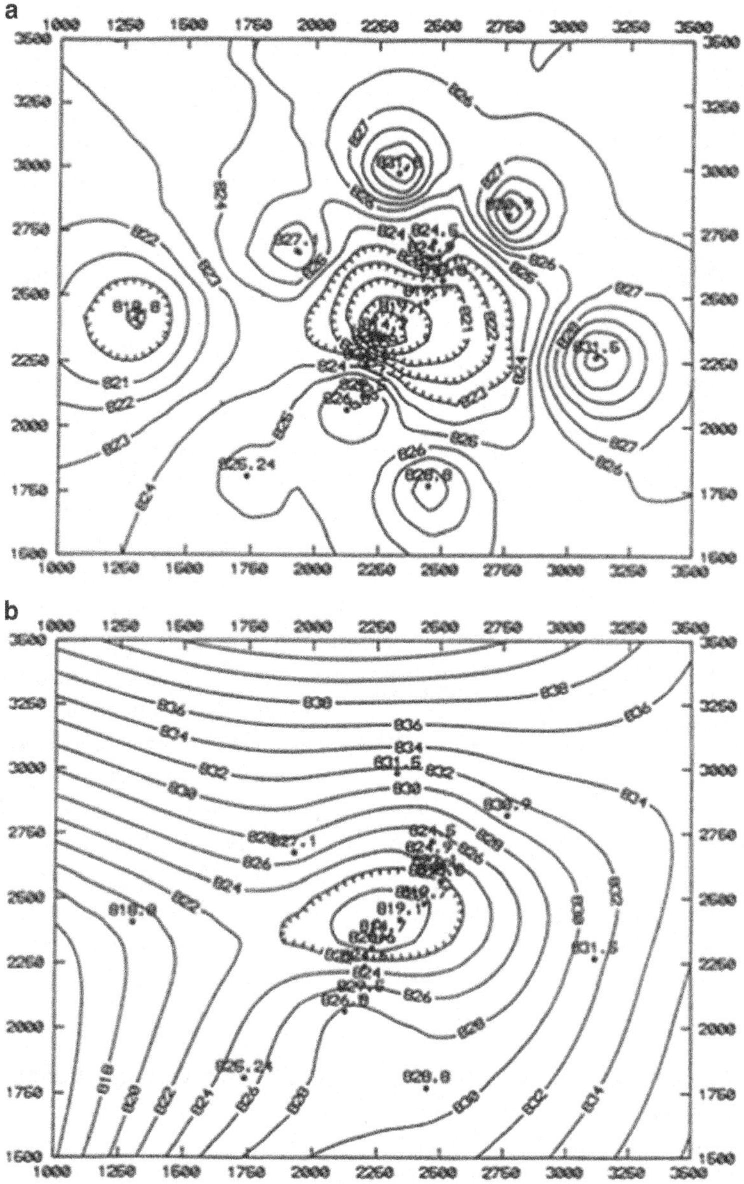

Fig. 23.2 Comparison of two different computer algorithms used for contouring the same data resulting in very different presentations

On the other hand, if it is found that the individual data sets are in disagreement or don't support the conceptual model for the site, something is likely wrong with our data, our interpretation or our conceptual model and additional data or analysis may be required along with revised interpretations.

The integration and correlation of individual data sets is essential for building a complete and accurate understanding of the site conditions. The three case histories presented in Part III utilize diverse data, which show spatial correlation in order to build a final conceptual model. Each individual set of data presents a piece of the puzzle. As an example, the karst investigation along a bridge alignment (Chap. 26) had seven distinct sets of data, which showed excellent spatial correlation (Fig. 23.4). These data included:

1. Spatial correlation of rod drops and fluid loss in the original set of borings logs,
2. An area of roadway that required frequent repaving due to settlement,
3. An aerial photo lineament passing through the center of the problem area,
4. A large microgravity anomaly indicating a low-density zone,

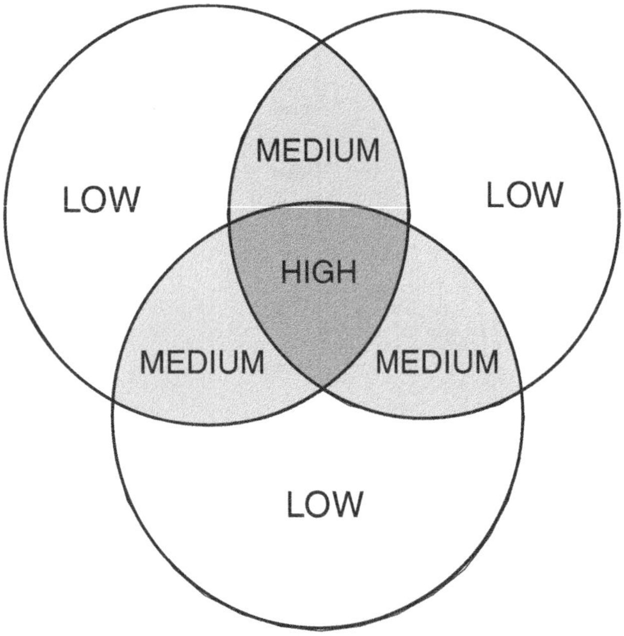

Fig. 23.3 The circles represent independent data sets. When two sets of data correlate, confidence levels increase from low to medium, when all three sets of data correlate, confidence levels are higher

5. Marine seismic reflection data indicating dipping strata at depth,
6. Additional borings confirming zones of rod drops and fluid loss
7. Geophysical logging showing areas of low-density materials.

Some of these independent data sets by themselves may not seem very significant. However, the spatial correlation of all data sets clearly increases our confidence level in both the individual interpretations as well as their correlation and integration. An additional piece of data included independent research by USGS off-shore that identified a 600 m diameter paleocollapse feature. The presence of such a feature shows what might be expected in the area and lends support to our interpretation of a large paleocollapse at the site.

23.5.2 Final Interpretation

While many of our engineering and hydrologic methods of measurement are highly quantitative, in many cases our final interpretations will be subjective in which the judgment of the investigator is used to quantify the results. Judgment is based on inductive reasoning that incorporates site-specific data, observations and professional opinion, experience and integrity of the site characterization team into an assessment. As part of the data analysis, subjective data and results must be supported by adequate site-specific data and observations to provide a convincing basis for the assessment.

An example of this occurred at a landfill in Missouri. The landfill had been constructed above an abandoned limestone mine. A review of hydrogeologic conditions and a groundwater monitoring plan was needed. The initial conceptual model of the site (Fig. 23.5a) showed that existing monitoring wells screened in the limestone and located closely around the perimeter of the landfill contained little, if any, water. The limestone mine under the landfill contained about a meter or so of water and was underlain by a impermeable shale layer. So how was there water in the limestone mine and not in the wells?

A detailed site characterization confirmed that the limestone was unsaturated, but found that the underlying shale layer was actually quite permeable. There was also a subtle dip within the mine to the west, and the water was found to be flowing into the mine through the shale layer immediately under the mine floor at this low area (Fig. 23.5b). The monitor wells were only drilled into the limestone and not into the deeper shale, which accounted for them being dry.

Again, it is the attention to details and the willingness to put the pieces of the geologic puzzle together that can clarify conditions. In this case, the facts were correct, there was water in the mine and the surrounding wells were dry, but the original conceptual model (Fig. 23.5a) was wrong. The assumption of shale always being impermeable was incorrect and the degradation by sulfides under the mine floor had been ignored even though it had been cited in the local literature (Hasan et al. 1988).

23.5.3 Final Conceptual Model

A preliminary conceptual model of geologic conditions is used to guide early work at the site. This preliminary conceptual model developed at the end of the desk study has been continually upgraded throughout the site characterization process as new data has become available Fig. 12.17. This process requires the conceptual model to be continually questioned, tested, verified, and modified as additional site-specific data is acquired to minimizing uncertainties and maximizing confidence levels (Yuhr 1998).

As each individual data set is evaluated and interpreted it is integrated into the conceptual model. Ideally each piece of data fits into the conceptual model supporting other data and expanding our understanding of conditions. But what happens when the data doesn't fit into the conceptual model?

The first step would be to re-check the data to confirm that it meets the criteria of appropriate, adequate and accurate data and make sure that the data's limitations are understood. If this step indicates that the data is solid then check the individual data interpretation. Is there more than one interpretation

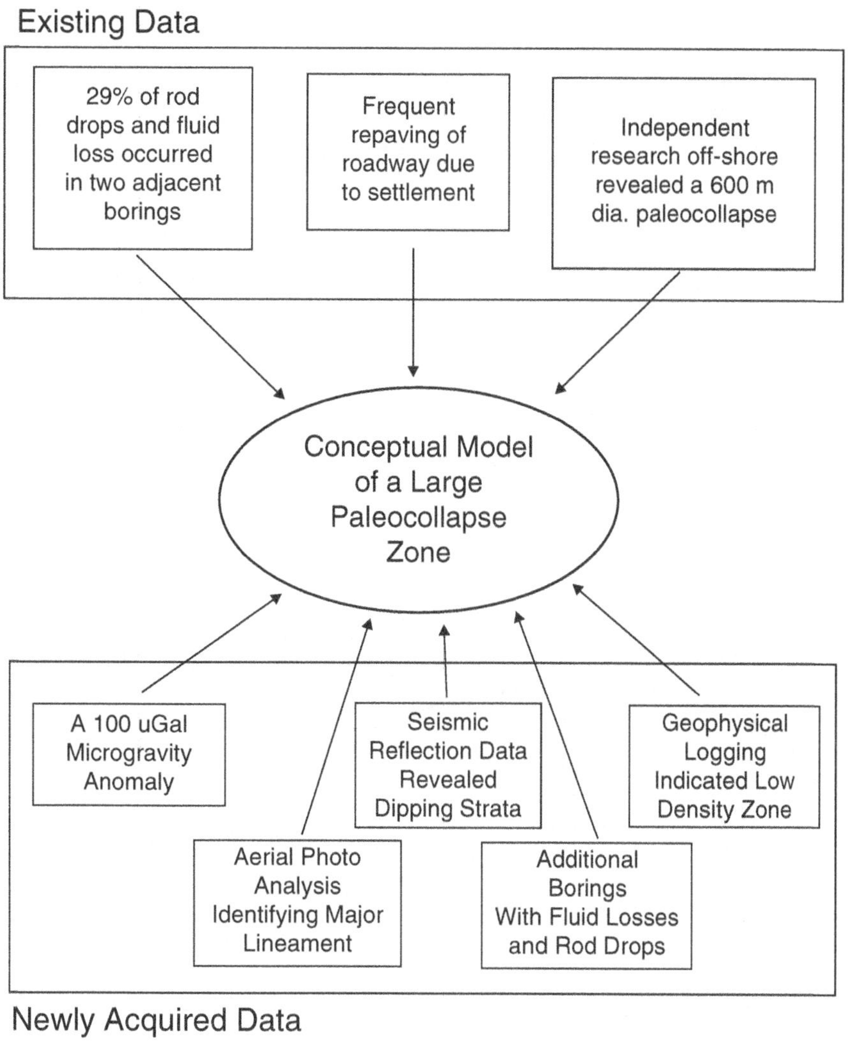

Fig. 23.4 A combination of existing and newly acquired data were used to develop and support the final conceptual model for the presence of a large paleocollapse

that can be made? Is there one that fits the conceptual model better? If the data and interpretation are not in question, then the conceptual model needs to be checked.

Initial conceptual models are often based upon early opinions rather than data or on data that has not yet been checked. While the conceptual models should be tested and modified throughout the site characterization process, early interpretation without sufficient data can easily lead to an erroneous understanding of site conditions and an incorrect conceptual model.

However, at this point, the integration of data over a wide range of scales and correlation of multiple data sets should have minimized the geologic uncertainties and increase the accuracy of the overall site characterization to an acceptable confidence level. By now we have the data to develop a reasonably, accurate regional and local conceptual model of site conditions which incorporates all of the essential features of the physical system. This final conceptual model must be supported by the available data including maps and cross sections of sufficient detail, specific properties, and calculations or models.

A final conceptual model of karst conditions is shown in Fig. 23.6 from the low-level radioactive disposal site (SLAPS) just north of the St Louis Airport. Supporting data for the conceptual model included:

- The geologic cross sections along with 81 borings over the site (Fig. 23.1)
- Geology and Geomorphology
- Analysis of topographic maps and aerial photos for lineament analysis
- Karst literature and cave maps

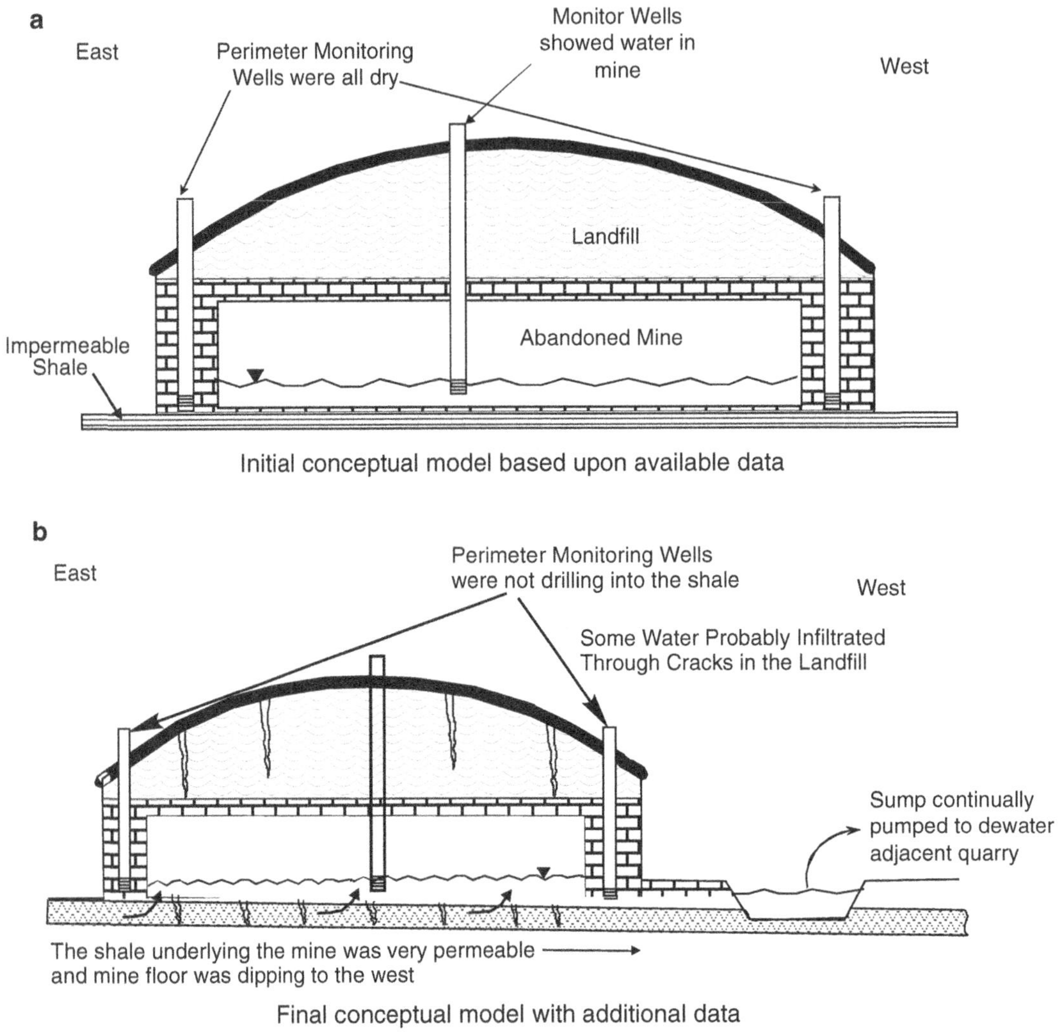

Fig. 23.5 The initial conceptual model for a site shows conflicting issues regarding the presence of water in the mine versus dry wells around the mine (**a**). An updated conceptual model illustrates the dip in the mine floor and permeability of the shale layer underlying the mine, both of which allows water to enter the mine while the monitoring wells in the limestone remains dry (**b**)

- Mapping outcrops
- Microgravity and TDEM resistivity measurements
- Geophysical logging

The final conceptual models included a plan view and cross section for the site. The buried valley was confirmed running south to north through the site. Within the Florissant Basin, which underlies the site the limestone had been exposed allowing extensive weathering and dissolution. The area was then filled with glacial material. Nearby outcrops confirmed open fissures were likely to be present in the upper most limestone. Significant weathering at the top of rock was confirmed in addition to preferential dissolutioning along joints and bedding planes. The eastern boundary of 3M clay layer had also been defined along with where it was missing from the eastern portion of the site. However, the lack of borings into the bedrock (Fig. 12.11) has limited any detailed assessment of conditions or contaminants within the limestone.

As complete as a conceptual model may be, it remains a simplified hypothesis of site conditions and must be understood as such. It is also expected that these conceptual models will be supported by more detailed technical data such as a contour map for a particular unit. However, we must constantly be aware that even a sophisticated conceptual model does not represent a totally complete or accurate description of a site.

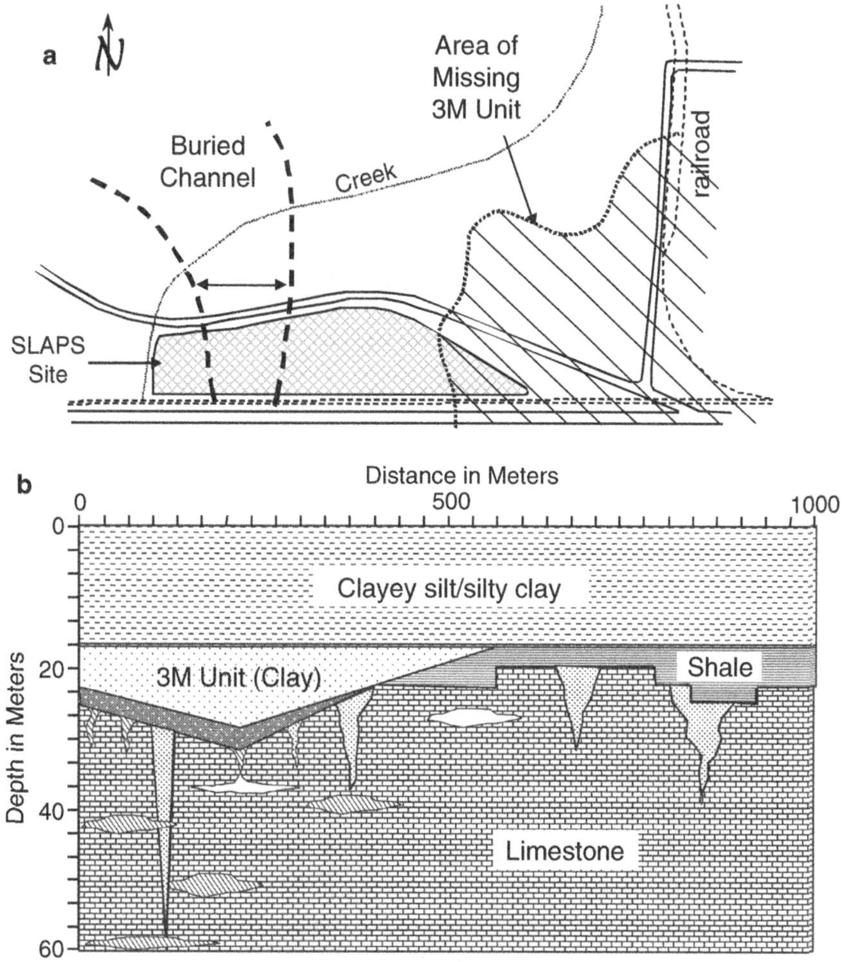

Fig. 23.6 The final conceptual model of geologic conditions for the SLAPS site. The extent of the buried channel and where the 3M clay unit is missing is shown using a plan view map (**a**), while the fissures in the upper most portion of the limestone is shown in a cross section (**b**)

23.6 Visualization and Presentation of Data

There are a wide range of graphics and presentation software that can be utilized for presentation of data along with many options for contours, cross sections and other formats. Even a simple sketch (Fig. 23.7) can be used to illustrate the conceptual relationship between observed surface fissures and a major mine collapse. This figure was the basis of our early conceptual model of conditions in our case history presented in Chap. 25, which was later supported by a wide range of additional data.

The use of software and particularly graphics software meets two critical project requirements; one is to aid the analysis and interpretation of data, the other is in the effective visual presentation of data. The graphics used for technical analysis, interpretation and reports may be different than graphics used for general presentations.

Graphics used for interpretation analysis and the report must contain supporting data upon which the figure is based and may be somewhat cluttered with this additional data. Graphics used for presentation purposes are intended to provide a summary of conditions and do not need to include the multitude of details needed for purposes of data interpretation, correlation and the report. Computer graphics can often be used to make complex geologic conditions clear to interested professionals and non-technical audiences.

Geology and hydrology is a three-dimensional system thus it is best understood and presented where possible in three-dimensions (3D). The use of 3D sketches is advocated by (Fookes 1997; Fookes et al. 2005). Such art skills seldom exist with most professionals today having been replaced by photographs and computer graphics. In many cases, simple lines of 2D graphics can be shown together as fence diagrams to provide a 3D representation of data. Figure 23.8 is an example of such data. This presentation illustrates the

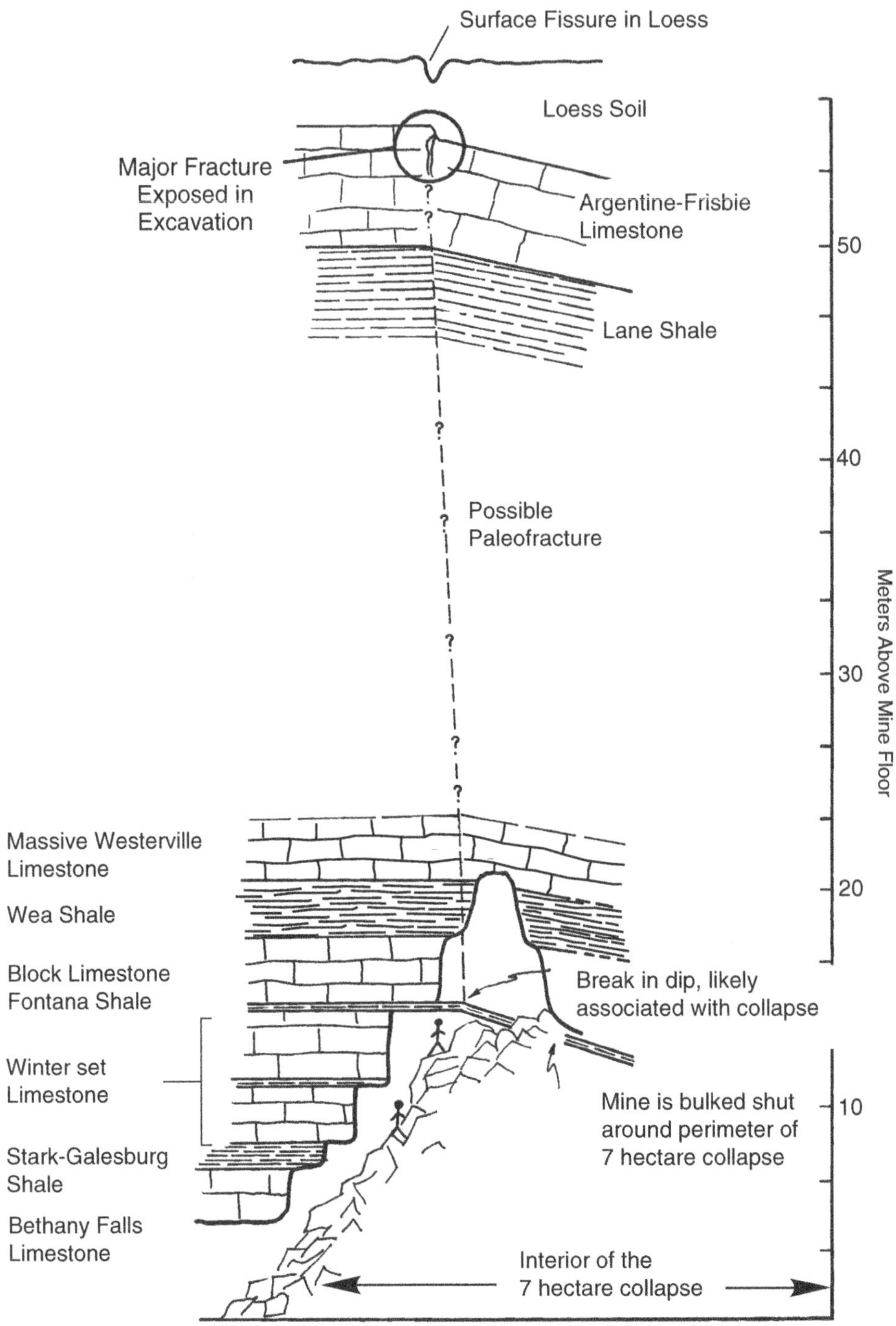

Fig. 23.7 A simple sketch of an initial conceptual model illustrates the possible connection between surface fissures seen in the loess soil, the uppermost rock and the major collapse within the mine

trends in the data from line to line and avoids unnecessary and unsupported extrapolation of data.

Three-dimensional graphics developed by computer may be pleasing to the eye but may be technically defective. Experienced professionals need to monitor the process and control results so that accuracy is maintained. As the use of computers and software continues to expand, there is a tendency to promote greater use of 3D graphics. In some cases the 3D imaging has been promoted to integrate many sets of measurements into one or more of the final images. The problem is that there are inherently increasing levels of interpolations and extrapolations from the raw data points in the process of developing such complex 3D images. These images often leave reality behind creating artifacts and possibly misrepresenting the actual data. Graphics should be used to convey information not to impress.

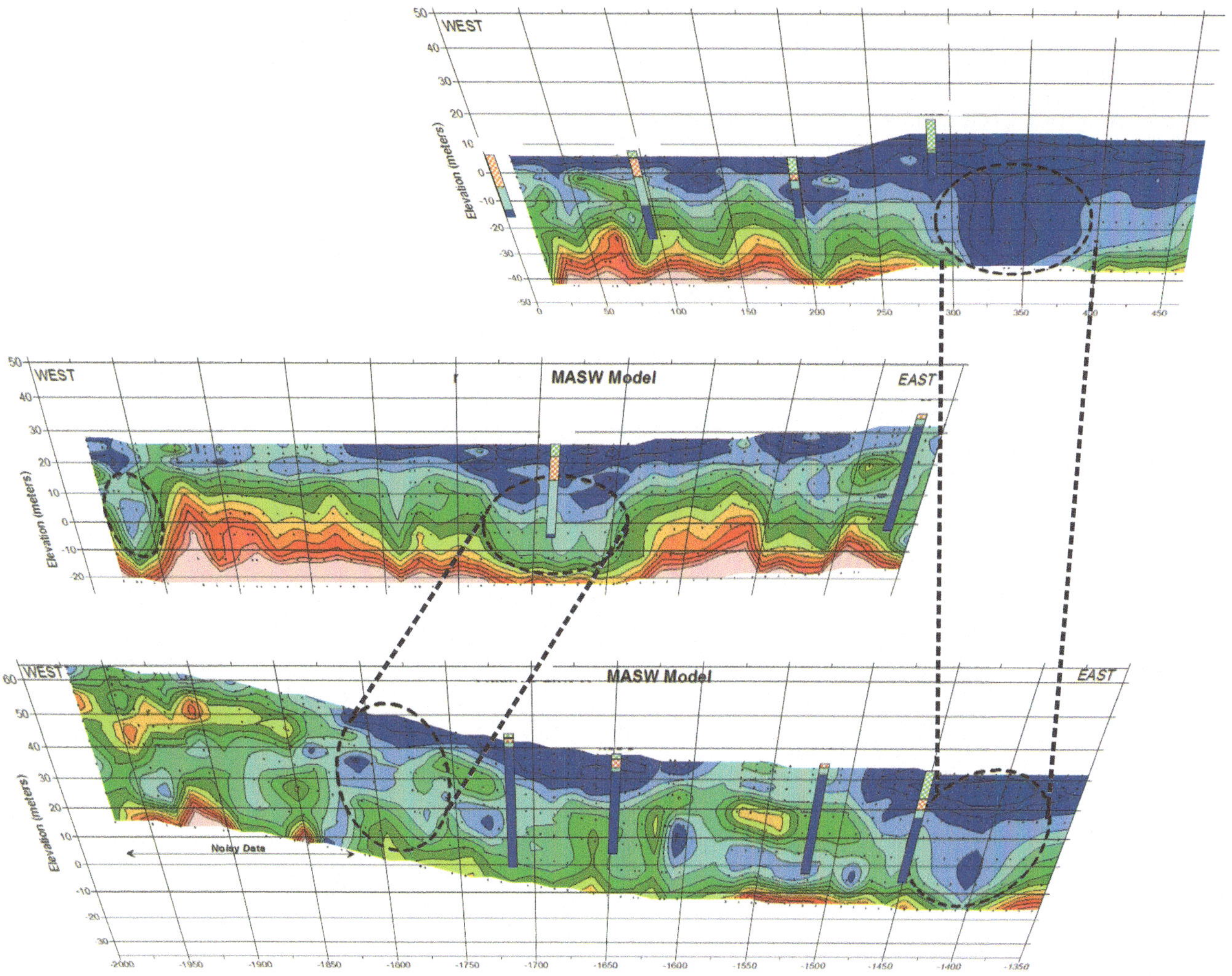

Fig. 23.8 Three lines of 2D MASW (surface wave) data show correlation between lines and illustrates their 3D relationship

23.7 Documentation: A Final Report

The final report of findings should be based upon a solid foundation of data and should include sufficiently complete information so that the owner and other professionals can understand the purpose, technical approach, and the results of the investigation along with any limitations. The report should include:

- The strategy for the site characterization effort along with its goals and with any limitations in scope,
- The methodologies used along with their advantages and limitations,
- Individual data sets should be discussed and then integrated with other data,
- A final conceptual model or models of site conditions along with a supporting data,
- The results of modeling or risk assessment (if needed), and
- Any inherent limitations of the site characterization effort.

The report should step a technical reader through the site characterization process so that the basis of the interpretations are clear. The report should be written so that other professionals can easily reach the same conclusions.

In most cases, it is recommended that a technical presentation of the site characterization and its results be made. All too often busy professionals may only read the executive summary and scan the report. A verbal presentation provides the focused review and opportunity for questions and discussion.

References

Benson RC, Sharma D (1994) The pitfalls in site characterization and how to avoid them. Short course presented by Environmental Education Enterprises, Salt Lake City

Benson RC, Sharma D (1995) Pitfalls of site characterization and how to avoid them. Short course presented by Environmental Education Enterprises, Charlotte, November

Fookes PG (1997) Geology for engineers: the geological model, prediction and performance. Q J Eng Geol Hydrogeol 30:293–431
Fookes PG et al (eds) (2005) Geomorphology for engineers. Whittles Publishing, Caithness
Hasan S et al (1988) Geology of greater Kansas City, Missouri and Kansas, United States of America. Bull Assoc Eng Geol 25(3):281–341
Schreuder WA (1997) On contouring surfaces. Principia Mathematica Inc. PM-TN-1997-1, Lakewood, Colorado
Schreuder WA (1999) Spatial interpretation and analysis of point data. Principia Mathematica Inc. PM-TN-1999-1, Lakewood, Colorado
Yuhr L (1998) Managing data and its integrity in the ESC process. Symposium on the application of geophysics to environmental and engineering problems, EEGS, 22–26 March 1998, EEGS, Denver, Colorado
Yuhr L et al (1996) Achieving a reasonable level of accuracy in site characterization in the presence of geologic uncertainty. In: Shackelford CD et al (eds) Proceedings of uncertainty '96, uncertainty in the geologic environment: from theory to practice. ASCE Geotechnical Special Publication No. 58(1), pp 195–209, ASCE Reston, Virginia

Risk Assessment

24

Abstract

Once the site characterization is complete, there is often the necessity of completing a risk assessment. Here the focus is twofold. The risk of sinkhole or collapse associated with natural voids in soluble rock, abandoned mines, other open structures within the ground or the vulnerability of groundwater in karst terrains. A brief introduction is provided to risk assessment including objective and subjective approaches. Risk assessment can be of a regional nature providing general trends or can be site-specific. A few examples are provided for both.

24.1 Definition of Risk

The terms hazard and risk are used in various ways and often interchangeably leading to confusion. The following definitions are derived from Sowers (1996) and Waltham et al. (2005), who specifically address karst as well as Vick (2002) who provides an overview of risk for engineering purposes.

Hazard – A condition with the potential for causing an undesirable consequence. Hazard is the event itself such as subsidence or collapse. Its magnitude or size is a description of its impact. For example, subsidence is generally slow and often occurs over a larger area while collapse can be quite rapid and is generally localized.

Risk – A measure of the probability and severity of an adverse effect to health, property or the environment. Risk is the combination of the likelihood or probability of the hazard occurring and its magnitude or consequences. Risk is usually described with terms such as unlikely, probable or highly likely or by a numeric scale of 1–10, with 10 being most likely.

There have been numerous approaches proposed for risk assessment. Beck (1991) had presented an approach for sinkhole risk assessment based upon available statistical data on sinkholes within the area. Engineers have proposed various approaches for dealing with general geologic risk assessment. Haneberg (2000) provides a mathematical approach using deterministic and probabilistic analysis in geotechnical engineering. Vick (2002) discusses both objective and subjective approaches to risk assessment. We have presented a non-mathematical approach to risk assessment, which is simple in concept but can be difficult in practice. Our approach is based upon having a complete and accurate site characterization in which to base any assessment of subsidence risk or groundwater vulnerability.

For further information there are a variety of books on geologic hazards that include: Ross (1984), Bolt et al. (1977), Scheidegger (1975), Shuirman and Slosson (1992), and Hunt (2007). Most of these books deal with earthquakes, landslides, or floods, only a few mention the hazard of subsidence, collapse or that of groundwater vulnerability.

24.2 Objective and Subjective Methods for Risk Assessment

Risk assessments are typically approached in one of two ways, either objective (quantitative) or subjective (qualitative). An objective approach relies on being able to quantify variables. A subjective approach relies on the professional

R.C. Benson, L.B. Yuhr, *Site Characterization in Karst and Pseudokarst Terraines*,
DOI 10.1007/978-94-017-9924-9_24

experience and the interpretation of conditions. Benson and Sharma (1994, 1995) have proposed:

> The assessment of risks in a quantitative manner is technically feasible for most man-made structures, because such structural and engineered materials under a wide variety of circumstances can be expressed through mathematical equations. Mathematical equations can range in complexity from simple algebraic equations to partial differential equations, but involve variables that can be analyzed through quantitative, statistical or stochastic means. Consequently, the influence of each variable upon the engineering decision or action can often be identified as a quantitative risk factor that may then be expressed in terms of some probability of occurrence.
>
> However, assessment of risks for most natural phenomena, such as the occurrence of sinkholes, is not tractable to systematic mathematical analysis in the same manner. While the broad range of factors that influence such phenomena are generally known, their specific inter-relationships in a given site or region are often the topic of conjecture. While the resulting subjective reasoning behind assigning risk factors may well be grounded in sound practical experience along with site-specific observations and measurements, they are not easily cast into mathematical expressions that then allow risks, or even probabilities of occurrences, to be quantified through analysis. For these reasons risk assessments of subsidence or collapse are commonly based upon subjective probabilities.

Risk can be developed using theory and numerical calculations (objective method). It is the process of using historical data (past occurrences) to predict future. Examples of this type of risk assessment include how the risk of earthquakes is assessed for a particular area at a particular magnitude. However, all too often, geologic hazards such as sinkhole collapse cannot be accurately quantified with such statistical methods. If the hazard does not have a defined history of occurrence, if the geologic conditions are modified by an engineered structure, or if localized anomalous geologic conditions are present but unknown, a subjective method of characterizing risk is needed. While these subjective methods are largely based upon the experience of the person assessing the risk, they must be based upon solid data and observations in order to support any conclusions. This is why a site characterization is needed for such efforts.

Vick (2002) points out that a subjective assessment of risk is equally as valid as an objective assessment of risk, and they are not mutually exclusive. Both the objective and subjective approach to risk assessments provides a valid approach if based upon a reasonable amount of data and experience. These approaches are often used in combination to develop a hybrid approach. For example, where data for one variable is in abundance and can be readily quantified while other variables have limited data and rely more on professional judgment.

24.3 Regional Risk Assessments

Regional risk assessments have been developed for subsidence and collapse as well as groundwater concerns. Subsidence and sinkhole collapse risk assessments are made based upon various geologic conditions and or historic trends from databases over an area of concern. For groundwater concerns a variety of programs have been developed which superimposed the presence of aquifers and their properties along with cultural factors to determine areas of groundwater vulnerability.

The following are some examples taken from work primarily in the State of Florida. However, other states or countries dealing with karst issues have developed similar databases, regional risk maps, or software programs to evaluate risks.

24.3.1 Sinkhole Databases

Many states have developed databases on sinkhole occurrence and have maps showing sinkhole prone areas. These data provide an indication of areas of sinkhole risk on a regional basis and provides a means of identifying potential risk in relation to a specific area of concern. This information is often used to develop the regional setting for a site-specific investigation. This has been discussed in Chap. 13 and an example of this is provided in Chap. 27. The sinkholes occurring within 1.6, 5 and 16 km from the site under investigation were identified in a sinkhole database on the west coast of Florida. Such data helps to set the stage for what might be expected at the site under investigation.

These databases typically include new sinkhole data as it is reported but not the older historic sinkhole data. The historic sinkholes provide the degree of maturity of the karst. The historic sinkhole data is obtained from a review of topographic sheets and older aerial photography, which is commonly available back to the 1930s (see Chap. 14). The presence of both historic sinkholes as well as recent sinkholes is needed to identify the complete nature of sinkhole prone areas.

In 1995, Bill Wilson, then with the Florida Sinkhole Research Institute developed a method to describe sinkhole occurrence based upon their statewide database. His results were presented as the number of new sinkholes per km^2 per year (NSH). This numeric evaluation provides an indication of the degree of active sinkhole development within an area thereby improving the regional risk assessment based upon the data in a sinkhole database (Wilson 1995).

24.3.2 Regional Sinkhole Risk Maps

Sinclair and Stewart (1985) developed one of the early sinkhole risk maps for the state of Florida (Fig. 3.10). It was developed based upon the type of overlying sediment (sand versus clayey material) and its thickness. Based upon these variables they categorized areas as follows:

- Thinly covered limestone (<10 m of sediment) – sinkholes are few, shallow broad and develop gradually

- 10–60 m of sand covered limestone – sinkholes are few, shallow, small diameter and develop gradually
- 10–60 m of clayey covered limestone – sinkholes are numerous, vary in size, develop abruptly
- Greater than >60 m of sediment covered limestone – sinkhole are very few, large diameter, and deep sinkholes do occur

This approach provided a reasonably accurate risk assessment on a regional basis.

Gao and Alexander (2003) used GIS mapping to develop sinkhole probability map for southeastern Minnesota. They used a risk-based decision tree, which considered factors such as type of bedrock, depth to bedrock, sinkhole density, and the distance to nearest sinkhole. Six zones of risk were characterized:

- No Sinkhole Risk (non-carbonate areas).
- Low Risk (areas of carbonate rocks with no sinkholes observed).
- Low To Moderate Risk (widely scattered sinkholes or clusters of sinkholes).
- Moderate To High Risk (areas where sinkholes are a routine part of the landscape).
- High Risk (minimum distance to the nearest sinkhole is 700 m and sinkholes are common with densities exceeding 10/km^2).
- Sinkhole Plains (where sinkholes are the dominant landscape feature, minimum distance to the nearest sinkhole is 400 m and sinkholes are common with densities exceeding 1/km^2).

The primary controls on sinkhole development were the type of bedrock geology and the thickness of sediment cover along with structure. A final risk factor was the nearest neighbor effect. The majority of sinkholes in southeastern Minnesota were found to form in concentrated zones. Gao (2002) found that 95 % of the sinkholes in the Devonian karst, and 99 % of the sinkholes in the Galena-Maquoketa karst, are less than 400 m away from their nearest neighbor. In the Prairie du Chien karst, sinkholes are less than 700 m away from their nearest neighbor.

Both of these examples have taken existing data on key variables and developed maps indicating what conditions may be expected for a given location. The example from Minnesota is also an excellent example of a hybrid approach to risk assessment effectively blending both objective and subjective data.

24.3.3 Groundwater Vulnerability in Karst

Karst terrains provide some of the more complex settings for evaluating groundwater, its flow and potential for contamination. Where the risk of sinkholes focuses on the variables that would cause or trigger a collapse, the groundwater vulnerability focuses on variables that would affect the transport of contaminants (thin soil cover, thick epikarst, high hydraulic conductivities). A number of approaches to assess the regional groundwater vulnerability have been developed. These are in essence risk maps for the potential of groundwater contamination.

LeGrand (1983) developed a standardized system for evaluating areas suitable for waste disposal sites. Later, a group with the National Water Well Association adapted the LeGrand System for broader hydrologic settings and not just waste disposal sites, called DRASTIC (Aller et al. 1987). This was a systematic means used to evaluate the groundwater pollution potential for any hydrogeologic setting within the United States. These were "screening tools" to be used with existing information for the assessment of potential groundwater vulnerability to pollution based upon a calculated index. The method could be applied to a wide range of geologic and hydrologic settings. DRASTIC recognized karst limestone settings, but did not utilize any site-specific details in its evaluation.

A modified version of DRASTIC was developed by the Florida Geologic Survey to accommodate karst conditions and is used as an Aquifer Vulnerability Model referred to as the Florida Aquifer Vulnerability Assessment (FAVA) (Arthur et al. 2005). This is a GIS-based modeling technique. The relative vulnerability is calculated based upon seven key factors that focus on aquifer thickness and hydraulic conductivity along with effective karst features.

$$\text{Relative vulnerability} = (Ts / Ks + Teg / Keg + Tconf / Kconf) + Kf$$

Ts – soil thickness
Teg – environmental geology (vadose) thickness
Tconf – confinement thickness
Ks – soil conductivity (weighted average)
Keg – environmental geology (vadose) hydraulic conductivity
Kconf – confinement hydraulic conductivity
Kf – effective karst features (the presence and proximity of closed depressions)

As a tool, FAVA provides a scientifically defensible water resources management and protection tool. It is based upon pre-development data (natural systems with no man-made alterations). Results are presented as three levels of site classifications, less vulnerable, vulnerable or, more vulnerable. The purpose is to facilitate regional planning and development to minimize adverse impacts on groundwater quality.

Another method called EPIK has been developed to define protection zones for karst aquifers in Switzerland (Doerfliger and Zwahlen 1995). Groundwater vulnerability maps based on this method can provide guidance for establishing groundwater protection zones within karst catchment areas. The EPIK vulnerability rating is based on four hydrogeological parameters:

1. E: The presence and character of the epikarst conditions,
2. P: The presence and character of the protective soil cover,
3. I: The rate of infiltration conditions and
4. K: The level of development of the karst network including the extent and interconnectability of the conduit system.

These parameters are used to assess specific aspects of the karst flow regime. The EPIK method is implemented in three stages. A semi-quantitative evaluation and field mapping of the four parameters is completed. Calculation of a protection index is completed by combining and weighting of the values of the four parameters for each unit area in the catchment. The cartographic representation is developed of the distribution of the protection index for the entire catchment.

24.3.4 Advantages and Limitations of Regional Methods

All of these regional risk assessments utilize a group of key variables that are evaluated individually then correlated spatially in order to assign a level of risk or vulnerability. These regional risk assessments have been used effectively for regional planning purposes, for insurance actuaries and even groundwater resource protection. In some cases, regional data can be used to improve the design of critical structures and infrastructure and determine set back distances from sinkholes, protected water bodies or wellfields. Regional data, risk maps or vulnerability maps can also be useful as a starting point for a site-specific risk assessment.

24.4 Site-Specific Risk Assessment

Risk assessment at the regional scale primarily utilizes historical data to predict future occurrences. A risk assessment for subsidence or collapse at the site-specific scale requires that more details be considered which impact the occurrence of subsidence or collapse. Therefore, when a site-specific risk assessment for subsidence or collapse is required, it should be an integral part of the site characterization process.

The risk assessment strategy summarized by Benson et al. (2003) is based upon having completed a site characterization and having a good understanding of site conditions along with a solid base of supporting data. However, a risk assessment is ultimately often very subjective and largely based upon professional experience. The authors have successfully applied the strategy for site-specific risk assessment for the past few decades in areas of collapse, subsidence and paleokarst (Benson et al. 2002, 2003; Benson 2001; Benson and Hatheway 2001). It has also been used to assess mine subsidence and collapse (Hatheway et al. 2000).

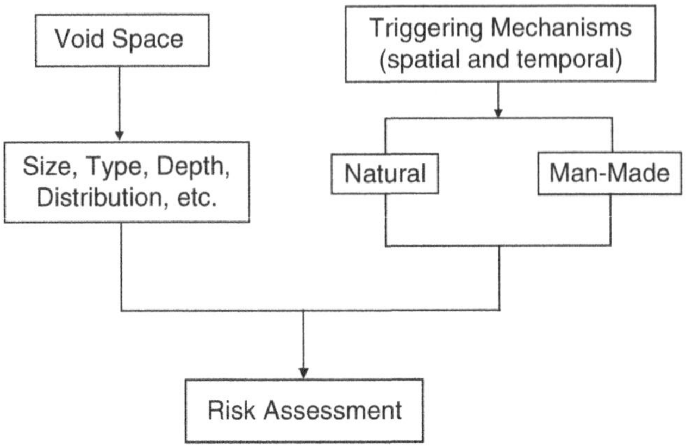

Fig. 24.1 Key factors in determining risk of subsidence or collapse

In concept you must have two conditions in order to have subsidence or collapse potential, a void space for material to move into and a triggering mechanism to precipitate the movement of the material (Fig. 24.1). The natural dissolution of limestone is a slow process, occurring at a rate of about 1–3 cm/1,000 years. As such, further natural dissolution of rock by itself is not generally of concern regarding the life of engineered structures unless acid spills or leaks are present. The concern then lies with the existing void space within the rock. Once the void space is identified, the possible triggering mechanisms must be identified. The triggering mechanisms can be natural, man induced or a combination of both. Without these two conditions (a void space and a triggering mechanism), subsidence or collapse would not necessarily be possible.

The strategy is summarized in the flow diagram (Fig. 24.2). A strong base of site-specific data must support the development of a risk assessment and begins with a site characterization. Therefore, with the completion of a site characterization the basic geologic and hydrologic data should be available to support the risk assessment.

From the site characterization effort we should have identified the types of sinkholes present (Fig. 3.1), the level of maturity at the site (Fig. 4.1) and the size or sizes of the sinkholes that might develop at the site (Fig. 10.3). This information should have been refined to be very site-specific. We should have the information in order to determine where subsidence or collapse may occur.

- The presence of geologic and hydrologic conditions at a site, which may be susceptible to subsidence or collapse. Where is the void space, how big is it and could it accept large quantities of materials? See Chap. 10 for the conceptual models of cover collapse sinkholes.

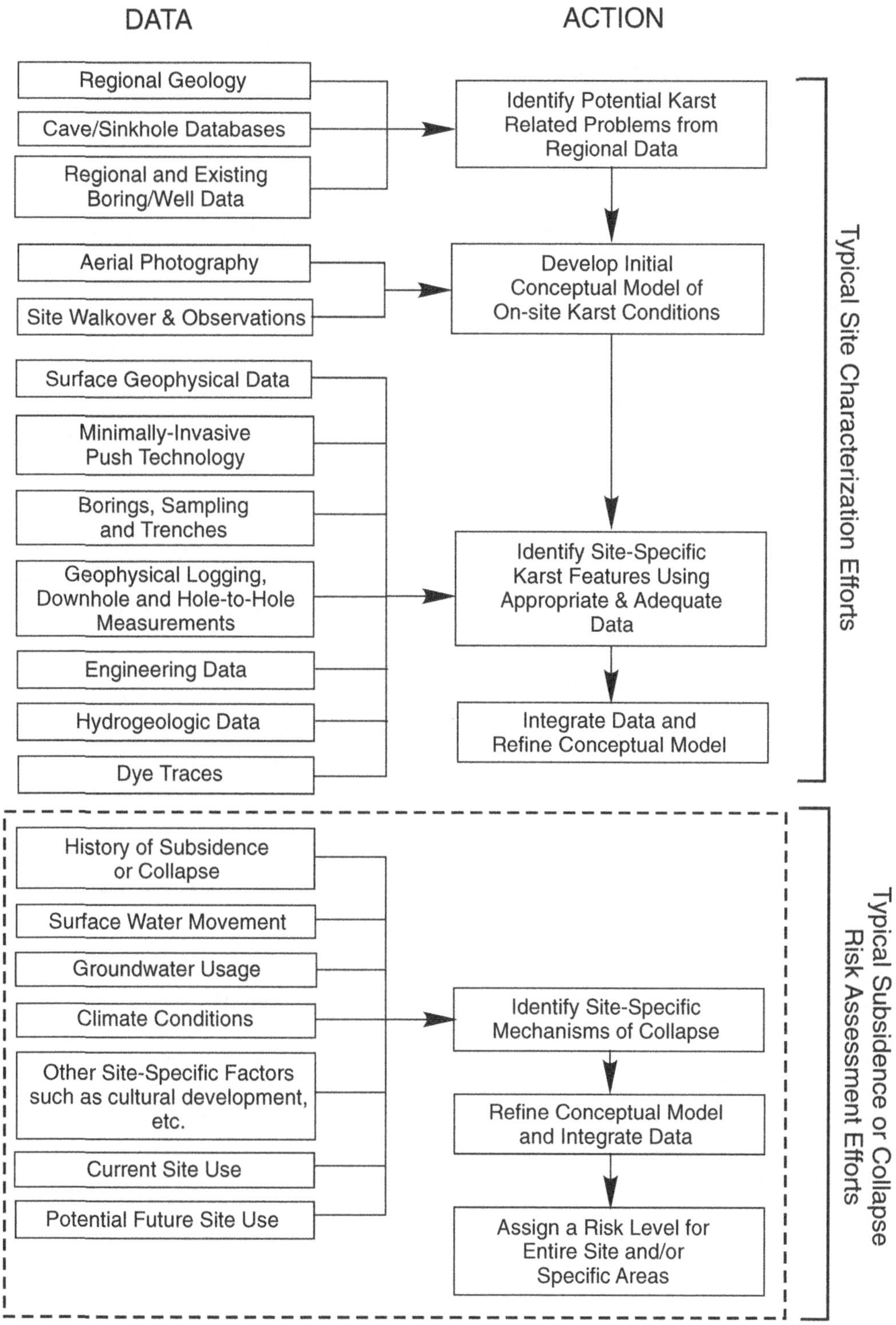

Fig. 24.2 A strategy for risk assessment in karst

- The spatial and temporal conditions at a site, which are likely to trigger subsidence or collapse (with a focus upon surface water flow and associated hydrologic conditions).

However, additional site-specific data will commonly be necessary to provide the additional details regarding the triggering mechanisms that may cause subsidence or collapse. These issues may not have been a part of the site characterization and are critical to a complete assessment of risk. The completeness and accuracy of a risk assessment is a function of our ability to identify and understand all of the risk factors that may cause subsidence or collapse at a particular site. This may include current and potential future use of the site such as development, construction, remediation, etc. that may change conditions and trigger subsidence or collapse.

Sinkholes do occur naturally without any man-made triggering mechanisms. However, in culturally developed areas, it is usually the interaction of man-made changes in site conditions that increase risk and trigger subsidence or collapse (see Chap. 8 for a discussion of triggering mechanisms). Conditions which are likely to trigger subsidence or collapse include areas of concentrated surface water runoff (large roof areas, paved parking lots, storm water basins, etc.), heavy rainfall or drought conditions, drilling, construction, and blasting, etc. or even broken utility lines (sewers or water lines). In some cases the triggering factors can be compounded, where two or more factors may interact to accelerate collapse. For example, heavy rains after a period of drought can often trigger soil raveling and ultimate collapse. It is the spatial distribution, interaction and correlation of all these factors, which will allow us to identify an area as having a higher or lower risk.

Assessment of risk should be relatively straight forward, in concept, if appropriate and adequate data has been acquired during our site characterization. This would include identifying the site-specific potential for weak rock or void spaces and triggering mechanisms. Then and only then can a realistic estimate of subsidence or collapse risk can be made for site-specific conditions. While the concept is simple, its execution requires extensive experience and extreme attention to details.

Risk assessments are commonly based upon subjective interpretations. The dilemma faced in subjective risk assessment is that there are often bold statements made (opinions) promoting either low or high risk with a minimum amount of site-specific data to back them up. In that case, the level of confidence in the risk assessment is more or less based upon the qualifications of the person or persons providing the risk assessment. Therefore, a risk analysis must clearly identify all interpretations and assumptions along with supporting data upon which it is based. This supporting data minimizes concerns of professional integrity and increases confidence levels. However, it should also be recognized that while the risk of sinkholes may be minimized by a proper site characterization along with management of triggering mechanisms, it is not likely to be totally eliminated.

Subjective risk predictions for subsidence or collapse must be made inherently general and limited, using terms such as:

- Very High; surface subsidence or collapse is underway or is eminent
- High: surface subsidence or collapse is actively occurring or is very likely to occur;
- Medium: surface subsidence or collapse may occur if further aggravated by on-site mechanisms of collapse;
- Low: surface subsidence or collapse is not likely to occur over lifetime of project.
- No Risk

Additional examples of risk assessment are included with each of the three case histories in Part III. There will be situations where some issues may not be clearly resolved by data. In these cases, subjective judgment and opinions will be supported by limited data. The examples provided within each of the case histories include characterization of paleocollapse features with a strong set of data. However, the assessment of risk of future subsidence or collapse related to these deep paleocollapse are supported by more limited data and subjective opinion.

Attempts to further quantify, model risks or predict time of collapse are generally inappropriate since detailed temporal data are commonly unavailable. However, if a site is properly instrumented and monitored, warnings of collapse may be reasonably estimated based upon a variety of data. Subtle precursors can often provide warnings of an eminent collapse. Examples of sites with monitoring instrumentation include:

- South Africa's use of telescoping bench marks
- Time domain reflectometry monitoring a road in Florida
- Tilt meters in boreholes monitoring a brine cavern in Carlsbad, New Mexico

Further discussion of these monitoring efforts is provided in Chap. 20, "Engineering Measurements and Monitoring".

24.4.1 Examples of Site-Specific Risk Assessment

Sowers (1996) offered a strategy for reducing risk for the foundation design of critical structures (hospitals, police and fire departments) in karst. "It may be prudent to design foundations to tolerate a large dome collapse, which is assumed to occur at any random location beneath the structure. Since, even in sinkhole prone areas simultaneous, closely spaced dropouts are rare. Therefore, a design for a one occurrence at a time at a random location beneath the structure is a reasonable presumption". This strategy is only effective if the nature of the event can be estimated in terms of its size so that an adequate design can be made.

The following examples illustrate approaches to site-specific risk assessment. The factors used to assess the risk vary and are site-specific.

24.4.1.1 Highway I-70 Near Frederick, Maryland

A risk assessment was made along a section of Highway I-70 near Frederick, Maryland (Zhou et al. 2003). The spatial distribution and density of 138 sinkholes along 8 km^2 of roadway were used to develop a risk assessment model. More than 70 % of the sinkholes were within 300 m of the highway. Eight factors were considered including:

- Topography,
- Proximity to topographic depressions,
- Proximity to existing sinkholes,
- Rock type,
- Soil type,
- The presence of geophysical anomalies,
- Proximity to geologic structure and
- Thickness of overburden.

A scoring system was then used to predict the occurrence of new sinkholes along the highway itself. A relative rating of 1–7 was developed for every 30 m segment of the highway with higher values indicating a relatively higher risk of new sinkholes occurring. Where multiple variables indicated risk, the risk was higher. Where fewer variables indicated risk, the risk was lower.

During an early assessment of this site by the authors in 1996, sinkhole activity had been observed along highway I-70 and to the north about 150 m (Fig. 24.3). As part of our investigation, a microgravity survey had been carried out along the highway. The microgravity data clearly indicated the boundary between the Lime Kiln and the Grove Formations (up to 800 microgals) due to a density contrast between the formations (Fig. 24.4a). In addition, superimposed upon the gravity data were two localized microgravity anomalies of 300–500 microgals. These two anomalies indicated significant mass deficits in the limestone along the highway, likely due to the presence of cavities and caves. Therefore, these areas have a high level of risk for sinkholes to occur. The areas of mass deficit were modeled and superimposed on the geologic cross section (Fig. 24.4b). The westernmost anomaly was estimated to be a lower density zone of about 180 m wide and 75 m deep. The easternmost anomaly was estimated to be a lower density zone of about 450 m wide and 30–45 m deep.

One boring was drilled within each of these anomalies along with a corresponding borings within adjacent background conditions. These borings confirmed our interpretation. The borings within the anomalies showed a distinct increase in the thickness of highly weathered and fractured rock as well as in the number of voids (Fig. 24.4b). A few months after completion of field work a large 30 m diameter and 18 m deep sinkhole occurred south of the roadway within the western most microgravity anomaly (Fig. 24.3). This is an example of how site-specific data can be used to improve site-specific risk assessment (Benson et al. 1997).

24.4.1.2 Spatial Distribution of Sinkholes

Both Gao and Alexander (2003) and Zhou et al. (2003) have commented upon the concentration or proximity of sinkholes. Zhou et al. (2003) suggested that new sinkholes tend to develop in the vicinity of previous ones because geologic

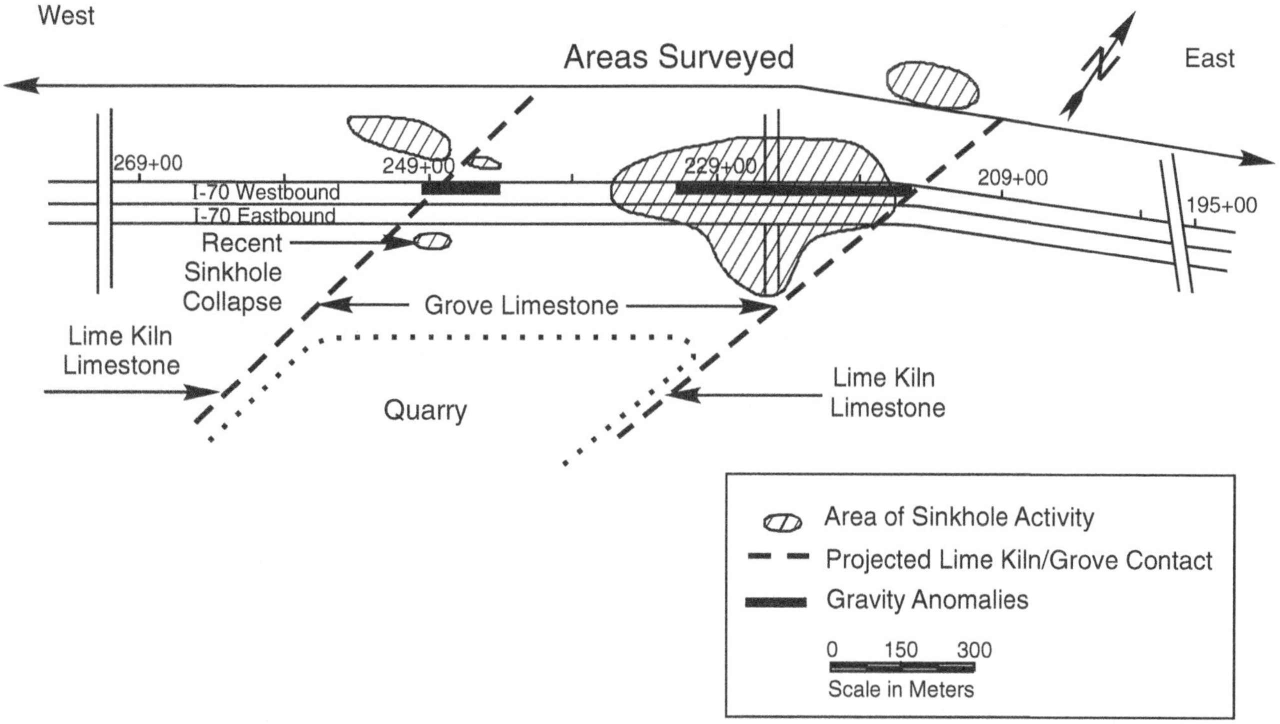

Fig. 24.3 Map of study area along I-70 in Maryland

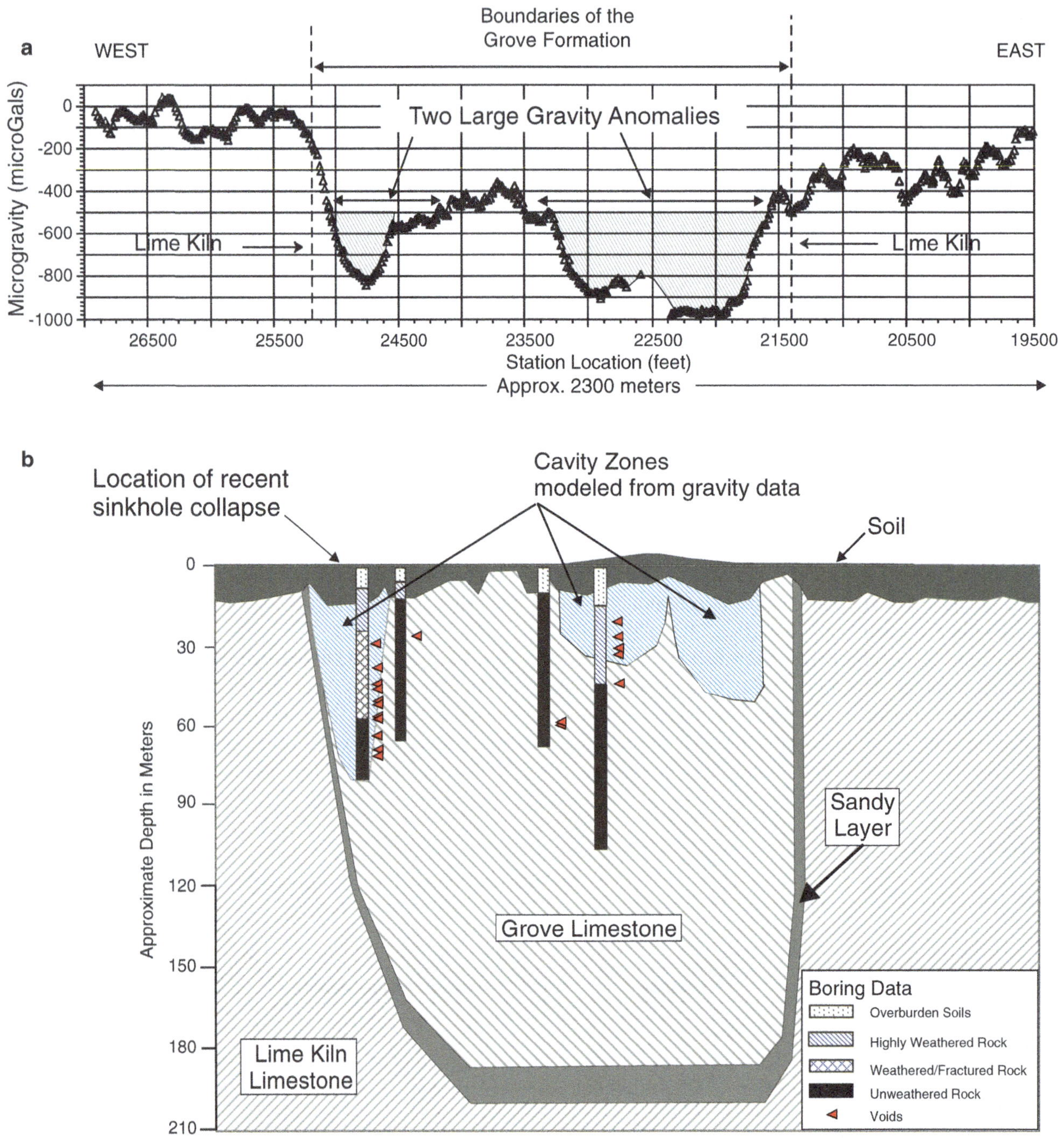

Fig. 24.4 Micrograyity data, geologic model and boring results along I-70 in Maryland (**a**) Microgravity data acquired along I-70 (**b**) Modeled microgravity data and boring results

and hydrologic conditions are favorable for sinkhole collapse in an area. Risk assessments have often looked at the spatial proximity of sinkholes and have deduced that there is a higher risk in closer proximity to existing sinkhole activity. Analysis of the sinkhole collapse along Interstate-70 near Fredrick, Maryland indicated that the radius of influence of a sinkhole was approximately 30 m at that site (Zhou et al. 2003). This indicated that the risk of additional sinkholes is very high within 30 m of an existing sinkhole.

This example illustrates that there is a spatial relationship to sinkhole density. However, looking at the spatial distribution of sinkholes without consideration of geologic conditions such as the depth of origin and the nature of the void space within the rock does not provide a complete picture.

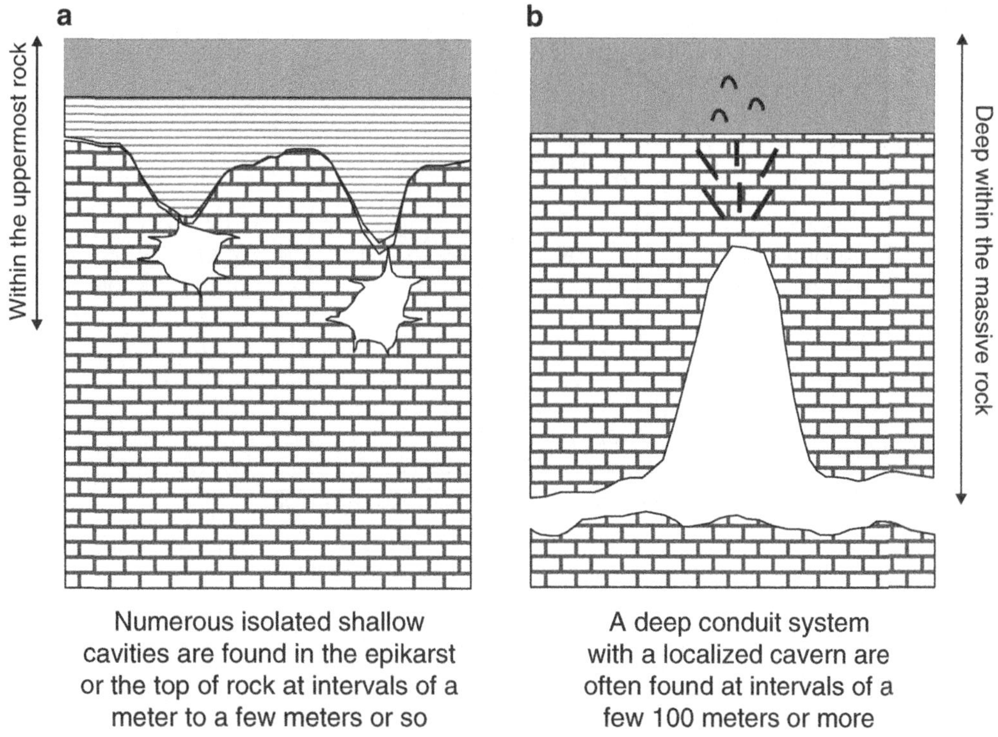

Fig. 24.5 Conceptual models of sinkhole development from epikarst or top of shallow rock are more pervasive (**a**), while large caverns formed deeper within the massive rock are more widely spaced (**b**)

This concept uses two of the five simplified conceptual models presented in Chap. 10. For example, a simplified conceptual model of cover collapse sinkhole development that incorporated the sinkhole size and spacing includes:

- Smaller sinkholes tend to originate from the epikarst or top of rock where many smaller isolated voids typically occur (Fig. 24.5a). This will result in the spacing between sinkholes to be relatively close (on the order of a few meters to a few tens of meters) and their locations are more random.
- Very large sinkholes are generally associated with the presence of a large cavern. A cavern will typically develop at weak points such as at the intersection of major fracture systems or at a critical structural feature (Fig. 24.5b). These sinkholes will typically occur at larger intervals of on the order of a few 100 m or so and result in very large sinkholes.

These conceptual models (Fig. 24.5) are very simplified models of two major factors (depth to rock and nature of the void space), which control the size and spatial distribution of sinkholes. This information can play a critical role in assessing risk. Such data may provide technical background for determination of setbacks from very large sinkholes or for engineering measures to improve the site conditions (Fig. 24.6).

24.4.1.3 EPA Superfund Site

An extensive site characterization was completed at the Superfund site in Tarpon Springs, Florida (Part III, Chap. 27). This site was a former phosphate ore processing area. The site is in an area of historic and current sinkhole activity. As part of this project, a risk assessment was completed at the site. The risk assessment was to address the site in general but also specifically address the planned remediation and its impact on potential sinkhole development.

The site characterization was completed at an unusual level of detail with shallow geologic data density approaching 100 %. As a result we knew the shallow geology and hydrology at the site extremely well. This level of data also provided a high level of confidence in which to base a risk assessment.

We were able to determine that there were three zones in which dissolution had occurred and where void space or caves had developed (Fig. 24.7).

- Zone 1: shallow, small voids occur within the top of the Tampa Limestone just below a semi-confining layer, generally less than 7.5 m deep.
- Zone 2: intermediate voids or conduits at the unconformity between the Tampa Limestone and the Suwannee Limestone at a depth of about 18 m.
- Zone 3: deeper conduits or caverns had developed at a depth of about 55–90 m at the lower third to half of the

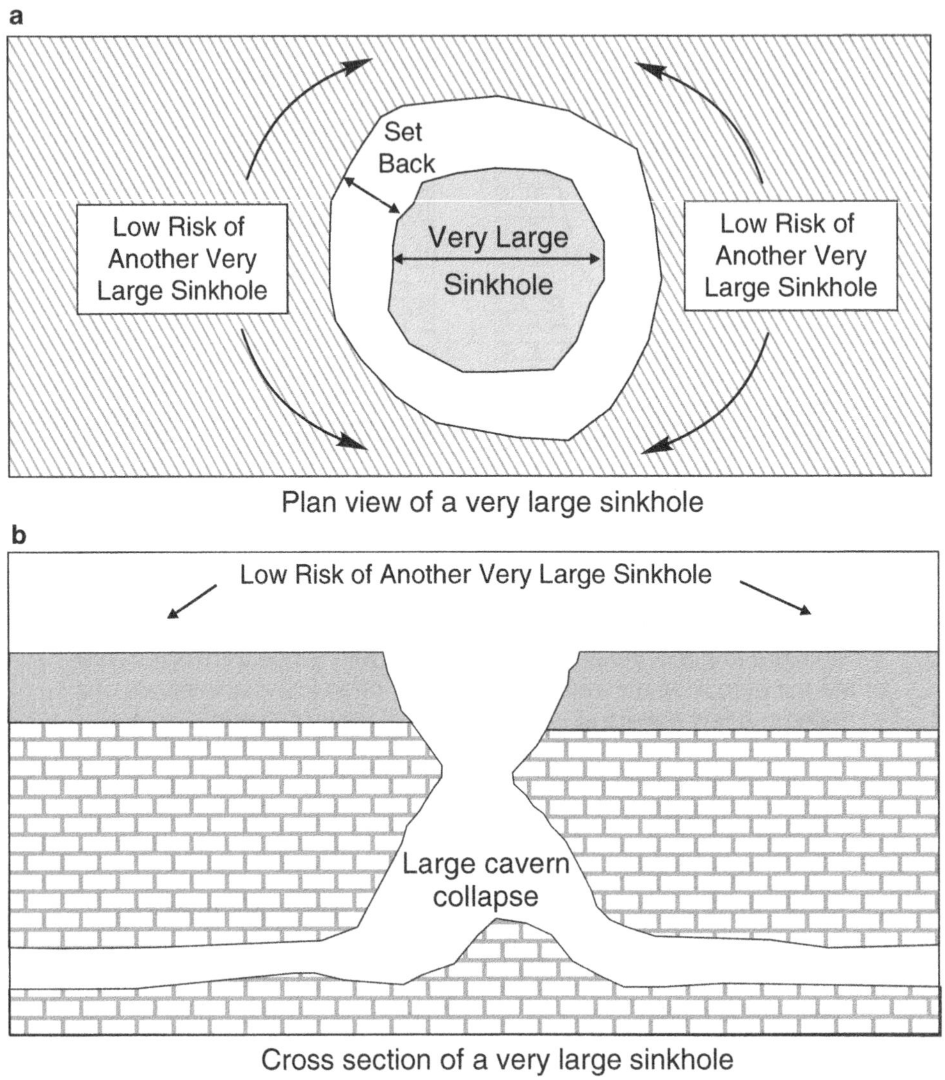

Fig. 24.6 Since large sinkholes are typically controlled by major fracture system structure within rock they occur further apart, therefore risk near the large sinkhole assuming a reasonable set back may be relatively low (**a**) Plan view of a very large sinkhole (**b**) Cross section of a very large sinkhole

Suwannee Limestone (this was the probable source of the large paleocollapse features on-site).

Groundwater levels at the site are controlled by the adjacent river and the Gulf of Mexico and remain relatively constant. There is only a small difference in head between the surficial and Floridan aquifers. Tidal changes are the most significant factor in changing groundwater levels and primarily influence the Floridan Aquifer. There is no excessive pumping of groundwater from either the surficial or Floridan aquifer within the property. The risk of sinkhole collapse at this site is very low, if the site were left by itself, unaffected by further activity.

Zone 1

All remediation efforts were expected to be fairly shallow, focused in the sands of the surficial aquifer, above Zone 1. No deeper excavation, drilling, etc. would or should be taking place in or below the semi-confining layer that would impact sinkhole development. However, triggering mechanisms for sinkhole development associated with remediation and construction include:

- Breaching of the semi-confining layer by excavation or drilling
- Creating an artificially higher head of water in the surficial aquifer which might tend to breach the semi-confining layer

	Zone 1 Shallow, Small Collapse	Zone 2 Deeper, Larger Collapse	Zone 3 Re-activation of Paleocollapse
Risk of Occurrence	Very Low, if SCL is not breached	Virtually Zero	Virtually Zero

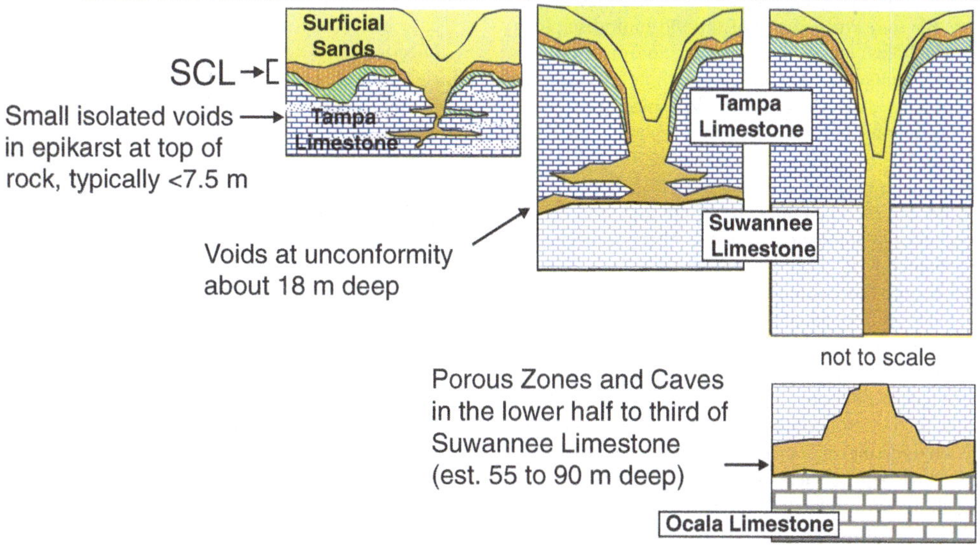

Fig. 24.7 Risk assessment at EPA Superfund site showing void development at three depths and their associated risk of sinkhole development or reactivation

- Over pressuring the Floridan aquifer by excessive drilling pressures when drilling into the Tampa Limestone, which might breach the semi-confining layer from below.

Any of these actions could cause a sinkhole to develop by breaching the semi-confining layer and allowing sands to ravel into the voids within the upper Tampa Limestone. However, this worse-case scenario would likely create a small (3–6 m in diameter) localized sinkhole.

Zone 2

The probability of development of small shallow cover subsidence sinkholes originating from 18 m deep (Zone 2) is considered to be virtually zero.

Zone 3

Approximately 8 % of the site (in two areas) contained paleocollapse features, which originated from a depth of 55–90 m bls (the dissolution at these depths occurred at lower sea level stands). The depth of origin is well beyond the depths of any remediation efforts. These two areas of the site were mapped in detail and were dated to 48,000 years ago using C-14 methods. These features were considered stable since there was no surface expression (topographic low), 40 years of facility operations occurring over or near them including loading from material piles, vibration due to trains and local traffic of heavy trucks. As a result, we concluded that the risk of subsidence or collapse due to reactivation of these features at their depths of origin is considered virtually zero.

References

Aller L et al (1987) DRASTIC: a standardized system for evaluating ground water pollution potential using hydrogeologic settings. US Environmental Protection Agency, Robert S Kerr Environmental Research Laboratory, Office of Research and Development EPA/600/2-87/035

Arthur J et al (2005) Florida aquifer vulnerability assessment (FAVA): contamination potential of Florida's principal aquifer systems. Florida Geological Survey Report for FDEP 21 Mar

Beck BF (1991) On calculating the risk of sinkhole collapse. In: Kastning EH, Kastning KM (eds) Proceedings of the Appalachian karst symposium. National Speleological Society, Radford, pp 231–236

Benson RC (2001) Strategies of site characterization and risk management presented to National Research Council, Committee on coal waste impoundment failures subcommittee on impoundment site characterization, Washington, DC

Benson RC, Hatheway AW (2001) A systematic approach for evaluating subsidence risk. AEG/AIPG Annual Meeting St. Louis

Benson RC, Sharma D (1994) The pitfalls in site characterization and how to avoid them. Short course presented by Environmental Education Enterprises, Salt Lake City

Benson RC, Sharma D (1995) Pitfalls of site characterization and how to avoid them. Short course presented by Environmental Education Enterprises, Charlotte, November

Benson RC et al (1997) Assessment and prediction of karst conditions. Presented at the Federal Highway Administration Southeastern Regional Meeting, 27–29 Oct 1997, Chattanooga

Benson RC et al (2002) Assessing the risk of karst subsidence AEG/AIPG Joint Annual Meeting Reno, pp 22–29

Benson RC et al (2003) Assessing the risk of karst subsidence and collapse. In: Barry BF (ed) Proceedings of the 9th multidisciplinary conference on sinkholes and the engineering and environmental impacts of karst, Huntsville, ASCE Geo-Institute Geotechnical Special Publication No. 122, ASCE Reston, Virginia, pp 31–39

Bolt BA et al (1977) Geologic hazards, 2nd edn. Springer, New York

Doerfliger N, Zwahlen F (1995) EPIK: a new method for outlining of protection areas in karst environment. In: Gunay G, Johnson I (eds) Proceedings of the 5th International symposium and field seminar on karst waters and environmental impacts, Antalya, September. Balkema, Rotterdam, pp 117–123

Gao Y (2002) Karst feature distribution in southern Minnesota: extending GIS-based database for spatial analysis and resource management. Ph.D. thesis, University of Minnesota

Gao Y, Alexander EC Jr (2003) A mathematical model for a map of relative sinkhole risk in Fillmore County, Minnesota. In: Barry BF (ed) Proceedings of the 9th multidisciplinary conference on sinkholes and the engineering and environmental impacts of karst, Huntsville, Alabama. ASCE Geo-Institute Geotechnical Special Publication No. 122, ASCE Reston, Virginia, pp 439–449

Haneberg WC (2000) Deterministic and probabilistic approaches to geologic hazard assessment. Environ Eng Geosci 6(3):209–226

Hatheway AW et al (2000) Geotechnical risk assessment of Route 169 over the Briarcliff Mine, Kansas City, Clay County, Missouri. Consultant Report to Kansas Dept. of Transportation

Hunt RE (2007) Geologic hazards, a field guide for geotechnical engineers. CRC Press, New York

LeGrand HE (1983) A standardized system for evaluating waste-disposal sites. National Water Well Association, Worthington

Ross SS (1984) Construction disasters, design failures, causes, and prevention, Engineering news record series. McGraw-Hill, Columbus

Scheidegger AE (1975) Physical aspects of natural catastrophes. Elsevier, New York

Shuirman G, Slosson JE (1992) Forensic engineering, environmental case histories for civil engineers and geologists. Academic, San Diego

Sinclair WC, Stewart JW (1985) Sinkhole type, development and distribution in Florida, Florida Geologic Survey, Map Series #110, Tallahassee

Sowers GF (1996) Building on sinkholes, design and construction of foundations in karst terrain. ASCE Press, Reston (out of print)

Vick SG (2002) Degrees of belief subjective probability and engineering judgment. ASCE Press, Reston

Waltham AC et al (2005) Sinkholes and subsidence, karst and cavernous rocks in engineering and construction. Springer-Praxis, Chichester

Wilson WL (1995) Sinkhole and buried sinkhole densities and new sinkhole frequencies in karst of northwest Peninsular Florida. In: Beck BF (ed) Proceedings of the 5th multidisciplinary conference on sinkholes and the engineering and environmental impacts of karst: karst geohazards, Gatlinburg, Tennessee. Balkema, Rotterdam, pp 79–91

Zhou W et al (2003) Sinkhole risk assessment along Highway I-70 near Frederick, Maryland. In: Barry BF (ed) Proceedings of the 9th multidisciplinary conference on sinkholes and the engineering and environmental impacts of karst. ASCE Geo-Institute Geotechnical Special Publication No. 122, ASCE Reston, Virginia, pp 591–601

Part III

Case Histories

Most site characterizations are carried out over a long period of time with a variety of objectives using different contractors, consultants and even owners. Even with the best of intentions on the part of owners and consultants it is difficult to complete such a site characterization in a meaningful way and to achieve an accurate understanding of site conditions. Such an approach results in a marginal site characterization effort at best.

The three case histories included here are those in which the authors have had complete cooperation of the owner throughout the site characterizations process. These case histories represent a range of conditions including the total time of the project, the strategy used, and the methods employed. Most important to the project was the fact that the senior staff was intimately involved from the beginning to the end of the project. This continuity has enabled us to achieve a high level of confidence in the site characterization effort for each of these three projects.

The Development of a Landfill over an Abandoned Limestone Mine

25

Abstract

The first case history deals with the expansion of a landfill over a 56 ha abandoned room and pillar limestone mine in Kansas City area. A massive 7 ha mine-roof collapse had occurred in the mine and fissures were discovered on the surface. The State of Kansas required that the mine be backfilled to avoid any surface subsidence. This project did not follow our standard site characterization strategy. Work at this site began with the single task of using a borehole video camera to evaluate the extent of mine collapse and the effectiveness of fly ash being used to backfill the mine. The project slowly evolved into a substantial site characterization effort that included an assessment of mine collapse, monitoring mine-roof collapse conditions over time, a groundwater monitoring plan for the landfill and a subsidence risk assessment along with many other tasks. The project was carried out over a period of 7 years from 1987 to 1993 with minor efforts in mine backfilling continuing through 2003. About half of the project effort was completed underground in the mine.

25.1 Background

There has been more than a century of extensive underground mining of the limestone within Greater Kansas City (Kansas and Missouri) area. The Bethany Falls Limestone provides an excellent limestone and is typically a one level room-and-pillar mine within the relatively flat-lying uniform geology. Kansas City limestone mines are almost always dry and stable giving the city the distinction of being number one in the world in terms of human use and occupancy of underground space after mining. More than 200 businesses are located underground occupying more than 484 ha as of 1983 (Hasan et al. 1988). In some cases, commercial buildings are developed over abandoned mines. State agencies now require that mines be backfilled prior to developing facilities over them to avoid possible subsidence.

The Tobin limestone mine is located in Wyandotte County, Kansas City, Kansas, west of highway 635 and north of interstate I-70 just north of the Kansas River (Fig. 25.1). The mine is overlain by a stratigraphic sequence of about 52 m of alternating shale and limestone (Pennsylvanian age) (Fig. 25.2). Up to 18 m of pleistocene-aged loess overlays much of the site. The mine operated from 1960 to 1980. Early mining began at the portal (Fig. 25.1) and extended eastward about 425 m, then mining extended west of the portal for about 730 m. The Bethany Falls limestone was mined to height of 4.2 m. The mine occupies an area of approximately 56 ha of which the westernmost 34 ha would underlie the proposed landfill (Fig. 25.1).

A major mine-roof collapse occurred in the western portion of the mine in the early 1970s. This collapse encompassed an area of up to 7 ha, which is referred to as the Central Collapse Area (CCA) (Fig. 25.3). The collapse is reported to have noticeably shook the ground, broke windows in the trucks within the mine and was recorded on seismographs in Lawrence, Kansas some 40 km away. After this large mine-roof failure of the CCA, inspection revealed a number of fissures within the loess at the surface (Fig. 25.3). The fissures occurred in a circular pattern within in the thick loess sediment and were centered over the CCA in the mine. The conclusion by the owner and a local consultant was that the mine-roof failure had resulted in the surface fissures. This was a reasonable conclusion based upon limited information available at the time.

R.C. Benson, L.B. Yuhr, *Site Characterization in Karst and Pseudokarst Terraines*,
DOI 10.1007/978-94-017-9924-9_25

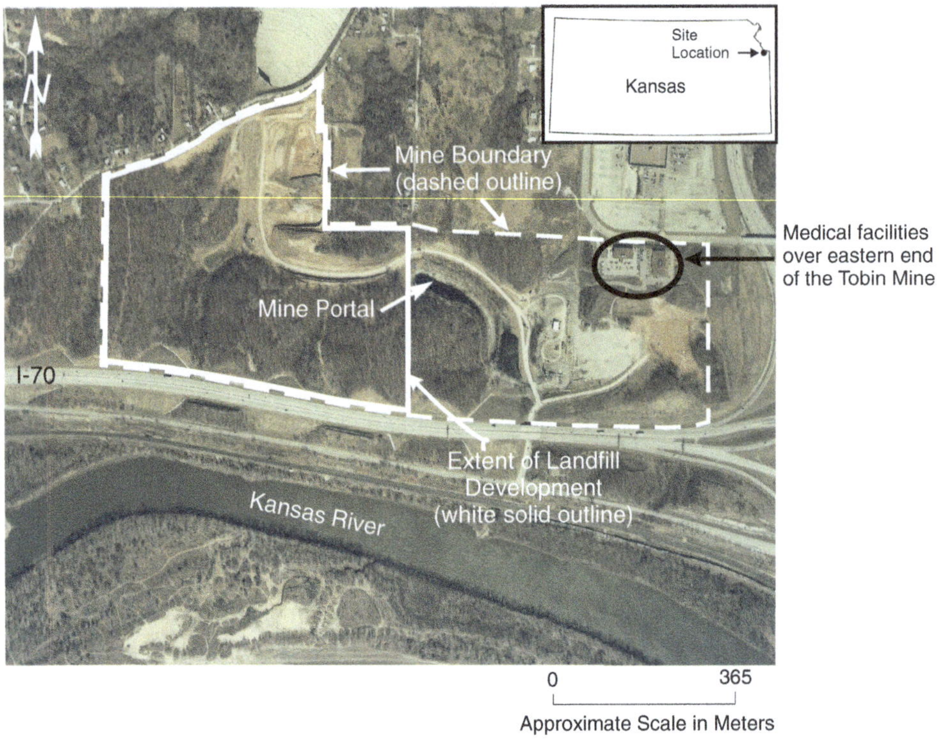

Fig. 25.1 A 1987 aerial photo shows the site just north of the Kansas River and the initial landfill can be seen in the northern most portion of the mine

Prior to the collapse of the CCA, pillar size and spacing were not uniform and distances were determined by pacing. After the CCA occurred, mining operations west of the CCA began using a uniform pillar layout of pillars of about 7.6 by 7.6 m on 18 m centers (Nicholson 1988), an extraction ratio of about 82 %.

As the mining ended, the owner at the time had started a landfill over the mine. As a requirement of the Kansas State permitting process, the mine had to be backfilled to prevent possible surface subsidence. Initially, fly-ash from local power companies was being used as backfill material. The efforts included dry fly-ash, cast in place columns of fly ash and a mixture of fly ash that was mixed in a cement truck and dumped into the mine via 20 cm uncased boreholes.

In July of 1986, an engineering firm's mining staff made a 1-day inspection of mine conditions. It was concluded, "the mine workings and overlaying rock are stable and not likely to cause significant movement or fracturing at the ground surface" (Golder Assoc. 1986). They evaluated the pillar stability in three ways:

- Estimated overburden stress of the rock and proposed landfill.
- Pillar strength based upon compressive strength of 82.7–137.9 MPa for the Bethany Falls Limestone
- Possible pillar punching into the weaker Hushpuckney Shale below the mine floor.

They concluded that failure of the pillars would be very unlikely even at 90 % extraction ratios.

In March of 1987, a localized but significant roof-fall of about 0.12 ha occurred about 60 m from the west portal entrance. The owners considered the mine unsafe and further access to the mine was denied based upon recommendations from their attorneys. The authors were then engaged to observe general conditions within the mine and evaluate the extent of fly-ash backfill using a downhole video camera. Observations by borehole camera were a very slow, incomplete, and a tedious process.

In June 1987, the authors gained physical access into the mine and the formal site characterization process began to characterize mine conditions and assess subsidence risk. This work extended over a period of 7 years from 1987 to 1993. Minor efforts continued through 2003 to support mine backfilling efforts. As time passed other tasks associated with the project were included such as developing and installing a groundwater monitoring plan for the landfill and providing quality control monitoring of the mine backfilling process. Work included hundreds of hours in the mine and endless hours of observations and measurements at the surface.

Figure 25.4 summarizes the main tasks for site characterization and monitoring associated with this project. Many of the tasks were carried out over time and interacted with one another.

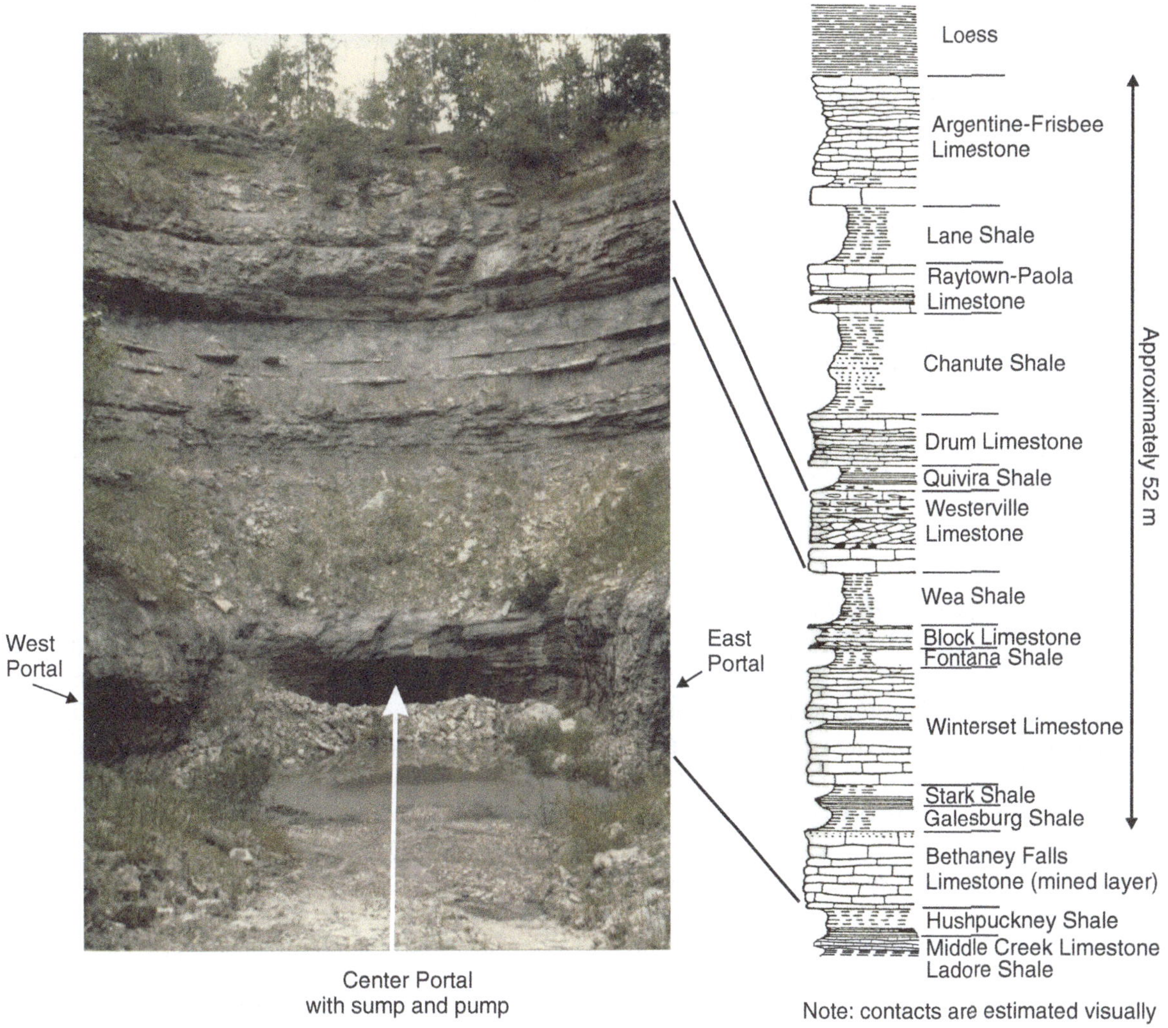

Fig. 25.2 The geologic section along with a photo of the mine portal entrances

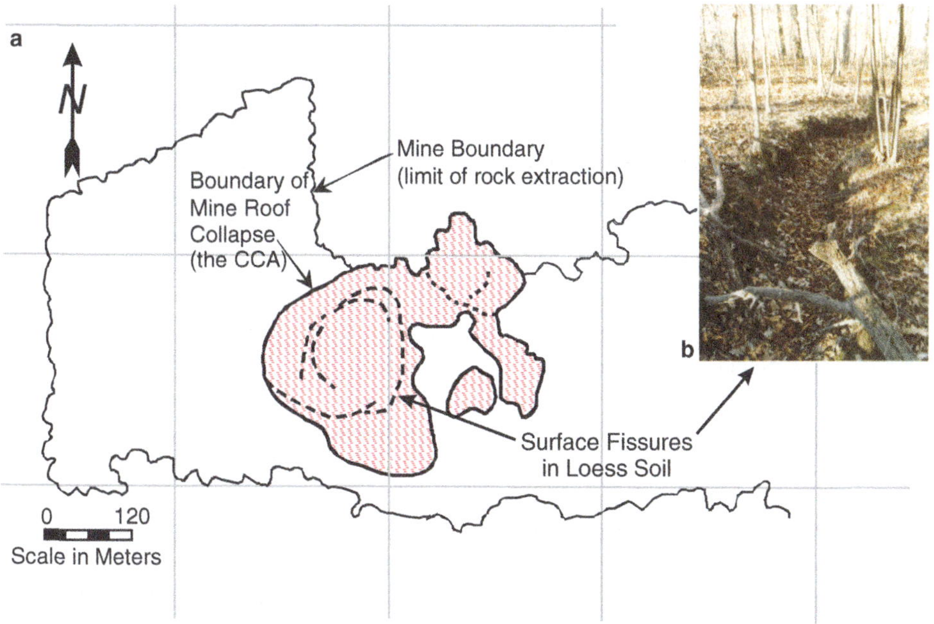

Fig. 25.3 The mine boundary along with the extent of the central collapse area (*CCA*) and location of the surface fissures shown in *dashed lines* (**a**). The photo shows a typical surface fissure (**b**)

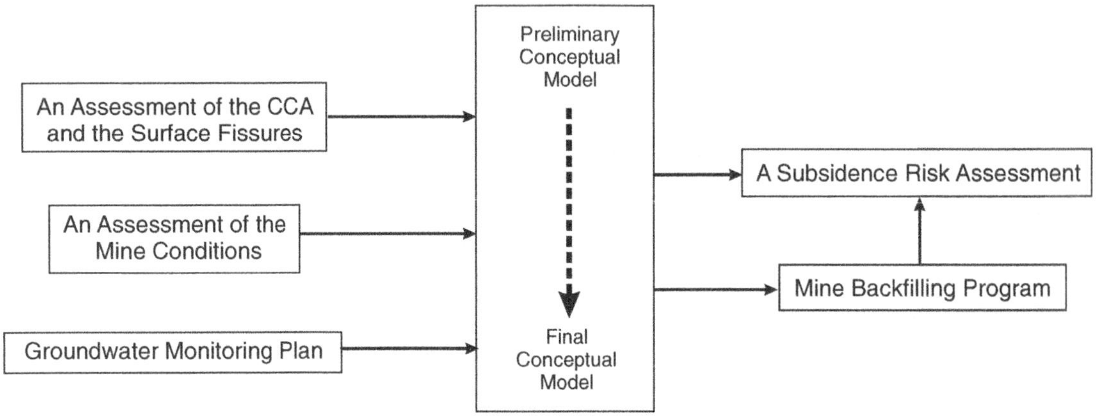

Fig. 25.4 The work tasks at this site included a wide range of activities over a 7 year period. These tasks supported the development of a conceptual model of site conditions along with a subsidence risk assessment

25.2 An Assessment of the CCA and the Surface Fissures

The initial site characterization effort focused upon developing an understanding of the mine-roof collapse and the relationship between the CCA and the surface fissures. The tasks completed to meet these objectives included:

- A review of regional and local geology and geomorphology
- Mapping of the surface fissures and selected trenching
- Developing a preliminary conceptual model of the CCA and their relation to the surface fissures
- An aerial photo analysis
- Site-specific geology (field mapping, limited drilling, and geophysical logging all new and existing boreholes)
- Confirmation of the preliminary conceptual model

25.2.1 Review of Regional and Local Geology

The geology of the area is well known and has been summarized in "Geology of Greater Kansas City" (Hasan et al. 1988), which also provides an excellent insight and overview of the limestone mines in the area along with a discussion on the commercial use of underground space. The subsurface strata are relatively flat-laying and uniform. The Pennsylvanian age bedrock overlaying the mine is made up of an alternating sequence of limestone and shale, about 52 m thick that is covered by a thick loess of up to 18 m or more (Fig. 25.2). The deeper Mississippian limestone and dolomite beds lie disconformably below the Pennsylvanian bedrock, and are over 120 m thick (Gentile 1984).

The Bethany Falls Limestone is considered the best quality limestone in the Kansas City area and is found in thickness of 4.2–6.4 m (Fig. 25.2). Two of the limestone strata above the mine are significant because of their thickness. The Winterset Limestone is about 8.8 m thick and is found about 2.4–3 m above the Bethany Falls Limestone. The Westerville Limestone is massive, about 6 m thick, and found about 18 m above the Bethany Falls Limestone (Hasan et al. 1988).

Two geologic aspects are known to impact the mine stability in the area (Hasan et al. 1988). The first is the presence of an anomalous zone of rock in the upper part of the Bethany Falls Limestone known locally as the "Rubble Zone", "Buckshot", or "Peanut Rock" by the local limestone miners. The second is the Hushpuckney Shale, which lies immediately beneath the Bethany Falls Limestone and contains sulfides which can lead to swelling and floor heave.

The "Rubble Zone" (Fig. 25.5) is a carbonate-nodular, and highly over-consolidated clay which is friable and typically 0.45–1 m thick. This zone appears to occur as channels that are much thicker at some locations and inadvertently exposed by mining operations. The rubble zone is known to deteriorate rapidly when exposed due to increased moisture, leading to localized roof failure (Hasan et al. 1988). This is the initial mechanism of most roof-failure of mines in the Kansas City area.

The Hushpuckney Shale occurs below the Bethany Falls Limestone (Fig. 25.5) and contains 5–6 % (by volume) sulfides in the form of pyrite, sphalerite and chalcopyrite. Conversion of sulfide minerals into gypsum results in a volume increase of six to eight times, generating high stresses resulting in swelling and floor heave. Noticeable heave of the floor can occur between 2 and 5 years after extraction of the Bethany Falls Limestone and total amount of heave (ranging from 2.5 to 20 cm), may occur over a 10 year period (Hasan et al. 1988).

Fig. 25.5 The two geologic conditions affecting mine stability include the rubble rock zone at the top of the Bethany Falls Limestone and the Hushpuckney Shale at the base of the Bethany Falls Limestone

25.2.2 Regional Geomorphology

The site itself is located on the northeastern flank of the Shawnee Syncline, which causes the bedrock at the site to dip to the southwest. As a result, the mine floor dips slightly to the southwest (up to a few degrees).

Gentile (1984) has discussed the extensive paleokarst that had developed in the deeper Mississippian rocks which has been recognized throughout the mid-continent. He has proposed that collapse occurring within the Mississippi limestone had propagated to the surface in the Kansas City area (Fig. 9.7a). These paleocollapse structures range in size from a few hectare or so to a couple of km^2 and include large blocks of bedrock that have moved downward a few meters along high angle, normal faults. They were formed during an earlier interval of geologic time and are covered by a loess regolith. Consequently, they are commonly unnoticed until uncovered in excavations. Observations of these paleocollapse zones have been made at road cuts in the Kansas City area by Gentile (1984) and by our site characterization team (Fig. 9.7b).

25.2.3 Mapping and Trenching of the Surface Fissures

A total of 23 surface fissures within the loess were located and surveyed. The main pattern of the surface fissures formed a set of large "concentric circles" overlaying the CCA (Fig. 25.3a). The larger set of fissures has a diameter of about 180 m. A partial set of concentric fissures was also located to the northeast. The surface fissures within the loess were 7.6–76 m long, 0.3–1.2 m wide and 0.3–1.2 m deep (Fig. 25.3b). Attention was focused upon the major set of concentric fissures that coincided with the center of the CCA.

Seven trenches up to 7.6 m long and up to 3.3 m deep were cut perpendicular to the surface fissures at different locations. Each was examined in detail and it was clear that the fissures had not occurred recently and in fact were quite old, likely occurring before the mine collapse in early 1970s. However, this was just an opinion based upon our observations and needed further verification. In addition to the fissures, there were numerous small depressions, sinkhole like in appearance.

These were quickly identified as being associated with the large open fissures within the Argentine Limestone where the loess cover was thin.

25.2.4 Developing a Preliminary Conceptual Model

The preliminary conceptual model (Fig. 25.6) for the CCA and surface fissures assumes that a paleocollapse zone originating within the deeper Mississippian rocks existed prior to mining. The mine intersected 7 ha of weakened rock due to the paleocollapse when it was extended to the west. The surface fissures were simply part of the original paleocollapse activity prior to the development of the mine. This preliminary conceptual model of the relationship between the mine collapse of the CCA and the surface fissures was developed based upon:

- Observations in the mine by a senior mining experts (Golder Associates 1986);
- Direct observations in seven trenches;
- Regional mapping and the conceptual model from Gentile (1984), and
- Observations of paleokarst collapse along road cuts noted by Gentile (1984) and subsequently observed by our site characterization team (Fig. 9.7b).

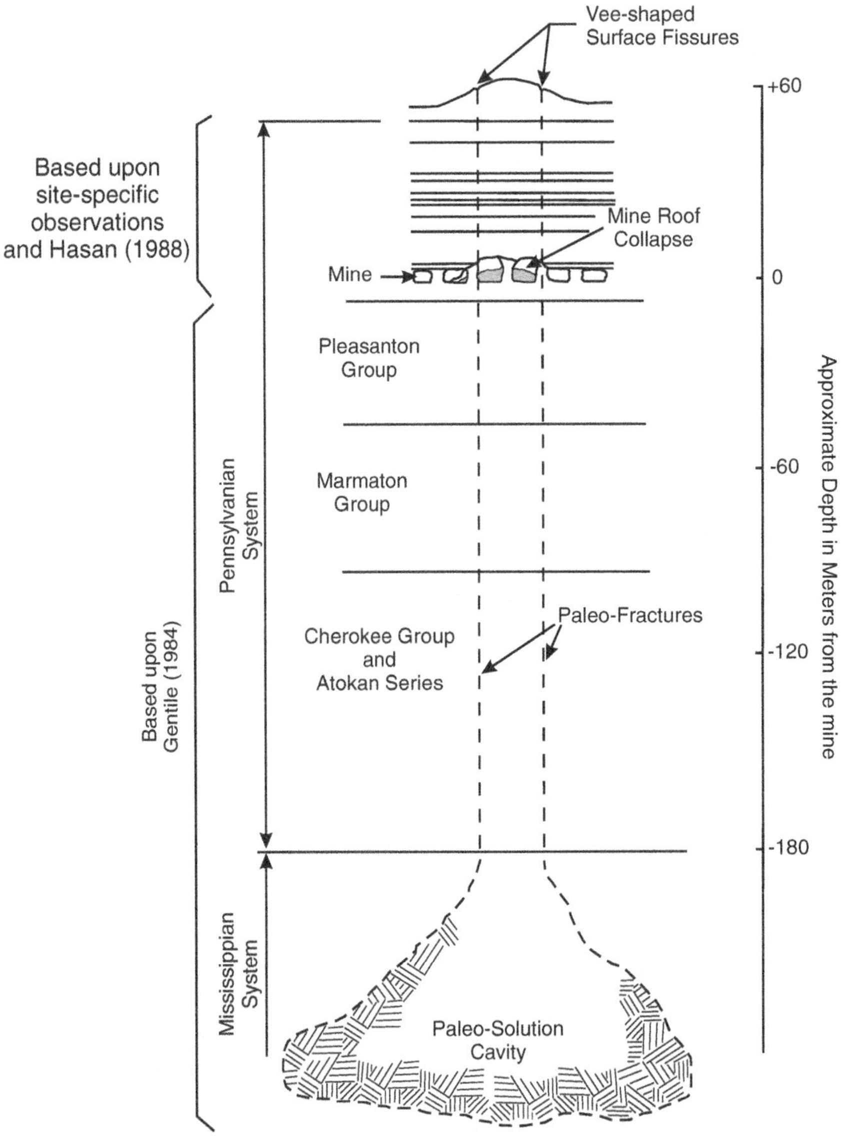

Fig. 25.6 A preliminary conceptual model illustrating the relationship between the surface fissures, the CCA and the paleocollapse within the deeper Mississippian bedrock

While we were reasonably certain of our preliminary conceptual model (Fig. 25.6) at this point it was merely an opinion and would need further verification.

25.2.5 Aerial Photo Analysis

An aerial photo analysis was carried out using both stereoscopic and monoscopic photos (Beccasio 1988). The aerial photography ranged from 1988, 1985, 1971, 1964, and 1954. This coverage provided information from pre-mining, throughout the years of mining (from 1960 to 1980) and post-mining. The objectives for the aerial photography analysis included fracture trace analysis, pre-development topography and observing the presence of surface fissures.

25.2.5.1 Fracture Traces

The 1954 air photos provided the best means for mapping fractures in the project area since there was minimal cultural development at that time. The primary fracture trend is north-northeast, which is reflected in the strong topographic and drainage alignments on-site. A secondary northwest fracture trend was also observed (Beccasio 1988).

25.2.5.2 Topography

While detailed topographic maps of the site had been developed, there had been extensive excavations and modification of surface topography as a result of on-going landfill development. The combination of a USGS topographic map and older aerial photography (1954) were used to evaluate original surface drainage and the depth of valleys over the site.

There is considerable topographic variation across the site. Figure 25.7 utilizes the USGS 7.5 min Shawnee Quadrangle (1975) to illustrate the variations in topography at the site. The elevation to the north is 292 m and decreases to the south to approximately 240 m, then decreases further to the Kansas River that is approximately 210 m south of the site. The east to west hill and valley terrain has elevation changes of up to 45 m.

There are two deeply incised north to south valleys over the mine (Fig. 25.7). The westernmost valley is located about 100 m west of the CCA and it extends downward to the top of the Westerville Limestone at the southern side of the mine (Fig. 25.8). The easternmost valley is located south of the CCA and it extends downward to the top of the Quivira Shale at the south side of the mine. These incised valleys are probably associated with fractures that could extend

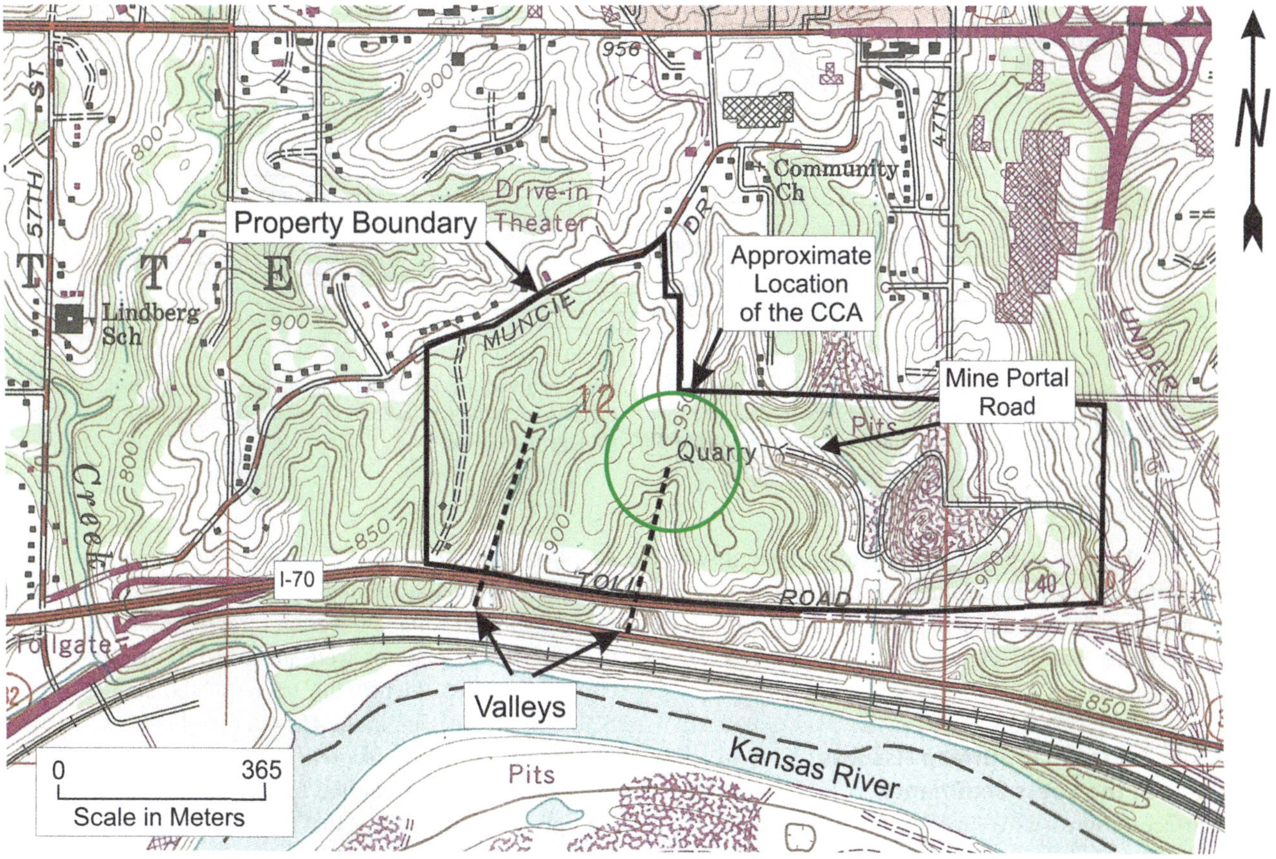

Fig. 25.7 The topographic map from the site reveals extensive topographic variation over the western portion of the mine. Two deep valleys can be seen across the site (USGS 7.5 Minute Quadrangle Shawnee, Kansas revised 1975)

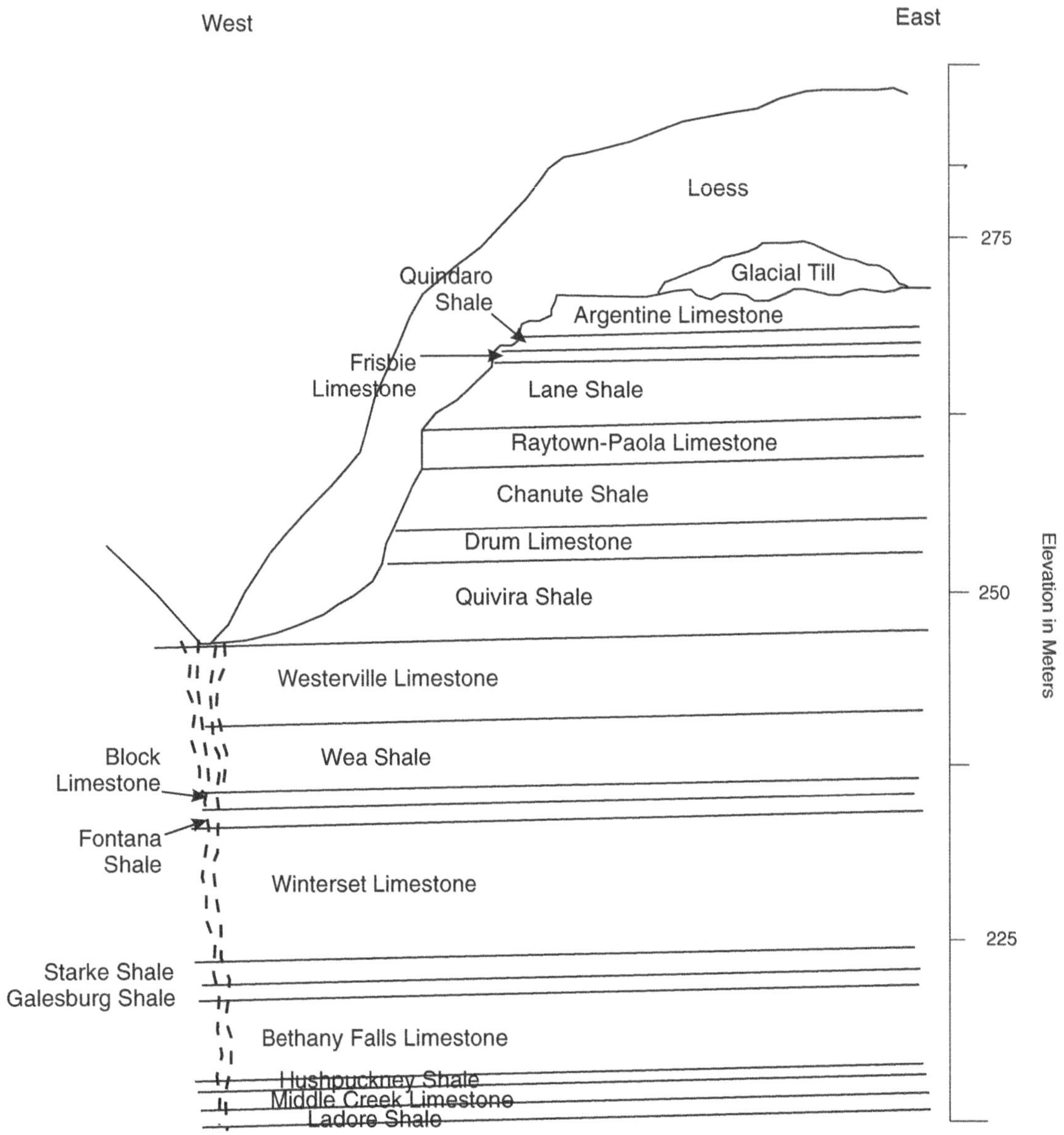

Fig. 25.8 A cross section through the westernmost incised valley (shown in Fig. 25.7) that extends to the top of the Westerville Limestone at the southern side of the mine. These surface features are probably associated with fractures that extend deeper into the rock, possibly to the mine resulting in a zone of weak rock and may be a potential geologic flaw at the site

deeper into the mine. They could provide a zone in which surface water can more easily enter the underlying rock, and over long periods of time will have weakened the rock. However, there was no evidence of such fractures from within the mine.

25.2.5.3 Evidence of Surface Fissures in the Loess Sediments

The surface fissures were first identified on the 1988 and 1985 air photos, characterized by a series of subtle furrows arranged in a concentric or curvilinear pattern based upon field mapping. Having identified the fissures in the recent aerial photos, the older 1954 pre-mining photos were then reviewed. Although limited by vegetation cover (mid-July with maximum foliage coverage) traces of similar concentric, curvilinear patterns were identified in the 1954 photos (Beccasio 1988). Identifying traces of the surface fissures before mining began, in the 1954 aerial photos, was a major factor in confirming our preliminary conceptual model and the relationship between the CCA and the surface fissures. This provided verification that the surface fissures were not a result of subsidence due to the mine-roof collapse. These fissures were present before mining and likely a result of a paleocollapse that had taken place long ago.

25.2.6 Site-Specific Geology

At the beginning of our investigation there were about 40 mine backfilling holes immediately to the north and west of the CCA. These boreholes were used for the purpose of injecting fly-ash backfill material into the mine. They were open boreholes that were destructively drilled. There were also five existing cased vent holes into the mine and a few older piezometers and monitor wells that had been installed by others.

While the regional geology was well described in the literature (Hasan et al. 1988) there were no core samples or recent site-specific geologic data available. To provide site-specific geologic data one new borehole was drilled as a geologic reference hole. NX-core was obtained during drilling and then the hole was geophysically logged with natural gamma, induction, gamma-gamma (density) and neutron (porosity) logs to provide a detailed reference for geologic conditions at this site. Since these logs can make measurements within open and PVC-cased boreholes, they were run in all available fillholes, vent holes and monitoring wells on-site. These logs provided very repeatable data indicating fairly uniform geologic conditions across the site.

Figure 18.5 shows the natural gamma logs from three boreholes at the site. The natural gamma logs were found to be very repeatable from hole to hole and provided distinct responses from three shale layers. From top to bottom they include:

- The Muncie Creek Shale a thin layer within the Raytown Paola Limestone sequence,
- The Quivira Shale over the massive Westerville Limestone, and
- The Stark Shale located just below the Westerville Limestone.

25.2.6.1 Excavation During Landfill Development

As the landfill was being developed, clearing of the land and excavation exposed a section of the Argentine Limestone (the uppermost limestone at the site) and the underlying Lane Shale across the western portion of the CCA. This unexpected window into the geology provided an opportunity to gain additional understanding of site conditions. This excavation was over 180 m long and extended over the western portion of the CCA. Figure 25.9a shows photograph of the excavation and a portion of the sketched excavation (Fig. 25.9b) to document conditions.

The top of the Lane Shale was marked at intervals along the entire excavation and the elevations surveyed. The location of each survey point along the excavation is shown in

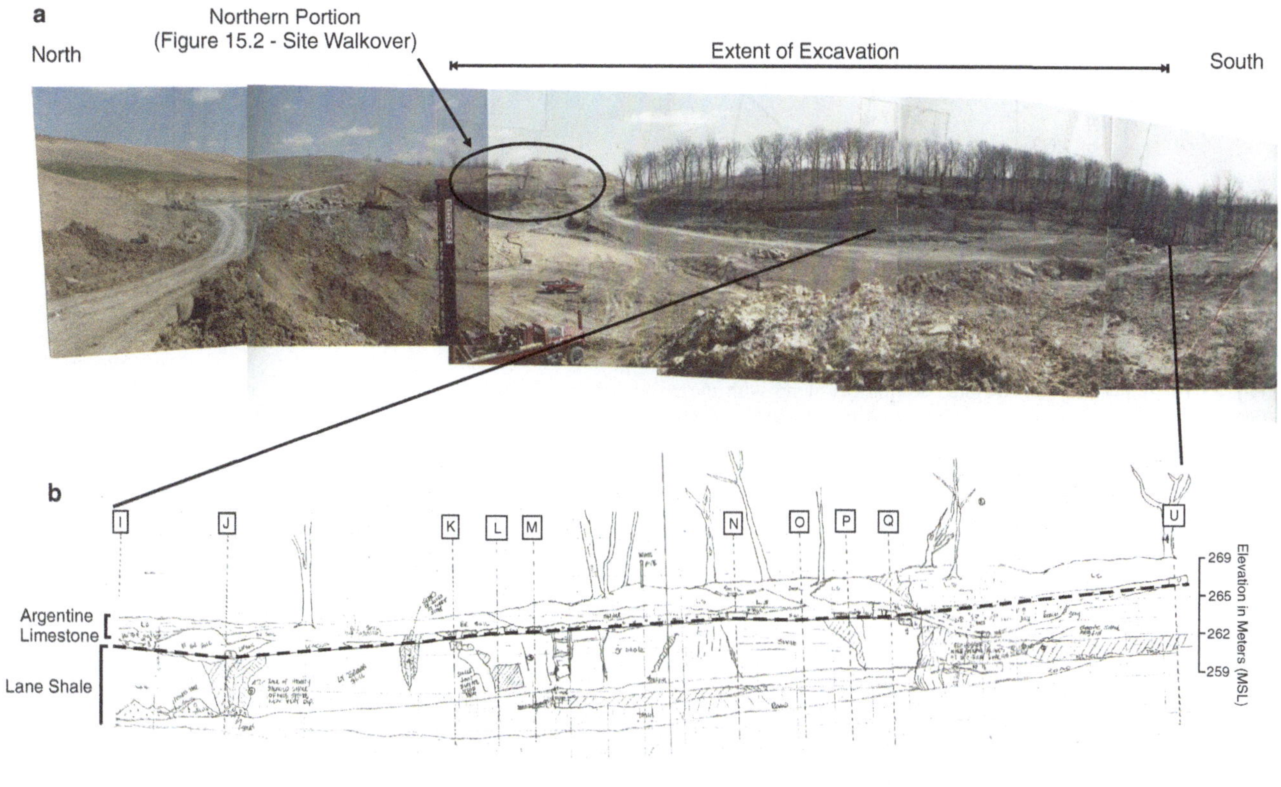

Fig. 25.9 A photograph of the north to south excavation that cuts across the western side of the CCA (**a**) and a partial sketch of the excavation (**b**)

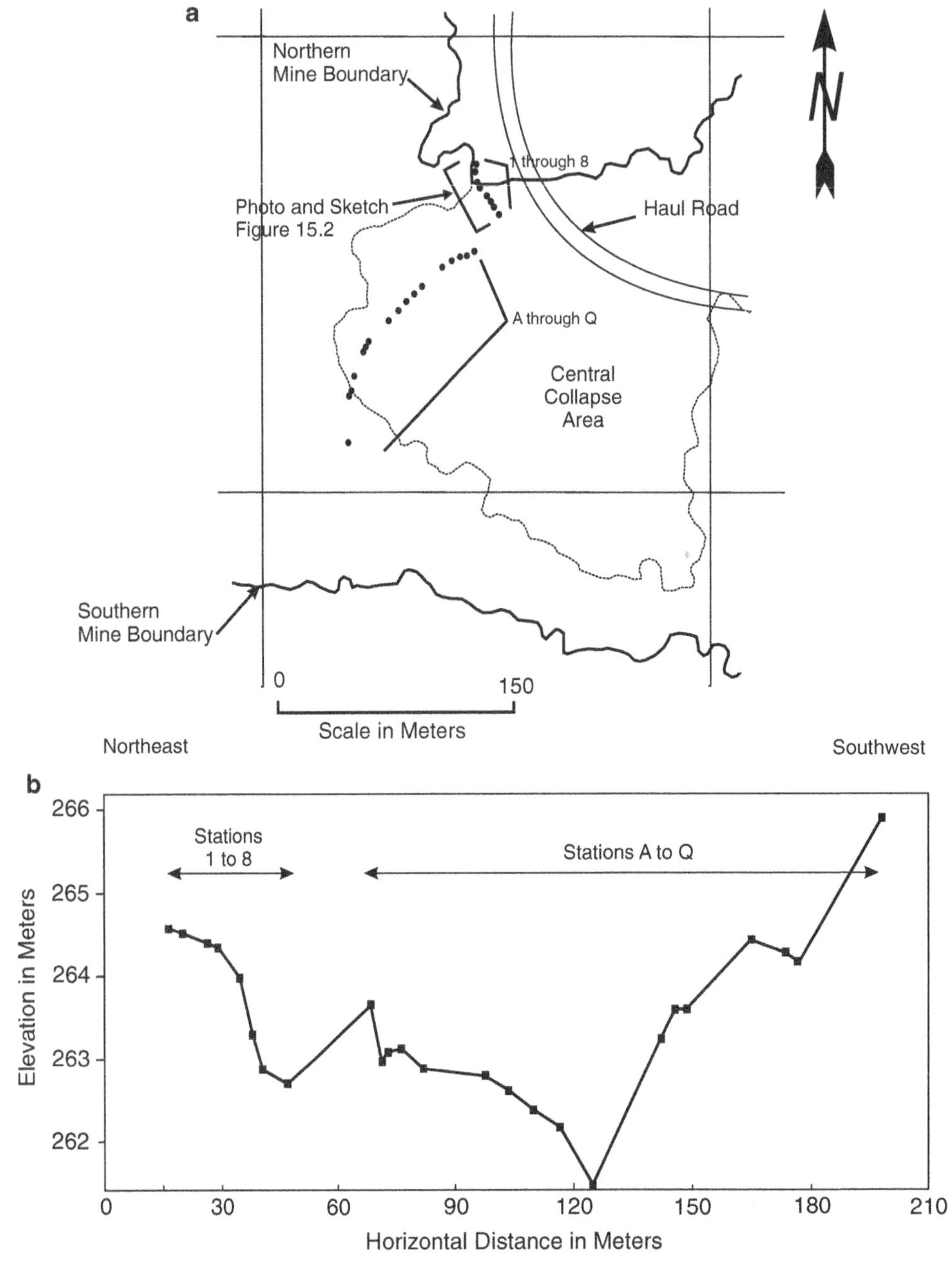

Fig. 25.10 A map of the excavated section (**a**) along with the profile of the top of the Lane Shale (**b**), which revealed the presence of a depression

Fig. 25.10a. The resulting elevation profile of the Lane Shale is shown in Fig. 25.10b. While the site appears to have flat-lying, uniform geology, the elevation profile along the excavation clearly indicates some unusual conditions. This exposure revealed the presence of a depression about 180 m wide with a displacement of as much as 4.5 m. Numerous local fractures, open voids and areas of high stress were seen in the Lane Shale along the face of this cut (Fig. 25.11). The extent of open fissures within the Lane Shale was a bit of a surprise. These features are quite weathered and appear much older than the time of the CCA less than 17 years earlier.

25.2.6.2 Identifying a Circular Depression Over the CCA

A contour map of the bedrock was developed over the CCA and its surrounding area. This contour map incorporated data from the 30 geophysical logs acquired over and around

Fig. 25.11 Local fractures and open voids along with areas of high stress are seen in the Lane Shale along the face of the excavation. These features are quite weathered and are thought to be much older than the CCA less than 17 years earlier (**a**) Open fracture (**b**) Evidence of stress along fractures (**c**) Typical evidence for water in joint of Lane Shale

the CCA. The Muncie Creek shale marker bed (the uppermost of the three key marker beds) identified in the natural gamma logs (Fig. 18.5) was used as a reference. In addition, the elevation measurements made along the cross section through the excavation western portion of the CCA (Fig. 25.10b) and strata within the mine portal (Fig. 25.2) were all used to develop the contour map of the area.

The contour map (Fig. 25.12) indicated the presence of a local circular area with up to 4.5 m of displacement. It is reasonable to assume that this local depression is related to the paleocollapse feature. Gentile (1984, 1988 personal communication) reported similar features in the area. These features can have 1–1.5 m of vertical displacement and extend a couple of hundred meters.

25.2.7 Confirmation of the Preliminary Conceptual Model

Many of our early opinions and assumptions presented in the preliminary conceptual model had now been supported by this additional data thereby improving our confidence level in the conceptual model. Additional data included:

- The results of the aerial photo analysis indicating that the surface fissures existed prior to the mining of the Tobin mine.
- The elevation profile developed along the surface excavation through the western portion of the CCA verified a collapse
- The local circular depression seen in the contour map over the CCA further verifying a collapse

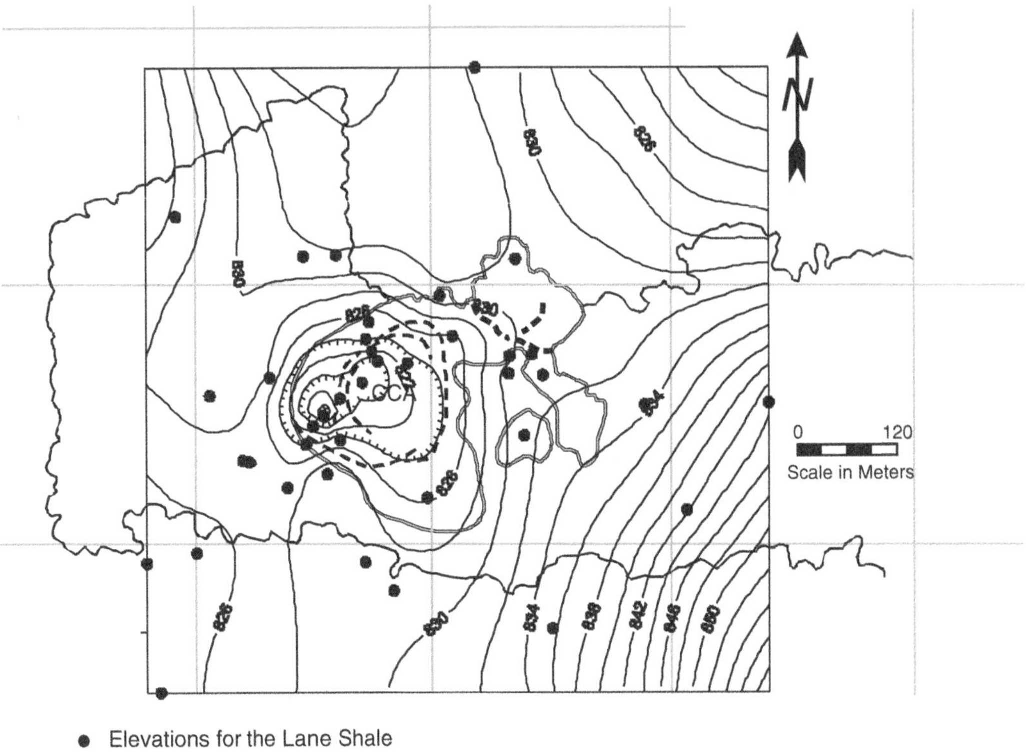

Fig. 25.12 A contour map was developed showing a local depression centered over the CCA (note contours in feet)

25.2.8 Hydraulic Connection of the Paleocollapse Fractures

As the conceptual model at the site developed, the hydraulic connection of the paleocollapse fractures extending from the surface fissures downward into the mine was of concern. The older portions of the landfill had used a clay liner at the base of the landfill. Now, a fabric liner and leachate collection system was being installed under newer portions of the landfill. However, It was unknown whether any fractures associated with the paleocollapse would provide a direct pathway for leachate migration from the landfill into the mine.

One of the surface fissures in the loess had a small hole at its base. More than 3,785 l of water was allowed to flow into this hole at a rate of more than 37 l/min without causing the hole to overflow. The loess was then excavated to expose the surface of the Argentine Limestone where a significant fissure in the rock was observed (Fig. 17.16a). This fissure is located over the highest mine-roof collapse found in the CCA and was suspected to be a fracture associated with the deeper paleocollapse.

Two angle borings were drilled through the exposed fracture. Observations were then made within the open boreholes with a borehole video camera (Fig. 17.16b). A number of fractures and open voids were observed near the projected location of the paleofracture all within the Lane Shale. These were probably similar to those observed in the excavation. In addition, it was found that these angle boreholes would not hold water.

A dye trace study to assess the potential for flow from the surface into the mine via the paleocollapse fracture was carried out utilizing this open joint in the Argentine Limestone as the dye injection point. Fourteen points were selected in the mine around the perimeter of the CCA to sample mine-water for the presence of dye using charcoal bugs (Fig. 25.13a). Jim Quinlan assisted with the design of the dye survey and the placement of charcoal bugs (Fig. 22.6b).

In addition, charcoal bugs were located within a number of piezometers and monitor wells and within two ponds on the surface. Background water samples were obtained from all stations within the mine as well as a number of piezometers, monitor wells and within two ponds on the surface. The city water and tanker truck used to provide water for the dye trace was also sampled. Quinlan (1991) and Aley (1991) provided analysis of all water and dye samples using a scanning spectrofluorometer.

Prior to injecting the dye into the large fracture (Fig. 17.16a) 15,140 l of water was injected to pre-wet the fracture system. Then 0.45 kg of fluorescein dye was mixed with 170 l of water and was allowed to flow into the injection site. The dye was followed by 90,840 l of water as a chaser at a flow of approximately 340 l/min.

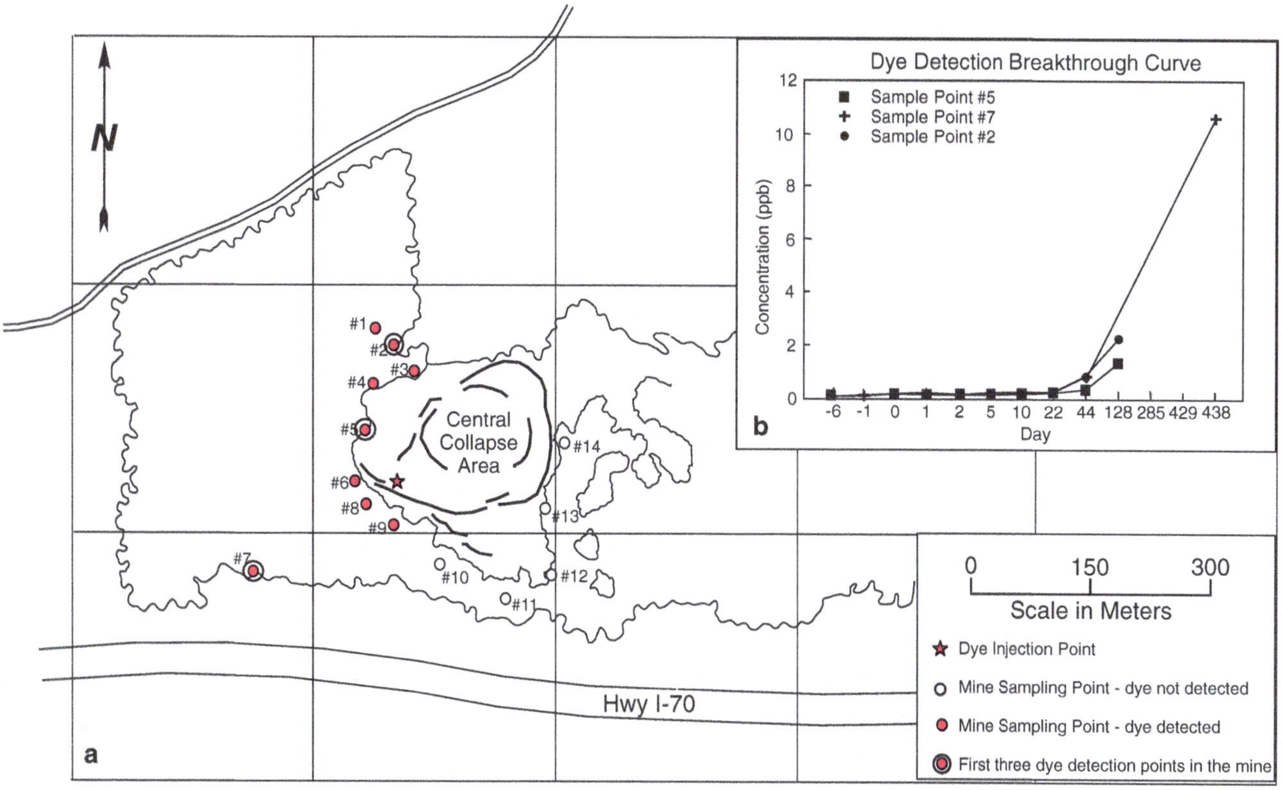

Fig. 25.13 Dye sampling was located at 14 points around the perimeter of the CCA (**a**) and results from the dye trace indicated the presence of dye in the mine at three locations between 22 and 44 days after injection (**b**)

At this point in the site characterization, the mine had flooded to the point that movement within the mine was accomplished by small boat. Prior to injection of dye and during the first 10 days after injection of dye, sampling was carried out in the mine using a Turner portable flourometer with continuous recording (Fig. 22.5). Water was pumped continuously through the flourometer as the boat traversed through a common course within the mine passing by each charcoal bug sample station. No dye was detected by the Turner flourometer during the first 10 days of sampling. The charcoal bugs in the mine were scheduled to be replaced and analyzed at approximately 1, 2, 5, 10, 20, 50, 100, and 500 days.

Samples of water with dye were recovered at two older monitor wells, about 22 and 152 m north of the injection site 1 day after dye injection. These wells were screened within the Raytown-Paola and Drum Limestone. This detection of dye was unexpected but indicated the potential for rapid lateral flow within these formations. This rapid flow may only occur within the stressed and fractured rock overlying the CCA.

Between 22 and 44 days, dye began to appear at three locations in the mine (Fig. 25.13b) and included:

- Sample station #5 at the west edge of the CCA and about 45 m north-northwest of the dye injection point,
- Sample station #2 north of the CCA and about 137 m north of the dye injection point,
- Sample station #7 at the south mine wall and about 180 m southwest of the dye injection point.

The presence of dye at station #5 suggests a possible hydraulic connection between the surface and the mine via the paleofracture. However, the other two sampling stations are located farther from a paleofracture suggesting that dye traveled there via means other than the paleofractures.

There are three possible pathways for dye to enter the mine (Fig. 25.14).

1. Via the paleofracture associated with the deep-seated collapse in the Mississippian Limestone (Fig. 25.6). The paleofracture was observed at three locations within the mine. The vertical fracture was so tight that thin knife blade could not be pushed into an opening of about 0.10 cm.
2. Via the open fissures, fractures and voids that had been observed visually in the Argentine Limestone and the Lane Shale (Fig. 25.11). It might be reasonable to assume that the entire column of rock over the CCA contains similar open fissures, fractures and voids. If so, these could be possible pathways for dye to have migrated into the mine.

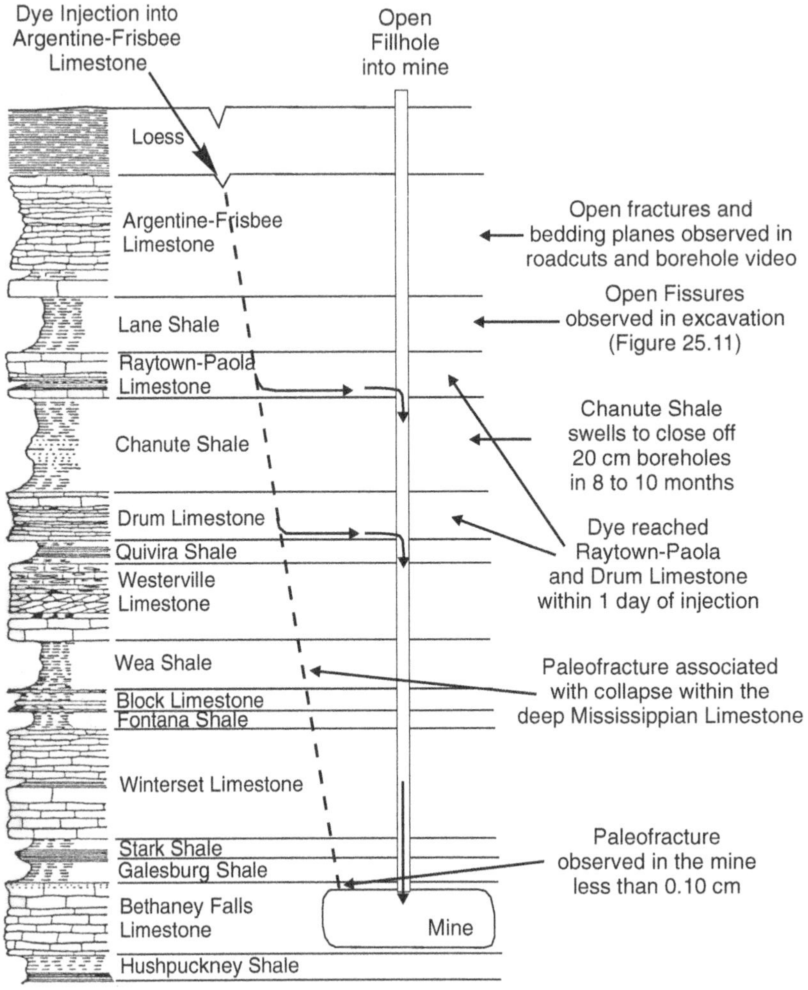

Fig. 25.14 Rapid lateral movement of dye within the Raytown-Paola and Drum Limestones could have intersected either fractured rock within the CCA or mine fillholes that had not been grouted and provided direct access to the mine

3. Via the mine fillholes that may not have been grouted prior to the dye injection. The lateral movement of water within the Raytown-Paola and Drum Limestones could have flowed down open boreholes being used to fill the mine.

The later two options are the most likely pathways into the mine: by the open fissures, fractures and voids that had been observed within the stressed column of rock over the CCA or via mine fillholes.

Additional backfilling of the mine by the rock slurry method began about 133 days into the dye tracer test. Backfilling water was being pumped from the mine and re-injected with the crushed rock backfill at about 5,677 l/min. After the mine backfilling began the amount of dye detected in the mine rapidly increased to 10 ppb or more and was found along the entire west side of the CCA where backfill material was being placed (Fig. 25.13a). Because of this further dye sampling was discontinued.

While the dye trace did not definitively answer the question of a direct connection for flow from the surface to the mine there was information gained. There was also the unexpected lateral flow within the Raytown-Paola and Drum Limestones within 1 day after injection of the dye. However, we do not know if this rapid flow was limited to the area over the highly fractured rock within the CCA or was common to the entire site. While this was not the focus of the dye trace study, it did ultimately impact the groundwater monitoring plan later in the project.

25.2.9 Subsidence Measurements

Through extensive characterization within the mine, it had been determined that the possibility of surface subsidence due to mine collapse was highly unlikely. Three outside experts Golder (1986), Tein (1988), and Brink (1989) who had inspected the mine conditions had also indicated that mine conditions were stable and would not result in surface subsidence. However, the question of mine stability still remained a concern by the state and subsidence measurements were undertaken.

Early on we recognized that such measurements were difficult and would require an extreme level of expertise along with quality assurance (QA) and quality control (QC). To accomplish this we engaged a survey firm who were experts in subsidence measurements to oversee the program and a local survey firm with experience in such measurements. The design and layout of the monument network was evaluated and it was estimated that the network could be expected to detect vertical displacements on the order of (0.021 cm using first order vertical measurements (Boston Survey Consultants (BSC) 1990).

The subsidence monument design was modified from that in (Dunnicliff and Green 1988) (Fig. 20.2). Monuments were located along a clear visual path for easy access of the survey crew. The monuments were anchored in the uppermost Argentine Limestone a minimum of 1.2 m. A network of 11 subsidence monuments were initially installed from south to north across the CCA (Fig. 25.15):

- Two monuments M-100 and M-9 were located off of the mine, one to the south and the other to the north to provide stable reference points.
- Two monuments M-2 and M-3 were located over the mine, but off of the CCA.
- Seven other monuments M-4, M-5, M-6, M-7, M-8, M-10, and M-11 were located over the CCA.

A series of baseline subsidence measurements were made 1 week and again 1 month after the monuments were installed to assess stability and provide baseline data. Then a sequence of subsidence measurements was carried out about every 3–4 weeks by Anderson Survey co. (1990).

Because of heavy truck and equipment traffic on-site during the week, survey measurements were made on weekends when there was little or no traffic present on-site.

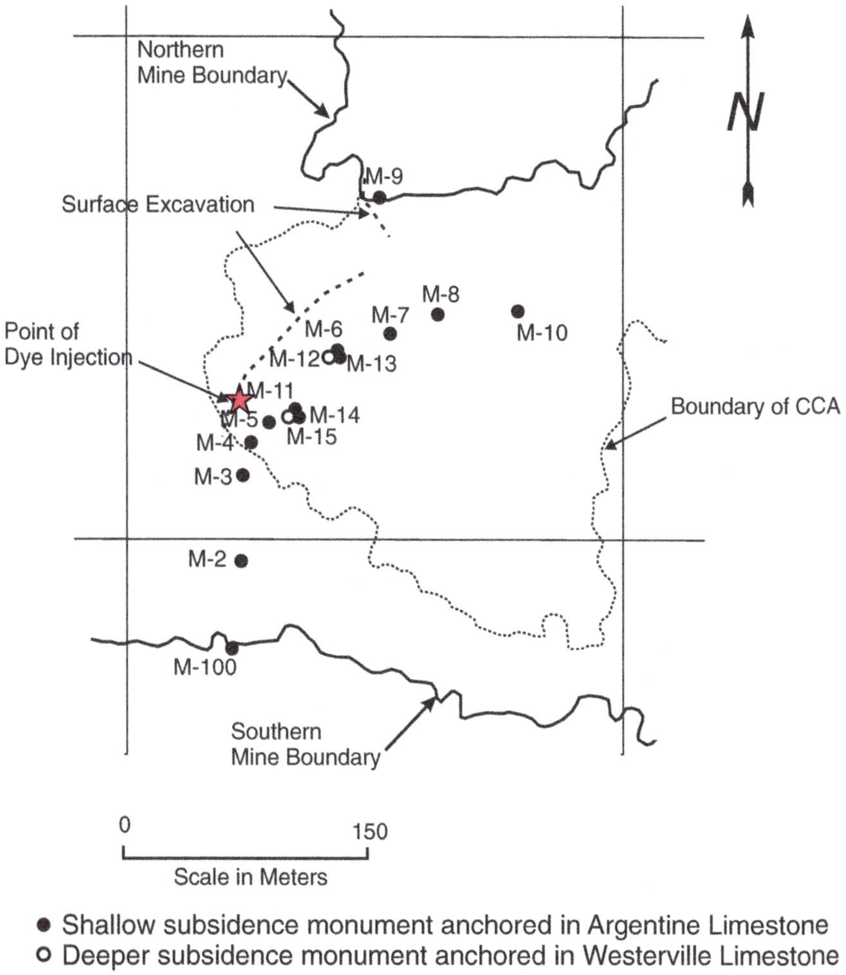

Fig. 25.15 The location of subsidence monuments that were installed across the CCA

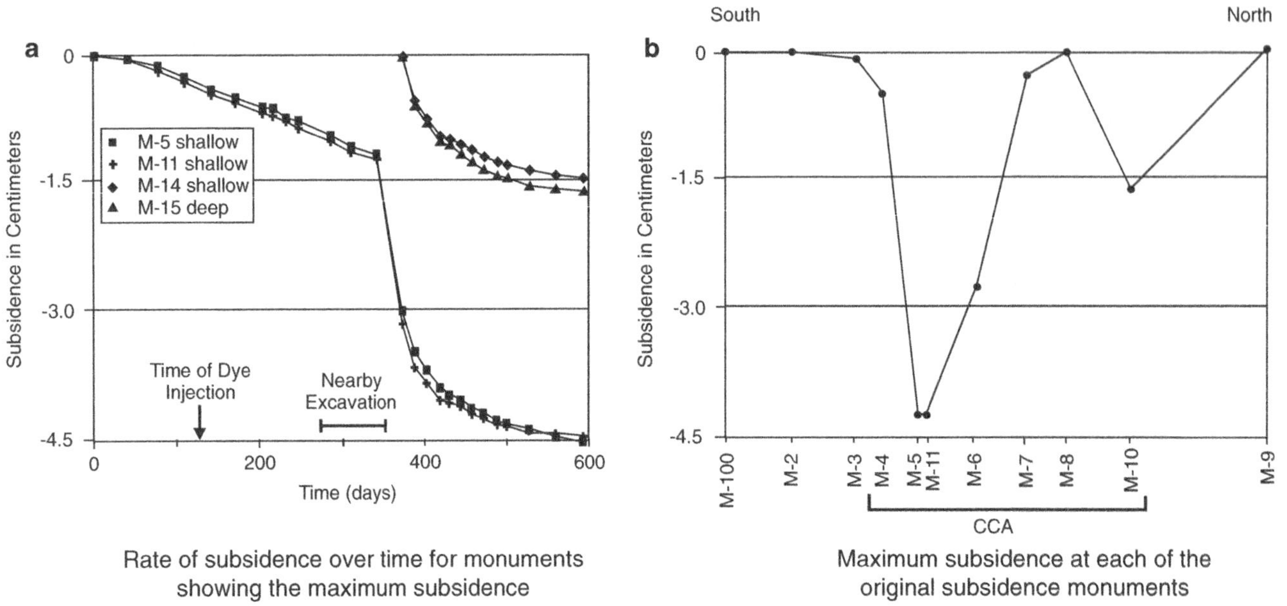

Fig. 25.16 Abrupt changes in subsidence are seen in two shallow monuments anchored in the shallow Argentine Limestone. A similar response is seen in the newly installed monuments one anchored in the shallow Argentine Limestone and the other anchored in the deeper Westerville Limestone. Note that subsidence seems to be slowing and stopping with time (**a**). The maximum subsidence of all initial 11 monuments is shown in profile (**b**)

BSC reviewed the monument design, field survey procedures and the data acquired by Anderson Survey Co. BSC also carried out an independent set of measurements as a means of quality control. The subsidence monument design and installation, the field measurements by two experienced firms, along with the QA/QC procedures have resulted in the subsidence data being the highest possible quality.

25.2.9.1 The Results of Subsidence Measurements

The initial subsidence monuments indicate little, if any, significant subsidence. Some of the movement is due to noise or error in measurement since some values indicate a movement upward.

Figure 25.16a shows the data from the two of the monuments showing the greatest subsidence (M-5 and M-11). The subsidence begins slowly, but after the 13th set of measurements (about day 340) an abrupt increase in subsidence occurs. As a result of this rapid subsidence, four additional subsidence monuments were installed near those monuments that showed an increased rate of subsidence. Two shallow monuments (M-12 and M-14) were installed in the Argentine Limestone and two deep monuments (M-13 and M-15) were installed in the massive Westerville Limestone at a depth of more than 30 m to evaluate whether the rapid subsidence was due to shallow or deeper cause. Figure 20.3 shows a cross section of all monuments both shallow and deep.

The newly installed monuments near M-5, both shallow and deep (M-14 and m-15), immediately began to show the same pattern of subsidence as the nearby monuments (Fig. 25.16a). Then the subsidence of all the monuments appears to decrease and subsidence seems to be stopping by day 600.

Figure 25.16b shows the maximum subsidence profile across the initial eleven monuments from south to north. The maximum subsidence measured was almost 4.5 cm. This was measured at two adjacent monument locations (M-5 and M-11). While these two adjacent monuments provided very similar data, there was considerable differential in both the rate and amount of settlement taking place along the line of monuments.

After 600 days and a total of 26 sets of measurements, further efforts were discontinued due to excavation for the expanding landfill impacting the monuments. No obvious changes in mine roof collapse were observed during this time.

Figure 25.15 shows the location of subsidence monuments along with the excavation west of the CCA and the point of dye injection. The excavation is location within 45 m of the monuments and was excavated a few months prior to the rapid change in settlement of shallow monuments. As a result, the monuments which were anchored in the shallow Argentine Limestone (a highly fractured rock) could have been affected by the lateral stress relief of the excavation which had removed as much as 10 m of rock.

The introduction of 105,980 l of water associated with the dye trace, 9 months before the rapid change in settlement has likely contributed to the movement of both shallow and deep monuments over the CCA. This was an exceptional amount of water rapidly injected into an otherwise dry system.

The combination of the excavation within 45 m from the line of monuments and the 105,980 l of water as part of the dye trace study seem to be the most likely cause of subsidence based upon known geologic and hydrologic conditions. We sometimes encounter results from our observations and measurements that are highly variable and are difficult to interpret with a high degree of certainty. While the subsidence data itself is considered quite accurate, our interpretations of the causes of subsidence still contain a level of assumptions and opinions that are not backed up by site-specific detailed data.

25.3 An Assessment of the Mine Conditions

Assessing mine conditions was focused upon the western portion of the mine over which a landfill was being developed. Observations in the eastern portion of the mine were limited to a 100 m or so around the portal entrance and under the mine portal road. The assessment of mine conditions was important to the project and addressed concerns of the state and ultimately aided in the assessment of subsidence risk. This work had been on-going since the early part of the project and included:

- Developing a detailed mine map
- Monitoring any changes in mine conditions for 3 years;
- Determining the mechanism and types of mine collapse;
- Acquiring bulking data
- Estimating the extent of upward collapse within the CCA;
- Identifying the source of water filling mine;
- Evaluating mine water quality and
- Assessing the potential subsidence over the CCA.

25.3.1 Development of a Detailed Mine Map

Access into the mine required a detailed mine safety plan. Part of the safety plan included a detailed working map of the mine and was one of the first priorities. The map provided a means of safely carrying out routine inspections within the mine as well as providing a means to document the location of conditions within the mine. A mine map was available from mining operations (Tuttle, Ayers, and Woodward Co. 1979) reference points had been marked on pillars and on the ceiling, but the map was quite cluttered and not easy to interpret. However, it was used as a basis for developing a new detailed mine map that incorporated the areas of mine-roof collapse (Fig. 25.17). The working mine map was started as soon as access to the mine was obtained in June of 1987. It focused upon documenting a detailed boundary of the CCA and other collapse areas, as well as identifying a routine pathway to be used in traveling within the mine. The working mine map evolved over time as we accessed various parts of the mine.

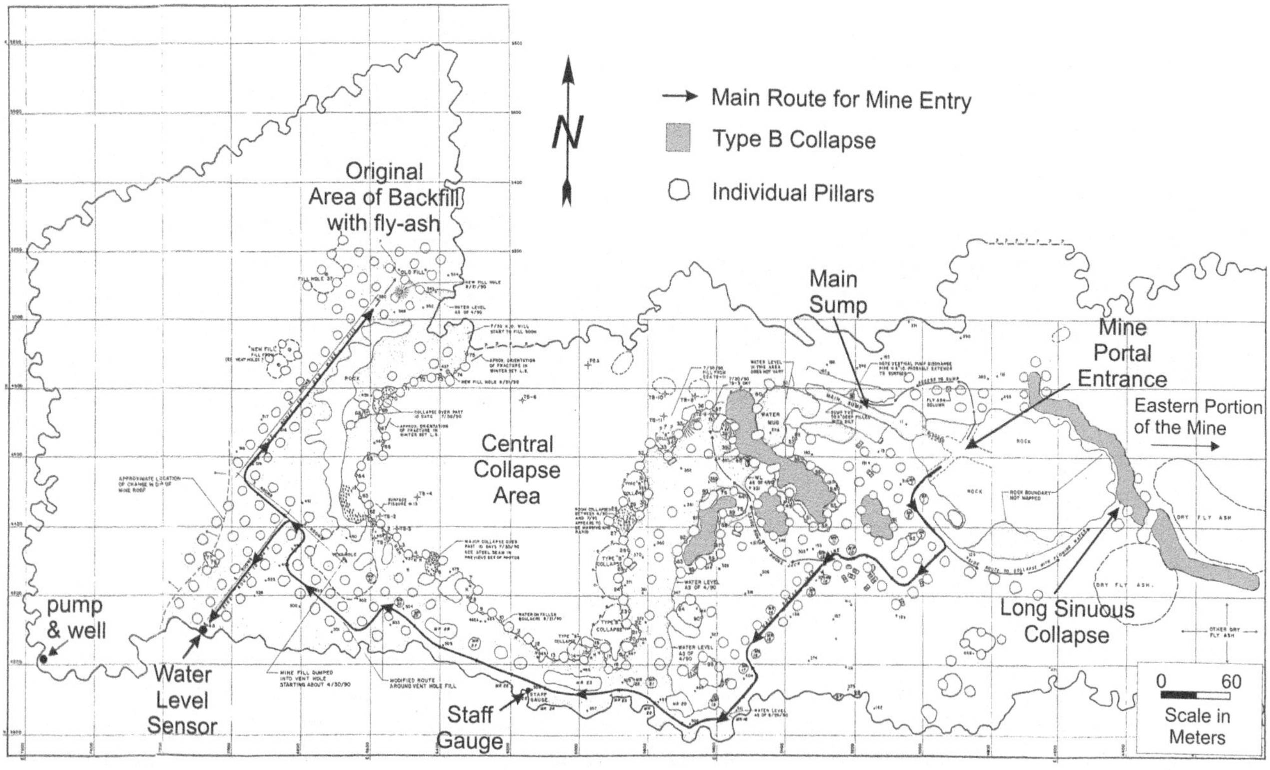

Fig. 25.17 A detailed map of the mine including the CCA and other collapse areas. The mine map also identifies the main routes used to travel within the mine

25.3.2 Monitoring of Changes in Mine Conditions over 3 Years

As part of developing the mine map, conditions within the mine were monitored. Monitoring of mine conditions was focused upon the CCA and other isolated collapse areas but also included general conditions of the pillars, floor and ceiling as well as water levels.

Monitoring of collapse areas consisted of visual inspections and photographing each collapse site every 3–4 months over a period of 3 years. Each area of collapse was identified using a large white sheet of plastic with black numbers and photographed over time providing long-term monitoring of any changes. A total of 125 collapse sites were identified for monitoring. The photographs from each inspection was reviewed and compared to the previous photographs. This inspection process provided a means to document any changing conditions at the large number of collapse sites. Note that the photos included here do not have numbers as these were the very first set of photos.

The CCA had not expanded laterally since the major collapse in the early 1970s. This was based upon roof stability warning devices and support beams that had been installed at the south edge of the CCA after the initial collapse. Furthermore over more than 3 years of monitoring the CCA there had not been any significant lateral or vertical expansion of the CCA.

About 60 % of the stations monitored (around the CCA and other isolated collapse areas) over the 3-year period have had some small level of additional collapse activity. The additional collapse were all small, localized failures, typically about 1 m^3 associated with further degradation of the rubble zone or edges of the Starke-Galesburg Shale.

Two independent mine experts were engaged to help evaluate mine conditions. Both experts, Tein (1988) from the Rolla School of Mines and Brink (1989) from South Africa provided similar and supporting observations and conclusions to that of an earlier consultant (Golder Associates 1986).

- The pillars show no signs of spalling, punching, or stress related failure.
- No evidence of roof failure due to cracking or spalling was apparent in the areas inspected around the perimeter of the CCA or in general.
- Generally the mine floor appears to be excellent, although floor heave was observed south of the CCA in an area where water was present on the mine floor. Floor heave is due to degradation of the Hushpuckney Shale as a result of the mine filling with water. About 15–30 cm of floor heave was noted at three locations.
- Engineered pillars of 7.6 by 7.6 m on 18 m centers would yield about 82 % extraction. However failure of pillars is unlikely even at an extraction of 90 %. Compression strength of the Bethany Falls Limestone is 82.7–137.9 MPa.
- No major roof collapse similar to the CCA had occurred within any of the oldest portion of the mine to the east where randomly spaced pillars were used.

The general conditions within the mine, outside the CCA, were considered good. The collective opinions of those experienced experts who had entered the mine strongly supported the fact that the surface fissures were a result of a collapse originating deep within the Mississippian Limestone and propagated to the surface, long before mining occurred. Recommendations by Brink (1989) included:

- Extension of the landfill over the CCA should be permitted provided that the surface paleocollapse fissures are exposed and sealed.
- That subsidence monitoring be made over the mine and
- A groundwater monitoring program should be pursued.

Although scattered roof failures can be expected in the future due to the exposure of the rubble zone. They will be of limited vertical extent and will not cause subsidence or fracturing at the ground surface. This was also the conclusion by Hasan et al. (1988). Furthermore, no other mine in the greater Kansas City area had encountered an instantaneous collapse of the size of the CCA.

25.3.3 Determining the Types of Mine Collapse

The mine-roof collapse areas were classified into four types to provide a convenient means of characterizing the nature of collapse.

25.3.3.1 Type A Isolated Collapse

Type A collapse are isolated, occur within individual rooms (centered between pillars) and are limited to about 0.04 ha (Fig. 25.18a). The Bethany Falls Limestone was mined to within 0.3–0.6 m from its top. Type A collapse extend upward into the Galesburg Shale, (about 2 m thick) and then into the Stark Shale (about 0.6 m thick) to the base of the Winterset Limestone. In a few cases, collapse would extend further upward into the thick Winterset Limestone (Fig. 25.18b). Type A collapse are the most typical of the roof collapse and are open so that the upward extent of collapse can be directly observed.

25.3.3.2 Type B Coalesced Collapse

Type B collapse are coalesced collapse, where adjacent isolated Type A collapse areas have merged to result in a collapse with a larger aerial extent between two or more

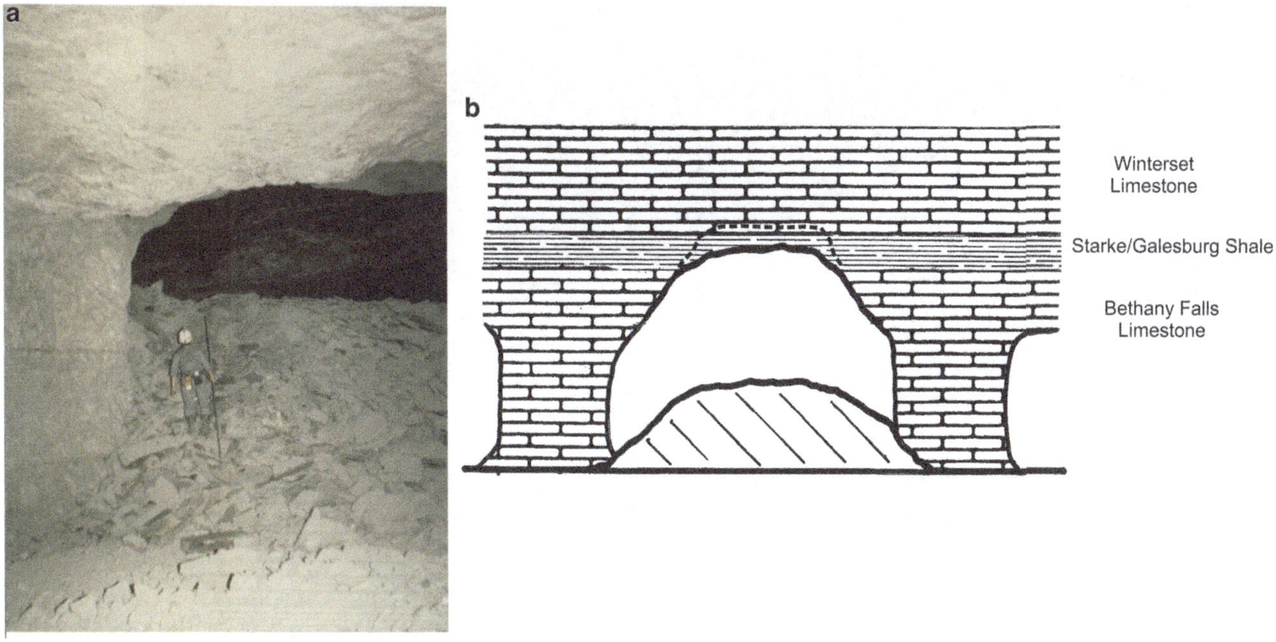

Fig. 25.18 Type A collapse are local, isolated collapse between pillars and would have a typical area of about 334 m^2 (**a**). Collapse would extend upward into the Stark Galesburg Shale and in some cases into the Winterset Limestone (**b**)

rooms with pillars intact. This type of collapse has occurred at an estimated 10–30 % of sites. They are open so that the upward extent can be directly observed (Fig. 25.19).

Two areas in particular illustrate extensive Type B collapse. They are located east and are separate from the CCA and occupy an area of about 0.2–0.4 ha (Fig. 25.17). In July 1990, a newly discovered but older long sinuous roof collapse was discovered in an area east of the mine portal road (Fig. 25.17). This roof collapse is more than 304 m in length. Here, the rubble zone is as much as 2.1 m thick compared to its typical 0.45–1 m thickness. Roof failure extended upward to the bottom of the Winterset Limestone. The collapse follows a sinuous path much like a meandering stream channel. This is a good example of a deep paleochannel of the rubble zone that had been inadvertently exposed by mining operations.

25.3.3.3 Type C Isolated High Collapse

Type C collapse are isolated high collapse which have extended upward beyond the Winterset Limestone (Fig. 25.20). Only three Type C collapse were discovered. However, there may be others that are blocked off and are inaccessible. These three areas around the perimeter of the CCA provided physical access up into the collapse zone.

Since these three mine-roof collapse were all located along the outer perimeter of the CCA they are thought to occur along the paleofractures associated with the CCA. A fracture was observed within the mine at each of these three locations within the Stark Galesburg Shale. No displacement was observed at the crack but strata dipped downward at an angle of 10–20° on the inside of the collapse. In each case, the paleofracture was within a few degrees of vertical and was tight and would not allow a thin knife blade to enter.

Access to one of these three sites was obtained by crawling upward on the pile of rubble (Fig. 25.20a). Upon reaching the top of the collapsed rock pile observations were made of what is believed to be the uppermost extent of rock failure within the CCA. Overhead we could see the base of the massive Westerville limestone (Fig. 25.20b) approximately 18 m above the mine-roof. The collapse tapered inward and became smaller as it extended upward (Fig. 25.20c). This is the uppermost extent of collapse observed within the CCA.

25.3.3.4 Type D Closed Collapse

Approximately 10 % of the collapse sites observed are Type D collapse. They are closed collapse (blocked off from access) in which the upward extent of collapse cannot be seen or accessed (Fig. 25.21). This type of collapse is typically found around the perimeter of the CCA.

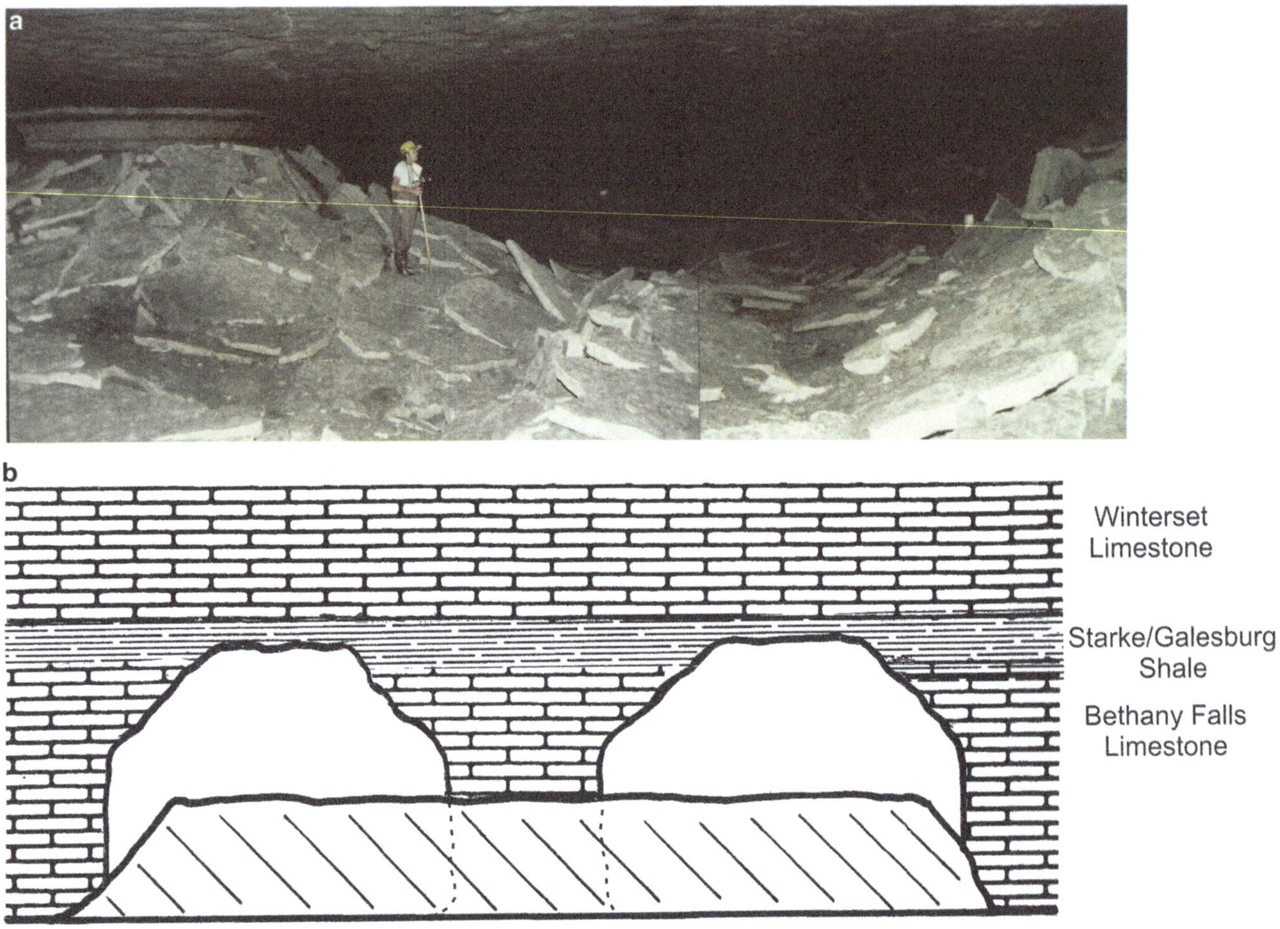

Fig. 25.19 Type B coalesced collapse extend latterly from room to room encompassing larger areas (**a**) and extend upward into the Stark Galesburg Shale similar to Type A (**b**)

25.3.4 Determining the Mechanism of Mine Collapse

Two geologic aspects appear to impact mine stability and cause roof collapse:

- The presence of paleofractures that are associated with the CCA and extend through the rock column from below the mine upward to the surface.
- The presence of a localized thicker zone of rubble rock in the upper part of the Bethany Falls Limestone.

The paleofractures have weakened the rock and are caused by the collapse of a large deep-seated cavern within the Mississippian Limestone which lies more than 180 m below the mine (Fig. 25.6). This deep-seated collapse had occurred well before mining had taken place. The paleofractures (Fig. 25.6) have had the greatest impact in causing mine-roof collapse. These features allowed the single large collapse event to occur forming the CCA (Fig. 25.3) and also appears to allow the highest upward migration of collapse, as much as 18 m above the mine (Fig. 25.20c). The paleofracture zone has resulted in a local circular depression with subsidence of up to 4.5 m over the CCA (Fig. 25.12). Rock over the CCA is extensively fractured and fissured as seen in the Lane Shale exposed at the excavation (Fig. 25.11) and also noted in various video logs.

The zone of rubble rock (Fig. 25.5) is often found in what appears to be channels that vary in thickness. In thicker channels, the channel bottom can inadvertently be exposed by mining operations. The rubble rock is the cause of the Type A, B and D collapse areas (Figs. 25.18, 25.19 and 25.20). One of the longest channels of rubble rock with thickness up to 2.1 m was found east of the mine portals (Fig. 25.17). It was more than 304 m long. In addition, the rubble zone is known to deteriorate rapidly when exposed to moisture, which leads to localized roof failure (Hasan et al. 1988). The outer perimeter of the CCA contains many collapse initiated by the rubble rock. In addition, all of the collapse areas outside of the CCA have been initiated by the rubble rock. This mechanism is responsible for the majority of roof failures in this and other mines in the area.

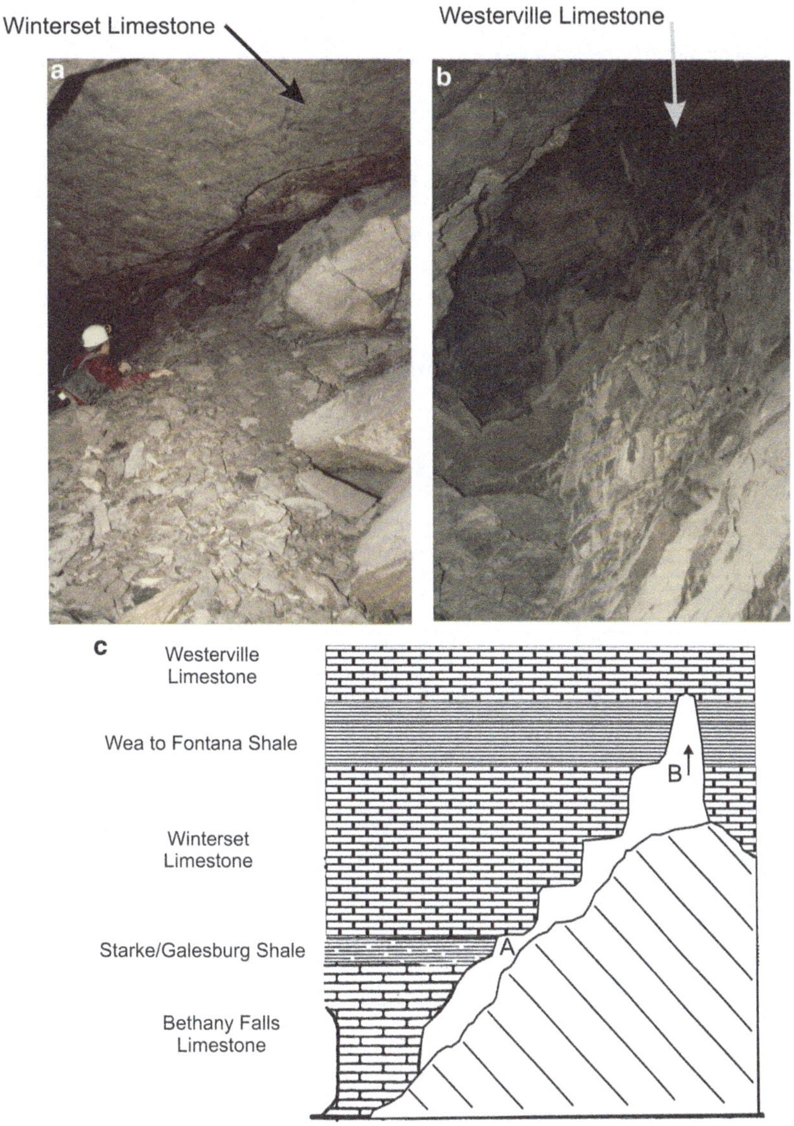

Fig. 25.20 Type C collapse are those in which roof collapse has extended upward higher than the Winterset Limestone up to the Westerville Limestone One of them allowed physical access upward to assess conditions (**a**). Looking up we could see the base of the massive Westerville Limestone some 18 m above the mine roof (**b**) the cross section shows the detail (**c**)

25.3.5 Bulking Measurements

Fallen rock will occupy a greater volume than it did when the rock was in place. This increase in volume is referred to as bulking or swell factor and has been discussed in Sect. 9.4.1. The bulking of fallen rock is one of the factors to be assessed when evaluating whether a mine collapse could potentially reach the surface. If the rock overlaying a mine (or cave) is thick enough, the volume of bulking rock will be more than the remaining void space and will eventually fill the void space preventing further upward migration of roof failure, before it reaches the surface (Fig. 9.5). Rock that falls in smaller pieces (Fig. 9.6a) result in a higher bulking factor than large intact slabs with little rotation (Fig. 9.6b). Hasan et al. (1988) indicate that bulking of failed roof rock over limestone mines in the Kansas City area is about 30 %.

The amount of bulking that occurs can vary considerably depending upon:

- The size and shape of the broken rock
- The relative strength of the rock
- The geometry of the open cave or mine.

Bulking will decrease as the opening becomes filled up and the fall distance is reduced, then there will be less rotation and breakage of the rock. The bulk rock volume may also decrease over time due to compaction of the rock fragments and the presence of water.

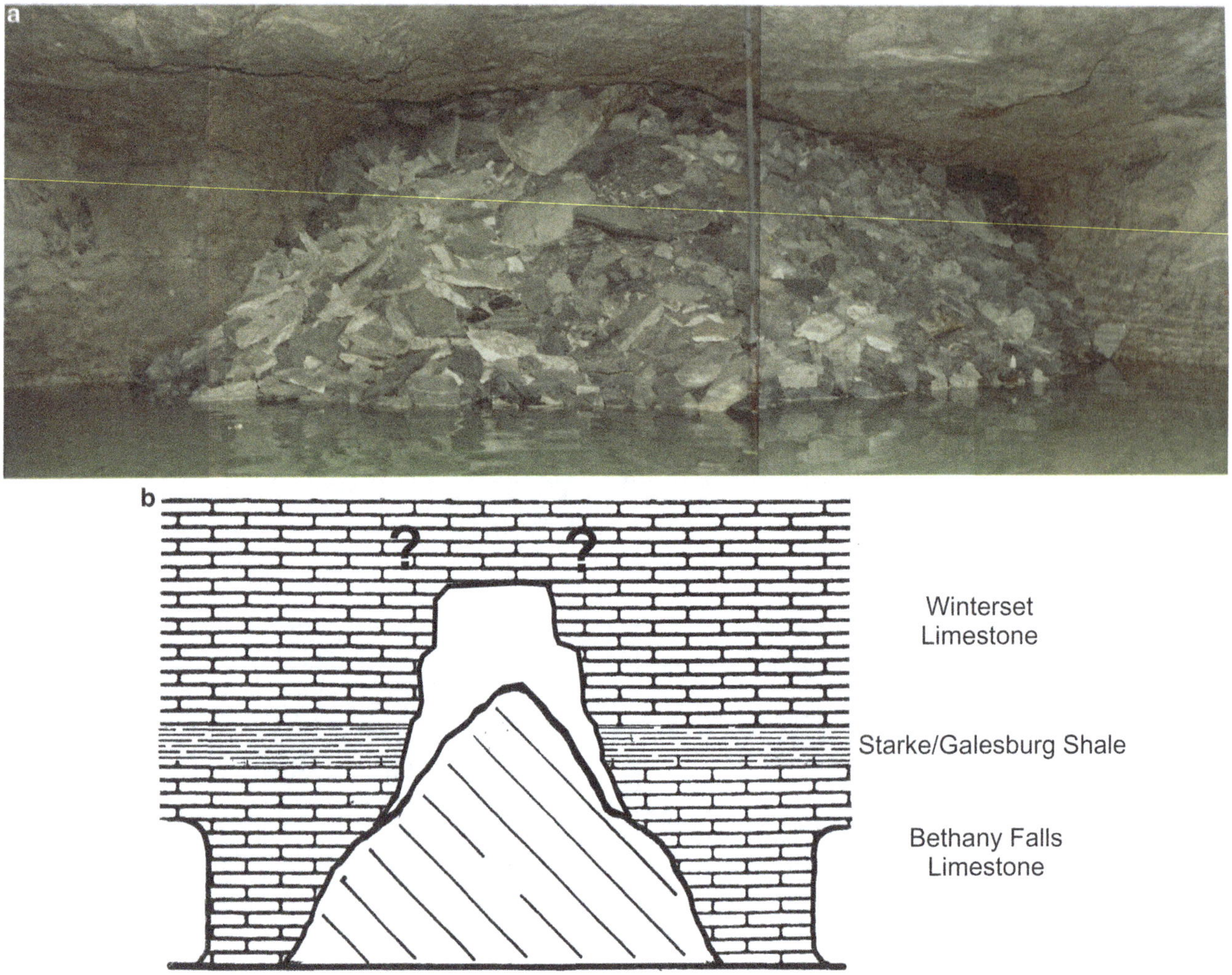

Fig. 25.21 Type D collapse are choked-off where fallen rock has prevented visual or physical access up into the collapse area (**a**). The vertical extent is unknown, but probably extends into the Winterset Limestone or higher (**b**)

Extensive direct measurements and photo documentation of bulking were made in the Tobin Mine. Measurements were made of the volume of the roof void created by the fallen rock and the volume of the rock rubble pile on the mine floor. At some locations the entire perimeter of the rock fall area was accessible enabling very accurate measurements to be made. In other locations access was only available on one side of the collapse requiring an estimate to be made. These bulking measurements resulted in an average bulking factor of 42 %. A conservative bulking factor as low as 30 % would limit upward collapse to below the base of the Westerville Limestone about 18 m above the mine (Fig. 25.2). Furthermore, the Westerville Limestone, which is massive and about 6 m thick, would likely limit any further upward collapse.

Because the interior of the CCA was inaccessible, a question remained as to the height of collapse and the degree of bulking within the CCA. Seven borings were drilled into the CCA and logged with a borehole video camera. These borings indicate open voids ranging from 0.6 to 2.4 m, averaging 1.5 m and estimated bulking factors averaging 36 % (Fig. 25.22). The highest extent of collapse from these seven borings was in TB-8, about 9 m above the mine roof into the Winterset Limestone. The result from these seven borings also suggests that the roof collapse over most of the CCA, is typically Type A and B collapse seen within the mine (having a vertical extent limited to the base of or within the Winterset Limestone) (Fig. 25.18 and 25.19). Based upon these borings into the CCA collapse an average bulking factor of 36 % was estimated.

25.3.6 Further Support of the CCA Conceptual Model

Our preliminary conceptual model of mine conditions and the relationship of the CCA and surface fissures has

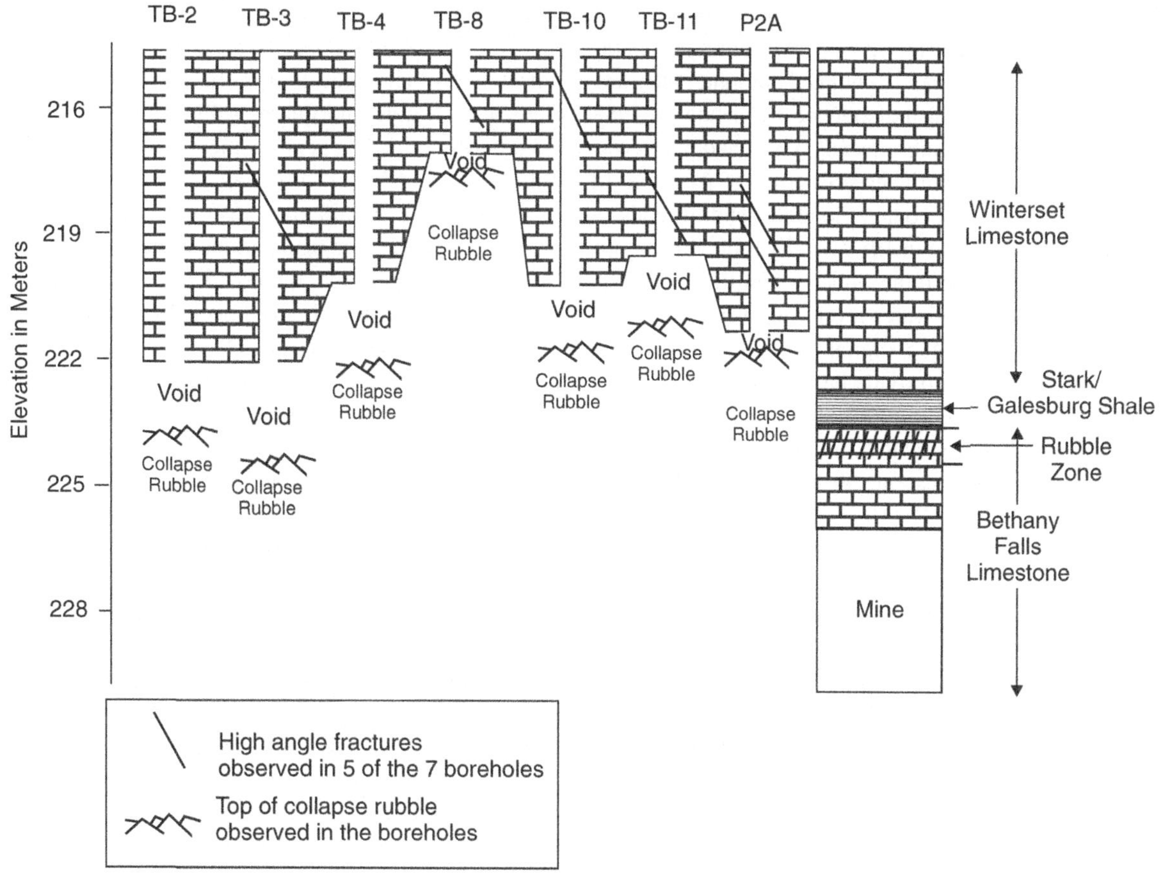

Fig. 25.22 Seven borings were used to evaluate the upward extent of collapse and bulking factor within the CCA. The results indicated an average bulking of 36 %

been supported by further direct evidence. A high level of confidence can now be placed on the conceptual model based upon:

- Our extensive direct observations and photo documentation around the perimeter of the CCA.
- Observations were supported by three independent mine experts, Golder (1986), Tein (1988), and Brink (1989) who also thought that the mine was stable. This was also the conclusion by Hasan et al. (1988).
- Extensive bulking measurements throughout the mine ranging from 35 to 42 % suggesting that worse case collapse scenario will not reach the surface.
- Seven borings into the CCA along with video logs to measure the height of roof collapse and the height of the rubble pile with an average bulking factor of 36 %.
- Direct observations of the vertical paleofracture at three locations within the CCA. The fracture was tight with a dip of 10–20° to the inside of the collapse.
- Direct observations at one point which exposed what was thought to be the uppermost zone of collapse within the CCA, about 18 m above the mine roof at the base of the massive Westerville Limestone.

25.3.7 Determining Sources of Water Filling the Mine

The former mine superintendent (Nicholson 1988), indicated that the mine was generally dry until mining extended southwest of the CCA by 90–120 m. Then a truck of water, about 3,700 l, was removed every few days. As the mine extended further westward, a number of sumps 0.9–1.5 m deep were installed below the mine floor and pumped as needed to control water. These sumps were installed within the Hushpuckney Shale, the Middle Creek Limestone, and the Ladore Shale, which is up to 3 m thick (Fig. 25.5). A large pump had been installed in the main sump at the center portal to remove runoff water from the portal road (Fig. 25.17). After mining operations ended in 1980 and without continued pumping, the mine began to slowly flood from the southwest (Nicholson 1988 personal communication). Subsequently a well and pump was installed in the extreme southwestern corner of the mine to provide water for daily dust control along the landfill roads.

Initial access to the mine in 1987 was by foot. During our first trip into the mine the mine floor around the CCA was mostly dry. Monitoring of water level within the mine was started in late 1987 using a digital water level recorder. A staff gauge was used to provide back-up data in case the electronic equipment failed and was read on each trip into the mine. Measurements indicated that water was increasing at about 3.3 cm/month or about 40 cm/year. At that time there was an estimated 227×10^6 l of water in the mine increasing at an estimated rate of approximately 1.3×10^6 l/month.

By April 1990 water levels had extended eastward to encircle the CCA. The extreme northernmost portion of the mine was dry at that time. As time passed, access to the northernmost portion of the mine to obtain water quality samples became more difficult due to raising water levels and finally became impossible. On our last trip into the northernmost portion of the mine to sample mine water in early 1991, the clearance between the water and the mine roof west of the CCA was about 45 cm requiring us to lie back in the boat and push on the mine roof by hand to obtain passage.

Possible sources of water entering the mine had been suggested by others and included:

- Seepage from the portal wall,
- Surface runoff into the mine from the portal road, and
- Seepage from the mine roof including the paleocollapse fracture zone.

The combination of water seeping out of the northeastern side of the portal wall and the water flowing down the portal road were both associated with intermittent rainfall and could not possibly account for the large volume of water entering the mine. In addition, the hundreds of hours of observations in the mine revealed no flow or significant seepage into the mine along fractures in the overlaying rock or along the paleofracture associated with the CCA.

These possible options for the source of water flowing into the mine could not account for the large quantity of water entering the mine, at an estimated 1.3×10^6 l/month. The focus now shifted to other possibilities for flow into the mine by a permeable zone below the Bethany Falls Limestone consisting of the Hushpuckney Shale, the Middle Creek Limestone, and the Ladore Shale (Fig. 25.5). This zone is up to 3 m thick (Hasan et al. 1988).

While no measurements were made in the Hushpuckney Shale at the Tobin mine, considerable work had been carried out at two other landfills and the large Missouri Portland Cement Co. mine and quarry some 29 km east in Jackson County Missouri. The geology there is identical to that of the Tobin mine.

Lateral flow within the Hushpuckney Shale had been observed at a number of locations at the Missouri Portland Cement Co. mine and quarry. A section of the exposed Hushpuckney Shale, Middle Creek Limestone and Ladore Shale can be seen in outcrops. These observations along with discussion with the geologist at the Missouri Portland Cement Co. indicated that the zone of rock immediately below the Bethany Falls Limestone was a permeable zone that transmitted water latterly quite rapidly (Camp 1988 personal communication).

Most of the mines in the Kansas City area, to the east, are well above the levels of the Kansas and the Missouri Rivers. At the Tobin mine the elevation of the mine floor in the southwest corner of the mine is about 7.6 m below that of the Kansas River, which is approximately 210 m to the south of the mine (Fig. 21.4). A piezometer was installed between the Kansas River and the southwest corner of the mine and showed that water from the Kansas River was flowing into the mine. Water from the Kansas River flows through a thickness of 12–21 m of alluvium capable of yielding from 567 to 3,785 l/min (O'Connor 1971). This water with a head of approximately 7.6 m or more, then flows through the permeable zone about 213 m into the mine. The issue of the source of mine water had now been resolved.

When the water level in the Kansas River and the mine are in equilibrium the mine will be flooded over most if not all of the CCA. Water level measurements were discontinued in 1995, but observations by downhole video camera in 2003 at the sump area near the portal entrance indicated water levels were above the mine floor level at the portal entrance.

25.3.8 Water Quality Measurements

As part of a groundwater monitoring plan a mine-water sampling program was initiated to characterize background quality of mine-water. Field parameters measured included pH, (9.7–12.4 near areas with fly-ash fill and 7.6–9.1 away from the fly-ash fill) specific conductance (2,260–7,340 umhos/cm) and temperature (12.9–16.4 °C). Laboratory analyses were made for those expected parameters to be required for Subtitle D – Phase I, U.S.-EPA solid waste landfill monitoring. In addition, the samples were analyzed for several pesticides and herbicides. Geochemical analyses for major ions indicate that the mine-water is predominantly sodium sulfate in character.

25.3.8.1 Possible Further Dissolution of Limestone

There was some concern by the state about the possible dissolution of the remaining mine pillars as well as the crushed-rock limestone that was being used as mine backfill material. Natural dissolution of limestone is caused by fresh rain water entering the limestone groundwater system. Since rain water is slightly acidic due to CO_2 in the atmosphere and humic acids, the slightly acid water slowly dissolves the limestone. Ford and Williams (2007) indicate that dissolution rates are on the order of 2.5 cm/1,000 years.

The water entering the mine is dominantly from the Kansas River, which is flowing through fractured rock, including limestone, into the mine. River water flowing through the limestone on the way to the mine becomes partially or totally saturated with calcium. The calcite saturation index (SI) was calculated for several mine water samples. The SI for a given sample is an indicator of whether the sample is saturated with respect to calcite (i.e. whether it has the capacity to dissolve any limestone). A positive SI value indicates that the water is oversaturated and will not dissolve calcite. The SI values were calculated using procedures developed by Hem (1989).

The high pH measured in the mine-water (7.6–9.1) and an average positive saturation index of (0.63) calculated from geochemical analyses, indicate that geochemical conditions are oversaturated and dissolution of the pillars or crushed limestone backfill will not occur. After water level in the mine is in equilibrium with that of the Kansas River, there will be no further flow into the mine and the water chemistry will come to equilibrium and any dissolution of limestone, however small, will cease.

25.4 A Groundwater Monitoring Plan

Very early portions of the landfill had been placed upon the top of thick loess or clay layers. Later when mine backfilling occurred the Argentine Limestone (the uppermost rock at the site) was removed and crushed for use as backfilling material. The landfill was then placed upon a plastic liner and a leachate collection system was installed. A detailed groundwater monitoring plan was developed for the site (Benson et al. 1991b) and is presented in Sect 21.7.1 and summarized here.

Sixteen of the open boreholes had been logged with a borehole video camera. The video logs were used to identify fractures and zones of water. Based upon the video logs, joints were common in the Raytown-Paola and Drum Limestone and limited quantities of water were encountered. Little, if any water, was found in the majority of bedrock overlying the mine. In addition, the Chanute Shale was seen to swell and seal off 20 cm diameter open boreholes that could then hold water in less than 10–12 months.

The results of the dye trace study indicated rapid lateral flow within the Raytown Paola and Drum Limestones (up to 150 m in 1 day). The Raytown Paola Limestone was also the first continuous layer of limestone under the landfill. These two zones were selected as the uppermost units to be monitored. In addition the mine water would be monitored (Fig. 21.4).

25.5 Subsidence Risk Assessment

All to often risk assessments are based upon an overview of conditions (opinions and assumptions) with little site-specific data to back them up. Because of the extensive work at this mine site and others within the area, the possible factors associated with mine collapse are well understood.

The strategy used to predict the risk of surface subsidence is simple, but highly dependent upon first identifying the key geologic, hydrologic and cultural factors which may possibly contribute to surface subsidence. Then, obtaining appropriate and adequate site-specific data to characterize these factors and carrying out a detailed assessment of them. This strategy also helps to minimize the use of opinions and assumptions. Subsidence risk prediction is then relatively straightforward in concept and a realistic estimate of subsidence risk can be made (Benson and Hatheway 2001).

25.5.1 Mine Stability Assessment

Mine conditions were assessed by Benson and Hatheway over a period of more than 3 years. In addition, three independent mine experts Golder (1986), Tein (1988), and Brink (1989) had made 1-day inspections of the mine. All concluded that the mine itself is stable and would not cause surface subsidence.

The key factors associated with possible mine collapse can be divide up into three zones:

- Zone 1 Conditions above the open mine,
- Zone 2 Conditions within the mine and,
- Zone 3 Conditions below the mine

Zone 1: Conditions above the open mine, include two potential problem areas:

1. The two deeply incised valleys (Figs. 25.7) one of which reaches the massive Westerville Limestone. These fracture zones may extend to the mine level allowing surface water to seep into the underlying strata thereby reducing its strength. However, there were no indications of fractures of mine roof failure within the mine at these two locations

2. A variety of data clearly shows the unique conditions of subsidence along with fractures and open fissures within the rock over the CCA. These conditions were further verified by the rapid lateral flow of dye trace water within the Raytown-Paola and Drum Limestones along with the results of subsidence measurements and the observations of high angle fractures (Fig. 25.22). These conditions of inherently weakened rock within the area of the CCA could result in further settlement. However, any such settlement would likely be small (probably only a few cm), see subsidence measurements (Fig. 25.16).

Zone 2: Conditions within the mine include the rubble rock on top of the Bethany Falls Limestone and the Hushpuckney Shale directly beneath the Bethany Falls Limestone (Fig. 25.5). Increasing water levels would negatively impact both of these conditions.

The mine water level will continue to increase until it comes into equilibrium with that of the Kansas River. As of 2003, water was over the mine floor at the portal entrance. Water will eventually fill the westernmost portion of the mine and the entire CCA. This will cause further degradation of the rubble rock, and the Hushpuckney Shale.

Further degradation of the rubble rock will initiate Type A and B collapse. These are localized collapse, which extend upward to or into the Winterset Limestone. Further degradation of the Hushpuckney Shale could also lead to pillar puncturing of a few centimeters so. However, this would be uniform over the mine. If this occurred it would impact the entire western half of the mine by uniform subsidence (on the order of a few cm).

Zone 3: Conditions below the mine would include reactivation of the original paleocollapse associated with the deep cavern system within the Mississippian Limestone some 180 m or more below the mine (Fig. 25.6). This is highly unlikely, since the original collapse of the cavern within the Mississippian Limestone occurred in past geologic time. These paleocollapse features are believed to be structurally stable and there have been no known reactivations of similar collapse zones in the Kansas City area (Gentile 1988 personal communication).

If the worse case conditions in Zone 1 and 2 were to occur, any subsequent surface subsidence would be small on the order of a few centimeters, and would be minimized by the presence of the mine backfill material.

25.5.2 Two Examples of Subsidence

The following four factors have been emphasized as an argument against subsidence reaching the surface at the Tobin mine site:

- A thick layer of rock over the mine (52 m) of limestone and shale)
- Arguments of mine stability by the authors and other mine experts
- A bulking factor of 30 % would have stopped collapse at or below the Westerville Limestone
- The presence of the massive and thick (6 m) Westerville Limestone

However surprising things do happen! The following are two examples of subsidence in which we have had the opportunity to make observations of the cause and effects of surface subsidence in the Kansas City Area. In both cases, the overlaying strata are identical to that in the Tobin mine, approximately 52 m of a limestone and shale. Under normal hydrogeologic conditions we would not have expected surface subsidence or collapse to occur at these two sites due to the thickness of rock over the mine and the associated bulking factor using a similar analysis to that of the Tobin mine.

25.5.2.1 The Inland Storage and Distribution Center

The Inland Storage and Distribution Center is located a few kilometers to the south of the Tobin mine and had railway access into a mine along with refrigerated storage. A localized portion of the mine-roof had collapsed and resulted in surface subsidence. The mine collapse had choked-off and was not accessible to determine the upper extent of roof failure similar to a Type D collapse in the Tobin mine (Fig. 25.21). Surface evidence of subsidence was seen as open fractures of a few centimeters. Surface observations along with topographic maps revealed that the collapse had occurred along a deeply incised valley. It is likely that this valley had focused surface water into a major fracture system and water had migrated downward into the rock over a long period of time resulting in a zone of weakened rock. This zone of weakened rock allowed the mine-roof collapse to propagate to the surface.

25.5.2.2 The Eastern Portion of the Tobin Mine

A medical facility consisting of two buildings and large paved parking lots had been constructed over the eastern end of the Tobin mine (Fig. 25.1). The area of the medical facility including the buildings and paved parking lots covered approximately 1.6 ha.

While this was an area of the mine that had not employed uniform engineered pillar design, no significant roof failure had been reported by the miners or noted by the surveyors in the eastern half of the mine. Furthermore, the mine superintendent, (Nicholson 1988) indicated that the Bethany Falls Limestone was generally better quality on the east side of the mine and that the eastern portion of the mine did not have

Fig. 25.23 Evidence of surface subsidence up to 2 m is seen in the paved parking lots of the two medical office buildings that were built over the easternmost extent of the Tobin Mine

a water problem. There had been a few small, localized roof collapse of Type A, similar to those in the western portion of the mine, but there was no large area of collapse such as the CCA. There was no reason to suspect surface subsidence would occur. However, subsidence did occur over much of the 1.6 ha, most obvious in the paved parking lots resulting in a undulating surface with as much 2 m of subsidence (Fig. 25.23). The medical offices were abandoned and removed.

An analysis of the aerial photos (Beccasio 1988) indicated the presence of a major north-northwest lineament crossing the northwest portion of the medical complex and a minor northeast lineament crossing the center of the medical complex. These lineaments intersected at the location of the medical facility buildings.

Three factors were identified that in combination likely contributed to triggering the collapse and subsequent surface subsidence at this location.

- First the site was located over or near the intersection of two photolineaments (fracture systems) which most likely resulted in a zone of weakened rock which could have extended downward to the level of the mine.
- Second the two buildings and their associated parking lots covered an area of approximately 1.6 ha. This resulted in a large amount of surface water runoff from the building roofs and paved parking lots being concentrated over these fracture zones of weakened rock.
- Third, there was little or no loess soil cover at this site to slow infiltration of rainfall and runoff. The highly fractured Argentine Limestone was at the surface providing a permeable zone for rainfall and runoff to penetrate.

The presence of the lineaments (fracture zones) alone may not have been significant. However, the combination of the lineaments along with the concentration of runoff from 1.6 ha created the mechanism to trigger collapse and subsidence at this site.

In cases where we have had the opportunity to review site-specific conditions, a geologic flaw is commonly found to be present along with possible cultural features which impact site conditions. When such conditions are identified, they can often be avoided or managed by proper engineering.

25.6 The Mine Backfilling Program

Although the mine had been assessed as being stable, the responsible State agency had required that the mine be backfilled to insure the long-term stability of the mine. Their specifications included:

- Filling the mine 30 m beyond landfill boundary.
- Filling the mine to at least 90 % full (or 3.8 m of the 4.2 m high mine)
- Fill material was to have a strength of 1.5 MPa.

25.6.1 Initial Efforts with Fly-Ash

A major effort of mine backfilling with fly ash began in late 1987. The mixture consisted of approximately 816 kg of type F fly ash, 90 kg of kiln dust, 90 kg of bottom ash, and 360 l of water. This slurry was created at an on-site facility using material waste from local power companies. It was mixed in cement trucks and dumped into the mine via 20 cm open boreholes.

While such a fly-ash mixture is an excellent backfilling material, it will harden in a short period of time much as cement. The fly-ash would set up and harden over evenings and weekends and would, at some fillhole sites, form a zone of conical fill under the borehole choking off the borehole and preventing further filling of the mine. A variety of alternatives were considered including a 24 h per day batch plant to mix and transport the fly ash to the mine via surface piping.

25.6.2 Crushed Rock Slurry Backfill

An experienced mine consultant, Goodson Associates, Inc., of Denver, Colorado recommended the process of using a crushed rock slurry. Based upon favorable economic factors, the owner elected to employ the crushed rock slurry backfill program, which provided three benefits:

- First, it removed the highly weathered and fractured uppermost limestone, the Argentine-Frisbie Limestone, from immediately beneath the landfill.
- Second, the on-site Argentine-Frisbie Limestone was an existing on-site resource of backfill material.
- Third, by removal of about 7.6 m thick Argentine-Frisbie Limestone, the volume available for the landfill was significantly increased and the cost benefits essentially paid for the backfilling of the mine.

The initial base of the landfill was designed to lie upon the loess soil or on a clay cover on top of the Argentine-Frisbie Limestone. Once the Argentine Frisbie-Limestone was removed, the landfill cells were then paced on a plastic liner over the 10.6 m thick Lane Shale (Fig. 25.2).

The upper layer of highly fractured and weathered Argentine-Frisbie Limestone (about 7.6 m thick) was removed using a D9-L caterpillar impact ripper with a single tooth. The rock was delivered to a crusher that provided 1.25 cm rock, which was then mixed with water and pumped via piping to a mine fillhole and injected as mine backfill material. The large volume of water used was obtained from the mine and was re-circulated as rock and water were injected into the mine.

The crushed rock pumped slurry method differs from open-gravity feed methods previously used with fly-ash. The pumping energy is used to achieve a dynamic suspension of solid crushed limestone in water and the system is essentially closed from the point of slurry mixing to the bottom of the injection borehole at 5,677 l/min. The energy provided by the pump and the static head in the borehole provided the velocity required to keep the solid particles in suspension and to transport them away from the borehole once they had entered the mine. Furthermore, the crushed rock slurry method did not set up and block the fillhole as did the fly ash. This allowed the filling operation to be conducted during a 10-h shift per day (Goodson Associates 1988).

As the slurry first enters the open mine from the injection hole, its velocity drops rapidly and solid particles settle out near the borehole, forming a doughnut-shaped mound on the mine floor (Fig. 10.1a). As the height of the mound approaches the mine roof, the velocity of the slurry increases through a narrow channel, and solid particles are transported to the outer diameter of the mound. Here the velocity decreases abruptly and solids are deposited.

As the mine backfill material begins to fill the mine to the roof, the flow of water and rock cut a new path to the perimeter of the fill. At some point friction begins to increase and a new channel is cut. This process of cutting a new channel occurs about every 20 min and continues in a random fashion around the perimeter of the fillhole placing rock to the roof of the mine until an area of about 22–30 m radius around the fillhole was backfilled to the mine-roof (100 % full) (Fig. 10.1). At this point, the friction loss of the rock slurry between the rock fill and the mine-roof became greater than the head of slurry being injected and further filling ceases.

Fillholes were spaced approximately 45–60 m apart in order to provide 90–100 % filling of the mine. The amount of fill placed in each hole would vary from 7,600 to more than 15,300 m^3 assuming a density of 1.4 tons/yd^3) depending upon the location of nearby mine walls and previous fill material. The slope of the fill (angle of repose) ranged from 4 to 9° and averaged 6°.

25.6.3 A QA/QC Program for Mine Backfilling

A Quality Assurance Audit program and Quality Control procedures were developed and included:

- Direct observation, measurements along with photos and video documentation of the early backfilling.
- Quality control monitoring included mapping the perimeter of the fill after each backfilling sequence. A contour map of fill was developed based upon these observations and measurements. Photographs were taken of the entire accessible perimeter of backfill to document conditions.
- Quantities of fill were recorded daily and
- Physical property measurements of the mine backfill material were made and included sieve analysis, Los Angeles (LA) abrasion test, minimum index density, maximum index density, and a California bearing ratio test.

Fig. 25.24 In-situ compression tests were made on the crushed rock fill material

25.6.3.1 In-Situ Compression Tests of Fill Material

In-situ compression tests were carried out on both fly-ash and crushed rock backfill which had been placed within the mine. The State required a minimum of 1.5 MPa for the backfill materials. Figure 25.24 shows the acquisition of compression tests over crushed rock. The average in-situ value for the tests on crushed rock was 24.7 MPa with an average vertical displacement of 0.0155 cm and no indication of failure. A strong limestone typically has unconfined compressive strength of 103.4–137.9 MPa. The average maximum in-situ values for the two tests on fly-ash was 5.5 MPa with an average displacement of 1.7 cm. These tests on fly-ash and crushed rock provided results well above the requirements of the 1.5 MPa by the State.

25.6.4 Summary of Mine Backfilling

The backfilling of the mine eliminated the State agency's concerns for long-term stability of the mine. While the state requirements were to backfill the mine to 90 %, the crushed rock slurry method provided essentially 100 % backfill to the mine roof over a radius of 22–30 m and fill to 90 % to the mine-roof over distances up to 68 m from the fillhole.

A fill material strength of 1.5 MPa was required to support the weight of a theoretically completely failed rock overburden (a maximum of 54 m of rock and a maximum landfill thickness of approximately 36 m).

- The crushed rock backfill strength is more than 15 times the amount required to resist compression in the theoretical worst-case instance of a completely failed rock overburden;
- The fly-ash backfill strength is more than 3.5 times that required.

In either case, the backfill material has adequate strength and an unusually high factor of safety against resistance to worst-case (but unanticipated) mine stability conditions.

The crushed rock slurry program proved successful and provided a dramatic increase in the degree of lateral and vertical filling to achieve closure of the open mine space. Once the permanent roof-void space was reduced to near-zero, the perceived risk of further subsidence was terminated. While both fly-ash and rock slurry proved to be good materials, from a strength point of view, the economics of the additional landfill air space gained by removal of 7.6 m of limestone made the crushed rock slurry not only a technical success, but an economic one as well. Further details are given in by Benson et al. (2000).

25.7 Conclusions

The work at this site was carried out over a period of 7 years and provided abundant site-specific assessment of geology, hydrology and mine conditions over time. Hundreds of hours were spent both above ground and underground in the mine.

At most sites we will not have nearly the time or data as we have at this site.

During this 7 year period there had been four different site engineers involved and the ownership had changed three times resulting in a loss of corporate memory as well as data files and reports. However, the same senior staff was involved with the site characterization over the entire 7 year period. This continuity was a key factor in acquiring the data and integrating the many pieces of the geologic puzzle at this site.

None of the issues identified in this case history required a high level of analysis. Simple observations and measurements along with an assessment and integration of a variety of data were used to piece together the geologic and hydrologic puzzle of the site conditions.

Existing literature from Hasan et al. (1988) provided a solid basis for understanding of the local geology hydrology of the region and specific mine conditions. Gentile (1984) had reported the presence of paleocollapse from within the deeper Mississippi Limestone, which became the basis of our conceptual model. The contributions of other mining experts Golder Assoc. (1986), Tein (1988), Brink (1989), Colaizzi (1990 personal communication), and Camp (1988 personal communication) also provided support for this work.

A preliminary conceptual model of what we feel caused the surface fissure and subsequent mine collapse is shown in Fig. 25.6. This conceptual model was developed based upon work by Gentile (1984) along with observations from trenching and within the mine. Subsequently the preliminary conceptual model was supported by

- Aerial photo interpretation that identified evidence of fissures in the 1954 photos indicating that the surface fissures were present prior to mining.
- The circular depression with up to 4.5 m of elevation change located approximately centered over the CCA (Fig. 25.12).
- Exposure of old fissures in the Lane Shale at an excavation that cut across the west portion of the CCA (Fig. 25.11).

There is sufficient data, based upon on-site and in mine measurements, observations, along with opinions of independent experts to provide a reasonable degree of confidence in the conceptual model of the mine and surface conditions. Two worse case scenarios are presented where subsidence has occurred under identical geologic conditions. When such subsidence occurs under circumstances in which it would not have normally been be predicted, it is a combination of a geologic flaw and possibly the impact of cultural conditions that have caused the subsidence. A risk assessment indicated that significant subsidence is highly unlikely and that the backfilling of the mine will prevent any significant surface subsidence.

This case history uses knowledge from extensive investigation of the Tobin mine and other mines in the Kansas City area. A series of papers have been presented on the work at this site and others in the area. They include:

- Hatheway et al. (1990)
- Benson et al. (1991a)
- Benson et al. (1991b)
- Benson et al. (2000)
- Benson (2001)
- Benson and Hatheway (2001)
- Hatheway et al. (2001)

The landfill operations have since been completed and the landfill has been covered over. In addition, the mine portal road has been filled with debris and the portals covered over.

References

Aley T (1991) Results of spectrofluorometer analysis of dye samples from charcoal bugs. Consultant report

Anderson Survey Co. (1990) Subsidence monument measurements. Consultant report

Beccasio D (1988) Photogeologic interpretation of the Forest View Landfill Site, Kansas City Kansas. Consultants report

Benson RC (2001) Strategies of site characterization and risk management. An invited presentation to the National Research Council, committee on coal waste impoundment failures, subcommittee on impoundment site characterization, Washington DC

Benson RB, Hatheway AW (2001) A systematic approach for evaluating subsidence risk. Presented at the AEG/AIPG annual meeting, St. Louis, MO

Benson RC et al (1991a) Site characterization for a landfill located over a mine with superimposed paleokarst collapse, a case history. In: Proceedings of the 3rd conference on hydrology, ecology, monitoring and management of groundwater in karst terranes, Nashville, 4–6 Dec, pp 131–145

Benson RC et al (1991b) Ground water monitoring in unsaturated and saturated zones at a site with paleocollapse structures. Presented at the annual meeting and 10th anniversary of the American Institute of Hydrology International conference on hydrology and hydrogeology, 3–7 November 1991, Orlando, FL

Benson RC et al (2000) Stabilization against geologic instability – underground documentation of a mine backfilling case history. ASCE Geo-Denver Conference, Colorado, 5–8 Aug

Boston Survey Consultants (BSC) (1990) Design and analysis of the subsidence monitoring network at Forest View Landfill. Consultant report

Brink T (1989) Inspection of the Tobin mine. Consultant report

Dunnicliff J, Green GE (1988) Geotechnical instrumentation for monitoring field performance. Wiley, New York

Ford D, Williams P (2007) Karst hydrogeology and geomorphology. Wiley, Chichester

Gentile RJ (1984) Paleocollapse structures: Longview Region, Kansas City, Missouri. Bull Assoc Eng Geol 21(2):229–247

Golder Associates (1986) Mine backfilling requirements for the Forest View Landfill Site, Kansas City, Kansas. Consultant report

Goodson Associates, Inc. (1988) Proposal for mine backfilling using crushed limestone. Consultant report

Hasan S et al (1988) Geology of greater Kansas City, Missouri and Kansas, United States of America. Bull Assoc Eng Geol 25(3):281–341

Hatheway AW et al (1990) Geologic factors responsible for mine collapse under a portion of the Forest View Landfill Site, Kansas City, Kansas. In: 33rd annual meeting of the Association of Engineering Geologists, Engineering Geology for the 90's, Pittsburgh, PA, 1–5 Oct

Hatheway AW et al (2001) Geotechnical risk assessment of US Route 169 over the Briarcliff Mine, Kansas City, Clay County, Missouri. Consultants report for the Missouri Department of Transportation, unpublished

Hem JD (1989) Study and interpretation of the chemical characteristics of natural water. US Geologic Survey, Water Supply Paper 2254

Nicholson M (1988) Former Tobin mine superintendent. Notes from personal interview with Hatheway AW at Forest View Landfill, July 12

O'Connor HG (1971) Geology and ground water resources of Johnson County, northeastern Kansas. Kansas Geological Survey Bulletin 203

Quinlan J (1991) Results of spectrofluorometer analysis of water and dye samples. Consultant report

Tein JC (1988) Mine stability at Forest View Landfill. Consultant Report

Tuttle, Ayers, and Woodward Co. (1979) Plat map of J.A. Tobin mine showing time development of mine boundaries. Consultants report. Kansas City, Missouri

Site Characterization Along Bridge Alignment

26

Abstract

The second case history involves a site characterization for a new bridge into the Florida Keys. There was concern that there may be paleosinkholes present because of the alignment of four circular shaped lakes. The presence of paleosinkholes could impact the foundation for the new larger and higher bridge over Jewfish Creek between the Florida mainland and Key Largo in the upper Florida Keys. The site characterization effort was limited to the existing two-lane road and the surrounding areas that were covered in water. A major drilling and sampling effort had already been completed as part of the Florida Department of Transportation (FDOT) geotechnical investigation. The project was carried out in 1994 over a period of 6 months in three phases following our standard site characterization strategy.

26.1 Background

A new bridge was being planned by the Florida Department of Transportation to ease traffic flow into and out of the Florida Keys, particularly during evacuations due to a pending hurricane. Access to the Florida Keys has been by two 2-lane narrow roads crossing from the lower tip of Florida's mainland across extensive bays and mangrove covered areas. Card Sound road to the northeast has a high bridge crossing the Florida Intracoastal Waterway and leads to the northern portion of Key Largo. The Florida Keys Scenic Highway, US1 crosses the bascule bridge constructed in 1944 at Jewfish Creek and then runs across Lake Surprise entering the middle of Key Largo (Fig. 26.1). The majority of traffic into or out of the Florida Keys uses US1. During a hurricane all traffic must exit the Keys by one of these two roads.

The planned new bridge must span Jewfish Creek, at the existing bascule bridge and cross Lake Surprise, a circular lake with US-1 passing though its center for about 1,066 m. The new bridge will begin northwest of the present bascule bridge at Jewfish Creek, then extend southeast to Key Largo for a distance of more than 2,400 m (Fig. 26.1). Drilled shaft foundations for the bridge will be placed over Jewfish Creek (the highest portion of the bridge) and through the center of Lake Surprise. The drilled shaft foundations will extend to depths of about 18 m.

There had been virtually no sinkhole activity reported in this area of the Keys, and initially there was no reason to suspect karst in the upper Florida Keys. However, a review of marine charts and aerial photos clearly shows Lake Surprise as one of four circular lakes in alignment along a northeast–southwest trend (Fig. 26.1). Such an alignment is common for sinkholes throughout central Florida often indicating the presence of an underground cave system. A project manager with the Federal Highway Administration (FWHA) had seen the linear trend of circular lakes on maps and had raised the issue of possible sinkhole problems at the site. However in the Florida Bay area and the Keys such patterns are often referred to as mangrove lakes and are not thought to be associated with sinkholes (Shinn 1994a).

US-1 is a narrow two-lane road constructed on fill through mangroves and Lake Surprise (Fig. 26.1). There are limited areas of developed land both sides of Jewfish Creek, which are associated with two marinas and a condominium. Other than the narrow two-lane US-1, the remainder of the site consists of either open water that is accessible by boat, or covered by mangroves with no practical access. Local wildlife, such as the saltwater crocodile, were also of concern.

R.C. Benson, L.B. Yuhr, *Site Characterization in Karst and Pseudokarst Terraines*,
DOI 10.1007/978-94-017-9924-9_26

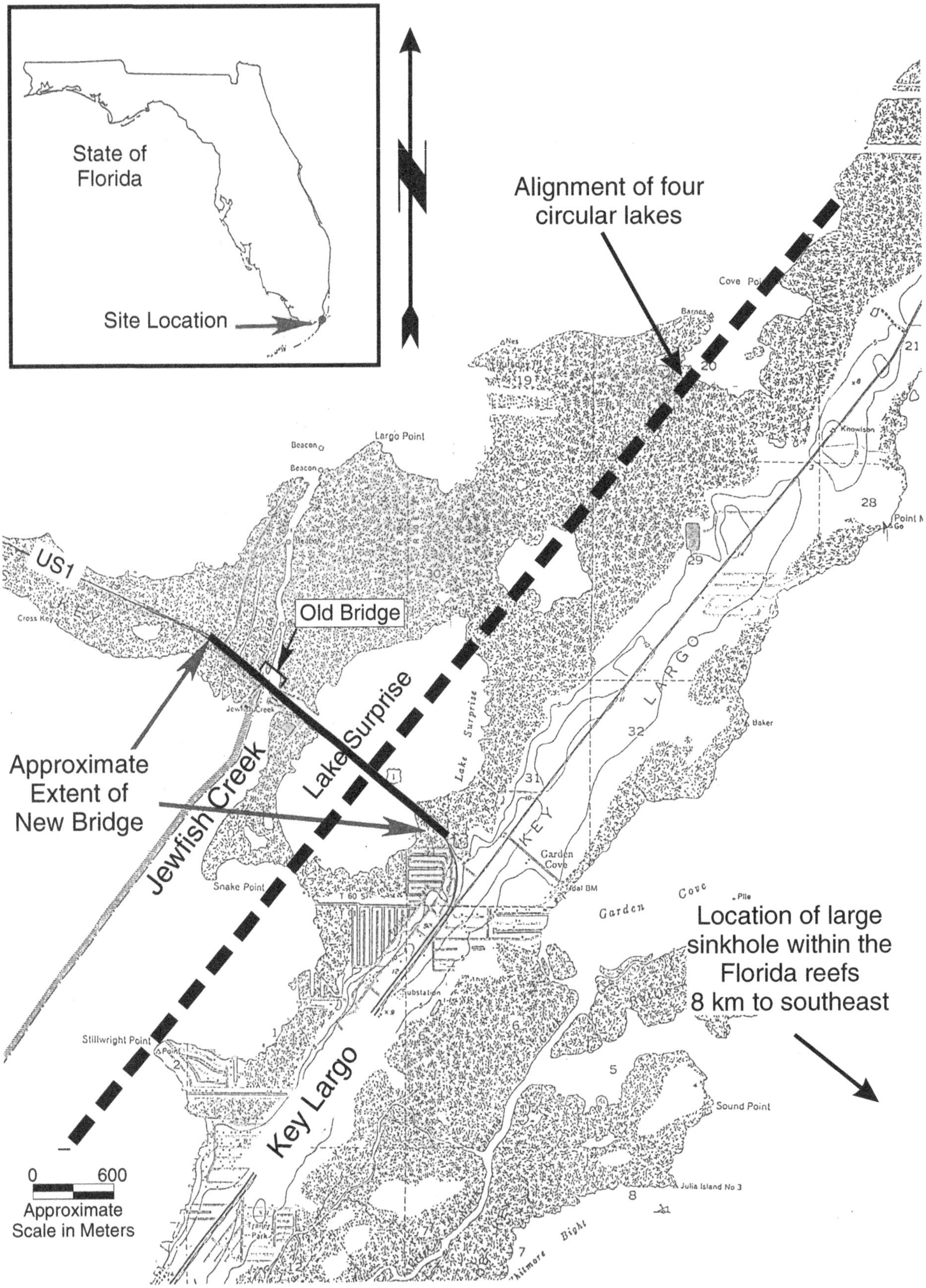

Fig. 26.1 Site location map showing the extent of new bridge construction over Jewfish Creek and Lake Surprise along with the alignment of four circular lakes

26.2 An Initial Site Assessment

Prior to beginning the site characterization effort there was a review of findings by USGS, and an initial site visit.

26.2.1 Findings by USGS

The United States Geological Survey (USGS) had located a large 594 m diameter paleokarst sinkhole collapse filled with unconsolidated materials among the reefs off of Key Largo, 8 km to the southeast of Jewfish Creek (Fig. 26.1). The sinkhole is located in about 6 m of water within the Key Largo National Marine Sanctuary. USGS has cored to a depth of 32 m and wash bored to a depth of 55 m finding only unconsolidated fill and no solid rock. The oldest C14 age from the jet probe was 5,650 +/−90 year before present (BP) (Shinn et al. 1994b). The 594 m diameter and minimum depth of 55 m suggest a very large paleosinkhole. The proximity of such a large paleosinkhole within the Florida Keys implies the possibility of other deep cave systems in the area.

26.2.2 Initial Site Visit

During the initial site visit to establish boat launching facilities, an obvious patch about 15 m long was noted in the southbound lane of US-1 about 243 m south of Jewfish Creek bridge (Fig. 26.2). The Jewfish Creek bridge tender and the nearby business owner at Gilberts Marina indicated that extensive road maintenance was routinely carried out by DOT at this location, (personal communication with the manager of Gilberts Marina and the Jewfish Creek bridge tender, 1994).

A meeting with the Florida Department of Transportation (FDOT) engineers indicated that extensive road maintenance due to continuing subsidence had been carried out at that location over a number of years. The FDOT engineers had explained the subsidence was due to decaying timbers, which had been encountered by shallow borings (Miro 1994 personal communication). The timbers were used in the construction of Flagler's overseas railway to the Florida Keys completed in 1912.

Given no other data or information one would consider this simply as a routine road maintenance problem. The problem with the explanation was that wooden piles and timbers had been used all along this area on US 1 to construct Flagler's overseas railway. However, this 15 m long area was the only location where such subsidence was occurring, there were no other such maintenance problems in the area along the old railway line form the Florida mainland to Key Largo.

26.3 The Approach

The FHWA, who provided guidance and funding for the project, initiated a karst investigation of the site. The purpose of this investigation was to evaluate the possible presence and impact of significant paleokarst (old sinkholes or cavities) along the alignment of the proposed bridge to be

Fig. 26.2 A patch about 15 m long was noted in the southbound lane of US-1. The Jewfish Creek lift bridge is seen in the top of the photograph. The view is towards the northwest

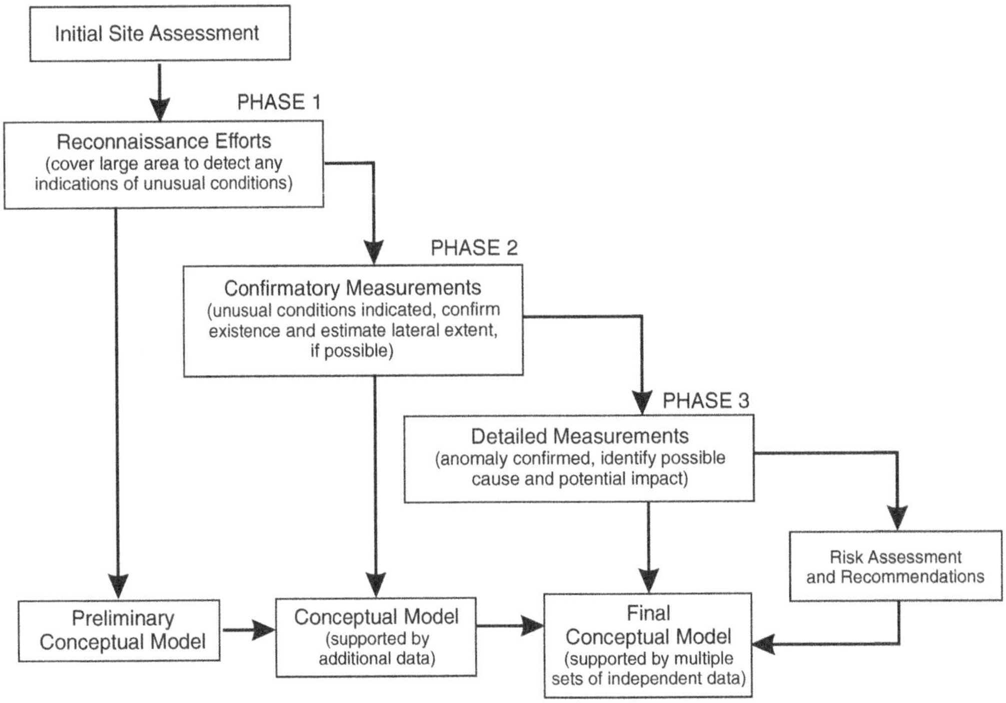

Fig. 26.3 The overall organization of tasks for the three-phased project

built over Lake Surprise and Jewfish Creek. The initial focus was upon Lake Surprise that was thought to be a possible old sinkhole.

Two critical questions were asked at the beginning of this study:

- Can an investigation fully identify unstable geology, such that bridge decks can be placed or lengthened to safely span the critical geologic conditions?
- Can the bridge as designed, be stable for at least 100 years in the given geologic environment?

Figure 26.3 shows the organization and flow of the overall project. A phased strategy was developed to investigate the possible karst conditions at this site. The investigation would consist of a three-phased approach:

- Phase I was a reconnaissance phase to assess whether or not potential karst problems existed;
- Phase II was a confirmation phase, to verify and expand findings of anomalous conditions that might be found in Phase I;
- Phase III was a detailed investigation to specifically assess the impact of any karst conditions found.

26.4 Phase I Reconnaissance Investigation

Phase I reconnaissance efforts consisted of:

- A review of geologic literature and site geomorphology;
- A site fly over to assess site access;
- Reconnaissance marine seismic reflection through Lake Surprise;
- A microgravity survey along US1 from Key Largo across Lake Surprise and extending northwest of Jewfish creek.
- An aerial photo lineament analysis; and
- A detailed review of existing boring data for possible indications of karst conditions.

26.4.1 Review of Regional and Local Geology

Sources of information for regional geologic conditions came from work by Parker et al. (1955), Lane (1981), and Scott (1988). Additional information was obtained from deeper regional investigations were being completed in southeast Florida and the Florida Keys by the University of Miami (McNeill et al. 1995).

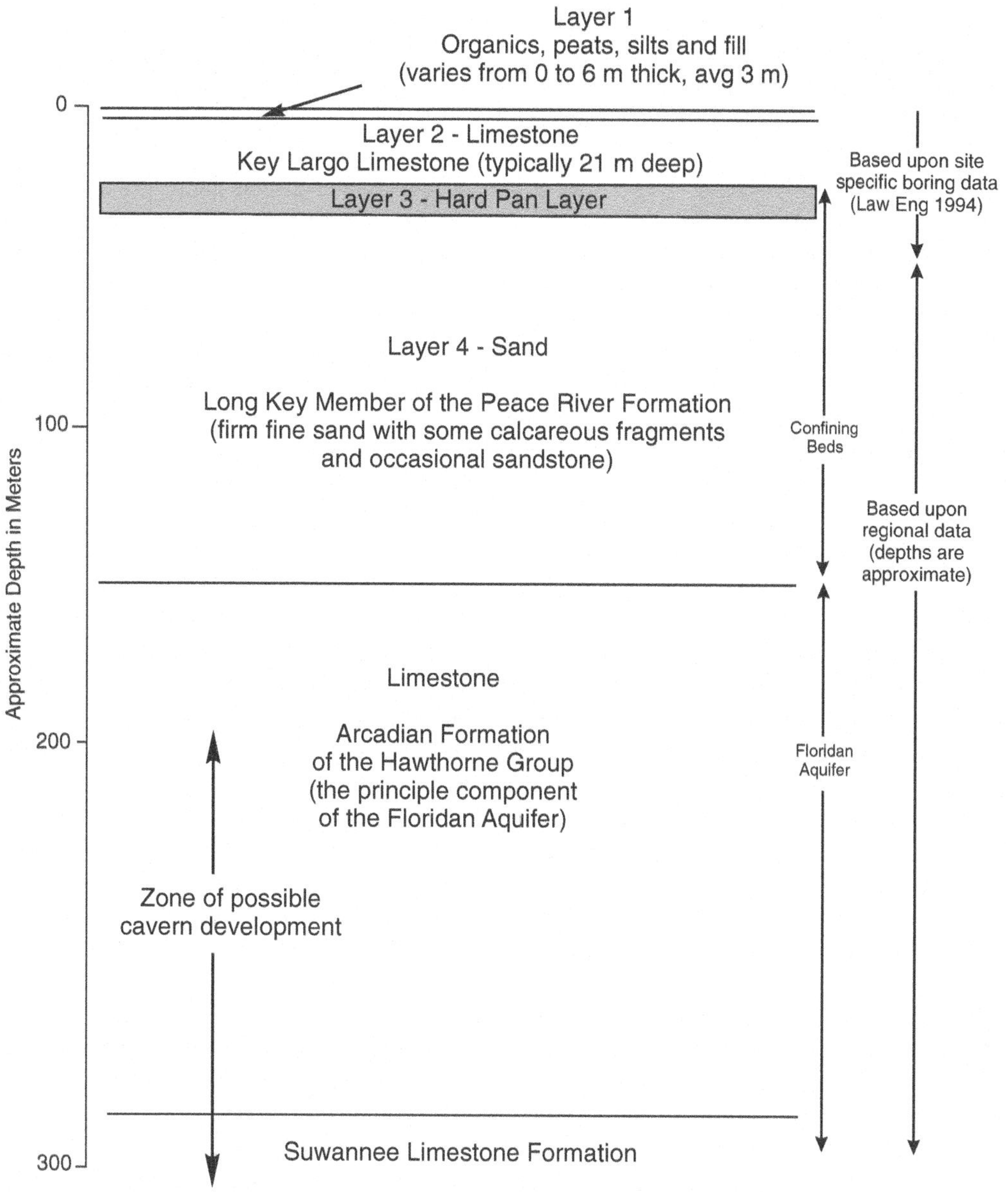

Fig. 26.4 The geologic setting for the site

Local geology was provided by the drilling data acquired for the bridge design. Prior to beginning our work, Law Engineering (1994) had completed a series of 34 borings over a distance of 2,621 m. The borings were spaced on average 76 m apart along the bridge right-of-way and had been drilled to depths of 20–47 m with the majority extending to 30 m.

The stratigraphic sequence at the site is shown in Fig. 26.4. Four geologic layers had been described as follows:

- Layer 1 consists of water, organics, peat, silt and fill materials that varies from 0 to 6 m thick, but is typically less than 3 m thick.
- Layer 2 is a vuggy limestone, which averages about 21 m deep over much of the site. This layer is equivalent to the Key Largo Limestone which is a coralline reef rock ranging from hard and dense to soft with open voids. The spaces between and around the coral heads are filled with limestone debris, cemented carbonate sands and uncemented carbonate sands.
- Layer 3 is a significant hard-pan layer present beneath the Key Largo Limestone at a depth of about 21–30 m. This hard-pan is the Caloosahatchee equivalent (Scott 1995 personal communication). It consists of sandy marls, clay, silt, sand, and shell beds and generally has low permeability.

- Layer 4 consists of a firm, fine quartz sand with some calcareous fragments and occasional sandstone. Only 4 of the 34 existing borings were drilled deep enough to encounter this layer. This layer extends from approximately 30–150 m deep; and is the Long Key Member of the Peace River Formation (McNeill et al. 1995).

Below Layer 4 between approximately 150 and 290 m, the Arcadian Formation of the Hawthorne Group limestone occurs (Scott 1988) (Fig. 26.4). Deep borings have encountered the Arcadia Formation both north and south of Jewfish Creek. This is the first limestone in which a large cave system could develop. The Suwannee (limestone) Formation lies below the Arcadian Formation (McNeill et al. 1995).

26.4.2 The Regional Geomorphology

Large surficial karst features or deeper caves have not been recorded in south Florida and in the Florida Keys as they are in central Florida where active subsidence and collapse is on-going and readily noticed. On a regional scale, the nearest large open sinkhole occurs in the Sarasota County 290 km northwest from Lake Surprise. Warm Salt Spring (in Sarasota County) is well known as an archaeological site (Ferguson et al. 1947). Other examples include the numerous Blueholes on Andros Island in the Bahamas 290 km to the east. Dean's Bluehole on Long Island about 442 km to the southeast in the Bahamas is at least 202 m deep and is described as the deepest blue hole in the world by Wilson (1994).

From a review of regional geomorphology we find that sea levels had been lower by as much as 130 m (Balsillie and Donoghue 2004). Dissolution and cave development is known to occur at the base of the fresh water lens (the fresh to salt water contact) (Ford and Williams 2007). The combination of much lower sea levels and a thick fresh water lens would have created an environment prone to the dissolution of limestone and development of caves at depths of 180–240 m or more below present sea level.

On a much more local scale, there are a number of factors which indicate that dissolution has occurred in the area.

- Numerous springs have been identified over 100 years ago within Biscayne Bay and offshore of Miami just 64 km north-northeast of Lake Surprise. The presence of springs indicates conduits with much greater porosity than the surrounding materials. Some of these appear to be in-filled sinkholes or blue holes (Wanless 1994 personal communication).
- Evidence of cavities and voids taking considerable grout has been encountered in the borings drilled for the geotechnical foundations of tall buildings and bridges in the Miami-Ft. Lauderdale area (Berkovitz 1994 personal communication).
- The proximity of the large sinkhole identified by USGS only 8 km from Lake Surprise (Shinn et al. 1994b) clearly implies the potential presence of other large deep cave systems and paleocollapse in the area.

While both regional and local evidence suggests dissolution has occurred in the area, the karst features typically expressed at the surface such as subsidence and collapse are not readily visible in South Florida. This is due to the infilling of karst features with sediments and their stability due to the very shallow water table in the area. Surface elevation on Key Largo is about 2.4 m above MSL.

26.4.3 A Site Fly Over

A site flyover was made prior to starting fieldwork. The aerial observations provided insight to access and limitations in and around the mangroves and identified any roads, trails, or areas of dry land that may be accessible which are not indicated on charts or topographic maps of the area. While standard marine charts provided a feel for the access by boat and topographic maps identified a few dirt roads and trails, the fly over provided a much better feel for the limitations and areas of access including dirt roads and trails within the mangroves as well as small channels not shown on the marine chart or topographic map. The oblique photos acquired were also useful for describing conditions to other members of the field investigation team and in planning fieldwork.

26.4.4 Reconnaissance Marine Seismic Reflection Survey

The Phase I reconnaissance efforts with seismic reflection were focused within Lake Surprise since it was thought to be a possible paleosinkhole lake. A total of 13.8 km of reconnaissance seismic reflection data was obtained from both north and south of US-1 on Lake Surprise (Fig. 26.1).

Data was obtained from a small boat using an EG&G Model 230 Uniboom system. Reflected signals were received by a single channel eight element hydrophone array. The data was recorded with a record length of 60 ms for a depth of about 63 m based upon an estimated velocity of 2,133 m/s. Although there is some lateral coverage to each side of the survey line, from a practical point of view the seismic reflection data can be considered to be from directly under the boat.

The seismic reflection data over Lake Surprise indicates horizontally bedded sediments. No obvious anomalies that might indicate the presence of karst conditions were identified in this data. While the 13.8 km of survey line provided a reasonable coverage to detect possible dipping strata associated with a sinkhole, the throat of a sinkhole is quite localized and

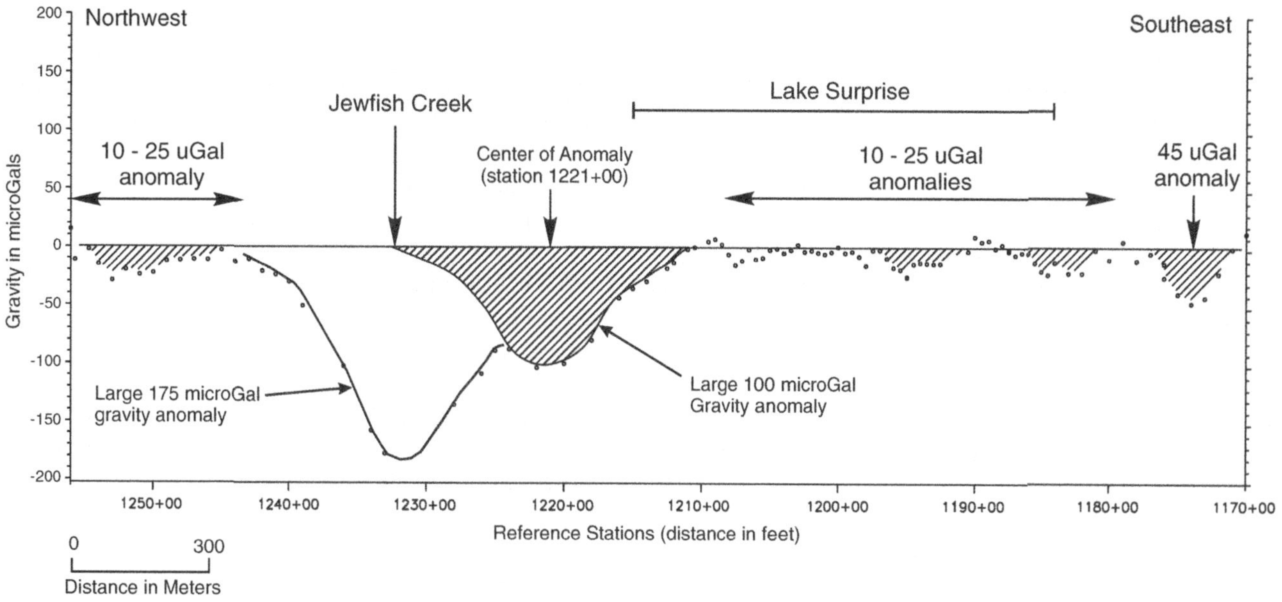

Fig. 26.5 The microgravity data along a 2,621 m survey line along US1 is shown in profile, with each station value shown as a dot

can lie anywhere within the sinkhole lake and might have been missed by this reconnaissance seismic survey.

26.4.5 Microgravity Data

A microgravity survey was run along the edge of the northbound lane of US-1. The survey extended from Key Largo across Lake Surprise and northwest of the Jewfish Creek Bridge. This covered the entire extent of the proposed new construction (a total of 2,621 m), however the focus was upon Lake Surprise. Gravity measurements were obtained at intervals of 7.6 m over Lake Surprise and intervals of 15 m beyond Lake Surprise for a total of 245 gravity measurements. Gravity measurements were made at night to minimize the noise from traffic along the narrow two-lane road. A Lacoste and Romberg Model G microgal gravimeter with electronic readout by EDCON was used to acquire the data.

The microgravity data for the Phase I survey is shown in Fig. 26.5. There is a level of geologic and vehicle noise present in the microgravity data. This noise level is estimated to be about ±10 microGals. Therefore, only anomalies above 20 microGals and with a consistent trend are considered to be significant.

There are two small gravity anomalies of 10–25 microGals over Lake Surprise and one negative anomaly of 45 microGals at the southeast end of Lake Surprise. Another small gravity anomaly of 10–25 microGals is seen at the northwest end of the survey line. Each of these can be explained by variations in the depth to rock seen in the boring logs and the presence of peat (a lower density material) within these low zones and as a result they can be ignored.

There is a large negative gravity anomaly of 175 microGals, that extends more than 900 m. This anomaly begins near the northwest edge of Lake Surprise and extends northwest of the Jewfish Creek bridge (Fig. 26.5). This anomaly consists of two independent anomalies, which are superimposed upon one another.

The large gravity anomaly of 175 microGals centered at Jewfish Creek can be explained by the increasing elevations of gravity stations along the built-up roadway approaching the bridge, plus the existence of the deeper channel at Jewfish Creek Bridge. Therefore, this large anomaly is a function of local changes in topography and can be disregarded. After removing the topographic effects, a smaller negative gravity anomaly of 100 microgals remains (Fig. 26.5). This very significant 100 microgal anomaly is centered at Station 1221+00 (FDOT station numbers in feet) and is the only significant anomaly identified by the microgravity survey.

The simplest possible cause of the gravity anomaly would be a localized increase in depth to rock with organics or loose sediments above rock. But ten boreholes within the large 100 microgal gravity anomaly do not indicate a deepening of the top of rock. In addition, these ten boreholes show no sign of a major void or other significant geologic feature that could account for a gravity anomaly of this magnitude, indicating that the source of the anomaly is deeper than 30 m, the depth of the borings.

26.4.6 Aerial Photo Analysis

While photo analysis normally uses aerial images over land, images over shallow water-covered areas can also be used, often to depths of 10 m (Finkl and Warner 2005). In this case, the aerial photo analysis was carried out over an area that is covered by a combination of shallow water, from 1.2 to 3 m deep, and by mangroves.

An analysis of aerial photos was carried out to detect the presence of photo-lineaments (Finkl 1994). The IDRISI image processing – raster GIS analysis program was used. Lineaments are significant linear features within the landscape that may be structurally controlled (Billings 1954) and are usually associated with one or a combination of joints, fractures, dikes, faults, geologic boundaries, or depositional history or geomorphology (Bagdley 1960, 1965). In the case of karst, a photo-lineament is a near surface indication of a possible fracture zone along which a cave system may be developed. Existing black and white stereo-paired aerial photographs (2-12-1991) were obtained from the Florida Department of Transportation for the general area. Two independent aerial photo analyses were carried out, one manually using a stereoscope by an experienced aerial photo interpreter and another by computer analysis. Both analyses identified a major NE-SW lineament centered at station 1224+00, about 100 m northwest of the center of the gravity anomaly. The focus now shifts to the narrow channel between Lake Surprise and Jewfish Creek based upon the microgravity anomaly, and the location of the photo-lineament.

26.4.7 A Detailed Review of Existing Boring Data

As a result of the microgravity data indicating anomalous conditions between Lake Surprise and Jewfish Creek, a detailed review of the existing boring data was made to assess indications of karst conditions. This data included the 34 existing borings by Law Engineering (1994). Core boxes were made available and selected cores were inspected. The geologic conditions seem to be reasonably well defined, but at this point the data had not been interpreted in terms of karst conditions.

26.4.7.1 Standard Penetration Tests

Standard penetration tests (SPT) were run in all 34 borings and were reviewed for patterns and trends within each of the geologic layers (Fig. 26.6).

SPT values within Layer 2, the Key Largo Limestone, range from 2 to 98 with an average of 28. They appear to be randomly distributed both vertically and laterally over the 34 borings. The base of the Key Largo Limestone is found at an average depth of about 21 m.

STP values within Layer 3, the hard-pan (Caloosahatchee equivalent) are typically 50 blows per about 7 cm of penetration defining a very hard layer. These SPT values are quite uniformly distributed both vertically and laterally over the 34 borings. The base of Layer 3 is found at an average depth of about 30 m.

STP values within Layer 4 (fine quartz sand) range from 5 to 70 and averaging about 30. Only 4 of the 34 borings had been drilled deep enough to encounter Layer 4.

26.4.7.2 Fluid Loss

Fluid loss was noted in the drill logs as it occurred in each borehole and indicated very porous zones or cavities. In their report, the drillers noted that circulation losses were generally abrupt. In many cases, total loss of circulation occurred

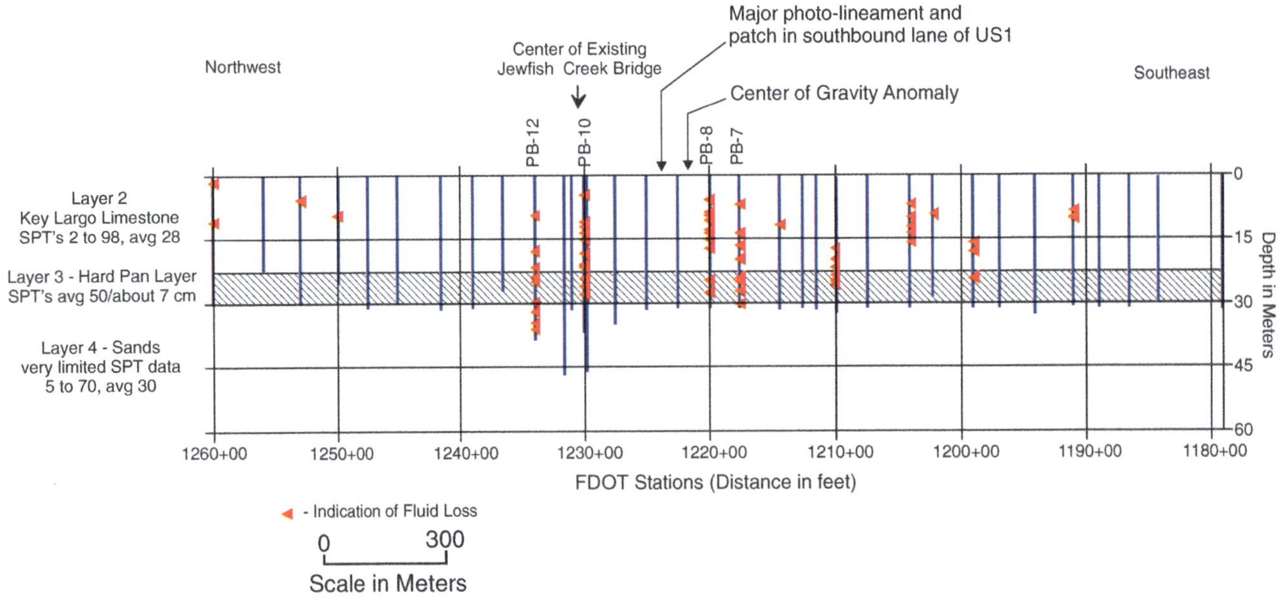

Fig. 26.6 The locations and depth of the original 34 borings along with averaged SPTs and the locations of fluid losses

even when using a thick drilling fluid (Law Engineering 1994). The fluid losses are shown in Fig. 26.6.

Two adjacent boreholes, PB-7, with eight fluid losses and PB-8, with ten fluid losses accounted for 27 % of all the fluid losses noted in the 34 borings. Two near-by borings PB-10 near the centerline of the Jewfish Creek Bridge and PB-12 just northwest of the Jewfish Creek Bridge account for an additional 34 % of all the fluid losses noted in the 34 borings. Sixty one percent of all fluid losses in the 34 borings are concentrated within 4 borings PB-7, PB-8, PB-10, and PB-12 that lie between stations 1217+50 and 1234+00. This is a distance of 503 m between the northwestern edge of Lake Surprise and the northwestern edge of the Jewfish creek bridge.

These fluid losses are distributed as follows:

- Layer 2 the Key Largo Limestone whose base occurs at a depth of about 21 m has 21 fluid losses.
- Layer 3 the hard-pan (Caloosahatchee equivalent) whose base occurs at a depth of about 30 m has 16 fluid losses.
- Layer 4 the sands of the Long Key Member of the Peace River Formation has three fluid losses in only one boring between 27 and 37 m.

26.4.7.3 Core Recovery and RQD Values

NX cores were obtained from 11 of the 34 borings at selected depths. Core recovery ranged from 0 to 100 % and averaged 53 %. RQD values ranged from 0 to 79 % and averaged 20. RQD values indicated very poor to good quality rock in those limited areas and depths sampled.

26.4.8 The Correlation of Anomalous Conditions

The shallow seismic reflection data and gravity data did not indicate anomalous conditions within Lake Surprise. However, the nature and spatial correlation of the following four independent sets of site-specific data (Fig. 26.7) provided strong evidence for potential karst conditions between

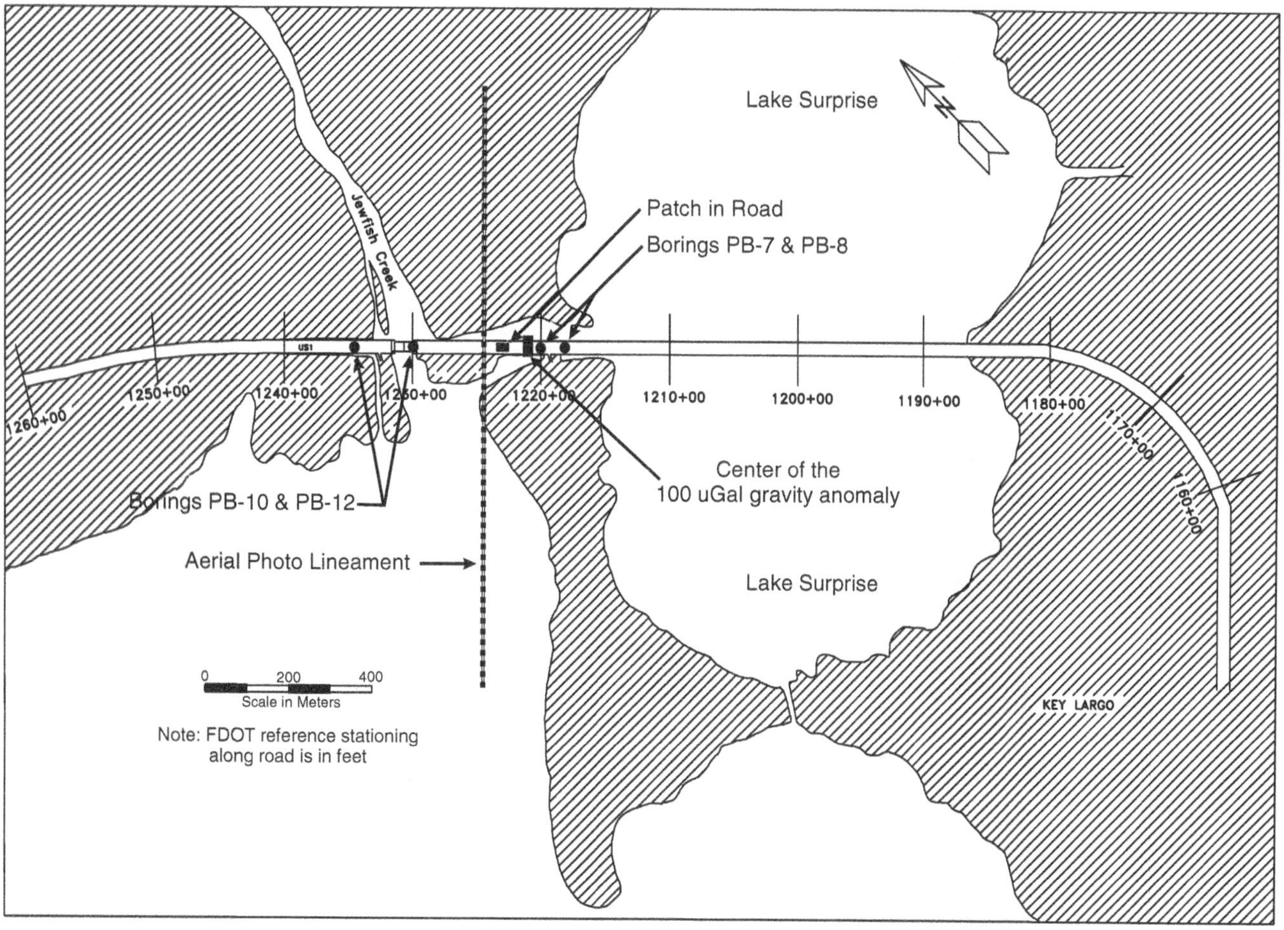

Fig. 26.7 The location of four anomalous conditions identified in Phase I show strong spatial correlation

Lake Surprise and Jewfish Creek. These four independent sets of data include:

- The wide 100 microGal gravity anomaly between Lake Surprise and Jewfish Creek. This was the first indication of possible significant karst conditions at the site because of the implied mass deficit (voids or low-density zones in the subsurface or the possibility of a paleocollapse cavern at depth).
- The lineament identified in the aerial photo interpretation that passes between the center of the gravity anomaly and Jewfish Creek.
- The two boreholes near the center of the gravity anomaly PB7 and PB8 and two boreholes PB10 and PB12 near Jewfish Creek that account for 61 % of all the fluid losses noted in the 34 borings.
- The patch in the road due to continued subsidence and maintenance along the southbound lane of US1 indicates anomalous conditions and lies about 45 m northwest of the center of the gravity anomaly.

This preliminary data begins to show spatial correlation of four independent sets of data focusing our attention upon the channel between Lake Surprise and Jewfish Creek. In addition, USGS had identified a large sediment-filled sinkhole just 8 km to the southeast (Shinn et al. 1994b). These preliminary indications of karst conditions lead to Phase II of the project.

26.5 Phase II Confirmation Phase

The purpose of Phase II was to verify and expand findings of anomalous conditions identified in Phase I. The area for additional measurements was focused upon the narrow channel about 38 m wide that connected Lake Surprise and Jewfish Creek (Fig. 26.8). Phase II confirmation phase consisted of:

- Marine seismic reflection data between Lake Surprise and Jewfish Creek which would aid in the interpretation of the microgravity data
- Microgravity data was obtained along the south side of US 1 parallel to the Phase I microgravity data. Additional gravity data was obtained to the south within accessible areas.

This data would confirm the original findings and help to develop a conceptual model of site conditions.

26.5.1 Additional Marine Seismic Reflection Data

Four closely spaced parallel lines (about 7.6 m apart) of seismic reflection data were obtained along the narrow channel between Lake Surprise and Jewfish Creek. Data was obtained

Fig. 26.8 Narrow channel between Lake Surprise and Jewfish Creek used for detailed marine seismic reflection measurements. The drilling barge is seen in the background. Lake Surprise lies beyond the drilling barge. The view is towards the southeast

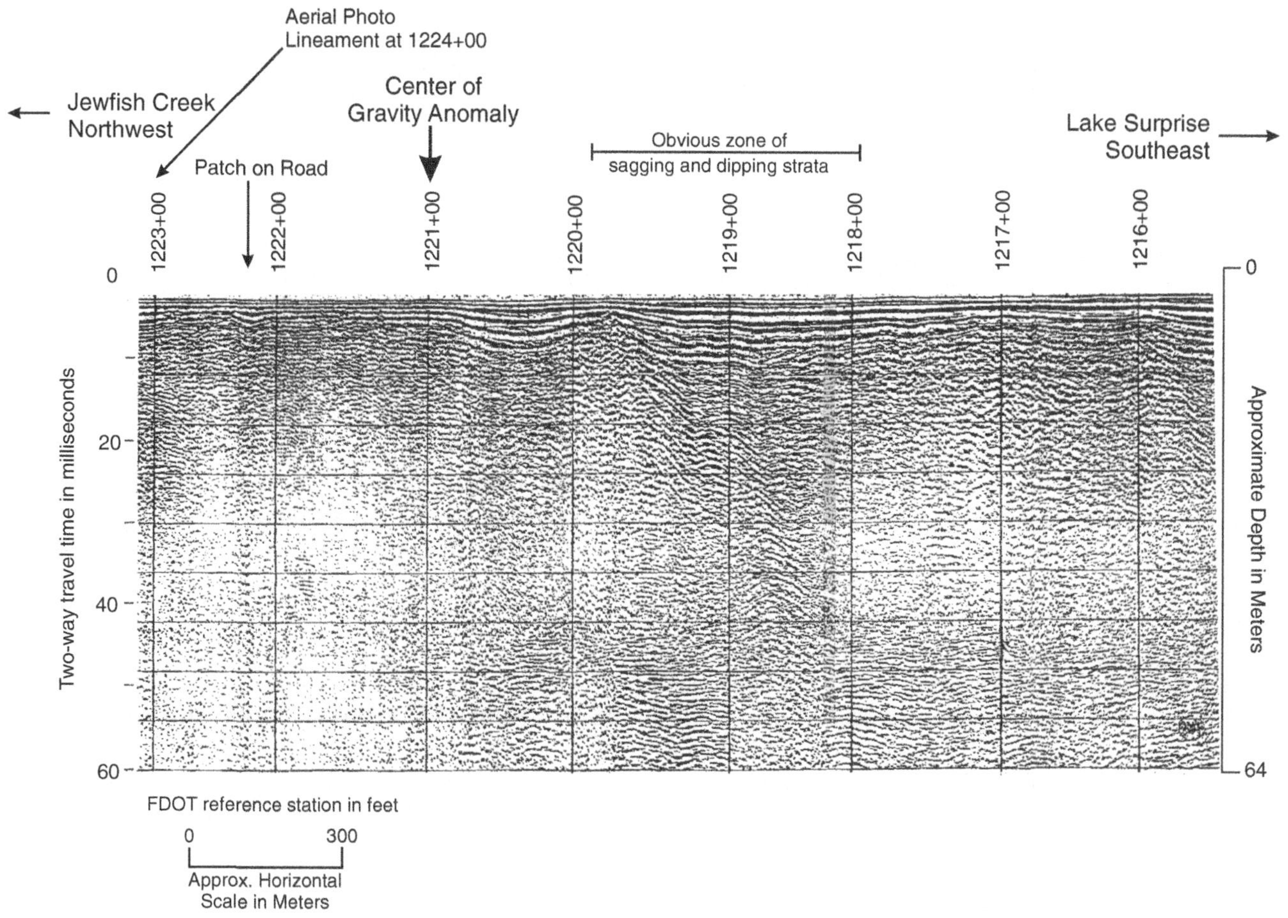

Fig. 26.9 A portion of one of the four shallow seismic reflection lines between Lake Surprise and Jewfish Creek showing dipping strata

with the same equipment and settings used in Phase I. Each survey line was about 396 m long and ran parallel to the microgravity survey line along US-1. The positioning of the survey boat was tightly controlled by an array of marker buoys that were referenced to the project survey grid. Each of the four lines clearly indicated the presence of an anomalous zone with sagging and dipping strata indicating subsidence (Fig. 26.9).

26.5.2 Additional Microgravity Data

During Phase II an additional 116 gravity measurements were made to confirm the anomaly and estimate its lateral extent. Gravity measurements were made parallel to the initial survey line within the anomalous zone along the southbound side of US1 and within the limited area of service roads and trails immediately southwest of US1 where there was solid ground.

Results from the gravity profile line on the south side of the road clearly repeat the data obtained from Phase I line, confirming the presence and magnitude of the gravity anomaly. Additional gravity data off of the road to the southwest extended the gravity anomaly at least 76 m off of US 1 to the southwest.

26.5.2.1 Analysis of the Microgravity Data

The 100 microgal gravity anomaly clearly indicates a deficiency in subsurface mass. A gravity anomaly can be caused by a number of geologic conditions and we must therefore, estimate possible causes based upon:

- Rules of thumb;
- Forward gravity models of simple geologic and karst geometries that are reasonable for the area; and
- Additional data which would constrain interpretations.

A simple rule of thumb that can be used to estimate the possible depth of the condition that is causing the gravity anomaly is the half-width rule (Butler 1980). The half-width depth for this gravity anomaly assuming a spherical cavity is about a depth of 172 m.

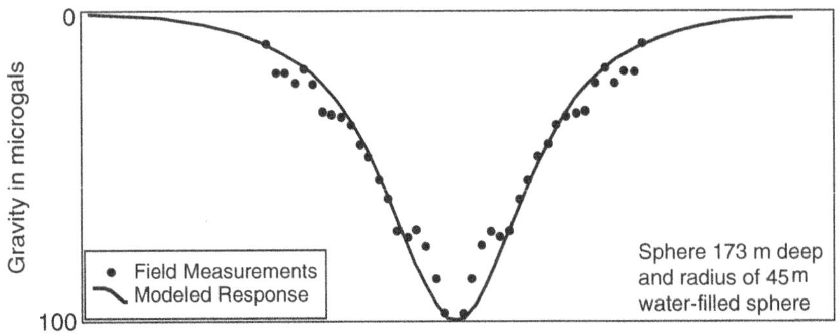

Fig. 26.10 Simplified geologic conditions can be modeled to fit the gravity data in order to estimate the cause and depth of the gravity anomaly

As a further refinement we can run inverse gravity models (matching modeled responses to actual field measurement). A model was calculated using a spherical water-filled cavity at depth (Fig. 26.10). The model indicates that a spherical water-filled cavity about 173 m deep with a radius of 45 m could be a possible cause of the mass deficit. Note that an open, water-filled sphere presents the worse case scenario in terms of potential karst features. A collapsed sphere, filled with broken rock and sediments would have to be of larger dimensions in order to match the gravity data acquired at the site. This is a more likely scenario based upon the overlying subsidence and dipping strata. No further models were run at the time.

Based upon regional geology a large cave system that had collapsed must be located in Layer 5, the Arcadian Formation (Scott 1988). This is the first limestone (Fig. 26.4) in which a large cave system could develop. The top of the enlarged cavern would occur at a minimum depth of 150 m (the top of the Arcadian Formation but could have developed at a depth somewhere between 180 and 240 m or more. This is consistent with the depth of other known cave systems, such as the Bluehole on Long Island in the southern Bahamas which is about 198 m deep (Wilson 1994) and lower sea levels of as much as 130 m (Balsillie and Donoghue 2004).

26.5.3 A Conceptual Model

A conceptual model was developed (Fig. 26.11) and is supported by the additional dipping strata to depths of 63 m seen in the shallow seismic reflection data between Lake Surprise and Jewfish Creek (Fig. 26.9) and the depths obtained from modeling of the gravity data (Fig. 26.10).

The seismic data indicated subsidence and dipping strata to a depth of 60 m or more indicating that the source or cause of this anomaly was probably deeper than 60 m. The combined estimated depth of 173 m from the gravity models (Fig. 26.10) along with geologic interpretations (Fig. 26.4) and regional geomorphology can be used to establish a reasonable conceptual model of the depth of origin of collapse at this site. Since the first limestone in which a cave could develop is the Arcadian Formation, it is reasonable to conclude that deeper geologic conditions should be investigated.

26.6 Phase III Detailed Investigation

As a result of the additional gravity and seismic reflection data in Phase II, further detailed work within the narrow channel between Lake Surprise and Jewfish Creek was pursued to provide additional data to constrain our interpretation. Phase III detailed investigation consisted of:

- Deeper seismic reflection data over the anomalous area,
- Measurements of seismic velocity
- Drilling of four boreholes within the anomalous zone to depths of 30–60 m with continuous sampling, and
- Geophysical logging in the four new boreholes

26.6.1 Deeper Seismic Reflection Data

The shallow seismic reflection data clearly indicated the presence of a paleokarst collapse zone by dipping strata, but the seismic data only extended to a depth of about 60 m. Seismic reflection data was obtained in Phase III to provide deeper data. A single survey line (approximately 600 m long) was run in the channel between Lake Surprise and Jewfish Creek.

Data was obtained with an EG&G model 267 three-element sparker array and the signal was received by a single channel eight-element hydrophone. A record length of 250 ms was used at a sample rate of 125 μs. The depth scale is based upon an estimated average seismic velocity of 2,133 m per second and indicates a maximum depth to the seismic data of 266 m.

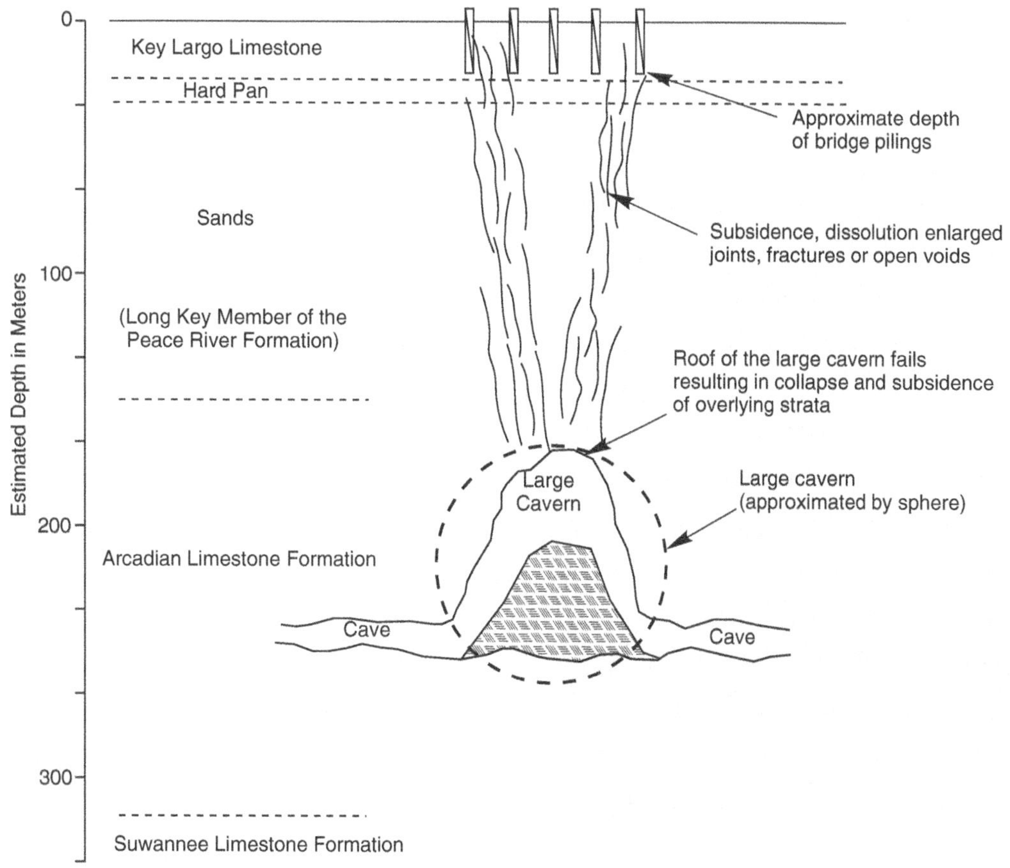

Fig. 26.11 A conceptual model can now be developed indicating a collapsed cavern probably within the Arcadian Formation at a depth greater than 150 m. This is the first section of limestone thick enough to support development of a cave system and an enlarged cavern

The deep seismic data clearly indicated subsidence and dipping strata over a distance of approximately 488 m (from stations 1213+00 to 1229+00). Relatively horizontal strata are seen on either end of the seismic record away from the paleocollapse zone. The deepest and most obvious feature of dipping strata is centered near 1215+00 that indicates possible collapse to a depth of 240 m or more. An example of the deep seismic data is shown in Fig. 26.12. This data suggests that the origin of collapse lies within the Arcadian (limestone) Formation or possibly the deeper Suwannee (limestone) Formation (Fig. 26.4). The depth of collapse is in keeping with dramatically lower sea levels (Balsillie and Donoghue 2004) and a thick fresh water lens.

26.6.1.1 Seismic Velocity Measurements

Up to now, the depth scales for the seismic data were based upon an assumed velocity of 2,133 m/s. To determine the actual depth of the seismic record the seismic velocity of the P wave was measured. Two independent velocity measurements were made, one by surface seismic refraction measurements over a distance of 300 m and another by downhole measurements to a depth of 45 m. McNeill et al. (1995) provided regional velocity measurements on other core samples from the area. These measurements provided an average velocity of 2,133 m/s that was used to establish the depth scale for the seismic data. Based upon this velocity, the shallow seismic reflection data (Fig. 26.9) has a maximum depth of approximately 63 m and the deeper seismic data (Fig. 26.12) has a maximum depth of approximately 267 m.

26.6.2 Drilling of Four Additional Boreholes

Four additional borings were drilled in Phase III. These were located 76–120 m southeast of the center of the gravity anomaly. Their locations were selected based upon the shallow seismic reflection data obtained in Phase II (Fig. 26.9). TB-1, and TB-2 were drilled to 30 m and TB-3, and TB-4 were drilled to 60 m. Drilling was done from a barge and continuous core was obtained from each of the four borings along with RQD values. The core showed considerable variation in RQD (Law Engineering 1995).

These four additional borings indicate similar overall geologic conditions to those found in the original 34 borings.

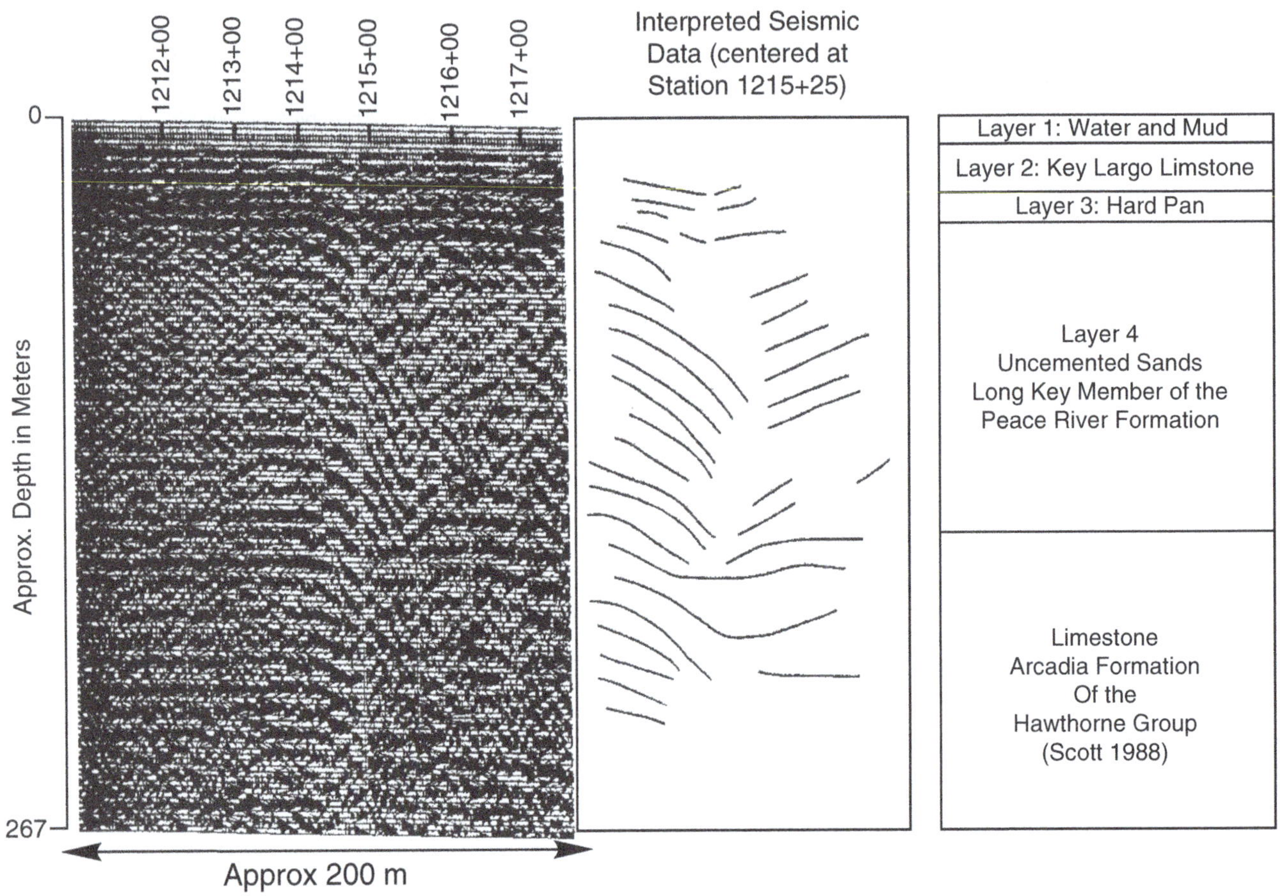

Fig. 26.12 A portion of the deep seismic reflection data to a depth of about 267 m is shown along with its interpretation and geologic section

However, dramatic differences in porosity were evident. Figure 26.13 shows two short lengths of core obtained from the Key Largo Limestone. One of the cores was obtained over Lake Surprise (outside of the paleocollapse zone) from PB-5 and is considered to be representative of the core from the initial 34 borings (Fig. 26.13a). The second core is from TB-4 and shows extensive dissolution and increased porosity. This core represents worse case conditions from the last four borings drilled within the paleocollapse zone (Fig. 26.13b).

26.6.3 Geophysical Logs from the Four Boreholes

A suite of three geophysical logs was run through the steel casing for each of the four boreholes. They included natural gamma, gamma-gamma (density) and neutron-neutron (porosity). For simplicity, we have shown only the density logs (Fig. 26.14). Higher measured values indicate lower densities. The density logs in all four borings show a number of zones of low density, some of which are open water-filled voids up to 2 m thick. Based upon these four logs, a dominant

Fig. 26.13 A sample of limestone core outside of the anomalous area within Lake Surprise (**a**) and within the anomalous area (**b**)

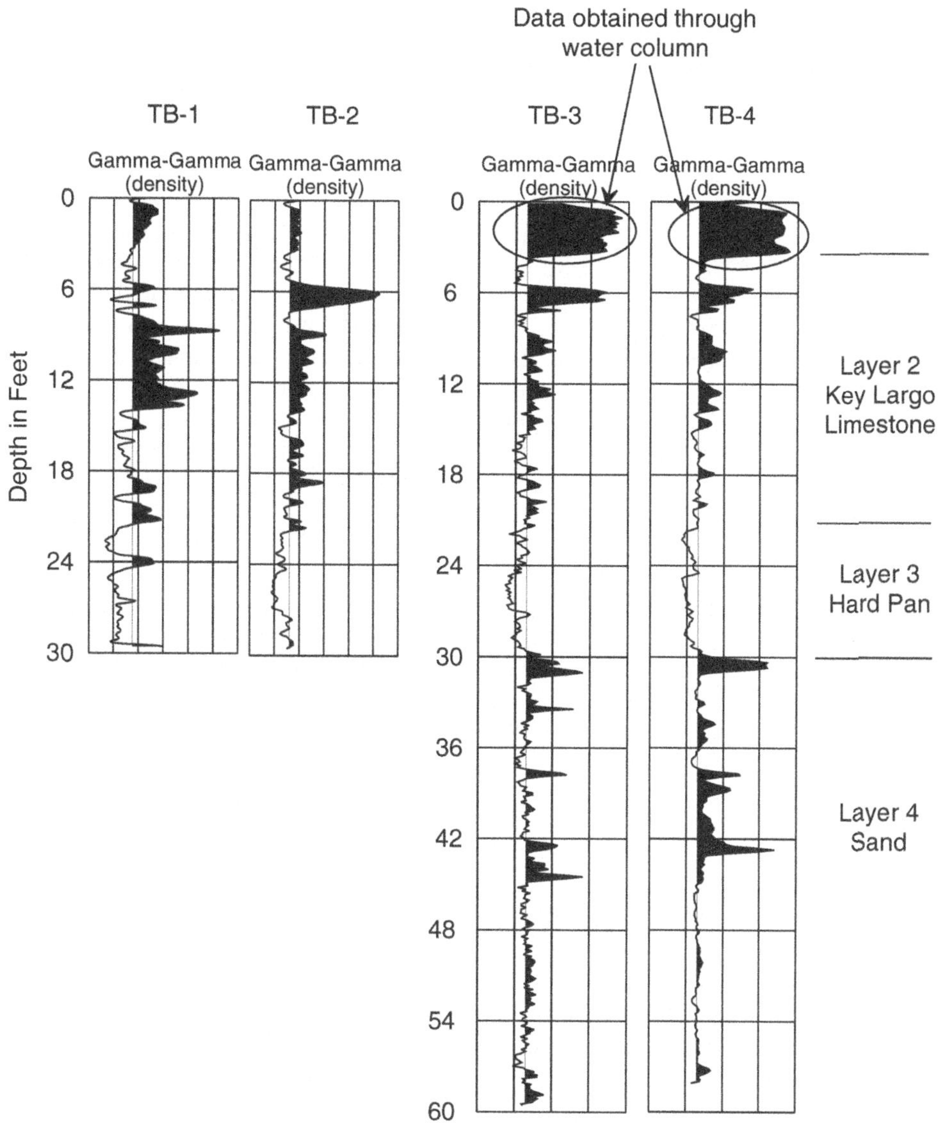

Fig. 26.14 The geophysical density logs from the four additional borings within the anomalous area (see Fig. 26.15 for locations)

zone of low-density occurs at a depth between 6 and 15 m (within Layer 2, the Key Largo Limestone). It is this zone that probably account for subsidence in US-1 (Fig. 26.2). However some low densities are seen to a depth of 21 m, the base of the Key Largo Limestone. The hard-pan layer (Layer 3) defined by high SPT values, is also seen as a high density zone (lower values) in the density logs.

Another deeper low-density zone is seen to occur in Borings TB-3 and TB-4 that extends to 60 m. These low-density zones occur in the upper portion of (Layer 4) the sandy Long Key Member of the Peace River Formation, between 30 and 45 m. These are possible zones of sediment piping.

It must be pointed out that only 5 of the 34 original borings were drilled to depths below 30 m. The four additional borings were all centered near PB-7 and drilled to 30 and 60 m deep. As a result, the available data regarding porous zones based upon the density logs and fluid losses is limited both laterally and vertically.

26.6.4 Support for the Final Conceptual Model

Our original focus was upon Lake Surprise itself as a possible paleosinkhole. However, there are no significant anomalies in the 14 borings, in the gravity data along US-1 or in the 13.8 km of reconnaissance subbottom profile lines through Lake Surprise. Based upon these three sets of data, Lake Surprise itself is probably not a large filled paleokarst collapse sinkhole. However, it is not uncommon for a large sinkhole lake to have a small sinkhole located off to one side of the lake, which could have been missed.

Table 26.1 Summary of site-specific findings

Phase I – reconnaissance phase	**Key findings**
A site visit	Noted limited access for investigation and identified patch in the southbound lane of US1
A site fly over	Provide further assessment of limited access
USGS findings	Provided evidence of deep-seated paleocollapse within the Florida Keys reef 8 km from site
Microgravity survey	Identified a 100 microGal anomaly between Jewfish Creek and Lake Surprise centered at station 1221+00 indicating mass deficit
Marine seismic reflection survey through Lake Surprise	Did not locate any indications of karst within Lake Surprise
Aerial photo-lineament analysis	Identified a major NE-SW lineament through site at station 1224+00
A detailed review of existing 34 original borings	Identified 61 % of all fluid losses concentrated within four borings which lie between stations 1217+50 and 1234+00
Phase II – confirmation phase	**Key findings**
Detailed seismic reflection survey in the channel between Lake Surprise and Jewfish Creek	Identified subsidence and dipping layers extending to a depth of 60 m suggesting causes are deeper
Additional microgravity data	Confirmed original findings and expanded anomaly to the southwest
Modeling of microgravity data	Suggested deep-seated cavern more than 150 m deep
Phase III – detailed investigation	**Key findings**
Deep seismic reflection data in the channel between Lake Surprise and Jewfish Creek	Identified subsidence and dipping strata between stations 1213+00 and 1229+00 extending to depths of 267 m or more
Drill four additional boreholes from 30 to 60 m with continuous core	Cores indicated highly variable porosity
Run geophysical logs in 4 boreholes	Multiple low-density and open zones identified between 6 and 15 m in the Key Largo Limestone and between 30 and 45 m in the sandy Caloosahatchee equivalent
Measure seismic P wave velocity	Provided seismic velocity to calculate depth of seismic reflection data

Multiple sets of site-specific independent data (Table 26.1) were used to develop the final conceptual model for this site. All sets of independent data have identified anomalous conditions between Lake Surprise and Jewfish Creek, all within a distance of approximately 480 m. The general location and spatial correlation of these anomalous conditions are shown in Fig. 26.15.

The 100 microgal gravity anomaly was the first piece of data which focused attention away from Lake Surprise and onto the narrow channel between Lake Surprise and Jewfish Creek. While rules of thumb and inverse modeling indicated a source depth of 150 m or more, it was the deep seismic reflection data (Fig. 26.12) that provided indications of subsidence and dipping strata to depths of 267 m or more that spatially correlated with the microgravity data. This indicated the presence of an old paleocollapse feature whose origin lies within the Arcadian Limestone Formation or possibly in the Suwannee Formation. Four additional borings and geophysical logs to depths of 30 and 60 m (Fig. 26.14) confirm that the anomalous area is more porous with zones of open voids.

Some of the individual sets of data identified in Table 26.1 such as loss of drilling fluid, repeated repair of subsidence in the road, and a major photo-lineament through the site by themselves do not provide conclusive evidence of a major karst feature. It is the spatial correlation of these independent sets of data that provide overwhelming evidence for the presence of a large paleocollapse originating at depth. Because multiple sets of independent data correlate, (Fig. 26.15) we can have a high degree of confidence in our interpretations and conceptual model (Fig. 26.12).

Additional support for deep-seated karst in the area is provided by:

- Lower sea levels by as much as 130 m (Balsillie and Donoghue 2004) and a fresh water lens; and
- The presence of the large sinkhole identified by USGS (Shinn et al. 1994b) in the reefs off of Key Largo just 8 km to the southeast.

26.7 Risk Assessment

When a karst feature does occur, it often can have significant impact upon structures. The area of concern extends from stations 1214+00–1226+00, which includes the location of the piers for the highest portion of the bridge over Jewfish Creek. Prior to this site characterization effort the only evidence of subsidence at the site was the small area of localized settlement seen on the southbound lane of US1 that required periodic maintenance (Fig. 26.2).

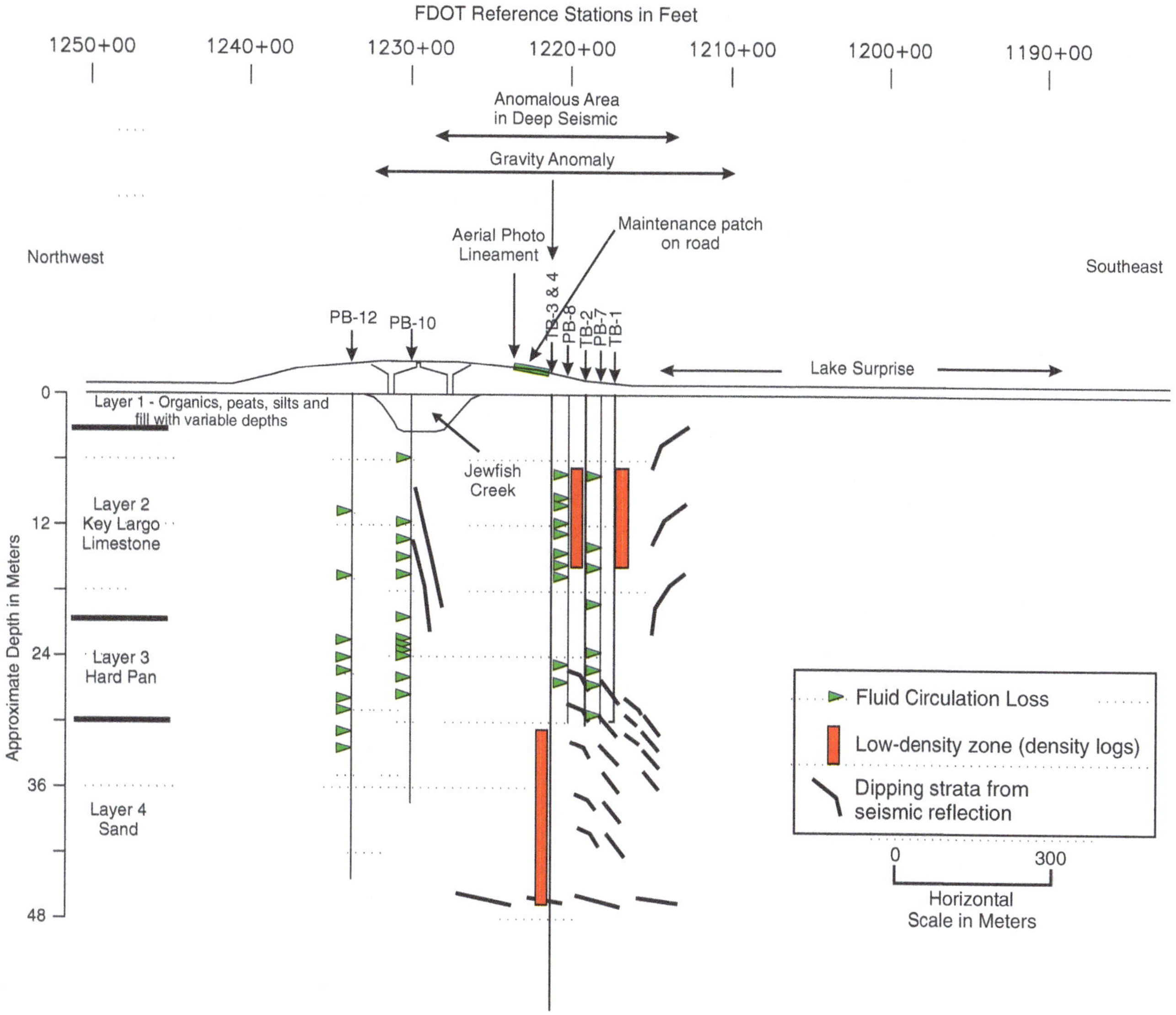

Fig. 26.15 Spatial correlation of anomalous conditions identified by independent data sets provides an increased confidence level in the interpretation and our conceptual model of geologic conditions (Fig. 26.11)

There are three geologic factors that could result in subsidence or collapse at this site, each are highly unlikely, but possible, and may impact construction and or long-term operation of the structure.

1. The most probable condition to impact the site is the localized low-density zones or open voids within the Key Largo Limestone at a depth of 6–15 m (Fig. 26.14)
2. Localized low-density zones or open voids within the unconsolidated sands beneath the Key Largo Limestone at a depth of 30–45 m (Fig. 26.14)
3. Larger open voids associated with a deep-seated cavernous system from which the paleocollapse is thought to occur at depths of 180–240 m or more (Fig. 26.11)

1. Some localized zones of increased permeability and open voids were found within the Key Largo Limestone (Layer 2) at depths of 6–15 m based upon the original 34 borings (Fig. 26.6) and the geophysical logs in the additional 4 borings (Fig. 26.14). These features could impact the construction or long-term maintenance.

The bridge piers will be founded on cast in-place drilled shafts, which will extend to depths of about 18 m in the Key Largo Limestone within the area of concern. By using drilled and poured shafts most of the load is carried by skin friction (compared to driven piles where most of the load is carried by the pile tip).

There is a hard-pan zone (Layer 3) at the base of the Key Largo Limestone between 21 and 30 m (Fig. 26.4). It has very high SPT values of 50 per about 7 cm. In areas where bridge piers are to be located within the anomalous zone, a pilot boring should be extended to establish the presence of this hard-pan layer beneath each bridge pier. Then grouting the pilot borehole to design depth.

Geologic conditions should be monitored closely during construction by careful inspection of each pier boring within the anomalous zone for evidence of any voids, monitoring of fluid loss and rod drop while drilling and monitoring of any concrete loss when pouring piers. If any unusual conditions are encountered an alternative would be to extend the bridge piers to rest upon the hard-pan layer at the base of the Key Largo Limestone.

Long-term maintenance concerns within this depth are associated with disposal of surface runoff from the bridge deck. The proposed approach for disposal of surface runoff from the bridge deck is by injection wells drilled into the first permeable zone. This practice should be avoided within the anomalous zone. The discharge of large quantities of surface water runoff directly adjacent or near bridge piles, within this zone, could possibly flush materials away creating larger voids. While the probability of inducing subsidence or collapse due to disposal of runoff within this depth is low, it should be avoided.

2. Localized low-density zones and open voids within the sands of Layer 4 at a depth of 30–45 m have been identified and could allow soil raveling or piping. The largest of those identified is about 1.2 m high and all lie below the hard-pan, Layer 3. The risk of these voids impacting the bridge piles is low since they tend to be relatively small, randomly located and occur below the hard-pan, Layer 3.

However, the lateral extent and depth of these low-density zones within Layer 4 is unknown. Only 4 of the original 34 borings and 2 of the additional borings were drilled to limited depths into the sands of Layer 4. All borings were located within the anomalous zone of the paleocollapse, which is 360 m wide. The lateral extent of these features is unknown.

3. The origin of the large, deep-seated cavernous system in which the paleocollapse occurred is believed to lie at a depth of 180–240 m or more below grade. The deep seismic data (Fig. 26.12) shows extensive subsidence and dipping strata throughout the zone of paleocollapse suggesting that the large cavern has collapsed and is likely filled in with sediments (Fig. 26.11). Further dissolution of the deeper cave systems has ceased and the cave system is now totally saturated by seawater.

Several factors exist to suggest that further collapse or reactivation of this deep zone is unlikely to virtually zero.

- Dissolution has ceased.
- The seismic data suggests that extensive subsidence has occurred over the collapse indicating that the large cavern has likely been filled with sediments
- These deep collapse are quite old and has been stable for a long time. Data from Shinn et al. (1994b) indicates C14 ages of 5,650±90 years before present in the Florida Keys and data from Yuhr et al. (2003) indicates ages of 40,000 and 48,800 years before present in the Tarpon Springs Florida area.
- In addition, all construction and other further activities at the site would not impact the geology at these depths.

26.7.1 Limitations

The limitation of this site characterization is the limited site-specific spatial data acquired at this site. The data that indicated increased porosity and open voids in Layer 2, the Key Largo Limestone was laterally limited to the line of borings (34 original and 4 additional). This data density along this survey line is extensive, but does not provide any data off of this line. The data that indicated increase porosity and open voids in Layer 4 (quartz sand) was very limited. Only four of the original borings extended less than 18 m into this layer. Two of the four additional borings extended 30 m into this layer. The limited number of borings and geophysical logs clearly provided an indication of the depth of the zones of increased porosity and open voids, the actual lateral and vertical extent of these zones has not been established. However, it is reasonable to assume that these zones extend laterally along the line of borings and off of the line of borings some distance within the anomalous zone of the paleocollapse.

26.8 Conclusions

The approach to this site characterization includes many of the key concepts for dealing with geologic uncertainty in the site characterization process. They include:

- A variety of site-specific measurements appropriate to site conditions;
- Measurements with a wide range of scale;
- Correlation of a number of independent measurements to improve confidence levels in the interpretation; and
- Development of a conceptual model supported by solid data.

It is interesting to note that prior to beginning the site characterization there were two independent pieces of data indicating potential karst conditions between Lake surprise and Jewfish Creek. These two pieces of data included the patch in the southbound lane of US-1 and the location of four borings that accounted for 61 % of all fluid losses in the 34 original borings. However, these were subtle indications that alone would not have indicated a serious problem. It was not until the gravity anomaly was detected in Phase I that we began to focus attention upon this area.

All of the data that has been used to solve the geologic puzzle and to develop an accurate conceptual model of geologic conditions at this site is summarized in Table 26.1. The data includes multiple sets of independent and diverse data with different scales of measurement and depth of measurement. No single set of measurements would have given a clear understanding of geologic conditions at this site. However, these independent sets of site-specific data correlated spatially (Fig. 26.15) to provide a solid basis for the final conceptual model (Fig. 26.11).

The data subsequently obtained indicated the presence of a large anomalous area (360 m) along the bridge alignment between stations 1214+00 and 1226+00. This anomalous zone is located at the highest portion of the new bridge. The mass deficit indicated by the microgravity data and overlaying subsidence and dipping strata seen in the deep seismic data and provided the evidence for a large cavernous zone originating deep within the Arcadian Formation at depths of 180–290 m or possibly deeper within the Suwannee Formation.

The possible presence of the deep cavern system was supported by conditions for dissolution created by a lower sea level of up to 130 m (Balsillie and Donoghue 2004) and a thick fresh water lens (Ford and Williams 2007) resulting in development of caves at significant depths below present sea level. Dean's Bluehole on Long Island in the Bahamas is at least 200 m deep and is an example of karst formed at these depths. Further support for the presence of deep-seated paleokarst is provided by the presence of a large paleocollapse sinkhole 8 km to the southeast within the Florida reefs (Shinn et al. 1994b).

Based upon the data acquired and conceptual model developed, the two critical questions that were asked at the beginning of this study can be addressed:

1. Can an investigation fully identify unstable geology, such that bridge decks can be placed or lengthened to safely span the critical geologic conditions?
2. Can the bridge as designed, be stable for at least 100 years in the given geologic environment?

The results of this study provided sufficient data to allow the bridge to be constructed with a minimum risk. This deep-seated paleocollapse feature appears to be stable and well below the depth of impact due to construction. While increased porosities and small voids could impact the project, design and construction recommendations are provided to minimize the potential long-term impact of geologic conditions and subsidence collapse of the bridge structure.

Technical reports and papers have been presented about this work include:

- Technos, Inc. (1994)
- Benson et al. (1995a)
- Benson et al. (1995b)

References

Badgley PC (1960) The analysis of structural patterns in bedrock. American Institute Mining and Metallurgical Engineers, S. M. E. Trans 225:381–389

Badgley PC (1965) Structural and tectonic principles. Harper and Row, New York

Balsillie JH, Donoghue JF (2004) High resolution sea-level history for the Gulf of Mexico since the last glacial maximum, Florida Geologic Survey Report of Investigation 103

Benson RC et al (1995a) Assessment of potential karst conditions for a new bridge in the Florida Keys. Symposium on the application of geophysics to engineering & environmental problems, Orlando, 23–26 Apr 1995, pp 22–26

Benson RC et al (1995b) Subsurface investigation of possible karst conditions at the Jewfish Creek bridge replacement, Key Largo, Florida. In: Beck BF (ed) Proceedings of the 5th multidisciplinary conference on sinkholes and the engineering and environmental impacts of karst: karst geohazards, Gatlinburg, Tennessee. Balkema, Rotterdam, pp 409–414

Billings MP (1954) Structural geology. Prentice-Hall, Englewood Cliffs

Butler DK (1980) Microgravimetric techniques for geotechnical applications. U.S. Army Engineer Waterway Experiment Station, Miscellaneous Paper GL-80-13, Vicksburg, Mississippi

Ferguson GE et al (1947) Springs of Florida. Florida Geological Survey, Geological Bulletin No. 31

Finkl C (1994) Report on FDOT Jewfish Creek Bridge. Consultants report, The Coastal Education & Research Foundation, CERF Remote Sensing Lab, Ft. Lauderdale, Florida

Finkl CW, Warner MT (2005) Morphologic features and morphodynamic zones along the inner continental shelf of southeastern Florida: an example of form and process controlled by lithology. J Coast Res 42:79–96, Special Issue

Ford D, Williams P (2007) Karst hydrogeology and geomorphology. Wiley, New York

Lane E (1981) Environmental geology series, Miami sheet, Map Series 101. Florida Geological Survey, Tallahassee

Law Engineering and Environmental Services (1994) Report of geotechnical exploration, Jewfish Creek Bridge Replacement, Volumes I and II

Law Engineering and Environmental Services (1995) Presentation of test boring records, Jewfish Creek – Supplemental Drilling, 8 Aug

McNeill D et al (1995) Core workshop: late tertiary-quaternary cores from the Florida keys and everglades. Sponsored by the Florida geological survey. University of Miami, RSMAS/MGG, Miami

Parker GG et al (1955) Water resources of southeastern Florida. USGS Geological Survey Water-Supply Paper 1255

Scott TM (1988) The lithostratigraphy of the Hawthorn Group (Miocene) of Florida. Florida Geological Survey, Bulletin No. 59

Shinn EA (1994a) Letter to Mr. David C. Miro, Florida Department of Transportation Dec 21

Shinn EA et al (1994b) A giant sediment trap in the Florida keys. US Geological Survey Center for Coastal Geology, St. Petersburg

Technos, Inc (1994) Subsurface investigation of possible karst conditions at Jewfish Creek bridge replacement Monroe County, Florida. Consultant Report for Stienman Boynton Gronquist & Birdsall and Florida Department of Transportation, 16 Nov

Wilson WL (1994) Morphology and hydrology of the deepest know cave in the Bahamas: Dean's blue hole, long island. Bahamas J Sci 2(1):10–14

Yuhr L et al (2003) A case history of a large karst investigation. Presented at the 3rd international conference on applied geophysics, Geophysics 2003 Orlando 8–12 Dec 2003

EPA Superfund Site

27

Abstract

This third case history was a 52 ha EPA Superfund site located in Tarpon Springs north of Tampa, Florida that had processed phosphate ore to produce elemental phosphorus. While the site had been investigated and a Record Of Decision (ROD) had been completed from EPA regarding the remediation, there remained concerns with the site. These concerns included the development of sinkholes on the site, as well as the presence of buried drums of elemental phosphorus waste material, and the potential for groundwater contamination. The project was carried out between 2001 and 2003 over a period of 30 months following our standard site characterization strategy. Because this was a superfund site, there were a number of stakeholders involved from beginning to end. The project included extensive meetings with local interests, county, state and federal agencies with EPA oversight. The client wanted this project completed right the first time with no questions left unanswered. Therefore, a large number of methods were utilized with heavy emphasis on surface, borehole and marine geophysical methods. Some of these methods were used to provide 100 % coverage of the property. The work at this site is probably one of the most complete karst site characterizations carried out.

27.1 Background

The site is located approximately 24 km northwest of Tampa, Florida, in the northwest corner of Pinellas County. The site is adjacent to the Anclote River, which flows into the Gulf of Mexico approximately 3 km west of the site. Figure 27.1 shows the regional setting for the Stauffer site from the Gulf of Mexico to the west, to alternate highway US-19 and US-19 to the east and the Anclote River to the south.

Production began in 1947 at the Stauffer Management Company site. The principal raw materials were phosphate ore (primarily tricalcium phosphate), along with coke and silica rock that are extensively mined in the west-central Florida area. Essentially all raw materials, products and byproducts were received and shipped by rail. A high temperature electric arc furnace melted the raw materials, and elemental phosphorus evolved as a gas, which was condensed under water and further processed to remove impurities. The site included production facilities, slag processing area, and a system of settling ponds. Figure 14.6 is a historical aerial photo that shows the plant and general operations at the site including storage piles of raw phosphate ore and railway operations. The view is toward the Anclote River to the southwest.

The site is divided into two areas, a North and South Parcel (Fig. 27.2). The production facility, a series of settling and disposal ponds were located on the South Parcel. The North Parcel was used for slag processing, storage and a waste pond. Two byproducts came from the furnace: slag and ferrophosphorus. The calcium silicate slag was drained periodically from the furnace into the adjacent slag pit. The slag was then cooled and solidified prior to being trucked to the slag processing area in the North Parcel for crushing to produce aggregate that was sold and used off-site. The ferrophosphorus was also removed from the furnace as a liquid, allowed to cooled and solidified for sale to the steel industry. Some of the waste phosphorus was placed in drums and covered with water then disposed of in on-site ponds on the South Parcel.

R.C. Benson, L.B. Yuhr, *Site Characterization in Karst and Pseudokarst Terraines*,
DOI 10.1007/978-94-017-9924-9_27

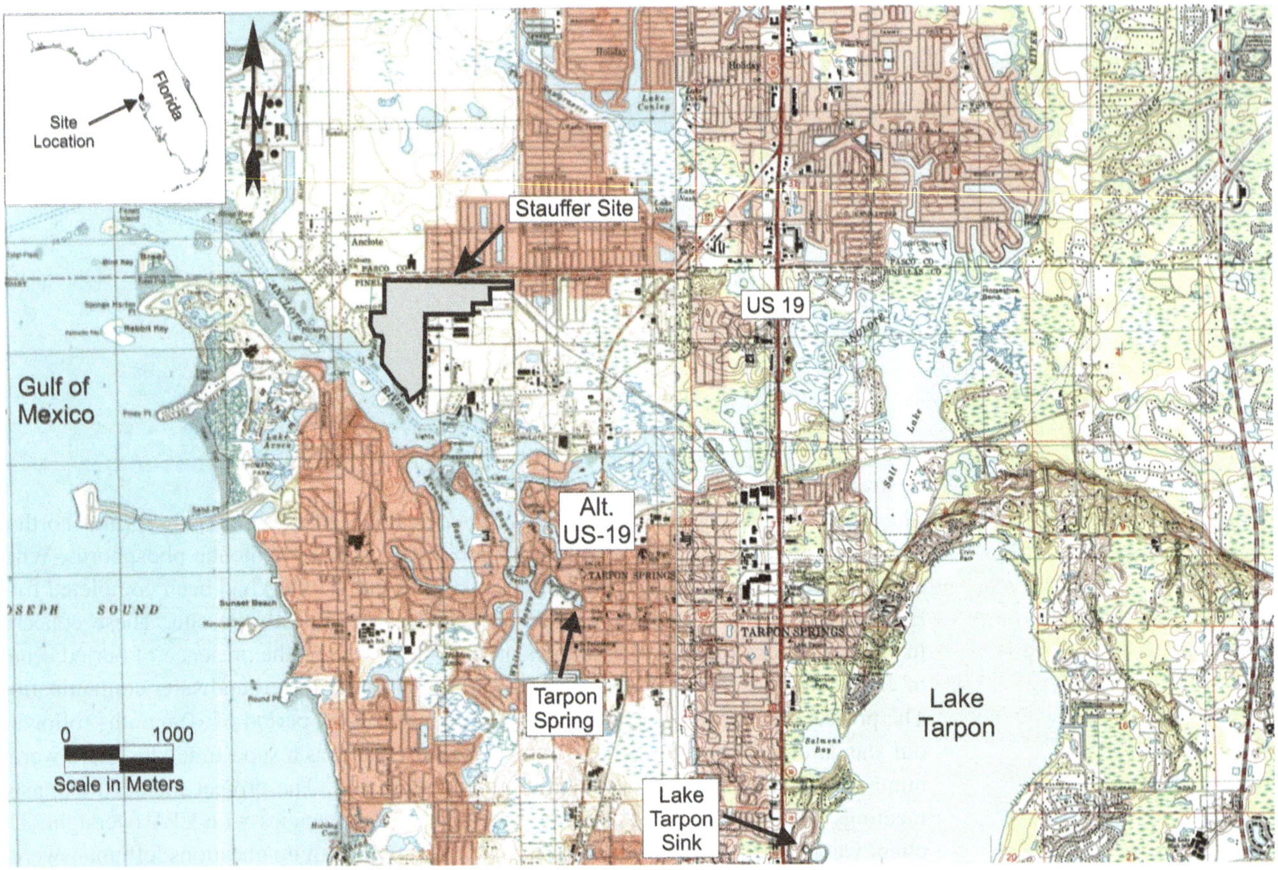

Fig. 27.1 Site map (USGS 7.5 min Series Tarpon Springs Quadrangle, 1995 Elfers Quadrangle, 1998)

After the plant was shut down in 1981, the plant production facilities and most of the buildings were decontaminated, demolished and removed from the site. The facility permanently closed in 1983. The only buildings remaining on-site at the time of this investigation were the main office building, an equipment building, a guard building, and a lunchroom building located on the northeast corner of the South Parcel (Fig. 27.2).

The southeastern most portion of the South Parcel was purchased in 1981 as a buffer along the southeastern edge of the ponds. The Corps of Engineers had previous used the property for placement of dredge spoils from the Anclote River in 1972 (Fig. 27.2). No production-related operations took place in this area.

The site topography is generally flat, with a slight slope southward towards the Anclote River. The average land surface elevation is about 3 m above sea level, but can be locally higher in the pond areas reaching as much as 10 m. These are areas where ponds have been filled with disposal materials.

27.1.1 Record of Decision (ROD)

A Record of Decision (ROD) was issued by US-EPA (1998) that outlined the remediation for the property. The remedy selected by the EPA includes the following components:

- Limited excavation of contaminated material and soil that exceeds residential cleanup standards.
- Consolidation of contaminated material and soil in the main pond area, slag processing area, and other areas on-site.
- In-situ solidification and stabilization of pond material and contaminated soil below the water table in the consolidation areas.
- Clay caps will be placed over the remaining consolidation areas.

In addition, institutional controls on the site including deed restrictions such as land use ordinances, physical barriers, and restrictions for water supply wells were put in place.

Fig. 27.2 Aerial photo of the site showing the North and South Parcels

27.2 Objectives of the Overall Investigation

On August 25, 2000 Stauffer Management Company and the EPA entered into an Agreement to conduct certain additional studies to evaluate whether the remedy selected for the source control will provide protection of human health and the environment throughout the life of the remedy (US EPA 2000a, b). Three major additional studies were undertaken to verify previous findings and resolve all remaining issues. These studies included:

- A geologic and karst characterization as well as a sinkhole risk assessment that was carried out by the authors and their firm, Technos Inc.
- A site-wide groundwater characterization and contaminant study that was carried out by Parsons Engineering Science, Inc. (Parsons).
- A solidification/stabilization treatability study on pond materials and soils was carried out by O'Brien & Gere Engineers, Inc. (OBG), the lead engineering team and responsible for the remediation of the site.

For this case history, the focus is upon the assessment of geologic and karst conditions, along with the site hydrogeology, as they impact the site and its proposed remediation.

27.2.1 Objectives of the Geologic and Karst Investigation

The objectives of the geologic and karst studies carried out by Technos, Inc. are summarized as:

- Evaluate whether or not sinkholes are present within the site and the probability that sinkholes or karst features will form over the required life of the remedy.

- Evaluate the nature and extent of the intermittent semi-confining clay layer (SCL) existing between the surficial aquifer and Floridan aquifers to the extent that it impacts the effectiveness of the selected remedy.
- Delineate the horizontal extent of the ponds and estimate the depths of the pond materials.
- Evaluate the presence of buried drums, storage tanks, or other potential sources of contamination at the site that have not already been identified.
- Evaluate the ability of the underlying geology to support the proposed remedy over the life of the remedy as required under 40 CFR Part 192 under reasonably anticipated site conditions.

27.2.2 Objectives of the Hydrologic Investigation

The objectives of Parson's hydrologic investigation were to evaluate the surficial aquifer (to approximately 6 m deep), the semi-confining layer at the base of the surficial aquifer, and the Upper Floridan aquifer (approximately 6–15 m deep). To accomplish this, the following work elements were completed:

- Assess the flow direction and estimate the lateral hydraulic gradients within the surficial and Upper Floridan aquifers and estimate the vertical gradient between these two aquifers.
- Evaluate the hydrologic nature of the semi-confining layer (SCL) between the surficial and Upper Floridan aquifers.
- Establish the locations and concentrations of groundwater constituents in both aquifers as well as the general geochemistry of the groundwater.
- Characterize the effects of tides on the aquifers and evaluate whether the tides in the Anclote River will affect the proposed remedy.

27.2.3 The Owner's Goals

Obtaining appropriate and adequate site-specific data is often a significant problem with budget and or time constraints along with many other factors will limit this effort. Numerous studies had already been completed at the site over a period of years, however, none had been comprehensive or had answered all of the questions about the sites geology, karst conditions, groundwater and contaminant issues. As a result the owners instructions at this site were clear, "Do whatever it takes to do it right the first time so that we don't have to come back and do more work or do things over".

27.2.4 Review and Oversight Committee

A technical oversight committee was formed to review and provide input on the work plan, as well as the data and its interpretation at key milestone meetings throughout the project. This committee consisted of representatives from the US-EPA, the Florida Department of Environmental Protection (FDEP), United States Geological Survey (USGS) and the Corps of Engineers as well as the Pinellas County Department of Environmental Management (DEM), Pinellas County Health Department, Dr. Mark Stewart of the University of South Florida as well as private sector interest groups.

The purpose of this committee was to establish consensus on the technical issues as the project moved forward and address any concerns or questions as they were encountered. They provided comments and approval of the project strategy and work plan. As the project proceeded, the oversight committee was presented each data set and its interpretation. Any concerns or questions were addressed and interpretations agreed upon. This allowed the project to move forward in a smooth and methodical manner with a consensus on all technical matters.

27.3 Technical Approach

The strategy used for the project is that presented in Chap. 12 which is based upon the ASTM D6235 Standard Practice for Expedited Site Characterization (ASTM 1998). Two key components of the strategy were utilizing a core team and diverse methodologies.

27.3.1 Core Team

One of the key aspects leading to the success of this project was the core group of companies and individuals that were used throughout the project from the beginning to end. This included key team members as well as subcontractors (surveyors, drillers, and laboratories). This consistency allowed the team to develop an understanding of the site and allowed the project to move forward in a coherent manner. The geologic and karst studies as well as the hydrologic studies were conducted concurrently in conformance with the Work Plans for the Additional Studies Program. In addition, field efforts and data were coordinated and shared, so that data was not duplicated and the geologic and hydrologic puzzle could be analyzed and integrated as efficiently as possible.

A quality control and quality assurance (QC/QA) program was an integral part of the site characterization effort because of the critical nature of the project from both a political and technical perspective. The fieldwork was initiated February 2002 and final reports were submitted June 2003.

27.3.2 Methodologies

The issue of resolving karst conditions is one of adequate spatial sampling, since it is quite easy to miss the spatially variable karst conditions. The measurements used and data acquired must not only be appropriate for the site-specific geologic and cultural conditions but must be spatially adequate, both laterally and vertically, to detect any geologic anomalies of interest. Therefore, to resolve the karst issues with a high level of resolution and confidence a wide range of surface and borehole geophysical surveys were utilized. These sets of data provided adequate and sometimes total site coverage and allowed background and anomalous conditions to be readily identified. Detailed invasive data could then be located and acquired using traditional drilling, sampling and laboratory analysis.

The methods selected provided a range of measurements with different independently measured parameters, along with a wide range of scale (regional, local and site-specific details) and depths of measurements. The combination of these diverse measurements and their integration provided a level of redundancy in the data obtained. If different measurement techniques result in a similar interpretation, their correlation, or redundancy, provides an increase in the level of confidence in the data and its interpretation. It is recognized that the range of measurements and their density is unusual and unique to this project. This was driven by the client's objective to "do whatever it takes to do it right the first time".

Table 27.1 summarizes the methods used at this site for the purposes of assessing geologic, hydrologic, and karst conditions; including the depth of measurement, and the site coverage provided by each measurement. This large set of on-site measurements provided a solid foundation of data from which to build the site conceptual model and reach conclusions that minimize opinions and assumptions about on-site conditions (Yuhr et al. 1996).

The geologic and karst issues were primarily divided into two zones, a shallow and deeper zone. Figure 27.3 summarizes these measurements and their depth:

- The shallow zone including the sands and the semi confining layer SCL which range from about 3 m on the north side of the property to about 6 m deep near the Anclote River and the surficial aquifer that lies within this zone; and
- The deeper zone below the SCL that included the Tampa and Suwannee Limestone that contains the upper Floridan aquifer.

The surface geophysical techniques utilized for assessment of the shallow zone included ground penetrating radar (GPR), frequency domain electromagnetics (EM), magnetics, and multi-frequency electromagnetics (MFEM). The assessment of the deeper zone utilized 2D resistivity imaging and microgravity measurements on-site and marine geophysical techniques in the adjacent river. In addition, the investigation included:

- invasive measurements (direct push electrical conductivity (DPEC), rotosonic drilling, hollow-stem augers and wash borings),
- geophysical logging in all borings and wells (existing and new);
- marine geophysical measurements to take advantage of data from within the adjacent Anclote river,
- in-situ monitoring, testing and laboratory analysis for a range of hydrologic and geologic parameters, and
- additional geophysical measurements to obtain compressional (P) and shear (S) wave velocities for calculating elastic modulus.

Figure 27.4 illustrates the overall strategy used on this project and follows the strategy presented in Chap. 12. The project started with getting the site ready for fieldwork and getting all the background data in order. This included a desk study, aerial photo analysis and site walkover. A preliminary conceptual model was developed based upon this information. When fieldwork began, each data set was independently analyzed and interpreted before being integrated with other data. Once integrated, it was checked against the conceptual model and the model was upgraded or modified as necessary. This iterative process allowed the project to address issues as they were identified and kept the project moving forward.

27.4 Site Preparation

Maximum site availability was required for this investigation to allow a high density of measurements to be made. This required debris piles to be removed or consolidated and overgrown brush and small trees to be cleared. An extensive clearing operation was completed to enable easy access and allow continuous survey lines of geophysical data to be acquired by a small four wheel drive vehicle. This site clearance was a major effort on-site and was completed by a subcontractor prior to any field measurements.

All new data acquired by the different team members required coordination and integration. Therefore, a standard reference grid was established for everyone to use. The use of a fixed survey grid facilitated the spatial integration of all old and new data. The grid also enabled anyone to quickly

Table 27.1 Summary of measurement techniques used

Method	Parameter measured	Approximate depth of measurement	Data quantity
Surface and borehole geophysical measurements			
Electromagnetics (EM31)	Conductivity and in-phase (metals)	0–6 m	Continuous 3 m line spacing, 177 km of data
Ground penetrating radar (GPR) with 100 MHz antenna	Complex dielectric constant	0–7.3 m	Continuous 3 m line spacing, 145 km of data
Multi-frequency electromagnetics (MFEM)	Conductivity and in-phase metals at multiple frequencies	Up to 6 m	Continuous 1.5 m line spacing over selected areas, 80 km of data
Magnetic	Magnetic susceptibility	Up to 6 m for a single steel drum	Continuous 3 m line spacing over selected areas, 18 km of data
2D resistivity dipole-dipole array	Electrical resistivity	0–24 m	6 m electrode spacing run along lines spaced 30 m apart, 16 km of data
Microgravity	Density	>30 m	30 m line spacing with 12 m station spacing. 6 m station spacing south portion of South Parcel, a total of 1,826 stations
Geophysical logging	Natural gamma, density, porosity, and electrical conductivity	Limited to depth of well or boring max 48 m	74 borings and wells
Invasive measurements			
Direct push electrical conductivity (DPEC)	Electrical conductivity	12 m	110 DPEC pushes
Direct push soil cores	Sediments (visual)	12 m	18 soil cores
Hollow stem auger	SPT values and soil and rock samples	3.6–15 m	23 shallow and 20 deep monitoring wells
Rotosonic test borings	Soil and rock samples	23–48 m	15 borings
Wash boring	Refusal of penetration	Up to 4.5 m	14 in sinkhole pond
Hydrologic measurements			
Slug test	Hydraulic conductivity	Over well screen Interval	9 tests
Pump test	Draw down hydraulic conductivity	Depth of pump and well screen	1 in surficial aquifer and 1 in Floridan aquifer
Water levels	Water level elevations	3.6–15 m	39 shallow and 20 deep wells monitored for 17 months
Tidal effects	Water level changes versus time	0–15 m	Every 15 min for 5 days in 21 wells plus tide gauge
Laboratory testing			
Water quality and testing	23 metals	3.6–15 m	28 shallow wells and 20 deep wells
	9 inorganics		
	5 radiological		
Permeability testing	Hydraulic conductivity	Depth of sample	15 samples from 7 locations
Particle size analysis – sieve analysis	Grain size	Depth of sample	18 samples from 9 test borings
Particle size analysis – hydrometer analysis	Grain size	Depth of sample	15 samples from 7 locations
C-14 dating	Age	15.8 and 24.6 m	3 samples from 2 boring locations
X-ray diffraction measurements	Mineral constituents	Within the SCL	6 samples from 6 locations within the SCL
Marine geophysical measurements			
Bathymetry	Bottom profile	River bottom	1.6 km in Meyers cove
Sidescan sonar	Bottom image	River bottom	16 km of data in the Anclote River
Subbottom profiling (seismic reflection)	Seismic coefficient of reflection	45 m below river bottom	19 km in the Anclote River
Geotechnical field measurements			
Seismic refraction	Velocity of P-waves	12 m	1.5 m geophone spacing, 2 test lines
Hole to hole seismic measurements (crosshole)	Velocity of P and S-waves	Up to 23 m	Between 3 boreholes 4.5 m apart, 2 sets of measurements
Vertical seismic profiling (VSP)	Velocity of P and S-waves	23–48 m	9 test borings

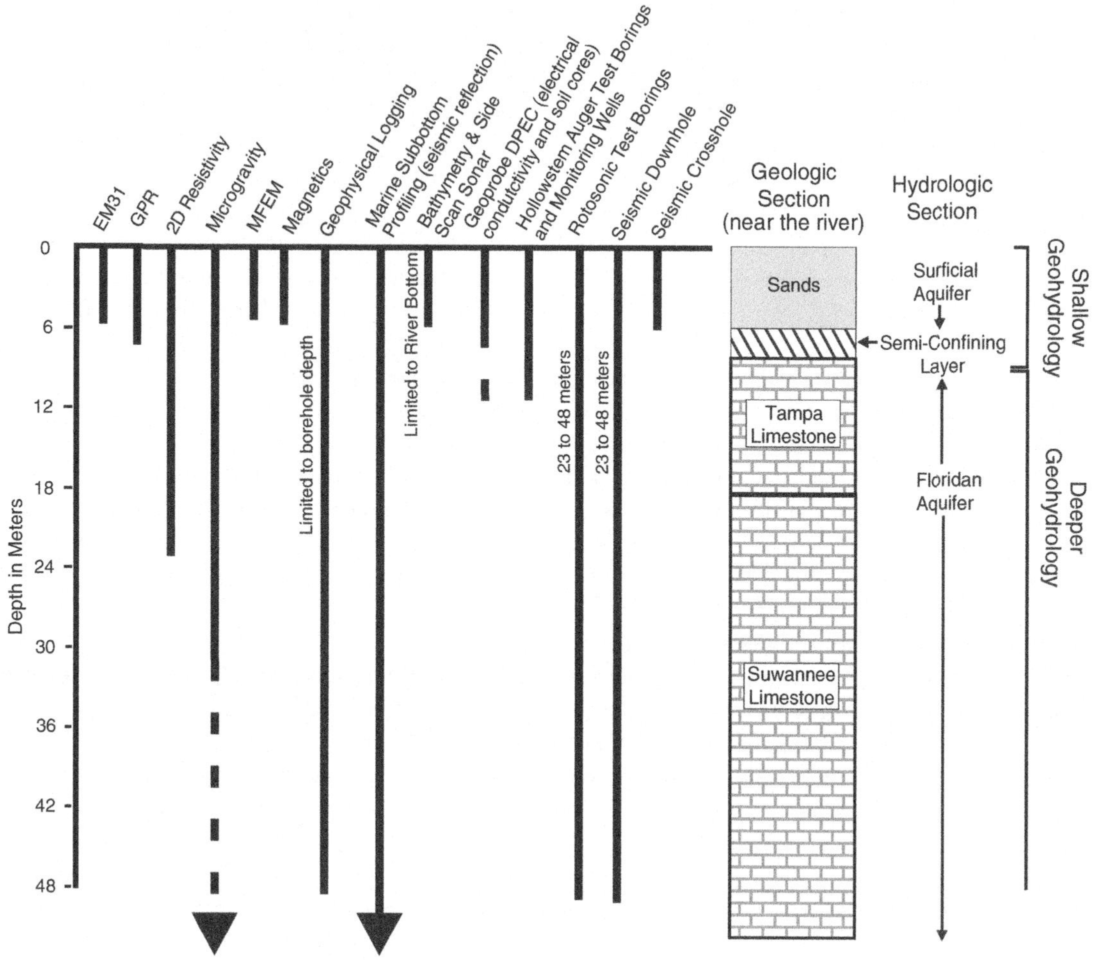

Fig. 27.3 Type and depth of measurements used to meet project objectives

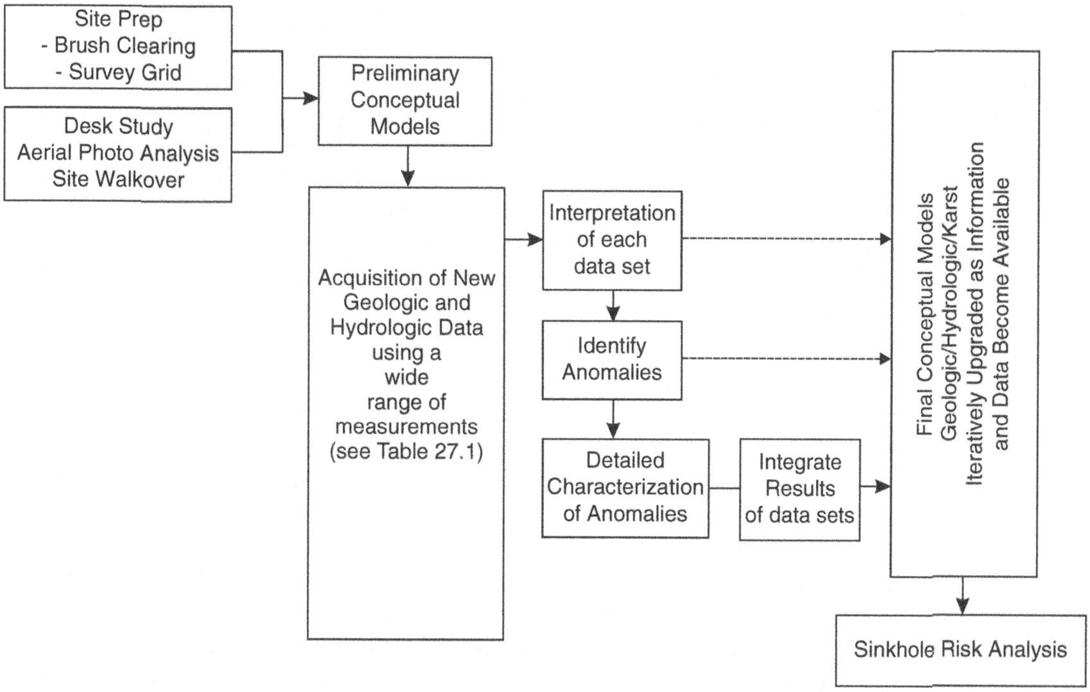

Fig. 27.4 Overall project strategy

walk to and reoccupy any location within the entire 52 ha without use of a GPS system.

All existing site features including buildings, borings, wells, disposal ponds and piles of waste materials were referenced to this standard grid. Current aerial photography of the site was also obtained after brush clearing and was used to develop a topographic map with a 0.3 m contour interval.

27.5 The Desk Study

While the site was being prepared for fieldwork (brush clearance, surveying and a topographic map) a desk study was carried out. An abundance of geologic and hydrologic information was available from the Florida Geologic Survey bulletins and reports. Karst data included cave maps, spring surveys, county sinkhole databases, state sinkhole databases and general karst reports provided by a variety of sources. These reports provided information on the stratigraphy, hydrologic setting, sinkhole distribution and their frequency of occurrence, lineaments, depth of known cave systems in the area, and distribution of springs. The information provided in these reports formed the basis for the initial understanding of the regional site conditions and included general geologic and hydrologic information.

27.5.1 The General Site Setting

The Stauffer site is located near the town of Tarpon Springs, in Pinellas County, on the west central coast of the Florida peninsula. The site itself lies on the north side of the Anclote River (Fig. 27.1) and is underlain by hundreds of meters of limestones and dolomites, which compose the Floridan Aquifer. For purposes of evaluating sinkhole hazards and the karst topography of Pinellas County, our focus was the geology of the uppermost limestones (about 60 m) and the overlaying clastic strata.

27.5.2 Regional Geology

A regional geologic section for the Tarpon Springs area was constructed using five existing borings. Two deep borings, W-16609 located approximately 4.8 km north-northeast of the site, and W-12943 located approximately 5.6 km south-southwest of the site. These two borings are referenced in cross-sections included in publications by the Florida Geologic Survey (Arthur et al. 2001; Green et al. 1995) and show a slight southward dip to the formations. Three additional borings in Pinellas County east of the site (TR15-2, 2560 and 1072) are added to the regional cross-section and provide more localized depth and thickness of the Tampa Limestone and the depth of the top of Suwannee Limestone at the site (Beck and Sayed 1991). These three borings indicate some variation in depth and thickness of the Tampa and Suwannee limestone, but generally support the regional trends.

The geologic section (Fig. 13.3) defines four lithostratigraphic units. From shallow to deeper, these units are:

1. Post-Hawthorn Group Undifferentiated Sands and Clays
2. Hawthorn Group consisting of the Tampa Member of the Arcadia Formation, herein referred to as the Tampa Limestone
3. Suwannee Limestone
4. Ocala Limestone.

Only the Tampa and Suwannee Limestone are of interest here. The deeper Ocala Limestone is not expected to have karst features that would impact the remediation at the site. The following descriptions are primarily from Green et al. (1995).

27.5.2.1 Post-Hawthorn Group Undifferentiated Sands and Clays

The Post-Hawthorn Group sediments are primarily composed of varying proportions of quartz sand, shell and clay with variable amounts of organics and naturally occurring re-worked phosphate. Much of the county is covered by unconsolidated sediment referred to as sand. Elevations at the site are approximately 3 m MSL with a thin 3–6 m cover of sand (Beck and Sayed 1991).

27.5.2.2 The Hawthorn Group

The Hawthorn Group Consisting of the Tampa Member of the Arcadia Formation and is referred to as the Tampa Limestone. In Pinellas County, the Hawthorn Group sediments lie unconformably above the Suwannee Limestone. Only the Tampa Member (Tampa Limestone) of the Arcadia Formation is present in the study area.

The Tampa Limestone is white to yellowish gray in color, and ranges from a limestone, quartz sand and clay (Green et al. 1995; Scott 1988). The Tampa Limestone commonly contains 10–30 % quartz sand. Based upon projections of the regional cross-section to the site (Fig. 13.3), the top of the Tampa Limestone at the Stauffer site was expected to be about 5–15 m below land surface (bls) and approximately 10–15 m thick.

27.5.2.3 Suwannee Limestone

The Hawthorn Group sediments, which include the Tampa Limestone, overlie the Suwannee Limestone unconformably. In some cases, the upper Suwannee lithologies appear to grade upward into the Hawthorn Group (Green et al. 1995; Beck and Sayed 1991). The lithology of the Suwannee Limestone ranges from a light-gray to yellowish-gray packestone to grainstone. These carbonates are variably moldic with trace amounts of sand and clay within the upper portions. Trace amounts of chert and organics occur throughout the unit. A dolostone or dolomitic limestone layer, approximately 3–6 m thick, commonly occurs within the lower one-third of the unit in the study area (Green et al. 1995). Based upon projections of the regional cross-section to the site (Fig. 13.3) the top of the Suwannee Limestone at the Stauffer site is expected to be about 15–24 m bls (Beck and Sayed 1991) with a thickness of about 60 m.

27.5.3 Regional Hydrology

The hydrogeologic section in Pinellas County consists of an unconfined surficial aquifer in the surficial sands and the karstified limestones of the Floridan Aquifer, consisting of the Tampa and Suwannee Limestones (Fig. 13.3). The Tampa Limestone is hydrogeologically connected with the underlying Suwannee Formation and is included in the Upper Floridan aquifer (Beck and Sayed 1991; Wetterhall 1964).

The surficial aquifer is thin, heterogeneous, has low values of transmissivity and low yields to pumping wells. The surficial aquifer in Pinellas County has only limited use as a supplemental or alternative source of water (Southwest Florida Water Management District 1988). The thin nature of the surficial aquifer in the vicinity of the site limits its usefulness as a drinking water supply; however, the aquifer is used for small irrigation purposes. The surficial aquifer at the site flows toward the south and discharges to the Anclote River. The surficial aquifer is separated from the Floridan aquifer by a semi-confining layer (SCL) of clay, sandy clay, marl, and limestone from the base of the Post-Hawthorn Group and upper weathered surface of the Tampa Limestone.

Regionally, the water table in the surficial aquifer is generally 1.5–3 m below the land surface and the potentiometric surface of the confined Upper Floridan aquifer is also relatively flat and is also generally 1.5–3 m above sea level. In the Tarpon Springs area, where surface elevations are low, the head difference between the surficial and Floridan aquifers is very small (Beck and Sayed 1991; Wetterhall 1964).

Further inland to the east, beneath higher ridges, the piezometric surface of the Upper Floridan aquifer is much higher. There is a significant head difference between these two aquifers in these areas. The regional flow direction within the Upper Floridan aquifer is shown to be westward to the Gulf of Mexico. However, within the areas north and south of the Anclote River, flow is toward the river indicating that the river serves as a discharge point for the Upper Floridan aquifer (Coble 1973).

The most productive aquifer in Pinellas County is the Upper Floridan aquifer and the most productive zone of this aquifer is in the uppermost limestones, the Tampa and Suwannee Limestones (Hickey 1982). The high productivity is due to the high primary porosity along with the karst character of these limestones. The Floridan aquifer provides some of the public water supply for Pinellas County, although it has limited use within the western portion of the county due to saltwater intrusion along the coast.

Saltwater intrusion is known to occur along the Anclote River. In 1963, chloride content in and around the Anclote River near the site was shown as more than 250 mg/L (maximum for drinking water) at depths of 0–38 m (Cherry 1966). The chloride concentration at selected locations within the Anclote River near the site in 1969 ranged from 3,100 mg/L at low tide to 15,200 mg/L at high tide (Reichenbaugh 1972).

27.5.4 Regional Geomorphology

The karst character of the limestone aquifer in the county is well-demonstrated by the existence of paleokarst sinkhole lakes on land, numerous springs and numerous bayous within the Anclote River. A bayou is a circular body of water within the Anclote River thought to be associated with a paleocollapse. The karst character of the limestone aquifer in the immediate area is well documented by the cavernous connection between Lake Tarpon and Tarpon Spring (Fig. 27.1). Lake Tarpon drained through a sinkhole, via a deep cavern and discharged at Tarpon Spring located in Spring Bayou (Wetterhall 1965).

From a review of regional geomorphology it was found that sea levels had been lower by as much as 130 m (Balsillie and Donoghue 2004). The combination of much lower sea levels and a thick fresh water lens would have resulted in the dissolution of limestone and the development of caves at significant depths of 240 m or more. This model can account for the many large and therefore deeper-seated paleokarst seen in the area.

27.5.4.1 Aerial Photo Analysis

Three types of aerial photos were obtained for this project and they include stereo-pairs, individual aerial photographs, and various historic oblique aerial photographs of the site. Stereo-pair photos were acquired and reviewed for the following years 1926, 1942, 1951, 1962, 1967, 1997 and 2001. These photos typically covered an area of 1.6 km radius or more around the Stauffer site.

A review of these aerial photos was made to identify obvious sinkholes in and around the site area. Each sinkhole identified in the latest 2001 aerial photography was also seen in the aerial photography from 1926, prior to the construction of the plant. A number of obvious paleosinkholes due to their circular shape are seen in the aerial photos including some bayous within the adjacent Anclote River and its branches (Fig. 27.5). Two of these were located along the

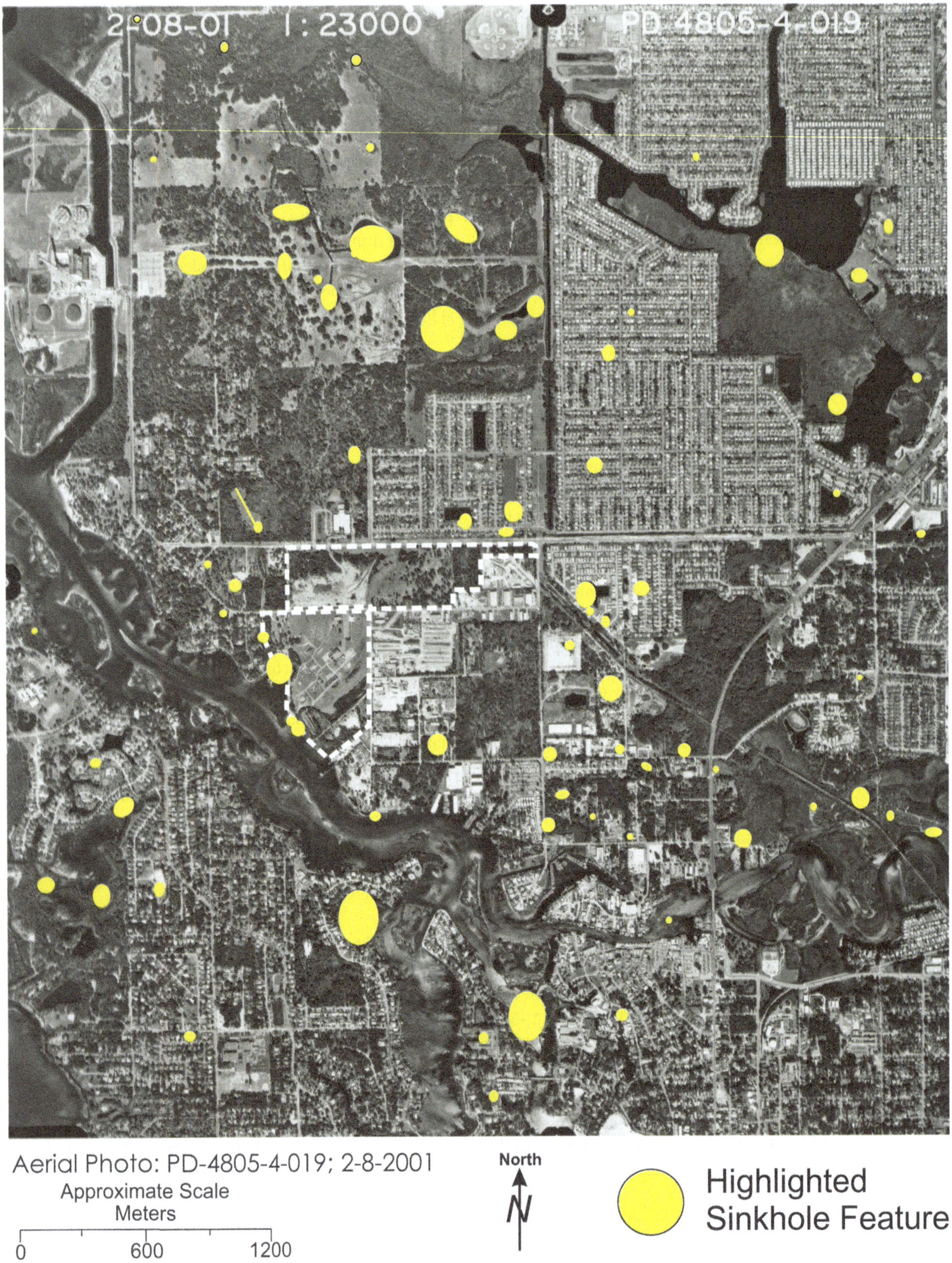

Fig. 27.5 Aerial photo of the site setting including numerous sinkholes and the Anclote River with its bayous (FDOT PD 4805-4-019; 2-18-2001)

western perimeter of the South Parcel of the site (Fig. 27.2). They are the Sinkhole Pond and Meyers Cove. All of these features appear in the oldest aerial photography from 1926 indicating that they are old and not due to recent sinkhole activity.

Photo lineaments are seen on aerial photographs as alignments of structure, joints, sinkholes, sinkhole lakes, streambeds, vegetation trends, and other surface features thought to be controlled by subsurface conditions. Lineaments that are manifestations of subsurface geologic conditions in limestone are often the result of dissolution of rock along joints, with major karst depressions centered where two sets of joints intersect (Parizek 1976).

Regional lineaments in the State of Florida generally trend northeast-southwest and northwest-southeast (Vernon 1951). A number of local lineaments were identified from the aerial photos and USGS topographic maps. The Anclote River and its bayous are to a large extent fracture controlled and contain a number of circular bayous, which are paleosinkholes. Figure 27.6 shows three significant lineaments in proximity to the Stauffer Site along with other features of interest.

- The most obvious local lineament is the stretch of the Anclote River south of the Stauffer Facility that forms a long lineament (oriented approximately 45°W). This is coincident with the connection between Tarpon Spring in Spring Bayou and Lake Tarpon sink in Lake Tarpon about 5 km to the southeast (Fig. 27.1) and described by Wetterhall (1965).
- Another local lineament is seen extending 10°W through the eastern part of the Stauffer site and Kreamer Bayou. This lineament becomes very subtle north of site, where it is only based upon a linear trend in vegetation.
- A possible lineament along the west side of the site extends through Meyers Cove and the sinkhole pond extending 10°W.

27.5.4.2 Cave Systems Within the Region

The caves in the area are all under water and are based upon maps developed by cave divers. In addition to the cave connecting Lake Tarpon and Tarpon Spring. A number of caves frequented by cave divers exist to the north of the site and include:

- Beaconwoods Cave System is located about 19 km north of the site underlying a large area of Bayonet Point, Florida. This is an extensive cave system about 3.2 km in length at a depth of approximate 38–45 m bls (about midway within the Suwannee Limestone at that location) (Wilson, 2001, personal communication).
- A large open paleosinkhole along US19 in Hudson, Florida is located about 22.5 km north of the site. The cavity system that collapsed to form this paleosinkhole is at a depth of 40–43 m bls within the lower third of the Suwannee Limestone at that location.

These examples indicate that large cave conduits would be expected to occur within the lower half to lower third of the Suwannee Limestone at depths of approximately 30–60 m below grade and even deeper. Wetterhall (1965) indicates that within Pasco County (the county immediately north of Pinellas) most domestic and many irrigation wells produce from the lower part of the Suwannee Limestone. This data can be used as an indication of where significant dissolution could be expected at the Stauffer site.

27.5.4.3 Springs

Springs in the area are discussed in Wetterhall (1965) and Scott et al. (2004). There are numerous springs located in Pinellas county and Pasco county to the north. The largest spring in the area is a tourist attraction called Weeki Wachee Springs about 45 km to the north. This is a first magnitude spring (Scott et al. 2004). The spring flows from near the contact between the Suwannee and the Ocala Limestone at a depth of 42 m deep. Its source of water is probably below the Ocala in the Avon Park Limestone.

Tarpon Spring is located about 2.7 km southeast of the site and is located within Spring Bayou at a depth of about 38 m deep (Fig. 27.1). Lake Tarpon Sink is located on the western shore of Lake Tarpon further to the southeast (Fig. 27.1). The sink has a depth of 35 m below MSL and is the eastern end of a deep conduit system providing a connection between Lake Tarpon Sink and Tarpon Springs. The connecting conduit lies within the Suwannee Limestone (Wetterhall 1965).

Based upon local knowledge, (discussions with fishermen and divers) there are several springs within the bayous of the Anclote River as well as in the area near shore in the Gulf of Mexico. The assessment of possible local springs was made visually at low tide. The nearest spring to the site is located in the Port Tarpon Marina about 0.5 km southeast of the site (Fig. 27.6) and is in the Tampa Limestone.

These local springs, such as the one identified in the Port Tarpon marina are of smaller magnitude (probably less than a few hundred GPM) and flow only for a short duration at low tide. This indicates a very small head difference between the river and the groundwater source for the spring. A small head difference would account for the low flow noticed at many local springs only at low tides.

27.5.5 Reported Sinkholes and Sinkhole Trends in the Region

A wide range of databases, reports, and newspaper clippings were available to provide sinkhole data. While none are

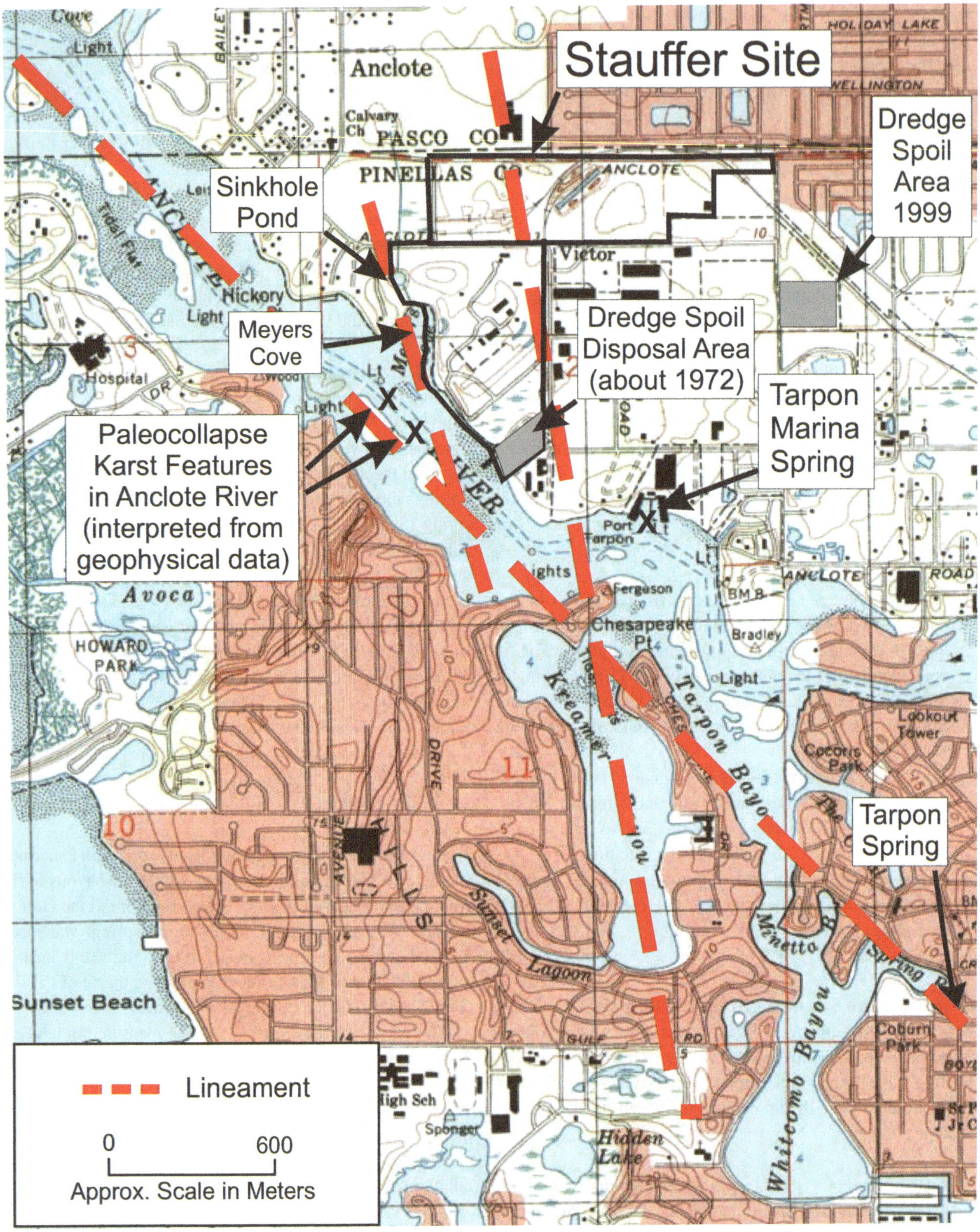

Fig. 27.6 The Stauffer site with locations of three photo-lineaments, karst features, two dredge disposal sites and two springs shown (USGS 7.5 min Series Tarpon Springs Quadrangle, 1995)

complete and there is redundancy, they are representative of reported sinkhole activity. The following summarizes the data obtained from these sources. Unfortunately, collapse caused by broken water or sewer lines are also often reported as naturally occurring sinkholes.

27.5.5.1 Sinclair and Stewart's Sinkhole Map

Sinclair and Stewart (1985) used the concept that the thickness of cover exerts a significant control on sinkhole development and have used this concept to develop their sinkhole probability map showing the expected types of sinkholes based upon cover thickness (Fig. 3.10). The Stauffer site lies within an area classified as "Bare or thinly covered limestone where the sinkholes are few, generally shallow and broad, and develop gradually". They are referred to as cover subsidence sinkholes where the sandy soils (about 3–6 m thick) migrate downward into local cavities. Sinkholes reported by newspapers in the vicinity are typically 3–6 m in diameter and 3–6 m deep, which supports the Sinclair and Stewart (1985) sinkhole map.

27.5.5.2 Corps of Engineers Dredge Disposal Sites

In 1972 the Corps of Engineers used a parcel of land immediately southeast and adjacent to the Stauffer site for dredge disposal (Fig. 27.6). The elevation at this site was about 3 m above sea level and the top of rock averaged about 6 m below grade and was covered by a layer of clay (the SCL). The site was adjacent to the Anclote River and both the surficial and Floridan groundwater levels are shallow with little difference in head. Both geologic and hydrologic conditions provided a relatively stable environment. No sinkholes developed at this dredge disposal site that is directly adjacent to the southern series of settling ponds at the Stauffer site (Fig. 27.6). Stauffer purchased this property in 1981 as a buffer to the settling ponds.

In 1999, the US Army Corps of Engineers utilized a parcel of land for disposal of Anclote River dredge material located about 915 m due east of the Stauffer site (Fig. 27.6). The area had been excavated to the top of rock removing the surficial sands and the semi-confining clay layer and a berm was built around the perimeter. Dredge spoil was pumped into the disposal area with large volumes of water increasing the local head of water.

A few days after dredging started, four small sinkholes (6–9 m across) reportedly developed in and around the site. Figure 27.7 shows two of these sinkholes. These sinkholes most likely developed as a result of sediment raveling into localized voids within the upper portion of the Tampa Limestone due to the missing semi-confining layer and the large quantities of water associated with the dredged materials. These are typical of the small, shallow sinkholes that occur in the area (Sinclair and Stewart 1985).

27.5.5.3 The Sinkhole Database

Bill Wisner and Steve Denahan began a sinkhole database in the 1960s, who were with the Florida Department of Transportation (FDOT) at the time. Bill Wilson then expanded this database, when he was with the Florida Sinkhole Research Institute, and was later continued by Bill at his own firm Subsurface Exploration Inc. (SEI). This database has been used as a means of estimating sinkhole locations, densities, and their temporal trends, new sinkholes per km^2 per year (NSH) and to provide sinkhole risk assessments (Wilson 1995).

A plot of historic sinkhole occurrence was provided by SEI for the sinkhole occurrence within 1.6, 8 and 16 km radius from the Stauffer site (Fig. 27.8). At the date of the study (June 2002):

- Seventy-eight sinkholes have occurred within the 16 km radius
- Thirty-five sinkholes have occurred within the 8 km radius
- Three sinkholes have occurred within the 1.6 km radius

Note that only three sinkholes were identified in the database within 1.6 km of the site and are located approximately 1.4 km east-southeast of the site where elevations start to become higher and there is considerable development and concentration of surface runoff.

Based upon the results of the database (Fig. 27.8) alone one could easily reach the conclusion that the site was susceptible to sinkhole activity. Yet, on site, and within the immediate surrounding area there had been no recent sinkhole activity based upon aerial photo analysis back to 1926, corporate records and discussion with former workers at the facility.

The database had recorded sinkholes that were all located east, inland and upgradient of the site by almost a kilometer. They occurred in the more developed areas at slightly higher elevations. Many of the sinkholes in the database have been caused by cultural conditions such as those associated with concentrated drainage of surface water, leaky sewers, water line breaks or where construction was taking place. Note that this database did not include the sinkholes from the Corps of Engineers dredge disposal site in 1999 just 914 m east of the site (Figs. 27.6 and 27.7).

27.5.6 Review of Corporate Files and Interviews with Previous Workers

On-site existing data and information was obtained from Stauffer's project files and included some site-specific geologic data, information on the ponds/disposal areas (locations and depths) and possible drum disposal areas. The results of work by previous consultants all had a very specific

Fig. 27.7 Sinkholes at Corps of Engineers dredge disposal site in 1999. The location of these sinkholes is due east of the Stauffer site and noted on Fig. 27.6. (**a**) Sinkhole north of bermed dredge spoil area. (**b**) Repaired sinkhole within bermed dredge spoil area (Photos courtesy of B. Bergen, Pinellas County Health Department)

focus and were not intended to be comprehensive. Where possible, this information was used to develop the beginning of an on-site database as well as preliminary conceptual model and included site-specific geologic data from drilling.

Communication with retired workers from the site indicated that there were no incidences of subsidence or sinkholes during the period of plant operations (1947–1981). This is a strong indication of geologic stability over the site under operating conditions which included variable static and dynamic loading from piles of stored materials and railway traffic (Fig. 14.6).

27.6 The Preliminary Conceptual Models

The general geologic and hydrologic conditions were reasonably well defined by the regional literature and site-specific reports prior to beginning any work at the site (Fig. 27.9). Subsequent efforts would be focused upon verifying these data and providing further detailed information of site-specific conditions. Then, as additional data were acquired, the conceptual model was verified or revised.

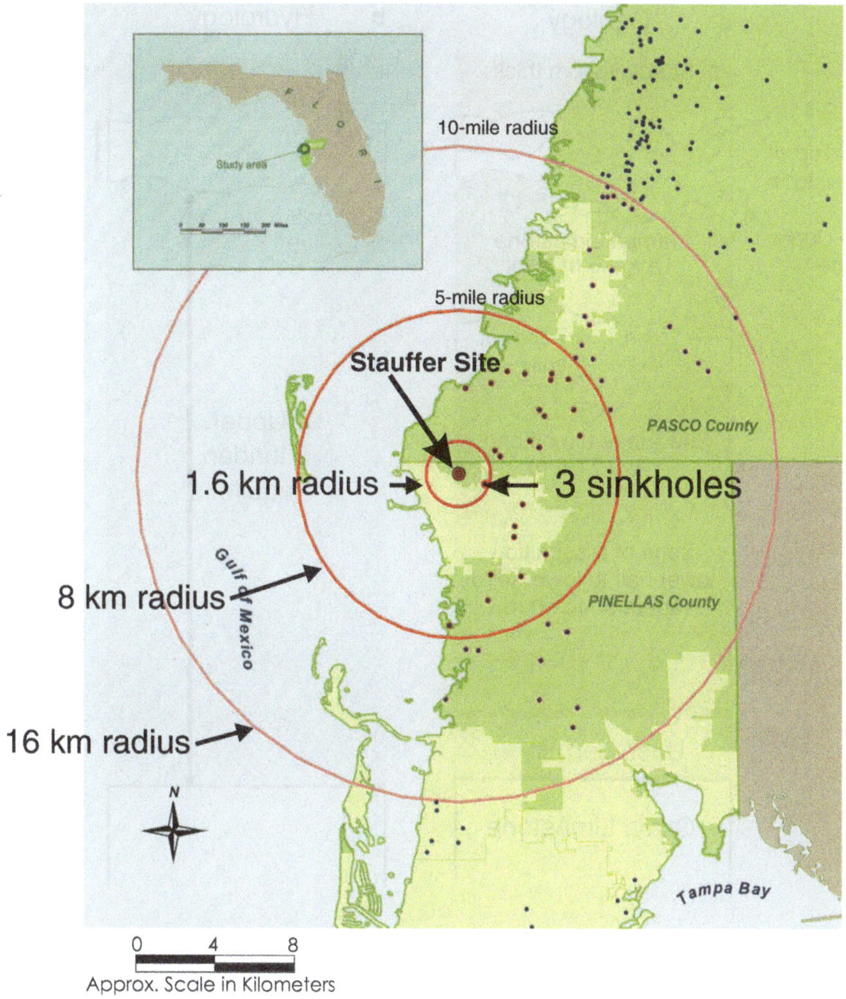

Fig. 27.8 The locations of sinkholes within a 1.6, 8 and 16 km radius of the site based upon a sinkhole database (SEI 2002)

27.6.1 Geologic Conceptual Model

The preliminary conceptual model of site geology (Fig. 27.9a) was based upon five regional borings extending 24–137 m below grade. These borings located the depths of the Hawthorn Group, the Tampa Formation and the contact with the Suwannee Formation. Site-specific data came from on-site existing borings and site-specific reports and indicated:

- a surficial layer of sand was found to be 3–6 m thick becoming thicker toward the Anclote River,
- a semi-confining clayey layer (SCL) ranges from 0.3 to 2.4 m thick,
- the Tampa Limestone was found to be about 15 m thick over the site and lies unconformably over the Suwannee Limestone. These formations were found to be hard, cemented layers interspersed with layers of soft, lime clay and silt with limestone fragments. Although the limestone contains some clays and silts, the overall lithology is competent and difficult to drill through.
- Zones of dissolution, high permeability and cave development were also cited in the literature

27.6.2 Hydrologic Conceptual Model

The site hydrogeology is a relatively flat, low flow system flowing to the south or southwest with ultimate discharge to the Anclote River by both the surficial and Upper Floridan aquifers (Fig. 27.9b). The flat gradients, in combination with the relatively low hydraulic conductivities of both aquifers, are indicative of a low flow velocity groundwater system. Water levels in both aquifers are maintained relatively constant due to the adjacent Anclote River and nearby Gulf of Mexico.

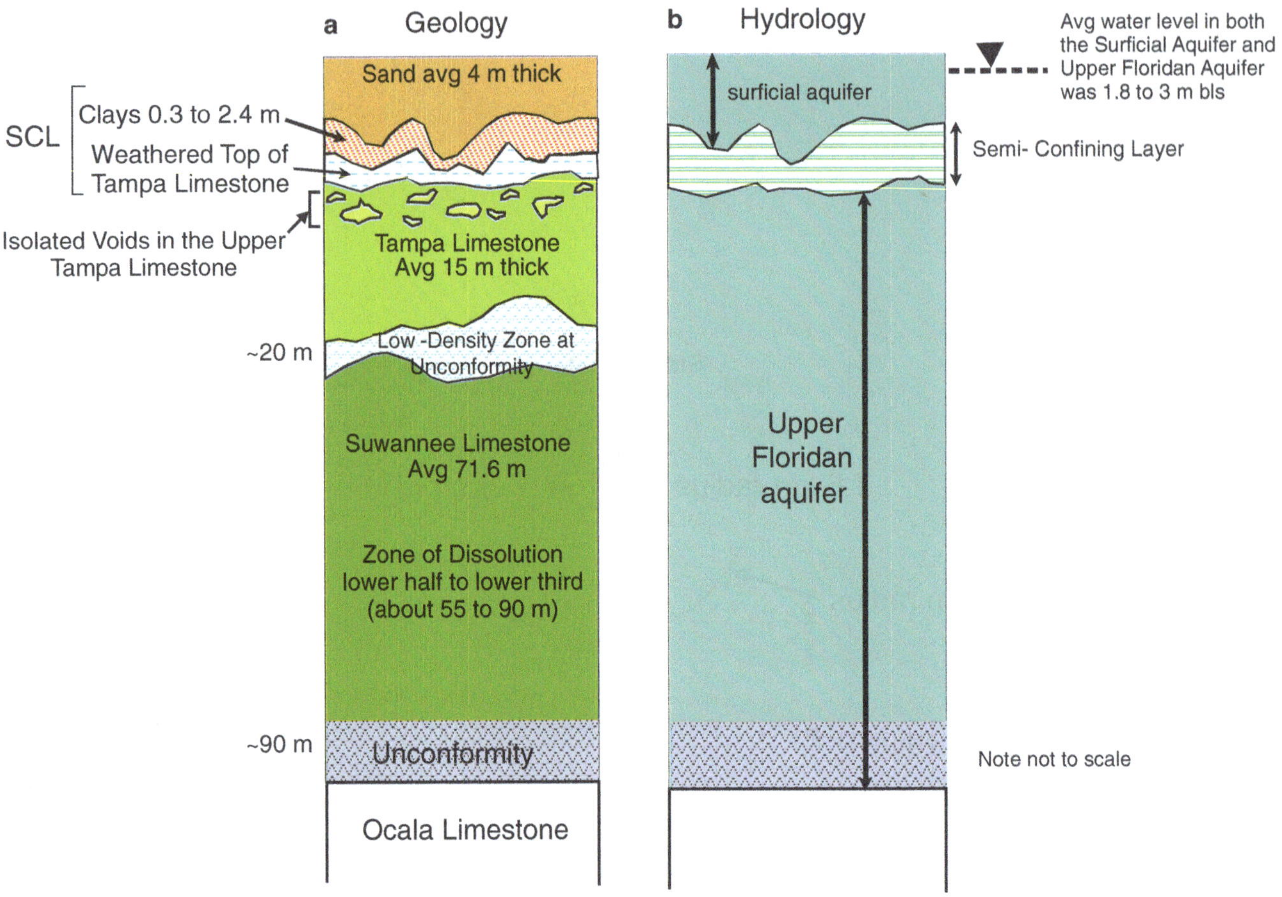

Fig. 27.9 Preliminary conceptual model of geologic conditions (**a**) and hydrologic conditions (**b**)

27.6.3 Karst Conceptual Model

There was clearly sinkhole activity in the region both old and new. Paleosinkholes were numerous in the aerial photo analysis with two identified along the west side of the South Parcel, the Sinkhole Pond and Meyers Cove (Fig. 27.2). Both of these features were identified in the earliest aerial photos (1926) indicating that they were old and stable.

The new sinkhole activity as reported in the databases and local newspapers is relatively small and likely due to shallow voids in the top of rock. In addition, this activity lies to the east at a somewhat greater elevation, with thicker sand cover, in an area with extensive development, and away from the stabilizing influence of the high water table of the nearby Anclote River and the Gulf of Mexico. No active sinkholes were noted near the Stauffer site in any of the data including the aerial photos dating back to 1926. Furthermore discussions with individuals who had worked at the Stauffer site for years had indicated that there had never been any sinkhole occurrence at the site.

The nearest sinkhole activity occurred off-site 914 m to the east at the Corps of Engineers dredge disposal site in 1999 (Fig. 27.6). This appears to be due to the removal of the SCL and the disposal of dredge material containing large quantities of water. Furthermore, consider that the Corps of Engineers dredge disposal site used in 1972, immediately adjacent to the property, did not have any sinkholes develop where the SCL was left intact.

Two photo lineaments had been identified extending through the eastern and western sides of the South Parcel, both at about 10°W (Fig. 27.6). The lineament to the west aligns with Meyers Cove and Sinkhole Pond, both of which are old features. In addition, there are no recent sinkholes or current sinkhole activity associated with the lineament to the east.

27.7 Shallow Geohydrologic Conditions

The shallow hydrogeologic conditions consist of the surficial sandy aquifer and the semi-confining layer (SCL). The SCL between the surficial aquifer and the Upper Florida aquifer (Fig. 27.9) is critical to the site for two main reasons:

- Limiting contaminant transport downward into the Floridan Aquifer and
- Limiting the downward movement of sands into local voids within the upper portion of the Tampa Limestone resulting in sinkholes.

In addition, the head or pressure difference [static and dynamic (tidal)] between shallow and deeper aquifers could impact a weak zone within the SCL by breaching this layer. For example, an upward head may lead to a local spring, and a downward head may aid in moving unconsolidated material downward through weakened zones in the SCL resulting in subsidence and possibly allow contaminants to flow into the Floridan Aquifer. In any case, significant (man-induced) changes in head or breaching of the SCL could result in contaminant transport and sinkhole collapse.

In addition, the lateral and vertical extent of the ponds were assessed as well as the potential for buried drums, tanks or other sources of contamination. These additional objectives fall within the shallow hydrogeologic conditions and will be the focus of the remediation.

27.7.1 The Sands and Surficial Aquifer

The shallow geology was densely sampled using two continuous surface geophysical techniques, electromagnetic conductivity measurements and ground penetrating radar (GPR). The electromagnetic measurements were completed using an EM31 which provide a composite conductivity value to a depth of approximate 5.5–6 m. In addition, a secondary measurement (in-phase data) indicative of the presence of buried metals was obtained simultaneously with the EM31. The GPR measurements were completed using a 100 MHz antenna and time window of 150 ns, which provided a 2D image of the subsurface to a depth of approximate 7.3 m. Both of these measurements were obtained along east to west parallel survey lines spaced 3 m apart. This provided approximately 100 % coverage over the entire site with the exception of buildings, pond areas and piles of debris materials.

Electromagnetic measurements at the site (EM31) indicate that the surficial sands are low conductivity materials, about 10 mS/m, that are generally uniform both laterally and vertically. Existing geologic logs as well as subsequent measurements and sampling on-site indicated that the surficial sands at the site are fine to medium-grained, tan to orange in color with traces of silt or clays and are found consistently across the site ranging in thickness from 2 to 8.5 m thick, averaging 4.5 m. Figure 27.10 is a contour map of the electrical conductivity across the site. Values of electrical conductivity higher than background can be accounted for due to utilities, building foundations, slag debris, settling ponds or buried metals. The GPR data did not provided indications of any anomalous conditions within the sands over the SCL such as dipping strata or piping within the sands that might indicate sinkhole activity.

In addition, slug test and pumping test results in the surficial aquifer indicated a range of hydraulic conductivities that are relatively low and consistent with that of a fine sand aquifer. The hydraulic conductivity of the surficial aquifer sands ranged from 0.3 to 37.8 m/day based on both slug and pumping tests, and the surficial aquifer transmissivity ranged from 75 to 85 m^2/day based on the pumping test results (Parsons 1999, 2003).

The general groundwater geochemistry in the surficial aquifer (sodium, chloride, salinity levels, etc.) indicates that with the exception of proximity to the Anclote River, the groundwater in the proposed remediation area is considered fresh water and there appears to be little geochemical variability that would affect the proposed remedy (Parsons 1999, 2003).

27.7.2 The Semi-confining Layer

Various techniques were used to characterize the continuity, thickness, and content of the semi-confining layer. Ground penetrating radar typically mapped the lateral extent and variations in depth of this layer, but not its thickness. While resistivity imaging was primarily used for assessing deeper geologic conditions, it also provided data on the SCL showing its continuity, depth and relative thickness across the site (Fig. 27.11). Detailed information on the SCL was provided by direct push electrical conductivity (DPEC). This Geoprobe technique was used throughout the site and locations were guided by the surface geophysical data. This technique provided information on the top of the SCL, the variability within the SCL (clay content) and the bottom of the SCL based upon electrical conductivity measurements as well as allowed core samples to be collected at select locations. Figure 17.3 shows an example of GPR over a low area in the top of the SCL along with DPEC data. The DPEC data measures variations in electrical conductivity, higher conductivity indicates more clay content. In addition, geophysical logs (natural gamma and induction) were obtained in all borings and provided very detailed profiles through the SCL showing its depth, variability in thickness and variability of clay content within the SCL.

The semi-confining layer is composed of a clayey layer that grades into a weathered, clayey limestone that is a part of the upper Tampa Limestone. A zone of clay and weathered limestone marks the top of the Tampa Limestone and is considered to be the bottom of the SCL. The weathered zone is typically harder and dryer, with limestone fragments that distinguish it from the clayey layer above. The clayey layer

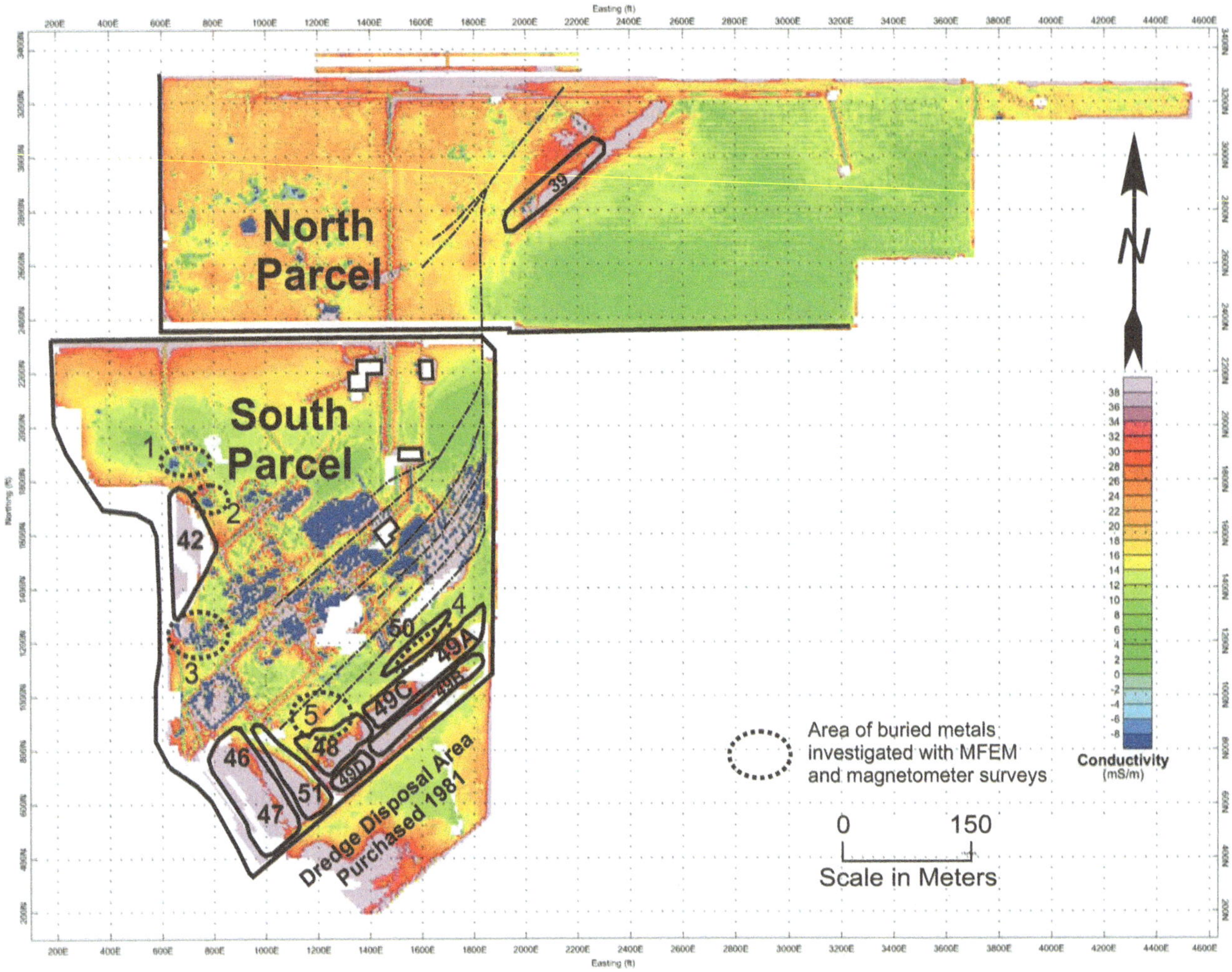

Fig. 27.10 EM conductivity contour map

is thought to be an unconformity at the top of the Tampa Limestone, which is an erosional surface.

Figure 27.12 shows a contour of the top of the SCL based upon DPEC data, all new and existing borings and geophysical logs. The semi-confining layer:

- is described in the geologic logs as moist to wet sands, silty clay grading with depth to very dry, very hard sandy, silty clay and limestone
- ranges in depth from 3 to 8 m bls, dipping to the south and has a range of elevations from almost +1.5 m MSL in the north to −3 m MSL in the south near the river
- has a variable thickness which ranges from 0.3 to 7.3 m, with an average of about 2.3 m
- varies in clay content based upon variations in 110 DPEC electrical measurements and 74 geophysical logs
- is variable, but generally high in silt and clay content based upon X-ray diffraction, sieve analysis and hydrometer analysis
- Localized lows in the top of the SCL were identified and are thought to be erosional surfaces
- The SCL is laterally continuous throughout the entire site, with the exception of one area along the eastern edge of the South Parcel.

The nature of the semi-confining layer was further defined by permeability testing, that indicate hydraulic conductivities of 2.84×10^{-4} to $<1.0 \times 10^{-8}$ cm/s which are typically two or more orders of magnitude lower than the hydraulic conductivities of the surficial aquifer 10^{-2} to 10^{-4} cm/s. The contrast in hydraulic conductivities indicates that the SCL would minimize the cross-connection between the surficial and Upper Floridan aquifers at the site (Parsons 2003).

The chemical data collected directly beneath or adjacent to settling ponds confirm that the semi-confining layer is impeding the flow between the surficial aquifer and Upper Floridan aquifer. Chemical data show that no primary drinking

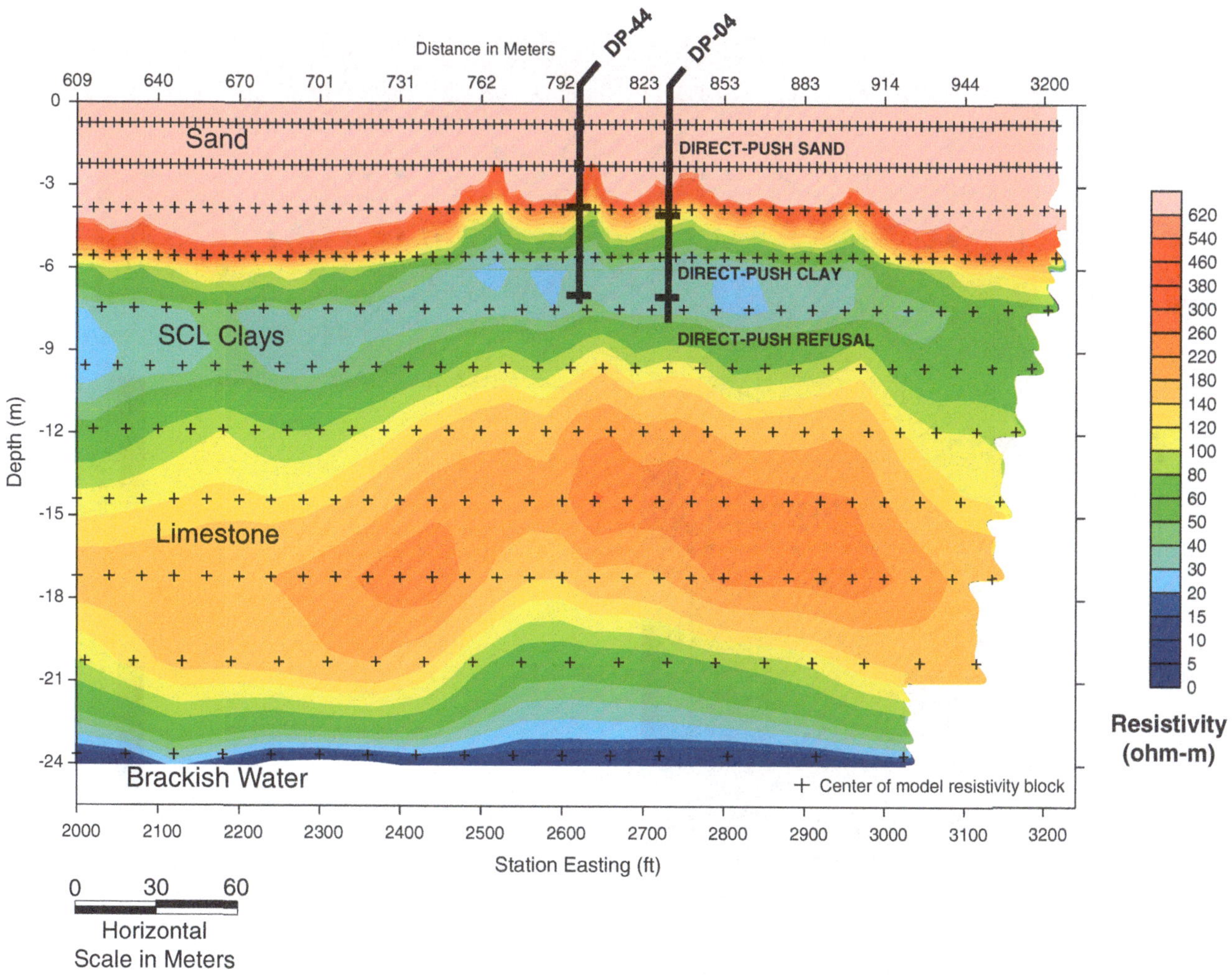

Fig. 27.11 Example of 2D resistivity data along with two DPEC pushes identifying the top and bottom of the SCL

water standards were exceeded in the Upper Floridan aquifer (Parsons 2003).

Pumping tests conducted in both aquifers showed no measurable effect on the non-pumping aquifer. The lack of measurable drawdown in the non-pumping aquifer during the aquifer pumping tests indicates that there is not a strong hydraulic connection between the two aquifers. The vertical gradient between aquifers is generally flat and varies from slightly downward to slightly upward. The lack of a strong vertical gradient would tend to minimize the potential for a weak zone or breach within the SCL layer. Therefore, the potential for contaminants in the surficial aquifer to migrate toward the Upper Floridan aquifer or to enable the downward movement of sands into local voids within the upper portion of the Tampa Limestone resulting in sinkholes is also minimized by the SCL (Parsons 2003).

Water level data was collected over several tidal cycles. Tidal influence in the surficial aquifer was hardly discernible beyond 150 m from the shoreline. Therefore, tides should have little or no effect on the proposed remedy in the surficial aquifer. The presence of the adjacent Anclote River and the nearby Gulf of Mexico provide consistency to the water levels at the site. In addition, the tidal change is relatively small (about 0.76 m semi-diurnal tide from tidal tables) and the head between the surficial aquifer and the Floridan aquifer is very small (fractions of a centimeter). These factors provide a high degree of stability and reduced stress on the SCL. This stability minimizes the opportunity for rupture of the SCL leading to any contaminant flow from the surficial aquifer into the Floridan aquifer or the flow of sand into voids in the uppermost Tampa Limestone.

27.7.3 Pond Boundaries

The pond disposal areas (Fig. 27.3) were characterized to determine their boundaries, depths, and possible migration of contaminants. Because the material in the ponds was elec-

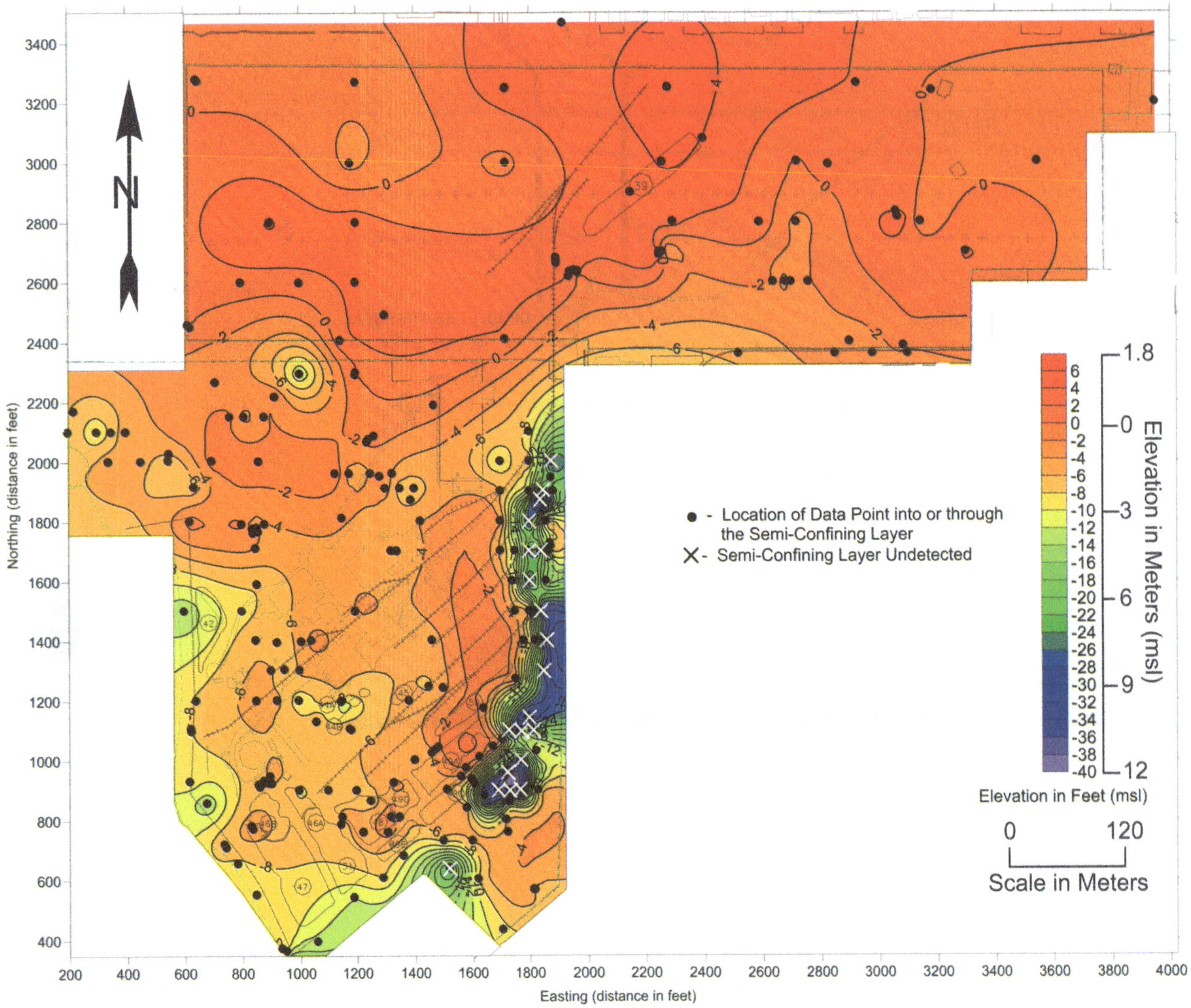

Fig. 27.12 Elevation *contour* map of the top of the SCL based upon DPEC data, existing and new borings as well as geophysical logs

trically conductivity, the electromagnetic survey was the primary technique used to meet this objective. Figure 27.10 shows the conductivity contour map from the electromagnetic survey.

A sharp electrical conductivity contrast (from 10 to greater than 30 mS/m) between the natural soil and pond materials allowed the pond boundaries to be mapped using electromagnetic techniques (EM31). Other areas of elevated conductivities are due to buried building foundations, utilities and remnants of slag piles. Based upon electromagnetic measurements, the locations and boundaries of the ponds are all in general agreement with previous maps produced by other site investigators, with the exception of Pond 39 located in the center of the North Parcel (Fig. 27.10).

Pond 39 appears to be approximately 30 % longer, extending to the north, and may be divided into two sections. A combination of DPEC and soil cores were used to confirm the depth of the pond materials which range from 1.2 to 6.4 m bls. Based upon this data the depths of the pond materials do not appear to breach the semi-confining layer.

All ponds were found to lie within the surficial aquifer sands and did not breach the SCL and no contamination was found below the SCL in the upper Floridan Aquifer.

27.7.4 Buried Drums

Concern was built around the rumor of a large cache of drums being buried in one or more location at the site. Both electromagnetic and ground penetrating radar provided essentially 100 % site coverage and included the areas where buried drums were suspected. These techniques were use to focus more detailed surveys for buried drums using multi-frequency EM system (MFEM) which responds to both ferrous and non-ferrous metal and magnetometer data which responds to only ferrous metal.

Based upon electromagnetic and MFEM measurements, five areas were identified that potentially contain buried metal debris (Fig. 27.10). While some ferrous metal was indicated in the magnetometer data there was no indications of the presence of a large cache of drums in one or more locations at the site.

The corrosion of buried steel drums is highly variable depending upon soil type, moisture, and chemistry, as well as the condition of the drum and its coatings. The drums of phosphorous waste would have been disposed of in water filled containers then disposed of below the water table to prevent ignition of the phosphorous.

The wall thickness of a 55-gal drum can range from 50 to 75 mils. Based upon uniform corrosion rates of 1 to 10 mil per year (Uhlig 1965; Romanoff 1957) it would take as long as 50–75 years or as short as 5–7.5 year for a drum to totally corrode. However, localized pitting corrosion can be much more rapid leading to leakage in a much shorter time. Data from Tuthill and Schillmoller (1966) indicates a rate of 15–30 mils per year or more for carbon steel in quiet seawater. At this rate pitting could penetrate a drum in 2.5–3.3 years.

Drums that were disposed of in the ponds have likely corroded, if not totally, then nearly so over time (more than 20 years since the plant closed in 1983). This would reduce the probability of their detection by magnetometer measurement and make a successful barrel recovery program unlikely.

27.7.5 Contaminants

Groundwater contamination in the surficial aquifer appears as "hot spots" at pond areas in the North and South Parcels. Metals of concern at the site that exceed the Primary Drinking Water Standards (PDWS) (EPA 2000) include antimony, arsenic, cadmium, nickel and thallium. Most of the wells containing these constituents only had one or two constituents each, while one South Parcel monitoring well had four of the constituents and another had three. These localized "hot spots" are closely associated with source areas and large areal plumes of these constituents do not exist at the site.

Metal constituents exceeding Secondary Drinking Water Standards (SDWS) (EPA 2000) in either aquifer include aluminum, iron, manganese, and zinc. Inorganic constituents exceeding the SDWS include chloride, fluoride and sulfate. Fluoride is the most widespread constituent at the site, being found in 18 surficial and 4 Upper Floridan wells at levels above the SDWS. Elevated sodium and chloride levels were found in wells near the shoreline, a result of the proximity of the river and saltwater intrusion. One surficial well in a source area in the South Parcel contained elemental phosphorus and two contained gross alpha above the PDWS.

27.8 Deeper Geohydrologic Conditions

The deeper geology consists of the Tampa and Suwannee Limestone, and the Upper Floridan aquifer. While no remediation will be taking place within this deeper zone, it was included in the investigation to determine the presence or risk of contaminants as well as the risk of sinkholes occurring within this zone.

Below the weathered zone that is considered part of the SCL, the Tampa Limestone consists of layers of fine grained limestones, fine to coarse-grained sands, gravelly limestones and silty/sandy stiff clays. The contact between the Tampa Limestone and the Suwannee Limestone is an unconformity and difficult to identify. However, there is typically a high production zone at this unconformity between the Tampa and the Suwannee (Hickey 1982). Below this zone, the Suwannee Limestone is highly variable, ranging in composition from fine grained to sandy to gravelly limestone, with layers of silty/sandy clays and clayey marls. Cave development is known to occur within the lower half to lower third of the Suwannee Limestone.

The deeper geologic conditions were characterized using 2D resistivity imaging and microgravity measurements. These two sets of geophysical data provided the capability of detecting changes in geologic conditions such as open cavity systems and paleocollapse within the deeper strata. Both of these measurements integrate a larger volume of the subsurface and were acquired using survey lines spaced 30 m apart.

The 2D resistivity imaging data was obtained using a 6 m electrode spacing and a dipole-dipole array providing data to a depth of 24 m. The resistivity imaging measurements provide a 2D cross section of variations in electrical resistivity. The data at this site consisted of three geologic layers, the shallow sands, the SCL and the upper Tampa Limestone which all had difference electrical properties. Figure 27.11 is an example of the resistivity data from the site showing these layers. A fourth layer, at the bottom of the data represents brackish water intrusion from the Anclote River.

Microgravity measurements provided a means of characterizing possible karst features such as large voids, a cavern system or areas of paleocollapse originating within the deeper Tampa and Suwannee Limestone due to changes in density. While the microgravity measurement results in a single value of relative gravity for each station measurement, it is the trend of these values along the survey line that indicate the presence of karst conditions based upon changes in subsurface density.

Microgravity measurements were made at station intervals of 12 m across most of the site, but a spacing of 6 m was used on the south half of the South Parcel, for a total of 1,826 measurements over the 52-ha. The tighter station spacing was requested by Dr. Stewart, from the University of South Florida, who was part of the project oversight committee, and wanted to see more detailed data in the southern half of the South Parcel where the critical site features were located.

In addition, drilling, sampling and geophysical logging added details to the deeper geohydrologic assessment. Geophysical logs were acquired in all existing and new wells, piezometers and test borings. The suite of geophysical logs included natural gamma, induction (electrical conductivity), gamma, gamma (density), and neutron (porosity) logs. The geophysical logs provided continuous vertical measurements of in-situ properties and were obtained through PVC or steel casing. These logs provided high-resolution vertical data at the site.

The transmissivity of the Upper Floridan aquifer ranged from 11 to 26 m^2/day, based on the pumping test. Pumping tests conducted in the Florida aquifers showed no measurable effect on the non-pumping aquifer. The lack of measurable drawdown in the non-pumping aquifer during the aquifer pumping tests indicates that there is not a strong hydraulic connection between the two aquifers.

Tidal influences were noted in all Upper Floridan aquifer wells including those in the North Parcel. Lag times in response to tidal fluctuations were generally shorter in the Upper Floridan aquifer than in the surficial. This shows that the tides in the Anclote River had a greater affect on the Upper Floridan aquifer.

27.8.1 Anomalous Areas

The microgravity data identified changes in subsurface density and areas of possible deeper karst conditions (Fig. 27.13). This method was not susceptible to cultural interference like the 2D resistivity data and provided high quality data throughout the site. The microgravity data indicated some distinct variations in subsurface density across the site. Three anomalous areas of low gravity values occur and include:

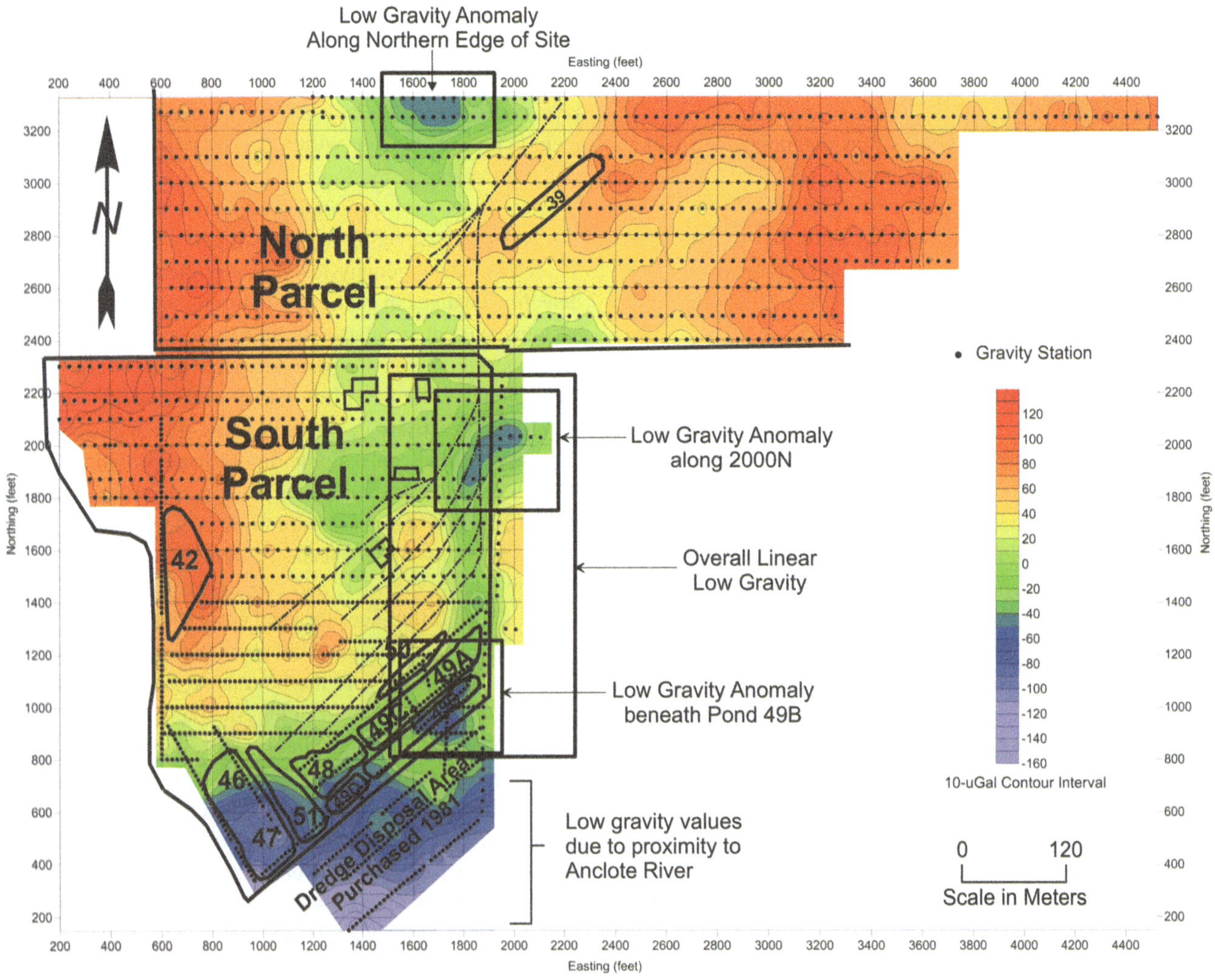

Fig. 27.13 Contour map of microgravity data indicating variations in subsurface density

- An area of low gravity values occurs along the northern edge of the North Parcel.
- A broad band of low gravity values is observed across the eastern side of the South Parcel.
- Extremely low gravity values occur adjacent to the Anclote River and can be ignored since they are due to the mass deficit associated with the river.

27.8.1.1 Northern Edge of the North Parcel

A high electrical conductivity anomaly (Fig. 27.10) correlates spatially with the low gravity anomaly detected along the northernmost fence line of the site in the North Parcel (Fig. 27.13). A review of historical aerial photographs indicated that previous on-site activity was conducted in this area. In addition, the location of the west-east road that runs along the northern edge of the site was actually moved south by about 30 m in the 1970s. DPEC measurements were completed in the area and a localized area of high conductivity materials were detected within the upper 2.4–3 m over the SCL layer. A deep test boring was completed within the area of the low-gravity values and geophysically logged. Low-densities were detected between 6 and 18 m based upon the gamma-gamma (density) logs. Additional microgravity data was acquired north of the site and indicated that the low-gravity values extended further to the north.

It appears very likely that the high electrical conductivity and low gravity values are spatially coincident, but may have separate causes. The high conductivity area along the northern fence line appears to be due to the localized presence of shallow, high conductivity materials associated with waste materials from plant operations. The area of low-gravity values is related to a dissolutioned zone or paleocollapse feature that is centered to the north, off of the Stauffer site.

27.8.1.2 Eastern Edge of the South Parcel

The eastern edge of the South Parcel contains a large linear south-to-north trend of lower gravity values (about −20 to −30 microgals). Within this linear trend are two localized areas of even lower gravity values (−60 to −80 microgals) (Fig. 27.13). This linear trend coincides with a major photolineament through the area (Fig. 27.6) and the SCL is missing in the GPR data throughout this area. DPEC measurements were used to evaluate whether this gravity low was due to a deeper SCL and top of bedrock.

Figure 27.14 shows the linear low gravity values along the eastern edge of the site and the corresponding top of SCL

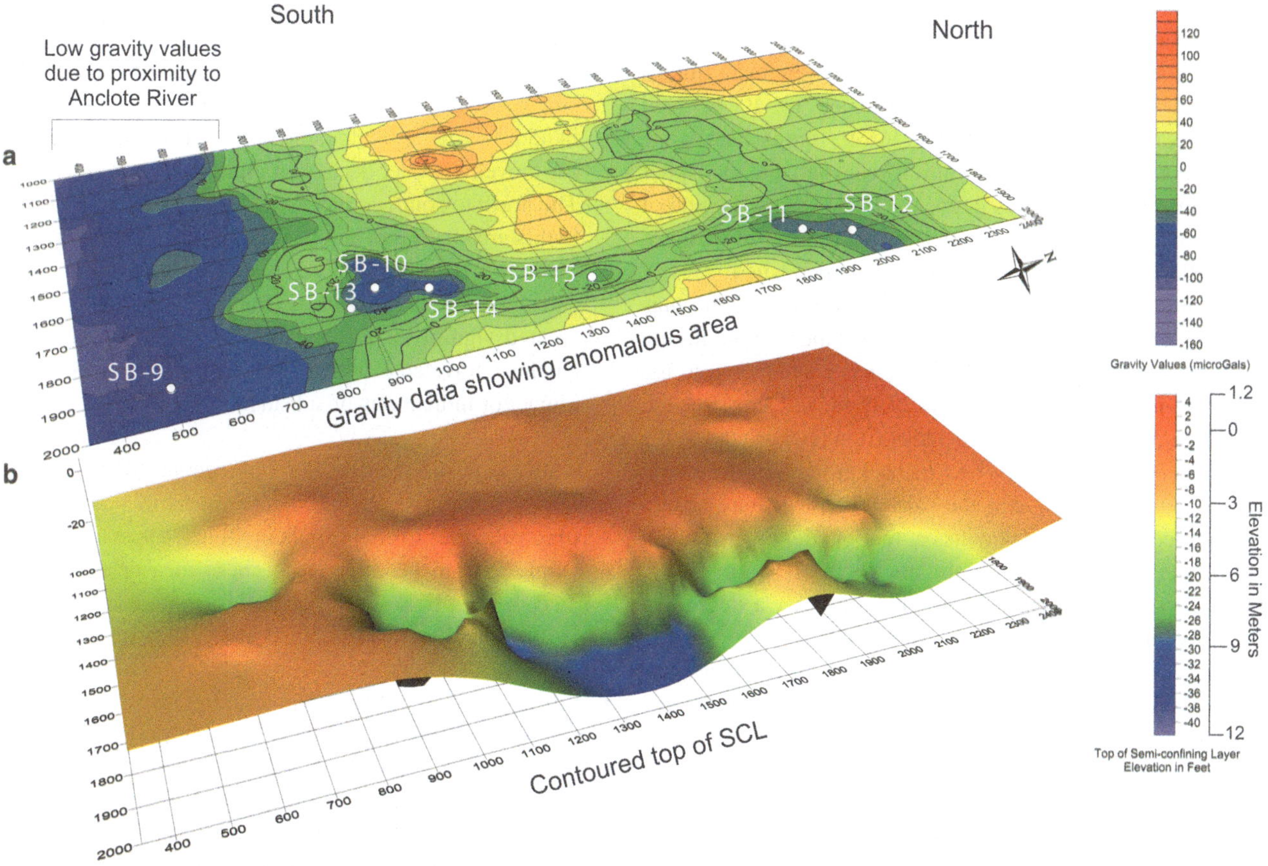

Fig. 27.14 Correlation between gravity anomaly along eastern edge of the South Parcel (**a**) and *top* of semi-confining layer (**b**). Seven rotosonic boring locations are also shown in (**a**)

contour based upon DPEC data that had a maximum depth of penetration of 12 m. The DPEC data indicated that the SCL dips inward around the perimeter of the low gravity values and is absent within the center of the low gravity values. There was a direct correlation between the missing SCL layer and the low gravity values.

Rotosonic borings were then used to assess deeper geologic strata across the site. Borings to depths of 23–48 m were made at 22 locations with the rotosonic drilling method. Continuous core samples were obtained and geologically logged. Rotosonic borings were made in both background and anomalous areas defined by the microgravity measurements.

Six rotosonic borings (SB-10 through SB-15) were located within the areas of lowest gravity values along the eastern edge of the site (Fig. 27.14). Boring SB-9 was located in a background area south of this low-gravity area. The borings within the area of low gravity indicated that the SCL and underlying limestone was missing. The materials encountered were generally unconsolidated, interbedded sand and clays with some limestone fragments. In addition, thin bands of organic materials (wood, peat, etc) were observed intermittently over a large depth interval. These unconsolidated layers along with organic materials were encountered where the Tampa and Suwannee Limestone should have been.

Samples of organic material from test borings were submitted for Carbon-14 age dating. These organic samples from 15.8 to 25 m were carbon dated to 40,000 and 48,800 years before present. The lower Tampa Limestone was deposited more than 24 million years ago and the Upper Suwannee Limestone was deposited before that. The age of the organic material indicates that the paleocollapse occurred long after the Tampa and Suwannee were deposited.

Geophysical logs obtained in all of the rotosonic borings within the gravity anomaly confirmed that the semi-confining layer was missing and no limestone was encountered to the maximum depth of borings. Figure 27.15 shows a cross section of the gamma-gamma (density) logs from background in the south through the gravity low to background to the north. These data also show that the paleocollapse feature has an abundance of low-density zones at random depths. Note that all of the low-density zones are at a depth of 12 m or more within the paleocollapse area. This suggests a degree of stability within the upper 12 m of sediment.

The integration and correlation of these independent data (microgravity, DPEC pushes, core obtained from rotosonic borings, geophysical logs, and carbon dating) clearly provide a solid case for interpretation of the area as a paleocollapse feature. The spatial extent, geologic character, and age of the paleocollapse have been defined (Fig. 27.16) along with its depth of origin at more than 48 m below grade (the limit of drilling data acquired).

Additional supporting data include:

- The cave systems and springs in and around this region all indicate that major dissolution has occurred within the Suwannee Limestone some 45–60 m bls.
- The lower sea levels of up to 130 m (Balsillie and Donoghue 2004) support the development of cave systems at depth.
- An obvious lineament was noted extending from the Kreamer Bayou in the Anclote River northward through the east side of the Stauffer site at about 10°W (Fig. 27.6). This lineament passes through the linear paleocollapse area and crosses through the microgravity anomaly at the north edge of the north parcel.

These supporting data are not conclusive on their own, but when combined with the other data they provide a coherent understanding of geologic conditions at this site.

The areas of low gravity values (Fig. 27.13) exhibit no visible surface evidence of settlement, nor are there any indications on historical aerial photos. Due to the age of the fill materials, it is believed that these paleocollapse zones are and will remain stable for the following reasons:

- Their age from the carbon dating suggest that they have been in place for >40,000 years
- Large piles of materials have been in place on-site at one time or another resulting in long periods of static loading over much of this paleocollapse area.
- In addition, dynamic loading has occurred due to railroad car movement over decades with no impact.
- The north to south county Anclote road over and adjacent to this area has been in place for decades. Continued traffic on this road includes heavy cement trucks and trailers carrying reinforced concrete structures. These trucks have created significant loads with low frequency vibrations and have traveled over this area for many decades with no impact or indications of subsidence.

Additional surface and borehole geophysical measurements were made to provide in-situ values of P and S wave seismic velocities to enable bulk modulus values of the sands and limestone to be calculated. These measurements were located in both background and anomalous areas for comparison and included:

- Surface seismic refraction measurements to provide representative P-wave velocities within the Tampa Limestone.
- Vertical seismic profiling (VSP) measurements to depths of 23–48 m to provide representative P and S wave seismic velocities within the Tampa and Suwannee Limestone.
- Hole to hole (P and Shear wave) measurements to 23 m deep were made in background conditions to provide

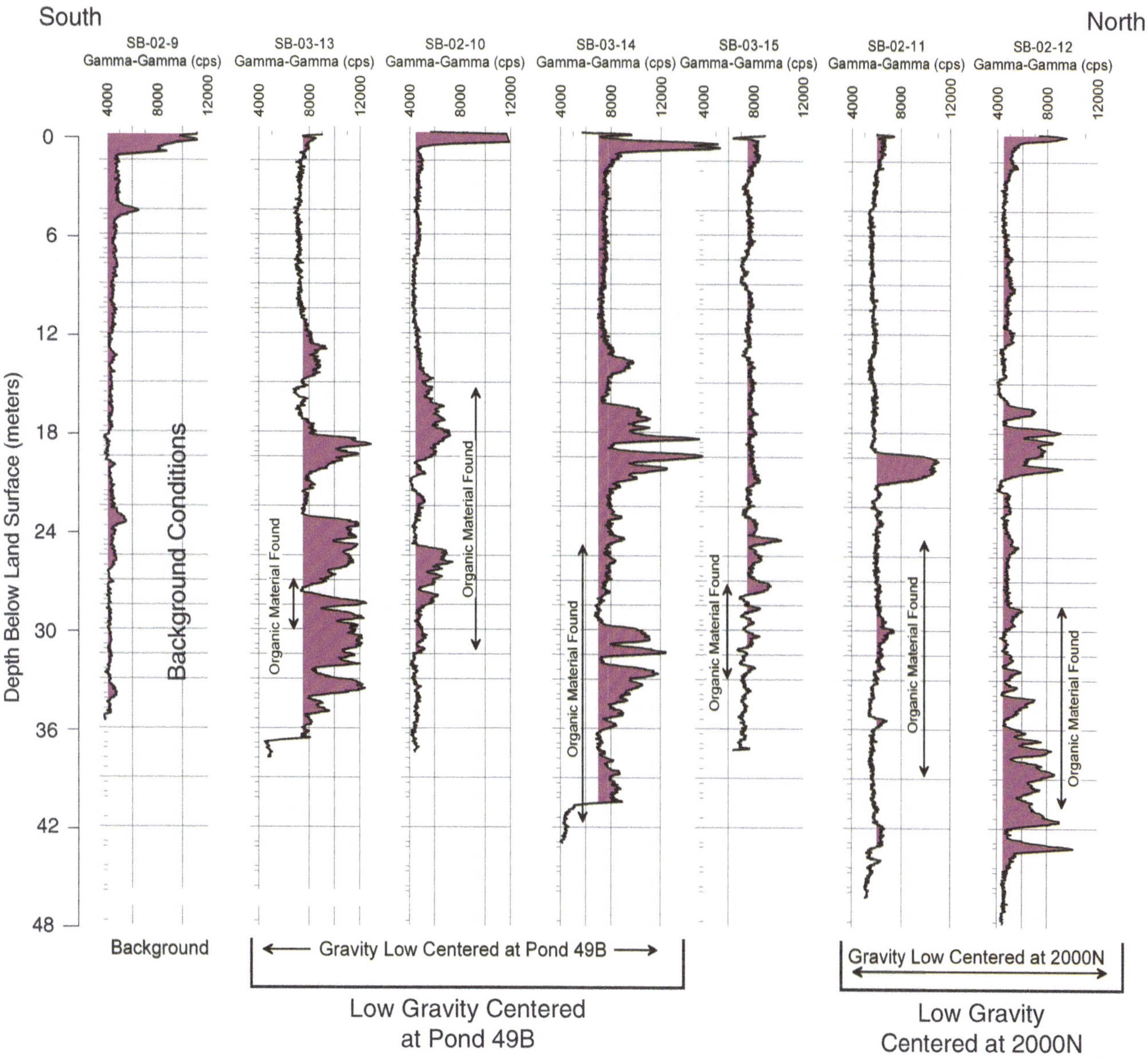

Fig. 27.15 South to north cross-section of seven geophysical density logs through gravity anomaly along eastern edge of the South Parcel. Location of the borings is shown on Fig. 27.14

representative P and S wave seismic velocities within the surficial sands and the upper Tampa Limestone.

It was found that the P-wave velocities across the site were quite similar in both the background and anomalous areas. The shear-wave velocities from the background areas were 40–60 % higher than shear-wave velocities measured within the unconsolidated materials of the paleocollapse zones (see Sect. 18.4, Figs. 18.17 and 18.18). This implies the presence of weaker, less consolidated materials within the paleocollapse features. However, no surface evidence of unstable conditions has been noted over the paleocollapse area throughout the plant operations of 40 years or during our site characterization.

27.8.2 The Sinkhole Pond

The sinkhole pond located on the northwestern edge of the southern parcel is a shallow circular depression of approximately 42 m in diameter (Fig. 27.2). It is water-filled much of the year but becomes mostly dry at times. Ground penetrating radar data indicate that the top semi-confining layer becomes deeper towards the pond (dipping to the west). This dip in the semi-confining layer was confirmed with DPEC data at the edge of the sinkhole. The data indicate that the SCL is approximately 4.5–5.5 m bls immediately east of the sinkhole pond.

A series of wash borings were used in the pond to detect the presence of the SCL and top of rock. Wash borings indi-

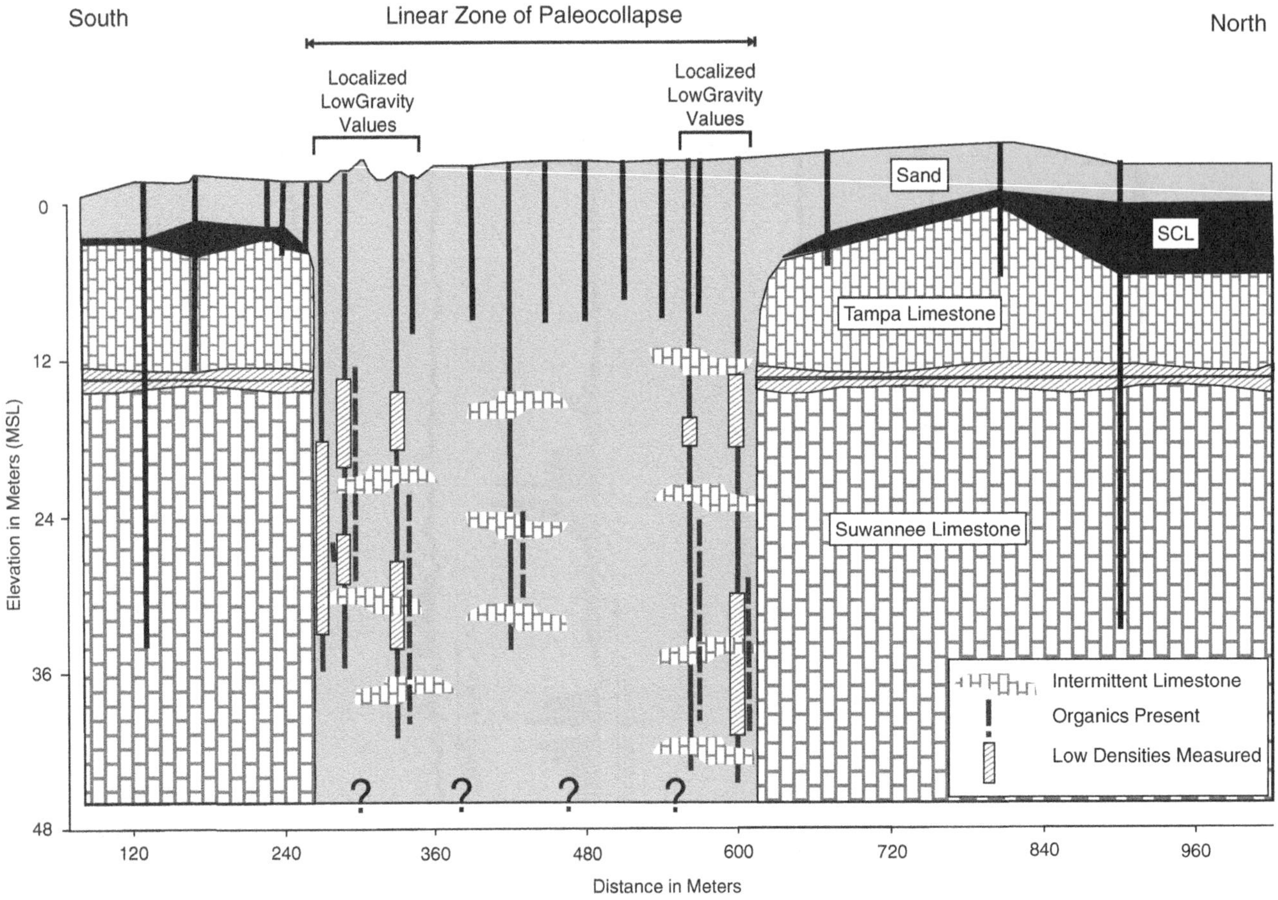

Fig. 27.16 South to north conceptual model of the paleocollapse feature based upon microgravity, DPEC data, borings, geophysical logs, the presence of organic materials and C-14 dating

cate that the sediment thickness increases toward the center of the pond indicating the presence of a sinkhole. The sinkhole pond is not in an area of potential remediation and should not have any impact on the proposed remediation activity.

27.8.3 Meyers Cove

Meyers Cove is located on the southwestern edge of the southern parcel just south of the sinkhole pond (Fig. 27.2). It is about 100 m east to west and 210 m north to south, contains boat docks for the adjoining housing development and is one of many large bayous seen within the Anclote River. Seismic reflection data were acquired in Meyers Cove along multiple parallel survey lines spaced 6 m apart. Several features in the seismic reflection data (hyperbolas and dipping strata) indicate past paleocollapse activity in this area as would be expected (Fig. 27.17). Based upon the seismic reflection data, the depth of origin for this paleocollapse feature is at least 45–60 m below MSL placing it in the mid to lower Suwannee Limestone.

27.8.4 Within the Anclote River Adjacent to the Site

The Anclote River lies immediately south of the site and has numerous circular areas (bayous) indicating the presence of paleokarst activity (Fig. 27.1). Marine geophysical surveys were run in the river to assess karst conditions immediately adjacent to the site. The measurements included bathymetry, side scan sonar and seismic reflection data. Positioning was accomplished by GPS.

Bathymetry and side scan imaging were obtained to look for indications of local depressions in the river bottom, which may indicate the presence of sinkhole activity. Seismic reflection data provided 2D cross sections and was used to evaluate the presence of paleokarst sinkhole activity below the river bottom.

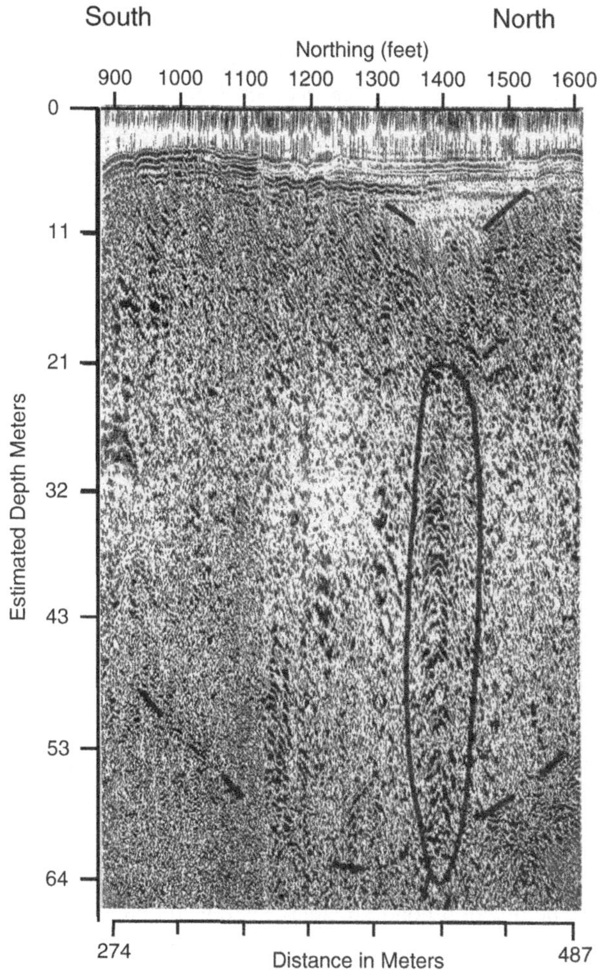

Fig. 27.17 Marine seismic reflection data from Meyers Cove showing dipping strata and voids (hyperbola) evidence of paleokarst to depths of 60 m or more

Seismic reflection data from the Anclote River southwest of the Stauffer property identified two paleokarst collapse features. Figure 27.6 shows the location of these anomalies. These features were observed in data collected on both the south and north sides of the channel, which is more than 30 m wide. This data was similar to that found in Meyers Cove (Fig. 27.17) suggesting a depth of origin of about 45–60 m.

27.8.5 Summary of Karst Conditions

The assessment of subsidence, sinkholes or paleocollapse activity ranged from a regional assessment to a detailed site-specific investigation using a wide variety of observations and measurements. The Stauffer site lies in an area where karst conditions are known to exist. Many karst features were identified in the area surrounding the site including; larger, paleosinkholes, which are seen throughout the Anclote River as circular features (bayous), along with Meyers Cove and the sinkhole pond. These features can be seen on the earliest aerial photos (1926) and therefore are quite old and appear to be stable (Fig. 27.5). Three photolineaments, a nearby spring, and small sinkholes at a number of locations were all found within the area.

Two additional karst features have been identified on site:

- One feature is located at the very north edge of the north parcel (Fig. 27.13). This appears to be the southern edge of a paleocollapse that extends further north off-site and was not of further interest to this site characterization due to its off-site location.
- The second feature was located along the eastern edge of the South Parcel. It was a large linear paleocollapse (about 4 ha). It extends along a linear south-to-north trend with a portion extending under an area of settling ponds and off-site to the east (Fig. 27.13). There was no evidence of this paleocollapse seen at the surface and it does not show up on the oldest aerial photos from 1926.

The photolineament (N10°W) along the eastern side of the site (Fig. 27.6) intersects both of these paleocollapse areas.

27.9 The Conceptual Model for the Site

Key features such as detailed lithology and karst conditions along with the vertical and horizontal hydraulic gradients, hydraulic conductivities, the direction of groundwater flow, the effects of tides, and contaminant distributions have been established. Based on these factors, a conceptual model of the geologic, hydrologic and karst conditions has been developed and is illustrated in Figs. 27.9, 27.16 and 27.18.

- These studies found that 92 % of the Stauffer site (about 48 ha) do not contain areas of deep-seated paleocollapse karst or any obvious indications of shallow subsidence activity and will easily support the proposed remedy.
- Two buried paleokarst features were found over the site (about 8 % or 4 ha) (Fig. 27.13). One is located on the north edge of the North Parcel and extends off site to the north. It is not of concern since only its southern edge extends onto the site. Furthermore, no remediation will take place there. Only one large paleokarst feature on the eastern side of the site is of concern (Fig. 27.16). The origin of the paleocollapse features appears to be in the lower portion of the Suwannee Limestone based upon rotosonic borings. Based upon C-14 dating this karst feature collapsed more than 40,000 years ago and then begin

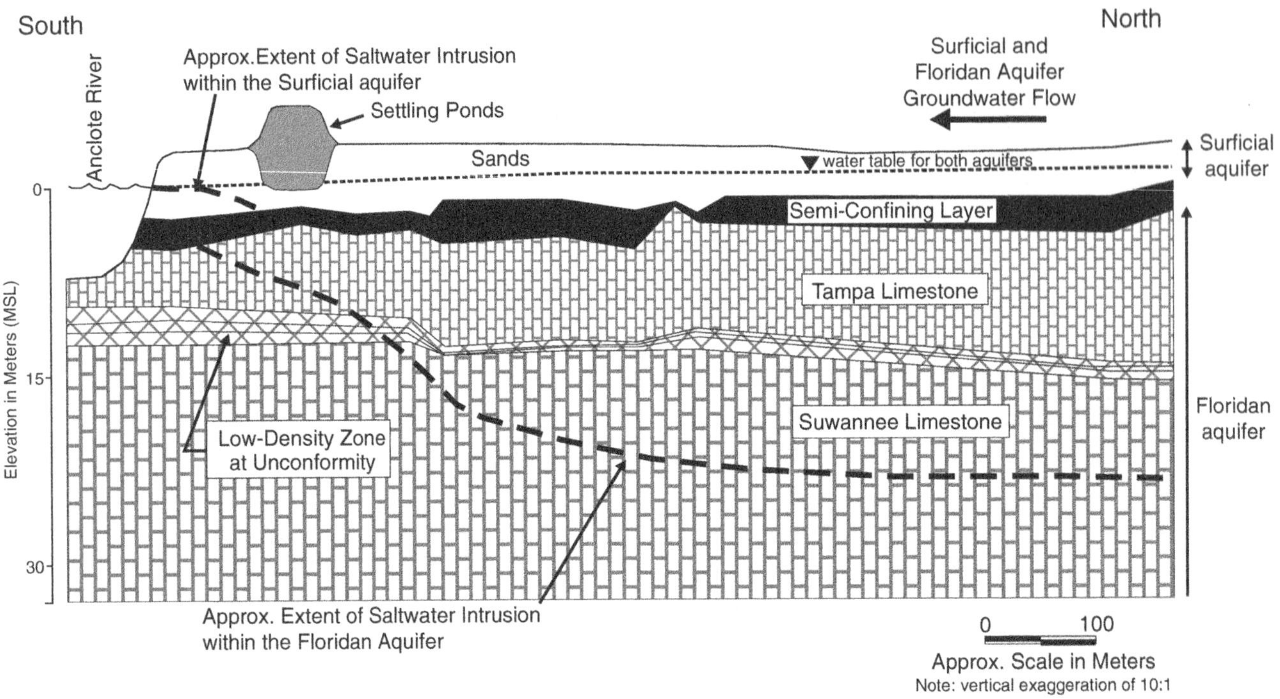

Fig. 27.18 South to north conceptual cross section of site geology and hydrology

to fill with sediments. This paleocollapse area appears to be very stable with an extremely low risk of reactivation over the life of the remedy.

- The semi-confining layer is continuous over about 92 % of the site. Its thickness ranges from 0.3 to 7.3 m, with an average thickness of 2.3 m. This semi-confining layer is minimizing cross contamination between the surficial and Upper Floridan aquifers and preventing sinkhole development caused by voids in the upper Tampa Limestone.
- The site hydrogeology is nearly flat, low flow system with ultimate discharge to the Anclote River by both the surficial and Upper Floridan aquifers.
- The vertical head between the Surficial aquifer and Upper Floridan aquifer is minimal; therefore there is little driving force for the downward migration of contaminants, even in those areas of the site where the SCL is thin or absent. In addition there is little driving force for the downward migration of sands and development of shallow sinkholes.
- Groundwater contamination for metals above the Primary Drinking Water Standard is generally localized near source areas in the surficial aquifer.
- The areas of brackish water intrusion in the surficial aquifer are generally restricted to the shoreline.
- Tidal affects on the surficial aquifer are minimal and should have little or no effect on the proposed remedy.
- Off-site drinking water wells are not impacted by the site because the groundwater flow in both aquifers discharges into the Anclote River.
- There are no groundwater factors at the site that should affect the implementation or the long-term effectiveness of the remedy presented in the Record of Decision.
- The disposal pond boundaries are well defined and there is no evidence that the pond material extends beyond the confines of the pond basins. The depth of pond materials does not appear to breach the semi-confining layer.
- Areas of buried metal debris have been identified; however, the data does not support the presence of a large cache of drums remaining in one or more locations at the site. It is very likely that the steel drums, if they were present, have corroded in the two or more decades since the plant has shut down.

27.10 Sinkhole Risk Assessment

The natural dissolution of limestone is a slow process, occurring at a rate of about 2.5 cm/1,000 years. As such, further dissolution of rock by itself is not a problem to the life of engineered structures (Sowers 1996). The concern then lies with existing dissolution features and the risk of sinkholes developing due to these dissolution features. Three zones of dissolution were identified at the site and briefly discussed in Sect. 24.4.1 Risk Assessments (Fig. 24.8).

Zone 1 is a shallow zone of dissolution and weathering that exists at the top of the Tampa Limestone just beneath the SCL. This zone contains small isolated voids or loose materials. This zone lies approximately 3–9 m bls (dipping from north to south). The presence of the semi-confining layer acts as the barrier, preventing the overlying sands or contaminants from migrating into these localized voids within the upper portion of the Tampa Limestone.

Zone 2 is the unconformity between the Tampa and the Suwannee Limestone (about 18 m bls) that was identified in both the regional literature and the geophysical logs as typically being a low-density zone.

Zone 3 is a deeper zone within the Suwannee Limestone some 45–60 m bls in which cave systems and springs have developed throughout the region based upon regional literature, and is further verified by the seismic reflection data in Meyers Cove and the Anclote River.

Only Zone 1 at the top of the Tampa Limestone just beneath the SCL is of any concern (Figs. 24.7 and 27.9a), and then only if the SCL is breached by remediation or subsequent activity on site. A worse case scenario would result in a small, localized sinkhole 6–9 m in diameter, similar to those at the Corps of Engineers dredge disposal site (Figs. 24.7 and 27.9a). The other two deeper zones of potential karst development, Zones 2 and 3 (Figs. 24.7 and 27.9a) are at depths well below any possible influence on remediation efforts at the site.

In addition, there is the paleocollapse (Zone 3), which is located on the eastern side of the South Parcel a portion of which is under the pond areas (Pond 49A and 49B) (Fig. 27.13). Concern here lies with the potential for reactivation of this paleocollapse feature.

Many factors may trigger sinkhole collapse including both natural and man-made activities. The dominant factors in triggering collapse are water related such as heavy prolonged rainfall and or drought, concentration of surface water run-off, or changes in groundwater levels by heavy pumping. In addition changes in surface loads, vibration, blasting, and excessive drilling fluid or grouting pressures may trigger collapse (Benson and Hussin 2002; Benson et al. 2003).

Groundwater levels at the site are controlled by the adjacent river and the nearby Gulf of Mexico and remain relatively constant. Tidal changes are the most significant factor in changing groundwater levels and primarily influence the Floridan aquifer. There is only a very small difference in head between the surficial and Floridan aquifers (thousandths of a meter). There is no excessive pumping of groundwater from either aquifer within the property.

Dynamic loading and vibrations are potential triggering mechanisms that have existed at the Stauffer site during plant operations (1947–1981). They were associated with railway operations along the eastern portions of the site, as well as static loading from large piles of coke, ferrophosphorous and slag (Fig. 14.6). These conditions existed directly over or near the area of the linear paleocollapse feature along the eastern boundary of the South Parcel with no effect.

27.10.1 Summary of Risk Assessment

The water levels in both the Surficial and Upper Floridan aquifers are shallow with flow towards the Anclote River and there are no significant vertical gradients. These conditions are moderated by the presence of the river and will aid in preventing any extreme groundwater conditions to occur on-site. If the site were left by itself, unaffected by further activity, the probability of development of small shallow cover subsidence sinkholes below the SCL in Zone 1 or redevelopment of a paleocollapse sinkhole or development of a new deep-seated cover collapse sinkhole within the Tampa or Suwannee Limestone (Zones 2 and 3) is considered to be virtually zero (Fig. 24.8).

27.11 The Ability of Geology to Support the Proposed Remedy

One of the objectives of this investigation was to evaluate the ability of the sites geology and hydrology to support the proposed remedy over the life of the remedy as required under reasonably anticipated site conditions. The following comments were provided at the completion of the site characterization investigation and are based upon the proposed remediation of soils at that time.

Most of the Stauffer site, about 92 % or 48 ha, was found to consist of background geohydrologic conditions. Evidence of past sinkhole activity is provided by the sinkhole pond and Meyers Cove along the western edge of the South Parcel. These two features were present in the 1926 aerials indicating that they are quite old. In addition, two areas of buried paleocollapse were identified, one along the northern edge of the site and one along the eastern edge of the site. Only the eastern edge paleocollapse is of concern due to its location at the edge of a remediation area.

The proposed remedy consists of:

- Excavation of certain materials
- Consolidation of materials
- Capping of materials with an impermeable layer
- In-situ stabilization of pond materials

Figure 27.19 illustrates the conceptual cross section of the proposed remedy and its potential impact on the geologic hydrologic and karst conditions. All remediation will occur above the semi-confining layer and above Zone 1 (the epikarst and top of Tampa Limestone). It is critical to avoid breaching the SCL during remediation or by long-term site use.

Excavation of materials will decrease loading within the excavated area. This is not considered to be a major risk factor. However, any topographic low created due to excavation will tend to concentrate surface water run-off. Storm water will have to be managed during and after construction so that

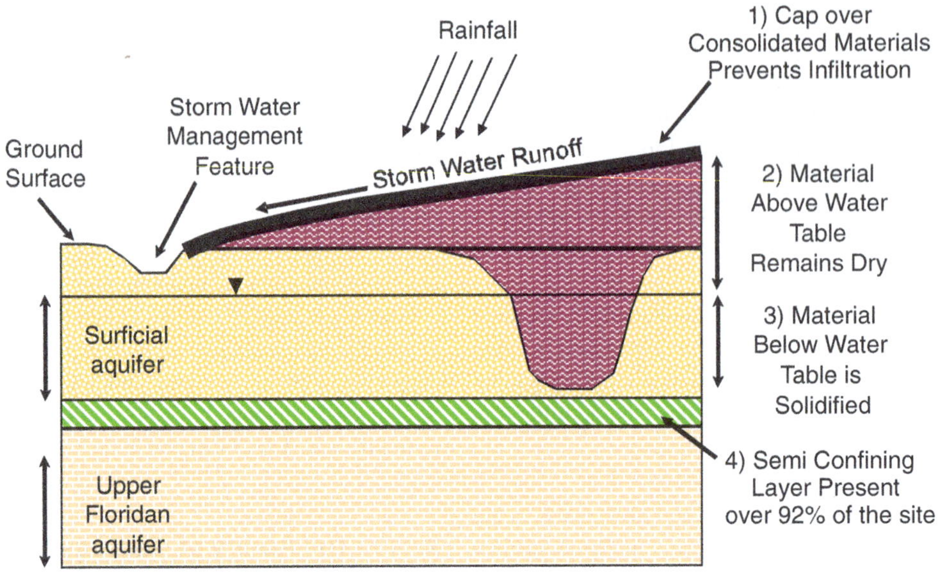

Fig. 27.19 Conceptual model of remediation factors and their impact to the shallow geology

large quantities are not concentrated in a local area. The materials to be excavated may extend to the top of the semi-confining layer. Therefore, care will be necessary to avoid breaching the semi-confining layer during excavation. Recall that two sinkholes 914 m east of the site resulted when the Corps of Engineers excavated the SCL and used the area for disposal of dredge material.

Consolidation of materials will increase loading in selected areas. Both static loading as well as dynamic loading has occurred at the site during plant operations and was much greater than that expected due to consolidation. The consolidation itself and its effects are not considered to be a risk factor.

Capping of materials with an impermeable layer will potentially change the surface water run-off characteristics. Storm water run-off will have to be managed so that it is not concentrated in a local area. The capping system will be designed to control run-off and avoid concentration or build up of surface water run-off. Therefore, capping is not considered to be a major risk factor.

In-situ stabilization is to be focused on pond materials and is carried out by blending a cement or binding mixture with in-place soils. Some vibrations and/or pressures are created during the mixing associated with the stabilization process. The amplitude of vibration and its frequency or pressures is not known at this time. Excessive vibrations and or pressures could breach the SCL allowing sands to ravel into small isolated voids within the surface of the Tampa Limestone and contaminants to migrate into the Florida aquifer. Measures will be incorporated into the design and implemented during construction to monitor and control vibrations and pressures. The depth of stabilization must also be controlled to avoid breach the semi-confining layer.

Portions of Ponds 49A and 49B (Fig. 27.13) that require remediation lie over the southern portion of the paleocollapse area covering a total of less than 0.2 ha. The SCL as well as the Tampa and Suwannee limestones are missing within this area of paleocollapse. Geophysical logs within the paleocollapse zone indicate localized low-density zones likely associated with the random sediments filling the paleocollapse. These low-density zones (Fig. 27.15) are all deeper than 12 m and deeper than remediation activities. The paleocollapse feature itself appears to be stable from its origin at depth within Zone 3 or deeper. Regardless of the very low risk, monitoring of conditions (vibration and or pressure) during remediation of Ponds 49A and 49B will be necessary to minimize any potential impact.

Once the stabilized material hardens groundwater flow patterns may be changed slightly. It is expected that the stabilized material would have reduced permeability, and will act as an obstruction to groundwater flow. Because no pond material appears to extend below the semi-confining layer, this change in groundwater flow should be limited to the surficial aquifer and is expected to have negligible impact to the sites hydrology.

27.12 About the Site Characterization Strategy

This project was unique in several ways, but the two that had the greatest impact was the involvement and technical understanding by the client to get the project done right the first time as well as a budget to allow the team to get things completed in a timely manner. In addition, the project followed many of the core site characterization strategies imbedded within the ESC process that included:

- An active team of experienced professionals who were involved from the beginning to the end of the project in the data acquisition, interpretation and integration. This experience and continuity of the project team created a high level of confidence in the project results.
- Accurate data was obtained by having hands-on experienced professionals in the field, who developed standard operating procedures (SOP) for the project and utilized a strong quality control program throughout the project.
- Because no single method or measurement will uniquely define subsurface conditions, a wide range of independent measurements responding to different parameters and volume of measurement (scale) were used and integrated.
- The correlation of independent measurements provided a significant increase in the level of confidence in the interpretation and final conceptual model.
- Active involvement and consensus building of all stakeholders from the beginning and throughout the project allowed technical questions or concerns to be addressed and resolved as they occurred.

While this project was unique, the basic approach can be scaled down and successfully applied to other karst site characterization.

The final investigation results were submitted by Parsons Engineering Science (2003) and Technos, Inc. under the cover of O'Brien and Gere (2003). Papers describing this project were presented by Yuhr (2003); Yuhr et al. (2003).

The once EPA Superfund site has been cleaned up and awaits redevelopment. It is one of the largest undeveloped tracts of land along the Anclote River left in Pinellas County. The contaminated soil was capped beneath a layer of dirt and sod. More than $21 million was spent for remediation efforts that were completed in 2011. Because polluted soil remains on site homes can never be built there. Local legislators are trying to attract industrial and manufacturing business to the site. The state of Florida has budgeted funds for dredging, wharf stabilization and road improvements at the site. One manufacturing client is seriously considering the property (Telford 2014).

References

Arthur JD et al (2001) Lithostratigraphic and hydrostratigraphic cross-sections through Levy-Marion to Pasco Counties, southwest Florida. Florida Geological Survey, Tallahassee. Open File Report 81

Balsillie JH, Donoghue JF (2004) High resolution sea-level history for the Gulf of Mexico since the last glacial maximum. Florida Geologic Survey Report of Investigation 103

Beck BF, Sayed S (1991) The sinkhole hazard in Pinellas County: a geologic summary for planning purposes. Florida Sinkhole Research Institute and Pinellas County Department of Public Works. Report 90-91-1

Benson RC, Hussin J (2002) Assessment of karst, pseudokarst and its remediation. Workshop presented at Geophysics 2002, 2nd annual conference on the application of geophysical and NDT methodologies to transportation facilities, Federal Highway Administration, 15–19 Apr 2002, Los Angeles, CA

Benson RC et al (2003) Assessing the risk of karst subsidence. In: Proceedings from the AEG/AIPS 2002 joint annual meeting, 22–29 Sept 2003, Reno, NV

Cherry RN (1966) Chloride content of ground water in Pinellas County, Florida in 1950 and 1963. Florida Geologic Survey, Map Series No. 20

Coble RW (1973) The Anclote and Pithlachascotee Rivers as water-supply sources. USGS, Map Series No. 61

Environmental Protection Agency (2000) National primary and secondary drinking water regulations 40 CFR Parts 9, 141, 142 and 143. Fed Regist Vol. 65 No. 87, 4 May

Green R et al (1995) Lithostratigraphic and hydrostratigraphic cross-sections through Pinellas and Hillsborough Counties, southwest Florida. Florida Geological Survey, Tallahassee, Open File Report 61

Hickey JJ (1982) Hydrogeology and results of injection tests at waste-injection test sites in Pinellas County, Florida. USGS, Water Supply Paper 2183

Parizek RR (1976) On the nature and significance of fracture traces and lineaments in carbonate and other terranes. In: Yevjevich V (ed) Karst hydrology and water resources, Vol. 1, karst hydrology, Ft. Collins, CO, Water Resources Publication 3-1/3-108

Parsons Engineering Science, Inc. (1999) Baseline groundwater monitoring report. Stauffer Management Company, Tarpon Springs, Consultants report

Parsons Engineering Science, Inc. (2003) Draft groundwater investigation of the Stauffer Site. Stauffer Management Company, Tarpon Springs, Consultants report

Reichenbaugh RC (1972) Sea-water intrusion in the upper part of the Floridan aquifer in coastal Pasco County, Florida, 1969. Florida Geologic Survey, Map Series No. 47

Romanoff M (1957) Underground corrosion. National Bureau of Standards. Circular 579

Scott TM (1988) The lithostratigraphy of the Hawthorn Group (Miocene) of Florida. Florida Geological Survey Bulletin No. 59

Scott TM et al (2004) Springs of Florida. Florida Geologic Bulletin No. 66

Sinclair WC, Stewart JW (1985) Sinkhole type, development, and distribution in Florida. US Geological Survey, Map Series No. 110

Southwest Florida Water Management (SWFWMD) (1988) Groundwater resource availability inventory: Pinellas County, Florida. Southwest Florida Water Management District, Brooksville

Sowers GF (1996) Building on sinkholes, design and construction of foundations in karst terrain. American Society of Civil Engineers, New York (out of print)

Technos, Inc. (2003) Geophysical studies report for Stauffer Management Company, Tarpon Springs, Florida. Submitted to O'Brien & Gere Engineers, Inc. June, Consultant Report

Telford S (2014) Cleaned up Stauffer Chemical plant site awaits redevelopment. Florida Specifier, September, Goldenrod, Florida

Tuthill AH, Schillmoller CM (1966) Guidelines for selection of marine materials. International Nickel Company, New York

Uhlig HH (1965) Corrosion and corrosion control an introduction to corrosion science and engineering. Wiley, New York

US Environmental Protection Agency (1998) Record of decision, the decision summary, operable unit 1. Stauffer Chemical Tarpon Springs Site, Tarpon Springs

US Environmental Protection Agency (2000a) Agreement to conduct studies to include: geophysical studies, treatability studies, and groundwater characterization studies. Stauffer Chemical Tarpon Springs Site, Tarpon Springs

US Environmental Protection Agency (2000b) Scope of work, agreement to conduct studies to include: geophysical studies, treatability studies, and groundwater characterization studies. Stauffer Chemical Company Superfund Site, Tarpon Springs

Vernon RO (1951) Geology of Citrus and Levy Counties, Florida. Florida Geological Survey, Bulletin No. 33

Wetterhall WS (1964) Geohydrologic reconnaissance of Pasco and southern Hernando Counties, Florida. Florida Geological Survey, Report of Investigations No. 34

Wetterhall WS (1965) Reconnaissance of springs and sinks in west-central Florida. Florida Geological Survey, Report of Investigations No. 39

Wilson WL (1995) Sinkhole and buried sinkhole densities and new sinkhole frequencies in karst of northwest Peninsular Florida. In: Beck BF (ed) Proceedings of the 5th multidisciplinary conference on sinkholes and the engineering and environmental impacts of karst: karst geohazards, Gatlinburg, TN. Balkema, Rotterdam, pp 79–91

Yuhr L (2003) A case history of a comprehensive karst investigation in Tarpon Springs, Florida. Presented at the 18th annual regional phosphate conference Lakeland, FL, 16–17 Oct 2003

Yuhr L et al (1996) Achieving a reasonable level of accuracy in site characterization in the presence of geologic uncertainly. In: Shackelford CD, Nelson PP, Roth MJS (eds) Proceedings of uncertainty 1996 – uncertainty in the geologic environment: from theory to practice. ASCE Geotechnical Special Publication No. 68

Yuhr L et al (2003) A case history of a large karst investigation. Presented at the 3rd international conference on applied geophysics, geophysics 2003 Orlando, FL, 8–12 Dec 2003

Index

R.C. Benson, L.B. Yuhr, *Site Characterization in Karst and Pseudokarst Terraines*,
DOI 10.1007/978-94-017-9924-9

The manufacturer's authorised representative in the EU is Springer Nature Customer Service Centre GmbH, Europaplatz 3, 69115 Heidelberg, Germany. If you have any concerns regarding our products, please contact ProductSafety@springernature.com

Printed and bound by CPI Group (UK) Ltd, Croydon, CR0 4YY
30/09/2026
02225477-0003